BIOGEOGRAPHY AND ECOLOGY OF SOUTHERN AFRICA

MONOGRAPHIAE BIOLOGICAE

Editor

J. ILLIES
Schlitz

VOLUME 31

Dr W. Junk bv Publishers The Hague 1978

BIOGEOGRAPHY AND ECOLOGY OF SOUTHERN AFRICA

Edited by

M. J. A. WERGER

with the assistance of
A. C. VAN BRUGGEN
for the zoological chapters

Dr W. Junk bv Publishers The Hague 1978

To Elke
I wish to dedicate my part in this work

M.J.A.W.

ISBN 90 6193 083 9

© 1978 Dr W. Junk bv Publishers The Hague

Cover design M. Velthuijs

No part of this book may be reproduced and/or published in any form, by print, photoprint, microfilm or any other means without written permission from the publishers.

Contents

Preface vii
Authors' addresses xi
List of selected synonymous toponyms xv

Environment, present and past

1. The geomorphology of central and southern Africa, by Lester King .. 1
2. Climatic indices and classifications in relation to the biogeography of southern Africa, by R. E. Schulze and O. S. McGee 19
3. Rainfall changes over South Africa during the period of meteorological record, by P. D. Tyson 55
4. Schematic soil map of southern Africa south of latitude 16° 30′S, by H. J. von M. Harmse 71
5. Late Cretaceous and Tertiary vegetation history of Africa, by D. I. Axelrod and P. H. Raven 77
6. Quaternary vegetation changes in southern Africa, by E. M. van Zinderen Bakker Sr 131

Biogeography and ecology

7. Biogeographical division of southern Africa, by M. J. A. Werger .. 145
8. Capensis, by H. C. Taylor 171
9. The Karoo–Namib Region, by M. J. A. Werger 231
10. The Sudano–Zambezian Region, by M. J. A. Werger and B. J. Coetzee 301
11. The Afromontane Region, by F. White 463
12. The Afro-alpine Region, by D. J. B. Killick 515
13. The Indian Ocean Coastal Belt, by E. J. Moll and F. White 561
14. The Guineo–Congolian transition to southern Africa, by F. White and M. J. A. Werger 599
15. Primary production ecology in southern Africa, by M. C. Rutherford 621
16. Megadrilacea (Oligochaeta), by R. W. Sims 661
17. Onychophora, by G. Newlands and H. Ruhberg 677
18. Arachnida (except Acari), by G. Newlands 685
19. Acari, by Magdalena K. P. Smith Meyer and G. C. Loots 703
20. Myriapoda, by O. Kraus 719
21. Odonata, by Elliot Pinhey 723
22. Orthoptera, by David C. Rentz 733

23. Isoptera, by J. E. Ruelle 747
24. Lepidoptera, by Elliot Pinhey 763
25. Diptera, by John Bowden 775
26. Coleoptera, by S. Endrödy-Younga 797
27. Hymenoptera, by A. J. Prins 823
28. Land molluscs, by A. C. van Bruggen 877
29. The Herpetofauna, by J. C. Poynton and D. G. Broadley 925
30. Birds, by J. M. Winterbottom 949
31. Mammals, by R. C. Bigalke 981
32. Patterns of man–land relations, by Bill Puzo 1049

Biogeography and ecology of special habitats

33. Freshwater plants, by D. S. Mitchell 1113
34. Freshwater invertebrates (except molluscs), by A. D. Harrison 1139
35. Freshwater molluscs, by D. S. Brown 1153
36. Freshwater fishes, by A. P. Bowmaker, P. B. N. Jackson and R. A. Jubb 1181
37. Mangrove communities, by E. J. Moll and M. J. A. Werger 1231
38. Coastal marine habitats, by A. C. Brown and N. Jarman 1239
39. High termitaria, by F. Malaisse 1279
40. The vegetation of heavy metal and other toxic soils, by H. Wild 1301

Conservation

41. Ecosystem conservation in southern Africa, by B. J. Huntley 1333

Subject index 1385
Systematic index 1395

Preface

Southern Africa is certainly not a naturally bounded area so that there are several possibilities for delineating it and concepts about its extent. Wellington* discussed the various possibilities for delineation and suggested that one line stands out more clearly and definitely as a physical boundary than any other, namely the South Equatorial Divide, the watershed between the Zaïre, Cuanza and Rufiji Rivers on the one hand and the Zambezi, Cunene and Rovuma Rivers on the other. This South Equatorial Divide is indeed a major line of separation for some organisms and is also applicable in a certain geographical sense, though it does not possess the slightest significance for many other groups of organisms, ecosystems or geographical and physical features of Africa. The placing of the northern boundary of southern Africa differs in fact strongly per scientific discipline and is also influenced by practical considerations regarding the possibilities of scientific work as subordinate to certain political realities and historically grown traditions. This is illustrated, for example, in such works as the Flora of Southern Africa, where the northern boundary of the area is conceived as the northern and eastern political boundaries of South West Africa, South Africa and Swaziland. Botswana, traditionally included in the area covered by the Flora Zambesiaca, thus forms a large wedge in 'Southern Africa'. On the other hand, most maps include this country in 'Southern Africa' and in several publications 'Southern Africa' is broadly defined as 'the area south of the Cunene and Limpopo Rivers'. For this book I have decided not to stick to a fixed line as the northern boundary, but with Wellington's considerations in mind, I suggested to the contributing authors that they make their own choice as to the northern boundary of the area to be discussed as either: (1) the southern political boundaries of Zaïre and Tanzania; (2) the ninth or tenth parallel of southern latitude; or (3) the South Equatorial Divide. In this way of arbitrarily deciding on the northern limit there was no artificial rigidity in suggesting a real boundary where none exists. Thus the following areas are covered in this book: Angola, part of Shaba, Zambia, Malawi, Moçambique, South West Africa, Botswana, Rhodesia, South Africa, Swaziland and Lesotho. Nevertheless, in a few cases authors found it impossible or undesirable to stick to this suggestion for good technical reasons and discussed either a smaller (e.g. Chapters 3 and 4) or a larger area (e.g. Chapter 5).

Southern Africa then, as conceived here, covers an enormous area of some 6,100,000 sq km and is extremely varied both in biotic and abiotic aspects. It varies in altitude from sea level to 3482 m, though a very large part of the area consists of a fairly flat plateau over 1000 m above sea level (Fig. 1). The pattern of altitudinal variation as well as of the other abiotic environmental factors shows a strong correlation with the biogeographical patterns as demonstrated in the following pages. This book concentrates on these patterns and their explanation in a historical and ecological way. It is a review of what is known about the biogeography and ecology of the southern African organisms and ecosystems. It has been attempted to integrate this knowledge which exists in printed form almost exclusively in either of five languages allochthonous to the area, i.e. English, Por-

---

* J. H. Wellington. 1955. Southern Africa. A geographical study. Vol. 1. Cambridge Univ. Press, Cambridge.

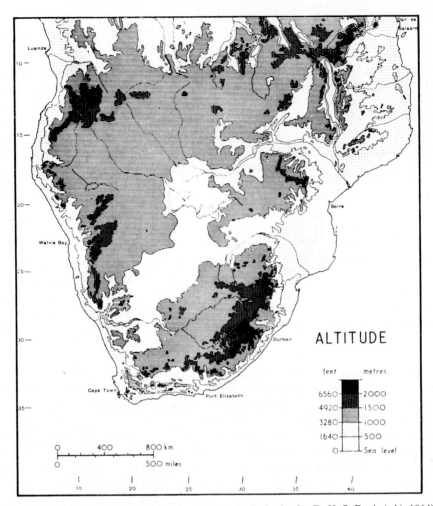

Fig. 1. Altitudinal map of southern Africa (from H. B. S. Cooke, In: D. H. S. Davis (ed.), 1964).

tuguese, French, German and Afrikaans. Not infrequently the knowledge is available only from publications which are difficult to obtain or from academic theses with a very limited distribution. This information has been largely integrated, together with that from more readily available sources, in the chapters that follow. The ecological aspects covered in this book are mainly of a descriptive nature, concentrating on diversity and variation in species and on a structural–functional evaluation; physiological aspects of the ecology and energy and nutrient flow systems remain virtually untouched. In fact, coordinated research in these fields has only just started in southern Africa, and it is hoped that the present review of biogeographical patterns and descriptive ecology is helpful in facilitating a justifiable choice of problems and objects on which ecophysiological and ecosystem research in southern Africa should concentrate.

Not only in its contents but also in its structure the present book reflects our present state of knowledge and the way we have assembled and still gather this knowledge: the botanically oriented Chapters 8 to 14 each deal with a major biome, while the zoologically oriented Chapters 16 to 31 each cover a major taxonomical group. Apart from these chapters arranged according to this dual scheme there are those (Chapters 33 to 40) on organisms and communities living in special habitats, where one environmental factor or factor complex is of overriding significance. Then there are the Chapters 15, 32 and 41 dealing with special aspects of southern Africa in general, and, of course, the introductory background-providing Chapters 1 to 6.

Though the book has grown rather voluminous it is, of course, far from a complete account on the biogeography and ecology of southern Africa. Several more chapters on special habitats could have been included in the botanical section, while chapters on many more animal groups with equally interesting biogeographical patterns and ecological features could have been included. I had to make a choice in preparing a general scheme for the book, and the one I made was partly based on practical considerations. All chapters, whatever their length, presented to their authors the same problems of how to select the main features from an overwhelming mass of facts and how to fit these into a clear general picture within the necessarily limited number of pages available. To achieve this, different authors opted for different approaches as the following pages will show. It is obvious that with so many authors contributing to one work there easily results a problem because of differences in concepts and opinions among these authors. I have attempted to keep the technical terminology as homogeneous as possible, and if differences in opinion between authors did necessitate a heterogeneity in terminology, I have tried to prevent it from conflicting with opinions and terms expressed elsewhere in the text. That such is possible is illustrated, for example, by the fact that two different systems of biogeographical subdivision of southern Africa have been used as the basis of different botanical chapters (e.g. Chapters 11 and 14 versus 9, 10 and 12). This could be done because these systems, though different in details, do not lead to conflicting conclusions when applied to the basic data. Similarly, strong preferentials of authors led, e.g., to the use of the terms 'wooded grassland' or 'savanna' for the same formation in different chapters. I personally feel that such differences are not at all disadvantageous to the integral character of the book.

Perhaps the synoptic tables included in the botanical chapters need a short explanation. They show the floristic composition of different communities and indicate the number of relevés on which they are based. The figures 1 to 5 following the species represent degrees of presence in intervals of 20 per cent, so that 5 means a presence of 81–100 per cent in the number of relevés indicated, etc.

Since southern Africa is one of those areas in the world where place-names are still changing rapidly, I add a short list of the most important previous, present and some possible future synonymous toponyms in the area, while throughout the text the current legally valid names (January 1, 1977) are used, sometimes, in first instances, followed by the old name in brackets.

There has been an earlier volume on southern Africa in the series Monographiae Biologicae.* The present book is no revision of or supplementary volume to this earlier publication, however, but should be viewed as an entirely independent endeavour altogether. A brief inspection of the two books will immediately confirm that, even though some of the contributors to the present book also wrote in the earlier publication.

* D. H. S. Davis (ed.). 1964. Ecological Studies in Southern Africa. Monogr. Biol. 14. Junk, The Hague.

This is a work of many authors, and I am grateful to all of them for their contribution and the pleasant cooperation. I received much appreciated help with various technical preparative matters from Th. Kuijper, who also prepared the index, K. L. Tinley, J. W. Morris, J. Eysink, W. Holzner, P. Smeets, C. de Groot, S. Premer, Mrs. A. van de Zand-Barten, Mrs. G. E. Thomas, Mrs. K. E. Werger-Klein, several members of staff of the illustration and photographic departments of the University of Nijmegen, and several other people, whom I hope I may be excused from mentioning by name. I am very grateful to all of them. I would further like to acknowledge gratefully the support I received in many instances from the director, B. de Winter, and several members of staff of the Botanical Research Institute, Pretoria, and from V. Westhoff, head of the department of geobotany, Nijmegen.

A very special word of thanks is addressed to A. C. van Bruggen who efficiently and continually helped me in many questions concerning the zoological chapters, both in the planning and in the review stages. My thanks are also due to all those persons who provided illustrations for the various chapters, and to the publishers who allowed me to reproduce illustrations from their publications. In each instance their names or the sources are mentioned in the captions. I am grateful for the financial support which I received in 1976 from the Netherlands Foundation for the Advancement of Tropical Research (WOTRO) and from the Faculty of Sciences, Nijmegen, thus enabling me to revisit southern Africa in a well-developed stage of preparation for this book so that I could personally discuss various matters with many authors. Finally, I thank Dr. W. Junk Publishers for their invitation to make this book and for the ample facilities provided.

Marinus J. A. Werger
Nijmegen, February 1, 1977

Authors' addresses

D. I. Axelrod,
Department of Botany, University of California, Davis, Ca. 95616, U.S.A.

R. C. Bigalke,
Faculty of Forestry, Department of Nature Conservation, University of Stellenbosch, Stellenbosch 7600, South Africa.

J. Bowden,
Entomology Department, Rothamsted Experimental Station, Harpenden, Herts. AL5 2JQ, Great Britain.

A. P. Bowmaker,
Zoology Department, University of Rhodesia, P.O. Box MP. 167, Mount Pleasant, Salisbury, Rhodesia.

D. G. Broadley,
Umtali Museum, Victory Avenue, Umtali, Rhodesia.

A. C. Brown,
Zoology Department, University of Cape Town, Rondebosch 7700, South Africa.

D. S. Brown,
Medical Research Council Project, P.O. Box 1971, Kisumu, Kenya.
Present address:
Medical Research Council, Experimental Taxonomy Unit, Zoology Department, British Museum (Natural History), Cromwell Road, London SW7 5BD, Great Britain.

B. J. Coetzee,
Department of Nature Conservation, Kruger National Park, Private Bag X 404, Skukuza 1350, South Africa.

S. Endrödy-Younga,
Transvaal Museum, P.O. Box 413, Pretoria 0001, South Africa.

H. J. von M. Harmse,
Department of Soil Science, Potchefstroom University for C.H.E., Potchefstroom 2520, South Africa.

A. D. Harrison,
Department of Biology, University of Waterloo, Waterloo, Ontario, Canada N2L 3G1.

B. J. Huntley,
Serviços de Veterinária, C.P. 527, Luanda, Angola.
Present address:
South African Savanna Ecosystem Project, C.S.I.R., P.O. Box 395, Pretoria 0001, South Africa.

P. B. N. Jackson,
J. L. B. Smith Institute of Ichthyology, Rhodes University, Grahamstown 6140, South Africa.

N. Jarman,
Seaweed Research Laboratory, c/o Department of Botany, University of Cape Town, Rondebosch 7700, South Africa.

R. A. Jubb,
Albany Museum, Grahamstown 6140, South Africa.

D. J. B. Killick,
Botanical Research Institute, Private Bag X101, Pretoria 0001, South Africa.

L. C. King,
7 Ribston Place, Westville 3630, South Africa.

O. Kraus,
Zoologisches Institut und Zoologisches Museum, Universität von Hamburg, Martin-Luther-King-Platz 3, D-2000 Hamburg 13, West Germany.

G. C. Loots,
Institute for Zoological Research, Potchefstroom University for C.H.E., Potchefstroom 2520, South Africa.

O. S. McGee,
Department of Geography, University of Natal, P.O. Box 375, Pietermaritzburg 3200, South Africa.

F. Malaisse,
Laboratoire de Botanique et Écologie, Faculté des Sciences, Université Nationale du Zaïre, B.P. 3429, Lubumbashi, Zaïre.

M. K. P. Smith Meyer,
Plant Protection Research Institute, Private Bag X 134, Pretoria 0001, South Africa.

D. S. Mitchell,
Botany Department, University of Rhodesia, P.O. Box MP. 167, Mount Pleasant, Salisbury, Rhodesia.
Present address:
C.S.I.R.O., Division of Irrigation Research, Private Bag, Griffith, NSW 2680, Australia.

E. J. Moll,
Department of Botany, University of Cape Town, Rondebosch 7700, South Africa.

G. Newlands,
Department of Entomology, South African Institute for Medical Research, P.O. Box 1038, Johannesburg 2001, South Africa.

E. Pinhey,
The National Museum, P.O. Box 240, Bulawayo, Rhodesia.

J. C. Poynton,
Department of Biological Sciences, University of Natal, P.O. Austerville, Durban 4005, South Africa.

A. J. Prins,
South African Museum, P.O. Box 61, Cape Town, South Africa.

B. Puzo,
Department of Geography, University of Botswana and Swaziland, Private Bag 0022, Gaborone, Botswana.
Present address:
Department of Geography, California State University, Fullerton, Ca. 92634, U.S.A.

P. H. Raven,
Missouri Botanical Garden, 2345 Tower Grove Ave., St. Louis, Mo. 63110, U.S.A.

D. C. Rentz,
Department of Entomology, California Academy of Sciences, Golden Gate Park, San Francisco, Ca. 94118, U.S.A.
Present address:
C.S.I.R.O., Division of Entomology, P.O. Box 1700, Canberra City, ACT 2601, Australia.

J. E. Ruelle,
Plant Protection Research Institute, Private Bag X134, Pretoria 0001, South Africa.

H. Ruhberg,
Zoologisches Institut und Zoologisches Museum, Universität von Hamburg, Martin-Luther-King-Platz 3, D-2000 Hamburg 13, West Germany.

M. C. Rutherford,
Botanical Research Institute, Private Bag X101, Pretoria 0001, South Africa.

R. E. Schulze,
Department of Agricultural Engineering, University of Natal, P.O. Box 375, Pietermaritzburg 3200, South Africa.

R. W. Sims,
Zoology Department, British Museum (Natural History), Cromwell Road, London SW7 5BD, Great Britain.

H. C. Taylor,
Botanical Research Unit, P.O. Box 471, Stellenbosch 7600, South Africa.

P. D. Tyson,
Department of Geography and Environmental Studies, University of the Witwatersrand, 1 Jan Smuts Avenue, Johannesburg 2001, South Africa.

A. C. van Bruggen,
Department of Systematic Zoology and Evolutionary Biology of the University, c/o Rijksmuseum voor Natuurlijke Historie, Raamsteeg 2, Leiden, the Netherlands.

E. M. van Zinderen Bakker Sr,
Institute of Environmental Sciences, University of the Orange Free State, Bloemfontein 9300, South Africa.

M. J. A. Werger,
Department of Geobotany, University of Nijmegen, Toernooiveld, Nijmegen, the Netherlands.

F. White,
Department of Botany and Forestry, University of Oxford, South Parks Road, Oxford OX1 3RB, Great Britain.

H. Wild,
Department of Botany, University of Rhodesia, P.O. Box MP. 167, Mount Pleasant, Salisbury, Rhodesia.

J. M. Winterbottom,
Percy Fitzpatrick Institute of African Ornithology, University of Cape Town, Rondebosch 7700, South Africa.

List of selected synonymous toponyms

Present name	Old name
Botswana	Bechuanaland
Bié	Silva Porto
Chipata	Fort Jameson
Dalatando	Salazar
Huambo	Nova Lisboa
Kabwe	Broken Hill
Kinshasa	Léopoldville
KwaZulu	Zululand (partly)
Lake Malawi	Lake Nyasa
Lesotho	Basutoland
Lubango	Sá da Bandeira
Lubango Province	Huíla Province
Lubumbashi	Elisabethville
Malawi	Nyasaland
Maputo	Lourenço Marques
Menongue	Serpa Pinto
Moxico	Luso
Ngiva	Pereira d'Eça
Ngunza	Novo Redondo
Rhodesia	Southern Rhodesia
Saurimo	Henrique de Carvalho
Shaba	Katanga
Uíge	Carmona
Zaïre	Congo Republic
Zaïre River (only within Zaïre)	Congo River
Zambia	Northern Rhodesia

Present name	Possible future name
Rhodesia	Zimbabwe
South West Africa	Namibia

1 The geomorphology of central and southern Africa

Lester King

1.	The birth of Africa	3
2.	Africa today	3
3.	Topographical detail	4
3.1	Rock constitution	4
3.2	Processes of denudation	6
3.3	Differential vertical uplift	7
4.	Geomorphic history of central and southern Africa	9
4.1	Relicts of the Gondwana landsurface in south-central Africa	10
4.2	The post-Gondwana (early Cretaceous) denudation of southern Africa	11
4.3	The great African planation	12
4.4	The Miocene 'rolling' surface	14
4.5	The Pliocene basins	14
4.6	Quaternary changes in African scenery	15
5.	Some effects of Quaternary climatic change in Africa	16
6.	A note on soil mapping	16
	References	17

1 The geomorphology of central and southern Africa

1. The birth of Africa

Africa is about 100 million years old. Prior to that time it was a part of the great southern supercontinent called Gondwanaland, and when that landmass broke up Africa was one of the daughter continents created by the disruption. New seas girdled its east, south and west coasts as they were formed, and as the earliest marine sediments in these locations are mid-Jurassic to earliest Cretaceous on the east, and early Cretaceous on the west, the Indian Ocean basin is slightly older than the Atlantic.

At the time of its birth, therefore, the African continent inherited a landscape that earlier had belonged to Gondwanaland. During the ensuing 100 million years this ancient (Gondwana) landscape has been attacked, and all but been destroyed, by the agencies of erosion: weathering, running water and wind; or has been buried in basins like the Kalahari and the Congo by sediment transported by those same agencies. But to this day, upon the highest plateau terrains in Africa, recognizable remnants of the original Gondwana landscape still survive. They are small, and negligible in area compared with the whole vast expanse of Africa; but their age and history render them specially worthy of study and record (p. 10).

Geomorphologists in Africa must therefore take full cognizance of:

(a) the manner of African landscape development under subaerial conditions, noting the dominance of scarp retreat and the lateral spread of pedimented lowlands. In relatively basined terrains they will note conversely the record preserved by continental deposition, including the use of fossils, if any, for dating.

(b) Very important too are the several tectonic deformations that have affected different regions (moving up, down or differentially) and interrupted the denudational activity.

(c) We shall correlate the information from these two major fields of study to derive a general history of landscape development in Africa from the mid-Jurassic to the present day.

2. Africa today

The continent of Africa is usually described under three major regions: west Africa, north Africa and central-southern Africa. We are now concerned only with the last of these. It is a region of high, interior plateaus indicative of broad subcontinental uplift, with two large interior basins (Congo and Kalahari) each affording a focus for widespread intra-continental-type deposition. These basins are not depressions, they too have been elevated, but they have failed to rise as much as the encompassing watersheds.

The accidents of nineteenth century exploration and annexation have divided the territory of central and southern Africa into a number of very unequal states, with unequal facilities for travel and study within their respective borders. But in the realm of geomorphic study sufficient is known to indicate that all this vast region has evolved since the Precambrian with a truly remarkable uniformity. One does not have to study the topography of each state individually and then build up a complex amalgam of diverse data. Any student familiar with the relatively simple principles and types of landform development over one area will find that other regions, however remote, conform to the same principles and develop similar suites of landforms.

Every territory of southern Africa has, for instance, elevated areas whereon is a feeling of spaciousness. In every direction no higher land is seen, there is only the sky above. So is the Highveld of the Republic of South Africa. All these areas tend to have ancient impoverished soils, no trees, and sparse dry grassland is the common vegetation. As a skyline, it appears to be ruled flat. This is the early Tertiary planation, planed over so long a span of time (at least eighty million years from the Late Cretaceous to the Miocene Period) that it is quite unmistakeable on its morphology alone. This is, indeed, the first geomorphic datum that an observer seeks to identify on entering a new territory, to ensure 'where he is' in the geomorphic history.

Again, about the high interior is a coastal selvedge from 300 to 1000 km wide, the broad structure of which is monoclinal outward, to and beneath the ocean. Because of strong Quaternary uplift of the interior coupled with this outward monoclinal tilting of the marginal regions to a hinge line near the position of the present shoreline, all the major rivers are deeply engorged with waterfalls and rapids relatively low in their courses and none are navigable for any great distance from the sea.

Also, transition from the landforms of one historic group to another is almost invariably abrupt and scarped. The observer, having climbed the scarp of a mountain with an ever-expanding view behind, finds from the summit no corresponding view upon the other side. He will find instead only the gently undulating landscape belonging to an earlier chapter of the geomorphic history. Ascent of the wall-like Drakensberg leads up in this way from the early Tertiary planation at its foot to the Mesozoic landsurfaces which make smooth summit plateaus (Fig. 2).

Rock structure has nothing to do with these Mesozoic summit bevels. Even the Cape ranges (of Triassic orogenic structure) bear a few tiny summit bevels that can be seen from a passing aeroplane.

So uniform is the geomorphic history throughout central and southern Africa that we shall not describe the area region by region, but state the history seriatim, giving with each denudational chapter a few examples from different territories to assist readers in identifying features of the geomorphic history in their own areas.

3. Topographical detail

Topographical detail of any major geomorphic region differs locally according to the variables of (a) rock constitution, (b) the denudational processes operative from place to place, (c) the amount of local uplift in the general differential plan (p. 7), and (d) effects due to climatic change.

3.1 *Rock constitution*

In the history of the earth three major different geologic regimes have prevailed over the area now known as southern Africa.

These are:

(1) Precambrian assembly of the Gondwana super-continent of which this region formed approximately the centre (Fig. 1). The picture at this stage shows an aggregation of numerous primitive cratons (Kaapvaal, Rhodesia, etc.) welding together along orogenic lines with much emission of juvenile basalt from below the crust. Upon a major scale this annealing took place at least five times: c. 3000 million years ago, 2500–2800 m.y., 1850 ± 250 m.y., 1100 ± 200 m.y., and 550 ± 100 m.y. ago respectively (Clifford 1970, Haughton 1963). None of the topographic features of those early events makes any considerable feature in Africa today; though the rocks themselves sometimes govern details of modern topography by the relative hardness of the formations which are etched into relief by the denudation of Tertiary and Quaternary times. For example, the hard ironstone ridges overlooking strike valleys eroded along weaker schists, or the Great Dyke of Rhodesia, or the domed 'bornhardt' mountains of Tanzania. In each of these the topographical expression is a direct result of rock composition and structure.

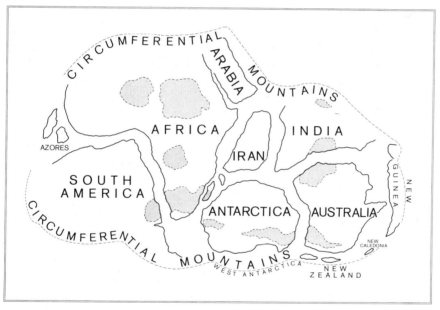

Fig. 1. A reconstruction of the ancient super-continent of Gondwanaland. Areas of correlative Gondwana-type sedimentary rocks stippled. The super-continent broke up in late Jurassic–early Cretaceous times and the parts then dispersed to their present positions.

(2) The second phase (late Palaeozoic to mid-Mesozoic) was marked by broad subsiding basins in which were laid thick sequences of freshwater sandstones and shales with continental fossil floras and faunas. Called the Karoo system in South Africa, these have equivalents in all the other southern continents, and demonstrate the former unity of Gondwanaland. Limited marine incursions are also known. The sediments and their contained fossils record a progressive change of climate through polar glacial to warm desert that is deemed nowadays to have

been due to a drift of Gondwanaland across the South Pole to subtropical regions. From the distribution and succession of these rocks the conclusion has been drawn that the major girdles of world climate were the same then as now (King 1958).

Karoo-type rocks crop out widely across the face of central and southern Africa. They preserve in the fossil state records of two earlier landscapes of Gondwanaland (one sub- and the other intra-Karoo, late Palaeozoic and Triassic in age respectively). Each is known in all the southern territories of Africa, and resurrected examples of the former reveal glacial pavements splendidly polished by the ice. In the modern topography Karoo rocks generally make flat plains with small table-topped hills.

The terminal landscape of Gondwanaland, combining all the southern continents with Iran and India in a single enormous land mass (Fig. 1), was very flat, being composed almost equally of denudational aspects worn down into plains, and still flatter depositional plains accumulated in wide basins, the whole being ringed around by circumferential mountains. As recorded by aeolian sandstones of Triassic age, vast regions were desert or semi-desert. It represented the end of an age, a long age wherein the forces of denudation had triumphed over tectonic disturbances. It was due for change and a new regime.

(3) The change came with drama. In every quarter of Gondwanaland huge emissions of basalt from the subcrust poured out over much of the surface. In the South African region alone, du Toit (1954) has estimated that the volume of basalt involved may have exceeded 200,000 cubic kilometres. This magnificent prelude was followed by the dismemberment of Gondwana into the present southern continents (with Iran and India), followed by the centrifugal drift of these continental-sized fragments to their present relative positions around the globe.

The third phase, from late-Mesozoic to present day, therefore deals with the African continent as an entity. We accordingly state its geomorphic evolution without reference to other continents, though these too have been shown to have pursued similar denudational histories (King 1962).

3.2 *Processes of denudation*

The most active denudational agents in central and southern Africa are running water and the mass-movement of earth materials downhill under gravity. These operate almost everywhere. Wind action has been dominant latterly in and about the Kalahari, obliterating much evidence of the stream activity formerly widespread in the region. Glaciation is known only upon the equatorial peaks Ruwenzori, Kenya and Kilimanjaro.

The evolution of African landscape is along orthodox lines. Incision of rivervalley patterns is followed by the establishment of stable slope gradients and scarp recession. Such hillslope recession is at constant angles determined by qualities of the bedrock and the potency of downslope denudational processes. In certain rock systems there exists over wide districts a close uniformity of mid-slope gradients. This indicates that once a stable gradient has been achieved it is maintained throughout the history of slope recession. Parallel scarp retreat is indeed the dominant form of landscape evolution in southern Africa: it applies equally to the recently formed sides of 'dongas' as it does to the 50 million year old Nyika of

Malawi or the Drakensberg of Natal. The latter, a huge mountain wall over 1000 metres high, is deemed to have receded more than 150 km westward since it was initiated during the late Mesozoic.

The retreat of scarps leaves lowlands and basin plains at their bases. These characteristically exhibit broad pediments sweeping from the scarp-foot towards the lowland rivers. In section, these pediments show at the scarp-foot a slope of 5° to 7° but this flattens basinward in an hydraulic curve to $\frac{1}{2}$° or less. Running water (often in sheets after thunderstorms) thus fashions the pediment surface. Gully-head formation by rill-wash, and mass movement under gravity govern the scarp-face and its retreat.

Between the tops of retreating scarps, upon interfluves, remnants of earlier landscape surfaces remain without significant alteration until opposing scarps meet and the interfluve is rapidly lowered. Such remnants of ancient surfaces retain their ancient soil profiles, e.g. laterite, a demonstration that direct down-weathering has been minimal. In contradistinction to the pronounced denudational activity upon steep scarp faces, no natural process exists to cause any marked change upon a landscape of almost negligible declivity.

Finally, after prolonged denudation, pediplains result. Africa is remarkable for the vast extent and smoothness of its pediplains, the most extensive of which was formed during the early Tertiary.

Here we arrive at one of the most important canons of landscape development, especially well exemplified in Africa. That, with the dominance of relatively rapid scarp-retreat and the absence of a universal active agent of downwearing (for residual soils on smooth near-horizontal landscapes are themselves end-products which change little with time) continental landscapes usually include one or more ancient surfaces at levels above the latest (modern) developing basins and lowlands. Geomorphic history is determinable, and is preserved for a very long time in multi-cyclic landscapes. The several stages of this history may, with experience, be readily distinguished by geomorphologists (p. 10).

3.3 *Differential vertical uplift*

In the beginning of phase 3, when Africa assumed independent continental status, gentle undulations appeared over the face of the interior plateau; and the new continental margins were strongly monoclinal, flexed down to and beneath the sea. This is the key to African geotectonics (Fig. 2). By it the major river catchments were prescribed. The continental divide which trends from the southwestern Cape Province eastwards to the great escarpment of the Drakensberg, which in turn carries the divide northwards around the headwater systems of the Orange and Vaal Rivers, is succeeded by a watershed passing west along the Witwatersrand to the Kalahari country, whence the divide reaches the Benguela highlands to pass round the catchments of the Limpopo and Zambezi Rivers. Thence along the Lunda axis of northern Angola it passes east once more between the Zambezi and affluents of the Congo, until south of Lake Victoria it bifurcates, the branches passing northwards on either side of Lake Victoria to enclose between them the northerly drainage of the Nile. This primitive divide probably dates from the earliest days of Africa, and some of it may have been inherited from Gondwanaland.

During Cretaceous to Recent time, several tectonic movements affected the face of Africa. These were vertical only in expression; but the amount of uplift differed very much from place to place and the deformation (which is recognized by the tilting of planed Tertiary landsurfaces) was gently differential. A common angle of tilt is one degree or less, but in the great spaces of Africa the vertical displacements often amount to several hundreds of metres. In general, the major uplifts were directed along linear axes, while the areas of lesser uplift form tectonic basins. The former make interfluves, the latter make centres for continental-type sedimentation either alluvial (Congo) or arid (Kalahari).

Many of the Tertiary movements did not coincide in position with the earlier (primitive) deformations. Some of the rivers are then antecedent; thus the Orange River traverses the Griqualand–Transvaal axis below Prieska. On the other hand the Lunda axis of northern Angola (though very flat) made a new divide between the present Zambezi headwaters and the affluents of the Congo. Older headwaters and source ranges of the Zambezi now appear 200 km north of the present divide. The rivers between, back-tilted, now flow in reversed direction through swamps to the Congo drainage.

Fig. 2. Section through Windhoek and Durban to show the present disposition of the 'Gondwana' and 'African' planations; and the Mesozoic and Tertiary warpings of Africa with coastal monoclines east and west, mountainous divides, interior plains and the central Kalahari basin with Tertiary sediments (Kalahari Marls below, and late Tertiary–Quaternary formations above).

Projections of the Gondwana surface are shown by broken line. The African surface is underlined or projected by dot and dash.

1. Atlantic coast; 2. Namib; 3. Brandberg; 4. Omaruru Flats; 5. Erongo Mt; 6. Windhoek; 7. Damaraland Plains; 8. Kalahari marls overlying Gondwana surface and topped by calcrete of the African planation (III), covered by late Tertiary and Quaternary Sands; 9. Kaap Plateau with crossing of the Gondwana and African surfaces to pass (a) below and (b) within the Kalahari sediments; 10. Vaal River incised below the Highveld (African surface); 11. Gondwana summit bevels of Lesotho highlands; 12. Drakensberg escarpment with Mesozoic bevels above and African planation on the 'Little Berg' below; 13. Natal Monocline; 14. Durban. Within the triangular section after the crossing, marine Cretaceous and Tertiary sediments are visible on the KwaZulu and Moçambique coastal plains to the north.

Crosswarping on the Kalahari–Rhodesia axis has likewise diverted the lower Okavango from the Limpopo to the Zambezi drainage and developed the swamps of Makarikari.

Some of the major uplifted axes and domes of East Africa have cracked open at the crest to form the Rift Valleys. This has happened several times, with maximal activity during the Quaternary era, so that the resulting scarps and associated landforms are comparatively fresh.

These morphotectonics do not imply stretching of the earth's crust. They are best explained as a function of vertical laminar displacements of the continental crust (cymatogeny), which also readily explains the abundant volcanicity characteristic of the rift valleys.

Within the major basins of the Congo and the Kalahari, Tertiary river systems deposited widespread sedimentary series of continental type. These are alluvial in the Congo, alluvial and later desert in the Kalahari. While most of these are clastics derived from surrounding areas, they vary much according to the agencies of weathering and transportation, being coarse and in poorly-assorted beds at the beginning of each stage and becoming sandy, silty or clayey in nature (e.g. Yangambi Series of the Congo or the Kalahari Marls) as each cycle of denudation proceeds. The end of each series is marked by unconformity, possibly over duricrust which is coeval with that over the surrounding landscape of advanced denudation. Such is the early to mid Tertiary calcrete which overlies the Kalahari Marls and corresponds with widespread calcrete overlying rocks in neighbouring parts of the northern Cape and the western Transvaal. The lateritic early Tertiary soils of Natal also correspond in time and mode of origin.

Earlier we have noted that when Africa was formed its new continental margins were strongly monoclinal, flexed down towards the sea (Fig. 2). This is the most important morphotectonic feature of all, for when Africa came under denudation the resulting detritus was shed into the sea about the new continental margins. Offshore, therefore, accumulated a sequence of Cretaceous to Recent marine sedimentary formations, most of which are highly fossiliferous and of prime importance for the dating not only of the sediments but also of the corresponding denudations upon the lands which provided renewed supplies of detritus from time to time. This provides the key to a geomorphic history which follows shortly in the text.

4. The geomorphic history of central and southern Africa

The central and southern parts of Africa have been uplifted episodically during Cretaceous, Tertiary and Quaternary time. Notable (but gentle) differential warping and tilting accompanied each uplift so that all pre-existing landscapes were lifted up into new elevations and attitudes where they could be operated upon anew by the forces of denudation.

Between these tectonic episodes were relatively prolonged quiet intermissions when denudation and deposition were free to operate to the new levels induced by the warping. These chapters, or cycles, of denudation have been identified by geomorphologists, and serve as a means for classifying the multitudinous landforms of the subcontinent into a relatively simple system of historically contemporaneous landscapes. The landscapes of different cycles are commonly separated by major scarps or zones of dissection. As each scarp or zone is in active retreat, the upper (and older) landscape is being progressively consumed while the lower (and younger) landscape is correspondingly extended.

All the cycles may thus be co-existent at any one time, though each has been initiated individually at times differing by millions of years. Moreover, where tilted uplifts have been great, and exceed (perhaps by multiples) the critical height for the local rock systems, more than one new landsurface may be generated following the tectonic episode. This was particularly so during the Pliocene denudation which sometimes displayed a multiplicity of related and essentially synchronous facets amid the corresponding landforms.

Such a history of denudations following episodes of tectonic uplift reduces the chaos of landforms over the subcontinent to a simple, clear system. Moreover, the fact that the several surfaces bend down in coastal monoclines to pass as unconformities into an offshore sequence of fossiliferous sediments enables each denudational chapter to be dated with tolerable accuracy.

The cyclic landsurfaces that have been identified are:

I — Gondwana landscape (Jurassic or older; pre-African continent).

II — Post-Gondwana landscape (Early and middle Cretaceous).

III — African landscape (Late Cretaceous to mid-Tertiary. Identifiable by its extreme planation).

IV — The post-African or 'rolling' surface (Miocene).

V — Valleys and basin plains younger than IV. The coastal plain round much of Africa belongs to this cycle (Pliocene).

VI — Deep river valleys with waterfalls, coastal dissection, tectonic arches and rift valleys (Quaternary).

These chapters of the geomorphic history are now discussed seriatim (cf. King 1967).

4.1 Relicts of the Gondwana landsurface in south-central Africa (I)

The oldest landscape that could feasibly be found in Africa would be a survival of the pre-African landscape of Gondwanaland.

At the time when Gondwanaland broke up, giving birth to the several continents of the southern hemisphere, with Iran and India, most of its topography was a flat plain of mixed denudational and depositional origin ringed about by a peripheral girdle of mountain ranges, outside which descent to the sea was steep (Fig. 1). The several daughter continents therefore inherited a rather featureless terrain with mountains upon one side only according with its aspect in Gondwana. The other coastal regions were of fractured origin and all were apparently monoclinal in form.

As the continents have since remained in great part above the sea, the original Gondwana landscape has been attacked ever since by the agents of denudation which have cut new landscapes at lower levels. It may be expected that after so great a lapse of time (over 100 million years), and such wide denudation, few relicts of the Gondwana surface should remain identifiable in modern landscapes, but they do exist and though the remnants are small, their importance in the geomorphic history is such that they warrant special notice. All these relicts illustrate a basic principle of geomorphology — that when a landscape is reduced to a flat plain, with low stream gradients it may remain in this form for an indefinite period until it is uplifted, and perhaps tilted, by tectonic movements which create new base levels to which the streams and rivers can operate. Most of these relicts preserve their original flatness, which sometimes is most striking. They always appear as the highest landscapes in any vicinity especially where they stand on or near continental divides or upon more recent axes of tectonic elevation. Nonetheless, closer inspection often reveals a 'microrelief' etched into them by continuous exposure to the weather for 100 million years, and no original Mesozoic soil is known to have survived. But some of the relicts bear small residual relief which still defines the original Mesozoic watersheds. This is so in Lesotho, where Ntabantlenyana, the highest and oldest point in southern Africa, stands upon a small ridge above the usual level of the Gondwana plateau.

Summit plateaus whereon Gondwana relicts have been identified are: in

Lesotho behind the Drakensberg; on the Windhoek and Benguela highlands; and upon the flanks of the East African rift valleys, notably the Nyika Plateau of Malawi. All the ancient relicts are indeed in large measure bounded by scarps.

Where the Tertiary history of the sub-continent has on the contrary involved subsidence or a minimal rise compared with adjacent uplifts, the Gondwana surface disappears from sight as an unconformity beneath younger sedimentary deposits. Internal basins such as Kalahari, Congo and perhaps Chad, have as a floor the Gondwana surface cutting across yet older rocks. Few dateable Jurassic sediments occur in south and central Africa to record the passage of Jurassic time. However, the Lualaba series of Zaïre has fossil fishes allied to the Jurassic marine fish of East Africa, and the Gokwe Series of Rhodesia also belongs here.

At the coasts, repeated outward monoclinal tilting carries the Gondwana planation downward in the continental shelf to form the floor upon which is laid a succession of late Mesozoic and Tertiary marine formations. The oldest of these defines the youngest possible age for the Gondwana landscape – late Jurassic or earliest Cretaceous.

4.2 The post-Gondwana (early Cretaceous) denudation of southern Africa (II)

Following the disruption of Gondwanaland and the roughing out of the modern outline of Africa, new base levels of denudation came into operation and a new cycle of landscape development was initiated. Operating by river incision, scarp retreat and the development of wide pedimented plains, this denudation gave a new aspect to Africa; but it did not everywhere destroy the older Gondwana landsurface which survived in the southern part of Africa in the form of extensive plateaus standing as much as 500 metres above the level of the new 'post-Gondwana' landscape.

The best region for study of this landscape is in high Lesotho where it makes most of the high plateau country. It appears atop the mountain wall of the Drakensberg where it makes the undulating crest of the escarpment. In this region, too, it bears many planed relicts of the older Gondwana landsurface. Both surfaces decline southward through the territory: on the older (Gondwana) surface from a maximum exceeding 3300 metres down to about 2600 metres. On the opposite side of South Africa both the Mesozoic planations are found upon the Windhoek and other highlands where they dip gently southeastwards towards the Kalahari basin. Topographic relations are thus similar to those in Lesotho. Followed northwards into Angola, the same conditions prevail in the Moçâmedes and Benguela highlands where very smooth Gondwana relicts stand above more extensive areas of the Cretaceous planation which themselves rear above great scarps separating the mountains from the great Tertiary plains of Africa.

Relicts of both the 'Gondwana' and 'post-Gondwana' planations continue into central Africa where they are generally aligned along the flanks of the rift valley system. They occur about Lake Tanganyika and Lake Rukwa, near Mbeya, above the Livingstone Mountains and about Iringa in Tanzania whence they continue northward into Kenya. Of particular interest are the Nyika (Gondwana) and Vipya (Cretaceous) plateaus of Malawi standing thousands of metres above Lake Malawi, for it was these relicts which first led Frank Dixey to the cyclic denudational interpretation of landscape in Africa in 1938 (cf. Dixey 1942).

Areas in the Republic of South Africa where Cretaceous summit bevels are known without the Gondwana surface overlooking them are: the Dullstroom area, Somerset East and the plateau above the Nuweveld Escarpment. The Gamsberg in South West Africa is similar, and on the farm Kangnas in Namaqualand, where a relict of the Gondwana planation is known, deposits in valleys incised into this surface have yielded bones of large Cretaceous dinosaurs.

Northwards through Africa, the two Mesozoic planations converge and through north and west Africa merge into a single planation (Cretaceous) of marvellous smoothness. This is so also in the northern Cape Province where the surfaces converge towards the Kalahari depression, where the surface acts as the floor for late Cretaceous and Tertiary continental sediments. A similar configuration appears at the southern end of the Congo basin.

The intermontane valleys of the Cape ranges were affected at this time by strong vertical faulting (e.g. Worcester Fault 480 km long with a vertical throw exceeding 3000 metres), and the depressions were part filled with extensive outcrops of early Cretaceous, Enon conglomerate followed in the east by the Wood Beds and the marine Sundays River Beds. At Empangeni in KwaZulu, too, was faulting with 3000 to 3750 metres of vertical throw. This great scarp, being near the coast, was early obliterated by erosion, for Upper Cretaceous marine strata transgress both sides evenly.

4.3   The great African planation (III)*

The landscape features described so far exist now only within small areas. Their historical significance is large; but their actual occurrence is small.

The great plains and savannas for which Africa is famous, however, are younger, and are widespread over vast areas. In the interior of Africa they stand often around 1300–1500 metres but they rise commonly towards the watersheds where they may attain an elevation of as much as 2000 metres. Most of this country is grassland, savanna and woodland and in the days before European occupation it was the habitat for vast herds of game animals.

The Highveld of South Africa, the uplands east of the mountains of South Africa and Angola, the watershed region of Rhodesia, much of the high country around the head of the Congo drainage, high wide plains in eastern Zambia and adjacent Malawi with southern Tanzania, the lands to either side of Ruwenzori, and the Kenya Dome all show this surface extensively, usually as broad interfluves. It affords a truly marvellous example of natural planation.

The fashioning of this great 'African' planation occupied an immense lapse of time, 80 million years from the mid-Cretaceous until the mid-Tertiary; and there is no evidence of noteworthy tectonic disturbance in Africa during this time.

Under the agencies of denudation the southern lobe of Africa was worn down to a single vast pediplain which in its wide development destroyed (except in a few favoured localities) all traces of the earlier Mesozoic surfaces, so that it appears as the oldest demonstrable land surface for the continent of Africa.

Because of the immense span of time during which it was developing under scarp retreat (with almost a negligible downwearing) the 'African' surface un-

---

* This widespread planation, late Cretaceous to mid-Tertiary in age, is known from all seven continents, usually as plateaux covered by poor, aged soil and equally poor grassy vegetation. From its usual appearance and vegetation it may well be called, world-wide, 'the Moorland cycle'.

derwent extreme soil evolution. Practically all the soluble constituents were removed, and there remained a most characteristic hard and thick carapace of ferricrete, calcrete or occasionally bauxite, according to the chemical composition of the bedrock from which the soils were derived. Lateral transport of soils seems to have been minimal and these duricrusts, as they are often termed collectively, are almost wholly residual. The presence of this deposit serves frequently to identify the smooth 'African' planation. Later deposits of lateritic type, thinner and commonly pisolitic, appear also locally in association with the next surface to be described. In part, they are often derived from debris of the earlier duricrust. Still younger iron pisolites are forming even to the present day where iron-rich solutions are being oxygenated under the atmosphere. These latest deposits may sometimes be distinguished because they contain Stone Age artifacts.

The first endeavour of a geomorphological surveyor should be to ascertain whether in his area any fragment of the 'African' planation exists upon the high ground, for this surface serves as a datum. Its presence informs the observer at what stage in the geomorphic history his activities are directed, whereas its absence indicates that the landscape he is studying is younger. Moreover, by its altitude and attitude the 'African' surface records the sum of all local vertical movements and deformations of the surface during late Tertiary and Quaternary times. Familiarity with the early Tertiary 'African' planation is all important in the study of the continental landscape. And from it all subsequent landforms have been carved.

In the two major sub-continental basins, Kalahari and Congo, prolonged sedimentation reigned during the time the rest of the continent was being reduced by erosion and shed its detritus into those basins. The Kalahari Marls vary in thickness according to irregularity of the floor and locale, but are commonly 100 metres (and may exceed 300 metres) thick. Drusy chalcedonic depositions are often characteristic of this series. At the top is a thick calcrete like the duricrust which covers the surrounding erosional surface and affords a terminal time 'marker' upon the erosional and depositional surfaces alike.

In the Congo basin also sedimentation reigned during the early Tertiary, and the 'grès polymorphe', equivalent to the Kalahari Marls, accumulated. It, too, is covered by a thin, lateritic gravel, like that found over a large area of the erosional 'African' landsurface between the Lualaba and the rift zone of eastern Zaïre.

Of outstanding importance is the marginal down-flexing of the 'African' landsurface towards the sea on both the east and west coasts so that it passes, in Zululand and Moçambique, in South West Africa and Angola, as an unconformity beneath lower to mid-Miocene marine strata, richly fosiliferous and providing a terminal date for the great planation.

At a few localities in East Africa (Lake Albert, Rusinga Island) also the surface passes below local sediments containing Miocene mammal remains. Subsequent to the mid-Tertiary uplift the early Tertiary planation soon underwent incision by wide-floored valleys so that nowadays more of the interior plateau belongs to these valleys than to the smooth early Tertiary planation. This planation may still be identified, however, on the flat interfluves and as occasional flat-topped hills, both of which commonly preserve the typical duricrusted soil profile of the early Tertiary. On the Highveld of South Africa, in Zambia, and on the central highland of Tanzania around Tabora these conditions are extensive.

4.4 The Miocene 'rolling' surface (IV)

About the early Miocene the attitude of the African landsurface was widely disturbed. Throughout southern Africa a number of axes of uplift came into existence between which lay lowland areas which were not necessarily of subsidence but had merely risen by a lesser amount than the axial uplifts. On either side of the axes, rivers and streams were accelerated and incised themselves so that while the flat African landscape remained on the crests of the arches it was destroyed upon their flanks and a new, much more irregular landscape developed under the new cycle of erosion.

This Miocene landscape is a characteristically rolling surface sometimes bearing flat-topped relicts of the earlier African planation. Over large areas relief is generally about 300 metres with more in the highlands. However, towards (a) the coasts, and (b) the interior tectonic basins, relief decreases and the early Tertiary and Miocene surfaces converge.

Most of the Miocene axes have remained as watersheds since mid-Tertiary time, and some have later been rejuvenated. In some of the basins sedimentary deposits accumulated complementary to the Miocene denudation about their margins. The Kalahari Marls of the early Tertiary, topped off by calcrete, were overlain by the plateau sands of presumed Miocene age, and in Zaïre the 'sables ocres' succeed the early Tertiary 'grès polymorphe'.

Some of the arches and domes developed crestal rifts. This was the beginning of the great rift valley systems of central Africa, though some of the rifts were sited along older structures of this kind which had been in existence even in Precambrian times. The rifts are commonly 40 km wide; but the crustal arches on which they occur are often as broad as 400 km. Rift depressions of this epoch sometimes filled up with gravels and sands containing early or mid-Miocene mammalian bones. On both east and west coasts, where monoclinal tilting occurred, extensive early to mid-Miocene sediments are often richly fossiliferous, and serve to date the land denudation which supplied the sediment.

The aspect of this landsurface is generally uneven, or 'rolling'. It stands at lower level than nearby remnants of the early Tertiary planation, but generally conforms with the tectonic deformation of that planation. The two surfaces rise and fall together, and generally occur in the same localities. Both stand above the younger basins of Pliocene age and well above the deeply incised gorges of the main rivers.

The soils of this 'rolling' Miocene surface are less extremely differentiated than the duricrusts of the early Tertiary surface, but nonetheless are old, considerably differentiated into sands and heavy clays and are generally infertile, which is not unexpected in soils of approximately 20 million years in age.

4.5 The Pliocene basins (V)

About the end of the Miocene, Africa was widely and greatly uplifted once more. The longer rivers and their tributaries were rejuvenated over great distances, and the retreat of valley-side scarps from these waterways created large numbers of local basins at various altitudes. A single river measuring not much longer than 150 km from the coast might have two or three of these basins along its length. Although the basins appear at different elevations, all are of essentially similar age and they are very alike in appearance.

The basins were formed by powerful scarp retreat and the resulting pediments which form the basin floors are wide and sweeping. There are many such basins in Zambia and Rhodesia.

The landforms of these Pliocene (V) basins cover a greater area of Africa than those of any other cycle. They are the landforms most actively developing at the present time. Scarps are retreating, pediments are being re-graded, 'dongas' are alternately being incised or filled up, and all the time soils are being destroyed and new soils constantly being made so that no soils are of great age, and a map of soil types looks like a patchwork quilt. Soil types generally reflect bedrock composition, but also are well enriched by the products of plant decay and are fertile. They are the best soils for agriculture provided that the bedrock type be suitable. Moreover, they are situated in the valleys adjacent to water supplies, unlike the soils of the ancient landsurfaces which are upon the highlands and watersheds.

River terraces, drainage changes, landslides and other minor times of the scenery often provide subjects for detailed research.

The Kalahari and Congo basins were once more the sites of extensive deposition by rivers which overflowed in flood to cover vast areas. The corresponding sandstones of the Kalahari are called the Pipe Sandstone and in the Congo the Plio-Pleistocene Sands.

Around the coasts monoclinal tilting permitted encroachment by the Pliocene sea, which then fashioned on the sea floor the coastal plains which have emerged in Quaternary time.

By the end of the Tertiary the aspect of southern Africa as a whole was recognizably ancestral to the present.

4.6 Quaternary changes in African scenery (VI)

The closing stages of Pliocene landscape-making were interrupted by vast tectonic upheavals which raised the interior plateaus by perhaps 1000 metres and also steepened the marginal monoclines to the coast by hundreds of metres. Many of the old axial uplifts were enhanced until the arches cracked open at their crests to complete the present system of rift valleys extending through central and east Africa. Local drainage patterns were often disrupted and areas of centripetal drainage were created. The ensuing lake bed deposits have preserved a fascinating record of the fossil African fauna, including early man. New tectonic axes sometimes created new watersheds at which some of the old river courses became antecedent, others were defeated (e.g. Okavango River at Makarikari).

Where the marginal monoclines steepened the courses of the rivers large and small, deep gorges were incised through which the rivers now run in magnificent scenery. Many of these gorges begin upstream at mighty waterfalls, several of which on the major rivers have advanced far upstream into the interior (e.g. Victoria Falls on the Zambezi; Aughrabies Falls on the Orange and Rua Cana on the Cunene).

This was the most severe tectonic deformation that Africa has undergone, and geomorphologically it is very young. Much of it took place during the occupation of the region by the *Australopithecines* and their successors Stone Age Man.

Over the Kalahari basin, from the Orange river northwards through Botswana into Angola, Zambia and Rhodesia, the chief Quaternary activity was the spread of an almost uninterrupted mantle of red to grey desert sand varying in thickness from 15 to 60 metres. The Congo basin is a vast centre of Quaternary alluvial deposition, not quite filled in as shown by Lakes Leopold and Tumba. Present rivers flow entrenched into these alluvials.

5. Some effects of Quaternary climatic change in Africa

In many countries and more books much emphasis is laid upon the topographic effects of Quaternary climatic change. Africa is no exception, and much has been written of arid and pluvial periods within its Quaternary history. But no single coherent scheme applicable even throughout the area of central and southern Africa has yet been forthcoming (see Chapter 6). Changes of drainage, or the waxing and waning of centripetal lakes, have nonetheless afforded useful concepts in explaining the evolution of mammalian faunas and the archaeology of Stone Age Man. There was no Pleistocene Ice Age in Africa, so except upon the trio of equatorial peaks – Ruwenzori, Kenya and Kilimanjaro – there are no recently glaciated landforms.

Apart from desiccation, no great topographic changes appear due to Quaternary climatic changes. Desert areas have contracted or expanded, lakes have waxed or waned, and 'dongas' have been either incised or re-filled with land waste accordingly with 'wet' or 'dry' periods. Thus over the central Transvaal 'dongas' were incised during the Palaeolithic Age. Towards the end of the Old Stone Age aridity set in and the 'dongas' began to re-fill with land waste. The lowest strata contain the last of the large stone hand axes. Aridity was long continued and the succeeding layers, to a total thickness of eight metres, contain a developing Middle Stone Age culture. Later, as the climate ameliorated, these deposits were trenched by the revived 'dongas' and now afford many interesting archaeological sites. Africa is rich in such occurrences, which help in the interpretation of landscapes by dating certain features. Greater topographic changes are due to tectonic activity, e.g., along the rift valleys.

But plants are more sensitive barometers than rocks, and great changes may ensue in the plant cover of the lands from even a few degrees of temperature or a critical change of rainfall. Thus, with only very minor changes of geography, great tracts of land in Africa have changed from forest to savanna or from savanna to desert during Quaternary time (see Chapter 6). In sympathy with the changes of climate and food supplies, large migrations have taken place among browsing and grazing animals and some may have changed their habits. Of none is this more striking than that of the ancestry of man.

Several of the primates were forest dwellers, surrounded by food supplies throughout their lives. But one of the smaller members, *Australopithecus*, was not vegetarian, he was omnivorous and especially on occasion carnivorous. When the forests shrank and the grassveld spread he found he could stand on his hind legs, look over the grass and see his next meal at a distance. At first a carrion-eater, he became a hunter and developed skills of his own, especially the invention of tools and weapons which enhanced his natural prowess. Amid the changing environment of African flora, hand and brain improved in co-operation, though the scenery of hills and valleys, scarps and pediments evolved but slowly.

6. A note on soil mapping

When soils were first mapped (in Russia) over a century ago, the conclusion was reached (by comparison) that soil maps resembled climatic maps for the same region, and therefore climatic agencies were dominant in soil making. This general conclusion overlooked two valid points – (a) that European Russia has very little relief, and (b) that it consists widely of uniform types of sedimentary rock.

Only much later, in Britain, was it found that soil maps in that country resembled geological maps, and the conclusion was drawn that bedrock types exercised the governing influence on soil formation.

Later still, Australian soils scientists making maps found that neither of these factors afforded a coherent and complete explanation of the pattern of Australian soils. Further research showed that geomorphic history explained much that was present in the soil maps, and the work of the Division of Soils, C.S.I.R.O. in their series of reports on Soils and Land Use conducted by teams of four men – a geologist, a chemist, a soils expert, and a geomorphologist, harmonized much that was necessary to an understanding of Australian soil types – on a basis of geomorphic history.

Some work in South Africa, too, has shown that geomorphic history should be considered when field mapping of soils is undertaken, and that it frequently affords useful and even decisive explanations of the distribution of soil types. I am reminded of the story of the expert who, when he compared soil maps of Uganda and neighbouring Congo (Zaïre), found a serious lack of correlation on opposite sides of the frontier. Going into the field himself to check, he found the facts correct; but whereas the Belgian administration had exercised a policy of nature conservation, the British Protectorate had permitted over-grazing to the extent that in twenty or thirty years nearly two metres thickness of top soil had disappeared!

Soils improve so long as rock breakdown furnishes inorganic plant nutrients and there is addition of organic breakdown, humus and humic acids, Under Australian or African conditions a soil can form in a few hundred years and be at its best at a thousand years or more. Between 1000 and 5000 years it may reach its optimum, after which the soluble nourishing materials become progressively more leached and the clay constitution increases and becomes denser. Fertility slowly declines to about 30,000 years and by 100,000 years a very stiff, almost unworkable clay, may result. Mottling, with separation of ferricrete, signals the beginning of laterization; but as soil removal often goes on pari passu with soil formation the appearance of this stage may be long delayed. On flattish terrains (old planations) it will be most in evidence, and there will be found the main duricrusted soils whose age can be measured by millions of years. Such infertile soils therefore sometimes help to identify ancient planation surfaces.

References

Clifford, T. N. 1970. The Structural Framework of Africa: In: African Magmatism & Tectonics, ed. Clifford & Gass, Edinburgh.
Dixey, F. 1942. Erosion cycles in Central and Southern Africa. Trans Geol. Soc. S. Afr. 45:151–158.
Du Toit, A. L. 1954. Geology of South Africa. Oliver & Boyd, Edinburgh. 3rd Ed.
Haughton, S. H. 1963. Stratigraphic History of Africa South of the Sahara. Oliver & Boyd, Edinburgh.
King, L. C. 1958. Basic Palaeogeography of Gondwanaland during the late Palaeozoic & Mesozoic Eras. Geol. Soc. Lond. Quart. Journ. 114:44–70.
King, L. C. 1962. The Morphology of the Earth. Oliver & Boyd, Edinburgh. 2nd ed. 1967.
King, L. C. 1967. South African Scenery, 3rd ed. Oliver & Boyd, Edinburgh.

2 Climatic indices and classifications in relation to the biogeography of southern Africa

R. E. Schulze and O. S. McGee*

1.	Introduction	21
2.	Light	21
2.1	The importance of light (solar radiation) in biogeographical studies	21
2.2	Radiation patterns in southern Africa	22
2.3	The influence of topography on solar radiation	24
3.	Temperature	25
3.1	The importance of temperature in biogeographical studies	25
3.2	Temperature distributions in southern Africa	26
4.	Moisture	29
4.1	The importance of moisture in biogeographical studies	29
4.2	Precipitation patterns over southern Africa	29
4.3	The rainy seasons of southern Africa	33
4.4	The dry seasons of southern Africa	33
4.5	Rainfall interception by vegetation	35
4.6	Fog – its importance and occurrence in southern Africa	35
4.7	Snow	36
5.	Climatic indices and classifications	37
5.1	Background	37
5.2	The Köppen climates of southern Africa	37
5.3	The Holdridge Life Zone System applied to southern Africa	40
5.4	Vegetation distribution and the Thornthwaite indices: a brief review	43
5.5	Potential evapotranspiration, thermal efficiency and thermal regions	44
5.6	Annual water surplus	46
5.7	Annual water deficiency	46
5.8	Moisture regions	48
6.	Conclusions	49
	References	50

* The authors wish to thank Mrs. H. Smithers and Mr. B. Martin, both of the University of Natal, Pietermaritzburg, for their assistance with the compilation of the maps. Many of the data were collected in the U.K. by one of the authors (R.E.S.) while a C.S.I.R. Overseas Grant holder.

2 Climatic indices and classifications in relation to the biogeography of southern Africa

1. Introduction

For land-based plant communities four divisions of potential restraints on growth exist, viz. climatic, topographic, edaphic and biotic restrictions (Watts 1971). Of these four, climatic restrictions are usually the most important – certainly for biogeographical and ecological studies on a subcontinental scale – for plants depend directly or indirectly on the atmosphere for certain fundamental materials, for their successful growth and reproduction. Different species, for instance, vary in their minimum requirements for, and in their tolerance of, particular climatic conditions and these conditions therefore play a major role in determining where a particular plant can or cannot exist (Tivy 1971).

The climatic factors of greatest importance in vegetation development are light, temperature and moisture, all of which vary subcontinentally as well as on a meso- and micro-scale.

There is a vast literature dating from the past century testifying to attempts made at climatic classifications, primarily by biologists – for instance by de Candolle (1855), Grisebach (1866) or Linsser (1867) – who recognized the existence between vegetation distribution or physiological response and certain climatic elements. In much of this and other earlier work, however, climatic parameters were treated separately, little attention being paid to the way in which they might interact. Today, however, the climatic interactions and the fact that climatic parameters operate in combination to produce homogeneous environments in which certain plant communities can attain importance are stressed.

This basic concept of climatic interaction, in terms of which all biologically vital processes are activated by light, but can only take place in the presence of water, also underlies the approach adopted in the present treatment (which is essentially a review) of the climatology of southern Africa. The ecologically important climatic parameters are first discussed individually in their southern African context before proceeding from a delineation and discussion of the Köppen climates of the subcontinent to an application of the Holdridge Life Zone classification of southern Africa and finally to an evaluation of various Thornthwaite moisture balance indices in the study area. Throughout the text an attempt has been made to relate the climatic factors to their ecological significance or response. Also, wherever possible, maps and diagrams illustrate the text, facilitating the extraction and interpretation of climatic data on a subcontinental scale.

2. Light

2.1 *The importance of light (solar radiation) in biogeographical studies*

The energy resources of all ecosystems are ultimately dependent upon the quantity of incoming solar radiation intercepted. As already mentioned, most vital biological processes (such as photosynthesis, photoperiodism, phototropism, deciduous plant leaf shedding, vertical zonation of plant groupings or sucrose formation) are activated by light. As with temperature (discussed in the following section) the light factor is most important in the details of plant morphology and ecology, and some caution should therefore be exercised in suggesting too close a degree of interaction between solar radiation patterns and vegetation distribution on a regional scale (Van Riper 1971, Watts 1971). Biogeographically it is more meaningful to view solar energy receipt in conjunction with other climatic indices and in relation to the effect of varying topography.

As the most fundamental climatic parameter that is present in the total environment of plants, an overall review of solar radiation receipt in southern Africa is therefore given, followed by an examination, also on a subcontinental scale, of the influence of topography on the light factor.

2.2 Radiation patterns in southern Africa

Systematic solar radiation measurements in southern Africa commenced in the early 1950s and pioneering research on the spatial patterns of radiation, using very little data, was first published in 1957 by Drummond & Vowinckel. For the present study, however, radiation maps have been redrawn using SI units and subsequent data from the South African Weather Bureau (S.A.W.B. 32 of 1968 for data up to 1962), supplemented by data from the S.A.W.B. Annual Radiation Reports of 1963–1973 as well as from Torrance (1972a) for Rhodesia, from Spain (1971) for Zambia and from Griffiths (1972) for Moçambique and Malawi.

The interpretation of maps of incoming radiation is sometimes rather difficult because altitude, sunshine duration, atmospheric moisture, cloud cover and dust content all exercise an influence on the amount of solar radiation received at a particular location. Of these factors generally only sunshine duration is known with sufficient accuracy for a successful investigation into the variability of incoming radiation.

The main factor determining the radiation patterns of southern Africa in winter (June–August) is the decrease in cloudiness south of 30°S (Drummond & Vowinckel 1957). Accordingly, in Fig. 1a the maximum solar radiation occurs in a zone extending from northern South West Africa, northeastwards to northern Zambia, with radiant flux densities exceeding 190×10^5 J m^{-2} day^{-1}. South of this zone of maximum radiation the western sector of the subcontinent receives more radiation than the eastern sector, principally because of the extremely dry subsiding air of the South Atlantic Ocean Anticyclone. Along the western coastal belt radiation decreases markedly in a sharp discontinuity to a mean of 130–140 $\times 10^5$ J m^{-2} day^{-1} (Fig. 1a). The large amount of fog (qv) caused by the cold Benguela Current as well as the cloud systems of the westerlies in the winter rainfall region of the southern Cape are the main factors responsible for the change in radiation.

With the sun in its southernmost position in summer (December–February) the zone of maximum radiation intensity moves southwards from its winter position to southern South West Africa (Fig. 1b), where the continued presence of the South Atlantic Ocean anticyclone results in small amounts of cloud in summer as well. Radiation flux densities there are of the order of 300×10^5 J m^{-2} day^{-1}. It is of interest to note that this zone of maximum radiation intensity is close to the

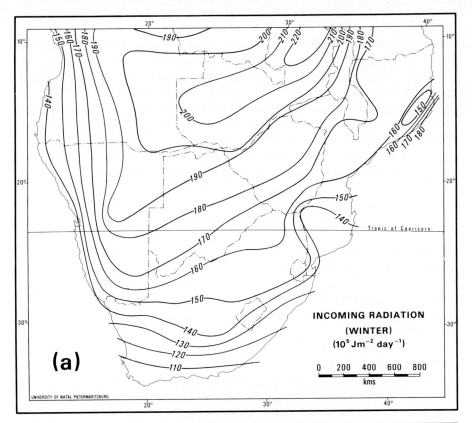
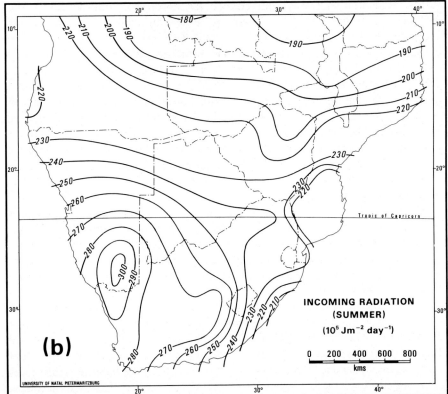

Fig. 1. Incoming radiation patterns.

coast. According to Drummond & Vowinckel (1957) Alexander Bay's radiation fluxes 'are probably among the highest ever to be recorded at sea level, amounting to 76 per cent of that available extraterrestrially'. North of about 18°S the daily incoming radiation values decrease steadily and this line of latitude may be regarded as constituting the boundary between equatorial air masses in the north and the drier subsiding air in the south. In both the summer and winter patterns of radiation the relatively low values along the Natal and Moçambique coasts south of 20°S indicate that the high atmospheric water vapour content associated with the warm Moçambique Current is the probable cause of radiation attenuation.

2.3 *The influence of topography on solar radiation*

In recent years ecologists in southern Africa have shown great interest in solar radiation budgets as an important indicator of plant communities' differences in meso-scale studies (for instance Edwards 1967, Granger 1975). Daily incoming radiant flux densities on sloping terrain as a function of slope, aspect and season have been presented for cloudless days in southern Africa for the latitudinal range 20°S–35°S (Schulze 1975b). The results are based on radiation data from southern Africa stations. Seasonal variations of radiation income on slopes are illustrated in Fig. 2 using December 22, March/September 22 and June 21 data as being representative of midsummer, the equinoxes and midwinter respectively.

In midsummer, on cloudless days, radiation flux densities in southern Africa generally exceed 250×10^5 J m^{-2} day^{-1}, the higher values being on the flatter slopes. Radiation increases with latitude for north, northeast/northwest and east/west aspects (Fig. 2). There is little variation with latitude on southeast/southwest slopes and only on south aspects is a decrease of radiation evident with increasing latitude, especially on the steeper slopes.

On cloudless days at the equinoxes slopes intercept about 250×10^5 J m^{-2} day^{-1} on northerly aspects in southern Africa, decreasing to 170–210×10^5 J m^{-2} day^{-1} on southern aspects. The north and northeast/northwest slopes exhibit relatively little variation of radiant energy interception with different gradients and with latitude. On the north aspects the steeper slopes receive most energy, but the more usual decrease of radiation on slope increases prevails on aspects deviating 90° or more from north. Around the equinoxes the influence of slope is most marked for the steeper southerly aspects and this influence is further accentuated as latitude increases (Fig. 2).

The midwinter influence of latitude on incoming radiation fluxes on slopes is most noticeable (Fig. 2). The lowest differences due to gradient throughout the latitudinal range graphed are found on east/west aspects. The most extreme effects of latitude and aspect again occur on the steep slopes which intercept considerably more radiation than flat slopes on northerly aspects and considerably less on southerly aspects.

For microscale studies techniques have recently been developed for actually mapping radiation patterns in topographically rugged areas (for instance Garnier & Ohmura 1968, Schulze 1975a), and, applying these techniques in the Natal Drakensberg, Granger (1975) has presented convincing evidence that topographically induced radiation – and consequently water balance – differences may largely account for vegetation successional changes from, for instance,

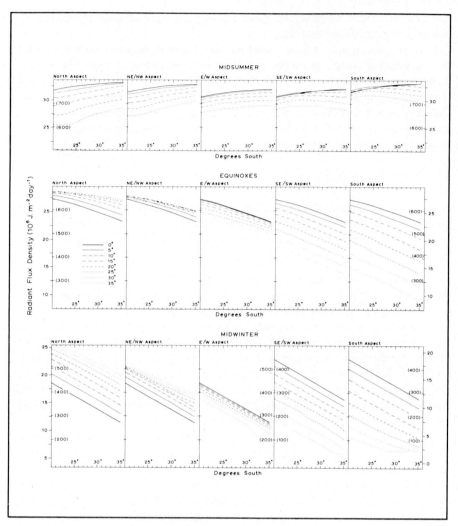

Fig. 2. The influence of topography on solar radiation (after Schulze 1975b).

Themeda triandra grassland to *Pteridium aquilinum* or from the latter to *Philippia evansii* in upland areas protected from the effects of fire and grazing.

3. Temperature

3.1 *The importance of temperature in biogeographical studies*

Temperature alone, or rather the availability of sensible heat, is not a significant factor in determining major regional vegetation formations, although its indirect influence on water availability through its effects on, for instance, evapotranspiration rates, is of primary importance (Van Riper 1971). On a meso- and microscale it does, however, play a major part in determining floristic variations. Within plant communities or associations the direct influences of temperature affect inter alia rates of growth, plant stature, seed germination, time of flowering and maturing of tissues, while certain plant mechanisms have adapted to seasonal or diurnal temperature fluctuations.

Critical temperature indices therefore, like summer maxima, winter minima (and associated frosts) or ranges are of more significance to plant distribution than means, which differentiate only the broad thermal divisions into megathermal plants (requiring mean monthly temperatures $\geqslant 20°C$ for $\geqslant 4$ months), microthermal plants ($\geqslant 8$ months with means $\leqslant 10°C$) and the mesothermal plants of the mid-latitudes and the greater part of southern Africa, whose physiology is adapted to strong seasonal climatic rhythms.

3.2 Temperature distributions in southern Africa

As a general guide and for application of annual biotemperatures in the Köppen and Holdridge Life Zone (vegetation) classifications (discussed in later sections) mean annual temperature patterns are depicted in Fig. 3a. The isotherms of Fig. 3a reveal four major characteristics, viz.

(i) an expected overall equatorwards temperature increase,
(ii) isotherms parallel to the coast over most of the area, which exhibit decreasing values with distance inland, reflecting the effects of continentality,
(iii) the effects of the cold Benguela and warm Agulhas/Moçambique Currents moving northwards and southwards on the west and east coasts respectively, and
(iv) the temperature irregularities induced by topographic variation on the subcontinent, for instance the lower temperatures along the escarpments on the perimeter of southern Africa ($<14°C$) or the higher temperatures along the Zambezi and Luangwa valleys ($>22°C$). The highest mean annual temperatures are found along the low-lying coastal plains of northern Moçambique.

The annual range of temperature (Fig. 3b) shows matching characteristics. Smallest ranges ($<6°C$) are evident towards the equator, as may be expected from radiation considerations (cf. Fig. 1) and also along parts of the west and northeast coasts. The greatest values (in excess of $16°C$) are over the southern Kalahari and northern Karoo, where the ameliorating effect of cloud cover is generally absent, and these values decrease towards the coast with sharp discontinuities, and, more gradually, towards the equator.

Fig. 4a illustrates mean daily summer maximum temperatures, January having been selected as being representative of summer, although it is known that in parts of southern Africa temperatures are actually higher in months preceeding or following January. Three broad features are apparent, viz.

(i) the influence again of the cold and warm ocean currents along the west and east coasts, the east coast generally recording mean daily maxima exceeding those of the west coast by $7-10°C$,
(ii) the high summer maxima in the Kalahari, which go hand in hand with the summer incoming radiation maxima there (Fig. 1b), and
(iii) the high January mean maxima in the lower reaches of the Zambezi and Limpopo valleys.

Mean daily minimum temperatures for winter are represented for the month of July in Fig. 4b. The highest values, $>15°C$, are recorded on the Moçambique coast north of 20°S while most of Moçambique and Malawi have mean daily

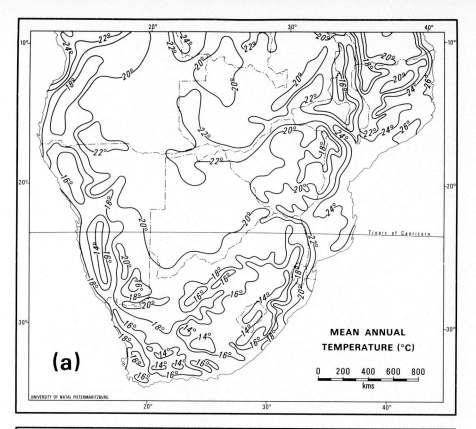
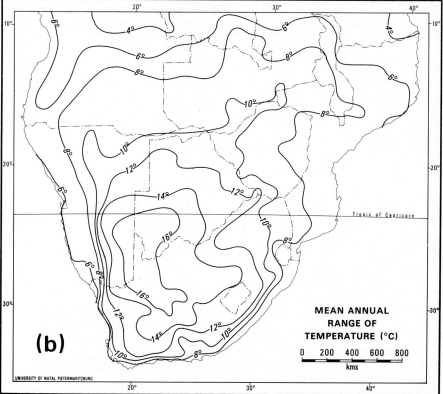

Fig. 3. (a) Mean annual temperature (after Knoch & Schulze 1957). (b) Mean annual range of temperature (after Knoch & Schulze 1957).

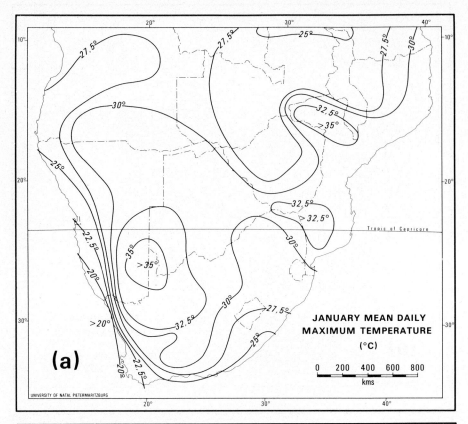
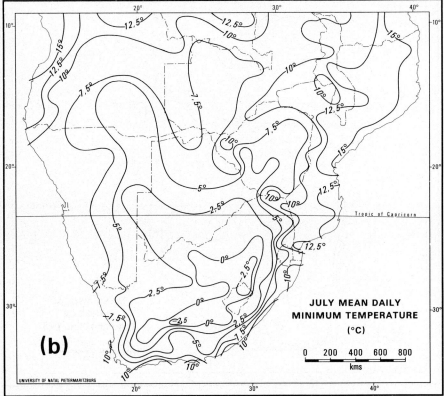

Fig. 4. Mean daily January maximum (a) and July minimum (b) temperatures (after Thompson 1965).

minima in excess of 10°C. Other areas with relatively high July mean minima are found in narrow zones along the southern and eastern coasts of South Africa, along the Angolan coast up to 300 km inland and equatorwards of 12°S. Vegetation in these areas is unlikely to suffer from frost damage (cf. Fig. 13 and Chapter 10, Fig. 2). The critical zero isotherm encompasses most of Lesotho, the interior plateaux above 1800 m in South Africa as well as the Karoo regions. The winter mean minimum temperature gradation from the 0°C isotherm is gradually north- and westwards but rapid south- and eastwards, being primarily a function of altitude and latitude.

Although less important biogeographically, the mean daily (actual) January and July temperature maps, given in Fig. 5, show considerably more detail than the minimum and maximum maps (Fig. 4) respectively, because more data were readily available for these statistics. Patterns similar to those described above emerge, but the influence of topography and the ocean currents on temperature conditions is more marked.

4. Moisture

4.1 *The importance of moisture in biogeographical studies*

Among the various individual climatic parameters which influence the gross features of vegetation differences on earth Van Riper (1971) and Walter (1972) consider the most important to be water. Limitations in water availability are frequently a restrictive factor in plant development, and water is essential for the maintenance of physiological and chemical processes within the plant (including germination, growth and reproduction), acting as an energy exchanger and carrier of nutrient food supply in solution.

The reservoir of soil water on which land plants draw is derived from precipitation mainly in the form of rainfall, fog and snow, of which the first two are considered important in southern Africa. Not all precipitation is, however, freely available to the vegetation through the soil, as some is intercepted by the plant before reaching the soil, some runs directly into streams as surface flow after storm events without being utilized by the plants, some percolates into the deep soil layers beyond the root zones and some is evaporated directly from bare ground without being transpired through the plant.

Most major subdivisions of vegetation formations on a subcontinental scale reflect the annual and seasonal soil moisture balances rather than gross precipitation income. Before aspects of the moisture balance of southern Africa are evaluated in a subsequent section, however, the overall patterns of annual and seasonal rainfall will be discussed as background, together with sections on the importance of rainfall interception, fog and snow in the biogeographical context of southern Africa.

4.2 *Precipitation patterns over southern Africa*

The distribution of mean annual precipitation is given in Fig. 6. Two overall features of the distribution are apparent:

(i) South of the tropic precipitation decreases uniformly westwards from the escarpment across the plateau. Between the escarpment and the sea, in both the

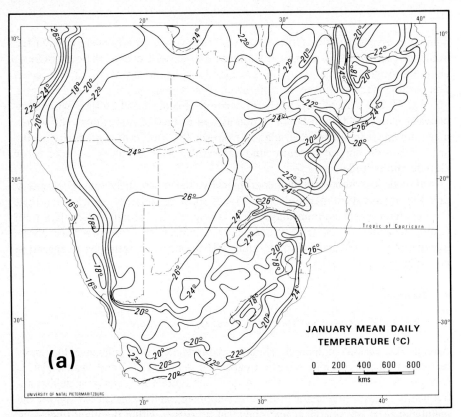
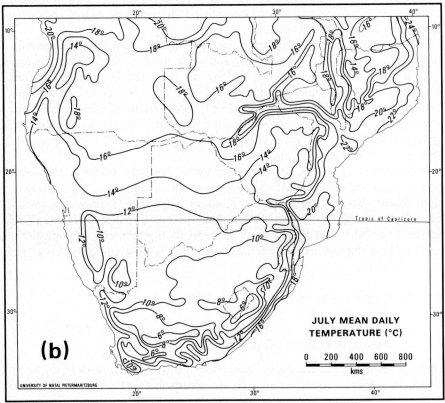

Fig. 5. Mean daily January (a) and July (b) temperatures (after Knoch & Schulze 1957).

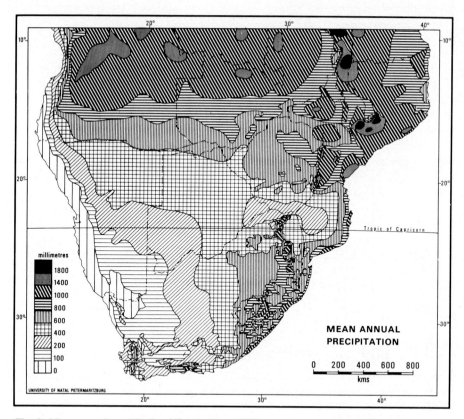

Fig. 6. Mean annual precipitation (after Jackson 1961).

southern and eastern coastal margins there is the expected complexity induced by topographical irregularities.

(ii) The second marked feature is the general west–east trend of isohyets north of the tropic, showing an increase from south to north except in a narrow coastal region of Angola and in Moçambique. Local variations again occur and are the result of topography, moist on-shore winds, shifts in the trajectories of the tradewinds, the location of the Inter Tropical Convergence Zone boundaries, the occasional visitations of tropical cyclones and the presence of large lakes.

The distribution of mean January precipitation (Fig. 7a) is in most respects similar to that for the year. Over almost the entire region it is in the December/January period that rainfall maxima occur. Only in the southwest Cape region of South Africa is there a winter maximum of rainfall, with the so called Mediterranean climate (Köppen's Csa/Csb). In general the July distribution of rainfall (Fig. 7b) therefore shows only vestiges of any pattern. Values from 10–50 mm occur in the south and east, but for the major part of the area the July rainfall totals average less than 10 mm.

Considered much more meaningful in terms of the distribution and characteristics of plant cover than the patterns illustrated in Figs. 6 and 7 are the duration, time of occurrence and the degrees of intensity of the rainy season and the dry season of southern Africa; particularly when viewed in the perspective of plant available soil water being able to meet the evaporative demand of the at-

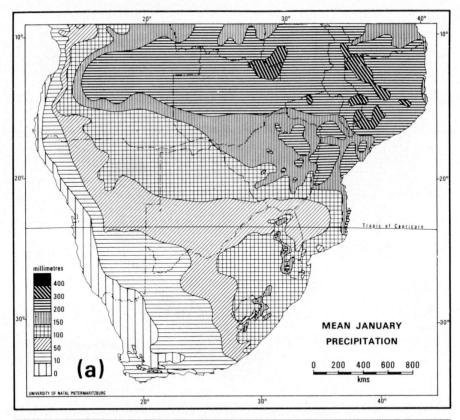
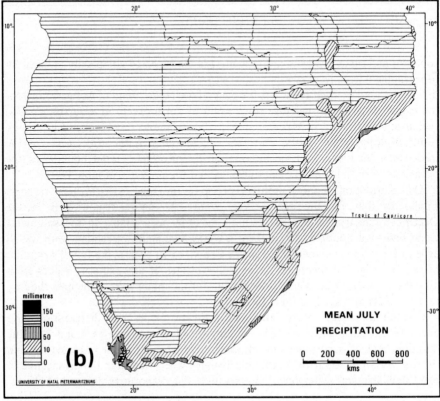

Fig. 7. Mean January (a) and July (b) precipitation (after Jackson 1961).

mosphere, with which is related plant physiological response and growth. Aspects of these two seasons are therefore discussed below.

4.3 The rainy seasons of southern Africa

In a designation of rainy seasons in southern Africa the definitions and the map used by Knoch & Schulze (1957) have been adhered to in this review. All months with $\geqslant 50$ mm of precipitation have been included in the rainy season, while months with precipitation $\geqslant 125$ mm, which signify a certain degree of intensity of the rainy season, are shown separately in Fig. 8a.

The dry west coast area of southern Africa exhibits a zone with no defined rainy season, the zone extending some 400 km inland in the south, but narrowing north of the tropic. This area is bordered by a zone with a late summer rainy season of short duration. The winter rainfall area of the southern Cape shows up clearly with the longest rainy season there towards the extreme southwest of the Cape Province. The only area with a distinct double rainfall maximum is in the 'all year' rainfall region around Port Elizabeth, where over 50 mm per month is recorded from September–November and again in May. South of 20°S the rainy season zones generally trend north–south with the rainy season becoming progressively longer from west to east. The longest rainy seasons in southern Africa are in northeastern Moçambique. Areas with monthly means exceeding 125 mm, demarcated by the stippled line in Fig. 8a, occur as an east coast and tropical feature east of 30°E and north of 20°S.

A more objective description of the temporal–areal variation of rainfall in southern Africa may be obtained from the harmonic analysis of rainfall undertaken by McGee & Hastenrath (1966).

4.4 The dry seasons of southern Africa

The definition of the dry season was again achieved by analysing mean monthly precipitation totals, months with $\leqslant 25$ mm being classified as dry. Two further degrees of drought intensity were established by Knoch & Schulze (1957), viz. months with rainfall totals $\leqslant 10$ mm and with < 0.1 mm (i.e. months with no 'measurable' precipitation). This classification of dry seasons is depicted in Fig. 8b.

Characteristic of the dry seasons of southern Africa is the general southwest-northeast trend of the varying degrees of drought as defined above. No dry season is experienced along a narrow coastal strip of the southern and eastern Cape and Natal. The drought zone in which mean monthly rainfalls $\geqslant 10$ mm extends from the southwestern Cape in a strip some 100–400 km wide to north-central Moçambique. Several seasonal variations are discernible in this zone, viz.

(i) the summer droughts of the southwestern Cape,
(ii) the virtually permanent drought in the Little Karoo and
(iii) The 3–4 month winter drought northeast of 27°E which, like the rainy season in summer in this area, commences progressively later in the season as latitude decreases.

The intensive drought zone (0.1–10.0 mm per month) extends over most of South Africa, eastern Rhodesia, Malawi and northern Moçambique, showing

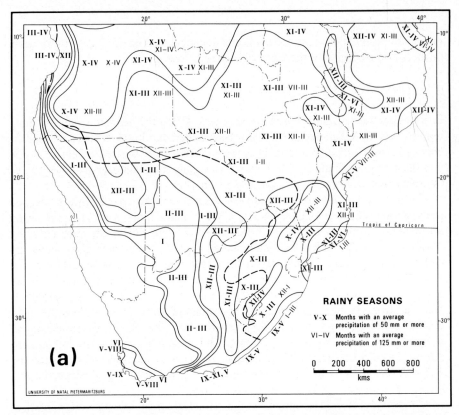
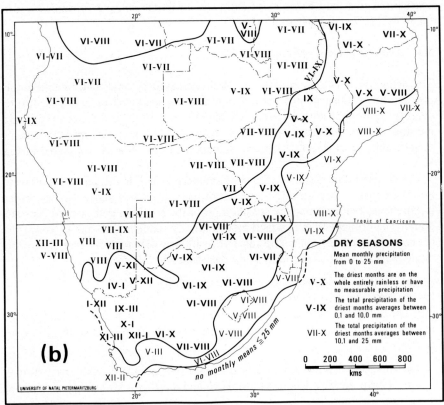

Fig. 8. The rainy (a) and dry (b) seasons (after Knoch & Schulze 1957).

drought occurrence trends similar to those of the previous zone. A very intense dry season (<0.1 mm per month) covers most of South West Africa, Zambia and Botswana, being most severe in the Namib desert region, where up to eight months of the year may, on average, receive no measurable rainfall. For most of the rest of this very intensive drought zone 2–4 months of the year normally receive no measurable precipitation.

Figs. 8a and 8b should be viewed together with the maps of water surplus and deficiency (Figs. 14 and 15 respectively) and should be used to supplement any evaluation of those maps, since they illustrate the times and degrees of intensity of the rainfall and drought phenomena as defined above.

4.5 Rainfall interception by vegetation

The interception of rainfall by plant foliage and plant litter is frequently one of the least considered aspects in biogeographical studies at any scale. Rainfall interception is a function of vegetation volume per unit area and the number of wetting cycles it undergoes, i.e. the number of raindays. In the southern African context, research from mature exotic forests indicates an 8–12 per cent 'loss' of mean annual precipitation to the soil by plant interception (Wicht 1971), while on the assumption of Whitmore (1970, 1971) that the tall grassveld of South Africa intercepts the first 1.5 mm of rain on each rainday, interception by vegetation would amount to 11.6 per cent of South Africa's rainfall and to 18.3 per cent of the rainfall in Natal. Research in mixed bushveld/open savanna vegetation in the northern Transvaal where the main tree species is *Acacia caffra* has shown that following a shower of 5 mm, between 0.8 and 2.4 mm may be intercepted, while for a shower of 15 mm these figures increase to 2.5–4.4 mm (De Villiers 1975).

4.6 Fog – its importance and occurrence in southern Africa

On foggy days and/or in zones where advective sea fog is prevalent, moisture may be intercepted by the vegetation even though standard rain-gauges do not necessarily record any precipitation during the same period. This deposition of fog droplets on the foliage is generally accepted to be beneficial to the plant and it may have a profound effect on the growth, development and distribution of plants (Kerfoot 1968).

Observations in southern Africa and elsewhere have shown that a relationship exists between fog incidence and the presence of vascular epiphytes. The *Welwitschia* zone of Angola and South West Africa, for instance, is near the sea and it has been suggested that, although fog is not entirely a limiting factor, the bulk of moisture required for current growth could be derived from sea fogs (Airy Shaw 1947).

Since the amount of fog is directly proportional to the liquid water content of the air, fog frequencies and amounts in the winter rainfall regions of South Africa are highest in winter. Similarly, on the subcontinent, summer fog predominates in the summer rainfall areas.

The systematic fog precipitation measurements of Nagel (1962) in the mountainous areas of winter rainfall area have shown inter alia that over a 5 year period at the highest point of Table Mountain (Cape Town), viz. Mclear's Beacon, the 5664 mm of fog intercepted per annum was 3 times as high as the annual rainfall, and that no month recorded less than 311 mm of fog. These may be extreme results, but at the Jonkershoek Mountains of Stellenbosch orographically induced moisture from fog (not recorded by standard gauges) exceeded 600 mm per annum.

Along the west coast of South Africa, radiation and advection fog is formed when warm inshore surface water mixes with upwelling cold water of the Benguela Current. The afternoon sea breeze blows this fog inshore. For Swakopmund Nagel (1962) cites 121 fog days per annum with an amount intercepted in 1958 equivalent to 130 mm of rainfall – more than 7 times the mean annual rainfall! Nagel estimates that along a 3 km coastal strip of the west coast around Swakopmund fog precipitation equivalent to 150 mm may be intercepted by plants, while between the sea, latitude 32°S and longitude 20°E, i.e. an area of some 50,000 km^2 in the southwestern Cape, the fog precipitation is equivalent to 300 mm per annum.

Observations of the contribution of fog to precipitation in the summer rainfall area are somewhat less scientific than those of Nagel's (1962). In Rhodesia Kreft (1972) has found the incidence of fog (generally of the radiation type, except on the eastern highlands), to be about 10 days annually in the west and about 40 days in the east, occurring most commonly during the main rains in February–April. One station, however, viz. Chisenga, records an average of 146 fog days per annum with a minimum incidence in July (8.1 days) and a maximum in March (16.6 days).

Preliminary results from 18 standard 'fog catchers' (as specified by Nagel 1956) in northern and eastern Transvaal have yielded precipitation figures in excess of those recorded by standard gauges of between 105.2 and 280.1 per cent (Fabricius 1969).

For central Natal Whitmore (1970) cites fog occurring on some 10 days per month in summer while Schulze (R.E., unpubl.) calculated that at Kranskop in the so-called Natal Mistbelt there was an average of 4 fog-days per month from November to February. In the higher lying areas in the foothills of the Natal Drakensberg the orographic fog contribution at 1800 m altitude is an additional 403 mm per annum – one third of the mean annual precipitation – with fog 'catch' values in each of the 6 months from October to March exceeding 40 mm and 68 mm in November (Schulze 1975c).

This brief review underlines the importance in southern Africa of fog as an ecological agent, indicating that considerable amounts of moisture not recorded conventionally may in fact be intercepted and utilized, directly or indirectly, by vegetation cover.

4.7 *Snow*

Snow occurs only spasmodically in southern Africa, mainly on the higher mountain ranges which constitute the 'great escarpment' (Schulze 1965), falling most often on the southwestern ranges of the Cape Province (with an annual snow frequency of 5.4 snowfalls, 3.2 of which may be expected from June to August) and, on account of the greater elevation, along the Drakensberg escarpment and in Lesotho (annual frequency 8.3 with an average of 5.2 falls from May to August). Snowfalls are rare on the eastern highlands of Rhodesia, and elsewhere north of 23°S. The ecological effects of snow in southern Africa are thus thought to be minimal.

5. Climatic indices and classifications

5.1 Background

Already in the nineteenth century many biologists had recognized the association between vegetation and climatic elements and had attempted climatic classifications. It was not, however, until the turn of the century, that Köppen (1900) was able to develop this idea sufficiently to produce a useful and practical climatic classification employing readily available climatic statistics.

To biogeographers, ecologists and others interested in southern Africa some indices and schemes of classification are better known and more familiar than others. In addition, some of the more recent classifications have not been applied as widely as others in southern Africa or they may not always have been related to vegetation patterns. For that reason the classifications and indices of Köppen, Holdridge and Thornthwaite are evaluated and discussed in their southern African context.

5.2 The Köppen climates of southern Africa

Because of its simplicity the Köppen classification has been widely applied in southern Africa (for instance by Coetzee & Werger 1975), and the climatic patterns shown in Fig. 10 have been derived from the work of Schulze (1947) for South and South West Africa, Botswana, Lesotho and Swaziland, of Torrance (1972b) for Rhodesia, Zambia, Malawi and Angola, and of Gonçalves (1970, 1971) and Griffiths (1972) for Moçambique. Further detailed work on this classification has been undertaken by Driscoll (1959) in relation to boundary fluctuations in Moçambique, by Buys & Jansen (1970) in the Cape and by others.

A simplified system used to compile Fig. 10 was adopted to render compatible the modified Köppen systems used by the authors listed above. Problems occurred on occasions in overlap regions where conflicting classifications were given; this also had to be resolved. The final system used for Fig. 10 is summarized in Table 1 and Fig. 9.

The A climates of southern Africa (Fig. 10) are represented mainly in Moçambique, Angola and in the Luangwa valley of Zambia, the majority of the area having Aw climates, with small areas of Am occurring. Most of the western interior of South Africa and South West Africa as well as a coastal strip of Angola has, as would be expected from Figs. 3a and 6, a BW climate which is divided into the k and h sub-types in Fig. 10, but which Schulze (1947) has subdivided further into 8 sub-types. Köppen's BS climates occupy large tracts of central southern Africa with extensions into the southwestern Cape. Except in isolated pockets the subdivision into h and w is essentially a latitudinal one.

The Mediterranean Cs climates are restricted to the extreme southwestern Cape while the Cf climatic province, as expected from Fig. 8, occupies the coastal belt from Mossel Bay to the border of Moçambique. Temperate climates with winter dry seasons, i.e. the Cw type, are found mainly in the eastern interior of South

Table 1. A simplified Köppen classification relevant to southern Africa (after Köppen & Geiger 1936).

1st letter	2nd letter	3rd letter
A, C, D Sufficient heat and precipitation for forest vegetation		
A Equatorial climates	f Sufficient precipitation during all months	
Mean temperature above 18°C for all months	m Monsoon climate (forest-vegetation despite dry season)	
	w Winter dry season	
B Arid zones	S Steppe climate	h dry-hot, mean annual temperature over 18°C
(limits: see Fig. 9)	W Desert climate (limits: see Fig. 9)	
		k dry-hot, mean annual temperature below 18°C
C Warm temperate climates Coldest month 18°C to −3°C	s Summer dry season	a warmest month over 22°C
	w Winter dry season	
	f sufficient precipitation during all months	
		b warmest month below 22°C, but at least 4 months above 10°C

Africa and Rhodesia as well as over central Angola, Zambia and Malawi with the sub-type b prevalent in plateau areas of higher altitudes.

Regarding the adequacy of the Köppen system of classification in delimiting vegetation boundaries it may be observed that in only very broad terms is there agreement in South Africa between, for instance, Köppen's BSh climate and Acock's (1953) delineation (of his vegetation map in A.D. 1400) of the 'bushveld', the BWh and 'succulent karoo', Cwb and 'sweetveld', Cs and 'fynbos' and Cf/Cw

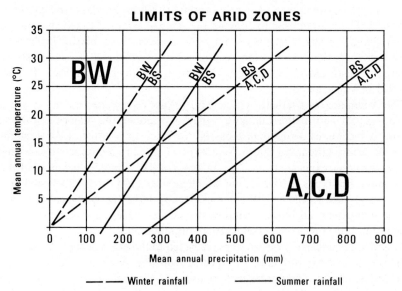

Fig. 9. Limits of arid zones in the Köppen climatic classification (after Köppen & Geiger 1936).

and 'forest and scrubforest'. Certain measures of agreement have also been pointed out by Schulze (1947) between the Köppen climatic regions and vegetation boundaries of older vegetation maps by Pole-Evans (1935) and Adamson (1938).

One of the disturbing aspects of Köppen's representation in southern Africa is the large coverage of the BSh climatic province, and in the opinion of Schulze (1947), both from vegetational and purely climatic reasons the eastern Transvaal lowveld and interior Moçambique should not be grouped with the rest of the climatic region in question, which includes the Kalahari. The reason why they are

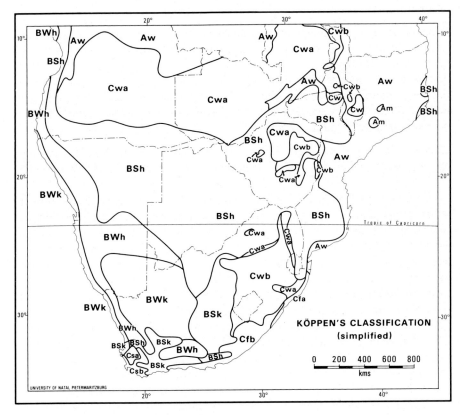

Fig. 10. Climates according to Köppen (after Schulze 1947, Gonçalves 1970, 1971, Griffiths 1972, Torrance 1972b).

classified together is because only gross annual rainfall and temperature are used as criteria.

The Köppen system still leaves much to be desired in its attempts to relate vegetation to broad climatic parameter values without adequate concern for such vegetation-limiting variables as water need in relation to supply, soil moisture storage or evapotranspiration (Mather 1974). Therefore, despite its still widespread use by ecologists, other additional climatic indices and classifications have to be assessed as well.

5.3 The Holdridge Life Zone System applied to southern Africa

One of the latest efforts in the classification of vegetation and climate is the Life Zone classification system proposed and modified by Holdridge (Holdridge 1959, et al. 1971) and based on a detailed knowledge of environmental relationships in the tropics and subtropics. Although the system consists of 3 levels of classification, only the primary level can be applied in this study because of the subcontinental scale of working.

Holdridge, finding that boundaries between the major vegetation units could be defined in terms of logarithmic increases of temperature and rainfall units, utilizes, in a triangular co-ordinate graph illustrated in Fig. 11, three criteria for vegetation delimitation, viz.

(i) the mean annual biotemperature (defined as the mean unit period temperature, with the substitution of zero for all unit period values below 0°C, and of 30°C for values above 30°C),
(ii) the mean annual precipitation and
(iii) the potential evapotranspiration (PE) ratio (derived by multiplying mean annual biotemperature by 58.93 to give a PE estimate, which is then divided by the mean annual precipitation to obtain a PE ratio yielding humidity provinces).

Diagonals at preselected values drawn from these criteria result in 30 interlocking hexagons representing the major vegetation types according to Holdridge (Fig. 11).

This Life Zone scheme having been evolved on the basis of the distribution of natural vegetation in response to climate, and thereby becoming a bioclimatic rather than merely an arbitrary climatic classification (Holdridge et al. 1971), should be of particular interest to biogeographers and ecologists, despite inherent shortcomings attributable to the use of annual climatic values. No detailed application of the System on a southern African scale being known to the authors, it was decided to prepare a preliminary Life Zone Map for the study area (Fig. 12) using the guidelines set out by Holdridge et al. (1971).

A total of some 750 climatic stations' data was analysed. Of these stations 54 were in Angola, 20 in Botswana, 8 in Lesotho, 22 in Malawi, 130 in Moçambique, 76 in Rhodesia, 357 in South Africa, 9 in South West Africa, 14 in Swaziland and 57 in Zambia. With the exception of parts of South West Africa, Botswana and the interior of Lesotho the station coverage was considered adequate. Using the hexagon-defined Life Zones (also where transitional zones were concerned) a total of 9 major Life Zones and 57 sub-zones, was delineated in Fig. 12. These zones are listed in Table 2.

Having been presented as a climatological exercise because of its particular relevance to plant geography, the interpretation of the map by botanists will be interesting (for instance, comparing the vegetation discontinuities in Fig. 12 with those of Coetzee & Werger (1975)). To the climatologist the relationship between the Life Zones and the vegetation maps compiled, for instance, by Acocks (1953) or by Rattray & Wild (1960) is only very broad. As in the case with the Köppen approach, annual temperature and precipitation values are not considered the most active factors resulting in the growth and development of vegetation. Furthermore by not taking into account seasonal patterns of temperature and moisture, which are necessary to differentiate between zones having similar annual values, the Holdridge Life Zones do not separate, for instance, the Mediterra-

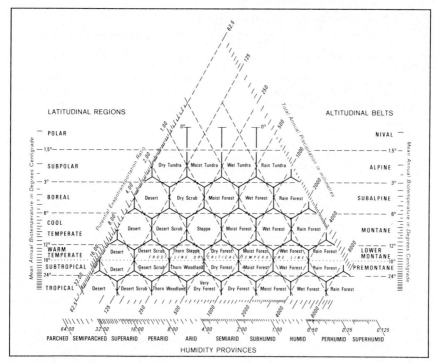

Fig. 11. The Holdridge Life Zone scheme (after Holdridge et al. 1971).

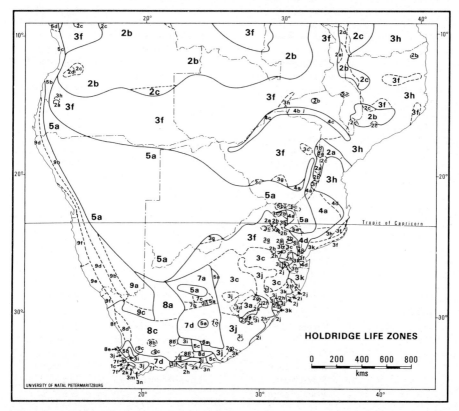

Fig. 12. Holdridge Life Zones in southern Africa.

Table 2. The Holdridge Life Zones in southern Africa.

		Holdridge life zones		
Wet forest				
	1a	Tropical	Lower Montane	Perhumid
	1b	Tropical	Premontane	Perhumid
	1c	Warm Temperate	Premontane	Perhumid
Moist forest				
	2a	Tropical	Lower Montane	Humid
	2b	Tropical	Premontane	Humid
	2c	Tropical	Premontane	Subhumid
	2d	Tropical	—	Humid
	2e	Tropical	—	Subhumid
	2f	Subtropical	Montane	Humid
	2g	Subtropical	Lower Montane	Perhumid*
	2h	Subtropical	Lower Montane	Humid
	2i	Subtropical	—	Humid
	2j	Subtropical	—	Subhumid
	2k	Warm Temperate	—	Humid
Dry forest				
	3a	Tropical	Montane	Subhumid
	3b	Tropical	Lower Montane	Humid
	3c	Tropical	Lower Montane	Subhumid
	3d	Tropical	Lower Montane	Semiarid
	3e	Tropical	Premontane	Humid
	3f	Tropical	Premontane	Subhumid
	3g	Tropical	Premontane	Semiarid
	3h	Tropical	—	Subhumid
	3i	Subtropical	—	Subhumid
	3j	Subtropical	Lower Montane	Humid
	3k	Subtropical	Lower Montane	Subhumid
	3l	Subtropical	—	Semiarid
	3m	Warm Temperate	—	Humid
	3n	Warm Temperate	—	Subhumid
Very dry forest				
	4a	Tropical	Premontane	Semiarid
	4b	Tropical	—	Subhumid
	4c	Tropical	—	Semiarid
	4d	Subtropical	—	Semiarid
Thorn woodland				
	5a	Tropical	Premontane	Semiarid
	5b	Tropical	Premontane	Arid
	5c	Tropical	—	Semiarid
	5d	Tropical	—	Arid
	5e	Subtropical	—	Semiarid
	5f	Subtropical	—	Arid
Steppe				
	6a	Subtropical	Lower Montane	Subhumid
Thorn steppe				
	7a	Tropical	Lower Montane	Semiarid
	7b	Tropical	Premontane	Semiarid
	7c	Subtropical	Lower Montane	Subhumid
	7d	Subtropical	Lower Montane	Subhumid
	7e	Subtropical	Lower Montane	Semiarid
	7f	Warm Temperate	—	Semiarid

Desert scrub
8a	Tropical	Premontane	Arid
8b	Subtropical	Lower Montane	Semiarid
8c	Subtropical	Lower Montane	Arid
8d	Subtropical	—	Arid
8e	Warm Temperate	—	Semiarid
8f	Warm Temperate	—	Arid

Desert
9a	Tropical	Premontane	Perarid
9b	Tropical	Premontane	Superarid
9c	Subtropical	Lower Montane	Perarid
9d	Subtropical	—	Superarid
9e	Warm Temperate	—	Superarid
9f	Warm Temperate	—	Parched

nean macchia of the southwestern Cape and thornveld and associated grasses of Natal (Fig. 12).

Many relationships between the distribution of vegetation and the annual values of the two commonly used climatic parameters are therefore thought by Mather (1974) to be largely fortuitous and based on the fact that the two factors are related to other more active climatic indices like those of Thornthwaite's (1948), which are discussed in the following section.

5.4 *Vegetation distribution and the Thornthwaite indices: a brief review*

The basis of Thornthwaite's (1948) climate classification is the climatic water budget, derived by the now well known bookkeeping procedure utilizing monthly values of precipitation and potential evapotranspiration which yield, inter alia, monthly values of actual evapotranspiration, water surplusses and deficiencies, and from which indices of humidity and aridity give moisture and thermal efficiency indices.

Thornthwaite indices have been applied in varying degrees of detail and by using different units in regional studies by a number of researchers in southern Africa, for instance by Carter (1954) on a continental scale, Howe (1953) for Rhodesia, Zambia and Malawi, Schulze (1958) for South Africa, South West Africa, Lesotho, Swaziland and Botswana, Gonçalves (1970, 1971) for Moçambique and Poynton (1971) for the subcontinental area under review. Mather (1962) has also presented statistics of the relevant water balance components on a monthly basis for 30 stations in Angola, 4 in Lesotho, 11 in Botswana, 98 stations in Moçambique, 79 in the former Central African Federation (Rhodesia, Zambia and Malawi), 153 in South Africa, 4 in South West Africa and 6 in Swaziland.

Relationships between vegetation distribution and Thornthwaite indices have been sought elsewhere by numerous workers. Thus Hare (1954) and Thornthwaite & Hare (1955) found that in temperate humid climates the moisture index could roughly distinguish forest types while in colder climates PE served as a better index to distinguish vegetation subdivisions. Major (1963), suggesting that vascular plant activity and growth might be related to actual water loss by vegetation (since this factor expresses the real rather than the potential water activity of the plant), considered actual evapotranspiration derived from the Thornthwaite water balance system for a vegetation delimitation. Mather & Yoshioka (1968), on the other hand, in plotting climatic data from 20 principal vegetation types of North America using the moisture index as the ordinate and PE as the abcissa variable, found it possible to locate discrete and largely non-overlapping vegetation areas defined by these two indices. Furthermore both Thornthwaite (1952) and Mather (1959) have shown that vegetation associations are also related to the two factors that make up the moisture index, viz. the aridity index and the humidity index. However, Mather (1974) concedes that in the high precipitation (>2500 mm) tropical areas the role of climate may be somewhat less important in delimiting vegetation associations than other factors. It would appear, nevertheless, that for most of southern Africa south of 10° latitude the Thornthwaite indices should, from work done elsewhere, provide significant background information for biogeographical and ecological distributions. Information on PE, water surplusses, water deficiencies and the moisture index are therefore presented below for southern Africa.

5.5 *Potential evapotranspiration, thermal efficiency and thermal regions*

Thornthwaite (1948) observed that a decision on whether a climate was moist could not be made if only precipitation data were available – it had to be known whether the precipitation was greater or less than the optimum quantity of water needed for evapotranspiration by a vegetation covered soil viz., the PE. This optimum water need by plants also serves as an index of thermal efficiency because Thornthwaite derived PE empirically from temperature, with consideration also being given for day length. 'It is not merely a growth index, but expresses growth in term of the water needed for growth. Given in the same units as precipitation, it relates thermal efficiency to precipitation effectiveness' (Thornthwaite 1948).

A number of writers have been critical of, or have suggested modifications to, Thornthwaite's thermal regions as defined by PE intervals on an annual basis. Thus Schulze (1958) pointed out that anomalies occur when thermal regions are based solely upon PE, remarking that it did not seem right that the west coast of South Africa and interior of Natal, for instance, or the Transkei coast and the interior of South West Africa at 1800 m altitude should fall in the same thermal efficiency region, as the respective climates certainly differed widely. Also from a biological standpoint, Poynton (1971) maintained, it is not always enough to know the thermal efficiency of a climate based on a period of a whole year, particularly in mesothermal climates in which temperature range as well as incidence and severity of frost may play 'a decisive part in determining the ability of many plants and animals to survive'.

In adopting Poynton's (1971) map of thermal regions and efficiency for southern Africa in this review (Fig. 13), it should be noted that he adhered to the broad thermal efficiency types of Thornthwaite but that the mesothermal zone, with the exception of the warmest of Thornthwaite's subdivisions, is subdivided on the basis of mean July minimum temperatures (Fig. 4b) and not according to PE.

In southern Africa Thornthwaite's tropical (megathermal) zone (Fig. 13) covers most low lying areas of Moçambique (with extensions along the Zambezi and Luangwa valleys) and also the low-lying coastal and inland areas of northern Angola. Subtropical thermal regions are found over most of the rest of Moçambique with smaller areas bordering the tropical regions. The warmer temperature

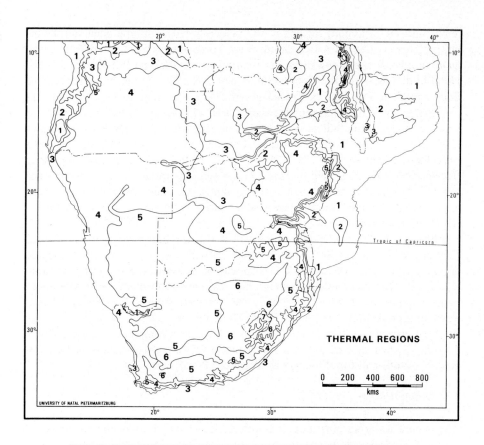

Fig. 13. Thermal regions (after Poynton 1971).

ZONE	CLIMATIC TYPE	THERMAL EFFICIENCY INDEX	FROST	APPROXIMATE MEAN MONTHLY MINIMUM TEMPERATURE FOR THE COLDEST MONTH
1	Tropical (Megathermal)	>1140 mm	none	—
2	Sub-Tropical (Mesothermal)	997 to 1140 mm	Occasional	5° to 10°C
3	Warmer-Temperate (Mesothermal)	570 to 997 mm	Very Light	5° to 10°C
4		570 to 997 mm	Light	0° to 5°C
5	Cooler-Temperate (Mesothermal)	570 to 997 mm	Moderate	-5° to 0°C
6		570 to 997 mm	Severe	< -5°C
7	Sub-Alpine (Microthermal)	427 to 570 mm	Very severe	"

regions with very light frost extend along the northern coast of South West Africa and Angola and along a narrow strip along the Cape and Natal coasts in addition to covering low-lying areas of northern Botswana and west and east of Zambia. Inland of these regions where mean monthly minimum temperatures for the coldest month vary between 0°–5°C, a 'cooler' warmer-temperate thermal region covers the northern parts of South West Africa, the Angolan plateau areas, most of Rhodesia, south Botswana and central/western areas of Zambia. The cooler-

45

temperate (mesothermal) regions of the moderate frost subtype occur on the interior plateaux of South Africa, South West Africa and the Kalahari desert as well as in an inland strip along the great escarpment in the Cape Province and in the Natal Midlands, while the higher altitude plateaux of the Transvaal, Orange Free State and Cape Province experience severe frosts according to Poynton (1971).

There is some disagreement between Carter (1954) and Schulze (1958) as to whether the microthermal zone occurs in southern Africa. Poynton (1971) follows Carter in ascribing to those parts of Lesotho lying above 3000 m altitude a microthermal climate (Fig. 13, zone 7), based on the premise that certain hardy tree species, while not killed by frost at such high elevations, fail to 'make normal growth'. His reasoning is on climatic grounds, viz. that the effective growing season is unduly short or that temperatures are too low to sustain a high rate of metabolic activity.

Although the Thornthwaite estimation of PE has been more widely adopted than any others and evidence has justified its application in almost every climatic region of the world (Ward 1975), it should be stressed that criticism has been levelled at Thornthwaite's extremely empirical derivation of PE. Probably more realistic patterns of PE per se may be obtained from the physically based Penman-type formulae (Penman 1956, Monteith 1965) using solar radiation as the most important variable, and such maps have been compiled for South Africa, Lesotho, Swaziland and parts of Botswana and South West Africa by Louw & Kruger (1968) and for Rhodesia by Prentice (1965).

5.6 Annual water surplus, S

The annual water surplus map of South Africa shows most of the subcontinent south of 17°S as well as coastal strips north of that latitude as having a meagre surplus of < 100 mm (Fig. 14). The largest region of appreciable annual surplus occurs within the tropics, extending from southwest Angola eastwards (with annual surplusses uniformly of the order of 100–600 mm per annum) to northern Moçambique (which shows more diversity in its surplus).

An annual water surplus of more than 200 mm occurs in higher lying areas of the northern and especially eastern parts of Rhodesia, while areas with similar surplusses in South Africa are found along the eastern escarpment and fold mountain areas of the southern Cape.

Throughout most of the area the surplusses occur in summer months within the period December–April, the exception being the southern Cape which has winter surplusses.

5.7 Annual water deficiency, D

A water deficiency, according to Carter (1954), is equivalent to drought – it is the need for moisture required by vegetation that soil storage and precipitation have failed to meet. Therefore, biogeographically, only regions without water deficiencies are free from drought by this definition.

A glance at the average annual water deficiency map (Fig. 15) already indicates that drought to some degree of severity occurs all over southern Africa except along the eastern fringe of South Africa. This region of low deficiencies extends from the vicinity of East London to Empangeni then inland to include Lesotho,

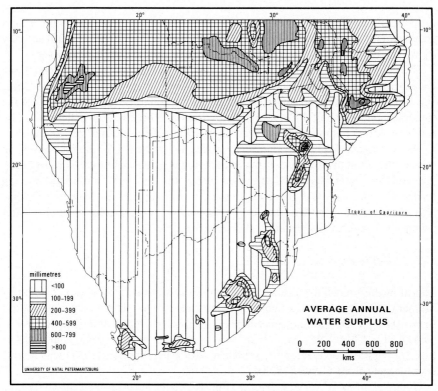

Fig. 14. Annual water surplus (after Carter 1954).

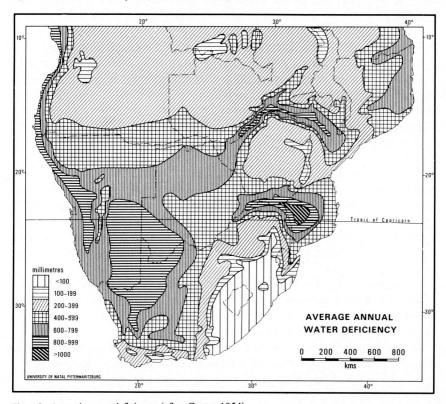

Fig. 15. Annual water deficiency (after Carter 1954).

with continuations radiating outwards from the main area along the ridges northwards and westwards towards Mafeking. A map by Schulze (1958) actually shows the eastern portions of this region to have no deficiencies whatever.

The greatest deficiencies, i.e. between 800–1000 mm per annum, occur in the Kalahari desert of southeastern South West Africa and the northwestern Cape, and also in the Limpopo and to some extent the Zambezi valleys. Towards the equator, again north of 17°S, and in large areas of Rhodesia, the deficiency is, by southern African standards, relatively low, being below 400 mm per annum.

The average annual water deficiency is less along the Namib desert coast than it is on the mountains inland of the Namib because, according to Carter (1954), the need for water is greatly depressed along the coast by pronounced cooling associated with the cold Benguela current. The coastal area from the Orange River northwards to Moçâmedes has negligible precipitation and so water deficiency is almost equivalent to the water need.

5.8 Moisture regions

Thornthwaite's moisture regions are derived from a moisture index, Im, defined by Mather (1974) in terms of a humidity index (based on a relative surplus of water), Ih = 100 (D/PE), and an aridity index (based on a relative annual deficiency of water), Ia = 100 (D/PE), so that Im = Ih − Ia. The moisture regions depicted in Fig. 16 are those delineated for southern Africa by Poynton (1971) which were taken largely from Carter (1954).

Humid climates of varying degrees (Fig. 16) cover most parts of the study area north of 14°S as well as the uplands and mountains along the eastern and southern perimeter of the subcontinent from Lake Malawi to the Cape of Good Hope.

The subhumid climates everywhere in southern Africa have some deficiency and some surplus (Carter 1954). In South Africa, apart from patches along the southern Cape coast, the subhumid climates extend westwards up to 25–26°E, the moisture indices decreasing with continentality. In Angola and Zambia 15–17°S latitude delimits subhumid areas, with moisture indices decreasing southward. In Moçambique the subhumid areas along the coastal tracts with their higher PEs are generally of the dry type (zone 4) becoming moist subhumid (zone 3) inland (Fig. 16).

The semi-arid areas of southern Africa are almost devoid of a significant water surplus (Carter 1954). The largest area of semi-arid climate on the subcontinent extends in an arc around the southern Kalahari from southern Angola to the Karoo, including most of Botswana, with extensions eastwards to Moçambique along the dry Limpopo and Zambezi valleys and an isolated area along the coast in northern Moçambique where PE is in excess of 1500 mm per annum (Mather 1962).

An extensive area of arid climates stretches southward along the Angolan and South West African coasts from 13°S to 32°S, extending inland to encompass the southern Kalahari and the Karoo. Moisture indices are mainly < −40 in Angola and < −50 in South West Africa and South Africa.

The biogeographical and botanical interpretations of the various Thornthwaite indices presented largely remain to be undertaken for most of southern Africa. As in the case of the Köppen or Holdridge classifications, only broad correlations

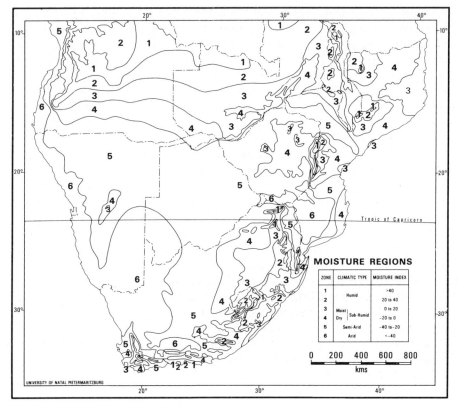

Fig. 16. Moisture regions (after Poynton 1971).

between vegetation types and these indices may be detected from available vegetation maps, as Howe (1953) has testified for Rhodesia and Zambia. When, however, some of the Thornthwaite indices are examined more closely in conjunction with the component values of the monthly water balances from which they are derived (and which are given by Mather (1962) or in diagrammatic form for some southern African stations by Schulze (1958) and Howe (1953)), little doubt exists as to the potential usefulness of these indices in biogeographical research on a subcontinental scale.

6. Conclusions

The response of vegetation to climate in southern Africa, as elsewhere, is both direct and indirect – direct through the role that the factors of radiation, temperature and moisture, independently or in some combination, play on the growth and development of the vegetation, and indirect through the influence of climatic factors on, for instance, soil conditions, competing botanic associations or cultural practices (Mather 1974). In addition – and on a scale not considered in this study of subcontinental southern Africa – the reciprocal influence of vegetation on the microclimate of an area and on the other factors of the microenvironment creates another level of influence that must be considered in evaluating those factors contributing to the distribution of vegetation (Mather & Yoskioka 1968).

Treatment of climatic factors in this review has been from both the individual parameter and the 'combined' index point of view, but has been by no means exhaustive. Thus, for instance, the Walter & Lieth (1960) climate-diagrams, much favoured by botanists, have been omitted because they are felt to be superseded by other indices. The effects of wind have also not been dealt with because they are considered to be largely local. Monthly water balances of individual stations have also not been presented because they are frequently not representative of larger areas and because they are, in any event, given for 385 stations in southern Africa by Mather (1962).

In conclusion it should be emphasized again that the degree to which the distribution of vegetation can be explained on the basis of climatic parameters and indices depends largely on the proper selection of the climatic factors. In this review it is therefore considered that for southern Africa no one factor or index should be used to the exclusion of others, although some indices dealt with are inherently more suitable than others. Thus, temperature and precipitation by themselves describe climate poorly, as precipitation does not really indicate whether a climate is moist or dry unless one is able to compare it with the water need of a place, and the temperature per se does not really reveal the energy that is useful for plant development unless the moisture condition of the soil is also known at the time. However, both temperature and precipitation serve as most important inputs in determinations of more 'active' factors of climate in relation to vegetation, like water surplusses or deficiencies, plant water usage (evapotranspiration), or the moisture index, which have been discussed in their southern African context.

References

Acocks, J. P. H. 1953. Veld Types of South Africa. Bot. Survey of S.A. Memoir 28:1–192.
Adamson, R. S. 1938. The Vegetation of South Africa. Brit. Emp. Veg. Com., London.
Airy Shaw, H. K. 1947. The vegetation of Angola. J. Ecol. 35:23–48.
Buys, M. E. L. & Jansen, J. P. 1970. Die praktiese aanwending van die Köppen-klassifikasie. Special publication of Society for Geography, Stellenbosch, S. Africa.
Carter, D. B. 1954. Climates of Africa and India according to Thornthwaite's 1948 Classification. Laboratory of Climatology, Publ. Climatol. 7:455–474 (and maps).
Coetzee, B. J. & Werger, M. J. A. 1975. A west–east vegetation transect through Africa South of the Tropic of Capricorn. Bothalia 11:539–560.
De Candolle, A. 1855. Geographie botanique raissonnée, ou exposition des faits principaux et des lois concernat le distribution geographique des plantes de l'epoque actuelle. 2 vols. Masson, Paris.
De Villiers, G. du T. 1975. Reënvalonderskeppingsverliese in die Republiek van Suid-Afrika – 'n Streekstudie. Unpubl. Ph.D. Thesis, Univ. Orange Free State, Bloemfontein.
Driscoll, D. M. 1959. Temperature and Rainfall Distribution in Mozambique. Thesis, Penns. State Univ., Univ. Park, Pa.
Drummond, A. J. & Vowinckel, E. 1957. The distribution of solar energy throughout Southern Africa. J. Meteor. 14:343–353.
Edwards, D. 1967. A plant ecology survey of the Tugela Basin. Mem. Bot. Surv. S. Afr. 36:1–285.
Fabricius, A. F. 1969, cited in Schutte, J. M. 1971. Die onttrekking van water uit die newellaag en lae wolke by Mariepskop. S.A. Dept. Water Affairs, Div. Hydr. Res., Techn. Note, 20:1–21.
Garnier, B. J. & Ohmura, A. 1968. A method of calculating the direct shortwave radiation income of slopes. J. Appl. Meteor. 7:796–800.
Gonçalves, C. A. 1970. Contribuição para o estudo do balanço hídrico e da caracterização climática da Provincia de Moçambique. Servico Meteorologico de Moçambique Mem. 55:1–11.
Gonçalves, C. A. 1971. Balanço hídrico e caracterização climática da Provincia de Moçambique. Servico Meteorologico de Moçambique Mem. 69:1–19.

Granger, J. E. 1975. The plant succession and some related factors in Catchment IX, Cathedral Peak Mountain Catchment Research Station. Unpubl. Ph.D. Thesis, Univ. of Natal, Pietermaritzburg.
Grisebach, A. 1866. Die Vegetations-Gebiete der Erde, übersichtlich zusammengestellt. Petermanns Mitt. 12:45–53.
Griffiths, J. F. (ed.) 1972. Climates of Africa. World Survey of Climatology. Vol. 10. Elsevier, Amsterdam.
Hare, F. K. 1954. The Boreal Conifer Zone. Geog. Studies 1:4–18.
Holdridge, L. R. 1959. Simple method for determining potential evapotranspiration from temperature data. Science 130:572.
Holdridge, L. R., Grenke, W. C., Hatheway, W. H., Liang, T. & Zosi, J. A. 1971. Forest Environments in Tropical Life Zones. Pergamon Press. Ch. 2:4–17.
Howe, G. M. 1953. Climates of Rhodesia and Nyasaland according to the Thornthwaite classification. Geogr. Rev. 43:525–539.
Jackson, S. P. (ed.) 1961. Climatological Atlas of Africa. CCTA/CSA, Lagos–Nairobi.
Kerfoot, O. 1968. Mist precipitation on vegetation. For. Abstracts 29:8–20.
Knoch, K. & Schulze, A. 1957. Niederschlag, Temperatur und Schwüle in Afrika. World Atlas of Epidemic Diseases. Vol. 2. Heidelberger Akademie der Wissenschaften, Heidelberg.
Köppen, W. 1900. Versuch einer Klassifikation der Klimate, vorzugsweise nach ihren Beziehungen zur Pflansenwelt. Geogr. Z. 6:593–611; 657–679.
Köppen, W. & Geiger, R. 1936. Handbuch der Klimatologie. Borntraeger, Berlin. 5 vols.
Kreft, J. 1972. Fog in Rhodesia. Clim. Infor. Sheet 41. Dept. Meteor. Serv., Salisbury.
Linsser, C. 1867. Die periodischen Erscheinungen des Pflanzenlebens in ihrem Verhältnis zu den Wärmeerscheinungen. Mem. Acad. Imp. Sci. St. Petersbourg, Series 7:11.
Louw, W. J. & Kruger, J. P. 1968. Potential evapotranspiration in South Africa. Notos 17:3–14.
McGee, O. S. & Hastenrath, S. L. 1966. Harmonic analysis of the rainfall over South Africa. Notos 15:79–90.
Major, J. 1963. A climatic index to vascular plant activity. Ecology 44:485–498.
Mather, J. R. 1959. The moisture balance in grassland climatology. In H. B. Sprague (ed.): Grasslands, pp. 251–261. Am. Assoc. Adv. Sci., Washington.
Mather, J. R. 1962. Average climatic water balance data: Part 1, Africa. Laboratory of Climatology, Publ. Climatol. 15:115–270.
Mather, J. R. 1974. Climatology: Fundamentals and Applications. McGraw-Hill, London.
Mather, J. R. & Yoshioka, G. A. 1968. The role of climate in the distribution of vegetation. Ann. Assoc. Am. Geogr. 58: 29–41.
Monteith, J. L. 1965. Evaporation and environment. Proc. Symp. Expl. Biol. 19: 205–234.
Nagel, J. F. 1956. Fog precipitation on Table Mountain. Quart. J. Roy. Met. Soc. 82:452–460.
Nagel, J. F. 1962. Fog precipitation measurements of Africa's southwest coast. Notos 11:51–60.
Penman, H. L. 1956. Evaporation: An introductory survey. Neth. J. Agric. Sci. 4:9–29.
Pole-Evans, I. B. 1936. A vegetation map of South Africa. Bot. Surv. of S.A. Memoir 15:1–23.
Poynton, R. J. 1971. A silvicultural map of Southern Africa. S. Afr. J. Sci. 67:58–60 and map.
Prentice, A. A. 1965. Potential evaporation in Rhodesia. Notes on Agric. Met. 13:1–7, and maps, statistics.
Rattray, J. M. & Wild, H. 1960. Vegetation map of the Federation of Rhodesia and Nyasaland. In: Atlas of the Federation of Rhodesia and Nyasaland. Fed. Govt. Printer, Salisbury.
Schulze, B. R. 1947. The climates of South Africa according to the classifications of Köppen and Thornthwaite. S. Af. Geog. J. 29:32–42.
Schulze, B. R. 1958. The climate of South Africa according to Thornthwaite's Rational Classification. S. Af. Geog. J. 40:31–53.
Schulze, B. R. 1965. Climate of South Africa. Part 8, General Survey. S.A. Weather Bur. 28. Govt. Printer, Pretoria.
Schulze, R. E. 1975a. Mapping potential evapotranspiration in hilly terrain. S. Afr. Geog. J. 57:26–35.
Schulze, R. E. 1975b. Incoming radiation on sloping terrain: A general model for use in southern Africa. Agrochemophysika 7:55–61.
Schulze, R. E. 1975c. Catchment evapotranspiration in the Natal Drakensberg. Unpubl. Ph.D. Thesis, Univ. of Natal, Pietermaritzburg.
Spain, E. S. 1971. Calculated radiation in Zambia. Clim. Data. Publ. 16:1–18. Dept. Meteorology, Lusaka.

South African Weather Bureau. 1968. Solar radiation and sunshine. S.A.W.B. 32. Govt. Printer, Pretoria.
South African Weather Bureau. 1963–1973. Annual radiation reports. Govt. Printer, Pretoria.
Thompson, B. W. 1965. The Climate of Africa. Oxford Univ. Press, Oxford.
Thornthwaite, C. W. 1948. An approach towards a rational classification of climate. Geog. Rev. 38: 55–94.
Thornthwaite, C. W. 1952. Grassland Climates. Laboratory of Climatology. Public. Climatol. 5:1–14.
Thornthwaite, C. W. & Hare, F. K. 1955. Climatic classification in forestry. Unasylva 9:50–59.
Tivy, J. 1971. Biogeography: A Study of Plants in the Ecosphere, Oliver & Boyd, Edinburgh.
Torrance, J. D. 1972a. Radiation over Rhodesia. Notes on Agric. Meteor. 22. Dept. Meteor. Serv., Salisbury.
Torrance, J. D. 1972b. Malawi, Rhodesia and Zambia. Ch. 13, 409–460, In: Griffiths, J. F. (1972).
Van Riper, J. 1971. Man's Physical World. McGraw-Hill, London.
Walter, H. 1972. Der Wasserhaushalt der Pflanzen in kausaler und kybernetischer Betrachtung. Ber. Dtsch. Bot. Ges. 85:301–313.
Walter, H. & Lieth, H. 1960. Klimadiagramm-Weltatlas, Fischer, Jena.
Ward, R. C. 1975. Principles of Hydrology. 2nd ed. McGraw-Hill, London.
Watts, D. 1971. Principles of Biogeography. McGraw-Hill, London.
Whitmore, J. S. 1970, The Hydrology of Natal. Symp. Water Natal, Durban.
Whitmore, J. S. 1971. South Africa's water budget. S. Af. J. Sci. 67:166–176.
Wicht, C. L. 1971. The influence of vegetation in South African mountain catchments on water supplies. S. Af. J. Sci. 67:201–209.

3 Rainfall changes over South Africa during the period of meteorological record

P. D. Tyson*

1.	Introduction	55
2.	Ideas concerning climatic change in South Africa	55
3.	Circulation and precipitation patterns	56
4.	Methods of analysis	56
5.	Regional rainfall changes	58
5.1	Progressive changes	58
5.2	Oscillatory variations	58
6.	The quasi twenty-year regional oscillation	59
7.	The quasi ten-year oscillation	60
8.	Manifestations of the quasi twenty-year oscillation	61
9.	Discussion	65
10.	Conclusions	67
References		68

* Since 1969 research into recent climatic change in South Africa has been a continuing project supported in part by the CSIR National Programme for Environmental Science. The work has been done under the direction of the author and those contributing to the project have been T. G. J. Dyer, M. N. Mametse, C. S. Keen and M. A. Abbott. Tree ring research is being carried out as part of the Dendroclimatological Research Project, with the author and T. G. J. Dyer as principal investigators, and with the financial support of the National Programme for Environmental Sciences.

Thanks are due to J. G. Rouse and P. Stickler for preparing the figures contained in this paper and to the CSIR for funding the research.

3 Rainfall changes over South Africa during the period of meteorological record

1. Introduction

That desert and semi-desert conditions are encroaching towards the northern and eastern parts of South Africa, and that Karoo vegetation is replacing grassland and in turn is being replaced by dryland woody species has been suggested by many research workers (Phillips 1938, Acocks 1953, Shantz & Turner 1958, Donaldson & Kelk 1970). Various reasons have been advanced from time to time to explain the phenomenon. It has been suggested that progressive climatic desiccation has taken place (Barber 1910, agricultural correspondents Agric. Jour. Union S. Afr. 1912–1914, Schwartz 1919, 1923, Kokot 1948, Vorster 1957). Secondly it has been suggested that man's activity in general and bad farming in particular have caused deterioration and allowed invasion of Karoo species to take place (Acocks 1953, South Africa (Union of) 1951, South Africa (Republic of) 1968–1972, Talbot 1961, Werger 1973). Yet another suggested cause is that the general ecological balance of marginal areas is so delicate as to be easily, or possibly semi-permanently, reversed in extended dry spells, and so allowing serious soil erosion. Proponents of soil erosion as a cause of desertification are Bayer (1955) and Tidmarsh (1966).

The most obvious possible cause of desertification is climatic and this contention needs to be examined carefully before the hypothesis can be either accepted or rejected.

2. Ideas concerning climatic change in South Africa

For years the debate as to whether South Africa is drying up or alternatively is undergoing some sort of cyclic rainfall variation has continued (Kokot 1948). The observations of early explorers and missionaries suggest that in former times periods of heavy rainfall and expanses of standing water occurred in areas now semi-arid (Moffat 1842, Livingstone 1857). Wilson (1865) was firmly of the belief that the Kalahari had become drier. In analysing meteorological records for 10 stations in Botswana (Bechuanaland), Wallis (1935) came to the same conclusion. In a general review Barber (1910) suggested South Africa as a whole was drying up, a view that was widely held by the public, both lay and informed (Agric, J. Un. S. Afr. 1913), but not by the Union Parliament Senate Select Committee on Droughts, Rainfall and Soil Erosion appointed at the time (Union of S.A. 1914). Nonetheless, in 1918 Schwarz reiterated the idea that South Africa as a whole was undergoing progressive desiccation (Schwarz 1919). His ideas were criticized by Cox (1926), Schumann & Thompson (1934) and Thompson (1936). The fundamental difficulty of refuting the assertion of desiccation lay in the limited length of rainfall record available for analysis.

Ideas concerning the possible cyclic nature of rainfall over South Africa were put forward as early as 1888 (Hutchins 1888, Tripp 1888). Using records extending back to the middle of the nineteenth century, Nevill (1908) found evidence for an 18-year periodicity in the rainfall of Natal. Cox (1925) found some evidence for a 14-year period for Cape Town over an 83-year set of observations. This was later refuted by Loor (1948) in the analysis of a 100-year record for the city. Periodicities in South West African rainfall for the interval 1771–1925 were analysed by van Reenen (1925). Likewise Peres (1930) analysed the available record for Lourenço Marques and found evidence of a 20-year periodicity.

More recently, Vorster (1957) concluded from linear regression analysis of data for 17 stations in the southwestern Cape Province that a general decline of rainfall occurred between 1881 and 1950. This finding was not corroborated by Brook & Mametse (1969), who did, however, support the spatial dependence of regression coefficients suggested by Hofmeyr & Schulze (1963).

In order to test the earlier assertions that South Africa has been undergoing a systematic decline in rainfall, and in order to make use of longer annual rainfall series than have hitherto been available, the evaluation of changes in annual rainfall over South Africa during the period of meteorological record has recently been the subject of detailed investigation (Tyson 1971, 1972, Tyson et al. 1975, Dyer 1975a, b, 1976, Tyson & Dyer 1975).

In this chapter the findings of these studies will be reviewed and extended.

3. Circulation and precipitation patterns

Consequent upon its location between 22° and 35°S most of South Africa is dominated by anticyclonic circulation patterns and attendant subsidence. This is particularly so in winter when the frequency of anticyclonic systems exceeds 70 per cent (Vowinckel 1956). In both summer and winter the mean circulation over South Africa is anticyclonic (Fig. 1, upper left). Rainfall decreases from east to west, except along the southern coast (Fig. 1, upper right). The monthly distribution of rainfall varies regionally, depending on the nature of the controlling precipitation mechanisms. In the northeastern parts of the country summer is the rainy season; in the southwest it is winter. By contrast, the southern coastal region receives an all-seasons rainfall and the arid interior experiences a weakly bimodal distribution (Fig. 1, lower left). The spatial extent of the core areas of these regions is indicated in Fig. 1, lower right.

4. Methods of analysis

The occurrence of trend, the delimitation of regions of like climatic change and the determination of spatial gradients of climatic change have been determined from data at 157 stations for the period 1910–1972 and for a decreasing number of stations with records extending back to 1880. The techniques used in these analyses included the Mann-Kendall test for trend, serial correlation, low-pass numerical filtering, Fourier analysis, spectral analysis (using the Parzen window), principal components analysis and the analysis of variance.

The way in which these techniques have been used and the manner in which the actual classification of regional climatic change has been effected fall beyond the scope of this chapter. Details are to be found in Tyson et al. (1975), Dyer (1975a,b, 1976) and Dyer & Tyson (1975).

Within regions of like climatic change, space mean rainfall series have been synthesized using a mixed two-way analysis of variance model (Tyson & Dyer 1975, Dyer 1975a, 1976). The resultant series define the temporal regional effect for a

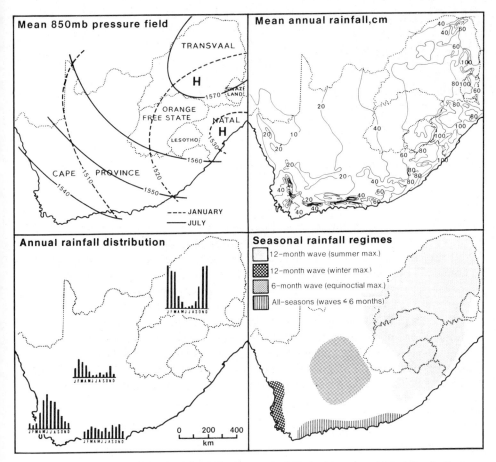

Fig. 1. Mean contours (g.p.m.) of the 850 mb pressure fields in January and July (after Jackson 1961); mean annual rainfall (after Jackson 1961); monthly distribution of rainfall in the summer, winter, all-seasons and equinoctial maxima regions (Pretoria, Cape Town, and Victoria West respectively); and core areas of seasonal rainfall regions (after Keen & Tyson 1973).

region as a whole. In this model the annual rainfall X_{ij} for the ith year (in a series of N years) and the jth station (for p stations) is given by

$$X_{ij} = \mu + \alpha_i + \beta_j + \varepsilon_{ij} \qquad (1)$$

where the best estimate of the space mean rainfall total for a region is defined as

$$\hat{\mu} = \frac{1}{Np} \sum_{j=1}^{P} \sum_{i=1}^{N} X_{ij} \qquad (2)$$

for the $N \times P$ matrix of annual rainfall totals (N rows) and stations (P columns). The best estimate of the mean series constituting temporal regional effect, $\hat{\alpha}_i$, is given by

$$\hat{\alpha}_i = \frac{1}{P} \sum_{j=1}^{P} (X_{ij} - \hat{\mu}) \qquad (3)$$

and the spatial regional effect, β_j, by

$$\beta_j = \frac{1}{N} \sum_{i=1}^{N} (X_{ij} - \hat{\mu}) \qquad (4)$$

The error variable ε_{ij} is determined by the difference in equation (1). The model is based on the assumption that the interaction between α_i and β_i is negligible and that the data are normal.

5. Regional rainfall changes

5.1 Progresseive changes

Non-parametric tests for trend at individual stations throughout South Africa show that the occurrence of trend is highly dependent on the length of series and the particular period being analysed. Of all the 157 series analysed, 24 showed negative trend and two positive trend over the period 1910–1972. Within the spatial distribution of these 24 stations no regional clustering occurred. Extending the record back to 1880, of 30 stations with this length of record, only three showed trend. Eliminating the independence between the data and using the two-way analysis of variance to separate regional from a local station effect, Dyer (1975a) has shown that trend does not occur on a regional scale. Whatever trend occurs, does so only at a local station level. The earlier hypothesis that South Africa has undergone progressive drying up must consequently be rejected as untenable.

5.2 Oscillatory variations

Analysis of unsmoothed data by both spectral and Fourier analysis reveals that in the high frequency range the whole of South Africa experiences a quasi triennial to quadrennial fluctuation in annual rainfall. These variations accounted for about 20 per cent of the observed variance in rainfall over the period 1910–1972. In the dry interior of the Plateau, centred approximately on Prieska, quasi biennial oscillations in rainfall predominate to a greater extent and account for up to 50 per cent of the observed variance (Tyson et al. 1975).

These short-period changes are superimposed on remarkably regular oscillations that exhibit a strong regional distribution. Throughout South Africa annual rainfall totals show a spectrum of variations. Examination of the peaks in rainfall spectra reveals what changes predominate at any particular place. Mapping the amount of variance associated with particular rainfall spectral peaks enables regions of the climatic change to be mapped (Tyson et al. 1975). These regions are distinctively distributed across the country (Fig. 2), and show a clear correspondence with the major rainfall regimes shown in Fig. 1. Summer rainfall areas have been affected predominantly by a quasi 20-year oscillation, the southern Cape coastal all-seasons rainfall belt by a weak 10-year oscillation and the Mediterranean southwestern Cape by complex fluctuations with periods greater than 20 years. Finally the arid interior that experiences equinoctial rainfall maxima has been affected predominantly by a quasi-biennial oscillation. The two most widespread long-period changes of rainfall that have been experienced are

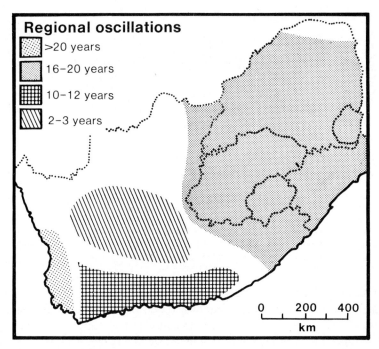

Fig. 2. Regional distribution of areas experiencing like predominant climatic changes during the period 1910–1972 (after Tyson et al. 1975).

those associated with the quasi 20- and 10-year oscillations. These can be considered in more detail.

6. The quasi twenty-year regional oscillation

Taken as a whole, the summer rainfall region of South Africa (as outlined in Fig. 1) has experienced a surprisingly regular pattern of rainfall fluctuations since 1910. Space mean percentage deviations show clearly how the decades 1916 to 1925, 1936 to 1944, and 1954 to 1963 experienced above-average rainfall, whereas the decades in between experienced below-average conditions. The series consisting of the temporal regional effect, and synthesized by the analysis of variance model, shows an almost identical variation. It is the series illustrating the temporal region effect which is given in Fig. 3. The spectra of the mean percentage deviation and temporal regional effect series are likewise almost identical and peak strongly with a period of about 20 years. The series given in Fig. 3 are only for the period 1910 to 1972. Throughout most of South Africa, and particularly in the summer rainfall region of the country, 1973, 1974, 1975 and 1976 have been, on the whole, years of above-normal rainfall. Thus inclusion of data for the years after 1972 shows the development of a fourth set of positive deviations, confirming the continuation of the oscillatory changes up to mid 1977.

In the case of the space mean series of percentage deviations for the summer rainfall region as a whole, the mean range of the fluctuation is 12.4 per cent and the absolute range 19.2 per cent. Owing to the smoothing effected by space averaging, these figures are conservative. The modulation about total annual rain-

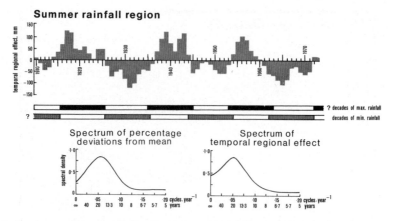

Fig. 3. Yearly temporal regional effect for the summer rainfall as a whole. Data have been smoothed using a 5-term binomial low pass filter. Spectra are for space mean deviations and temporal regional effect and have been determined by spectral analysis of smoothed data.

fall is somewhat higher when based on the analysis of the mean range of the quasi 20-year fluctuation at individual stations. Based on the crude and conservative estimate of the range of the fluctuations as twice the relative variability of the series, it appears that throughout the summer rainfall region the quasi 20-year oscillation is about 20 per cent of the mean (Fig. 4). Using the combined amplitudes of the Fourier waves with amplitudes between 16 and 22 years, it would appear that the range of the oscillation is in most areas nearer 30 per cent of total annual rainfall (Fig. 4).

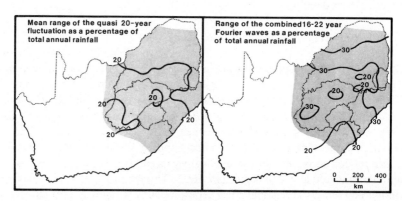

Fig. 4. Range of the quasi 20-year rainfall fluctuation as determined by, left, the ratio of twice the relative variability to total annual rainfall and, right, twice the amplitude of the 3rd and 4th Fourier waves expressed as a percentage of total annual rainfall. The methods of determining range are given in more detail in Tyson & Dyer (1975).

7. The quasi ten-year oscillation

Taken as a whole, the all-seasons rainfall area of the southern Cape has also experienced the effects of the 20-year oscillation. Superimposed on this wave, and somewhat more strongly developed, is the quasi 10-year oscillation (Fig. 5). The

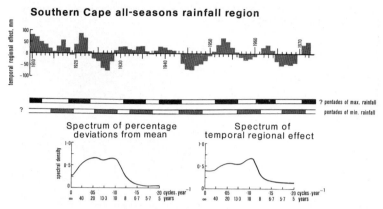

Fig. 5. Yearly temporal regional effects for the all-seasons rainfall as a whole. Data have been smoothed using a 5-term binomial low pass filter. Spectra are for space mean deviations and temporal regional effect and have been determined by spectral analysis of smoothed data.

10-year wave is much more irregular and is consequently associated with a much greater variance than the 20-year wave of the northeastern parts of the country. Maximum and minimum percentage deviations of the space mean series for the region are higher than those for the summer rainfall series. The temporal regional effect is, however, less pronounced. The spectra of the two southern region series are almost identical and clearly show double peaks, one at 10 and the other at 20 years.

Mapping the percentage range of the quasi 10-year fluctuation at individual stations falling within the region experiencing the fluctuations, it is clear that throughout the region the range of the quasi 10-year oscillation is about 20 to 35 per cent of the mean at any one place (Fig. 6). Based on the percentage deviations about the space mean for the region as a whole, the mean range of the oscillation is 16 per cent and the absolute range 45 per cent.

8. Manifestations of the quasi twenty-year oscillation

Further support for the reality of the quasi 20-year oscillation in summer rainfall is to be found in similar oscillations in temperature (Keen 1971), run-off (Abbott

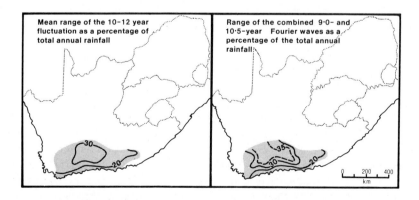

Fig. 6. Range of the quasi 10-year rainfall fluctuation of the all-seasons rainfall region.

61

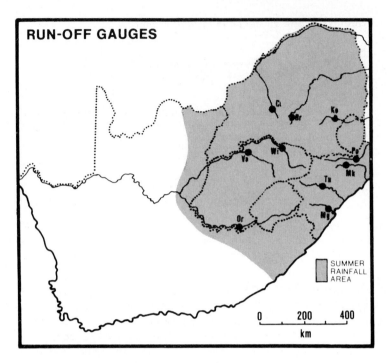

Fig. 7. The location of temperature and run-off stations in the summer rainfall region.

& Dyer 1976) and tree ring widths (Hall 1976). The locations of stations for which mean annual temperature series were determined, and of gauging points (for which catchment run-off data were available for suitable lengths of record) are given in Fig. 7. Tree ring data are only available for Kaarkloof, Natal.

(i) *Temperature changes*

Over the period of analysis (i.e. 1910–1972) temperature and rainfall have shown similar but inverse fluctuations (Table 1). Every station falling within the summer rainfall region experienced a quasi 20-year oscillation in temperature that was 180° out of phase with the rainfall oscillation at that station. Even stations in the southern Cape experienced the 20-year temperature oscillation.

The fact that the quasi 20-year temperature and rainfall fluctuations vary inversely suggests that the fluctuations in the two parameters reflect similar persistent changes in the general circulation of the atmosphere over the sub-continent, since similar but inverse fluctuations may be expected if the controlling mechanism is the general circulation. Periods of higher temperature and lower rainfall are likely to be brought about by persistent anticyclonic conditions favouring subsidence, clear skies, dry air and optimum insolation. Similarly, lower temperature and higher rainfall are likely to be due to persistently greater cyclonic activity, increased convergence and cloud cover accompanied by decreased insolation (Schulze 1965). Such a relationship holds over a timescale of months, when a close correlation exists between seasonal anomalies of precipitation and

Table 1. Amplitudes of rainfall and temperature changes for the stations shown in Fig. 7. Also given are the phase lags between the two series. (Modified after Keen 1971).

	Station number	Rainfall amplitude (mm)	Temperature amplitude (°C)	Phase lag (years)
	157	25	1.7	9
	153	35	1.6	9
summer	140	43	2.2	10
region	136	36	1.2	10
	135	44	1.1	11
	86	58	1.6	10
	67	11	0.4	9
	53	11	0.8	10
	36	14	1.5	11
	29	16	0.9	10
non-summer	23	21	0.8	9
region	17	31	1.1	9
	49	16	2.7	7
	38	17	2.5	15
	79	8	2.0	4

pressure departures over southern Africa (Rubin 1956). Thus an abnormally wet summer over the eastern part of the sub-continent is associated with a general negative departure of pressure, stronger cyclonic activity in the westerlies and a deeper continental trough. By contrast, dry summers coincide with positive pressure anomalies over and to the east of the continent.

(ii) *Run-off changes*

Obtaining continuous run-off data for the period 1910 to 1972 is not easy. However, some are available for river catchments in the summer rainfall area (Table 2).

Table 2. Some river catchments in the summer rainfall area. (The abbreviations given refer to the locations of the gauges shown in Fig. 7.)

River catchment	Length of record
Komati, Ko	1909–1972
Orange, Or	1914–1965
Mgeni, Mg	1936–1972
Tugela, Tu	1927–1966
Mkusi, Mk	1928–1967
Pongola, Po	1929–1967
Vaal, Va	1913–1962
Wilge, Wi	1913–1965
Crocodile, Cr	1922–1971
Bronkhorstspruit, Br	1904–1965

The run-off data have been analysed by filtering each series with a 5-term binomial low-pass filter prior to determining, for successive 5- and 9-year cumulative deviations from the mean, those 5- and 9-year periods of time in which deviations are either highest or lowest. By graphing the periods of maximum and minimum rainfall for each run-off series it is easy to assess the nature of the oscillations present in each series (Fig. 8). In the case of each catchment, except that of the Vaal River, the 9-year periods of maximum and minimum river flow correspond to the decades of maximum and minimum rainfall (shown by the heavy bars in Fig. 8).

The only run-off series covering the 1910–1972 period is that for the Komati River in the eastern Transvaal. Spectral analysis of this data clearly shows the evidence for the quasi 20-year fluctuation.

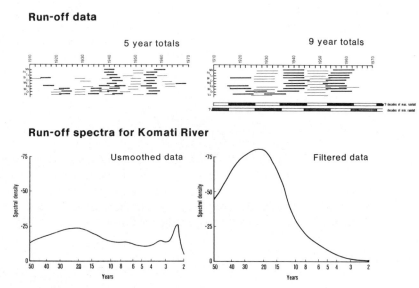

Fig. 8. Years showing maximum and minimum observed run-off from some catchments in the summer rainfall region. The abbreviations of the rivers are defined in Table 2. Decades of maximum and minimum rainfall are given for the summer rainfall region as a whole beneath the 9-year run-off bars. Spectra have been determined by spectral analysis of both raw and smoothed data. Heavy bars indicate above-normal run-off; light bars denote below-normal conditions. (After Abbot & Dyer 1976).

(iii) *Tree ring changes*

The growth habits of many species of trees respond to changes in environmental moisture and temperature. These changes are reflected in the variability of tree ring widths which can be used to provide information on past climates (Fritts 1965, 1971, 1974, LaMarche 1975). Apart from the studies of Walker (1940), Guy (1970) and Storry (1975) in South West Africa and Rhodesia, little work has been done on southern African dendroclimatology. Currently a major dendroclimatological study is being undertaken at the University of the Witwatersrand in Johannesburg. To date (mid 1977) the longest tree ring series available is that produced by Hall (1976) for a *Podocarpus* tree felled in 1910 in the Karkloof, Natal. Preliminary analysis suggests the section of the tree is at least 600 years old. Exact dendrochronological dating is not possible, however, owing to the occurrence of missing rings and incomplete ring sequences in many sectors of the cross-section. Nonetheless, it is possible to determine a reasonably good ring sequence for the last 150 years of the tree's life. A quasi 20-year oscillation was clearly discernible between 1760 and 1910 (Fig. 9).

By coring different species of trees throughout South Africa, and particularly within the summer rainfall region, and analysing the ring width sequences it should be possible to infer the manner in which regional changes in environmental conditions have taken place over the last few centuries.

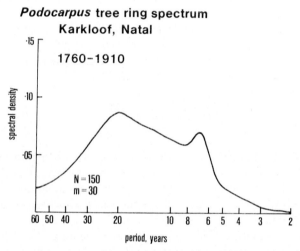

Fig. 9. *Podocarpus* tree ring spectrum from a tree felled in the Karkloof, Natal near Pietermaritzburg in 1910. The data were supplied by Hall (1976) and smoothed with a 5-term binomial filter before being submitted to spectral analysis.

9. Discussion

The quasi 20-year oscillation in rainfall that has been present since 1910 exhibits a distinctive regional occurrence almost coincident with that of the summer rainfall region of South Africa. Decades in which cumulative rainfall has been above normal have centred on about 1921, 1940 and 1958. Decades with cumulative rainfall below normal have centred on about 1911, 1930, 1949 and 1967. Although sparse, evidence does exist in early records to suggest that in the summer rainfall areas of South Africa regular spells of wet years occurred in the nineteenth century. Peak rainfall was reported by the Natal Government Astronomer for the colony in the eighteen fifties, eighteen seventies and eighteen nineties (Nevill 1908). It is quite definite that the eighteen nineties were wet over most of the country. Around 1910 the rhythmic quasi 20-year variation of wet years appears to have faltered to become re-established from the nineteen twenties onwards. Throughout the region of its occurrence the oscillation has a range between 20 and 30 per cent of total annual rainfall.

Given that the rhythm and pattern of change over the last sixty years or so continues to repeat itself, then it can be expected that the wet seasons the country is experiencing presently will continue until towards the end of the decade. Likewise the beginning of the next decade may well be drier than normal. This is not to say that for the rest of the seventies one or two years will not experience droughts or that the early and middle eighties will not experience occasional floods. On the basis of past experience all that can be said is that there is a good chance that the cumulative total rainfall in the summer rainfall areas will be above normal for the rest of this decade and below normal for much of the next.

Qualitative prediction of near-future wet and dry pentades in the southern Cape is far more difficult, owing to the mixed and more irregular nature of the oscillations in that area. From Fig. 6 it appears that the pentades 1971 to 1976 and 1981 to 1986 might be wetter than those immediately prior to or after and that 1976 to 1981 will be drier.

Quantitative estimation of future rainfall changes has been attempted by Dyer (1975a) using the temporal regional effect series defined by equation (3) and given in Figs. 3 and 5. Dyer's forecasting model is a trigonometric regression type in which predicted rainfall R_i in any year i is given by

$$R_i = A_0 + \sum_{m=1}^{k} \left[A_m \cos \frac{2\Pi i}{\lambda_m} + B_m \sin \frac{2\Pi i}{\lambda_m} \right] \qquad (5)$$

for $0 \leqslant i \leqslant N-1$ and $1 \leqslant m \leqslant k$ where A_m and B_m are regression coefficients determined by least squares; A_o is the regression constant; N is the number of observations of R_i; k is the number of prominent peaks in the spectrum (as determined by spectral analysis) and λ_m the periods of the oscillations at each of the $m = 1 \ldots k$ peaks.

The prediction equation differs from its Fourier counterpart in that the values of λ_m are determined from the spectrum and do not necessarily have to be integer values. Furthermore, they are not constrained to be multiples of the fundamental period (i.e. length) of the series. Regression coefficients used to predict future changes in the summer and all-seasons rainfall regions are given in Table 3.

Table 3. Regression coefficients used for predicting future long-period rainfall changes (after Dyer 1975a).

		Summer rainfall region		
λ_m (Years)	A_o (mm)	A_m (mm)	B_m (mm)	SE (mm)
18.60		−20.77	−66.03*	
3.60		40.76*	51.62*	
2.34		60.44*	10.58	
	2.92			105.27
		All-seasons rainfall region		
λ_m	A_o	A_m	B_m	SE
10.40		16.7	34.49*	
3.50		37.87*	4.88	
3.00		32.16*	2.45	
2.10		24.39*	9.41	
	−0.32			63.09

* significant at the 5 per cent level.

The wavelengths given in Table 3 refer to the peaks in the spectrum of raw regional temporal effect series. The forecast annual rainfall totals given by equation (5) and Table 3 have been smoothed with a 5-term binomial low-pass filter to give the predicted series given in Fig. 10. It is interesting to note that the positive deviations from the mean predicted by Dyer for 1973, 1974, 1975 and 1976 have all been realized for the summer rainfall region.

In order to refine forecast models such as that given in equation (5) it is vital to extend the series from which the regression is determined as far back in time as possible. It is for this reason that the dendroclimatological study now being undertaken is so important. Not only will it provide insights into past climatic changes but a means for estimating those of the future.

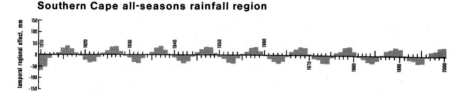

Fig. 10. Predicted rainfall changes in the summer and all-seasons rainfall regions to the year 2000. The annual forecasts are based on equation (5) and have been smoothed using the 5-term binomial low-pass filter. (After Dyer 1975a.)

10. Conclusions

Space mean rainfall series for the northeastern summer rainfall region and for the southern Cape all-seasons rainfall region show that the predominant oscillation in the former region is a quasi 20-year wave; in the latter a quasi 10-year wave. In neither case is there any evidence of progressive trends of either a positive or negative nature. In both cases the major changes in rainfall since 1910 have been oscillatory and, particularly in the case of the summer rainfall region, have been remarkably regular.

The implications of such changes, should they continue to occur in the future as they have in the past may be profound, particularly concerning food production in arid or semi-arid areas. Likewise, the general environmental implications need careful consideration. The absence of clear evidence for steadily declining rainfall shows that desertification in South Africa cannot have been the result of a progressive drying up of the country since the turn of the century. This does not mean that lack of rainfall has not been a major factor in desert encroachment. Now that the oscillatory character of South African rainfall has been demonstrated beyond all reasonable doubt for the period 1910 to 1972, the causal link between extended dry spells and changing vegetation must be examined carefully for different ecosystems in the country. Only when this has been done will it be possible to reconsider the complex interaction between man, his activities, and his environment with a view to explaining desert encroachment in South Africa.

References

Abbott, M. A. & Dyer, T. G. J. 1976. On the temporal variation of rainfall run-off over the summer rainfall region of South Africa. S. Afr. J. Sc. 72:276–278.
Acocks, J. P. H. 1953. Veld types of South Africa. Mem. Bot. Surv. S. Afr. 28:1–192.
Agric. J. Union S.A. 1912–1914. Is South Africa drying up? v. 4, 662–666; v. 6, 693–694, 1027–1029; v. 7, 47–60, 113–114, 150–154, 256–261, 243–424, 523–538, 732–739, 894–898; v. 8, 107–113, 262–265.
Barber, F. H. 1910. Is South Africa drying up? Agric. J. Cape G.H. 86:167–170.
Bayer, A. W. 1955. The ecology of grasslands. In: Meredith, D. The grasses and pastures of South Africa. Pp. 539–550. Central News Agency, Johannesburg.
Brook, G. A. & Mametse, M. N. 1969. Rainfall trend patterns in South Africa. S.A. Geog. J. 52:134–138.
Cox, G. W. 1925. Periodicity in rainfall. Trans. Roy. Soc. S. Afr. 12:295–299.
Cox, G. W. 1926. Some notes on the circulation of the atmosphere over Southern Africa. S.Afr.J. Sc. 23:103–167.
Donaldson, C. H. & Kelk, D. M. 1970. An investigation of the veld problems of the Molopo area: I. Early findings. Proc. Grassl. Soc. S. Afr. 5:50–57.
Dyer, T. G. J. 1975a. Secular variations in South African rainfall. Ph.D. thesis, Univ. of Witwatersrand, Johannesburg. (Unpubl.)
Dyer, T. G. J. 1975b. The assignment of rainfall stations into homogeneous groups: an application of principal components analysis. Quart. J. R. Met. Soc. 101:1005–1013.
Dyer, T. G. J. 1976. On the components of time series: the removal of spatial dependence. Quart. J. R. Met. Soc. 102:157–165.
Fritts, H. C. 1965. Tree ring evidence for climatic changes in Western North America. Mon. Weath. Rev. 93:421–443.
Fritts, H. C. 1971. Dendroclimatology and dendroecology. Quater. Res. 1:419–449.
Fritts, H. C. 1974. Relationships of ring widths in and site conifers to variations in monthly temperature and precipitation. Ecol. Monog. 44:411–440.
Guy, G. L. 1970. Adansonia digitata and its rate of growth in relation to rainfall in South Central Africa. Proc. & Trans. Rhod. Sci. Assoc. 54:68–84.
Hall, M. J. 1976. Dendroclimatology, rainfall and human adaptation in the later Iron Age of Natal & Zululand. Ann. Natal Mus. 22:693–703.
Hofmeyr, W. L. & Schulze, B. R. 1963. Temperature and rainfall trends in South Africa during the period of meteorological records. Proc. Rome Confr. on Changes of Climate with special reference to Arid Zones. Arid Zone Res. 20:81–85.
Hutchins, D. E. 1888. Cycles of drought and good seasons in South Africa. Times Office, Wynberg.
Jackson, S. P. 1961. Climatological Atlas of Africa. CCTA/CSA, Nairobi.
Keen, C. S. 1971. Rainfall spectra and the delimitation of rainfall regimes in South Africa, M.Sc. thesis, Univ. of Witwatersrand, Johannesburg. (Unpubl.)
Keen, C. S. & Tyson, P. D. 1973. Seasonality of South African rainfall: a note on its regional delimitation using spectral analysis. Arch. Met. Geoph. Biokl., Ser. B., 21:207–214.
Kokot, D. F. 1948. An investigation into the evidence bearing on the recent climatic changes over Southern Africa. Union of S.A., Irrigation Dept. Memoir. Govt. Printer, Pretoria.
LaMarche, V. C. 1975. Climatic clues from tree-rings. New Scientist 66 (943):8–11.
Livingston, D. 1857. Missionary travels and research in S.A. London.
Loor, B. de. 1948. Die ontleding van Kaapstadse reënval (1938–1946), Tydskr. Wet. en Kuns, New Series 8:34–36.
Moffat, R. 1842. Missionary labours and scenes in South Africa. London.

Nevill, E. 1908. The rainfall in Natal. Agric. J. Natal 11:1531–1533.
Peres, M.A. 1930. Preliminary investigations on the rainfall of Lourenço Marques. S. Afr. J. Sc. 27:132–135.
Phillips, J. 1938. Deterioration in the vegetation of the Union of South Africa, and how this may be controlled. S. Afr. J. Sc. 35:476–484.
Rubin, M. J. 1956. The associated precipitation and circulation patterns over Southern Africa. Notos 5: 53–63.
Schulze, B. R. 1965. General Survey, Climate of South Africa. Part 8. Govt. Printer, Pretoria.
Schumann, T. E. W. & Thomson, W. R. 1934. A study of South African rainfall, secular variations and agricultural aspects. University of Pretoria, Series No. 1, 28:1–46.
Schwarz, E. H. L. 1919. The progressive desiccation of Africa: the cause and the remedy. S. Afr. J. Sc. 15:139–190.
Schwarz, E. H. L. 1923. Kalahari scheme as the solution of the South African drought problem. S. Afr. J. Sc. 20:208–222.
Shantz, H. L. & Turner, B. L. 1958. Photographic documentation of vegetational changes in Africa over a third of a century. Univ. of Arizona.
South Africa (Republic of) 1968–1972. Commission of enquiry into Agriculture, Reports. Interim Report, R.P. 61 of 1968, 100 pp., chairman, M. D. Marais; second report, R. P. 84 of 1970, 234 pp., chairman, M. D. Marais; third (final) report, R.P. 19 of 1972, 221 pp., chairman, S. J. du Plessis.
South Africa (Union of) 1951. Desert encroachment committee, Report. Government Printer, Pretoria, U.G. 59 of 1951, 27 pp., chairman, C.R. van der Merwe.
Storry, J. G. 1975. Preliminary dendrochronology study in Rhodesia. S. Afr. J. Sc. 71:248.
Talbot, W. J. 1961. Land utilization in the arid regions of Southern Africa, Part 1: South Africa. Arid Zone Res. 17:299–331.
Thomson, W. R. 1936. Moisture and farming in South Africa. South African Agricultural Series, 14:1–260.
Tidmarsh, C. E. M. 1966. Pasture research in South Africa. Proc. Grassl. Soc. S. Afr. 1:21–26.
Tripp, W. B. 1888. Rainfall of South Africa, 1842–1886. Quart. J.R. Met. Soc. 14:108–123.
Tyson, P. D. 1971. Spatial variation of rainfall spectra in South Africa. Ann. Assoc. Am. Geogr. 61: 711–720.
Tyson, P. D. 1972. Rainfall spectra and recent climatic variation in Southern Africa. International Geography 1972. Pp. 202–204. Toronto Univ. Press.
Tyson, P. D. & Dyer, T. G. J. 1975. Mean fluctuations of precipitation in the summer rainfall region of South Africa. S. Afr. Geogr. J. 57:104–110.
Tyson, P. D., Dyer, T. G. J. & Mametse, M. N. 1975. Secular changes in South African rainfall: 1910–1972. Quart. J. R. Met. Soc. 101:817–833.
Union of S.A. 1914. Report from the Select Committee on droughts, rainfall and soil erosion. The Senate, Union Parliament, S.C.2., Pretoria.
van Reenen, R. J. 1925. Note on the apparent regularity of the occurrence of wet and dry years in South West Africa. S. Afr. J. Sc. 22:94–95.
Vorster, J. H. 1957. Trends in long range rainfall records in South Africa. S. Afr. Geog. J. 39:61–66.
Vowinckel, E. 1956. Ein Beitrag zur Witterungsklimatologie des suedlichen Mozambiquekanals, Miscelânea Geofisica Publicada Pelo Servico Meteorológico de Angola em Comemoracâo do X Aniversário do Servico Meteorológico Nacional, pp. 63–86.
Wallis, A. H. 1935. Is our rainfall getting less? The 1820 Magazine 6:35–37.
Walter, H. 1940. Die Jahresringe der Bäume als Mittel zur Feststellung der Niederschlagsverhältnisse in der Vergangenheit, insbesondere in Deutsch-Südwestafrika. Die Naturwissenschaften 38:607–612.
Wilson, J. F. 1865. Water supply in the basin of the River Orange or 'Gariep South Africa', J. R. Geog. Soc. 35:106–129.
Werger, M. J. A. 1973. Phytosociology of the Upper Orange River valley. V & R, Pretoria, 222 pp.

4 Schematic soil map of southern Africa south of latitude 16°30′S

H. J. von M. Harmse

1. Introduction 73
2. Description of soils 73

References 75

4 Schematic soil map of southern Africa south of latitude 16°30′S

1. Introduction

The nature and composition of the soil mantle of regions as large as southern Africa, could be represented in many different ways. To avoid lengthy descriptions, however, and to show the geographical distribution, interrelationships and salient features of the soil in a comprehensive manner, a schematic map was chosen as the most suitable form of data representation. Such a map of southern Africa, south of latitude 16°30′ was compiled from available data on soils, geology, geomorphology and climate (Du Toit 1964, F.A.O. 1973, Haughton 1969, King 1962, Soil Survey Staff 1973, Thompson 1965, Weather Bureau 1957).

Soil, such as other natural objects, may be classified by any suitable attribute or set of attributes. A special classification for each purpose is possible (Smith 1965). For this particular objective, the legend had primarily been designed to group the major soils of the area into categories which not only would have relevance to the distribution of the natural vegetation, but would also reflect the genetic factors of soil formation. The degree of extrapolation and abstraction to construct such a map, however, increases in an inverse ratio to the scale of the map.

Mapping units in the form of associations, representing the most commonly occurring soils were chosen. The first of the set of symbols refers to the dominant soils in a particular mapping unit, whereas the remainder refer to other constituents in order of incidence. A detailed description of the geology, climate and topography of each association is not intended. For this the reader is referred to the works of Haughton (1969), King (1962) and the Weather Bureau (1957) (see also Chapters 1 and 2).

2. Description of soils

The following description of soils corresponds more or less to the group level of abstraction. The criterion used for differentiation is mainly the degree of alteration of the original parent material. The materials from which a soil was formed has often been subjected to one or more cycles of weathering. In some instances the minerals of the deposits, for example those of aeolian sand, are so resistant to weathering that clay formation is practically non-existent. In this grouping the processes have been ignored, only the end product has been considered.

A. Ferrallitic soils

These soils are formed in areas where the mean annual rainfall is more than 800 mm. In southern Africa they are usually found in areas of relatively high altitudes. Horizons are weakly differentiated and boundaries are diffuse. There is little or no reserve of weatherable minerals. Clay minerals, of the 1:1 lattice types are mixed with significant quantities of free iron oxides and occasionally with aluminium oxides. The cation exchange capacity of the clay fraction is usually less than 20 me/100 g and the degree of base saturation in the A and B horizons is less than 40 per cent. In the mist belt some of these soils have humic epipedons.

B. *Fersiallitic soils*

In southern Africa these soils are usually found in a broad zone where the mean annual rainfall is between 500 and 800 mm. Transition between horizons are diffuse, but there may be a marked degree of segregation of iron oxides and hydroxides, especially in those solums with yellow B horizons. Some of the profiles may have incipient A_2 horizons. B-horizons are usually textural and structures are weakly to moderately developed.

The reserve of weatherable minerals may vary from very low, in soils developed on aeolian deposits, to appreciable in soils developed on basic rocks. The clay minerals are usually kaolinitic with or without sub-ordinate quantities of 2:1 lattice clays. The amount of 2:1 lattice clays increases with decreasing rainfall. The base saturation in the B-horizon is generally more than 40 per cent.

C. *Black and red montmorillonitic clays*

These soils are formed mainly on basic and ultrabasic igneous rocks and on calcareous argillaceous sediments. They are generally dark coloured and their morphology is characterized by a varying degree of argillo-pedoturbation. They occur in both upland and bottomland positions, especially in hot, dry areas. This category contains soils with vertic and melanic epipedons.

D. *Solonetzic and planosolic soils*

All the soils with textural B-horizons, with or without A_2-horizons, are included in this category. B-horizons have strongly to moderately developed pedocutanic or prismacutanic characteristics. Some B-horizons are 'sodic' or 'magnesic'. Also included in this category are soils with textural B-horizons which do not qualify as natric, i.e. planosols. Planosols may have been former solonetz which have lost most of the sodium by leaching. Colours vary from greyish brown to reddish brown. Soils of this category usually occur extensively in bottomland positions and lower slopes on granite and argillaceous sediments.

E. *Halomorphic soils*

All soils containing appreciable amounts of free salts, including soils which have been described as solontchaks in the literature, are included in this category.

F. *Arenosols*

The salient feature of the soils included in this category is their low reserve of weatherable minerals and their low silt/clay ratio. Apart from their colour, red in upland sites and on dune ridges, yellow in level areas and grey in bottomland sites and/or areas with impeded drainage, all their characteristics appear to be inherited from the parent material. The parent material is aeolian sand or sands derived from aeolian deposits. Shifting sands were classified as regosols. Soils associated with stabilized littoral or near littoral deposits were also included.

These soils become progressively acid with an increase in rainfall and their properties then approach those of the fersiallitic and ferralitic groups.

G. *Alluvial and other weakly developed soils of low lying areas*

Marshlands, young immature soils flanking the major rivers and dark coloured soils between sief dunes, comprise the major components of this category. Incipient soil formation is manifested in signs of clay movement and faunal activity.

H. *Weakly developed shallow soils*

Soils belonging to this category are located in a zone which comprises the transition between fersiallitic soils and soils of the arid regions. In higher rainfall areas they occur on dissected upland sites. Reserves of weatherable minerals are appreciable, base saturation and CEC are usually high. These soils are mainly reddish-brown to greyish-brown, sandy loams and loams, overlying reddish-brown to yellow-brown, sandy loams and clays. The sub-soil is often mottled but not hydromorphic. The soils are shallow to moderately shallow, and stonelines are common. Parent material is usually granite, gneiss, or metamorphosed siliceous rocks.

Solonetzic soils in bottomland sites, are extensively associated with soils of this category.

I. *Weakly developed soils of arid regions*

These include shallow brown to greyish-brown soils with or without B-horizons. Their properties and characteristics depend largely on the composition of the parent material. These soils often contain free lime. Gypsum may occur in areas where the rainfall is less than 300 mm.

J. *Lithosols*

Shallow soils with weak profile differentiation, containing coarse fragments and solid rock at depths of 30 cm and less. Topography is the dominant soil forming factor. Removal of the products of weathering exceeds their formation.

References

Du Toit, A. L. 1954. The geology of South Africa. Oliver & Boyd, London.
F.A.O. 1973. Soil map of the world; scale 1:5 000 000. Unesco, Rome.
Haughton, S. H. 1969. Geological history of Southern Africa. Geological Soc. of S.A., Johannesburg.
King, L. C. 1962. The morphology of the earth. Oliver & Boyd, London.
Smith, G. D. 1965. Soil Classification. Pedologie, Nr. 4.
Soil Survey Staff 1973. Soil map, Republic of South Africa; C. N. MacVicar (ed.). National Research Institute for Soils and Irrigation, Pretoria.
Thompson, J. G. 1965. The Soils of Rhodesia and their classification. Government Printer, Salisbury.
Weather Bureau 1957. Climate of South Africa. Government Printer & Weather Bureau, Pretoria.

5 Late Cretaceous and Tertiary vegetation history of Africa

D. I. Axelrod and P. H. Raven*

1.	Introduction	79
2.	Cretaceous breakup of Gondwanaland	80
3.	Late Cretaceous–Paleogene	84
3.1	Tropical rainforest	84
3.2	Temperate rainforest	87
3.3	Montane rainforest	92
3.4	Subtropical evergreen forest	93
3.5	Subtropic woodland-scrub	93
3.6	Summary	95
4.	Neogene	95
4.1	Terrain	95
4.2	Climate	97
4.3	Tropical rainforest	99
4.4	Montane rainforest	102
4.5	Subtropical evergreen forest	106
4.6	Savannas and woodlands	106
4.7	Temperature rainforest	109
4.8	Sclerophyll vegetation	110
4.9	Cape vegetation	110
4.10	Deserts and semideserts	114
4.11	Summary	116
5.	Conclusions	116
6.	Epilogue	118
References		119

* We are grateful to the U.S. National Science Foundation for a series of grants to each of us independently, and to H. P. Bailey, R. Estes, E. P. Plumstead, J. P. Rourke, A. J. Tankard, and H. Wild for useful suggestions and information. Peter Goldblatt has reviewed the entire manuscript and aided materially in its preparation.

5 Late Cretaceous and tertiary vegetation history of Africa

1. Introduction

Twenty-five years have elapsed since Moreau (1952) completed his outstanding essay, 'Africa since the Mesozoic: with particular reference to certain biological problems'. This valuable synthesis resulted from his interest in a major problem raised by the avifaunas. They reveal great differences between the four principal habitats of the continent: the arid regions, savanna, lowland evergreen forest, and montane evergreen forest. Since two or more of these ecosystems regularly are contiguous over wide areas, the coexistence of distinct avifaunas in them for a long period of time requires explanation, as does the existence of the different vegetation zones themselves.

Moreau's collation of geologic and climatic data for analysis of the problem can now be modified and supplemented in the light of data from the new field of plate tectonics, which clarifies certain aspects of African environments since the mid-Cretaceous (e.g., Bird & Isacks 1972). In addition, there are now available broad syntheses of its geologic history (DuToit 1954, Haughton 1963, Truswell 1970), its ancient landscapes (King 1962, 1963), and its land life especially as it relates to distribution and migration before the breakup of Gondwanaland. The older vertebrate faunas have been reviewed by Cox (1974), Keast (1972), Corydon & Savage (1973), and Cracraft (1973, 1975), and Cooke (1972) has analysed the history of the Tertiary mammal faunas of Africa.

Its plant life from the later Paleozoic (Plumstead 1973, Chaloner & Lacey 1973) into the early Cretaceous (Barnard 1973, also DuToit 1954) is also well outlined. But when we turn to the later Cretaceous and Tertiary, we find – as did Moreau – that the record of plant life in Africa is exceedingly incomplete. Apart from a few older Tertiary records in the Saharan region, only a few sites have yielded small, mostly inadequate samples of fossil floras that are younger than the middle Cretaceous. They are chiefly near the coast in South Africa, and in the rift system that penetrates into it from the north. The incompleteness of the plant record reflects in large measure the uplift of the continent and the progressive restriction of lowland areas following the breakup of Gondwanaland, and hence of basins in which the record of plant life might be preserved in accumulating fine-grained sediments. Although the broad Tertiary downwarps within the continent (Burke & Wilson 1972) no doubt have fossiliferous sedimentary deposits, they are largely inaccessible because they are covered by broad alluviated plains. Despite the poor representation of plant life following the middle Cretaceous, we can nonetheless draw some provisional inferences about vegetation in southern Africa from data in North Africa, and also from fossil sites on land areas which were formerly united with Africa, and have since been rafted away by ocean-floor spreading. The following analysis of vegetation history therefore relies chiefly on inferences drawn from three major lines of evidence:

(1) The geologic setting of the continent in the Cretaceous when it was connected with, or close to, lands that are now in distant regions. This provides an indication

of the ages of allied taxa that are now on remote continents, and affords a reliable basis for inferring the general environment in southern Africa itself.

(2) The general isolation of Africa following the later Cretaceous, and the effect of its geomorphic history on climate and terrain, and hence on the history of vegetation and of life dependent upon it.

(3) The topographic and accompanying climatic changes during the middle and later Tertiary that restricted older adaptive (vegetation–climate) zones, and opened up new expanding ones for the more modern ecosystems.

2. Cretaceous breakup of Gondwanaland

By Triassic time, Gondwanaland was separated from Laurasia on the north by the broad Tethys Sea (Bird & Isacks 1972, Tarling & Runcorn 1973, Hughes 1973, Tarling & Tarling 1971, Monod 1975, Hallum 1973). Early Mesozoic Gondwanaland had low relief (King 1962, also Chapter 1), being composed of broad plains and basins, and scattered Precambrian highs. As Gondwanaland was rifted in the Triassic and Jurassic, basalts totaling thousands of cubic kilometres in volume welled up and spread widely over the peneplaned surface (Burke & Dewey 1974). This was followed by separation of the fractured segments into the present southern continents (as well as Arabia and India), and thence centrifugal rafting to their present positions (Fig. 1). It is important to recall the times when the major fragments parted company with the African nuclear area for they bear directly on the times of separation of taxa that were formerly linked, and their later evolutionary history.

(1) The South Atlantic gradually widened as active sea-floor spreading commenced along the developing mid-Atlantic Ridge (Dietz & Holden 1970, Smith & Hallum 1970, Smith et al. 1973). The Falkland plateau at the tip of South America separated from the Agulhas escarpment of South Africa in the earliest Cretaceous (Larson & Ladd 1973), but they were still in proximity well into the middle Cretaceous (Dingle 1973, Bolli et al. 1975, as shown in Fig. 2). Saline deposits accumulated in the Aptian Sergipe–Angola basin (Allard & Hurst 1963) and elsewhere prior to the development of open circulation in the South Atlantic (Bolli et al. 1975, Burke 1975). As the Cretaceous progressed, continuing volcanism along the mid-Atlantic Ridge gradually widened the South Atlantic. It joined with the North Atlantic when the area of Pernambuco (Brazil) and Guinea (Africa) gradually moved apart in the mid-Turonian (Reyment 1969, Reyment & Tait 1972). During this time the Walvis thermal center had become active, developing the Rio Grande Rise and the Walvis Ridge which were carried generally northwesterly and northeasterly, respectively, by the moving plate (Dietz & Holden 1970). Their trace provides evidence of the latitudinal movement of Africa and South America since the Walvis thermal center was activated in the early Cretaceous and that subsidence of the eastern segment of the Walvis Ridge occurred after mid-Cretaceous (110 m.y.) time (Pastoret & Goslin 1974, Goslin & Sibuet 1975, Ladd et al. 1973). By the beginning of the Tertiary, Liberia was separated from Paraiba (Brazil) by about 800 km. Nonetheless, the Atlantic was relatively shallow and populated by numerous islands that provided stepping

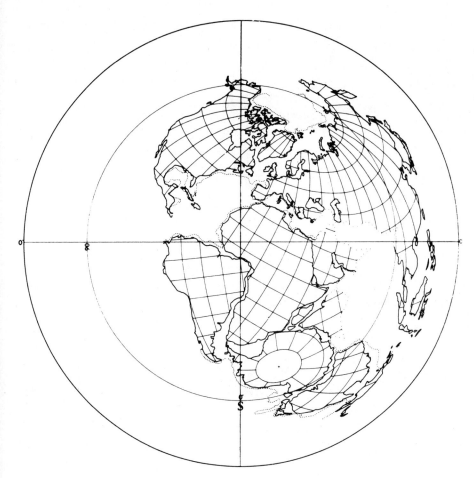

Fig. 1. Arrangement of the continents in the medial Cretaceous (from Briden et al. 1974).

stones across the region which have since been carried subsea (Tarling & Tarling 1971, Axelrod 1972, Fig. 25, Carr & Coleman 1974, Fig. 2).

(2) Madagascar was connected with Africa into the mid-Cretaceous, when it was situated about 15°N of its present area, against Tanzania–Kenya (Smith & Hallam 1970, Embleton & McElhinney 1975, review by P. J. Smith 1976). Madagascar was then a part of the now largely submerged Mascarene Plateau which joined India on the east into the latest Cretaceous. It was this Malgasy–Mascarene subcontinent that provided a route for the Cretaceous dinosaurs common to Africa–India–Madagascar and other regions, as well as for the rich angiosperm floras of Madagascar and the Seychelles that have numerous relicts. The exact time of separation of Madagascar–India from Africa remains to be determined, but could have occurred at any time between the mid- and late-Cretaceous. An earlier date might explain the absence of primary freshwater fish from Madagascar, if they originated later—after separation.

(3) India separated from the Madagascar–Mascarene subcontinent in the early Paleocene (65 m.y.; Molnar & Francheteau 1975). It moved rapidly north to meet

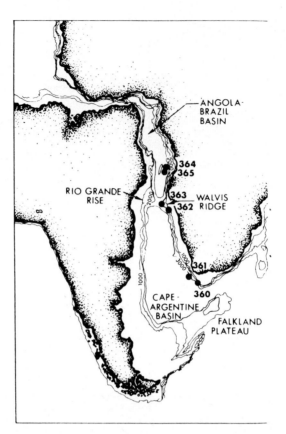

Fig. 2. Africa and South America in the middle Cretaceous (from Bolli et al. 1975). Sites 360–365 are from Leg 40 of the Deep Sea Drilling Project by the Glomar Challenger.

the Asian landmass by the middle Eocene (McKenzie & Sclater 1973, Sclater & Fischer 1974, Powell & Conaghan 1973) a date consistent with the occurrence of Laurasian mammals in the middle Eocene of the Murree Hills (Ranga Rao 1972).

(4) To the south (Fig. 3), Antarctica formed a bridge between South America (Fuegia) and Australia into the Eocene (McKenzie & Sclater 1971, Smith et al. 1973). As rifting and thence spreading along the Indian Antarctic Rise commenced in the middle and later Eocene (50 m.y.), Australia moved to warmer middle latitudes. However, Antarctica did not separate from the South Tasman Rise until the middle or late Oligocene (30–25 m.y.). As the opening was complete between Antarctica and the South Tasman Rise, effective circum-Antarctic circulation now commenced (Raven & Axelrod 1972, Kennett, Houts et al. 1974, Jenkins 1974). As Drakes Passage opened in the Eo-Oligocene (Foster 1974), westerlies rapidly gained strength as the global temperate gradient increased. To judge from the pollen floras on Antarctica (see below), Oligo-Miocene (30–25 m.y.) sea-level temperatures probably were like those now in southern cool temperate New Zealand or Chile. With a low annual range of temperature, firn limit probably was close to 1800–2000 metres, with ice tongues reaching down to lower levels. As temperatures were progressively lowered, the spread of ice and the increased speed of the circum-Antarctic circulation system now had a

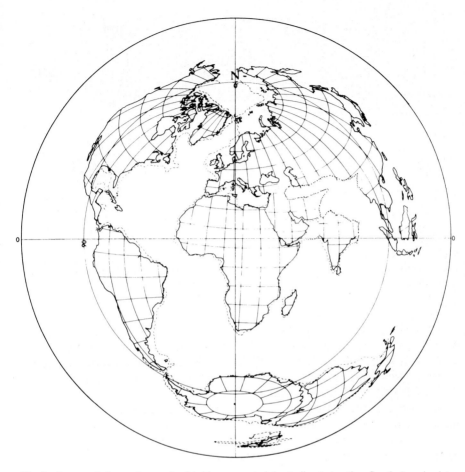

Fig. 3. An essentially continuous land-bridge connected Australia–Antarctica–South America into the middle Eocene (from Briden et al. 1940).

profound effect on life, both on land (Raven & Axelrod 1972, Raven 1973a) and in the sea (Kennett et al. 1972, 1974).

(5) Union of Africa with Eurasia was not completed until the middle Miocene ($\sim$17–18 m.y.), prior to the opening of the Red Sea (Dewey et al. 1973). The movements that brought northeast Afro-Arabia against south-central Asia in the middle Miocene effectively closed the Tethys, and hence the long-persistent latitudinal circulation system which brought warm moist climate into the entire mediterranean region and southern Asia as well (Fig. 4). In addition, it enabled for the first time a broad intermingling of the Eurasian and African mammalian faunas (Van Couvering & Van Couvering 1975, Berggren & Van Couvering 1974). Proboscideans, giraffoids, pigs and other characteristic members of the early Miocene mammalian fauna now moved north, whereas equids, perissodactyls, mustelids and others moved south. Many modern genera of birds likewise appear in Africa for the first time in the mid-Miocene (Rich 1974). Van Couvering & Van Couvering (1975) raise the possibility that some of the intermittent interchange that had occurred earlier in the Tertiary may owe not to regular

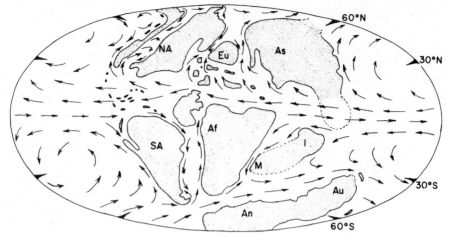

Fig. 4. A broad Tethyan seaway separated Africa and Eurasia in the Cretaceous and Paleocene (from Gordon 1973). Note the major latitudinal current arrangement in the inner tropics. Although numerous islands dotted the Tethys, the floras show important differences on its opposite shores (see Fig. 5).

sweepstakes dispersal (sensu Simpson 1943), but to transport via microplates detached from Europe that joined Africa (see Boccaletti & Guazzone 1974, Alvarez et al 1974, Channell & Tarling 1975, P. J. Smith 1974). If microplates acted as 'Noah's Arks' (sensu McKenna 1973) for transport to (or from) Africa, to judge from their size, one might expect much larger mammalian samples would have debarked when the arks landed. On the other hand, the problem of interchange between Africa and Eurasia before the Miocene is a serious one (Raven & Axelrod 1974, Schuster 1976, p. 116–126), and continuing studies of Mediterranean geology will be important in this respect.

3. Late Cretaceous–Paleogene

In view of the incomplete record, we follow Moreau (1952) and shall consider the major ecosystems which are defined by the principal vegetation belts insofar as they can be recognized. Some of these have been segregated into related subzones that are now widely separated by the central rainforest belt, and we shall note them at appropriate points.

3.1 *Tropical rainforest*

Climatic belts have always been arranged symmetrically about the equatorial region, with cooler and drier climates away from the central, warm and moist rainforest belt. Critical pollen studies in tropical West Africa and South America (e.g., Hoeken-Klinkerberg 1966, Roche 1974, Jardine et al. 1974, Herngreen 1974, 1975) reveal a change from essentially identical floras in the Albian–Cenomanian when these areas were still connected, to gradual differentiation during the latest Cretaceous–Paleocene, to distinctly different floras in the Eocene. A similar conclusion has been inferred from distributional evidence of living taxa that provide links across the Atlantic (Camp 1947, Boughey 1957, 1965, Axelrod 1970, 1972a). Particularly noteworthy, as emphasized by

Herngreen (1974, 1975) and Brenner (1976) is the similarity of taxa in tropical South America (Guiana–Venezuela, Brazil) and West Africa (Nigeria), and their marked difference from floras north of the Tethys Sea in Eurasia (Fig. 5), and also in North America which was then widely separated by sea from South America (see Fig. 4). The distribution of the later Cretaceous microfossil floras is thus similar to that of the early Tertiary megafossil floras which show a broad central Tropical Geoflora flanked to the north and south by the temperate Arcto-Tertiary and Antarcto-Tertiary Geofloras, respectively and with indications of savanna climate in the lower-middle latitudes (Axelrod 1960, 1975, Fig. 5). In each case, the similarities in floras, and hence in vegetation, are greatest around the North Pacific and the North Atlantic basins, with the present differences having developed chiefly since the middle Miocene (Axelrod 1960).

Within the inner tropics, the ties between the South American–African tropical floras are now chiefly at the level of families, but include some tribes and genera (Camp 1946, Axelrod 1970, 1972a, Raven & Axelrod 1974). This is expectable for these areas have been separated since the later Cretaceous (~85 m.y.), and the taxa have been subject to different environmental influences, both biotic and climatic, as well as to continuing genetic change. Among the tropical African links with South America that appear to have developed as sea floor spreading continued are numerous alliances, including Annonaceae, Apocynaceae, Araceae, Araliaceae, Bombacaceae, Chrysobalanaceae, Flacourtiaceae, Marantaceae, Menispermaceae, Myrtaceae-Myrtoideae, bambusoid Poaceae, Sapotaceae, and Simaroubaceae (for others, see Hutchinson 1973).

That northern Africa was covered with forest during the late Cretaceous is apparent from older records reported by Couyat & Fritel (1919) from the vicinity of Aswan. The plants referred to are palm, fig, and other taxa that are chiefly temperate today, which were no doubt misidentified owing to their comparison with illustrations in monographed floras from the United States and Europe. This collection should be restudied.

That tropical rainforest covered much of North Africa, which was then situated near the equator, is shown by fossil leaves from the Paleocene on the Red Sea coast (26°N) southwest of Quesir in Wadi Zeraib (Seward 1935). Referred to various genera, including a presumed dipterocarp, the leaves are quite large (mesophyll) and obviously lived under a warm, humid climate without any dry season. Engelhardt (1907) illustrates a group of taxa from the early Tertiary of Fajûm. The well-preserved leaves are chiefly notophyll in size, and on this basis imply savanna or woodland vegetation. Numerous taxa were identified, but the collection obviously needs restudy. Engelhardt illustrates *Artocarpidium*, *Ficus* (5 spp.), *Tetranthera*, *Maesa*, *Litsea*, *Securidaca*, *Cinnamomum*, *Pterocarpus*, *Cassia*, and a member of the Myrtaceae. This flora clearly is not younger than middle Eocene, and indicates a well watered region, probably with a dry season.

Abundant fossil woods (often large logs) scattered over the present Saharan region have been allied with tropical families chiefly. These were monographed by Kräusel (1939), and added to by Boureau (1958). Among the taxa recorded by Kräusel (1939) are woods representing Arecaceae (6 spp., including *Nypa* fruits), Moraceae (*Ficus*, 7 spp.), Proteaceae, Monimiaceae (*Atherospermoxylon*), Lauraceae (*Litsea*), Fabaceae (6 spp.), Rutaceae (*Evodia*, 3 spp.), Polygalaceae (*Securidaca*), Sterculiaceae (4 spp.), Clusiaceae (2 spp.), Ternstroemiaceae, Dipterocarpaceae (2 spp.), Rhizophoraceae, Myrsinaceae, and Ebenaceae. In ad-

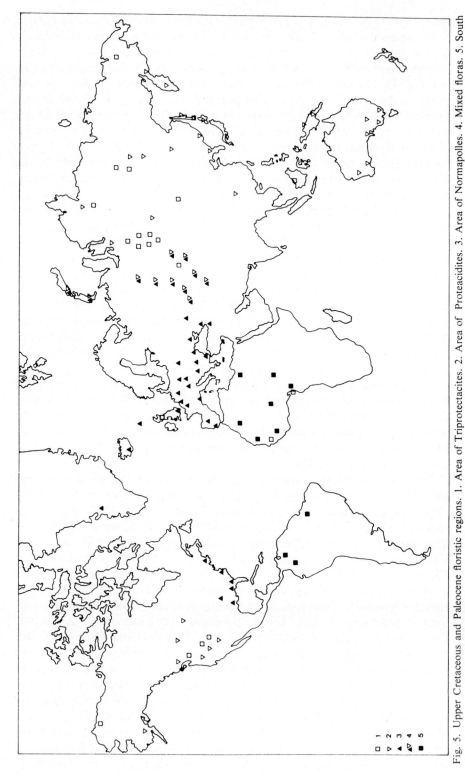

Fig. 5. Upper Cretaceous and Paleocene floristic regions. 1. Area of Triprotectacites. 2. Area of Proteacidites. 3. Area of Normapolles. 4. Mixed floras. 5. South America–African floristic region (from Roche 1974).

dition, Chiarugi (1929) described fossil woods from Cyrenacia, Libya, which show relationship to those in Egypt, notably species of palm, fig and laurel. More recently, Koeniguer & Louvert, whose extensive studies of the fossil woods of Egypt–Libya have been reviewed (Aubréville 1970, Axelrod 1975), record numerous taxa from the present desert that represent chiefly those of tropical rainforest, or of the bordering savanna–woodland in the later Eocene. Among the families recorded are Annonaceae, Combretaceae, Dipterocarpaceae, Ebenaceae, Euphorbiaceae, Fabaceae, Lauraceae, Moraceae, Myrtaceae, Sapindaceae and Sterculiaceae. Inasmuch as the highlands of East Africa had not yet been elevated, rainforest probably extended from coast to coast in Paleocene time (Fig. 6A). This is consistent with the large leaves recorded by Seward (1935), and occurrence of fruits and seeds of tropical taxa (Annonaceae, Euphorbiaceae, Icacinaceae) in the early Paleocene of the Red Sea Hills (Chandler 1954), some of which are also in the London Clay flora (Reid & Chandler 1933, Chandler 1964), a perhumid rainforest. To the south, fossil wood of Monimiaceae is recorded from the late Cretaceous rocks of Pondoland (Mädel 1960), an area in which the genus *Xymalos* of this family still occurs.

It is appropriate to note (see below) that the Deccan flora of India, situated near Lat. 10°S in the early Paleocene (McKenzie & Sclater 1973), has numerous tropical alliances that represent moist rainforest taxa (Lakhanpal 1970). At that time India was connected with the Mascarene Plateau–Madagascar–Africa via numerous islands. Hence we infer a commonality of tropical rainforest taxa for Africa–India in the Paleocene, a relation evident today in the numerous genera and related taxa that are still common to these areas (Good 1974, Wild 1965).

There are numerous links between the floras of East Africa and Madagascar. These may be understood in the light of evidence that the areas probably were linked by a basaltic plateau during the middle Cretaceous (Kent 1972, Kent et al. 1971, Tarling & Kent 1976). Such a plateau may have served as a pathway for numerous plants, and also for the large terrestrial dinosaurs recorded there. It may also have hindered the passage of primary fresh-water fish, which are absent from Madagascar, because of the high concentration of toxic ions (e.g. Cr^{+++}, Ni^{++}, Sr^{++}, Cu^{++}, S^{++}, F^-, BO_4^{---}) in the waters issuing with the magma. As noted above, however, they may have originated after separation. Southward movement of Madagascar has been postulated by Wild (1968, 1975) on the basis of plant distribution. He notes that 80 species are common to Madagascar and eastern Africa, distributed from Ethiopia to the south coast rainforest. Since the proportion of taxa common to southeast Africa–Madagascar is about 4 times that for the northern (Somalia–Kenya–Tanzania) area, Wild (1968, 1975) suggests the stocking of Madagascar by overseas dispersal as it moved south. However, it seems likely that there were more numerous links between Madagascar and the Somali–Kenya region into the Miocene, following which they were eliminated there as aridity spread over the region (see below). Furthermore, Wild's analysis demands the presence of modern species, including a number of Asteraceae, in the later Cretaceous, which is scarcely probable.

3.2 *Temperate rainforest*

Africa was about 15°–18° farther south in the late Cretaceous–Paleocene, a time

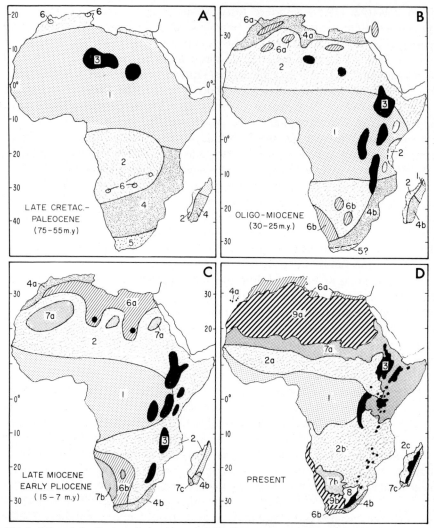

Fig. 6. Inferred distribution of vegetation. A. Late Cretaceous–Paleocene. B. Late Oligocene–early Miocene. C. Middle–Late Miocene. D. Recent (simplified from Greenway 1970).

Late Cretaceous–Paleocene
1. Lowland rainforest
2. Savanna–woodland
3. Montane rainforest
4. Subtropic rainforest
5. Temperate rainforest ('austral affinities')
6. Sclerophyll woodland

Oligo-Miocene
1. Lowland rainforest
2. Savanna–woodland and thorn scrub
3. Montane rainforest
4. Subtropic laurel forest
   a. Canarian
   b. Natal

5. Temperate rainforest? ('austral affinities')
6. Sclerophyll vegetation
   a. Tethyan
   b. Cape

Late Miocene–Early Pliocene
1. Lowland rainforest
2. Savanna–woodland
3. Montane rainforest–afroalpine
4. Subtropic laurel forest
   a. Canarian
   b. Cape
5. (eliminated)
6. Sclerophyll vegetation
   a. Tethyan
   b. Cape

when temperate austral lands were covered with a dense *Podocarpus–Nothofagus-*evergreen dicot forest of moist, warm temperate to cool temperate requirements. That such a forest was widespread on southern lands has been recognized for many years (e.g., Hooker 1853, Berry 1938, Skottsberg 1936, Couper 1953a, b, 1960a, b, Axelrod 1960). More recently, Penney (1969) reviewed the diverse spore–pollen assemblages from Australia and New Zealand in the late Cretaceous and Paleocene–Eocene. Taxa allied to *Podocarpus, Dacrydium,* and *Nothofagus* are regularly recorded as well as *Symplocos, Elaeocarpus* and members of Olacaceae, Sapindaceae and others that range into warmer climates. Such mixtures still occur in the mild temperate, equable parts of New Zealand and southeastern Australia. Temperate rainforest persisted into the later Eocene in southwest Australia, near Albany (Hos 1975), where Podocarpaceae (cf. *Dacrydium, Phyllocladus, Microcachrys*), *Nothofagus* (*brassii* group), Myrtaceae, Bombacaceae, Malvaceae, and Sapindaceae (Cupanieae) are recorded. Some 17 taxa extend down into the Paleocene, and 13 into the late Cretaceous (Hos 1975, Table 1). At that time, the fossil site was close to Lat. 50° as compared with Lat. 35°S today.

That the character of this temperate austral forest may have been quite different from anything existing at the present day is suggested by the presence of tropical mangrove communities at Perth, Western Australia, during the middle to late Eocene (Churchill 1973). Conditions there may be compared with those obtaining at the time of deposition of the Eocene London Clay flora at about the same time.

To the west, Kemp & Harris (1975) record pollen from deep-sea drilling sites in Paleocene (and Oligocene) sediments on the Ninetyeast Ridge, Indian Ocean. They suggest that the samples represent floras from oceanic islands, transported there by long-distance dispersal. Reconstructions of Indian Ocean paleogeography indicate that the Paleocene sample probably inhabited a larger land area, one now split into Broken Ridge and Kerguelen Ridge (see Sclater & Fischer 1974, Fig. 13). Similarity to nearby pollen floras of Australia does not suggest major floristic imbalance, as long-distance dispersal requires. Site 214 of Paleocene age, which accumulated near Lat. 45°S, includes pollen provisionally

---

7. Thorn scrub–succulent woodland
   a. Sahelian
   b. Kalaharian
   c. Malgasan

Present Vegetation (simplified from White, in Greenway 1970)
1. Lowland rainforest
2. Savanna–woodland
   a. Sudanian
   b. Zambezian
   c. Malgasan
3. Montane rainforest and afroalpine
4. Subtropic laurel forest
   a. Canarian
   b. Natal
5. (eliminated)
6. Sclerophyll vegetation
   a. Tethyan
   b. Cape

7. Thorn scrub–succulent woodland
   a. Sahelian
   b. Kalaharian
   c. Malgasan
8. Grassland
9. Desert and semidesert
   a. Saharan–Libyan
   b. Namib–Karoo

> NOTE: The names of subtypes (listed under a, b, or c) indicate only a general, ancestral relation to modern vegetation. All Tertiary vegetation types differed importantly from those of the present: they were far more diverse in composition. The modern communities arose as climates gradually changed, and especially rapidly in the later Pliocene and Pleistocene.

assigned to Araucariaceae (*Araucaria*), Podocarpaceae (cf. *Dacrydium, Microcachrys*), Arecaceae, Chloranthaceae (*Ascarina*?), *Casuarina*, Myrtaceae, Proteaceae, Didymelaceae (*Didymeles*?), and Gunneraceae (*Gunnera*). The Oligocene Site 254, then situated near Lat. 35°S, had many of these taxa, as well as Sapindaceae (Cupanieae), Loranthaceae, Restionaceae, Moraceae, Sapotaceae, and Asteraceae, implying a warmer climate consistent with displacement northward. There is no evidence here for a seasonally dry climate of any significant degree (see below), yet the area clearly is now under anticyclonic influence.

Pollen records of a *Nothofagus–Podocarpus* forest occur in the late Cretaceous of West Antarctica (Snow Hill and Seymour Island, Cranwell 1959), the Paleocene–Eocene of East Antarctica (Kemp 1972), and the Eocene of Black Island, McMurdo Sound (McIntye & Wilson 1966). The absence of warm temperate taxa gives these floras a more temperate aspect consistent with their higher latitude. For example, the Black Island assemblage includes grains referred to *Podocarpus, Dacrydium, Phyllocladus, Araucaria* and *Libocedrus* among the conifers, and dicots representing Myrtaceae, Proteaceae, and *Nothofagus* of three groups: *brassii* group, now in New Guinea–New Caledonia; *fusca* group, now in Chile–New Zealand–Tasmania; and *menziesii* group, now in Chile–New Zealand–Australia–Tasmania. Similar taxa are recorded also from the late Cretaceous and Paleocene of Patagonia (Archangelsky & Romero 1974, Romero 1973), not only the conifers but all three groups of *Nothofagus* as well.

Inasmuch as the *Nothofagus–Podocarpus*-evergreen dicot forest had a wide distribution on all southern lands during the late Cretaceous–Paleocene, we infer that it probably also covered the tip of southern Africa (Fig. 6A). It was then situated in similar latitudes, it had a similar climate, and was connected to them by scattered islands (e.g., Perchneilson et al. 1975). The very restricted Cretaceous deposits in southern Africa should be examined for records of the forest, but for the present the best evidence is that afforded by the distribution of living organisms. In and near the temperate rainforest that now extends in disjunct patches across the south coast of South Africa, from near Cape Town to the vicinity of Grahamstown, are a number of animals and plants that are of austral derivation (Levyns 1964). Many other characteristic ones, such as *Nothofagus* itself, are absent; their absence might be explained by the relatively low latitude of southern Africa at present, the early date at which its connections with Antarctica were severed (before the evolution of modern groups of angiosperms), and the limited area of the forest, which has contracted sharply as a result of human activities even within the past 500 years (Acocks 1953).

Some of the austral plants and animals of southern Africa may have dispersed across a mid-Cretaceous Indian Ocean that was limited in size, via Madagascar–India to Australasia (Raven & Axelrod 1974). Levyns (1962) has provided a review of such links amongst the seed plants, listing the Cunoniaceae, *Brabeium* (Johnson & Briggs 1975), and *Curtisia* as possible examples. We believe that *Acaena, Gunnera, Metrosideros*, and the inuloid Asteraceae, among others, probably reached southern Africa by long-distance dispersal more recently, judged from the age of their respective groups and their extra-African patterns of distribution (Raven & Axelrod 1974). The amphi-Indian Ocean distribution of the genus *Villarsia*, a mainly Australian one of about a dozen species with one, which is distylous and strongly self-incompatible (Ornduff 1974), is most unusual. Other possible examples of more direct migration across the southern hemisphere

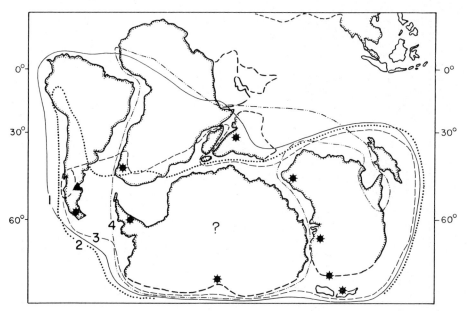

Fig. 7. Distribution of several austral families assembled on a pre-drift world. 1. Proteaceae, with ✹ marking some Paleocene and Late Cretaceous fossil localities. 2. Philesiaceae. 3. Restionaceae. 4. Aponogetonaceae, with ▲ marking a Cretaceous site in Argentina, its only known occurrence in South America.

are listed by Adamson (1948, Tables 1, 2), Wild (1968, Fig. 8a, 6), and Raven & Axelrod (1974, p. 616) (see Figs. 7, 8, also).

The greater similarities between the plants and animals of temperate South America and temperate Australasia than between either and South Africa is understandable on the presence of land connections between South America–Antarctica–Australasia into Eocene–early Oligocene time (Fig. 3), and the severance of the African connections much earlier, as noted by Adamson (1948). The distributional patterns of chironomid midges in the southern hemisphere led Brundin (1966) to deduce such a history independently of geological evidence, and a similar pattern is exhibited by the mayflies (Edmunds 1972). Both groups are much older than the flowering plants and clearly dispersed directly between Antarctica and southern Africa, as almost certainly did the gymnosperm *Araucaria*, now extinct in Africa (Seward 1903, Florin 1963). The other southern African genera of gymnosperms, *Podocarpus* and *Widdringtonia*, as well as the cycads and *Welwitschia*, may likewise reflect this pattern. Another example of a southern African animal with austral connections is the frog *Heleophryne* (Lynch 1971). Angiosperms are, however, more recent, and the limited number of obvious austral affinities amongst South African representatives of this group led Bews (1922) correctly to conclude that there had been no 'Antarctic continent' in existence since the time of origin of the angiosperms (or at any rate, their modern groups).

The fossil flora at Banke, Namaqualand (30°24'S, 18°31'E), preserved in a thick sedimentary infill of a diatreme which seems similar to those in the nearby area dated at 38.5 m.y. (Kroner 1973), appears to be like that now in the Cape Region, including *Podocarpus*, *Gunnera*, *Myrica*, *Ficus* (Adamson 1931), and a

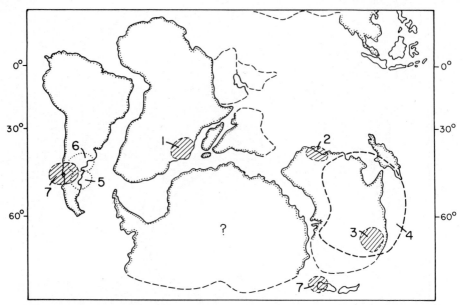

Fig. 8. General distribution of modern genera of Philesiaceae, assembled on a pre-drift world. 1. *Behnia*. 2. *Elachanthera*. 3. *Eustrephus*. 4. *Geironoplesium*. 5. *Philesia*. 6. *Lapageria*. 7. *Luzuriaga*.

probably proteaceous genus similar to *Brabeium* (Rennie 1931, Kirchheimer 1934, Cranwell 1964, J. R. Rourke, pers. comm.), as well as leaves of xerophilous appearance. Such a mixture is typical of that growing in the hills and flats of the Cape today. At this locality, where the pipid frog *Eoxenopoides* occurred (Haughton 1931, Nevo 1968, Estes 1975), temperate rainforest was certainly depleted by the end of the Eocene and mixed with more xeric elements ancestral to those so well developed in the southwestern Cape Province at present.

3.3 Montane rainforest

Fossil evidence regarding the early Tertiary nature of this vegetation zone is not now available in Africa. The history of the taxa that comprise it must have commenced whenever highlands had sufficient altitude. The principal relief on the pre-Miocene surface was provided by the ancient crystalline highs of the Saharan region (Hoggar, Air, Tibetsi massifs of Precambrian rocks chiefly), which assuredly were larger than at present. In addition, the East African and Ethiopian warps, which preceded volcanism (Burke & Wilson 1972, Baker et al. 1972), had summit levels near 1500 to 1600 meters at the close of the Paleogene (Baker & Wohlenberg 1971). This altitude approximates that of the lower margin of montane rainforest in equatorial regions today, decreasing in altitude toward middle latitudes. All Paleocene–early Neogene climatic zones, however, reached to lower levels than they do at present, owing to a moister and highly equable climate (Axelrod 1965, Axelrod & Bailey 1969). Hence, during the Paleogene the higher mountains of tropical North Africa, the swells in East and Southeast Africa, probably were sufficiently high and cool to enable certain tropical rainforest taxa to radiate upward into a temperate montane zone, and presumably also for north temperate and south temperate plants and animals to extend their ranges towards or even past the equator.

3.4 Subtropical evergreen forest

Subtropical evergreen forest now occurs in southeast Africa and extends south to Pondoland (Acocks 1953). Since many of the taxa range widely to the north, contributing to tropical forests there (Wild 1968), a wider, more continuous distribution is implied for it prior to the spread of dry climate. This finds a modern parallel in the distribution of subtropical rainforest that blankets the east coast of central and southern Australia, merging southward into the *Podocarpus–Nothofagus-*evergreen dicot forest. Also, in Brazil the subtropical rainforest extends from near Rio de Janeiro to 30°S, and to somewhat higher latitudes in the mountains. It reached well into Argentina in the Eocene, under a regime of moister climate (Berry 1938, Menéndez 1969, 1972), giving way to a *Podocarpus–Nothofagus-*evergreen dicot forest of temperate requirements farther south.

A former continuous distribution of subtropical rainforest into South Africa can be inferred also from the paleobotanical record in India (reviewed in Axelrod 1974), which was situated against the Mascarene Plateau (now largely submerged) into the latest Cretaceous (Molnar & Francheteau 1975, Sclater & Fischer 1974, Kutina 1975, Flores 1970). In the late Cretaceous (70 m.y.), Ceylon was at 45°S (Molnar & Tapponier 1975). The rich flora from the Deccan Traps, dated at 65 m.y. (Molnar & Francheteau 1975, McElhinney 1970), was by then near 30°S, at the southern edge of the tropics. The families represented are preponderantly those of tropical regions, notably Anacardiaceae, Arecaceae, Burseraceae, Combretaceae, Datiscaceae, Elaeocarpaceae, Euphorbiaceae, Flacourtiaceae, Lecythidaceae, Rutaceae, Sapindaceae, Simaroubaceae, Tiliaceae, Vitaceae and many others (see Lakhanpal 1970, Prakash 1965, 1972). However, there are also Deccan records of austral temperate taxa that are no longer in Peninsular India, notably Araucariaceae, and the Podocarpaceae which occur now only in southern and eastern India. In addition, the now-largely austral Casuarinaceae are recorded there, and pollen of Proteaceae is reported from the Eocene of Kutch (Mathur 1966). The frog *Indobatrachus* from the Deccan Traps represents the austral temperate Leptodactylidae, also no longer in India, but in South Africa–South America–Australia. In view of its position in the late Cretaceous–Paleocene, it seems likely that in Africa tropical rainforest graded southward into subtropic rainforest, as outlined in Fig. 6A.

3.5 Subtropic woodland-scrub

Areas of truly dry climate were highly restricted in the Paleocene–early Eocene, as judged from records of rainforest in north Africa (Seward 1935, Boureau 1958, Aubréville 1970). This agrees with evidence reviewed for the history of Tethyan sclerophyll vegetation, for taxa representing it entered the record in the later Eocene in southwestern Asia and also in southwestern North America where its Madrean counterpart is recorded (Axelrod 1975). Both appear to have been derived chiefly from subtropical alliances that adapted to drier climate.

There have always been high pressure cells at the edge of the tropics, but not of the intensity of those of today. We infer that by the later Eocene rainforest graded into open savannas, with evergreen subtropic woodland and scrub occupying the driest part of the climatic belt, inhabiting areas that have since been converted into desert. This is not only true for northern Africa and southern Eurasia, as well as southwestern North America, but it is indicated also by the small flora from Banke, Namaqualand (Rennie 1931), now considered transitional Eo-Oligocene (~38 m.y.) in age (see above). The evidently coarse, sclerophyllous nature of most of the leaves in this fossil flora suggests the existence of a dry season in the area by the late Eocene, when it was near 35°S.

This was also a time when dry climate was spreading over southwestern Asia and North America. In this regard it is noteworthy that an Oligocene flora from a deep-sea drilling site situated on Ninetyeast Ridge, which was then situated at 35°S (Kemp & Harris 1975), provides no evidence of arid climate in an area where there is now a moderate dry season. Also, arid climate had not yet appeared on the west coast of South America, to judge from the floras known there (Berry 1938, Menéndez 1969). The data suggest that in the Southern Hemisphere anticyclonic circulation systems did not have the intensity of the present ones, an inference consistent with evidence that they increased in strength and severity during the Neogene, as outlined below.

To judge from the history of sclerophyllous vegetation elsewhere, dry woodland and scrub probably originated early in the southwest Africa, wherever dry sites were available. Initially, these included broad structural lows which were in semi-rainshadows, and especially the abundant and widespread dry sites provided by broad tracts of extremely resistant Archean basement rocks that stood as essentially soil-less inselbergs and rocky ridges surrounded initially by mesic vegetation. From an origin in scattered edaphic dry sites, which occur today even in the wet tropics, dry-adapted (or preadapted) taxa could then spread as regional climate became drier (Axelrod 1970, 1972b).

This may well explain the antiquity that must be postulated for many of the taxa of arid requirements in the region. In this regard, Van Zinderen Bakker (1975) infers that the high degree of endemism in the Namib Desert indicates great antiquity for it. The peculiar endemics may have survived in coastal dunes present in the area since the South Atlantic first opened in the early Cretaceous, under conditions of highly equable climate (Axelrod 1965, 1967). Thus the desert itself may not be especially old.

Even during the Pleistocene pluvial cycles, temperate rainforest expanded greatly in southern Africa as shown by records from near Cape Town (Schalke 1973). At such a time, sclerophyllous vegetation must have shifted northward, into the present Namib and a good part of the western Karoo, where some relict patches exist today, as well. Such a shift parallels that of the pinyon pine–juniper woodland down into the present deserts of western North America during the last glacial (Wells 1966, Wells & Berger 1967). Inasmuch as there probably were four shifts of this order during the past 2.7 m.y., adaptation to the present desert environment need not be attributed to great antiquity. Semidesert regions commenced to spread in the late Tertiary (7–5 m.y.), their areas previously being occupied by sclerophyllous woodland, grassland and thorn scrub vegetation. Desert taxa have in large measure been derived by adaptation of taxa in woodland, savanna and thorn scrub vegetation to the somewhat drier conditions (~250 mm) that developed over the areas of their former occurrence (Axelrod 1950, 1958, 1970), chiefly since the Pliocene. The complex shifts in climate and vegetation in southern Africa in the late Quaternary have been well discussed by Van Zinderen Bakker (1976).

3.6 *Summary*

By the close of the Oligocene, when the African plate had moved north to essentially its present position, the vegetation of southern Africa had assumed a near-modern aspect. However, the composition and distribution of vegetation differed from that of today in several important respects (Fig. 6B). Subtropical rainforest had contracted somewhat, though patches may still have reached the east coast locally (cf. Aubréville 1970, Andres & Van Couvering 1975). The higher parts of East Africa, where the rift system was about to develop, evidently had sufficient altitude to support lower montane rainforest which is inferred to have also inhabited the higher mountains of the Saharan region. Inasmuch as climate was more humid as compared with that today, montane rainforest probably had a wider distribution, grading at lower elevations and southward along the east coast into subtropical rainforest. Its counterpart in the north was an evergreen laurel forest ancestral to the present Canarian forest. Flanking the rainforest over the interior was a savanna that had expanded with spreading dry climate. Bordering it in the driest climatic areas was sclerophyllous vegetation, fynbos in the southwest and related woodland (with macchia shrubs) in the Tethyan region to the north.

4. Neogene

4.1 *Terrain*

Southern Africa was transformed into a near-modern world after the late Oligocene–early Miocene (25–23 m.y.). As reviewed by King (Chapter 1, also 1962, 1963, Haughton 1963, DuToit 1954), the African landscape (Stage III) of the Cretaceous to the middle Tertiary (sub-Miocene or pre-Miocene surface of some investigators) was flexed into a number of broad warps and basins during the middle Tertiary (Burke & Wilson 1972) (Fig. 9). On the east and west coasts the sub-Miocene surface passes beneath mid-Miocene marine strata, and in the interior of East Africa (Lake Albert, Rusinga Island) it is covered with non-marine sediments and volcanogenic detritus containing early Miocene (23–25 m.y.) mammalian fossils (Van Couvering & Van Couvering 1975). As warping commenced, rifts developed on the arches, volcanism was initiated, and the rift valleys were now developing (Baker et al. 1971, 1972, Kahn 1972).

Africa was uplifted epeirically at the end of the Miocene (King, Chapter 1, also 1962). Rivers were rejuvenated over long distances, and as valley walls retreated numerous new local basins were carved at different altitudes. The close of the Pliocene and early Pleistocene (Stage V) was characterized by major uplifts which raised the interior plateaus more than 1,800 m (Baker & Wohlenberg 1971, Fig. 2) above the Miocene level, and steepened the coastal monoclines by hundreds of meters (King, Chapter 1, also 1963, DuToit 1954). The great rift valleys developed in phases, in the early Pliocene, and especially in the late Pliocene–early Pleistocene (Baker & Wohlenberg 1971), and new high volcanos were built upon their margins.

There also was major tectonism during the later Tertiary in South Africa, elevating the central Highveld and tilting the marginal regions outward (King 1962, 1963, Gough 1973, Newton 1974). Maximum uplift took place along a line outside the Great Escarpment, while the central Karoo plateau developed as a basin over which aridity was soon to spread. Superimposed on this were minor

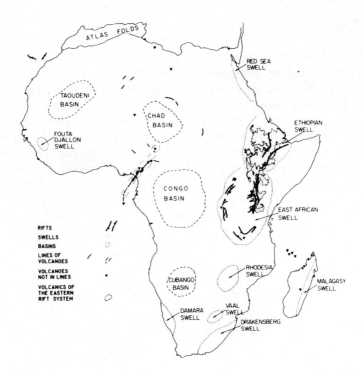

Fig. 9. Neogene basins, swells, volcanic uplifts and lines of volcanos in Africa (from Burke & Wilson 1972).

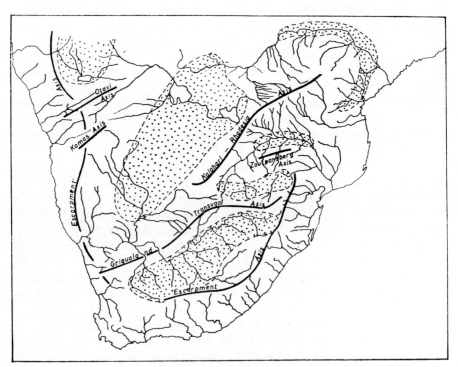

Fig. 10. Upwarped axes and basins of depression resulting from the Plio-Pleistocene deformation in southern Africa (from King 1963).

upwarps and basins, and the central plateau was elevated far above its former level. The local uplifts are in the form of long ridge-like axes, separated by relatively broad depressed basins (Fig. 10).

It is critical to realize that the area of the Plio-Pleistocene upwarps and downwarps trends northeasterly into the East African rift system that reaches southward to the Zambezi River. Further, the seismicity of the Afro-Arabian rift system continues down to the Cape (Kahn 1975, Fig. 16), and the major carbonatite intrusions have the same trend (Haughton 1963, Fig. 40). Since the Pliocene rocks of the rift valley are folded along with the older rocks, and the entire area was then elevated fully 2000 meters or more, uplift and warping in South Africa seems reasonably correlative with it. The problem of whether the seaward-dipping erosion surfaces owe in part to renewed tensional movement along the great Agulhas scarp can only be raised. Its structure is highly suggestive (see Scruton & du Plessis 1973) and it is the African counterpart of the Falkland Plateau scarp (Larson & Ladd 1973). The evidence implies that following a long period of quiet, when the relatively smooth African pre-Miocene surface was developed, major deformation occurred in the Miocene, and again in the Pliocene–Pleistocene culmination of rifting. These importantly modified not only the topography, but the climate of the region, and hence the history of life.

4.2 *Climate*

Major regional differences in climate, and hence in vegetation, developed following the early Miocene. Old, relatively stable ecosystems were restricted and were replaced by new ones that were expanding and diversifying as a broad belt of progressively drier climate spread over areas previously covered with forest and/or savanna or woodland.

The gateways through which moist, maritime air enters Africa at present fail by far to include the entire coastline of the continent. As a result, major rainfall gaps are on the Atlantic coastal strip from Morocco to Senegal, and from Cabinda to St. Helena Bay near Cape Town, as well as the entire coast north of Kenya. In addition, the Mediterranean coast of Libya and Egypt is quite dry. In the late Cretaceous and Paleocene, the circumstances of paleogeography suggest that maritime air had much better access to the continent (Fig. 4), and that dryness gradually spread as its geography was modified. Recall that in the late Cretaceous–Paleocene Africa was 15° south of its present position, that the continent was not yet in contact with Asia, that the Tethys separated it from Europe–Asia, and that the highlands of East Africa and the Eurasian Alpine axis had not yet emerged. The Atlantic was much narrower, and its water was warmer. Africa, then an island continent, reached into the southern westerlies which contributed abundant moisture south of the present 20th parallel. Northward of that parallel, and extending to the present equator was the southern subtropical region which received then, as now, more moisture from the Indian Ocean than from the Atlantic. Lowered contrast between sea-surface and land-surface temperatures would have resulted in a less intense and not as stationary a high pressure center over the ancestral South Atlantic as that of today. The difference made possible in the Paleocene the entry of Atlantic moisture across the west coast, thus supplementing precipitation from the Indian Ocean to subtropical latitudes.

A major difference in Paleocene climate was the fact that the equator then crossed Africa near the modern 15th parallel north, allowing abundant moisture to enter north Africa across coasts now extremely arid (Fig. 6A). Toward the northern margin of the continent, descending air of the subtropical high would have imposed episodes of fair weather, but particularly in the northern summer monsoonal inflow of air would have occurred from the broad, warm sea then forming the northern boundary of the continent. Similarly, in the southern subtropics there was a broad monsoonal area with drier savanna conditions resulting from the descending air.

Factors that contributed to the trend toward increasing drought may be found in changes in paleogeography which are known to have occurred in the Miocene and later. First was the destruction of the Tethys Sea (see Fig. 4) by the union of Africa–Arabia with Iran during the middle Miocene. This broad seaway had earlier controlled the major latitudinal system of circulation (Crowell & Frakes 1970). At the same time the rift zone was developing, bringing increasing altitude and active volcanism, and hence blocking the ingress of moist air across Africa. Somewhat later and far to the west, the Panamanian portal was closing, the Mexican plateau was being elevated, and the Andes were rapidly increasing in altitude. Furthermore, the great alpine system of southern Eurasia was also being elevated, and seaways were retreating from the continents. The net result was that the present great anticyclones in the lower-middle latitudes became progressively stronger, more persistent, more stable in position, and dry climates increased in area. Evidence of such change is well documented in India, where humid forest taxa rapidly retreated eastward to areas of moister climate following the middle Miocene (Prakash 1972). The spread of dry climate is reflected also in the abrupt appearance in the Miocene of numerous woody Fabaceae taxa which earlier were poorly represented (Lakhanpal 1970). During the middle Tertiary the change was from an ancient (Cretaceous–Paleogene) broadly latitudinal circulation with well distributed moisture over wide tracts, to more restricted north–south cellular circulation systems that were increasingly intensified as colder climates developed in high latitudes, leading to the progressive expansion of more arid climates over the lower latitudes as the thermal gradient increased.

In this regard, recall that Antarctica parted company with the South Tasman Rise in the middle to late Oligocene (Kennett et al. 1974), and thence shifted to full polar position. A small pollen flora of early Miocene age (26 m.y.) from Ross Sea indicates climate was cool temperate (Kemp & Barrett 1975). The sample suggests an environment like that now in the evergreen forest of South Island, New Zealand or southern Chile, near 43°–44°S. This implies that glaciers were in the uplands, with the firn limit near 1800 to 2000 m, and with ice tongues descending into an open Ross Sea. As the 'proto'-Benguela Current commenced to bring colder water to the west coast of Africa, a drier season now spread gradually over the southwest coast, aided by the increasing strength of the summer anticyclone offshore. As a major ice sheet commenced to spread after 5 m.y. (Shackelton & Kennett 1975, Kennett & Brunner 1973, Blank & Margolis 1975), the Benguela Current increased in strength and became progressively colder, and the south Atlantic anticyclone increased in intensity and became more stable. Inasmuch as the Tethys was blocked, and the Panamanian portal was closing (or closed), the Canary high pressure system also increased in strength, stability, and area. The subsidence of dry air during the warm season now brought increased drought to the west coast of moist tropical Africa. Also, adaptation was now increasingly for taxa that could withstand drought, whether in rainforest, savanna, sclerophyll, scrub, or semidesert communities. Selection for drought resistance was intensified considerably over that of the early Miocene. During the times of 'ice-age aridity' there was even more intense selection for taxa that could withstand drought, thus restricting forest and savanna–woodland even further at the expense of grassland, scrub, and other communities. Even in the past 20,000 years there has been major climatic changes over much of Africa (Geyh & Jakel 1974, summary in Livingstone 1975, Van Zinderen Bakker 1976 and Chapter 6), making a direct connection between present and past vegetation very difficult to establish. Some of the results of the climatic deterioration may be inferred from the nature of a number of small fossil floras that have been described.

4.3 *Tropical rainforest*

A small flora in the western rift zone from South Kivu on the Mobale River (Lakhanpal 1966) appears to represent rainforest. The fossils occur in a claystone in a subbasaltic river system that overlies river gravel. Most of the specimens represent the palm *Sclerosperma*, a monotype confined now to swampy areas near the Gaboon River in lower Guinea. Associated dicot leaves are referred to *Tetracera*, a liana or scandent shrub; *Garcinia*, a small tree; a species of *Strychnos* like one in the Congo; and several unidentified leaves and a leguminous pod. The flora evidently represents rainforest in a swampy drainage basin, presumably one ponded by the initial basalts. Structural evidence implies that the region was situated at a lower elevation in the Miocene, possibly near 500 m according to Lakhanpal (1966).

Andrews & Van Couvering (1975) suggest that tropical rainforest probably extended to the east coast during the Miocene, interrupted locally by woodland and grassland. This is based on evidence that prior to rifting and volcanism in the early Miocene, the continental divide was low, with drainage toward the Atlantic over most of Uganda, western Kenya and Tanzania. Mammals preserved in the early Miocene deposits just west of the divide are chiefly forest dwellers, implying climate in the eastern rift region was moister than at present. They point out that this is consistent with plant distribution in the eastern rift region today, for the area has a number of taxa that occur in the Congo rainforest, or have their nearest allies there, implying continuity prior to the spread of drier climate. In addition, there are genera that are well developed in the west that have only one or two species in the east coast patches of rainforest on the upper slopes of the plateau facing the Indian Ocean, implying former connection with the western forest.

The plant species disjunctions probably date at least from the early Miocene, as judged from the nature of the Rusinga flora, and other evidence (see below). The fruit and seed flora from Rusinga Island, Lake Victoria, was studied only in provisional manner by Chesters (1957). Most of the fossils are similar to living tropical African genera, though a few show a more distant affinity with modern forms, which is consistent with their age (19.5 m.y.). Families represented include Anacardiaceae, Annonaceae, Burseraceae, Capparaceae, Connaraceae, Cucurbitaceae, Euphorbiaceae, Fabaceae, Meliaceae, Menispermaceae, Oleaceae, Rhamnaceae, Rutaceae, Sapindaceae and possibly Apocynaceae and Lauraceae. Others that were identified tentatively included Combretaceae, Icacinaceae,

Olacaceae, Rubiaceae, Sterculiaceae, and Tiliaceae. Most of the taxa have close living allies and the majority are trees. One from Mfwanganu Island was referred to *Entandrophragma*, which is a component of montane rainforest but also contributes to mixed semideciduous forests (Lind & Morrison 1974). In Uganda, three species occur in the Bugoma Forest at elevations near 1200–1500 m, where rainfall is from 1250–1675 mm at a minimum. Climbers are prominent in the Rusinga flora and are common today in tropical forests especially along riverbanks and natural openings where there is ample light. The abundance of endocarps of Apocynaceae also suggests riverbank vegetation, with gallery forests along the watercourses in a region of broken rainforest.

Hamilton (1968) gives preliminary notice of an apparently rich flora from Bukwa, on the northeast slope of Mt. Elgon at an elevation of about 1,800 m. He notes leaf impressions are abundant, and illustrates flowers tentatively identified as *Bersama* (Melianthaceae) and *Cola* or *Pterygota* (Sterculiaceae). Species of the latter taxa are represented today chiefly in the Congo rainforest. Associated in the nearby area is a rich mammal fauna, preserved with volcanic rocks dated at 20 m.y.

That the general region was not a continuous rainforest, but one of interrupted rainforest and gallery forest, with woodland or savanna in the region, is consistent with two lines of evidence. First, the Miocene surface was already warped and volcanism had commenced by 25 m.y., giving the Rusinga flora (19.5 m.y.) a position to the west of the continental divide (see Burke & Wilson 1972, Fig. 2). Second, the older Bugishu flora (25 m.y.) from the lower slopes of Mt. Elgon (Chaney 1933) is represented chiefly by small-leaved (microphyll, nanophyll) taxa, woody legumes are abundant, and the plants accumulated in sediments deposited on a lateritic surface covering the Archean basement, all indicating the seasonal rise and fall of the watertable.

In this regard, Moreau (1933) noted that on the escarpment of the African plateau, rainfall in the mountains (e.g., Usambaras, Ulugurus) on the coast opposite Zanzibar Island may total 2000 mm, which exceeds that of most other areas in East Africa. As a result, there is a well developed montane rainforest at altitudes above 900 m, facing the ocean. The rainforest is rich in endemics, many with West African affinities. He suggested that an increase in precipitation of 500 mm would be adequate for rainforest to bridge the continent from coast to coast. However, this could result only if precipitation was well distributed through the year. Since changes in terrain and in air circulation systems were establishing a definite dry season during the early Miocene, the continuous rainforest of the early Tertiary probably was being disrupted by woodland and seasonal deciduous forests at that time.

In his discussion of the phytogeography of south–central Africa, Wild (1968) notes that a number of species and genera in Rhodesia–Tanzania provide links with the principal forest areas of the Congo and West Africa. He suggests that during the Quaternary pluvials a 50 per cent increase in rainfall may have enabled presently riverine species to form a more or less continuous forest at lower altitudes through much of tropical Africa. This seems unlikely unless considerable rainfall was distributed also in the present dry winter season (May through August), to provide a more nearly even distribution for rainforest. Such an event seems unlikely in view of the strength of the anticyclonic circulation systems at this late date. The present links between the Congo–West Africa closed rainforest and the relict stands in Rhodesia and border areas may therefore date from the early Miocene, at which time woodland apparently was spreading at the expense of rainforest.

The African rainforest was progressively impoverished as aridity spread during the Miocene, and notably in the Pliocene as the ice cap developed on Antarctica (after 5 m.y.), bringing the driest part of the Tertiary to low–middle latitudes (Axelrod 1948). This was followed by several episodes of 'ice-age aridity' that must have had a disastrous effect on the moisture-sensitive rainforest taxa, eliminating many from Africa. Chapin (1923: 111) and Richards (1973) have emphasized that the richest part of the flora of tropical West Africa occurs today in areas with least drought. Richards has also noted that the flora of tropical West Africa is very poor in Annonales, palms are rare, bamboos are poorly developed, orchids are not abundant, nor are epiphytes and lianas as conspicuous as in the rainforests of South America, Malaysia, or Madagascar.

Furthermore, the total number of taxa is low in tropical Africa as compared with other areas. Lind & Morrison (1974: 4) give comparative figures that illustrate this relationship. Whereas in the Mpanga forest of the Lake Victoria belt only 50 species exceed 30 cm in girth, the total is 109 species on 1.5 ha in the Cameroons, and rises to fully 200 in plots of only 1.6 ha in Malaya. Whereas the upper canopy trees in the Bugoma forest, Uganda, total not more than 80 taxa in an area of 45 km^2 (including diverse slope relations), in Malaya 375 taxa form the canopy in 23 ha in the lowland forest. It seems clear that under the moister climates of the middle and late Tertiary the African rainforest was probably very rich in taxa, especially into the Miocene, following which progressively spreading dry climate eliminated many of them (Axelrod 1972a, Raven & Axelrod 1974). Their nearest allies survive now in the rainforests of Madagascar, southeast Asia or in tropical South America, in areas where precipitation remained more evenly distributed during Miocene and later times. This impoverishment of the African tropical forest by climatic changes over the past 20 m.y. has led Aubréville (1955) and others to overstress the isolation of tropical Africa from other tropical areas.

With respect to the degree of impoverishment that may have occurred in the later Cenozoic, attention is directed to a rich, largely uninterpreted late Tertiary or early Quaternary flora from five localities in the Cameroon mountains of West Africa (Menzel 1920). Over 230 species of tropical West African plants distributed in 48 families are listed. The leaves, which are of excellent preservation, occur in basalt tuff and are associated with volcanic flows that probably can be dated radiometrically. The fossils, reportedly very similar to living species, are in Annonaceae, Araceae, Arecaceae, Bombacaceae, Burseraceae, Capparaceae, Connaraceae, Dichapetalaceae, Dioscoreaceae, Euphorbiaceae, Fabaceae (Mimosoideae, Caesalpinioideae, Faboideae), Flacourtiaceae, Hippocrateaceae, Lauraceae, Malpighiaceae, Marantaceae, Meliaceae, Menispermaceae, Moraceae, Octoknemaceae, Olacaceae, Piperaceae, Rosaceae, Sapindaceae, Sterculiaceae, Tiliaceae, and others. Manifestly, a comparison of this flora with the present rainforest can provide a real measure of the degree of floristic change in tropical West Africa during the later Cenozoic, and certainly after the later episodes of 'ice-age aridity' had affected it. The tropical rainforests of Africa are continuing to contract rapidly at present under the influence of human activities and probably also climate (Aubréville 1949).

4.4 Montane rainforest

The only African fossil flora that appears to represent montane rainforest was a large pollen flora recovered from the basal part of an 82-meter core in Ruizi valley, Burundi. The site is 35 km north of Lake Tanganyika, in the graben that leads north to Lake Kivu (Sah 1967). The strata, which underlie younger alluvium that fills the middle of the graben, may have been deposited in a larger, more extended Lake Tanganyika during the late Pliocene or early Pleistocene. Sah recognized some 43 genera and 64 species, but most were assigned to family only. Among these are Acanthaceae, Adiantaceae, Anacardiaceae, Aquifoliaceae, Bombacaceae, Brassicaceae, Buxaceae, Cyatheaceae, Dennstedtiaceae, Dicksoniaceae, Ericaceae, Fabaceae (Mimosoideae), Geraniaceae, Meliaceae, Olacaceae, Oleaceae, Osmundaceae, Pedaliaceae, Podocarpaceae, Polypodiaceae, Proteaceae, Rhamnaceae, Rubiaceae, Simaroubaceae, Solanaceae, Tiliaceae, and Vitaceae. The abundance of ferns, the rarity of grasses, and the presence of several alliances of temperate montane environments, including Ericaceae and Podocarpaceae, implies a temperate montane rainforest environment, evidently with ample rainfall.

The interrupted nature of the chain of volcanos extending down East Africa may help to explain why there are many fewer north temperate groups of plants represented in southern Africa than in either South America or Australasia (Raven 1973a). In addition, the limited number of high mountains in southern Africa and its relatively low latitude may be factors. Despite this scant direct evidence, it is clear that montane rainforest, which may have been in existence in Paleogene time, expanded from the Miocene onward as East Africa was elevated and volcanos were built up along the rift zone.

Taxa may have contributed quickly to montane rainforest (and the higher Afroalpine zone) as volcanos were built up along the rift zone during early Miocene and later times. Since strato-volcanos with elevations in excess of 2,000 to 3,000 meters regularly are constructed in less than 1 million years, alpine conditions may have appeared in the summit sections of the larger volcanic edifices early in the Miocene (25–23 m.y.), reaching down the rift belt into Moçambique, near Lat. 16°S. Inasmuch as these volcanos have always been discontinuous and separated by montane rainforest, the populations at their higher altitudes were fragmented and isolated from the beginning. The discontinuous patterns exhibited by pachycaulous species of *Senecio* and *Lobelia*, however, have probably originated since the late Pliocene. This would be consistent with the age of the genera themselves and the drastic geological and climatic changes in the area during the past few million years (Moreau 1963, Hedberg 1965, 1969, Mabberley 1975).

The high volcanos of East Africa probably have provided an interrupted pathway for the migration of the plants and animals of temperate regions between the Northern Hemisphere and southern Africa since the later Miocene, and continue to do so at present (Hooker 1885). Examples of northern genera that have reached East or even South Africa by this route are *Anemone, Arabis, Astragalus, Cardamine, Carduus, Carex, Cerastium, Delphinium, Galium, Heracleum, Hypericum,* and *Luzula* (Hedberg 1965). Southern African genera that have extended their ranges northward along the mountains of East Africa include *Dierama, Disa, Erica, Euryops, Gladiolus, Hesperantha, Kniphofia, Pen-*

taschistis, Protea, Stoebe and *Thesium* (Hedberg 1965, compare also Chapters 11 and 12).

At lower elevations there are numerous other genera characteristic of the Cape flora which range northward in the East African mountains, some extending to as far as Ethiopia. They include some sclerophyllous scrub genera (*Cliffortia, Muraltia, Philippia, Passerina, Protea, Stoebe*) as well as others, including *Aristea* and *Restio* (see Chapter 8). This Afro-montane-Cape flora has a discontinous distribution, chiefly in highlands which are separated by more arid valleys with savanna vegetation.

Inasmuch as the discontinuous areas receive high rainfall (150 to 200 cm yearly), the connections between the now-discontinuous patches probably date back into the Miocene. This corresponds with the principal time of disruption of forests in western North America and in southwestern Asia. In both areas relict forests are confined now to the moister, higher mountains separated by broad tracts of drier lowland climates and subhumid vegetation. Whether the genera have come from East Africa into the Cape region and produced many species there, or have migrated northward into East Africa, is a matter to be considered for each genus independently. A few Cape genera (e.g., *Aristea, Philippia, Phylica, Restio*; Adamson 1958) are even discontinuous to Madagascar, which we believe they reached by long-distance dispersal, contrary to the view of Adamson (1958) and Levyns (1962, 1964). The time of separation of Madagascar from the mainland of Africa remains to be established. Any notion of a continuously favourable habitat for such genera between the mountains of the Cape region and those of Madagascar at any time must be rejected out of hand, which makes the time of separation of Madagascar irrelevant to the time of differentiation of these genera. Such an assertion is likewise consistent with the fact that the shared genera constitute a very small proportion of the respective floras.

The relatively low representation of endemics in the alpine zone of the mountains of East Africa and the patterns of distribution observed would be consistent with the arrival of their ancestors during the past several million years. Miocene and Pliocene volcanos, long eroded, may have provided earlier pathways for migration between north and south, but there are probably no remnants of their floras still living at present.

The altitudinal profiles presented by Bader (1965) for the genera *Juniperus, Podocarpus* and *Rhamnus* in eastern and southern Africa, stretching from the Abyssinian highlands to the Drakensberg (excl. *Juniperus*), reinforce this conclusion. His illustrations clearly show that many of the taxa of these elements could only have attained their present distribution during the late Pliocene and Quaternary, after the highlands had formed (Fig. 11). This closely parallels the time and mode of invasion of other boreal taxa into the southern Andes via the American cordillera, and into New Zealand via the highlands of the Alpine axis–East Indies during the Quaternary (Raven 1963, 1973a, b), and again suggests an evolution of the Afro-alpine taxa within the past several million years. Just as in Central–South America and Southeast Asia–Australasia, the pathway appears to have functioned to the greatest extent during the cool 'pluvial' periods.

Vegetation allied to montane rainforest occurs today in the Canary Islands, inhabiting a volcanic archipelago no older than middle Miocene. This implies that the adjacent coast of northwest and northern Africa may have supported rather similar vegetation and taxa in the middle Tertiary (Axelrod 1975). The relation

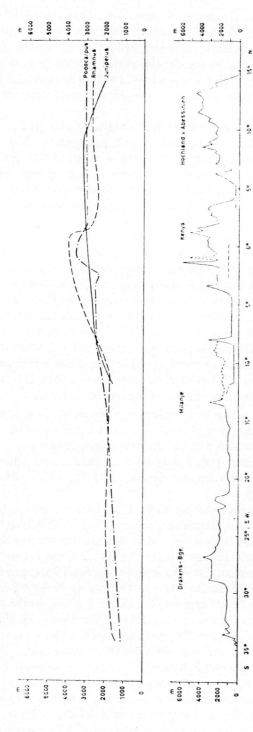

Fig. 11. Altitudinal distribution of selected taxa in Africa, and a topographic profile from Abyssinia to the Cape region (from Bader 1965). Building up of this high relief by warping, faulting and volcanism provided a north-south corridor for numerous plants during the Neogene, including taxa from Eurasia in the Pleistocene.

may be inferred also from the composition of the Neogene floras presently known from the Tethyan region, stretching from Portugal eastward across southern Europe into the Caucasus.

This inference is supported by the present occurrence in the Tibetsi, Ennedi, Hoggar, Marra and Air massifs of a group of taxa that are now disjunct from the East African montane region (Miré & Quezel 1959, Quezel 1965). Among these are species of *Abutilon, Acacia, Commicarpus, Ehretia, Erica, Ficus, Hermannia,* and *Parietaria* that occur in the Saharan massifs chiefly at altitudes between 1800 and 2400 m, on southwest warmer slopes. Elsewhere in the region at lower levels is relict, sclerophyllous Mediterranean vegetation, composed of *Capparis, Ceratonia, Cupressus, Helianthemum, Juniperus, Nerium, Pistacia,* and others, together with associated herbs and perennials. In both groups there are endemic species that differ from their nearest relatives, either to the north in the Mediterranean region, or to the east in the mountains of tropical Africa.

At altitudes in the Saharan region above 2800 m there are a few taxa that link the high Tibetsi and Hoggar with the eastern highlands of Ruwenzori, Kenya and Ethiopia. These also include related endemics, as well as a few identical species in the following herbaceous genera: *Agrostis, Avena, Crassula, Dichrocephala, Festuca, Fimbristylis, Helichrysum, Nepeta* and *Silene,* and doubtless their disjunct patterns of distribution are not older than the late Pleistocene. Numerous taxa of African montane affinity are recorded from Neogene floras of the Mediterranean region, and some have persisted into the early Pleistocene. For example, species of *Cassia, Pittosporum* and *Sapindus,* which no longer occur in northern Africa, lived with a typical Mediterranean sclerophyllous flora (*Ceratonia, Laurus, Olea, Quercus* spp., *Rhus*) in Tunisia (Arambourg et al. 1954). There also are a number of living Canarian species, distributed in *Aeonium, Ceropegia, Dracaena, Euphorbia, Kleinia, Myrsine, Notelaea, Prunus, Rhamnus* and *Visnea,* that appear to have their nearest relatives in East and South Africa (Lems 1960). The links were not across the central rainforest belt, but more probably by highlands (Hoggar, Air, Tibesti massifs) in the central Saharan region which provided an east–west corridor to the south-trending older warps of East Africa (see Figs. 6B, 6C).

These three very different floristic groups which occur above the Saharan desert region where there are very few endemics, and which attest to its recency, obviously entered the region at different times and have survived in favorable sites. No doubt many other trees and shrubs of the East African montane forest disappeared from the uplands of the Saharan region as precipitation was lowered following the Miocene. That a rise in rainfall of 40–50 cm would be adequate to support sclerophyll vegetation over much of the Saharan region is implied by the flora from Lake Ichkeul (Arambourg et al. 1953) which has several warm temperate to subtropical taxa that are no longer in North Africa. Such conditions are inferred to have been widely present into the early Pliocene (10–7 m.y.) at least, a belief consistent with the occurrence of wood of *Quercus* in the Pliocene south of Tindouf (Lemoigne 1967). During the Miocene, links between the Canarian laurel forest and the montane forest of East Africa doubtless were both more numerous and more continuous than at present (Figs. 6B, 6C).

4.5 Subtropical evergreen forest

There appears to be only one Neogene flora from within the area now occupied by subtropical evergreen forest, which extends south along the east coast to Pondoland. A particularly fine example is the Ngoya Forest, which is almost at the southwestward limit of this vegetation type (compare Chapter 13). Fossil plants from Fort Grey near East London (Adamson 1934) include *Podocarpus*, *Widdringtonia*, one or more ferns, fragmentary dicot leaves, and several unidentified fruits and seeds. Adamson noted that the forest was dominated by gymnosperms whereas they are rare in the area today, being found in moister climates. This fossil flora tells us little about the evolution or past occurrence of subtropical evergreen forest, although it must have contracted and expanded considerably during the past few million years like the other vegetation types with which it is in contact. Suggestive of the gradual breakup of subtropical evergreen forest into discontinuous patches is the present range of the cocosoid palm *Raphia*, which is represented by about 30 species in tropical Africa and Madagascar, and one on the coast of Natal (Adamson 1948, Moore 1973, Fig. 2H). Bews (1922) has provided an instructive discussion of the relationships of this forest as it occurs in South Africa, and the derivation of subtropical from tropical elements southward.

4.6 Savannas and woodlands

With a Neogene trend to drier climate, brought on by general uplift of the continent, changes in circulation, and the resultant decrease in moisture, closed forests retreated as savannas and woodlands replaced them. Savannas attained a progressively wider area as the trend to aridity increased during the Miocene and Pliocene, an inference consistent with the increasing numbers of grazing mammals in younger rocks. In this regard, attention is directed to the possible early evolutionary role of the developing rift system as rainforest retreated to the west. The development of rainshadows in the rift valleys favoured the early spread of savanna and then grassland, at first in local, small basins and then more widely as the basins lengthened and deepened. Such an environment may have created a small-population structure, favourable for rapid diversification not only of flowering plants, but of grazers (gazelles, antelopes, etc.) and their dependent carnivores (lions, cheetah, hyaena). This entire ecosystem spread rapidly as drier open environments (subzones, sensu Simpson 1944, 1953) expanded in the Pliocene and especially in the late Pliocene and Pleistocene as major renewed rifting commenced and as warping developed new dry lowland areas throughout the region.

The small Bugishu flora, from basal sediments that rest on Archean basement (Chaney 1933), is covered by the lower Miocene Mt. Elgon Volcanics, dated at 24 m.y. (Van Couvering & Van Couvering 1975, Fig. 3). The assemblage has taxa referred to *Bauhinia*, *Berlinia*, *Cassia*, *Dalbergia*, *Parinari*, *Pittosporum* and *Terminalia*, with the legumes especially common. The abundance of small (nanophyll–microphyll) leaves suggest dry savanna–woodland, not rainforest vegetation. Phillips (in Chaney 1933) indicated that the composition of the flora suggests seasonal climate with definite wet and dry seasons. The geologist E. J. Wayland (in Chaney 1933) concurred, for the leaf-bearing beds were deposited on a peneplaned basement mantled with lateritic ironstone, the formation of which is due to the rise and fall of ground water during the wet and dry seasons.

Other records from the Mt. Elgon volcanics include woods representing a cyatheoid tree fern (Bancroft 1932) and wood of Dipterocarpaceae (Bancroft 1933). Dipterocarps are represented today by some 600 species, centered in southeast Asia. In Africa, the family is represented by the isolated subfamily Monotoideae, composed of *Monotes* and *Marquesia*. Another genus, the monotypic *Vateriopsis*, is on the Seychelles. Environmentally, dipterocarps range from wet rainforests into savannas and other dry tropical environments.

The importance of the Bugishu flora is that it indicates independently that seasonal climate had already appeared west of the continental divide by the early Miocene. It implies that dense rainforest in this region probably was now being restricted, occurring chiefly on moister upland slopes and as gallery forests along waterways, as suggested also by the Rusinga flora. Nearby woodland and savanna was composed of taxa that had already adapted to drier climate. Regional geologic relations show that the Bugishu flora was in the lee of a low range where drier conditions would naturally result.

A Miocene tropical semideciduous forest appears to be represented by fossil woods from the Karugamania beds, situated at the base of the Bogoro scarp, Lake Albert (Lakhanpal & Prakash 1970). The flora has been dated at 19 m.y., making it a contemporary of the Rusinga (Van Couvering & Van Couvering 1975, Fig. 3). One species is allied to the Meliaceae (cf. *Entandrophragma* or *Carapa*), but all the others appear to represent members of the Fabaceae (cf. *Baphia, Brachystegia, Dichrostachys, Isoberlinia, Newtonia*). Judged from the author's comments, several of these have their nearest relatives in the nearby region. This implies that dry-adapted forests and savannas were now spreading in the young rift valleys as climate became drier, an inference also consistent with the occurrence of numerous woody legumes in the flora.

This analysis is consistent with fossil records now available from the Ethiopian plateau to the north which was developing as a volcanic upland in the early Miocene, 22–23 m.y. (Baker et al. 1971, Mohr 1968, 1971, Abel-Gawad 1969, Jones & Rex 1974). Fossil sites occur at several sites near Debre Libanos (alt. >2500 m) about 110 km north of Addis Ababa (Beauchamp et al. 1973). The taxa occur in tuffs overlying thick breccias that rest on an upper Cretaceous marine sequence. Fossil woods are referred to *Dombeyoxylon, Evodioxylon, Ficoxylon, Guttiferoxylon, Hibiscoxylon, Sapindoxylon,* and *Ternstroemioxylon,* and leaves identified as *Acrostichum, Cassalpinia, Ficus,* and *Stenochlaena*. Although the assemblage was referred to the Eocene on the basis of the similarity of the woods to those in the Egyptian Eocene (Kräusel 1939, Boureau 1958), the fossils may well be younger because wood structure is conservative. This agrees with evidence that the oldest volcanic rocks in the Ethiopian Plateau fall in the range of 23–27 m.y. (Jones & Rex 1974), implying that the flora is early Miocene, or younger. To judge from the small size (microphyll) of the few leaves illustrated, the flora suggests dry climate, possibly savanna–woodland, or a mixed semideciduous forest.

Similar evidence comes from other localities in the region (Lemoigne et al. 1974). The sites occur on both the Ethiopian plateau (Mush Valley, Molale) and Somali plateau (Wondo), as well as within the Ethiopian rift proper (Debre Sina). The Mush Valley flora (alt. ~3000 m) has leaf fossils referred to *Annonaphyllum*

(3 spp.), *Brideliophyllum*, *Leguminophyllum*, *Nymphaeophyllum*, 4 unidentified leaves, and wood referred to *Dialioxylon* and *Terminalioxylon*. The relatively moderate size of most of the leaves (microphyll–notophyll) suggests a semideciduous forest or savanna–woodland. The small Molale sample (alt. ~3000 m) has two species of *Leguminoxylon*, the Wondo sample has *Ficoxylon* and an unidentified wood. The Debre Sina flora (~3000 m) from the rift zone has wood referred to *Afzelioxylon*, *Entandrophragmoxylon*, *Euphorbioxylon*, *Ficoxylon*, *Sapotoxylon*, *Syndoroxylon*, and an unidentified species. This assemblage may represent a relict montane rainforest in the region, assuming *Entandrophragma* is correctly identified.

Podocarpus occurs in the Miocene of the Welkite region, about 150 km east of Addis Ababa, where it is recorded with woods referred to *Pahudia*, *Terminalia* and a leguminous tree of uncertain affinity (Lemoigne & Beauchamp 1972). *Podocarpus* is a rough indicator of elevation. In East Africa it occurs chiefly in the range of 1800–3000 m (Lind & Morrison 1974), though on the western shore of Lake Victoria it flourishes at about 1000 m in a seasonal swamp. Since the Welkite woods are associated with the imprints of reeds (*Phragmites*, *Typha*) that indicate swampy conditions, a somewhat similar setting may be inferred tentatively for the Welkite flora. A minimum altitude near 1000–1800 m seems indicated, whereas the region is now near 3000 m.

Building up of the Ethiopian highland during the Miocene, as a result of rifting and uplift of the plateaus bording the rift zone, seems reasonably indicated. That regional climate had a definite dry season is also apparent. Thus the spreading and strengthening anticyclone, as well as the topographic high, now tended to divert the SE trades northward, bringing a more seasonally dry climate sufficient to disrupt rainforest and restrict it farther west. By the middle Miocene, therefore, lowland rainforest probably had only a patchy distribution along the northern parts of the east coast, an area where it is now locally relict, as at the foot of the Moçambique escarpment. The taxa that represent the small samples presently known from the plateau region suggest seasonal tropical climate at the time of deposition. They occur now in the uplands under a temperate, equable climate at altitudes well over 2000 m. Uplift occurred in the late Pliocene and Pleistocene, as the rift deepened and its flanking plateaus were elevated.

The spread of seasonally dry climate not only favored savannas at the expense of lowland rainforest, it confined montane rainforest to highlands over the interior where there was adequate moisture. As a result, there was a progressive development of more discontinuous forests, and of taxa that comprised them. The present savannas and woodlands display marked differences as shown by the Sudanian and Zambezian floristic Domains (White 1965). Generic endemism in Sudano-Zambezian Region is slight, with about 15 per cent of the genera endemic, and about 61 per cent occurring in the moist Guineo-Congolian rainforest Region. Although 51 per cent of the genera and 19 per cent of the species are common to the Sudanian and Zambezian Domains, the Sudan has a much poorer flora, so the relations between the two savanna domains are asymmetric. That is, 85 per cent of the Sudanian genera and 46 per cent of its species occur in the Zambezian Domain. The Zambezian Domain, as exemplified by Zambia with 334 species, has a richer flora and three times as many endemics as the Sudanian (White 1965). These relations are understandable on the basis of the general trend to aridity during the Miocene and later. The Sudanian Domain is much drier than the

Zambezian. Its greater aridity has no doubt been the crucial factor in impoverishment of its flora, and in the asymmetry of savanna-type vegetation depicted on many vegetation maps (e.g., Bartholomew 1963).

In southern Tunisia, the lowlands in mid- to late Miocene time seem to have been occupied by a rather well watered open savanna (Rich 1974), much more mesic in character than the present desert character of the vegetation would suggest. Similar plant associations evidently had become common throughout the lowlands of Africa by the late Miocene.

4.7 Temperate rainforest

There are a few fossil records of temperate rainforest, which now extends in a series of disjunct areas from near Cape Town to near Port Elizabeth, but was much more widespread in historic times (Acocks 1953, compare Chapter 11). The Knysna flora (Phillips 1927) comes from a sequence of lignite, sandstone, and claystone that probably represent a lagoonal deposit, though it is now 400 m above sea level (DuToit 1954). It may be late Miocene or early Pliocene in age. Wood of *Widdringtonia* is abundant, leaves of *Podocarpus* are present, as are those of several dicots that live in the nearby region, notably the mesic endemic *Curtisia*, as well as *Gonioma*. According to Phillips, the flora represents a moister climate than that at the site today, for the taxa occur chiefly in the bordering mountains. More recently, Thiergart et al. (1963) conducted a pollen study of the Knysna lignites. They identified four taxa in the Podocarpaceae, one of which is referred to *Dacrydium* which is not now in Africa, but in the Fuegan–Tasman region. Other families represented are Restionaceae, Proteaceae with five presumed species, and Myrtaceae with one. A number of other grains were referred doubtfully to Anacardiaceae, Araliaceae (3 taxa), Fabaceae (6 taxa), Meliaceae, and Sapotaceae, all of which are represented in the area today. Spores allied to *Alsophila*, *Gleichenia* and *Schizaea* are also reported.

That some of the mesic forest plants like those which now grow near Knysna were more widespread under the moister climates of the Neogene may be inferred from the late Pleistocene pollen record studied by Schalke (1973), and reviewed by Livingstone (1975) and Van Zinderen Bakker (1976). The sample comes from an interfingering marine–nonmarine sequence that is slightly older than 40,000 years and ranges up to about 20,000 years. It accumulated on the Cape Flats just north of Cape Town when sea level was 18–24 m lower than at present, indicating contemporareity with the last glacial. The flora records two invasions of temperate rainforest elements which now occur several to tens of kilometers away. They grew with or displaced the sclerophyllous vegetation that dominated under mediterranean climate. The times when *Curtisia*, *Ilex*, *Podocarpus* and their associates lived there were separated by a return of sclerophyllous vegetation. During the early moist phase (~34,000 yrs. B.P.) a few grains of *Nothofagus* are recorded, and were considered contaminants, probably transported by the westerlies from South America (Schalke 1973).

From the Cape Flats, Adamson (1951) recorded the presence of trunks of *Podocarpus falcatus* (one 17.3 m long and 1.2 m in diameter) at depths of 4.9 m, or 14 m above present sea level. They appear to have lived in the nearby hills, judged from the rich pollen record of *Podocarpus* in the area, and therefore at lower elevations than at present.

4.8 Sclerophyll vegetation

Hardier sclerophylls gradually assumed dominance over the lowlands as the temperate rainforest retreated in response to increasing drought. The sclerophylls were derived from taxa in the retreating vegetation, and in the bordering subtropical evergreen forest and montane rainforest, a process that probably commenced in the later Eocene. Some of their ancestral forms may have occurred in locally dry or unusually well drained or rocky situations in a more continuous temperate rainforest. The Eo-Oligocene Banke flora of Namaqualand seems to represent an ecotonal area between temperate rainforest and sclerophyllous vegetation comparable to situations widespread from the vicinity of Cape Town to the vicinity of Grahamstown. The formation of comparable sclerophyllous associations seems to have occurred at about this time in southwestern Asia and in North America, which tend to lend credence to the notion of an Eocene origin for such vegetation in southern Africa.

The sclerophyll vegetation of Africa represents two disparate units. In the north, a mixed woodland of oak–juniper–laurel–pistacio–olive and associated macchia shrubs was replaced by spreading grassland and semidesert and then full desert in the drier phases of the Quaternary. The mixed woodland and macchia associations have survived only locally in North Africa in favorable sites, mostly near the Atlantic Ocean or the Mediterranean Sea. Except for a few structurally low basins dominated by grassland and semidesert scrub, much of the Saharan region probably was covered with sclerophyll woodland–savanna and scrub well into the Pliocene, and probably at least in part during the Pleistocene pluvials (Axelrod 1973). As noted above, this is consistent with the present occurrence of relict stands of oak–olive–juniper sclerophyll vegetation in the high mountains (Hoggar, Tibesti, Air) of the Sahara, where there is adequate moisture (Quezel 1965). As emphasized earlier (Axelrod 1973, 1975), sclerophyllous vegetation lived under a climate with summer rain into the early Pleistocene, and it still does in areas away from regions of mediterranean climate. Most of the woody taxa that occur in the latter areas are older than the mediterranean climate. They evidently adjusted to increasing summer drought during the Pliocene by shifting the time of establishment into the moister, cooler part of the year.

By analogy with the more complete record of plant history in the Tethyan region, we infer that sclerophyllous vegetation was more widespread in southwest Africa by the start of the Neogene, occupying areas presently semidesert and desert. At that time, it lived under a regime of summer rainfall, a relation shown also by floras of the Madrean–Tethyan region. The recurrent formation of sclerophyll vegetation on infertile soils has been discussed well by Johnson & Briggs (1975). Toward the interior and equatorward in Africa, this vegetation gave way to tropical woodland–savanna which lived under a climate with rain in the warm season.

4.9 Cape vegetation

That the flora of the southwestern Cape Province of South Africa is unique in comparison with those in other areas of similar climate is understandable: it has been isolated for well over 100 m.y. (Raven 1973b). This flora extends from the vicinity of Port Elizabeth, where rainfall is well distributed throughout the year, to the vicinity of Cape Town, with a progressive tendency toward the elimination of summer rainfall westward. The western part of the area has a mediterranean climate, but the Cape flora extends well beyond the area of summer aridity. The whole region can be said to have a highly equable climate, and it is one in which the products of many ancient and distinctive evolutionary lines have survived (e.g., see Carlquist 1975, for *Geissoloma*). As we shall discuss below, the development of a mediterranean climate within this region of sclerophyll-forest ecotone has been both secondary and geologically recent.

Among the families that are most characteristic of the Cape flora and richest in genera and species are Ericaceae, Fabaceae, Iridaceae, Proteaceae, Restionaceae, Rosaceae, Rutaceae–Diosmeae, and Thymelaeaceae (compare Chapter 8). Endemic to the region or nearly so are Bruniaceae (12 genera, 75 species), Penaeaceae (5 genera, 25 species), Geissolomataceae (1 species), Grubbiaceae (2 genera, 5 species), Stilbaceae (5 genera, 12 species), Retziaceae (1 species), and Roridulaceae (1 genus, 2 species). South African endemics not in the Cape vegetation or not centered there include Achariaceae (3 genera, 3 species), Greyiaceae (1 genus, 3 species), and Heteropyxidaceae (= Myrtaceae? 1 genus, 3 species).

The plants that contribute to the sclerophyllous fynbos, principal vegetation type of this area, are chiefly low shrubs, restioids and geophytes and trees are rare. Leaves of the woody plants are quite small, leathery, and often ericoid in type. Thus they provide a marked contrast with most of the taxa of the Mediterranean basin, and also of other mediterranean-climate areas. The large number of species of the important genera in the Cape vegetation (e.g., *Agathosma* 134 species; *Aspalathus* c. 250; *Cliffortia* 70; *Erica* c. 600 species; *Muraltia* 106; *Pelargonium* c. 150; *Phylica* c. 140; *Protea* 84; *Restio* 94) is a striking feature of the flora as compared with the nearby temperate rainforest to the east where genera have few species.

A few of these genera appear to have austral affinities and to have been derived from temperature antecedents (e.g., Proteaceae, Restionaceae, Rutaceae, Thymelaeaceae), as inferred from distributions shown in Figs. 7 and 8. Many others seem to have existed in subtropical or montane forests or other vegetation types in Africa, and to a certain extent, still exist there. Which of these areas represent the individual source for the various groups is a difficult question, and the place of origin of such taxa as Ericoideae or Proteaceae may never be known with certainty.

Proteaceae are fairly widespread in Africa outside of the Cape area, and Levyns (1958, 1962, 1964) and others have argued that the southern African representatives probably came from the north, with the probable exception of *Brabeium*. At a more fundamental level, Proteaceae are an austral family, with no authenticated fossils in the Northern Hemisphere; they may have dispersed originally between Africa and Australasia across a much smaller Indian Ocean (Raven & Axelrod 1974). Levyns (1964) has, in effect, hypothesized the same for *Phylica*. In ecological terms, to say that they came from the north is equivalent to saying that they are now better represented in the specialized and derivative vegetation types of the southwest than in regions of more generalized vegetation elsewhere. It is difficult to view the past through the eyes of the present, but Africa clearly lacks the more primitive rainforest Proteaceae of northern Australasia (Johnson & Briggs 1975). As demonstrated by the geological review provided earlier in this chapter, the pathways for migration across the Indian Ocean (via

Ninetyeast Ridge–India–Madagascar) between Africa and Australasia existed until mid-Cretaceous time, whereas those involving Africa and Antarctica – a true austral, temperate route – were severed tens of millions of years earlier, and clearly long before the origin of any existing groups of angiosperms.

The evidence reviewed above strongly suggests that summer-dry and desert climates are of recent origin in southern Africa. In the Miocene and Pliocene, under a climate of higher precipitation and milder temperatures, sclerophyllous woodland and patches of savanna seem to have covered much of the area now grassland, semidesert, and desert, judged from the shifts in vegetation reviewed above. Even in the late Pleistocene (last pluvial cycle), the climate was much wetter than at present, and xeric vegetation types must have been much restricted (e.g., Butzer et al. 1973). For example, at Alexandersfontein Pan, Kimberley (Butzer et al. 1973), at the east edge of the Cape province, precipitation is now annually 397 mm but is estimated to have been nearly twice that during the last pluvial (19,000 B.P., compare Chapter 6).

In particular, there is no evidence that the summer-dry (mediterranean) climates in southern Africa existed before the formation of a major ice sheet on Antarctica ($\sim$5 m.y.). Earlier, sea-surface waters were considerably warmer as judged from micropaleontological and sedimentological data (e.g., Kennett & Brunner 1973, Blank & Margolis 1975, Ciesielski & Weaver 1974, Fillon 1975, Kennett & Watkins 1974, Kennett & Vella 1975). The evidence from Antarctica thus implies that waters as cold as those at present did not exist off the coast of southwest Africa (contra Van Zinderen Bakker 1975 and Chapter 6), and also that sea surface temperatures fluctuated considerably during the past 5 m.y., from 'warm' to 'cold', with numerous oscillations (Blank & Margolis 1975, Fig. 5).

This agrees with evidence along the Namib coast itself. The rich molluscan fauna 3 km northeast of Bogenfels characterized by *Turritella* evidently is Miocene (Haughton 1963). Its warm-water implications militate against a cold-water Benguela Current at that time. Furthermore, the 'oyster-line' beach, dominated by shells of *Striostrea margaritacea* (= *Ostrea prismatica* of Haughton) near the mouth of the Orange River, is a warm-water fauna and is now regarded as early Pleistocene (Haughton 1932, Carrington & Kensley 1969, Martin 1973, Tankard, letter of 26 Feb. 1976). This indicates that the Benguela Current was not as strong or as cold as at present, and that the desert and mediterranean climates were not then in existence as pronounced regional features. Further, there is clear evidence of warmer water along the present Benguela-washed coast even in the last interglacial (Tankard 1975, Kilburn & Tankard 1975).

It probably was in the late interglacial ages that mediterranean-type climate (dry summer) appeared, giving way to biseasonal rainfall as the oceans warmed. This is consistent with evidence in southeastern Australia, where data now indicate that the transition from a humid temperate rainforest to the present dominant *Eucalyptus–Acacia* vegetation occurred at $\sim$4.5 m.y. (Gill 1975), a time when aridity also reached a peak in the western United States (Axelrod 1948), and well prior to the appearance of a mediterranean type climate. At 18,000 years before present, the Benguela Current was much stronger in winter than it is now, with sea ice reaching the southeastern African coast (CLIMAP Project Members 1976), and a great expansion of sclerophyllous vegetation and deserts is depicted at that time. They show savanna reaching the coast in the latitude of central Angola

(~20°S), with expanded deserts occupying the coast north of Cape Town: paleaobotanic and floristic data lend no support to this idea.

As in other areas of mediterranean climate, the adaptive structural features of those Cape sclerophylls that are most successful under conditions of summer drought evidently were established by the Oligocene, if not earlier. As the taxa that required summer rain gradually were restricted eastward, the environment was open for taxa with the tolerance to withstand summer drought, and many were able to do so. In effect, a new island habitat was being created within the area of an established vegetation type. Just as in the other areas of the world with a mediterranean climate, those groups that were able to survive had great opportunities for evolutionary radiation and the production of hundreds of new, localized species (Levyns 1952). As Levyns (1964) has put it, 'These unwieldy genera are the direct result of intensive speciation in the west'; i.e., within the area of summer drought, displaced northward during the Pleistocene pluvial cycles. Speaking of *Erica*, Adamson (1958) has rightly pointed out that the very high concentration of species in the southwestern Cape Province implies neither a center of origin there, nor survival from an early southward migration, but rather that the genus has found conditions especially favorable to differentiation in this area. He generalizes this argument to apply to other groups such as Proteaceae and *Stoebe* also. The southwestern Cape Province, and espcially its winter rainfall area, are not a center of survival, as Levyns suggested in 1938, nor are its mountains ancient; rather it is a zone of very active speciation owing to recent orogenic and climatic change. Some of the evolutionary patterns have been discussed very well by Dahlgren (1970, 1971). The ability to hybridize and consequently to produce a wide array of recombinants in response to the demands of a harsh marginal, and rapidly changing habitat seems clearly to have been important in many of these groups.

This does not mean that all the hundreds of species of genera such as *Aspalathus*, *Cliffortia*, *Erica*, and *Muraltia* originated in their present area. They most certainly did not. The evidence reviewed above indicates that the fynbos was severely limited within its present area by the expansion of temperate rainforest even during the most recent pluvial cycle. This implies that the present Cape vegetation was then displaced to the north, into the present semidesert and desert. On this basis, the sclerophyllous Cape flora was largely swept into its present area as dry climate expanded following the last pluvial, and further radiation has taken place even within the last 10,000 years (Weimark 1941, Stebbins 1974, Chapter 8, compare Van Zinderen Bakker, Chapter 6). In this sense, the present Cape flora with its numerous species of many genera may represent but a remnant of a much richer sclerophyllous flora that ranged over the present desert and steppe areas into the Pleistocene, as did comparable vegetation in the western United States, and in southwestern Asia as well. When it invaded its present area, it doubtless came in contact with isolated pockets of sclerophyllous vegetation, as on steep and rocky mountainsides, and these may have contributed directly and through hybridization to the overall diversity of the area.

In all of this, however, it must be borne in mind that the summer-dry area is only the western portion of that occupied by Cape vegetation. In that area, however, endemism at the specific level appears to be more pronounced, and speciation processes more active than in in those regions farther to the east with a more abundant supply of summer rainfall. It is likely that just as in California the notion that the Cape flora was once much more widespread in areas presently semidesert and desert, agrees with evidence that remnants of this flora occur also on mountains of sufficient height in Namaqualand and the Karoo generally at levels above 900–1200 m (Cannon 1924, Adamson 1938, Levyns 1938, see Chapter 8). The taxa in these scattered sites are largely identical with those in the Cape, although some have been isolated sufficiently so that a few endemics have developed (Levyns 1964). This suggests that as recently as the last pluvial period, the Cape flora may have been far more widespread (Levys 1938), a relation paralleled by the nature of the plants recovered from late Pleistocene wood rat middens of the Mohave Desert and border areas (e.g. Wells & Berger 1967).

Levyns (1964) suggested that the youthfulness of the flora in its present mediterranean-climate environment is implied also by the growth rhythms of some of the taxa. These are said to enter into a period of rapid vegetative growth toward the end of summer when water supply is lowest, a response scarcely explicable under present conditions, but one expectable in a region with summer rainfall. That the relation might imply evolution under a regime of summer rainfall is suggested also by the occurrence of the same phenomenon in South and West Australia (Specht 1969, Specht & Rayson 1957, Burbidge 1960, Johnson & Briggs 1975), where it has been interpreted in similar manner. This could be taken to mean that the mediterranean climate is not ancient, but is so recent in this area that the plants have not yet fully adapted to it. If the existence of this phenomenon can be substantiated by further studies in South Africa, it will be most interesting.

The fynbos itself, however, has greatly expanded its area to the eastward, into areas of summer rainfall during the past 500 years (Acocks 1953). While the temperate forest and scrub-forest have largely disappeared except around Knysna and in numerous smaller patches elsewhere during this relatively short period of time (Breitenbach 1972), the fynbos has expanded from small patches in edaphically favorable sites to become the dominant vegetation all the way to the vicinity of Grahamstown (Acocks 1953, maps 1 and 2). Most endemics and by far the highest concentration of species, however, are found within 200 km of the west coast (e.g., Levyns 1938, 1955, 1964), in summer-dry areas affected by the cold Benguela Current. From this region and its late Pliocene and Pleistocene antecedents farther north a relatively limited number of species, still however representing a rich flora, have migrated eastward in historic time to replace temperate rainforest as a direct consequence of its disruption by human activities (Acocks 1953).

4.10 *Deserts and semideserts*

As indicated above, the present desert and semidesert regions of southern Africa are not old geologically. A rise in precipitation of 250–300 mm would enable grassland and patches of sclerophyllous vegetation now in the region to descend into the lowlands and occupy areas where semidesert and desert now occur. This is substantiated not only by the evidence from fossil floras reviewed above, but also by the marine faunas preserved in older warped terraces along the west coast, which are probably Miocene in age (Haughton 1963, DuToit 1954). These deposits include animals which now occur in areas affected by the Moçambique Current on the east coast but not along the western shores washed by the cold

Benguela Current. This implies that in the Miocene sclerophyll vegetation probably occupied all the present semidesert and desert area, with relict patches of forest in favorable sites. As far back as Eo-Oligocene time (~38.5 m.y.), the Banke flora provides evidence of such vegetation in this area.

Concerning the development of the succulent Karoo, Levyns (1964) notes that near Ladismith the east–west trending Swartberg Range greatly affects local climate. As the area is approached, succulent Karoo vegetation is quickly replaced by a typical Cape sclerophyllous flora on the koppies facing south. Just north of Ladismith, where annual precipitation is 330 mm, a stream flows from the adjacent range and supports stands of the evergreen *Ilex mitis*, a common tree in the temperate rainforest near Knysna. Thus, a decrease in rainfall of from near 890 mm rather well distributed through the year (as at Knysna), to 380–500 mm (Cape flora), to less than 250 mm (the Karoo), involves a shift from a rich forest to succulent semidesert. Comparable changes in vegetation and climate are well documented for California–Nevada, and also for southwest Asia since the Miocene. There is every reason to assume that South Africa is similar in this respect.

One of the most extreme and unusual vegetation types in southern Africa is the succulent Karoo, which lies within 200 km of the west coast and is the vegetational counterpart of the succulent vegetation of Baja California, or northern Chile. Levyns (1964) has shown that many of the genera that are dominant in this vegetation type appear to have come from the savannas and woodlands that now lie in warmer and moister climates to the north: for example, *Acacia*, *Aloe*, *Buddleia*, *Euclea*, *Pappea*, and *Rhigozum*. On the other hand, Acocks (1953) has stressed the strong fynbos affinity of the Karoo, which is clearly a complex ecotonal region. *Rhus*, which is common in the Karoo, is frequent in drier vegetation types all over Africa. The succulent bushes and geophytes that are so prominent in the Karoo have often produced such overwhelming outbursts of speciation here that their extradesert affinities are difficult to determine. They include such genera as *Babiana* and *Oxalis*, each of which has its metropolis in the Cape region, together with other genera that are absent or less common there, including *Galenia*, *Mesembryanthemum*, *Pteronia*, and *Tetragonia* of the Aizoaceae; *Crassula* and *Zygophyllum*, which are also well developed in the Cape; and the Stapelieae (see Chapter 9).

In the remainder of the Karoo, a vast area where succulent shrubs are not conspicuous, some vegetation types contain woody genera and grasses related to woodland and savanna types characteristic of Africa farther north, but mixed with elements possibly derived from the Cape flora, including both woodland and fynbos types. A vast array of annuals and geophytes, mainly of temperate affinities, contributes to the beauty of this area, and especially of Namaqualand, after sufficient rainfall. The whole region is in many ways the analogue of the Sonoran–Mohave Desert system of North America, with plants derived from or related to the regions of sclerophyllous scrub more abundant on the latter.

These dissimilar dominants in the semideserts of South Africa obviously represent the hardy survivors of vegetation that formerly inhabited the region, notably savanna and sclerophyllous Cape vegetation. The taxa were already adapted structurally to some drought, and hence were preadapted to the trend to aridity that increased, especially after the ice sheets spread over Antarctica about 5 m.y. ago. When the semideserts and deserts expanded, many of these genera experienced rapid speciation and diversification into the patterns seen today.

The pulses of alternating wet and dry climate during the Quaternary probably account largely for the numerous species in the Karoo, as well as in the Cape flora. At times of moister climate taxa of the sclerophyllous Cape flora spread widely over the present Karoo region (see Chapter 6), speciating rapidly in many groups. As drier climate returned, the flora shifted coastward into its present area, bringing the new taxa with them and leaving relict stands in locally moist situations. During the dry phases the taxa representing the Karoo flora were able to spread out and differentiate new populations, aided no doubt by the dry edaphic sites already scattered in the region.

4.11   *Summary*

The modern aspects of African vegetation were outlined by the early Miocene. Vegetation was altered materially as climate became progressively drier, restricting forests at the expense of spreading savanna, thorn scrub and grassland. The later phases of this trend probably saw the appearance of local semidesert areas in the Pliocene, but widespread, regional semideserts and deserts are a phenomenon of the later periods of 'ice-age aridity'.

Links between the forest vegetation of tropical and southern Africa were more numerous and widespread in the past, and gradually became restricted by spreading drought. The flora of South Africa probably received increments from the tropics into the Miocene, and from boreal regions after highlands of sufficient elevation had been constructed, chiefly in Plio-Pleistocene times. Remnant links with distant austral lands, which are Cretaceous, have survived in the coastal strip of south Africa under mild equable climate, sheltered from the effects of aridity to the north by the east–west trending ranges.

5. Conclusions

The changing type and area of vegetation in Africa since the middle Cretaceous have been shaped by diverse physical factors: land connections, climate, and terrain. As their influence has waxed and waned, changing taxa have been selected by increasingly more complex climate–terrain environments to add to the diversity of vegetation in the region, and to breakup older vegetation zones into derivative ones of narrower area and lower diversity. In its broadest features, vegetation history in Africa has paralleled that of other continents which have remained relatively stable in latitude since the Cretaceous. The over-riding factors of a climatic trend to increasing aridity, coupled with increasing topography, have provided opportunities for taxa to contribute to new vegetation zones as older ones have become restricted. Among the major events the following deserve special emphasis.

(1) From the middle Cretaceous into the early Tertiary, Africa moved north about 15° Lat., from a near-union with Antarctica–India–Madagascar–South America–Australia. The links with austral lands were largely severed during the late Cretaceous, though plant dispersal via connecting islands was effective for some time.

(2) From the late Cretaceous into the early Tertiary, Africa was relatively low,

with a shallow seaway in the north covering parts of the Saharan region. The climatic belts were broad in extent, supporting a lowland equatorial rainforest that reached from coast to coast, and was flanked by savannas to the north and south. Sclerophyll vegetation had appeared at the outer, drier margins of each monsoon region by the later Eocene, in areas of subhumid climate characterized by summer rainfall. Local montane areas in the inner tropics were clothed with rainforests which descended to lower elevations to the north and south where taxa contributed to temperate rainforests.

(3) Uplift commenced in the late Oligocene–early Miocene, accompanied by warping that created broad swells and basins. Volcanism commenced on a major scale, accompanied by the initiations of the East African rift valleys and building of major volcanos. This provided greater topographic diversity, and rainforest taxa now commenced to invade higher, colder zones that probably included Afroalpine areas as well. Savannas were now commencing to spread more widely at the expense of rainforest, and sclerophyllous vegetation was attaining optimum development both in the north and south under the seasonal rainfall.

(4) The African plate (including Arabia) joined the Asian plate (at Iran) in the middle Miocene, bringing an end to the long-enduring Tethyan Sea and its control on broadly zoned climates. Conditions now became drier and extremes of temperature increased over lowland areas. Anticyclonic circulations in the bordering oceans now increased in area, intensity, and stability, and dry climate became more severe and spread more widely. This resulted in the spread of savanna, deciduous forest, thorn forest, and sclerophyllous vegetation at the expense of rainforests in particular, forcing them to higher, moister elevations as savanna and thorn forest replaced them.

(5) Antarctica had shifted to its present position by the early Miocene, and was the site of mountain glaciation. A full ice sheet did not appear until about 5 m.y., and it waxed and waned for 2–3 m.y. The introduction of cold water to the west coast by the Benguela Current accentuated the on-going trend to increased summer drought along the west coast. The strengthening high pressure systems now brought drier climate to the interior, restricting savanna–woodland and rainforest, lowering their diversity, and favoring the spread of grassland, thorn forest and semidesert. Sclerophyllous taxa that had lived earlier under summer and winter rain, were now adapting to increasingly drier summers.

(6) There was a symmetrical distribution of climate and vegetation with respect to the central moist tropical belt throughout the Tertiary. As the moist tropical belt was progressively narrowed, savanna, thorn scrub, grassland, and semidesert environments gradually increased in area, with full desert environment attaining maximum area and intensity at the times of 'ice-age aridity'.

(7) The present African vegetation shows considerably greater asymmetry than that of the early to late Tertiary. At present the Sudanian savanna contrasts with the Zambezian, the Saharan Desert with the Namib and Kalahari, the temperate Natal evergreen forest with the Canarian, and the Cape fynbos with the Mediterranean macchia vegetation. Basic differences were no doubt present early, for their areas were in proximity to lands that provided unique migrants to each region that did not penetrate the central rainforest barrier. Thus, the Tethyan

sclerophyll vegetation has greater affinity with the Madrean of western North America than with the fynbos, and the Canarian laurel forest lacks the austral relicts Podocarpaceae, Proteaceae, Restionaceae that occur in the temperate evergreen forest in South Africa. Nonetheless, some taxa managed to penetrate southward, chiefly in the Miocene and more recently, as relief increased down East Africa. The differences in composition of vegetation were accentuated largely in the Pliocene and later as dry climate spread more widely.

(8) There were at least two major episodes of rapid speciation in South Africa. The first evidently commenced in the Miocene, with the broad warping and uplift of Africa. The new intermontane basins favored the spread of open savanna and grassland at the expense of forest, and then steppe environments commenced to spread as aridity increased. Many new plant taxa probably originated in these expanding open areas. The large-mammal fauna no doubt responded as a new, open grazing zone spread. Restricted at first to local basins, the fauna may have proliferated rapidly as aridity increased and new subzones (food preference) appeared in more local areas.

The second burst of speciation in South Africa probably resulted from Plio-Pleistocene deformation and accompanying fluctuation of climate. As mountains were elevated appreciably around the rim of South Africa and broad basins developed over the interior, the low areas became drier and the mountains moister, and erosion sculptured many new small basins. The terrain was composed of diverse rocks, not only sedimentary, but varied crystallines (granite, gneiss, schist, ultrabasics). Climate was alternating between wet and dry phases so that populations were shifting continuously. At the times of moist climate sclerophyllous vegetation invaded the area of the present semidesert, only to return to the Cape as conditions became drier. Thus on several occasions large and diverse populations were swept into the Cape area which now includes a great concentration of species of numerous genera, as well as relicts that survive under a highly equable, near-maritime climate.

6. Epilogue

Some progress has been made in understanding the pre-Quaternary history of African vegetation during the 25 years since the problem was initially studied by Moreau (1952). Numerous sites have been discovered but in most cases only small, inadequate collections have been made. The greatest need is concentrated collecting so as to provide the largest possible representation of taxa. After all, a rainforest or a savanna can not be reconstructed on the basis of five or six species. As for pollen studies of Tertiary floras, these can be useful, but only if the taxa can be identified to genera at least: lists that indicate only family affinity of taxa are often not sufficiently indicative either of paleoecology or of age.

The earlier, provisional studies of the five largest Tertiary floras now known from Africa, each of which represents a different ecosystem, should be completed. These include (1) the Rusinga seed flora, the study of which can now be supplemented by associated leaf floras; (2) the well preserved Bugishu flora from Uganda that represents woodland vegetation is represented by abundant leaves on the tuffaceous slabs, but the sample is so small that it is not adequate to reconstruct the vegetation or climate; (3) the Cameroon flora, preserved in basalt tuff associated with flows that can be dated radiometrically, should be recollected and compared with the present rainforest so as to determine the magnitude of change that resulted from the stages of 'ice-age aridity'. (4) the Eocene flora from the Fajum which appears to represent woodland–savanna, or possibly semideciduous forest, needs intensive collecting and analysis. (5) the Banke site, which includes taxa representing plant associations similar to those found in the southwestern Cape at present, should be reopened with earth-moving equipment so that an adequate sample can be recovered.

The urgent need for larger, more complete samples is apparent also from a review paper by Werger (1973) which highlights the diverse ideas of floristic relations in South Africa. There is little agreement as to the affinities of taxa in terms of floral regions. This is expectable because during the later Tertiary and Quaternary climatic changes segregated taxa into more restricted environments, and greatly modified the composition of local floras in all parts of South Africa, as well as in the inner tropics. Clearly, the historical development of the floras in local areas in southern Africa will be clarified only when an adequate fossil record has been assembled and analysed.

References

Abdel-Gawad, M. 1969. New evidence of transcurrent movements in Red Sea area and petroleum indications. Amer. Assoc. Petrol. Geol. Bull. 53:1466–1499.
Acocks, J. P. H. 1953. Veld types of South Africa. Mem. Bot. Surv. S. Afr. 28:1–192.
Adamson, R. S. 1931. Notes on some petrified wood from Banke, Namaqualand. Trans. Roy. Soc. S. Afr. 19:255–258.
Adamson, R. S. 1934. Fossil plants from Fort Grey, near East London. South Afr. Mus. Ann. 31:67–96.
Adamson, R. S. 1938. Notes on the vegetation of the Kamiesberg. Mem. Bot. Surv. S. Afr. 18:1–25.
Adamson, R. S. 1948. Some geographical aspects of the Cape flora. Trans. Roy. Soc. S. Afr. 31:437–464.
Adamson, R. S. 1951. Buried trees on the Cape Flats. Trans. Roy. Soc. S. Afr. 33:13–24.
Adamson, R. S. 1958. The Cape as an ancient African flora. Advancem. Sci. 58:118–127.
Allard, G. O. & V. J. Hurst. 1963. Brazil–Gabon geologic link supports continental drift. Science 163:528–532.
Alvarex, W. T. Cocozza & F. C. Wezel. 1974. Fragmentation of the alpine orogenic belt by microplate dispersal. Nature 248:309–314.
Andrews, P. & J. A. H. Van Couvering. 1975. Paleoenvironments in the East African Miocene. In: F. S. Szalay, Approaches to Primate Paleobiology 5:62–103. Karger, Basel.
Arambourg, C. J. Arènes & G. Depape. 1953. Contribution a l'étude des flores fossiles Quaternairies de l'Afrique du Nord. Archiv. Mus. Nat. Hist., Paris, 2:1–81.
Archangelsky, S. & E. J. Romero. 1974. Polen de gimnospermas (Coníferas) del Cretaceo superior y Paleoceno de Patagonia. Ameghinoana 11:217–236.
Aubréville, A. 1949. Climats, Forêts et Désertification de l'Afrique tropicale. Sociéte d'Éditions Géographiques, Maritimes et Coloniales, Paris.
Aubréville, A. 1955. La disjonction africaine dans la flora forestière tropicale. Soc. Biogéogr. Paris Compt.-Rend. 278:42–49.
Aubréville, A. 1970. Le flore tropicale Tertiaire du Sahara. Adansonia n.s. 10:9–14.
Axelrod, D. I. 1948. Climate and evolution in western North America during middle Pliocene time. Evolution 2:127–144.
Axelrod, D. I. 1950. Evolution of desert vegetation in western North America. Carnegie Inst. Wash. Pub. 590:215–306.
Axelrod, D. I. 1958. Evolution of the Madro-Tertiary Geoflora. Bot. Rev. 24:433–509.
Axelrod, D. I. 1960. The evolution of flowering plants. In: S. Tax (ed.), Evolution after Darwin 1:277–305. Univ. Chicago Press, Chicago.

Axelrod, D. I. 1965. A method for determining the altitudes of Tertiary floras. Paleobotanist 14:144–171.
Axelrod, D. I. 1967. Quaternary extinctions of large mammals. Univ. Calif. Publ. Geol. Sci. 74:1–42.
Axelrod, D. I. 1970. Mesozoic paleogeography and early angiosperm history. Bot. Rev. 36:277–319.
Axelrod, D. I. 1972a. Ocean-floor spreading in relation to ecosystematic problems. Univ. Ark. Mus. Occas. Paper 4:15–76.
Axelrod, D. I. 1972b. Edaphic aridity as a factor in angiosperm evolution. Amer. Naturalist 106:311–320.
Axelrod, D. I. 1973. History of the Mediterranean ecosystem in California. In: F. di Castri & H. A. Mooney (eds.), Mediterranean Type Ecosystems – Origin and Structure, pp. 225–277. Springer-Verlag, Berlin.
Axelrod, D. I. 1974. Plate tectonics in relation to the history of angiosperm vegetation in India. B. Sahni Inst. Palaeobotany Spec. Pub. 1:5–18 (1971).
Axelrod, D. I. 1975. Evolution and biogeography of Madrean–Tethyan sclerophyll vegetation. Ann. Missouri Bot. Garden 62:280–284.
Axelrod, D. I. & H. P. Bailey. 1969. Paleotemperature analysis of Tertiary floras. Palaeogeogr., Paleoclimatol., Paleoecol. 6:163–195.
Bader, F. J. W. 1965. Some boreal and subantarctic elements in the flora of the high mountains of tropical Africa and their relation to other intertropical continents. Webbia 19:531–544.
Baker, B. H., P. A. Mohr & L. A. J. Williams. 1972. Geology of the eastern rift system of Africa. Geol. Soc. Amer. Spec. Paper 136.
Baker, B. H. & J. Wohlenberg. 1971. Structure and evolution of the Kenya rift valley. Nature 229:538–542.
Baker, B. H., L. A. Williams, J. A. Miller & F. J. Fitch. 1971. Sequence and geochronology of the Kenya rift volcanics. Tectonophysics 11:191–215.
Bancroft, H. 1932. A fossil cyatheoid stem from Mount Elgon, East Africa. New Phytol. 31:241–253.
Bancroft, H. 1933. A contribution to the geological history of the Dipterocarpaceae. Geol. Forhandl. 55:59–100.
Barnard, P. D. W. 1973. Mesozoic floras. In: N. F. Hughes (ed.), Organisms and continents through time. Paleontological Assoc. Spec. Papers in Palaeontology 12:175–188.
Bartholomew, J. 1963. The Advanced Atlas of Modern Geography Ed. 7. Oliver and Boyd Ltd., London.
Beauchamp, J., Y. Lemoigne & J. Petrescu. 1973. Les paléoflores tertiaires de Debré Libanos (Éthiopie). Ann. Soc. Geol. Nord., Lille 93:17–32.
Berggren, W. A. & J. A. Van Couvering. 1974. Neogene Geochronobioclimatopaleomagneto-stratigraphy: a Mediterranean synthesis. Geol. Soc. Amer. Abstr. 6:1022–1024.
Berry, E. W. 1938. Tertiary flora from the Rio Pichileufu, Argentina. Geol. Soc. Amer. Soec. Papers 12:1–149.
Bews, J. W. 1922. The south-east African flora: Its origins, migrations, and evolutionary tendencies. Ann. Bot. 36:209–223.
Bird, J. M. & B. Isacks, (eds.). 1972. Plate Tectonics: Selected Papers from The Journal of Geophysical Research. Amer. Geophys. Union. Washington, D.C.
Blank, R. G. & S. V. Margolis. 1975. Pliocene climatic and glacial history of Antarctica as revealed by Southeast Indian Ocean deep-sea cores. Bull. Geol. Soc. Amer. 86:1058–1066.
Boccaletti, M. &. G. Guassone. 1974. Remnant arcs and marginal basins in the Cainozoic development of the Mediterranean. Nature 252:18–21.
Bolli, H. M., W. B. F. Ryan et al. 1975. Basins and margins of the eastern South Atlantic (DSDP 40). Geotimes 20:22–24.
Boughey, A. S. 1957. The origin of the African flora. Oxford Press, Oxford.
Boughey, A. S. 1965. Comparisons between the montane forest floras of N. America, Africa and Asia. Webbia 19:507–517.
Boureau, E. 1958. Paléobotanique africaine. Evolution des flores disparues de l'Afrique nord-équatoriale. Bull. Scient. Com. Tr. Hist. et Scient, pp. 1–64 (not seen).
Breitenbach, F. von. 1972. Indigenous forests of the Southern Cape. J. Bot. Soc. S. Afr. 58:18–47.
Brenner, G. J. 1976. Middle Cretaceous floral provinces and early migrations of angiosperms. In:

Beck, C. B. (ed.), Origin and Early Evolution of Angiosperms. pp. 23–47. Columbia Univ. Press, New York and London.

Briden, J. C., G. E. Drewry & A. G. Smith. 1974. Phanerozoic equal-area world maps. J. Geology 82:555–574.

Brundin, L. 1966. Transantarctic relationships and their significance, as evidenced by the chironomid midges, with a monograph of the subfamily Podonomineae, Aphrotaeniinae and the austral Heptagiae. Kgl. Sv. Vetensk. Handl. IV. 11:1–472.

Burbidge, N. T. 1960. The phytogeography of the Australian region. Austral. J. Bot. 8:75–212.

Burke, K. 1975. Atlantic evaporites formed by evaporation of water spilled from Pacific, Tethyan, and Southern oceans. Geology 3: 613–616.

Burke, K. & J. F. Dewey. 1974. Two plates in Africa during the Cretaceous. Nature 249:313–316.

Burke, K. & J. T. Wilson. 1972. Is the African plate stationary? Nature 239:387–389.

Butzer, K. W., G. J. Bock, R. Stuckenrath & A. Zilch. 1973. Palaeohydrology of Late Pleistocene Lake, Alexandersfontein, Kimberley, South Africa. Nature 243:328–330.

Camp, W. H. 1947. Distribution patterns in modern plants and the problems of ancient dispersals. Ecol. Monogr. 17:159–183.

Cannon, W. A. 1924. General and physiological features of the vegetation of the more arid portions of southern Africa, with notes on the climatic environment. Carnegie Inst. Wash. Publ. 354:1–159.

Carlquist, S. 1975. Woody anatomy and relationships of Geissolomataceae. Bull. Torrey Bot. Club 102:128–134.

Carr, A. & P. J. Coleman. 1974. Seafloor spreading theory and the odyssey of the green turtle. Nature 249:128–130.

Carrington, A. J. & B. F. Kensley. 1969. Pleistocene molluscs from the Namaqualand coast. Ann. S. Afr. Mus. 52:189–223.

Chaloner, W. G. & W. S. Lacey. 1973. The Distribution of late Paleozoic floras. In: N. F. Hughes (ed.), Organisms and Continents Through Time. Palaeontological Assoc. Spec. Papers in Palaeontology 12:271–289.

Chandler, M. E. J. 1954. Some upper Cretaceous and Eocene fruits from Egypt. British Mus. Mat. Hist. Geol. 2:149–187.

Chandler, M. E. J. 1964. The Lower Tertiary floras of southern England. IV. A summary and survey of findings in the light of recent botanical observations. British Mus. Nat. Hist. Geol. 12:1–151, pl. 1–4.

Chaney, R. W. 1933. A Tertiary flora from Uganda. J. Geol. 41: 702–709.

Channell, J. E. T. & D. H. Tarling. 1975. Paleomagnetism and the rotation of Italy. Earth Planetary Sci. Letters 25: 177–188.

Chapin, J. P. 1923. Ecological aspects of bird distribution within the limits of West Africa. Ibis 13 (1):255–302.

Chesters, K. I. M. 1957. The Miocene flora of Rusinga Island, Lake Victoria, Kenya. Paleontographica 101B:30–71.

Chiarugi, A. 1929. Legni fossili (di Giarabùb). Resultati scientifici della Missione alla Oasi di Gairabùb. Paleontologia. Realc. Soc. Geogr. Ital. 3:397–430.

Churchill, D. M. 1973. The ecological significance of tropical mangroves in the early Tertiary floras of southern Australia. Geol. Soc. Austral. Sp. Publ. 4:79–86.

Ciesielski, P. F. & F. M. Weaver. 1974. Early Pliocene temperature changes in the Antarctic seas. Geology 2:511–516.

CLIMAP Project Members. 1976. The surface of the ice-age earth. Science 1911:1131–1144.

Cooke, H. B. S. 1972. The fossil mammal fauna of Africa. In: A. Keast, F. C. Erk & B. Glass (eds.), Evolution, Mammals and Southern Continents. pp. 89–140. State Univ. New York Press, Albany, N.Y.

Corydon, S. C. & R. J. G. Savage. 1973. The origin and affinities of African mammal faunas. In: N. F. Hughes (ed.), Organisms and Continents Through Time. Palaeontological Assn. London Spec. Pap. Palaeontol. 12:241–269.

Couper, R. A. 1953a. Distribution of Proteaceae, Fagaceae and Podocarpaceae in some Southern Hemisphere Cretaceous and Tertiary beds. New Zealand J. Sci. Tech. B. 35:247–250.

Couper, R. A. 1953b. Upper Mesozoic and Cainozoic spores and pollen grains from New Zealand. New Zeal. Geol. Surv. Paleontol. Bull. 22:1–77.

Couper, R. A. 1960a. Southern Hemisphere Mesozoic and Tertiary Podocarpaceae and Fagaceae and their palaeogeographic significance. Proc. Roy. Soc. London 152B:491–500.

Couper, R. A. 1960b. New Zealand Mesozoic and Cainozoic plant microfossils. New Zeal. Geol. Surv. Paleont. Bull. 32:1–87.
Couyat, J. & P. H. Fritel. 1919. Sur la presence d'imprintes végétales dans le grès nubian des environs d'Assouan. Compt. Rend. Hebd. Séanc. Acad. Sci. Paris 151:961–964.
Cox, C. B. 1974. Vertebrate paleodistributional patterns and continental drift. J. Biogeogr. 1:75–94.
Cracraft, J. 1973. Vertebrate evolution and biogeography in the Old World tropics. In: D. H. Tarling & S. K. Runcorn. (eds.), Implications of Continental Drift to the Earth Sciences. pp. 271–392. Academic Press, London and New York.
Cracraft, J. 1975. Historical biogeography and earth history: perspectives for a future synthesis. Ann. Missouri Bot. Gard. 62:227–250.
Cranwell, L. M. 1959. Fossil pollen from Seymour Island, Antarctica. Nature 184:1782–1785.
Cranwell, L. M. 1964. Antarctica: cradle or grave for its Nothofagus? In: L. M. Cranwell (ed.), Ancient Pacific floras. pp. 87–93. Univ. Hawaii Press, Honolulu.
Cranwell, L. M. 1969. Palynological intimations of some pre-Oligocene Antarctic climates. Palaeoecology of Africa 5:1–19.
Crowell, J. C. & L. A. Frakes. 1970. Phanerozoic glaciation and the causes of ice ages. Amer. J. Sci. 268:193–224.
Dahlgren, R. 1970. Parallelism, convergence, and analogy in some South African genera of Leguminosae. Bot. Not. 123:552–568.
Dahlgren, R. 1971. Multiple similarity of leaf between two genera of Cape plants, *Cliffortia* L. (Rosaceae) and *Aspalathus* L. (Fabaceae). Bot. Not. 124:292–304.
Dewey, J. F., W. C. Pitman III, W. B. F. Ryan & J. Bonnin. 1973. Plate tectonics and the evolution of the Alpine system. Geol. Soc. Amer. Bull. 84:3137–3180.
Dietz, R. S. & J. C. Holden. 1970. Reconstruction of Pangaea: breakup and dispersion of continents, Permian to Present. J. Geophys. Res. 75:4939–4956.
Dingle, R. V. 1973. Mesozoic palaeogeography of the southern Cape, South Africa. Palaeogeogr., Palaeoclimatol., Palaeoecol. 13:203–213.
DuToit, A. L. 1954. The Geology of South Africa. Ed. 3, S. H. Haughton (ed.). Oliver & Boyd, Edinburgh and London.
Edmunds, G. F., Jr. 1972. Biogeography and evolution of Ephemeroptera. Ann. Rev. Entom. 17:21–42.
Embleton, B. J. J. & M. W. McElhinny. 1975. The palaeoposition of Madagascar: Palaeomagnetic evidence from the Isalo group. Earth and Planet. Sci. Letters 27:329–341.
Engelhardt, H. 1907. Tertiäre Pflanzenreste aus dem Fajûm. Beiträge zur Palaont. und Geol. Österreich-Ungarns und des Orients 20:206–216.
Estes, R. 1975. Fossil *Xenopus* from the Paleocene of South America and the zoogeography of pipid frogs. Herpetologica 31:263–278.
Fillon, R. H. 1975. Late Cenozoic paleo-oceanography of the Ross Sea, Antarctica. Bull. Geol. Soc. Amer. 86:839–845.
Flores, G. 1970. Suggested origin of the Mozambique Channel. Trans. Geol. Soc. S. Afr. 73:1–16.
Florin, R. 1963. The distribution of conifer and taxad genera in time and space. Acta Horti Bergiani 20:121–312.
Foster, R. J. 1974. Eocene echinoids and the Drake Passage. Nature 249:751.
Geyh, M. A. & D. Jakel. 1974. Late glacial and Holocene climatic history of the Sahara Desert derived from a statistical assay of 14C dates. Palaeogeogr., Palaeoclimatol., Palaeoecol. 15:205–208.
Gill, E. D. 1975. Evolution of Australia's unique flora and fauna in relation to the plate tectonics theory. Roy. Soc. Victoria Proc. 87:215–234.
Good, R. 1974. The Geography of Flowering Plants. Ed. 4. Longman, London.
Gordon, W. A. 1973. Marine life and ocean surface currents in the Cretaceous. J. Geology 81:269–284.
Goslin, J. & J. C. Sibuet. 1975. Geophysical study of the easternmost Walvis Ridge, South Atlantic: Deep Structure. Geol. Soc. Amer. Bull. 86:1713–1724.
Gough, D. I. 1973. Possible linear plume under southernmost Africa. Nature Phys. Sci. 245:93–94.
Greenway, P. J. 1970. A classification of the vegetation of East Africa. Kirkia 9:1–68.
Hallum, A. (ed.). 1973. Atlas of Palaeobiogeography. Elsevier Sci. Publ. Co., Amsterdam, London, New York.
Hamilton, A. C. 1968. Some plant fossils from Bukwa. Uganda J. 32:157–164.

Haughton, S. H. 1931. On a collection of fossil frogs from the clays at Banke. Trans. Roy. Soc. S. Afr. 19:233–249.
Haughton, S. H. 1932. The late Tertiary and Recent deposits of the west coast of South Africa. Trans. Geol. Soc. S. Afr. 34:19–57.
Haughton, S. H. 1963. The Stratigraphic History of Africa South of the Sahara. Oliver and Boyd, Edinburgh and London.
Hedberg, O. 1965. Afroalpine flora elements. Webbia 19:519–529.
Hedberg, O. 1969. Evolution and speciation in a tropical high mountain flora. Biol. J. Linn. Soc. 1:135–148.
Herngreen, G. F. W. 1974. Middle Cretaceous palynomorphs from northeastern Brazil. Sci. Geol. Univ. Louis Pasteur de Strausbourg Bull. 27:101–116.
Herngreen, G. F. W. 1975. Palynology of the middle and upper Cretaceous strata in Brazil. Mededelingen Rijks Geologische Dienst n.s. 26:39–91.
Hoeken-Klinkerberg, P. M. J. van. 1966. Maastrichtian, Paleocene and Eocene pollen and spores from Nigeria. Leidse Geol. Medel. 38:37–48.
Hooker, J. D. 1853. Botany of the Antarctic Voyage of H.M. Discovery Ships Erebus and Terror in the years 1831–43, vol. 2: Flora Novae-Zelandiae, Pt. 1, Introductory Essay.
Hooker, J. D. 1885. Preface to Oliver, D., List of the Plants collected by Mr. Thomson F.R.G.S., on the mountains of Eastern Equatorial Africa. J. Linn. Soc. Bot. 21:392–406.
Hos, D. 1975. Preliminary investigation of the palynology of the Upper Eocene Werillup Formation, Western Australia. J. Roy. Soc. Western Austr. 58:1–14.
Hughes, N. F. (ed.). 1973. Organisms and Continents Through Time. Paleontological Assoc. London. Special Papers in Palaeontology 12:1–334.
Hutchinson, J. 1973. The Families of Flowering Plants. Oxford. 3rd ed.
Jardine, S., G. Kieser & Y. Reyre. 1974. L'individualisation progressive du continent Africain vue a travers les donnees palynologiques de l'ere secondarie. Sci. Geol. Univ. Louis Pasteur de Strausbourg Bull. 27:69–85.
Jenkins, D. G. 1974. Initiation of the proto circum-Antarctic current. Nature 252:371–373.
Johnson, L. A. S. & B. G. Briggs. 1975. On the Proteaceae – the evolution and classification of a southern family. Bot. J. Linn. Soc. 70:83–182.
Jones, P. W. & D. C. Rex. 1974. New dates from the Ethiopian plateau volcanics. Nature 252:218–219.
Keast, A. 1972. Continental drift and the biota of the mammals on southern continents. In: A. Keast, F. C. Erk & B. Glass (ed.), Evolution, Mammals and Southern Continents, pp. 23–194. State Univ. New York Press, Albany, N.Y.
Kemp, E. M. 1972. Reworked palynomorphs from the west ice shelf area, East Antarctica, and their possible geological and paleoclimatological significance. Marine Geol. 13:145–157.
Kemp, E. M. & P. J. Barrett. 1975. Antarctic glaciation and early Tertiary vegetation. Nature 258:507–508.
Kemp, E. M. & W. K. Harris. 1975. The vegetation of Tertiary islands on the Ninetyeast Ridge. Nature 258:303–307.
Kennett, J. P. et al. 1972. Australia–Antarctic continental drift, Palaeocirculation changes and Oligocene deep-sea erosion. Nature Phys. Sci. 239:51–55.
Kennett, J. P. & C. A. Brunner. 1973. Antarctic late Cenozoic glaciation: evidence for initiation of ice rafting and inferred increased bottom-water activity. Bull. Geol. Soc. Amer. 84:2043–2052.
Kennett. J. P. & R. E. Houts et al. 1974. Development of the Circum-Antarctic Current. Science 186:144–147.
Kennett, J. P. & P. Vella. 1975. 19. Late Cenozoic planktonic foraminifera and paleooceanography at DSDP Site 284 in the cool subtropical South Pacific. In: Kennett, J. P. & R. E. Houz et al., Initial Reports of the Deep Sea Drilling Project 29:769–799.
Kennett, J. P. & N. D. Watkins. 1974. Later Miocene–Early Pliocene paleomagnetic stratigraphy, paleoclimatology, and biostratigraphy in New Zealand. Geol. Soc. Amer. Bull. 85:1385–1398.
Kent, P. E. 1972. Mesozoic history of the East Coast of Africa. Nature 238:147–148.
Kent, P. E., J. A. Hunt. & D. W. Johnstone. 1971. The geology and geophysics of coastal Tanzania. Inst. Geol. Sci., Geophys. Paper 6:1–101. London.
Khan, M. A. 1975. The Afro-Arabian rift system. Sci. Progress Oxford 62:207–236.
Kilburn, R. N. & A. J. Tankard. 1975. Pleistocene molluscs from the west and south coasts of the Cape Province, South Africa. Ann. S. Afr. Mus. 67:183–226.

King, L. C. 1962. The Morphology of the Earth: a study and synthesis of world scenery. Hafner Publ. Co., New York.
King, L. C. 1963. South African Scenery: A Text Book of Geomorphology. Ed. 3. Hafner Publ. Co., New York.
Kirchheimer, F. 1934. On pollen from the Upper Cretaceous dysodil of Banque, Namaqualand (South Africa). Trans. Roy. Soc. S. Afr. 21:41–50.
Kräusel, R. 1939. Ergebnisse der Forschungsreisen Prof. E. Stromer's in den Wüsten Aegyptens. IV. Die fossilen Floren Aegyptens. Abh. Bayer. Ak. Wissensch. Mathem.-naturwissen Abt. N.F. 47:1–140.
Kröner, A. 1973. Comment on 'Is the African Plate stationary?' Nature 243:29–30.
Kutina, J. 1975. Tectonic development and metallogeny of Madagascar with reference to the fracture pattern of the Indian Ocean. Geol. Soc. Amer. Bull. 86:582–592.
Ladd, J. W., G. O. Dickson & W. C. Pitman III. 1973. The age of the South Atlantic. In: A. E. M. Narin & F. G. Stehli (eds.), The Ocean Basins and Margins, Vol. 1, The South Atlantic. pp. 555–573. Plenum Press, New York and London.
Lakhanpal, R. N. 1966. Some middle Tertiary plant remains from south Kivu, Congo. Musee Royal de l'Afrique centrale. Tervuren, Belgique Annales ser I–8°, Sci. Geol. no. 52:21–30.
Lakhanpal, R. N. 1970. Tertiary floras of India and their bearing on the historical geology of the region. Taxon 19:675–694.
Lakhanpal, R. N. & U. Prakash. 1970. Cenozoic plants from Congo. I. Fossil woods from the Miocene of Lake Albert. Musee Royal de l'Afrique Centrale – Tervuren, Belgique Ann. VIII. Sci. Geol. 64:1–20.
Larson, R. L. & J. W. Ladd. 1973. Evidence for the opening of the south Atlantic in the Early Cretaceous. Nature 246:209–212.
Lemoigne, Y. 1967. Reconnaissance paléobotanique dans le Sahara occidental (Région de Tindouf et Gara-Djebilet). Ann. Soc. Geol. Nord 87:31–38.
Lemoigne, Y. & Beauchamp, J. 1972. Paléoflores tertiares de la région de Welkite (Ethiopie, province d'Shoa). Compt. Rend. Somm. Séanc. Soc. Géol. France 1:35.
Lemoigne, Y., J. Beauchamp. & E. Samuel. 1974. Etude paléobotanique des dépôts volcaniques d'âge tertaire des bordures est et ouest du système des rifts éthiopiens. Geobios 7:267–288.
Lems, K. 1960. Floristic botany of the Canary Islands. Sarracenia 5:1–94.
Levyns, M. R. 1938. Some evidence bearing on the past history of the Cape flora. Trans. Roy. Soc. S. Afr. 26:401–424.
Levyns, M. R. 1952. Clues to the past in the Cape flora of today. S. Afr. J. Sci. 49:155–164.
Levyns, M. R. 1955. Some geographical features of the family Polygonaceae in southern Africa. Trans. Roy. Soc. S. Afr. 34:379–386.
Levyns, M. R. 1958. The phytogeography of members of Proteaceae in Africa. J. S. Afr. Bot. 24:1–9.
Levyns, M. R. 1962. Possible Antarctic elements in the South African flora. S. Afr. J. Sci. 58:237–241.
Levyns, M. R. 1964. Migrations and origin of the Cape Flora. Trans. Roy. Soc. S. Afr. 37:85–107.
Lind, E. M. & M. E. S. Morrison. 1974. East African Vegetation. Longman, London.
Livingstone, D. A. 1975. Late Quaternary climatic change in Africa. Ann. Rev. Ecol. Syst. 6:249–280.
Lynch, J. D. 1971. Evolutionary relationships, osteology, and zoogeography of leptodactylid frogs. Univ. Kansas Mus. Nat. Hist. Misc. Publ. 53:1–238.
Mabberley, D. J. 1975. The giant lobelias: Toxicity, inflorescence and tree building in the Campanulaceae. New Phytol. 74:365–374.
Mädel, E. 1960. Monimiaceen-Hölzer aus den oberkretizischen Umzamba-Schichten von Ost Pondoland (S. Afrika). Senckenbergia Lethaia 41:331–391.
Martin, H. 1973. The Atlantic margin of southern Africa between Latitude 17° South and the Cape of Good Hope. In: A. E. M. Narin & F. G. Stehli (eds.), The Ocean Basins and Margins. Vol. 1, The South Atlantic. pp. 277–300. Plenum Press, New York and London.
Mathur, Y. K. 1966. On the microflora in the Supra-Trappeans of western Kutch, India. Quart. J. Geol. Miner. Mettal. Soc. India 38:33–51.
McElhinny, M. W. 1970. Formation of the Indian Ocean. Nature 228:977–979.
McIntyre, D. J. & G. J. Wilson. 1966. Preliminary palynology of some Antarctic Tertiary erratics. New Zeal. J. Bot. 4:315–321.
McKenna, M. C. 1973. Sweepstakes, filters, corridors, Noah's Arks, and beached Viking funeral

ships in paleogeography. In: D. H. Tarling & S. K. Runcorn (eds.), Implications of Continental Drift to the Earth Sciences. 1:295–308. Academic Press, London and New York.

McKenzie, D. P. & J. G. Sclater. 1971. The evolution of the Indian Ocean since the Late Cretaceous. Geophys. J. 25:437–528.

McKenzie, D. P. & J. G. Sclater. 1973. The evolution of the Indian Ocean. Sci. Amer. 228:62–72.

Menéndez, C. A. 1969. Die fossilen Floren Südamerikas. Monogr. Biol. 19:519–561.

Menéndez, C. A. 1972. Estudios paleobotánicos en la Argentina, avances, problems y perspectivas. Mem. Sympos, I Congr. Latin-Amer. Bot., pp. 61–97.

Menzel, P. 1920. Über Pflanzenreste aus Basalttuffen des Kamerungebietes. Beiträge zur Geologischen Erforschung der deutschen Schutzgebiete 18:17–32.

Miré, P. D. de & P. Quezel. 1959. Sur la présence de la bruyère en arbre (*Erica arborea* L.) sur les sommets de l'Emi Koussi (Massif du Tibesti). Soc. Biogéogr. Paris Compt.-Rend. 315:66–70.

Mohr, P. A. 1968. The Cainozoic volcanic succession in Ethiopia. Bull. Volcanology 32:5–14.

Mohr, P. A. 1971. Ethiopian rift and plateaus: some petrochemical differences. J. Geophys. Res. 76:1967–1984.

Molnar, P. & J. Francheteau. 1975. Plate tectonics and paleomagnetic implications for the age of the Deccan Traps and the magnetic anomaly time scale. Nature 255:128–130.

Molnar, P. & P. Tapponnier. 1975. Cenozoic tectonics of Asia: effects of a continental collision. Science 189:419–426.

Monod, T. (ed.). 1975. Biogéographie et liasons intercontinentales au cours du Mésozoïque. Mem. Mus. Nat. Hist. n.s. A, Zoologie, vol. 88.

Moore, H. E., Jr. 1973. Palms in the tropical forest ecosystems of Africa and South America. In: B. J. Meggers, E. S. Ayensu & W. D. Duckworth (eds.), Tropical Forest Ecosystems in Africa and South America: A Comparative Review, pp. 63–88. Smithsonian Institution Press, Washington, D.C.

Moreau, R. E. 1933. Pleistocene climatic changes and the distribution of life in East Africa. J. Ecol. 21:415–435.

Moreau, R. E. 1952. Africa since the Mesozoic: with particular reference to certain biological problems. Zool. Soc. London Proc. 212:869–913.

Moreau, R. E. 1963. The distribution of tropical African birds as an indicator of climatic change. In: F. C. Howell & F. Bourliere (eds.), African Ecology and Human Evolution, pp. 28–42. Aldine Pub. Co., Chicago.

Nevo, E. 1968. Pipid frogs from the early Cretaceous of Israel and pipid evolution. Mus. Comp. Zool. Bull. 136:255–318.

Newton, A. R. 1974. Nature of South Africa's Cape fold velt. Nature 248:499–500.

Ornduff, R. 1974. Cytotaxonomic observations on *Villarsia* (Menyanthaceae). Austral. J. Bot. 22:513–516.

Pastoret, L. & J. Goslin. 1974. Middle Cretaceous sediments from the eastern part of Walvis Ridge. Nature 248:495–496.

Penney, J. S. 1969. Later Cretaceous and early Tertiary palynology. In: R. H. Tschudy & R. A. Scott (eds.), Aspects of Palynology, pp. 331–376. Wiley–Interscience, New York.

Perchneilsen, K., P. R. Supko. et al. 1975. Leg 39 examines facies changes in South Atlantic. Geotimes 20(3):20–28.

Phillips, J. F. V. 1927. Tertiary plants near Knysna, South Africa. S. Afr. J. Sci. 24: 188–197.

Plumstead, E. P. 1973. The Late Paleozoic Glossopteris Flora. In: A. Hallum (ed.), Atlas of Paleobiogeography, pp. 187–205. Elsevier Sci. Publ. Co., Amsterdam and New York.

Powell, C. McA. & P. J. Conaghan. 1973. Plate tectonics and the Himalayas. Earth Planetary Sci. Newsl. 20:1–12.

Prakash, U. 1965. A survey of the fossil dicotyledonous woods from India and the Far East. J. Paleontol. 39:815–827.

Prakash, U. 1972. Paleoenvironmental analysis of Indian Tertiary floras. Geophytology 2:178–205.

Quezel, P. 1965. La végétation du Sahara. Gustav Fischer Verlag, Stuttgart.

Ranga Rao, A. 1972. Further studies on the vertebrate fauna of Kalakot, India. Directorate of Geology, Oil and Natural Gas Commission Dehra Dun, India Spec. Paper 1:1–22.

Raven, P. H. 1963. Amphitropical relationships in the floras of North and South America. Quart. Rev. Biology 38:151–177.

Raven, P. H. 1973a. The origin of the alpine and subalpine floras of New Zealand. New Zeal. J. Bot. 11:177–200.

Raven, P. H. 1973b. The evolution of Mediterranean floras. In: F. di Castri & H. A. Mooney (eds.), Mediterranean Type Ecosystems – Origin and Structure, pp. 213–224. Springer Verlag, Berlin.

Raven, P. H. & D. I. Axelrod. 1974. Angiosperm biogeography and past continental movements. Ann. Missouri Bot. Gard. 61:539–673.

Raven, P. H. & D. I. Axelrod. 1977. Origins and relationships of the California flora. Univ. Calif. Publ. Bot. 72:(in press).

Reid, E. M. & M. E. J. Chandler. 1933. The London Clay Flora. British Mus. Nat. Hist., London.

Rennie, J. V. L. 1931. Note of fossil leaves from the Banke Clays. Trans. Roy. Soc. S. Afr. 19:251–253.

Reyment, R. A. 1969. Ammonite stratigraphy, continental drift and oscillatory transgressions. Nature 224:137–140.

Reyment, R. A. & E. A. Tait. 1972. Biogeographical data on the early history of the South Atlantic Ocean. Trans. Roy. Soc. London 264:55–95.

Rich, P. V. 1974. Significance of the Tertiary avifaunas from Africa (with emphasis on a mid to late Miocene avifauna from southern Tunisia. Ann. Geol. Surv. Egypt 4:167–210.

Richards, P. W. 1973. Africa, the odd man out. In: B. J. Meggers, E. S. Ayensu & W. D. Duckworth (eds.), Tropical Forest Ecosystems in Africa and South America: a Comparative Review. pp. 21–26. Smithsonian Institution Press, Washington, D.C.

Roche, E. 1974. Paléobotanique, paléoclimatologie et dérive des continents. Sci. Geol. Univ. Louis Pasteur de Strausbourg Bull. 27:9–24.

Romero, E. J. 1973. Polen fosil de 'Nothofagus' ('Nothofagites') del Cretaceo y Paleoceno de Patagonia. Rev. Mus. La Plata VII. Paleontologia 47:291–303.

Sah, S. C. D. 1967. Palynology of an upper Neogene profile from Rusizi valley, Burundi. Tervuren, Musee Royal de l'Afrique Centrale, Ann. Sec. Geol. 57:1–173.

Schalke, H. J. W. G. 1973. The upper Quaternary of the Cape Flats Area (Cape Province, South Africa). Scripta Geol. 15:1–57.

Schuster, R. M. 1976. Plate tectonics and its bearing on the geographical origin and dispersal of angiosperms. In: C. B. Beck (ed.), Origin and Early Evolution of Angiosperms, pp. 48–138. Columbia Univ. Press, New York and London.

Sclater, J. G. & R. L. Fisher. 1974. Evolution of the east central Indian Ocean, with emphasis on the tectonic setting of the Ninetyeast Ridge. Geol. Soc. Amer. Bull. 85:683–702.

Scruton, R. A. & A. du Plessis. 1973. Possible marginal fracture ridge south of South Africa. Nature 242:180–182.

Seward, A. C. 1903. Fossil floras of Cape Colony. Ann. S. Afr. Mus. 4:1–122.

Seward, A. C. 1935. Dicotyledonous leaves from the Nubian Sandstone of Egypt. Publ. Geol. Surv. Egypt.

Shackleton, N. J. & J. P. Kennett. 1975. Paleotemperature history of the Cenozoic and the initiation Antarctic glaciation: oxygen and carbon isotope analyses in DSDP Sites 277, 279, and 281. In: J. P. Kennett, R. E. Houtz et al., Initial reports of the Deep Sea Drilling Project. 29:743–755.

Simpson, G. G. 1943. Mammals and the nature of continents. Wash. Acad. Sci. J. 30:137–163.

Simpson, G. C. 1944. Tempo and Mode in Evolution. Columbia Univ. Press, New York.

Simpson, G. C. 1953. The Major Features of Evolution. Columbia Univ. Press, New York.

Skottsberg, C. 1936. Antarctic plants in Polynesia. In: T. H. Goodspeed (ed.), Essays in Geobotany in Honor of William Albert Setchell, pp. 291–311. Univ. California Press, Berkeley.

Smith, A. G. & A. Hallum. 1970. The fit of the southern continents. Nature 225:139–144.

Smith, A. G., J. C. Briden & G. E. Drewry. 1973. Phanerozoic world maps. In: N. F. Hughes (ed.), Organisms and Continents through Time. Paleontol. Assoc. London, Spec. Papers Paleontol. 12:1–43.

Smith, P. J. 1974. Sicily as a part of Africa. Nature 251:102.

Specht, R. L. 1969. A comparison of the sclerophyllous vegetation characteristics of mediterranean type climates in France, California and southern Australia. I. Structure, morphology and succession. Austral. J. Bot. 17:277–292.

Specht, R. L. & P. Rayson. 1957. Dark Island Heath (Ninety-Mile Plain, South Australia). I. Definition of the ecosystem. Austral. J. Bot. 5: 52–85.

Stebbins, G. L. 1974. Flowering Plants. Evolution above the Species Level. Belknap Press of Harvard Univ. Press, Cambridge, Mass.

Stebbins, G. L. & J. Major. 1965. Endemism and speciation in the California flora. Ecol. Monogr. 35:1–35.

Tankard, A. J. 1975. Thermally anomalous Pleistocene molluscs from the southwestern Cape Province, South Africa. Ann. S. Afr. Mus. 69:17–45.
Tarling, D. H. & P. E. Kent. 1976. The Madagascar controversy still lives. Nature 261:304–305.
Tarling, D. H. & S. K. Runcorn (eds.). 1973. Implications of Continental Drift to the Earth Sciences. 2 vols. Academic Press, London and New York.
Tarling, D. H. & M P. Tarling. 1971. Continental Drift: A Study of the Earth's Moving Surface. Doubleday, Garden City, N.Y.
Thiergart, F., F. Frantz & K. Baukopf. 1963. Palynologische unterschungen von Tertiärkohlen und einer Oberflächenprobe nahe Knysna, Südafrika. Adv. Front. Plant Sci. 4:151–178.
Truswell, J. F. 1970. An Introduction to the Historical Geology of South Africa. Purnell & Sons, Cape Town, South Africa.
Van Couvering, J. A. & J. A. H. Van Couvering. 1975. African isolation and the Tethys seaway. Proc. VI Congr. Internat. Union Geol. Sic., Comm. on Stratigraphy, Committee on Mediterranean Neogene Stratigraphy, pp. 363–367.
Van Couvering, J. A. H. & J. A. Van Couvering. 1975. Geological setting and faunal analysis of the African Early Miocene. In: G. L. Issac (ed.), Perspectives in Human Evolution. Vol. III.
Van Zinderen Bakker, Sr., E. M. 1975. The origin and paleoenvironment of the Namib desert biome. J. Biogeogr. 2:65–74.
Van Zinderen Bakker, Sr., E. M. 1976. The evolution of late Quaternary paleoclimates of southern Africa. Palaeocology of Africa 9:1–69.
Weimarck, H. 1941. The groups, centers and intervals within the Cape flora. Lund Univ. Årssk. N.F. II. 37:1–143.
Wells, P. V. 1966. Late Pleistocene vegetation and degree of pluvial climatic change in the Chihuahuan Desert. Science 153:970–975.
Wells, P. V. & R. Berger. 1967. Late Pleistocene history of coniferous woodland in the Mohave Desert. Science 155:1640–1647.
Werger, M. J. A. 1973. Phytosociology of the Upper Orange River Valley, South Africa. V & R, Pretoria (thesis Univ. Nijmegen).
White, F. 1965. The savanna woodlands of the Zambezian and Sudanian domains: an ecological and phytogeographical comparison. Webbia 19:651–671.
Wild, H. 1965. Additional evidence for the Africa–Madagascar–India–Ceylon land-bridge theory with special reference to the genera *Anisopappus* and *Commiphora*. Webbia 19:497–505.
Wild, H. 1968. Phytogeography in south central Africa. Kirkia 6:197–222.
Wild, H. 1975. Phytogeography and the Gondwanaland position of Madagascar. Boissiera 24:107–117.

Addendum: Additional Annotated References

A number of important papers that relate either directly or indirectly to the history of African vegetation were published, or came to our attention, after the manuscript was submitted. The following add particularly critical data that support our general inferences and conclusions.

Aubréville, A. 1949. Contribution à la Paléohistorie des forêts l'Afrique tropicale. Soc. d'Édit. Géogr., Maritimes et Coloniales. Challamel, Paris.
  Lists numerous disjunct taxa that provide evidence for a wider, pan-African Tropical-Tertiary flora. Also presents data for a southward shift of montane taxa during the Quaternary pluvials. Unfortunately no bibliography.

Aubréville, A. 1976. Essai d'interprétation nouvelle de la distribution des Diptérocarpacées. Adansonia sér. 2, 16:205–210.
  Useful summary of the occurrence of fossils, chiefly woods, previously referred to this family. Relates history of subfamilies Monotoideae (African) and Dipterocarpoideae (Asian) to plate tectonics.

Aubréville, A. 1976. Centres Tertiaires d'origine, radiations et migrations des flores angiospermiques tropicales. Adansonia sér. 2, 16:297–354.
  Discussion oriented to plate tectonics, with numerous useful maps of modern distributions; favours several small centers of origin in Southern Hemisphere before and during drift.

Barquin Diaz, E. & W. W. de la Torre, 1975. Diseminación de plantas canarias. Datos iniciales. Vieraea 5:38–60.

Important discussion of various forms of dispersal on the Canary Islands, with reference to saurochory (by *Lacerta*), ornithochory (by *Corvus*), and anemoballisty (in *Centaurea*).

Bonnefille, R. 1976. Implications of pollen assemblage from the Koobi Fora Formation, East Rudolf, Kenya. Nature 264:403–407.

Dated at 1.5 m.y., it includes a conspicuous montane element now in the mountains of Ethiopia, indicating a moister and cooler climate on the east shore of Lake Turkana (Lat. 4°N) during an early glacial-pluvial stage.

Bonnefille, R. & R. Letouzey. 1976. Fruits fossiles d'Antrocaryon dans le vallée de l'Ome (Éthiopie). Adansonia sèr. 2, 16:65–82.

Fossil fruits of *Antrocaryon* (Anacardiaceae) occur in 3.0 m.y. old deposits in the lower Omo River Valley, Ethiopia. They indicate at least a tropical gallery forest in this area which is now an *Acacia* thorn scrub or semidesert, a conclusion consistent with indications of fossil pollen, mammals and paleosoils. The increased precipitation at this time, which accompanied the building up of the first major ice sheets, was paralleled in California-Nevada by a comparable shift of forests into areas now steppe and desert.

Cifelli, R. 1976. Evolution of ocean climate and the record of planktonic foraminifera. Nature 264:431–432.

Notes progressive southward shift of warm water taxa in the Pliocene, which was intermediate between Miocene and present conditions. The shift seems to correspond to the increasing aridity over Africa.

Ferris, V. R., G. G. Goseco, & J. M. Ferris. 1976. Biogeography of free-living soil nematodes from the perspective of plate tectonics. Science 193:508–509.

The first biogeographical analysis based on the morphology and known distribution of a group of free-living soil nematodes. The data indicate a pre-Jurassic origin followed by West Gondwanaland radiation for some genera and a Laurasian radiation for others. Links between India and South Africa are rather numerous, probably because their sheltered habitat in the ground was protected from climatic change as the Indian plate moved north.

Floret, J.-J. 1976. A propos de Comiphyton gabonense (Rhizophoraceae-Macarisieae). Adansonia sèr. 2, 16:39–49.

Phytogeographic data on the monotypic genus *Comiphyton* and its relations to other genera of the tribe Macarisieae. Maps show pantropical and disjunct tropical genera (India, Madagascar, America–West Africa) that may express evolutionary events resulting from post mid-Cretaceous ocean-floor spreading.

Guillaumet, J. L. & G. Mangenot. 1975. Aspects de la speciation dans la flore malgache. Boissiera 24:119–123.

Discusses in preliminary manner some of the endemic areas in Madagascar.

Hedberg, O. 1969. Evolution and speciation in a tropical high mountain flora. Biol. J. Linn. Soc. 1:135–148.

Hedberg, O. 1970. Evolution of the Afroalpine flora. Biotropica 2:16–23.

Important summaries of Hedberg's work on the origin of this unique flora which dates chiefly from the Quaternary, a topic largely outside our area of discussion.

Hedberg, O. 1975. Studies of adaptation and speciation in the afroalpine flora of Ethiopia. Boissiera 24:71–74.

Comparison of adaptive trends and taxonomic differentiation in the Afroalpine flora of southern Ethiopia with that in tropical Africa. Same life forms and adaptations are in each region, but the Ethiopian alpine flora has more numerous ephemeral plants.

Herz, N. 1977. Time of spreading in the South Atlantic: information from Brazilian alkalic rocks. Geol. Soc. Amer. Bull. 88:101–112.

Refinement of times of initial rifting, the spread of the sea, and the separation of Brazil (Minas Gerais-Paraná) from Africa (Gabon-Angola).

Johnson, B. D., C. McA. Powell, & J. J. Veevers, 1976. Spreading history of the eastern Indian Ocean and Greater India's northward flight from Antarctica and Australia. Geol. Soc. Amer. Bull. 87:1560–1566.

Three periods of major movement are documented. From 80 m.y. to 64 m.y. Ceylon moved from

45°S to 30°S; from 64 to 53 m.y. it moved to 10°S, and it shifted to its present position at 10°N after 32 m.y., and probably by the Miocene.

Kunkel, G. (ed.). 1976. Biogeography and Ecology of the Canary Islands. Monographiae Biologicae 30. W. Junk, The Hague.

Excellent and up-to-date discussions of numerous critical topics, including geology (by Schmincke), endemic flora (Bramwell), laurisilva of Hierro (Schmid), lichen flora (Follmann), fungal flora (Gjaerum), fauna of the laurisilva (Machado), avifauna (Bacallado), herpetology (Klemmer).

Leroy, J.-F. 1976. Essais de taxonomie syncrétique. 1. Étude sur les Méliaceae de Madagascar. Adansonia sér. 2, 16:167–203.

Important biological analysis of Meliaceae of Madagascar. To be studied in conjunction with Pennington & Styles (1975), see below.

Lewalle, J. 1975. Endémisme dans une haute vallée du Burundi. Boissiera 24:85–89.

In every vegetation type, "new species belonging to various families, to monotypic or large genera, underline the endemic nature of the flora in a valley of the montane belt." The valley, with an average elevation near 1800 m, has a very equable climate and deserves further exploration, as do other valleys in the area near Bururi.

Lobreau-Callen, D. 1974. Problèmes de palynologie liés à la dérive des continents. Étude de quelques taxons tropicaux. Strausborg Sci. Géol. Bull. 17:147–168.

Cautions that continental drift is not the sole answer to distributional problems across the tropics. Reaffirms the notion that long distance dispersal, as well as migration over stepping stones, may also explain some of the patterns.

Maxson, L. R., V. M. Sarich, & A. C. Wilson. 1975. Continental drift and the use of albumin as an evolutionary clock. Nature 255:397–400.

Important exploratory analysis of value of proteins as an evolutionary clock, with change in amino acid sequence during evolution as a primarily time-dependent process. Results are encouraging when related to plate movements.

McElhinny, M. W., B. J. J. Embleton, L. Daly, & J-P. Pozzi. 1976. Paleomagnetic evidence for the location of Madagascar in Gondwanaland. Geology 4:455–457.

Madagascar moved from an earlier position off the Somalia-Kenya coast to its present area prior to the Middle Cretaceous (before 90 m.y.). This dating negates Wild's opinion (*in* Boissiera 24:107–117) that plant distribution in East Africa-Madagascar supports the southward movement of Madagascar. For the most part, the taxa that he considers were not yet in existence by the Middle Cretaceous, when Madagascar was in its present position.

Merxmüller, H. & K. P. Buttler. 1975. Nicotiana in der Afrikanischen Namib. Ein pflanzengeographisches und phylogenetisches Rätsel. Mitt. Bot. Staatssamml. München 12:91–104.

Remarkable find of a *Nicotiana* on isolated mountains in the middle Namib of South West Africa. This represents the first record of the genus in Africa. *Nicotiana africana* is a unique species, distantly allied to those that are otherwise native only to America, the South Pacific and Australia, and may be a relatively ancient (pre-Neogene?) relict in Africa, probably arrived following long-distance dispersal from South America.

Pennington, T. D. & B. T. Styles. 1975. A generic monograph of the Meliaceae. Blumea 22: 419–540.

Excellent monograph delimiting the subfamilies, tribes, and genera in this pantropic family. The family is now ripe for an evolutionary analysis in terms of sea-floor spreading and plate tectonics. Also see Leroy (1976), above.

Scrutton, R. A. 1976. Fragments of the earth's continental lithosphere. Endeavour 35:99–103.

Useful discussion of the continental fragments that lie within the ocean basin, includes a map and table listing them. Notes that the status of some is unresolved (e.g., Kerguelen).

Smith, P. J. 1976. So Madagascar was to the North. Nature 263:729–730.

Latest review of evidence indicating that Madagascar moved south to its present position from the Somalia-Kenya coast during the Early Cretaceous.

Smith, M. M. B. 1974. Southern biogeography on the basis of continental drift: a review. Australian Mammalogy (Jour. Austr. Mammal Soc.) 1:213–229.

Useful summary of varied biotic evidence, chiefly austral, that supports major continental movements during the past.

Straka, H. 1975. Palynologie et différentiation systématique d'une famille endémique de Madagascar: les Didieréacées. Boissiera 24:245–248.

This family, endemic to dry areas of southern Madagascar, constitutes an isolated group according to pollen morphology, a conclusion in agreement with the characters of its flowers and fruits. Straka constructs a geneological tree of the four genera in the family, and emphasizes its isolation in the Centrospermae.

Street, E. A. & A. T. Grove. 1976. Environmental and climatic implications of Late Quaternary lake-level fluctuations in Africa. Nature 261:385–389.

Clear evidence for markedly a moister climate across the present Saharan region in the last glacial maximum, and also in present arid southern Africa. At the same time more arid conditions were in the inner tropical belt.

Tankard, A. J. & J. Rogers, 1977. Progressive late Cenozoic dessication of the west coast of southern Africa. Unpubl. ms.

Outstanding review of evidence (plant, marine invertebrate, mammal paleosol) in South Africa for progressive desiccation, and for the recency of regional desert and mediterranean climates.

Van Campo, E. & J. Sivak. 1976. Présence de pollens de Dracaena dans le Néogene Meditérranéen. Rev. de Micropaléontol. 18:264–268.

Pollen of *Dracaena* is recorded from the Neogene of Tunisia. *Dracaena* is not known today from the mainland of Africa north of the Sahara desert, but does occur on the Canary Islands and in Socotra.

6 Quaternary vegetation changes in southern Africa

E. M. van Zinderen Bakker Sr

1. Present climate and vegetation 133
2. Quaternary climates and vegetation 134
2.1 The Cape coastal area 135
2.2 The plateau of the interior 139
2.3 The Austro-afroalpine area 141
2.4 The present savanna and woodland area 141
2.5 The Karoo 142
2.6 The Namib desert 142
3. Summary 142

References 142

6 Quaternary vegetation changes in southern Africa

1. Present climate and vegetation

In this section a sketch will be given in broad outline of the interrelation of climate and vegetation and the indications which can be implied from this correlation for the age of certain vegetation types.

The vegetation pattern of southern Africa gives an accurate image of the climatic regime which is characterized by an extreme aridifying influence penetrating the subcontinent from the western side and by rain bringing systems converging from the other coastal regions and from the tropical zone. This pattern dates from mid-Tertiary times (van Zinderen Bakker Sr 1976) (compare Chapter 5).

The atmospheric and oceanic circulation, which dominate the climate of southern Africa, originated as a consequence of the Antarctic glaciation. Under present-day conditions, which are comparable to the telocratic stage or second half of an interglacial the Westerlies penetrate far into the subcontinent and bring cyclonic rain only in winter to the most southwesterly part of the Cape. This region forms the nucleus of the old flora of Capensis with its great diversity of species. The ancestors of this flora migrated southward from the mountain areas in central Africa when the continent, after its isolation from Gondwanaland, moved northward (Wild 1968, van Zinderen Bakker 1970). The great wealth of species of Capensis must be the result of its isolation and the repeated fragmentations of its area which were caused by drastic climatic changes in late-Tertiary and Quaternary times (van Zinderen Bakker Sr 1976). This flora element is found in many elevated parts of the eastern escarpment (see Chapter 8).

Beyond the S.W. Cape, the rest of southern Africa receives summer rainfall, which is mainly concentrated in the eastern half, with the result that an impressive west–east humidity gradient exists (see Chapter 2). This sequence of increasing rainfall represents at the same time an altitudinal gradient ranging from sea-level on the west coast to the high mountains of nearly 3500 m in the eastern escarpment. At the western extreme of this gradient the hyper arid Namib desert gradually merges into open grassland, which is followed by a wide zone of savanna in the Kalahari Basin (see Chapter 9). The sequence then divides eastward into two directions, with the woodlands of Rhodesia and the Transvaal in the northern part and in the south the open *Cymbopogon-Themeda* grassland of the interior plateau which receives more erratic precipitation than do the northern regions (see Chapter 10). These woodlands and grasslands stretch as far as the foothills of the eastern escarpment, where they merge into patches of isolated montane and ravine forest which are related to the evergreen montane forest of the mountains of East and southern Africa (van Zinderen Bakker Jr 1973, Coetzee 1976, and Chapter 11). These forests are favoured by a lower temperature and more orographic rainfall. The final stage of this interesting and enormously elongated gradient is reached above the treeline where the austro-afroalpine grasslands dominate (van Zinderen Bakker Sr & Werger 1974, and Chapter 12). The lower limit of this alpine region depends on the latitudinal position, and in Lesotho is situated at c. 2250 m. These alpine grasslands receive an annual rainfall of over

1600 mm per year and as a consequence of the low temperature snow can fall here during every month of the year.

This impressive ecological gradient is dominated by the anticyclonic belt of mid-latitudes. The very stable and strong high pressure system over the south Atlantic Ocean is situated very near the coast of the continent and is centred around 30°S. The subsiding and divergent winds of this system together with the accompanying cold Benguela Current have an aridifying influence which deeply penetrates the continent. Over the continent itself at an altitude of 2000 m anticyclonic conditions also prevail throughout the year (Trewartha 1966, Jackson 1952), so that rainbringing winds of a northern origin can only reach higher latitudes during atmospheric disturbances. This rain of I.T.C. origin, which falls during the southern solstice in the woodland areas, diminishes rapidly southward.

At both ends of the ecological gradient under discussion, isolated biota with an ancient origin occur. The sand dunes of the Namib harbour some of the most remarkable adaptations in beetles, spiders and reptiles which exist on earth. Plants such as *Stipagrostis sabulicola* and *Acanthosicyos horrida* and certainly *Welwitschia mirabilis* of the flat parts of the desert are typical ancient endemics. At the other extreme of the gradient in the high eastern mountains alpine elements of a northern origin are found, which also point to entirely different palaeoclimatic conditions during early Tertiary times (van Zinderen Bakker 1970, Werger in van Zinderen Bakker & Werger 1974).

In the intermediate section of the ecological gradient, the grasslands, savannas and woodlands, endemism is less spectacular. The reason for this is certainly that these vegetation types occur widespread over vast areas to the north and that migrations in their area are not hampered by ecological barriers. This does not apply to the evergreen montane forest elements which have certainly for a long time been prevented from migrating in a meridional direction by the wide and hot Limpopo valley (Coetzee 1976, van Zinderen Bakker Jr 1973).

In this very broad sketch of the climatic pattern of the vegetation the eastern coastal region and the Karoo have not been mentioned. The dry semi-desert Karoo occupies an extensive area between the Cape Folded Mountains and the ecological gradient discussed above (see Chapter 9). The Karoo could be called a climatic no-man's-land as it lies between the two climatic systems which bring rain to South Africa. This vegetation of xerophytic dwarf shrubs and succulents, which shows such a high species diversity and interesting adaptations, must certainly be of great antiquity just as the climatic system of southern Africa.

2. Quaternary climates and vegetation

The world wide changes in temperature with their different amplitudes, which occurred during the Quaternary had a profound impact on the climates of southern Africa. During a glacial period (Fig. 1) the thermal equator was displaced southward as a consequence of the growth of the northern ice sheet. This displacement did, however, not result in a southward extension of the I.T.C. zone in southern Africa as the westerlies moved simultaneously northward. The summer rainfall area was not only diminished in size, but also received less precipitation (Butzer et al. 1972, van Zinderen Bakker Sr 1972, Hamilton 1972). This phenomenon is ascribed by Flohn (1953) to the lower evaporation of ocean water during glacial periods and consequently a diminished advection of moist air to the continents.

At higher latitudes conditions were entirely different during a glacial maximum as the northward shift of the westerlies brought cyclonic rain perhaps as far inland as 24°S (van Zinderen Bakker Sr 1976). The evidence for this shift is of varied nature, but does not yet give exact information on the northern boundary of this former rainbelt. It can, however, be assumed that a vast region which at present receives summer rain, was under the regime of winter rains during glacial times. The northward shift of cyclonic rains coincided with a considerable lowering in temperature. The depressions were accompanied by influxes of very cold polar air which deeply penetrated the subcontinent. These cold fronts aggravated the already cold climate.

The opposite development occurred during warm interglacial and interstadial periods when the climatic belts moved southward (Fig. 2). The I.T.C. could then extend its influence in the present savanna and woodland regions which received more summer rainfall. The zone of the westerlies shifted to higher latitudes with the consequences that the rain bringing depressions could reach a limited area of the S.W. Cape only in winter. This shift in the rainfall belts was accompanied by a considerable amelioration in temperature.

The above-mentioned climatic changes had manifold influences on the vegetation of different regions and it is at this stage impossible to reconstruct any former detailed topographic vegetation patterns. Not only do we not know enough about the ecological implications of the climatic changes but our knowledge of the former vegetation as such is far too limited. Pollen analytical studies have so far only been carried out with success in the Cape coastal region and on the interior plateau, where they have provided valuable dated information on vegetational changes (Figs. 1 and 2).

Throughout the complete glacial–interglacial cycle the changes in vegetation in each region consist of a great number of stages, which cannot all be distinguished in the fossil record. Some stages which have lasted several thousand years, however, were characterized by a particular fauna and vegetation. In southern Africa only some major episodes of the complex Late-Quaternary time scale have so far been recognized and it will therefore be sensible to describe the ecological changes for only some regions. For this purpose the following terminology of von Post (1946) and Iversen (1954) will be used. From the full-glacial or cryocratic stage climatic amelioration leads via the protocratic stage to the mesocratic temperature optimum of the interglacial. Our present-day conditions can be equated with the second half of the cycle, the telocratic stage, during which the climate deteriorates again to glacial conditions.

2.1   *The Cape coastal area*

Two pollen analytical studies, by Martin (1968) and Schalke (1973), provide valuable information on vegetation changes in this area. The investigation by Martin of the deposits of the Groenvlei coastal lake in the eastern part of the coastal belt covers the last 8,000 years of the time scale, while the work of Schalke was concerned with the southwestern coastal area and provides information on the long period of the last 50,000 years. Schalke is of the opinion that in his area of study *Podocarpus* forest occurred during the colder stadials, while fynbos

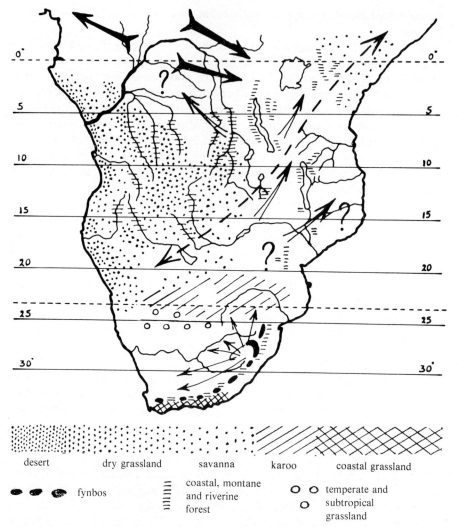

Fig. 1. Some trends of the vegetation pattern during a glacial maximum
Double shafted arrows = migration of woodland
Single shafted arrows = possible routes of alpine grassland
Thick arrows in Congo basin = possible migration of humid tropical forest
Broken line = position of 'arid corridor'

dominated during the warmer interstadials. Martin came to the conclusion that in the more eastern part of the coastal plain evergreen forest spread during the warmer period of the Holocene. This forest extension was, however, temporarily counteracted by the spread of dunes.

It may seem that these results cannot be reconciled with each other, but it should be realized that the climatic and ecological conditions of these two sites cannot be compared for totally different periods of time. It can further be argued that the counts of 5–15 per cent *Podocarpus* pollen are far too low to infer the growth of those forests in the S.W. Cape (Schalke 1973). If the radiometric dates are reliable the high counts of 40 per cent *Podocarpus* pollen shows, however, that

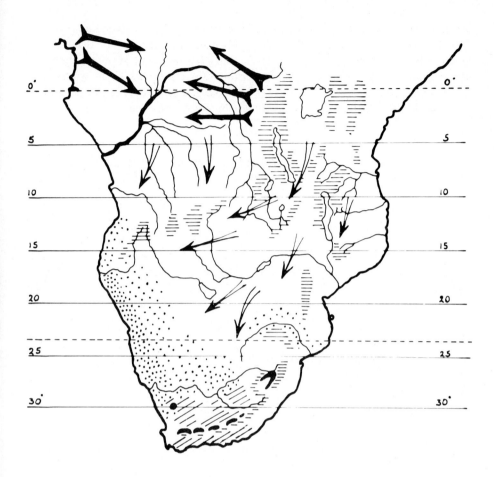

Fig. 2. Trends of vegetation shifts during an interglacial. See legend of Fig. 1.

Podocarpus forest must have been growing in the Cape Flats during the cold Salt River Interval of an age of 40,500–36,500 B.P. Only some pollen of *Ilex mitis*, which has been identified, supports the occurrence of this forest.

The *Podocarpus* forest which may have occurred in the S.W. Cape under cold conditions is not comparable with the 'Knysna' type of forest, which reached maxima during warm Holocene times, as these two forests have different ecological requirements.

Further interesting information on former vegetation is inferred by Klein (1972, 1974) from the fossil assemblages of grazing animals, which have been found in various caves in the Cape coastal area. The age of these assemblages has only been suggested on archaeological grounds. The general conclusions which can be implied suggest that in early MSA times, which could be of interglacial or interstadial age, forest and bush dominated, while in the youngest MSA levels a grazing fauna occurred. The open grassland must have been replaced by closed vegetation between 12,000 and 9000 years ago, as concluded by Klein (op. cit.) from the fossil faunas of the Nelson Bay, Elands Bay and Melkhoutboom Caves.

This very important palaeobotanical and palaeontological information can best be explained by the following sequence of events during the last glaciation (Fig. 3). The forest which dominated the coastal plain towards the end of the last interglacial was replaced by bush during the Early Glacial, while grassland covered the coastal region during the coldest periods of the Pleniglacial. In winter the windy climate must have been very wet and cold as a consequence of the regular penetrations of cold fronts. The sudden rise in temperature round 12,400 B.P. at

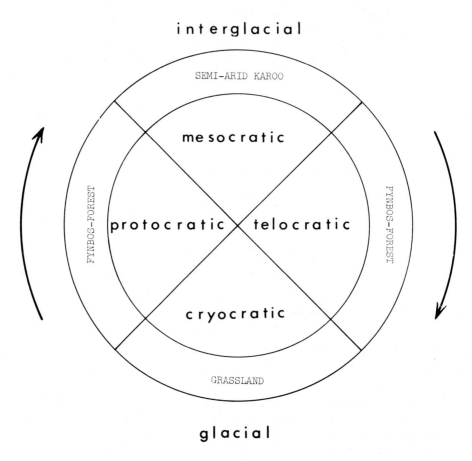

Fig. 3. Full glacial–interglacial vegetation cycle of the Cape coastal region.

the beginning of the combined Bölling–Alleröd interstadials of the European chronology, made it possible for the fynbos, which had survived in protected mountain areas, to expand and to invade the grassland. The grazing fauna was then replaced by browsing animals which were adapted to thick bush. During this protocratic period *Podocarpus* forest could perhaps compete with the dense fynbos in certain parts. These plant communities also competed for dominance during the cool recession which was coeval with the Younger Dryas Stadial of N.W. Europe (c. 10,900–10,100 B.P.). The gradual warming up of the climate will have been associated with a slightly drier climate, so that from 9000 B.P. onward (at Groenvlei from about 7000 B.P.) optimal conditions existed for the dominance of the evergreen coastal forest. The increase in temperature toward the optimum of the Hypsithermal Period and the associated drier climate, however, led to the encroachment of dunes into the retreating forest, while karroid vegetation occurred widespread in the coastal plain during this mesocratic ecological stage. A consequent lowering in temperature during telocratic times set in the revertence to forest conditions. Finally the influence of man caused the regression of most of the evergreen forest.

The glacial–interglacial cycle proposed here needs much more proof and amplification as only some stages of it have so far been verified by fossil evidence.

2.2 The plateau of the interior

The plateau of the Orange Free State and adjacent areas, which is at present covered by temperate *Cymbopogon-Themeda* grassveld, falls within the region which had a wet and cold climate during the last glacial maximum. It will have received winter rainfall, while the winter temperature may have dropped as much as 8–10°C (Harper 1969, Butzer 1973; and for the Transvaal: Talma et al. 1974). This drastic lowering in temperature will have depressed the upper limit of tree growth by about 1000 m with the consequence that the plateau under discussion must have been treeless under full-glacial conditions. The lowering of the vegetation zones and the climate suggest strongly that the plateau was then covered with an alpine grassland of the type which occurs at present in the austral part of the Afroalpine Region of the eastern escarpment (van Zinderen Bakker Sr & Werger 1974, and Chapter 12).

Very convincing proof for higher rainfall during this period is given by the studies of Butzer et al. (1973) of the high lake levels in the Alexandersfontein depression, east of Kimberley. Shortly before $16,010 \pm 185$ B.P. the extensive former lake reached levels of $+17$ to $+19$ m. Butzer et al. (op. cit.) showed that with a temperature depression of only 6°C, double, the present rainfall will be required to support the hydrology of a lake with the former high water levels.

The vegetation changes which occurred on the interior plateau during a full glacial–interglacial cycle are amplified by the palynological studies which have been carried out in the area (Fig. 4). The pollen analysis of the deposits around the Florisbad volcanic spring in the centre of the plateau gave the following results (van Zinderen Bakker Sr 1957). At a time more than 48,900 years ago (GrN – 4208) a dry and warm climate prevailed and the spring was surrounded by a dry type of Karoo vegetation. This mesocratic stage could have been of interglacial (Eemian) or interstadial age. The only other section of the profile which contained fossil pollen is bracketed in time between $28,450 \pm 2200$ B.P. (L – 271C) and $19,350 \pm 650$ B.P. (L – 271D). The lowest part of this section, which is of interstadial age, again shows the presence of the Karoo. At 25,000 B.P. (calculated age) a drastic change occurred as the Karoo was replaced by grassveld. This shift indicates a change in precipitation and/or temperature. Climatic considerations, which have been discussed already, indicate that both a higher precipitation and a decrease in temperature were responsible for the ecological change during this cryocratic stage. This information is supplemented by the detailed study of Coetzee (1967) of the pollen content of the deposits round the thermal spring of

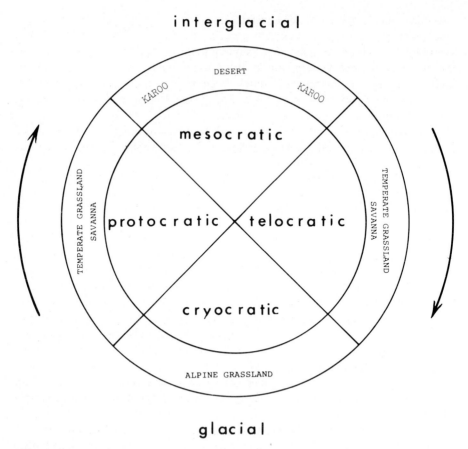

Fig. 4. Full glacial–interglacial vegetation cycle of the inland plateau.

Aliwal North, c. 250 km further southward, near the Orange River. The deposits date from 13,200 ± 180 B.P. (I – 2107) to 9650 ± 150 B.P. (GrN – 4012). Aliwal North is situated in a marginal area, between the Karoo to the south of it and grassveld to the north. The pollen diagrams show that six pollen zones can be distinguished during the episode of the European Late-Glacial. These pollen zones indicate alternations of 'wet–cool–grassveld' conditions and 'dry–warmer Karoo' conditions which were practically coeval respectively with stadials and interstadials of the stratigraphy of N.W. Europe.

The fossil pollen data from Florisbad and Aliwal North indicate that during warm conditions dry Karoo vegetation invaded the plateau. Geomorphological evidence suggest that during these mesocratic times windblown sand dunes covered large areas on the western part of the plateau.

Extremely interesting circumstantial evidence for dry conditions during warmer periods has been provided by a comparison made in South Africa by J. Deacon (1974) of the distribution in time and space of 223 dated archaeological sites, covering the Wilton and Smithfield cultures. The results show that no dates exist for occupation sites from the inland plateau for the period between 9500 and 4600 years B.P. During this mesocratic warm period the waterless semi-desert of the plateau apparently did not offer enough resources for the existence of prehistoric man.

During the proto- and telocratic stages of the ecological cycle the present day *Cymbopogon-Themeda* grassveld occupied the eastern part of the plateau, while the Kalahari savanna invaded the western part.

2.3 The Austro-Afroalpine area

During full-glacial times the alpine vegetation was unable to survive at the highest altitudes as it does at present. Radiocarbon dates of the age of the swamps, which occur at about 3000 m altitude in Lesotho, suggest that the vegetation only reached this altitude again after the onset of the Holocene warming up of the climate, about 8000 years ago (van Zinderen Bakker Sr & Werger 1974).

The full glacial–interglacial cycle of the alpine area may have consisted of the following sequence:

cryocratic stage: snow cover, open screes, erosion, inorganic sedimentation in tarns, patterned ground.

protocratic stage: pioneer vegetation.

mesocratic stage: growth of bogs, alpine grassland, encroachment of tarns by vegetation, rise of tree line above present level.

telocratic stage: regression of vegetation cover, erosion, accumulation of sand and gravel horizons on swamp deposits.

Vegetation shifts of the same nature but of a smaller magnitude must have occurred at lower latitude in the high mountains of eastern Rhodesia.

2.4 The present savanna and woodland area

For this large region no pertinent information is available on vegetation changes except that studies of fossil pollen (Hamilton 1972) and former lake levels (Butzer et al. 1972) have shown that rainfall was diminished during colder periods. The vast woodland and savanna region from the equator southward to the Transvaal is under the influences of the tropical high-sun rainfall system. It has already been discussed that during full-glacial times the rainfall was not only diminished, but its influence did not reach as far south as at present. At the same time the northward shift of the anticyclone over the south Atlantic and the associated northward movement of the Benguela Current displaced the aridifying influence of this system further north (van Zinderen Bakker Sr 1975). A study of the present arrangement of the isohyets reveals that only small decreases in rainfall will open up an arid corridor running from S.W. Africa in a northeasterly direction to northern Tanzania and Kenya (Knoch & Schulze 1956). This corridor has repeatedly been available during drier and colder times for the exchange of 'semi-arid elements' between these distant parts of Africa (van Zinderen Bakker Sr 1969, and Chapter 9). Current pollen analytical research in the Transvaal may before long supply valuable information on the changes which took place in this marginal area.

During warm mesocratic conditions the tropical rainfall will have been intensified and will also have reached higher latitudes.

2.5 The Karoo

This semi-desert vegetation with its xerophytic dwarf shrubs and great species diversity of succulents extended its area considerably during warm interglacial times. As has already been described the Karoo invaded the present grassland plateau and probably reached as far south as the coastal plain.

It is not known where the Karoo complex survived the cryocratic conditions. The semi-desert must probably have moved far northward to an area where neither the southern cyclonic rains nor the tropical summer rains could affect it. The most logical place might well have been the northern Cape Province or southern Botswana between the latitudes of 20 and 25°S.

2.6 The Namib Desert

The ancient Namib Desert has only in its extremities been affected by climatic changes. The core of the desert with its wealth of highly adapted elements has always remained hyper-arid since its origin in end-Tertiary times. These biota are closely correlated with the old climatic pattern which is dominated by the south Atlantic anticyclone. During cryocratic times the southern edge of the Namib will have received winter rainfall, while desertic conditions spread further north along the coast of Angola (van Zinderen Bakker Sr 1975). The reverse processes took place during warm mesocratic times. The extent of these movements and their ages offer interesting fields of palaeoecological research.

3. Summary

The vegetation changes which occurred during the Quaternary in southern Africa were considerable and were caused by changes in temperature, the amount of precipitation and seasonality of the rainfall. These changes especially affected the southern end of the subcontinent, perhaps as far north as the 24th parallel of latitude. During glacial periods a substantial lowering in temperature, the influx of polar air and winter rainfall caused major shifts in vegetation. The temperate grassveld of the inland plateau was replaced by alpine grassland, the Cape coastal plain was devoid of forest and covered by grassland, while the semi-desert Karoo moved far northward. During warm interglacial times Karoo and desert sand dunes moved into the present grassveld plateau and karroid vegetation even invaded the Cape coastal plain. At the same time the summer rainfall in the savanna and woodland region increased.

The limited data, which is available on these vegetation changes, is discussed for a glacial–interglacial cycle, using the terminology of von Post (1946) and Iversen (1954).

References

Butzer, K. W. 1973. Pleistocene 'periglacial' phenomena in southern Africa. Boreas 2:1–11.
Butzer, K. W., Isaac, G. L., Richardson, J. L. & Washbourn-Kamau, C. 1972. Radiocarbon dating of East African lake levels. Science 175:1069–1076.
Butzer, K. W., Fock, G. J., Stuckenrath, R., & Zilch A. 1973. Palaeohydrology of Late Pleistocene Lake, Alexandersfontein, Kimberley, South Africa. Nature 243 (5406):328–330.

Coetzee, J. A. 1967. Pollen Analytical Studies in East and Southern Africa. Palaeoecology of Africa. 3:1–146, 6 charts.
Coetzee, J. A. 1976. Phytogeographical aspects of the montane forests of the chain of mountains on the eastern side of Africa. Erdwissenschaftliche Forschung 12: (in press).
Deacon, J. 1974. Patterning in the radiocarbon dates for the Wilton/Smithfield complex in Southern Africa. S. Afr. Archaeol. Bull. 29 (113 + 114):3–18.
Flohn, H. 1953. Studien über die atmosphärische Zirkulation in der letzten Eiszeit. Erdkunde 7: 266–275.
Hamilton, A. C. 1972. The interpretation of pollen diagrams from highland Uganda. Palaeoecology of Africa 7:45–149.
Harper, G. 1969. Periglacial evidence in southern Africa during the Pleistocene epoch. Palaeoecology of Africa 4:71–101.
Iversen, J. 1954. The late-glacial flora of Denmark and its relation to climate and soil. Geol. Survey Denmark, ser. 2, 80.
Jackson, S. P. 1952. Atmospheric Circulation over South Africa. S. Afr. Geograph. J. 34:48–59.
Klein, R. G. 1972. The Late-Quaternary Mammalian Fauna of Nelson Bay Cave (Cape Province, South Africa): Its Implications for Megafaunal Extinctions and Environmental and Cultural Change. Quaternary Res. 2:135–142.
Klein, R. G. 1974. Environment and subsistence of prehistoric man in the Southern Cape Province, South Africa. World Archaeol. 5:249–284.
Knoch, K. & Schulze, A. 1956. Niederschlag, Temperatur und Schwüle in Africa. 12pp., Karten 72–77. Falk Verlag, Hamburg.
Martin, A. R. H. 1968. Pollen analysis of Groenvlei Lake sediments, Knysna (South Africa). Rev. Palaeobotan. Palynol. 7:107–144.
Post, L. von 1946. The prospect for pollen analysis in the study of the earth's climatic history. New Phytol. 45:193–217.
Schalke, H. J. W. G. 1973. The Upper Quaternary of the Cape Flats Area (Cape Province, South Africa). Scripta Geol. 15:1–57, 8 charts.
Talma, A. S., Vogel, J. C. & Partridge, T. C. 1974. Isotopic Contents of Some Transvaal Speleothems and their Palaeoclimatic Significance. S. Afr. J. Sci. 70:135–140.
Trewartha, G. T. 1966. The Earth's Problem Climates. Methuen & Co., London.
Van Zinderen Bakker Jr, E. M. 1973. Ecological investigations of forest communities in the eastern Orange Free State and the adjacent Natal Drakensberg. Vegetatio 28:299–334.
Van Zinderen Bakker Sr, E. M. 1957. A pollen analytical investigation of the Florisbad deposits (South Africa). Proc. 3rd Pan-Afr. Congr. Prehist. Livingstone 1955. Pp. 56–67.
Van Zinderen Bakker Sr, E. M. 1969. The 'arid corridor' between south-western Africa and the Horn of Africa. Palaeoecology of Africa 4:139–140.
Van Zinderen Bakker Sr, E. M. 1970. Observations on the Distribution of Ericaceae in Africa. Coll. Geographicum 12:89–97.
Van Zinderen Bakker Sr, E. M. 1972. Late-Quaternary lacustrine phases in the Southern Sahara and East Africa. Palaeoecology of Africa 6:15–27.
Van Zinderen Bakker Sr, E. M. 1975. The origin and palaeoenvironment of the Namib Desert biome. J. Biogeogr. 2:65–78.
Van Zinderen Bakker Sr, E. M. 1976. The Evolution of Late-Quaternary Palaeoclimates of Southern Africa. Palaeoecology of Africa 9:160–202.
Van Zinderen Bakker Sr, E. M. & Werger, M. J. A. 1974. Environment, vegetation and phytogeography of the high-altitude bogs of Lesotho. Vegetatio 29:37–49.
Wild, H. 1968. Phytogeography of South Central Africa. Kirkia 6:197–222.

7 Biogeographical division of southern Africa

M. J. A. Werger

1. Introduction 147
2. Phytochorology of southern Africa 147
3. Zoochorology of southern Africa 160
4. Conclusion 167
   References 167

7 Biogeographical division of southern Africa

1. Introduction

Scientific knowledge of the southern African flora and fauna began soon after the initiation of colonial rule in the area. Particularly showy Cape plants early attracted the interest of Europeans and the first Cape plants were already described and depicted before the establishment of a revictualling post there. The first description and illustration of a plant from southern Africa appears to be that of *Protea neriifolia* in a work by Clusius published in 1605. Some of the larger animal species occurring over much of the African continent, like lion, giraffe, black rhinoceros and elephant, were well-known in Europe before the beginning of the Christian era. Others were brought back to Europe with the early expeditions to the coastal regions of southern Africa. After the establishment of the outpost at the Cape, however, knowledge of southern African organisms quickly increased. Numerous specimens collected on expeditions penetrating ever further into the hinterland were sent to scientific centres in Europe (cf. Palmer & Pitman 1972). From these collections it became gradually understood towards the middle of the nineteenth century that the various species each had a specific distribution pattern that mostly correlated with roughly known patterns of climate or major landform features of southern Africa.

The early Portuguese explorers were only interested in plants useful for their nutritive or medicinal value, which were few in the coastal areas. Although some plants are named in their early narratives, the first herbarium specimens brought back to Europe by Portuguese from Portuguese territories in southern Africa date from the 18th century. Earlier, some British explorers had collected near Luanda (see Exell 1960, Mendonça 1962). Animal species were assiduously collected; first the larger and showy ones like mammals and birds, but from the beginning of the eighteenth century also species of lower vertebrates and invertebrates, such as insects, arachnids and molluscs. The diaries and accounts of the early travellers were also particularly important in the accumulation of knowledge and during the second half of the 19th century several accounts of journeys deep into southern Africa and reporting on flora and fauna were readily available. It was also then that an effort was made to divide the world, including southern Africa, into meaningful biogeographical units based on the natural occurrences of plants and animals. These divisions were carried out separately for plants and animals, establishing phytochoria* and zoochoria of various ranks and, as the distribution patterns in and modes of dispersal of plants and animals are not entirely similar, resulted in somewhat different phyto- and zoogeographical choria.

2. Phytochorology of southern Africa

Stimulated particularly by the works of Humboldt and Schouw, a geographical interest developed in the distribution of plant species and vegetation formations. The

---

*A 'chorion' or 'chorium' is a chorological unit of any rank, e.g., region, domain, etc. (compare the term 'generalized track' in Croizat et al. 1974); it should not be confused with the other meanings of this word in zoological anatomy.

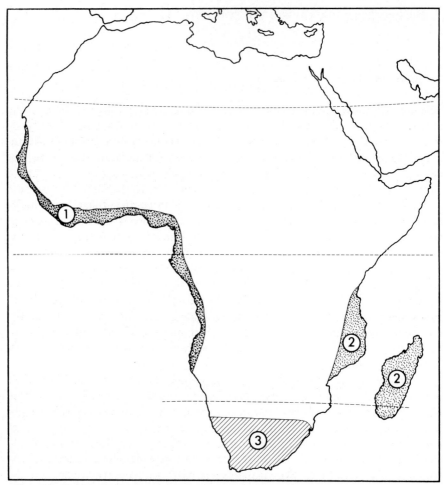

Fig. 1. The three floristic kingdoms in Africa distinguished by Schouw 1823.
1: Regnum Africae occidentalis; 2: Regnum Africae orientalis; 3: Regnum Mesembryanthemorum et Stapeliarum.

latter author already had published a work in 1823 in which he distinguished 22 floristic kingdoms in the world, including a Regnum Africae occidentalis (Atlantic coast of Africa), a Regnum Africae orientalis (East African coastal zone south of the equator, and Madagascar) and a Regnum Mesembryanthemorum et Stapeliarum (extra-tropical South Africa) (Schouw 1823, cf. also Möbius 1937, Schmithüsen 1968) (Fig. 1). After that date many explorers made their way through the African interior, collecting information in many fields and afterwards publishing their findings and experiences in often widely circulated accounts. The information from many such accounts was evaluated and summarized in Grisebach's famous treatise on the vegetation of the world, published in 1872. Based largely on Burchell's (1822–24) and Lichtenstein's (1811–12) observations that the Orange River formed a 'botanical limit', Grisebach reckoned it to be the border between the Cape and the Kalahari Floral Regions, the latter also including the Transvaal and most of the South African Highveld, while Lesotho and the area further east were included in the Sudan Floral Region covering the vast remainder of subsaharan Africa. Grisebach thus included the Karoo in the Cape Floral Region mainly because of the narrow ericoid leaves common in both Cape and Karoo plants and because of predominance of synantheroid families, particularly Asteraceae, in the Karoo (Fig. 2). Bolus (1875) disagreed and held that the only sharp phytogeographical boundary in southern Africa is the one between the true Cape flora and the flora of the interior, running along the Swartberge and the Hex River. He drew that boundary in his map published in 1886 (Fig. 3). Bolus (1875) also emphasized that the flora of the Karoo differs more from that of the Cape Region than from that of the Kalahari and that there is no sharp boundary between the Karoo and Kalahari floras. He suggested that a possible boundary could be drawn south of the Orange River, where patches of red Kalahari sand start to occur. Later, Bolus (1886, 1905) and Marloth (1887) held to the view that in the flora of southern Africa two chief types could be distinguished, a

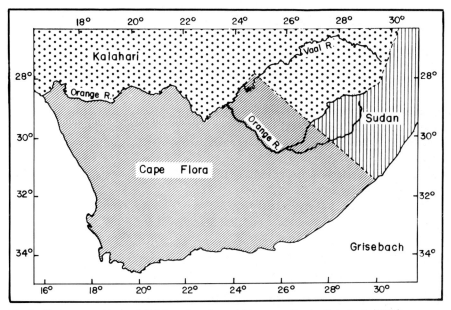

Fig. 2. Phytogeographical subdivision of southern Africa according to Grisebach (1872).

Southwestern or Cape type, occurring in the southwestern Cape, and an African type, which occurred in the rest of the area and could be subdivided into several different regions, including the first separation of a Western Coastal (Namib) Region from the Kalahari and Karoo Regions in 1905 (Fig. 4). Rehman (1880) agreed with Bolus (1875) on the major phytogeographical boundary and he constructed a remarkably useful phytochorological map of southern Africa (Fig. 5), even more meaningful than that of Bolus (1886). Drude (1887, 1890) only slightly modified Rehman's map. Engler (1882) also recognized the major phytogeographical boundary in southern Africa between the Cape flora and the 'Steppes of the palaeotropic flora', but unlike Rehman he did not subdivide this latter area properly (cf. Marloth 1908) (Fig. 6). Later authors mostly have followed Bolus (1875, 1886, 1905), Rehman (1880) and Engler (1882) in recognizing the Cape flora as very different from the other African flora, and they have usually

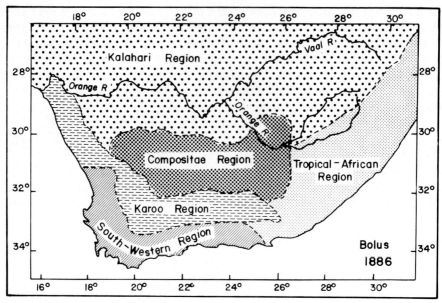

Fig. 3. Phytogeographical subdivision of southern Africa according to Bolus (1886).

recognized a Cape Floral Kingdom (Capensis) (Marloth 1908, Good 1947, Monod 1957, Volk 1966, Takhtajan 1969, cf. White in Chapman & White 1970).

A detailed and useful phytogeographical map of the southernmost part of Africa was presented by Marloth (1908). Here the extent of the Cape floral Kingdom, and of the two subdivisions of the Karoo is given with indications of the surrounding phytogeographical units (Fig. 7).

Pole Evans (1922) distinguished between the Cape Region, covering the area now recognized as Capensis, and the South African Region, covering the

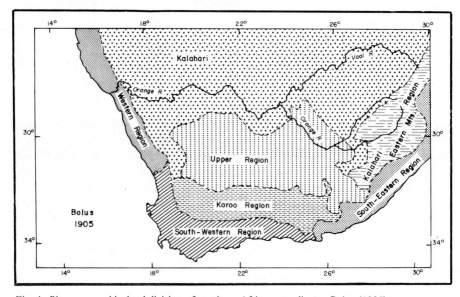

Fig. 4. Phytogeographical subdivision of southern Africa according to Bolus (1905).

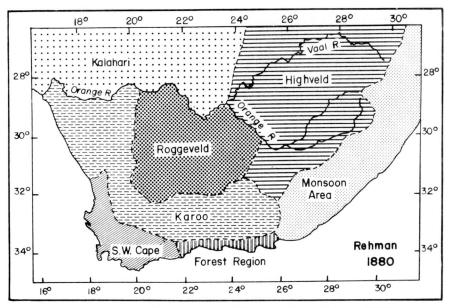

Fig. 5. Phytogeographical subdivision of southern Africa according to Rehman (1880).

remainder of South Africa. The latter was subdivided into four provinces, the Namaqualand Desert Province including the Namib, the Karoo Province, the Kalahari Park and Bush Province including most of the Transvaal, and the South African Steppe and Forest Province which was subdivided into four areas (Highveld, eastern mountains, eastern Cape and Natal grasslands, and east coast forests). Several phytogeographic boundaries on this map have persisted even in the present phytogeographical map of southern Africa. Acocks (1953) also recognized the Cape flora as entirely distinct from the savanna and grassland flora

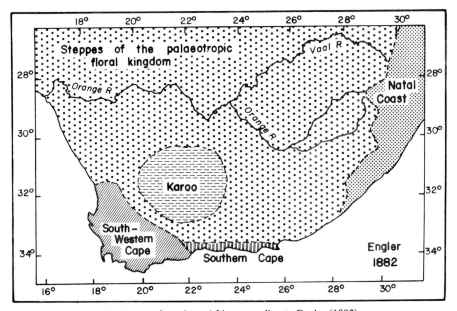

Fig. 6. Phytogeographical map of southern Africa according to Engler (1882).

151

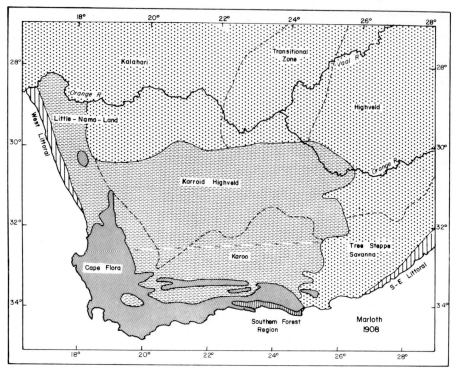

Fig. 7. Phytogeographical map of southern Africa according to Marloth (1908).

of South Africa, and regarded the Karoo flora as resulting from a mixture of the Cape and the tropical floras. Acocks did not present a phytogeographical map, however.

After Marloth (1908) and Pole Evans (1922) several botanists have been concerned with the phytogeographic subdivision of the African continent, but many of them concentrated virtually entirely on West and East Africa (cf. Monod 1957). The first authors to include again southern Africa were Hutchinson (1946), Good (1947) and Lebrun (1947). Hutchinson subdivided southern Africa into numerous units, without using definite chorological criteria, however. Good (1947) recognized the South African Kingdom comprising the Cape flora and the Palaeotropic Kingdom which he subdivided into eight regions, three of which affected southern Africa, namely: the West African Rain-forest Region, the East African Steppe Region and the South African Transition Region. These regions were subdivided into a number of smaller floristic units, which were not mapped. Lebrun (1947) published a phytogeographical map of Africa which has formed the basis of most subsequent phytogeographical subdivisions of that continent (Fig. 8). In Lebrun's scheme southern Africa belongs largely to the Sudano–Zambezian Region which is subdivided in a Zambezian, a Kalahari, a Namaqualand–Karoo, and a South African Domain. In the very north, southern Africa fringes on the Guinean Region, and in the very south, the Cape Region is found. Monod (1957) basically agreed with the pattern in Lebrun's map, but he adjusted the hierarchy and also recognized an Afro-alpine Region, originally proposed by Hauman (1955). In Monod's map (Fig. 9) the boundary between the

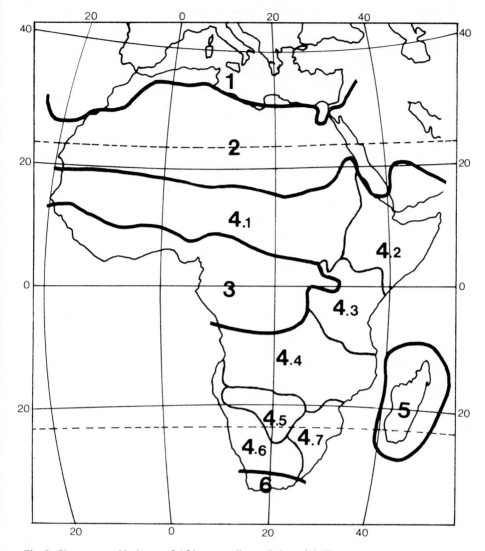

Fig. 8. Phytogeographical map of Africa according to Lebrun (1947).
1. Mediterranean Region
2. Saharo–Sindian Region
3. Guinean Region
4. Sudano–Zambezian Region
4.1 Sahelo–Sudanian Domain
4.2 Somalo–Ethiopian Domain
4.3 Oriental Domain
4.4 Zambezian Domain
4.5 Kalahari Domain
4.6 Namaqualand–Karoo Domain
4.7 South African savanna and forest Domain
5. Malagasian Region
6. Cape Region

Guineo–Congolian and the Sudano–Zambezian (Sudano–Angolan) Regions is somewhat more precise, while the East Coast Forest Domain also is included with the former region as previously suggested by Moreau (1952). Furthermore, the Karoo–Namib is recognized as a region, containing the Karoo, the Namaqualand and the Namib Domains. The Sudano–Zambezian Region in southern Africa consists of the (Angolo-) Zambezian Domain and the Southern Sahel Domain. Troupin's (1966) map differs in details from that of Monod (1957). With regard to

153

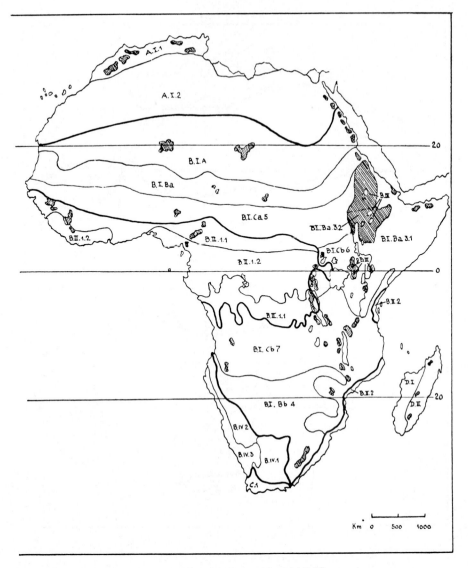

Fig. 9. Phytogeographical map of Africa according to Monod (1957).
A. Mediterranean Region
A.I.1. Eumediterranean Subregion
A.I.2. Saharo–Sindian Subregion
B.I. Sudano–Angolan Region
B.I.A.1. Saharo–African Domain
B.I.Ba.2. Atlantico–Nilotic Domain
B.I.Ba.3. Somalo–Ethiopian Domain
B.I.Bb.4. Southern Domains (Kalahari, etc.)
B.I.Ca.5. Senegalo–Nilotic Domain
B.I.Cb.6. Oriental Domain
B.I.Cb.7. Angolo–Zambezian Domain
B.II. Guineo–Congolian Region
B.II.1. Atlantico–Congolian Domain
B.II.1.1. Peripheral Subdomain
B.II.1.2. Central Subdomain
B.II.2. Eastern Forest Domain
hatched: mountain facies
B.III. Afro–alpine Region
B.IV. Karoo–Namib Region
B.IV.1. Karoo Domain
B.IV.2. Namaqualand Domain
B.IV.3. Namib Domain
C. Cape Region
D. Malagasian Region

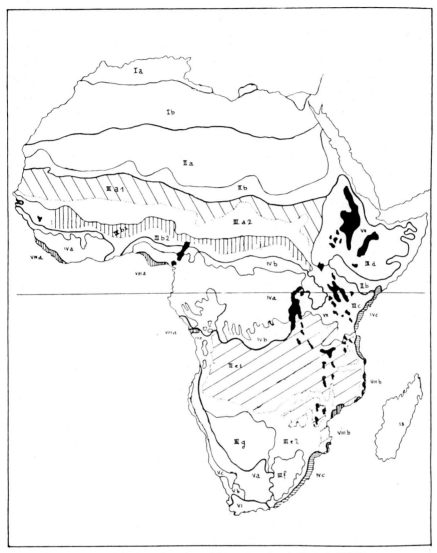

Fig. 10. Phytogeographical map of southern Africa according to Troupin (1966).

- I. Mediterranean Region
  - Ia. Afromediterranean Subregion
  - Ib. Saharo–Sindian Subregion
- II. Saharo–Sahelian Region
  - IIa. Saharo–African Domain
  - IIb. Sahelo–Ethiopian Domain
- III. Sudano–Zambezian Region
  - IIIa. Sudanian Domain
  - IIIa1. Northern Subdomain
  - IIIa2. Southern Subdomain
  - IIIb. Sudano–Guinean Domain
  - IIIb1. Northern Subdomain
  - IIIb2. Southern Subdomain
  - IIIc. Oriental Domain
  - IIId. Ethiopian Domain
  - IIIe. Zambezian Domain
  - IIIe1. Katango–Rhodesian Subdomain
  - IIIe2. Angolo–Zambezian Subdomain
  - IIIf. Austral Domain
  - IIIg. Kalahari Domain
- IV. Guineo–Congolian Region
  - IVa. Forest Domain
  - IVb. Postforest Domain
  - IVc. East Coast Domain
- V. Karoo–Namib Region
  - Va. Karoo Domain
  - Vb. Namaqualand Domain
  - Vc. Namib Domain
- VI. Cape Region
- VII. Afromontane Region (black)
- VIII. Intertropical Littoral Region
  - VIIIa. Atlantic Domain
  - VIIIb. Asiatic Domain
- IX. Malagasian Region

southern Africa Troupin's dissenting subdivision of the Karoo–Namib Region is relevant (Fig. 10). Aubréville's (1975) map modifies only slightly the picture presented by Troupin. White (1965, 1971, Chapman & White 1970) drew attention to the distinctness of the Afromontane flora from those adjacent, and parallel to the Afro-alpine flora he distinguished an Afromontane Region distributed as an archipelago mainly along the eastern African mountain ridges. White also did not recognize the Kalahari as a separate domain but included it in the Zambezian Domain mainly on the basis of its woody species. He referred to Monod's East Coast Forest Domain as the Usambara–Zululand Domain (Fig. 11). Volk (1966) and Werger (1973a) later modified this boundary and included the southern Kalahari into the Karoo–Namib Region. In the southeastern Cape, where the floras of Capensis, and four African palaeotropic regions meet (Karoo–Namib, Sudano–Zambezian, Afromontane and Tongaland–Pondoland), the boundaries are difficult to determine. The marked relief in this area provides a wide variety of habitats, which results in an intermingling of the various floras in some communities and a mosaic distribution in other cases.

As might be expected, an interesting, recent computer-based analysis of the distribution of West African grasses by Clayton & Hepper (1974) appeared to confirm the main phytochoria distinguished since Lebrun (1947).

The various phytochoria discussed so far have been analysed floristically in unequal proportions. For Capensis the work by Weimarck (1941) and Nordenstam (1969) contributed important information on the distinction of centres of speciation, or sectors, while workers such as Marloth (1908), Acocks (1953), Levyns (1958, 1962, 1964) and Beard (1959) discussed the origin of the Cape flora (compare Chapter 8). Volk (1964, 1966), Werger (1973a, b), Nordenstam (1974), and others, contributed to the analysis of the Karoo–Namib Region (see Chapter 9). Important analyses of the Zambezian Domain of the Sudano–Zambezian Region include the work of Lebrun (1961), White (1965) and Wild (1968) (see Chapter 10), while many studies dealt with the Guineo–Congolian Region, including that of Lebrun (1961). The Afromontane Region was particularly well studied by Chapman & White (1970) and parts of it also by Wild (1964), Van Zinderen Bakker (1970), Kerfoot (1975), Clayton (1976) and Hamilton (1976) (see Chapter 11), while the Afro-alpine Region was studied by Killick (1963), Hedberg (1965), Coetzee (1967), Van Zinderen Bakker & Werger (1974), and others (see Chapter 12). Coetzee (1967) drew attention to the differences between the southern Afro-alpine and central and eastern Afro-alpine floras and vegetation, and suggested an Austro–Afro-alpine zone, which may be recognized as a domain.

The present floristic knowledge thus results in the phytochorological subdivision of southern Africa as presented in Fig. 12. In the southwest one finds Capensis with its rich flora and very many endemics. Here the vegetation consists mainly of Fynbos and restioid formations with forest remnants containing species of Afromontane affinity in the protected ravines and gorges and on some screes. Insular outliers of Capensis occur somewhat further to the north on mountain ridges (see Chapter 8).

The entire remaining area of southern Africa belongs phytochorologically to the African Subkingdom of Palaeotropis, which is subdivided into regions. North of Capensis in a wide area that gradually narrows and tapers out in southwestern Angola, lies the Karoo–Namib Region to be subdivided into four domains. This

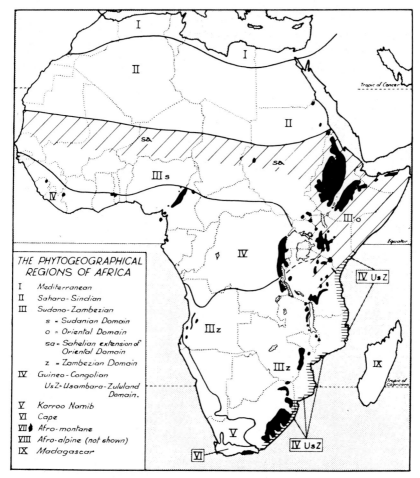

Fig. 11. Phytogeographical map of Africa according to White (1965, 1971).

region is characterized by a fairly rich and dry flora with many endemic genera and species. Particularly Asteraceae are rich in dwarf shrub species in the region. There is some floristic affinity between Capensis and the Karoo–Namib Region, especially in the winter rainfall parts, where many genera of the Mesembryanthemaceae, Aizoaceae, Geraniaceae, Oxalidaceae, and other families occur which are also common in Capensis. The eastern boundary of this region in common with the Sudano–Zambezian Region often is not very sharp and there is a more gradual transition in species of different affinity. The vegetation of the Karoo–Namib Region is largely an open dwarf shrub formation or a desert, with taller woody plants on the mountain slopes (see Chapter 9).

To the east and north of the Karoo–Namib Region the flora belongs to the vast Zambezian Domain of the Sudano–Zambezian Region. This floristically rich area (cf. Lebrun 1960) is characterized by a large number of species with a very wide range of distribution which extend beyond southern Africa. The flora also contains many endemic species. A fair number of Sudano–Zambezian species have closely related species in the rain and montane forests. The typical vegetation of the Zambezian Domain consists of woodlands and savannas (see Chapter 10).

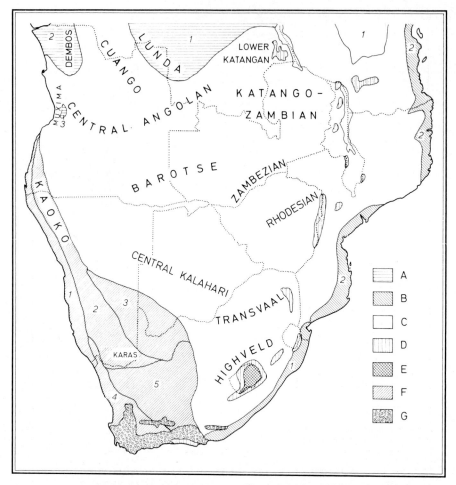

Fig. 12. Phytochorological subdivision of southern Africa (names of some centres are included in the map).

- A. Guineo–Congolian Region
  - A1. Congolian Domain
  - A2. Nigerian–Cameroonian Domain (including Littoral South Atlantic Domain)
  - A3. Amboim Section of A2
- B. Indian Ocean Coastal Belt
  - B1. Tongaland–Pondoland Regional Mosaic
  - B2. Zanzibar–Inhambane Regional Mosaic
- C. Sudano–Zambezian Region (Zambezian Domain)
  - C1. Oriental Domain
- D. Afromontane Region
- E. Afro-alpine Region (Austral Domain)
- F. Karoo–Namib Region
  - F1. Namib Domain
  - F2. Namaqualand Domain
  - F3. Southern Kalahari Subdomain
  - F4. Western Cape Domain
  - F5. Karoo Domain
- G. Capensis

Along the very northward margin of southern Africa a somewhat transitional part of the Guineo–Congolian Region, with three domains, is encountered. This region penetrates into the Zambezian Domain along the major river valleys and the Angolan escarpments. The flora of this region is rich and contains many endemics. Many species have closely-allied species in the montane forests and in the gallery forests of the Zambezian Domain. The characteristic vegetation of the Guineo–Congolian Region is dense forest, with tall grass savannas occurring in the transitional zone (see Chapter 14).

Along the coast of the Indian Ocean one finds patchily distributed the Indian Ocean Coastal Belt (previously recognized as the Usambara–Zululand Domain of the Guineo–Congolian Region), which can be divided into two chorological units of regional rank, namely the Zanzibar–Inhambane Regional Mosaic and the Tongaland–Pondoland Regional Mosaic (see Chapter 13). This belt, with a rich flora containing many endemics, shows a relict type of distribution and today continues to be further diminished as a result of anthropogenic activity. The flora shows some strong affinities with that of the Guineo–Congolian and the Afromontane Regions. The typical vegetation is forest (see Chapter 13).

Spread as an archipelago over southern Africa, mainly along the eastern escarpment, but in the south also reaching the coast, is the Afromontane Region. This region has a diverse flora with a number of endemics, and shows affinities to the Sudano–Zambezian and the Guineo–Congolian Regions and to Capensis. The Cape flora or related species radiate out northwards along the eastern escarpment where they are found at gradually increasing altitude. The vegetation of the Afromontane Region mainly consists of dense forests, but also of grasslands and savannas, which are possibly secondary (see Chapter 11).

On the high plateau of the Drakensberg in South Africa the Afro-alpine Region is represented by what possibly could be best called, the Austral Domain. This region has a fairly rich flora with several endemics, and with some affinities to Capensis. The affinities with Holarctis are also considerable as can be seen from such genera as *Poa, Festuca, Koeleria, Trifolium, Cerastium* and *Geum*. The vegetation characteristically consists of alpine grasslands and dwarf shrub communities (see Chapter 12).

Recently, White (1976) presented a new phytochorological map of Africa. Unlike the usual chorological maps which divide an area in mutually exclusive choria of a specific rank and subdivide these in a hierarchical manner, White's map (Chapter 11, Fig. 2) is quite pragmatic in that it emphasizes endemism by using four major categories of equal rank. These categories are regional centre of endemism, archipelago-like centre of endemism, regional transition zone, and regional mosaic. This map obviously maintains the salient structures of the other phytochorological maps discussed earlier, but wide or narrow transition zones separate the clearly characterized centres of endemism, which mainly correspond to regions or domains of earlier maps. Both this division and the hierarchical phytochorological division of southern Africa can be used compatibly. This is illustrated in the following chapters where one or the other system has been employed as a basis for phytochorological discussions. It should be realized, however, that the term 'region' as used in the system according to White (1976) is not entirely equivalent to a region in the hierarchical chorological system.

Obviously the phytochorological picture sketched above is based on data from vascular plants only. The taxonomic and distributional knowledge of bryophytes and lichens is very scanty, while that of Fungi, a group of organisms traditionally often included in the plant kingdom, is very poor indeed.

Due to insufficient collecting and taxonomical knowledge pronouncements on the geographical distribution of Hepaticae in southern Africa should be made with caution. In general the impression exists, however, that the hepatics show the same general patterns of distribution as the vascular plants, although probably with a far higher proportion of widely distributed species (E. W. Jones, pers. comm.).

For Musci the situation is equally difficult. Herzog (1926) suggested a clear distinction between the moss floras of the tropical rain forest zone (which he defined somewhat more broadly than the present Guineo–Congolian Region), the drier areas of eastern Africa, and the tropical high-mountain areas. At present, it is clear that there is a considerable affinity between the moss floras of the Cape and that of the eastern mountain ranges (A. Touw, pers. comm.) and the preliminary phytogeographical analysis of African *Bryoideae* by Ochi (1972–73) has not resulted in a chorological pattern seriously at variance with that of vascular plants.

A preliminary phytogeographical sketch of southern African lichens was provided by Almborn (1966), who pointed out that it is still premature to draw any far-reaching conclusions on the phytogeography of southern African lichens on the basis of present knowledge. Almborn has distinguished several groups of species on a geographic–ecological base: the ubiquitous species, the steppe and desert species, the montane species, the oceanic species, the tropical–oceanic species, and the maritime species. These groups all contain species with a worldwide distribution, although there are also some endemics. While the ubiquitous species do not show a restricted pattern of distribution, the steppe and desert species mainly occur in the Karoo–Namib Region; the montane species occur in the mountains of Capensis and in the Afromontane Region at higher altitudes; and the oceanic species on wet sites in Capensis and the Afromontane Region, also at lower altitudes; the tropical–oceanic species seem to be indicative of the Indian Ocean Coastal Belt while the maritime species are restricted to the spray zone along rocky sea coasts. Although the number of endemics in southern Africa recounted in the work by Doidge (1950) was first thought to be high, subsequent taxonomic studies have since revealed that it is rather small (Almborn 1966).

The present knowledge of the taxonomy and distribution of African Fungi is insufficient to such an extent that no patterns can be detected yet (R. A. Maas Geesteranus, pers. comm.).

3. Zoochorology of southern Africa

The earliest zoochorological publications on a world basis date from the first half of the nineteenth century and are, as in phytogeography, of historic interest only (cf. Schmidt 1954). However, as early as 1858 Sclater proposed a division of the world into six faunal regions. This work was based on his study of the distribution of birds and has subsequently proved significant for zoochorological division until the present day. One of these regions was the Ethiopian comprising virtually all of Africa (except the northwestern corner) and southeastern Arabia, Madagascar and the Mascarenes. With the Palaearctic, Indian and Australian Regions the Ethiopian Region was part of the Creatio Palaeogeana. Wallace (1876) provided substantial information on the distribution of other animal groups to prove the general validity of Sclater's zoochorological division. Most later zoogeographers (cf. De Beaufort 1951, Darlington 1957, De Lattin 1967, Udvardy 1969, Franz 1970, Thenius 1972, Briggs 1974) have accepted this division basically, although they have sometimes redefined the boundaries of the regions slightly. For the Ethiopian Region this usually implied a restriction to subsaharan Africa, and according to several authors, the separation of southwest Arabia, the Mascarenes and Madagascar from the region. Also Sclater's concept of the Creatio Palaeogeana was no longer regarded as useful and the Ethiopian Region was included in one of three realms, the Arctogaean or Megagaean Realm, together with the Oriental (Indian), Palaearctic and Nearctic Regions; the latter two are usually combined as the Holarctic Region. Müller (1974) distinguished five realms which agree more with the phytogeographical kingdoms and which correspond with Schmidt's (1954) regions. The Ethiopian Region (which is considered a subregion in Schmidt's scheme) is then included in the Palaeotropical Realm, together with the Madagascan and Oriental Regions. Bowden (1973 and Chapter 25) has conceived the Ethiopian Region as including the entire African continent and proposed to rename it 'African Region'. His views are based primarily on the distribution patterns in Diptera.

Wallace (1876) proposed a subdivision of the Ethiopian Region into four subregions: the East African Subregion which is largely identical to the phytogeographical Sudano–Zambezian Region; the West African Subregion, being comparable to the phytogeographical Guineo–Congolian Region; the South African Subregion, roughly comprising Africa south of the Tropic of Capricorn and the coastal zone of Moçambique; and the Malagasy Subregion (Fig. 13). In his description of these areas Wallace indicated that contrary to the boundary shown in his map the West African Subregion extends further eastward to include some of the great lakes of East Africa. This subdivision implies a distinction between forest species, savanna species and species of extratropical southern Africa, features which have been recognized by many authors working on various groups of animals (cf. Monod 1957).

Chapin (1932) presented a subdivision of the Ethiopian Region based on his studies of African birds that proved important and acceptable for other terrestrial animal groups. Chapin distinguished two subregions: the West African, coinciding with the phytogeographical Guineo–Congolian Region, and the East and South African Subregion, comprising all of the remainder of subsaharan Africa (Fig. 14). He recognized two provinces and six districts in the West African Subregion and four provinces and eleven districts in the East and South African Subregion respectively, but these only partly coincide with phytochorological units. Chapin thus rejected Wallace's South African Subregion as a separate zoochorion of high rank, but he remarked that the South African part of his map probably needed further subdivision. Koch (1944) on the other hand, basing his views on a study of Adesmiini, a group of tenebrionid beetles, distinguished again a South African Region next to an Ethiopian Region (Fig. 15), but his application of rank to the various zoochoria is not directly comparable to that of other authors. Koch also recognized the marked differences between the tropical forest area and the savanna area, and distinguished several provinces within the various regions. Moreau (1952) confirmed the zoochorological map by Chapin, and added that there are five main biotic types in Africa to which birds tend to be specific, namely the arid type, the lowland evergreen rain forest type, the montane evergreen rain forest type, the savanna type, and the subalpine moorland type. These types were then subdivided by Moreau into several geographic subunits. Moreau (1952) also distinguished the temperate winter rainfall area of the

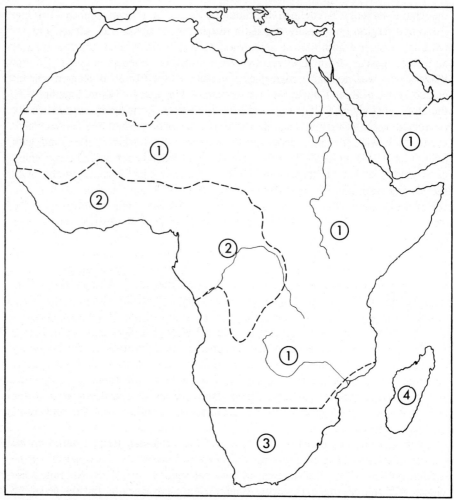

Fig. 13. Zoogeographical map of Africa (Ethiopian Region) according to Wallace (1876). 1: East African Subregion; 2: West African Subregion; 3: South African Subregion; 4: Malagasy Subregion.

southwestern Cape as a distinct unit. This was also emphasized by Davis (1962), Stuckenberg (1962) and Müller (1974). Stuckenberg (1962) demonstrated that the 'palaeogenic invertebrates' of southern Africa show that there is a clear affinity between the montane areas of southern Africa and the Cape area, but that the two are nevertheless clearly distinct. Van Bruggen's (1969) studies on land molluscs confirm this. Without reference to a hierarchical zoochorological division of Africa, Moreau (1966) discussed the differences in bird faunas of the various biomes in Africa. He concluded that the main differences lie between the bird fauna of the forest and the non-forest areas and noted that those of the lowland and montane areas are very distinct fron one another. His discussion of the bird faunas of the various biomes was based mainly on the floristic–physiognomical vegetation map of Keay (1959).

Southernmost Africa has been subdivided in some 19 districts on the basis of the bird fauna and some environmental factors by McLachlan & Liversidge (1958). Each district is characterized ecologically and most are typified by one or

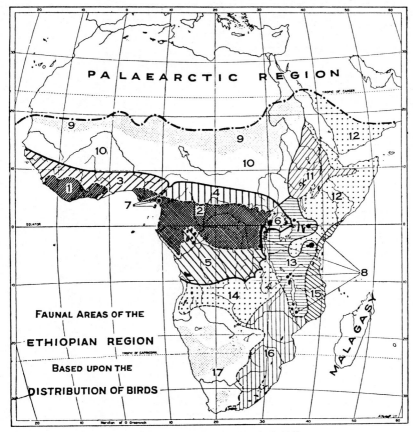

Fig. 14. Zoogeographical map of Africa according to Chapin (1932).
I. WEST AFRICAN SUBREGION— A. Guinean Forest Province:
1. Upper Guinea Forest District; 2. Lower Guinea Forest District. B. Guinean Savanna Province:
3. Upper Guinea Savanna District; 4. Ubangi–Uelle Savanna District; 5. Southern Congo Savanna District; 6. Uganda–Unyoro Savanna District.
II. EAST AND SOUTH AFRICAN SUBREGION—C. Humid Montane Province: 7. Cameroon Montane District; D. Sudanese Province: 8. Eastern Montane District. 9. Sudanese Arid District; 10. Sudanese Savanna District. E. Northeast African Province: 11. Abyssinian Highland District; 12. Somali Arid District. F. Eastern and Southern Province: 13. East African Highland District; 14. Rhodesian Highland District; 15. East African Lowland District; 16. Southwest Veld District; 17. Southwest Arid District.

more endemic species or races of birds. It is difficult, however, to compare this and Moreau's (1966) findings with other zoochorological schemes.

Based on a comprehensive study of African butterflies, Carcasson (1964) proposed a detailed subdivision of the Ethiopian Region (Fig. 16) into four subregions, the Sylvan Subregion, the Subregion of the open formations, the Cape Subregion, and the Malagasy Subregion. The two first subregions were subdivided into two divisions each, and within each unit several zones were distinguished. Carcasson's map agrees with Chapin's in strongly separating the forested areas from the more open ones, but Carcasson delineated the Guinean area differently and he included the montane forest areas into the Sylvan Subregion, although into a separate division. Carcasson also included the eastern coastal zone into the Sylvan Subregion, Lowland Forest Division, and pointed out that although the

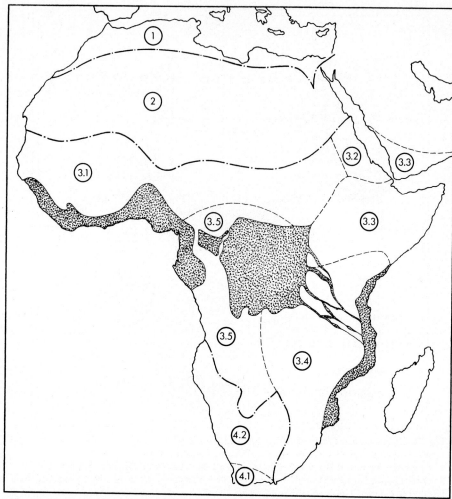

Fig. 15. Zoogeographical map of Africa according to Koch (1944). 1: Mediterranean Region; 2: Saharian Region; 3: Ethiopian Region; dotted: Forest Subregion; remainder: Savanna Subregion; 3.1: Sudanian Province; 3.2: Erithraean Province; 3.3: Somali–Abyssinian Province; 3.4: East African Province; 3.5: West African Province; 4: South African Region; 4.1: Fynbos Province; 4.2: Steppes and Deserts Province.

area is clearly distinct, it has unmistakable affinities with the western forest fauna. A similar feature exists in the floras of these two areas, as has been pointed out above. The Highland Forest Division which consists of seven zones distributed as an archipelago over the African continent is surprisingly uniform in its butterfly fauna, which suggests, according to Carcasson, that they once formed a continuous area. The delineations of the Subregion of the open formations and the Cape Subregion are also somewhat different from Chapin's East and South African Subregion particularly in southern Africa, where Carcasson showed more detail. Apart from these Subregions Carcasson distinguished four special habitats: the tropical montane grasslands which show, just as in their flora, affinities with the Cape and the Palaearctic; the highland and lowland swamps, and the littoral sand dunes.

Poynton (1964) made a comprehensive study of southern African amphibians

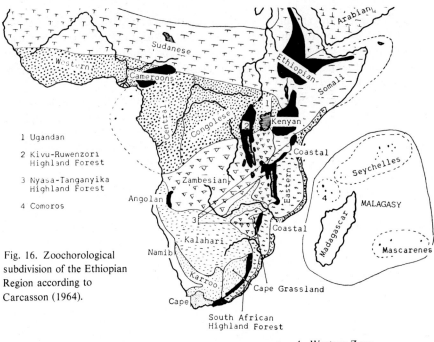

Fig. 16. Zoochorological subdivision of the Ethiopian Region according to Carcasson (1964).

1 Ugandan
2 Kivu-Ruwenzori Highland Forest
3 Nyasa-Tanganyika Highland Forest
4 Comoros

SYLVAN SUBREGION
- Lowland Forest Division
  - Western Subdivision
    1. Western Zone
    2. Central Zone
    3. Congolese Zone
    4. Ugandan Zone
    5. Coastal Zone
  - Eastern Subdivision
- Highland Forest Division
  1. Cameroons Zone
  2. Kiwu–Ruwenzori Zone
  3. Ethiopian Zone
  4. Kenyan Zone
  5. Tanganyika–Nyasa Zone
  6. Angolan Zone
  7. South African Zone

SUBREGION OF OPEN FORMATIONS
- Northern Division
  1. Sudanese Zone
  2. Somali Zone
  2. Arabian Zone
- Southern Division
  1. Eastern Zone
  2. Zambezian Zone
  3. Kalahari Zone

CAPE SUBREGION
1. Cape Zone
2. Karoo Zone
3. Cape Grassland Zone
4. Namib Zone

MALAGASY SUBREGION
1. Madagascar Zone
2. Comoros Zone
3. Mascarenes Zone
4. Seychelles Zone

SPECIAL HABITATS
1. Tropical Montane Grassland
2. Highland Swamps
3. Lowland Swamps
4. Littoral Sand Dunes

and concluded that the amphibians show a chorological pattern which justifies the separation of a Cape Region from the Ethiopian Region. Poynton emphasized that patterns of distribution in other animal groups confirm the separation of a Cape Faunal Region coexistent with the Cape Floral Region, and distinct from the Ethiopian Region.

Franz (1970) presented a zoochorological map of the Ethiopian Region very similar to that by Wallace (1876). The main differences are that, in contrast to Wallace's South African Subregion, Franz distinguished a Karoo–Kalaharian Subregion covering the dry country, and a South African Subregion covering the mesic zone from the Cape to the Limpopo between the coast and the escarpment, an area that appears as distinct in zoochorological maps on land molluscs (Van Bruggen 1969). Also, the West African Subregion protrudes much further south in Franz's map. Franz indicated, however, that the Sahel–East African Subregion should probably be subdivided into three separate units, a Sahel, an East African,

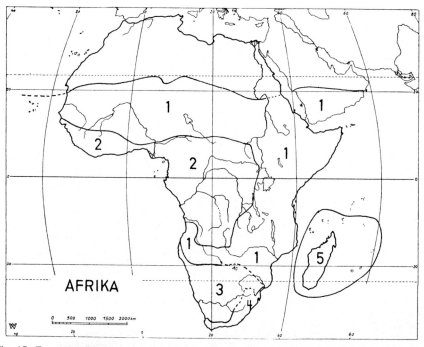

Fig. 17. Zoogeographical map of Africa according to Franz (1970). 1: Sahelian–East African Subregion; 2: West African Subregion; 3: Karoo–Kalaharian Subregion; 4: South African Subregion; 5: Madagascan Subregion.

and a Southern area. He also recognized the high African mountain areas as distinct, but did not include the southern African mountains within this unit (Fig. 17).

In the fresh water environment the situation is obviously different due to the different modes and possibilities of dispersal of the aquatic organisms in contrast to terrestrial ones. Blanc (1954) presented a simple chorological map of African fresh water systems, based on ichthyological studies. The boundaries in his map are very similar to those on Wallace's map of 1876 although the interpretation differs. Blanc distinguished a Berberic Region on the northwestern edge of Africa, a South African Region south of the Orange and Limpopo Rivers and an Ethiopian Region covering the vast remainder of Africa. Madagascar was also distinguished as a region of its own. Within the Ethiopian Region, Blanc made a clear distinction between the West and Central African forest area, the savanna and steppe area, and the large Central African lakes, these latter not being described as a homogeneous group, however. Recently, Roberts (1975) distinguished ten ichthyofaunal provinces in Africa.

4. Conclusion

It is thus apparent that reasonable unity in opinion on the zoogeographical boundaries in Africa was reached much earlier than was the case with the phytogeographical boundaries. Only the southern part, in particular extratropical southern Africa, has presented some difficulties as to the delineation of zoochoria. Basically, it is agreed among most zoogeographers, however, that all of southern Africa belongs to one region, the Ethiopian, and unlike the situation in phytogeography, the temperate winter rainfall area of the southwestern Cape, is not sufficiently distinct to justify recognition as a region in its own right. It is different from the remainder of southern Africa, however, and it could perhaps be recognized as a zoochorion of lesser rank. With regard to the subdivision of the other part of (southern) Africa it can be concluded that there is a major distinction between the West African Subregion, covering an area largely similar to the phytogeographic Guineo–Congolian Region, and the East and South African Subregion, similar in area to the phytogeographic Sudano–Zambezian Region. Also an Arid Southwest Subregion can be distinguished which coincides largely with the phytogeographic Karoo–Namib Region, although the eastern boundary of the Arid Southwest Subregion always is conceived as further inland than that of the Karoo–Namib Region. Also a Montane and an East Coast zoochorion can be distinguished, and perhaps even an Alpine, but predominantly these zoochoria, if distinguished, are not considered to be of subregion rank. These latter three zoochoria also correspond in general to phytogeographical regions.

It can therefore be concluded that (southern) Africa can be subdivided meaningfully into a number of well-defined biochoria. This subdivision can be made on the basis of both their floristic and their faunistic character since they both result in a mutually largely acceptable biogeographical map with the corresponding phyto- and zoochoria in general showing the same geographical pattern. The details that contribute to this pattern and inconsistencies within this pattern set by various zoological taxa are elaborated in Chapters 16 to 31.

Biogeographical features of the fresh and salt water habitats are reviewed in Chapters 33 to 36 and Chapter 38 respectively.

References

Acocks, J. P. H. 1953. Veld Types of South Africa. Mem. Bot. Surv. S. Afr. 28: 1–192.
Almborn, O. 1966. Revision of some lichen genera in southern Africa. I. Bot. Notis. 119: 70–112.
Aubréville, A. 1975. Essais sur l'origine et l'histoire des flores tropicales africaines. Application de la théorie des origines polytopiques des Angiospermes tropicales. Adansonia, Sér. 2, 15: 31–56.
Beard, J. S. 1959. The origin of African Proteaceae. J. S. Afr. Bot. 25: 231–236.
Blanc, M. 1954. La répartition des poissons d'eau douce africains. Bull. I.F.A.N., Sér. A. 16: 599–628.

Bolus, H. 1875. Letter from Mr. Bolus to Dr. J. B. Hooker. J. Linn. Soc. 14:482–484.
Bolus, H. 1886. Sketch of the flora of South Africa. In: Official Handbook of the Cape of Good Hope. Richards, Cape Town.
Bolus, H. 1905. Sketch of the floral regions of South Africa. In: Flint, W. & Chilchrist, J. D. F. (ed.), Science in South Africa. pp. 198–240. Maskew Miller, Cape Town.
Bowden, J. 1973. Studies in African Bombyliidae. VII. On Distichus Loew and related genera, and Bombylosoma Rondani, with some zoogeographical considerations. J. ent. Soc. Sth. Afr. 36:139–158.
Briggs, J. C. 1974. Operation of zoogeographic barriers. Syst. Zool. 23:248–256.
Burchell, W. J. 1822–24. Travels in the interior of southern Africa. 2 Vols. Paternoster Row, London.
Carcasson, R. H. 1964. A preliminary survey of the zoogeography of African butterflies. E. Afr. Wildl. J. 2:122–157.
Chapin, J. P. 1932. The birds of the Belgian Congo. I. Bull. Am. Mus. Nat. Hist. 65:1–756.
Chapman, J. D. & White, F. 1970. The evergreen forests of Malawi. Commonw. For. Inst., Oxford.
Clayton, W. D. 1976. The chorology of African mountain grasses. Kew Bull. 31:273–288.
Clayton, W. D. & Hepper, F. N. 1974. Computer-aided chorology of West African grasses. Kew Bull. 29:213–234.
Coetzee, J. A. 1967. Pollen analytical studies in East and Southern Africa. Palaeoecology of Africa 3:1–146.
Croizat, L., Nelson, G. & Rosen, D. E. 1974. Centers of origin and related concepts. Syst. Zool. 23:265–287.
Darlington, P. J. 1957. Zoogeography: The geographical distribution of animals. Wiley, New York.
Davis, D. H. S. 1962. Distribution patterns of southern African Muridae, with notes on some of their fossil antecedents. Ann. Cape Prov. Mus. 2:56–76.
De Beaufort, L. F. 1951. Zoogeography of the land and inland waters. Sidgwick & Jackson, London.
De Lattin, G. 1967. Grundriss der Zoogeographie. Fischer, Jena.
Doidge, E. M. 1950. The South African fungi and lichens to the end of 1945. Bothalia. 5:1–1094.
Drude, O. 1887. Atlas der Pflanzenverbreitung. Justus Perthes, Gotha. 8 maps.
Drude, O. 1890. Handbuch der Pflanzengeographie. Engelhorn, Stuttgart.
Engler, A. 1882. Versuch einer Entwicklungsgeschichte der Pflanzenwelt insbesondere der Florengebiete seit der Tertiärperiode. 2 Bd. Engelmann, Leipzig.
Exell, A. W. 1960. History of botanical collecting in the Flora Zambesiaca area. In: Flora Zambesiaca 1:23–34. Crown Agents, London.
Franz, H. 1970. Die gegenwärtige Insektenverbreitung und ihre Entstehung. In: Franz, H. & Beier, M.: Die geographische Verbreitung der Insekten. Handb. d. Zool. IV. 2–1/6, pp. 1–139. De Gruyter, Berlin.
Good, R. 1947. The geography of flowering plants. Longmans, London.
Grisebach, A. 1872. De Vegetation der Erde nach ihrer klimatischen Anordnung. 2 Bde. Engelmann, Leipzig.
Hamilton, A. 1976. The significance of patterns of distribution shown by forest plants and animals in tropical Africa for the reconstruction of palaeoenvironments: a review. Palaeoecology in Africa 9:63–97.
Hauman, L. 1955. La "Région Afroalpine" en phytogéographie centro-africaine. Webbia 11:467–469.
Hedberg, O. 1965. Afroalpine flora elements. Webbia 19:519–529.
Herzog, T. 1926. Geographie der Moose. Fischer, Jena.
Hutchinson, J. 1946. A botanist in southern Africa. Gawthorn, London.
Keay, R. W. J. (ed.). 1959. Vegetation map of Africa. Oxford Univ. Press, London.
Kerfoot, O. 1975. Origin and speciation of the Cupressaceae in Sub-Saharan Africa. Boissiera 24:145–150.
Killick, D. J. B. 1963. An account of the plant ecology of the Cathedral Peak area of the Natal Drakensberg. Mem. Bot. Surv. S. Afr. 34:1–178.
Koch, C. 1944. Die Adesmiini der tropischen und subtropischen Savannen Afrikas. Rev. Zool. Bot. Afr. 38:139–191.
Lebrun, J. 1947. La végétation de la plaine alluviale au sud du lac Édouard. Explor. Parc. Nat. Albert. Fasc. 1:1–800. Parcs Nat. Congo belge, Bruxelles.
Lebrun, J. 1960. Sur la richesse de la flore de divers territoires africains. Bull. Acad. Roy. Sc. Outremer 4:669–690.

Lebrun, J. 1961. Les deux flores d'Afrique tropicale. Mém. Acad. Roy. Belg., Cl. Sc. 32(6):1–81.
Levyns, M. R. 1958. The phytogeography of members of Proteaceae in Africa. J. S. Afr. Bot. 24:1–9.
Levyns, M. R. 1962. Possible Antarctic elements in the South African flora. S. Afr. J. Sci. 58:237–241.
Levyns, M. R. 1964. Migrations and the origin of the Cape flora. Trans. Roy. Soc. S. Afr. 37:85–107.
Lichtenstein, H. 1811–12. Reisen im südlichen Africa in den Jahren 1803, 1804, 1805 und 1806. 2 Bde. Salfeld, Berlin.
Marloth, R. 1887. Das südöstliche Kalahari-Gebiet. Ein Beitrag zur Pflanzengeographie Süd Afrikas. Bot. Jahrb. 8:247–260.
Marloth, R. 1908. Das Kapland, insonderheit das Reich der Kapflora, das Waldgebiet und die Karoo, pflanzengeografisch dargestellt. Wiss. Ergebn. Deutsch. Tiefsee-Exped. "Waldivia", 1898–1899. Bd. 2. T.3. Fischer, Jena.
McLachlan, G. R. & Liversidge, R. 1958. Roberts' Birds of South Africa. Central News Agency, Cape Town.
Mendonça, F. A. 1962. Botanical collectors in Angola. C.R. IV Réun, A.E.T.F.A.T.: pp. 111–121.
Möbius, M. 1937. Geschichte der Botanik. Fischer, Stuttgart.
Monod, T. 1957. Les grands divisions chorologiques de l'Afrique. C.S.A./C.C.T.A. Publ. No. 24:1–150. C.S.A./C.C.T.A., London.
Moreau, R. E. 1952. Africa since the Mesozoic: with particular reference to certain biological problems. Proc. Zool. Soc. London 121:869–913.
Moreau, R. E. 1966. The bird faunas of Africa and its islands. Academic Press, London.
Müller, P. 1974. Aspects of zoogeography. Junk, The Hague.
Nordenstam, B. 1969. Phytogeography of the genus Euryops (Compositae). Opera Botanica 23:1–77.
Nordenstam, B. 1974. The flora of the Brandberg. Dinteria 11:3–67.
Ochi, H. 1972–73. A revision of African Bryoideae, Musci. I+II. J. Fac. Educ. Tottori Univ. 23:1–126; 24:23–50.
Palmer, E. & Pitman, L. 1972. Trees of southern Africa. 3 Vols. Balkema, Cape Town.
Pole Evans, I. B. 1922. The main botanical regions of South Africa. Mem. Bot. Surv. S. Afr. 4:49–53.
Poynton, J. C. 1964. The Amphibia of southern Africa. Ann. Natal. Mus. 17:1–334.
Rehman, A. 1880. Geo-botaniczne stosunki potudniowéj Afryki. Pam. Akad. Umiej. w Krakowie (Denkschr. Akad. Wiss. Krakau) 5:28–96. also: Geo-botanische Verhältnisse von Süd-Afrika. Bot. Centralbl. 1:1119–1128.
Roberts, T. R. 1975. Geographical distribution of African freshwater fishes. Zool. J. Linn. Soc. 57:249–319.
Schmidt, K. P. 1954. Faunal realms, regions, and provinces. Quart. Rev. Biol. 29:322–331.
Schmithüsen, J. 1968. Allgemeine Vegetationsgeographie. 3. Aufl. De Gruyter, Berlin.
Schouw, J. F. 1823. Grundzüge einer allgemeinen Pflanzengeographie. + Pflanzengeographischer Atlas. 12 maps. Reimer, Berlin.
Sclater, P. L. 1858. On the general geographical distribution of the members of the class Aves. J. Proc. Linn. Soc. Zool. 2:130–145.
Stuckenberg, B. R. 1962. The distribution of the montane palaeogenic element in the South African invertebrate fauna. Ann. Cape Prov. Mus. 2:190–205.
Takhtajan, A. 1969. Flowering plants: origin and dispersal. Oliver & Boyd, Edinburgh.
Thenius, E. 1972. Grundzüge der Verbreitungsgeschichte der Säugetiere. Fischer, Stuttgart.
Udvardy, M. D. F. 1969. Dynamic zoogeography. Van Nostrand Reinhold, New York.
Van Bruggen, A. C. 1969. Studies on the land molluscs of Zululand with notes on the distribution of land molluscs in southern Africa. Zool. Verh. Leiden 103:1–116.
Van Zinderen Bakker, E. M. 1970. Observations on the distribution of Ericaceae in Africa. Colloq. Geogr. (Bonn) 12:89–97.
Van Zinderen Bakker, E. M. & Werger, M. J. A. 1974. Environment, vegetation and phytogeography of the high-altitude bogs of Lesotho. Vegetatio 29:37–49.
Volk, O. H. 1964. Die afro–meridional–occidentale Flora Region in Südwestafrika. In: Kreeb, K. (ed.): Beiträge zur Phytologie. pp. 1–16. Ulmer, Stuttgart.
Volk, O. H. 1966. Die Florengebiete von Südwestafrika. J.S.W.A. Wiss. Ges. 20:25–58.
Wallace, A. R. 1876. The geographical distribution of animals. Reprint. 1962. Hafner, New York.

Weimarck, H. 1941. Phytogeographical groups, centres and intervals within the Cape flora. Lunds Univ. Årsskr. N.F. Avd. 2, Bd. 37, No. 5:1–143.
Werger, M. J. A. 1973a. Notes on the phytogeographical affinities of the southern Kalahari. Bothalia 11:177–180.
Werger, M. J. A. 1973b. Phytosociology of the upper Orange River valley, South Africa. V & R, Pretoria.
White, F. 1965. The savanna–woodlands of the Zambezian and Sudanian Domains. Webbia 19:651–681.
White, F. 1971. The taxonomic and ecological basis of chorology. Mitt. Bot. Staatssamml. München 10:91–112.
White, F. 1976. The vegetation map of Africa. The history of a completed project. Boissiera 24:659–666.
Wild, H. 1964. The endemic species of the Chimanimani Mountains and their significance. Kirkia 4:125–157.
Wild, H. 1968. Phytogeography of South Central Africa. Kirkia 6:197–222.

8 Capensis

H. C. Taylor*

1.	Environment and phytogeography	173
1.1	The physical setting	173
1.2	General delimitation and characteristics	174
1.3	Earlier designation and description	176
1.4	Floristic characteristics and affinities	178
1.5	Migrations and origin	180
2.	Vegetation and ecology	184
2.1	Fynbos	184
2.1.1	Mountain Fynbos	185
2.1.2	Arid Fynbos	199
2.1.3	Coastal Fynbos	204
2.1.4	Fire in Fynbos	206
2.2	Strandveld	211
2.2.1	Pioneer littoral dune vegetation	212
2.2.2	Successional littoral dune communities	212
2.2.3	Coast scrub	213
2.2.4	Marsh and pan communities	214
2.2.5	Literature and further research	214
2.3	Coastal Renosterveld	215
3.	Conservation and utilization	218
3.1	Conservation	218
3.1.1	Shrinking ecosystems and vanishing species	219
3.1.2	Pest-plants	220
3.2	Utilization	221
References		223

* The help of colleagues, both in Pretoria and Stellenbosch, in giving unstintingly of their time to discuss contentious points, is gratefully acknowledged. All photographs in this chapter were taken by the author.

8 Capensis

1. Environment and phytogeography

Capensis comprises the distinctive temperate floral area of the southwestern and southern Cape Province found between the Karoo–Namib Region, some outliers of the Sudano–Zambezian and Afromontane Regions and the coast at latitudes between about 31° and 35° south and longitudes between 18° and 27° east.

1.1 *The physical setting*

This is the region of the Cape folded mountain belt, the mountains occurring for the most part in sub-parallel ranges with an average height of 1000 to 1500 m, individual peaks reaching over 2000 m. In the south these ranges strike from east to west while in the west the strike is more nearly north-north-west (vide King 1963, Fig. 55). The two series of folds meet near Ceres where the axes of folds strike east-north-east (Haughton 1969).

The major ranges are constructed of the Table Mountain sandstones and the minor ones of smaller sandstone folds or the Witteberg quartzites of the Cape System, while the Cape granites commonly form the foothills and lower slopes in the western part. The intervening valleys and parts of the coast belt are formed from the Bokkeveld shales and sandstones of the Cape System and the Malmesbury shales of the Archaean Complex. The coastal lowlands consist of sands, conglomerate and limestones of Tertiary to Recent origin.

The soils weathered from the sandstones in situ are generally acid and of low fertility, and on the mountain slopes, of course, very thin. In valleys the clayey soils derived from the Bokkeveld series are more fertile and of good texture especially where they are mixed with some sand from the mountain slopes (Wellington 1955). Coast belt soils derived from shales are often shallow and generally rather impervious, compact and clayey; those derived from granites are mostly sandy loams of moderate moisture retentiveness; while soils derived from sands have a low water retaining capacity and are either acid and relatively infertile or, near the coast, alkaline, with a distinct horizon of lime accumulation (Talbot 1947) (see Chapter 4).

Capensis has a predominantly Mediterranean-type climate with a rainfall in excess of 250 mm per annum, mostly from 300 to 2500 mm. The western part receives over 50 per cent of its rain in winter (Marloth 1929); on the higher peaks the summer drought is alleviated by moisture-bearing clouds of the southeast wind (Marloth 1904, 1907b, Stewart 1904). From Swellendam eastwards the rainfall is more evenly distributed throughout the year (see Chapter 2). During winter, snow falls regularly on the higher mountains, especially in the west, but the lowlands enjoy an equable climate and frost is rare except in some of the deep valleys of the interior.

The vegetation consists mainly of the sclerophyll types that Acocks (1953) has called Macchia (Veld Type 69), False Macchia (Veld Type 70) and Coastal Macchia (Veld Type 47), but includes Acocks' Coastal Rhenosterbosveld (Veld Type 46) and the portion of his Strandveld (Veld Type 34) receiving over 250 mm of rain per annum that extends southwards in a narrow belt down the west coast from the vicinity of Eland's Bay.

The sclerophyll types make up the Cape fynbos vegetation that is the home of the distinctive Cape Flora. The Coastal Rhenosterbosveld and Strandveld are transitional types containing elements of the Cape, the Karoo–Namib and the Afromontane Floras. These transitional types are included here because they occur within the geographical boundaries of Capensis and share common climatic, topographic, geological and some pedological features with the sclerophyll types and are to some extent floristically related. The emphasis in this account will, however, fall on the Cape floral element and the Cape fynbos vegetation.

Two other vegetation types that range partly within the boundaries of Capensis will be excluded from this chapter because they lack the typical elements of the Cape Flora and have different climatic requirements. They are dealt with in Chapters 11 and 9. These types are:

(1) Temperate Evergreen Forest of the Knysna Region (Acocks' Veld Type 4), concentrated on the southern coastal shelf and mountain foothills between George and Storm's River, with outliers in fire-free habitats extending westward as far as the Cape Peninsula; and

(2) Karroid Broken Veld (Acocks' Veld Type 26), a predominantly succulent vegetation that has its main distribution centre in the arid lowlands of the Karoo–Namib Region north of Capensis but intrudes into the broad, dry valleys and plains between the parallel mountain ranges of the eastern part of Capensis.

1.2 *General delimitation and characteristics*

Fynbos is an indigenous word probably used for the first time in a botanical work by Bews (1916) and now replacing older, ambiguous terms such as Sclerophyllous Bush (Schimper 1903, Pole Evans 1936, Adamson 1938a), Sclerophyllous Scrub (Riley & Young, 1966), Maqui (Warming 1909), Macchia (Phillips 1931, Acocks 1953, Roberts 1966) and Heath (Martin 1965). The word implies both the fine-leaved form of many of the shrubs and the bushy structure of the vegetation.

The extent and composition of the typical Cape Flora represented in the fynbos vegetation has been described by early writers such as Bolus (1886, 1905), Bolus & Wolley-Dod (1904), Marloth (1908, 1929) and Adamson (1929, 1938a). The mountain fynbos is concentrated in an irregularly L-shaped area with its centre in the angle of the L from Worcester to Stellenbosch to Caledon. Northward it tails out beyond the Cedarberg to the Van Rhyn's Pass escarpment, and eastward it extends as a thin double line along the major mountain chains, the Swartberg–Baviaanskloof ranges inland and the Langeberg–Outeniqua–Tzitzikamma ranges along and parallel to the coast, as far as Port Elizabeth. Major outliers occur in the Kamiesberg to the north, and near Grahamstown to the east.

The coastal fynbos occurs on part of the low-lying plains between the mountains and the sea.

Floristically, fynbos can be defined by one or two salient features: the lack of single species dominance, and/or the conspicuous presence of members of the family Restionaceae (Taylor 1972a). Physiognomically, fynbos is characterized by three elements, restioid, ericoid and proteoid. These elements comprise plants that resemble typical members of the Restionaceae, Ericaceae and Proteaceae respectively, in growth form, but do not necessarily belong to these families. The Restionaceae and some Cyperaceae, tufted plants with near-leafless tubular or wiry non-woody stems, give the vegetation its most characteristic physiognomic feature – the restioid element (Fig. 1). The only other constant physiognomic feature, the small, narrow, often rolled leaves of some of the shrubs, is the ericoid element (Fig. 2). Typical Cape plants representing the ericoid element belong to families such as Ericaceae (*Erica*), Rutaceae (*Agathosma*), Bruniaceae (*Brunia*), Polygalaceae (*Muraltia*), Thymelaeaceae (*Struthiola*), and to many species in genera like *Aspalathus* (Fabaceae), *Cliffortia* (Rosaceae), *Phylica* (Rhamnaceae) and a number of Asteraceae including *Metalasia* and *Stoebe*. Taller bushes with moderate-sized hard leaves with a dull surface, comprising the proteoid element

Fig. 1. Dry restioid community: Tussocks of *Cannamois* (with silvery inflorescence), *Hypodiscus* and many others, 40–50 cm tall. Skurweberg, Koue Bokkeveld, c. 1600 m.

(Fig. 3), belong mainly to the family Proteaceae, e.g. *Leucadendron, Leucospermum, Mimetes* and often *Protea* itself. In certain habitats this element may be absent. (See Sect. 2.1.1 for greater detail).

Coastal Renosterveld (Acocks' 'Coastal Rhenosterbosveld') occurs in two blocks, a western and a southern, on shales of the Malmesbury and Klipheuwel formations in the west and the Bokkeveld Series in the south, between the fynbos of the mountains and of the coastal plain. The landscape is rolling or undulating. Renosterveld is dominated over most of its range by the grey ericoid-leaved *Elytropappus rhinocerotis* (Asteraceae); typical Cape plants are absent or rare.

The Strandveld discussed in this chapter has a wider connotation than Acocks' 'Strandveld of the West Coast'. It refers to the non-fynbos vegetation of dune

Fig. 2. Predominantly ericoid community, 50–60 cm tall, of *Metalasia* (white flowers), *Salaxis* and others. Cape of Good Hope Nature Reserve, c. 150 m.

landscapes along both the south and west coasts and includes a range of communities within a broadly related successional sequence from pioneer vegetation of littoral dunes to closed scrub.

1.3 *Earlier designation and description*

In the last century, Capensis was described in works by travellers who dealt broadly with world vegetation; for South Africa these descriptions, based on scant information, are largely conjectural. Drege travelled throughout the Cape Colony from 1826 to 1834 and, with Meyer, published a map of the floral regions of the country (Drege & Meyer 1843). Subsequent maps and descriptions published by Grisebach (1872), Rehman (1880), Engler (1882), Drude (1887), Schimper (1898), Engler (1903) and others are reviewed in some detail by Marloth (1903, 1908) (see also Chapter 7).

The area covered by the Cape flora was designated by early authors in various ways. Rehman and Engler applied the term 'Southwestern region' to the country as far east as Mossel Bay. Drude used the term 'Evergreen scrub region' for the same area, and Schimper termed it 'Sclerophyllous Scrub Area.'

With the advent of resident botanists, description and delineation of the 'Region of the Cape Flora' became more accurate. Bolus (1886) was the first local author to define and describe the 'South Western Region'. His simple map was refined in a later publication (Bolus 1905) in which he noted the physiognomic similarity of Capensis vegetation to that of the Mediterranean Region. Marloth's (1906) map and classification were only a slight improvement on Bolus' revised version, but in his beautifully illustrated work (Marloth 1908) he produced the first full-scale ac-

Fig. 3. A semi-open (about 20%) cover of *Protea laurifolia*, c. 2 m tall, in predominantly restioid vegetation on rocky terraces in the northern Cedarberg. Near Heuningvlei, c. 1000 m.

count of Cape vegetation, describing subdivisions of fynbos based on physiognomy of prominent species in selected localities.

Bews (1916) described the successional aspects of Cape vegetation and enumerated, albeit briefly, the factors involved. Bews appears to have been the first botanical author to use the vernacular name 'Fijnbosch' for the Cape sclerophyll vegetation, defining it as 'any sort of small woodland growth which does not include timber trees' and stressing its lack of single-species dominance. In his subsequent work, Bews (1925) employed the term 'southwestern region' and called its vegetation the 'mountain and southwestern vegetation'.

Pole Evans (1920) called this vegetation the 'coast veld' and in a later publication, according to Marloth (1929, p. 155) referred to the 'Cape Region'. In his 'Vegetation Map of South Africa' (Pole Evans 1936) fynbos was depicted as 'Evergreen Sclerophyllus Bush'. Adamson (1938a) continued to use Pole Evans' term 'Bush (Sclerophyll) Vegetation' for fynbos, with which, like some previous writers, he included 'Rhenosterveld'. Hutchinson (1946), in a short chapter on 'The Floral Regions of South Africa', briefly described the 'Cape Region' (macchia), which he considered to be 'of the so-called bushwood type'.

Finally, Acocks (1953), in his detailed map of the 70 'Veld Types of South Africa', depicted fynbos as both 'Macchia' and 'False Macchia', the latter term indicating that, in his view, much of what is today fynbos has been derived from an unpalatable, wiry or 'sour' grassveld on the one hand and a 'transitional forest climax' on the other. Acocks specifically excluded 'Rhenosterbosveld' from the fynbos types. While noting the absence of grassveld in true Macchia, he stated that 'there are a few indications that more than one kind of grassveld is possible' and considered it probable that 'the Restionaceae have replaced a lot of grass, especially at higher altitudes'. In contrast with the detail in the rest of his work,

Acocks' treatment of fynbos was scant; while admitting its importance, he considered that it was not possible to deal adequately with such a complex and inaccessible vegetation in the time available to him.

1.4 *Floristic characteristics and affinities*

As a phytogeographical unit, Capensis is so distinct that it is generally recognized as a floral kingdom of its own (cf. Good 1974, and Chapter 7). Characteristics that justify such treatment are its richness in species, the high degree of endemism (especially of families) and the disjunct distribution of many of its members.

The Cape Flora is noted for its richness in species, both in small areas and over its whole range. For small areas, 121 species of flowering plants have been recorded in a single 100 m^2 quadrat in a homogeneous stand (Taylor 1972a) and this tally was probably not complete. In 1950, 2622 species of vascular plants were known to occur in the Cape peninsula, an area of 471 km^2 (Adamson & Salter 1950). This density of 5.6 species per km^2 is the second highest for the 43 selected areas listed by Good (1974).

Lack of up-to-date figures makes it difficult to compare the high concentration of species in Capensis with those of other Mediterranean-type floras. Marloth (1908) put the number of species occurring 'in the Cape' at about 6000, but even this is probably an underestimate. The figure nevertheless compares with Western Australia's 4400 species (Good 1974) and with the floras of Mediterranean lands which range from 1170 species for Cyprus to 6530 for the Balkan Peninsula (Good 1947). The Californian chaparral, by contrast, has only 900 species of vascular plants (Ornduff 1974).

Many authors have quoted examples of taxa that are endemic to or concentrated in the Cape Floral Kingdom (cf. Bolus 1886, 1905, Bolus & Wolley-Dod 1904, Marloth 1929, Walter 1968, Good 1974, Aubréville 1975). The fact that no less than seven families, Bruniaceae, Geissolomataceae (monotypic), Grubbiaceae, Penaeaceae, Retziaceae, Roridulaceae and Stilbaceae, are entirely confined to Capensis is a primary justification for its separation as a distinct floral kingdom. Among taxa with a distinct concentration in this region are:

Families: Restionaceae (with the endemic genera *Hypodiscus, Thamnochortus* and *Willdenowia*); Papilionaceae (20 endemic genera); Thymelaeaceae.

Sub-family: Persoonioideae of the Proteaceae (the African section of which, excluding the genera *Protea* and *Faurea,* has 204 species within Capensis and only five outside it).

Tribe: Diosmae of the Rutaceae (12 endemic genera with over 200 species).

Genera: Erica (with nearly 600 species in Capensis) and the minor genera of the Ericaceae such as *Salaxis, Scyphogyne, Acrostemon* and *Syndesmanthus; Cliffortia* (Rosaceae, 108 species); *Muraltia* (Polygalaceae, 115 species); *Pelargonium* (Geraniaceae, over 80 species); and in the Asteraceae, besides near-endemics such as *Metalasia* and *Stoebe,* even widespread genera have high Cape concentrations e.g. *Helichrysum* (over 76 spp.) and *Senecio* (over 100 spp.).

Weimarck (1941), in a detailed study of endemism and disjunctions, found that of 282 genera with their centre of origin in Capensis, 212 are confined to that region. This figure may be the highest rate of generic endemism in the world (Good 1974). Weimarck quoted evidence to show that the four Quaternary cold periods and four warmer periods in southern Africa may have been contemporaneous with the European Ice Ages. He distinguished six main 'intervals' or disjunctions within the Cape Floral Kingdom and considered the floras of the intervening 'centres' to be remnants resulting from past climatic fluctuations between dry and wet epochs, 'during which the Cape element was an inferior competitor with the Karoo element and the forest element' (p. 104). In examining local endemism Weimarck found that the two westernmost Cape centres together contained 45.5 per cent of the total number of endemics represented within the 'Cape proper'. He postulated that the more pronounced winter rains and summer droughts of the western centres favoured the Mediterranean-type Cape flora at the expense of the Karoo and forest floras which competed and intermingled with the Cape Flora to a greater or lesser extent in other parts of its range where the rainfall was more evenly distributed.

Some species, e.g. the shrubs *Erica fairii, Serruria florida, Protea aristata, Orothamnus zeyheri, Mimetes hottentotica* and *M. argentea,* and the geophytes *Sparaxis tricolor* and *Gladiolus aureus,* are confined to a few square kilometres or a single mountain range or summit. Other species like *Leucadendron argenteum, Homoglossum merianellum, Herschelia purpurascens* and *Cytinus capensis* have disjunct and limited distributions. Marloth (1929) and Goldblatt (1972), among others, have attributed this local endemism and disjunction partly to the diverse topography of the region, with its concomitant diversity of soils and local climates, and partly to the great age of the flora which was able to survive numerous climatic changes by retreating into more favourable localities. Indeed, climatic change still continues today, and Marloth (1915) has observed species that have disappeared from certain localities during periods of prolonged drought in his own lifetime, thereby creating new disjunctions.

Local endemism may also indicate youthfulness, where speciation has occurred in relatively recently disturbed habitats. This type of endemism is found along the southern Cape coastal flats which were subjected to marine invasions and regressions during Tertiary and Pleistocene times (see Sect. 1.5).

While the distinctive combination of taxa found in Capensis is unique, the flora does contain some elements found more commonly elsewhere. Genera such as *Anemone, Rubus, Scabiosa, Geranium* and *Dianthus,* with a few species at the Cape, have their main centres of concentration in the cool temperate or Mediterranean regions of the Northern Hemisphere. *Aloe, Euphorbia* and the Asclepiadaceae, uncommon in Capensis, are prominent members of the African flora to the north. *Gladiolus* has many members in the Cape but is widespread elsewhere in Africa and beyond. *Rhus* and *Euclea,* two of the few woody shrubs common in Capensis, have many more species in the subtropical forest, scrub and savanna. *Oxalis, Babiana* and *Osteospermum* appear to be typical members of the Cape Flora but their maximum concentration is in Namaqualand. Species of *Pteronia* and *Zygophyllum,* rather poorly represented in Capensis, also increase in the dry regions of Namaqualand and South West Africa to the north.

These similarities at the generic level suggest comparatively recent migrations and intermingling of floral elements in response to fluctuating climatic changes from wet to dry. Similarities at a higher taxonomic level suggest older affinities. Adamson (1958) considered that the following pairs of taxa provided evidence of a once widespread temperate flora that became fragmented as the climate changed, and then evolved in isolation:

	Capensis	Mediterranean including North Africa
Family	Selaginaceae	Globulariaceae
Genera	*Dimorphotheca*	*Calendula*
	Lobostemon	*Echium*
	Crassula	*Sedum*
	Widdringtonia	*Tetraclinis*

The affinity of the Cape Flora with that of southwestern Australia, though obvious, is far more remote. Except for widespread genera such as *Helipterum* and *Helichrysum,* the relationships are usually at a higher taxonomic level. The Thymelaeaceae, Haemodoraceae and Droseraceae, characteristic of the Australian vegetation, are common also in the Cape. The Diosmae, a large tribe of the Rutaceae endemic to Capensis, has its counterpart in the Australian tribe Boroniae (Bolus & Wolley-Dod 1904). The Cyperaceous genus *Tetraria* of the Cape is closely paralleled by the Australian genus *Gahnia,* and the Cape *Phylica* (Rhamnaceae) by *Cryptandra* in Australia (Adamson 1958). Both subfamilies of the Proteaceae, the Persoonioideae and Grevilleoideae, are well represented in Australia while the African Proteaceae, with one exception, all belong to the Persoonioideae. The exception is *Brabeium stellatifolium,* a member of the Grevilleoideae inhabiting streamsides within the region of the Cape Flora. It has a close parallel in the Australian genus *Macadamia* (Levyns 1964). The family Ericaceae, so characteristic of the Cape Flora, is replaced in Australia by the large and closely allied family Epacridaceae which is almost confined to Australasia. There are 293 species of Restionaceae in South Africa and 84 in Australia; the genera in the two countries are usually regarded as distinct except for *Restio* with 117 species in South Africa, 27 in Australia and one in Madagascar (Walter 1968).

1.5 Migrations and origin

The floristic characteristics and affinities of Capensis suggest a long and varied history of geology and climate. On these grounds and because dominance by one or more species is a rare phenomenon in mature fynbos, the flora has generally been regarded as an ancient one (cf. Marloth 1915, Bews 1925, Weimarck 1941, Levyns 1952, Adamson 1958, Dyer 1966).

Yet despite general agreement about its great age, there has been much controversy about the origin of the Cape Flora. One school has postulated an origin in the northern hemisphere, another in the southern, while a third contends that it originated somewhere in central Africa. The early proponents of a northern origin, among them Thiselton-Dyer (1909) and Guppy (in Bews 1925), considered that plants radiated southwards from the Arctic Circle at the onset of each northern Ice Age and retreated back there during each warm period, resulting in a mixing of floras. The richness of the Cape Flora is thus explained as an accumulation of species at the limit of one or more such migrations from the north (see also Chapter 6).

Proponents of the opposite view (cf. Bews 1925, Boughey 1957, Adamson 1958, Beard 1959, Plumstead 1961, 1969, 1974) maintained that the bonds of affinity between the present flora of Australia, the Cape and South America cannot be explained by the theory of a northern origin and considered that they represent fragments of an ancient, once continuous floral area in the south. They suggested that the present temperate floras of the southern continents, including the Cape Flora, arose from the distinctive pre-angiospermous fossil flora common to South Africa, Australia, South America, India and Antarctica, and explained the lack of close affinity between the present floras by the early break-up of Gondwanaland and the subsequent evolution of the floras in isolation (compare Chapter 5).

Today there are some Cape taxa with a predominantly southern distribution that have few or no records of their existence, either past or present, in the north. The case of Restionaceae and paired genera like *Tetraria–Gahnia* and *Phylica–Cryptandra* have already been quoted. Others that may be considered as partly of the Cape floral element are *Widdringtonia, Metrosideros, Brabeium, Acaena* and *Gunnera*; and in the forest element of the Cape there are *Podocarpus, Cunonia, Platylophus* and *Curtisia* (Levyns 1962). These may well have been derived directly from an austral source.

Levyns (1938, 1952, 1958, 1964) and Van Vuuren (1973) pointed out that very many members of the Cape Flora show another type of distribution pattern. Though they are concentrated in the Cape Floral Kingdom, there are clear traces of them scattered throughout Africa, mainly on mountains, as far north as Ethiopia. Weimarck (1941) listed 225 species of plants belonging to Cape genera, that are found on tropical African mountains. Patches of Cape-like vegetation, often separated from one another by hundreds of kilometers, occur within the present tropical African flora. Identical or closely related species are frequently found in these island refuges, especially in the heath zone on East African mountains (cf. Robyns 1959, and Chapter 12). Proceeding southwards, these islands become more frequent until, south of the Swartberg, all the scattered mountains of the Little Karoo have cappings of Cape plants while the flora of the lowlands is entirely different (cf. Levyns 1950).

From evidence of morphology, anatomy and cytology, Levyns showed that the more primitive members of Cape plants in such groups as the Ericoideae of Ericaceae, the Persoonioidae of Proteaceae and the larger genera like *Stoebe, Cliffortia, Aristea* and *Muraltia* are to be found on mountain outliers within the tropics, whereas in the southwestern Cape many of the species are advanced and occupy restricted geographical ranges. Rourke (1972) and Goldblatt (1972) found the same type of distribution in *Leucospermum* and the family Iridaceae, respectively. The tribe Diosmae of the Rutaceae consists of shrubs almost restricted to the Cape Flora, while *Calodendrum*, apparently a primitive member of the tribe, is a forest tree extending from Kenya in the north to George in the south. The shrubby family Penaeaceae, endemic to the southwestern Cape, bears close resemblance in its floral anatomy and morphology to the more primitive woody family Oliniaceae of central Africa, only one member of which, *Olinia ventosa*, reaches the forest remnants of the Cape Peninsula. These distribution patterns suggest that a flora of the Cape type was once widespread in central Africa. Such a conclusion is supported by the belief of palaeobotanists that a drought-loving flora occupied much of Africa soon after the flowering plants came into being (Levyns 1963). According to King (1962), an even land surface occurred over a large continuous area of southern Africa in the Early Tertiary. This area could have been uniformly clothed with a single flora which would have become fragmented during the Pliocene when folding of the land occurred.

Later, when moister conditions prevailed in central Africa, the members of this ancient flora yielded to competitors better adapted to a high rainfall, and concentrated in the south, where the temperate climate remained more favourable to them, leaving traces on the northern mountains. This led to a burst of evolutionary activity in the southwestern Cape and new species, new genera and even new families arose.

The two opposing theories, of a southern and a central African origin for the Cape Flora, are both supported by valid evidence. A consideration of the time scales involved may show how these two apparently discrepant views could both contribute to a holistic interpretation of the facts of present distribution. The following reconstruction of events explains the obvious but ancient affinities of the Cape Flora with those of other austral lands; its closer affinities with the isolated mountain floras of central Africa; the island-type distribution of the Cape Flora within Capensis itself; the apparent southward movement of the flora from a previous station in central Africa to its modern centre in the southwestern Cape; and the high rate of speciation, endemism and disjunction in its present position (compare Chapter 5).

The distinction between the fossil floras of the northern and southern hemispheres, apparent even in the first land plants of Devonian times, was accentuated by the marked climatic differences between north and south in the Carboniferous period. The Glossopteridophyta that flourished in the southern hemisphere at that time differed completely from the flora of the north and gradually developed morphological structures resembling those of primitive Angiosperms (Plumstead 1969, 1974). Before the Cretaceous period commenced, the southern continent of Gondwanaland slowly moved northward and fragmented, Africa being the first land mass to be severed from the rest. The subsequent elevation and erosion of the African continent caused a break in the fossil record, but today there exist many disjunct angiosperm families endemic to the southern continents; these families must have evolved to the family level before the break-up of Gondwanaland and may well have been derived from the Glossopteridophyta.

This old flora had been adapted to and was evolving in temperate conditions. As Africa moved north, the increasingly warmer and wetter climate would have eventually become inimical to the existence of this flora which was consequently overwhelmed in the north by tropical competitors but survived in the south where temperate conditions still prevailed. Of interest here is Raven's (1973: p. 211 and 214, and Chapter 5) view that the Mediterranean-type climate with its principal growing season in winter is a unique type, transient on a geological scale, that evolved only during the Pleistocene – perhaps less than a million years ago – from a summer-rain climate to which it will return when the polar ice-caps melt. An interesting feature in support of this view is that many Cape plants enter into a period of rapid vegetative growth towards the end of summer at a time when water supplies in the Mediterranean-type climate of the southwestern Cape are at their lowest. This strangely ill-adapted growth rhythm suggests that the ancestors of these plants 'evolved in some place having a summer rainfall. The same phenomenon has been recorded for south Australia, where a similar change in climate is postulated to account for the same, apparently abnormal, features of growth' (Levyns 1964). With successive climatic fluctuations from wet to dry during the Pleistocene (Van Zinderen Bakker 1967), the flora in the southwestern Cape would have repeatedly extended and contracted its range, leaving relics on the cooler, moister mountains with each retreat. Under these conditions speciation was accelerated by isolation and by the creation of new niches with each climatic shift.

Each advance and retreat of the coastline would have generated the same processes. Much of the coastal strip in the Divisions of the Cape, Caledon and Bredasdorp was subjected to marine invasions and regressions during the Tertiary and Pleistocene (du Toit 1966, Hamilton & Cooke 1965), and may consequently be regarded as a disturbed habitat where rapid and relatively recent evolution took place. Levyns (1954) has given clear evidence that a high proportion of *Muraltia* species occupying this coastal strip are local, youthful endemics that have not had time to spread beyond their present territories. Rourke (1972) quoted examples of several pairs of ecological vicariads of *Leucospermum*, each very specific in its edaphic requirements, growing in adjacent localities on different geological formations of the Bredasdorp coastal flats.

Some present disjunctions can probably also be explained by these geological disturbances. For instance, *Muraltia mitior* (Levyns 1954) and *Leucospermum hypophyllocarpodendron* subsp. *hypophyllocarpodendron* (Rourke 1972) occur on recent sands of the Cape Flats and reappear only on the similar substrate of the Bredasdorp Flats about 160 km to the east. They are separated by a predominantly rocky coast that provides few suitable habitats for these species. When the sea retreated during Pleistocene times, much of the coastal shelf was probably land which would have provided a suitable habitat and pathway for migration between the now disjunct localities.

The Gondwanaland theory accords with the views of Beard (1959) who maintained, largely from evidence of the distribution and morphology of *Protea*, that while the Proteaceous flora of the southwestern Cape appeared to be of southern origin, it was more likely to have been derived from a tropical montane flora that was thought to have originated in Gondwanaland. In his study of *Leucospermum*, Rourke (1972) found a considerable amount of evidence to support Beard's theory.

Though the migrations between central Africa and Capensis would have taken place mainly along the highlands and mountains in the eastern part of the subcontinent (see e.g. Bews 1917, Story 1952, Roberts 1961, Killick 1963, Edwards 1967 for descriptions of fynbos communities along this route), there is a well-recognized enclave of the Cape floral element on the Huila Plateau in Angola which may have been connected by a western pathway. About half way between the Huila Plateau and the nearest outliers of the Cape flora proper on the Kamiesberg in Namaqualand, and about 800 km from each, lies the Auas Mountains (c. 2460 m) near Windhoek in South West Africa. Both Rennie (1935) who found nine supposed Cape elements on the summit of the Auas Mountains, and Levyns (1952) thought that a western route via Auas was possible. Acocks (1953), however, pointed out that a western route would have been much less favourable than the eastern for as long as present climatic patterns and ocean currents existed. Nordenstam (1974) found no typical representatives of the Cape flora on the Brandberg, the highest mountain in South West Africa, and he considered that, in the light of modern distributional evidence, only *Stoebe plumosa* of Rennie's collections belonged to the true Cape flora, the others representing a

Karoo–Cape transitional element. Nordenstam felt that the Cape element in Angola could best be explained by a connection between Angola and the eastern pathway (cf. also Chapter 11, Fig. 4).

There is nevertheless firm cytological evidence for a migration, though perhaps a minor one, from Namaqualand to the southwestern Cape. Both *Lobostemon glaucophyllus* (Levyns 1952) and *Ferraria antherosa* (M. P. de Vos pers. comm.) have only their primitive diploid forms towards and in Namaqualand and only tetraploid forms in the southwestern Cape. The implied Namaqualand origin and subsequent southward migration of these plants may only serve to highlight the fact that the Cape flora contains, especially on its drier outskirts, an admixture of elements from adjacent floras like the Karoo (see Section 2.1.2).

2. Vegetation and ecology

2.1 Fynbos

Despite its uniqueness and scientific interest, fynbos has not yet been classified in structural or floristic units, partly because the main botanical effort has hitherto been directed towards solving the manifold taxonomic problems and partly because of the complexity of the vegetation. Also, since fynbos does not possess the economic potential of grassland or forest, the practical need for vegetation studies has not hitherto been felt.

Besides Marloth's (1908) monumental work, the only relatively complete account of fynbos is that of Adamson (1938a), whose broad categories have been retained here, with some changes in nomenclature. Rapidly increasing human pressures on the Cape environment have now underlined the urgent need for a broad reconnaissance of Capensis vegetation, especially as a basis for land use planning and management in both the lowland and mountain areas. Until such a survey is completed it would be unwise to attempt anything more than a generalized description based on Adamson's work and Acocks' (1953) three 'macchia' types, supplemented by the writer's experience and by published accounts of the vegetation of regional and local areas.

In Section 1, the general physiognomic features of fynbos were summarized and the high species diversity stressed. Since many species have narrowly circumscribed habitat preferences, the composition of the vegetation changes with every nuance of habitat variation, and most plant communities are not clearly defined. The prevalence of fire, and the consequent range of regeneration stages even within a comparatively uniform habitat, further complicates community recognition.

In the broadest terms, there are two major subdivisions, Mountain and Coastal Fynbos, both having typical fynbos families and physiognomy but differing in species composition. Adamson did not make this basic distinction but divided his 'Bush (Sclerophyll) Vegetation' into five types, the first of which, 'Sclerophyll Bush ... the type of the plains and lower slopes', apparently included part of Coastal as well as part of Mountain Fynbos. Acocks (1953) separated 'Coastal Macchia' from 'Macchia' and 'False Macchia'. The two last-mentioned comprise Mountain Fynbos. There are indications of two major categories of Mountain Fynbos ecosystems roughly east and west of the Hottentots Holland Mountain

Divide and its northward extension (Taylor 1969, Kruger 1974) but this boundary does not coincide with that of Acocks' 'Macchia' and 'False Macchia'.

2.1.1 Mountain Fynbos

Sometimes, as in the Kogelberg and parts of the Cape Peninsula, this vegetation approaches very close to the coast, but only where the mountains fall abruptly to the sea. The major ranges are separated from the sea by a coastal plain or platform 15–100 km wide, narrowest along the east coast at George and Storms River.

Mountain Fynbos, the largest and most important unit of Capensis vegetation, occurs in two blocks, a western and an eastern. The western block (Acocks' 'Macchia') extends from the region of Cape Agulhas northwards to beyond the Cedarberg, a distance of approximately 400 km. It covers the mountain ranges with a roughly south–north trend that receive at least half their rainfall during the six winter months. In the eastern block (Acocks' 'False Macchia') the west–east trending ranges run roughly parallel to the Cape south coast for nearly 600 km, and the proportion of summer rain gradually increases eastwards (see Chapter 2). The inland boundary of the region, where the mountains form an effectual check to the rain-bearing winds, is fairly clear-cut, but its northern and eastern extremities pass over to other regions by more gradual transitions. Grass and forest species gradually increase in number and importance as one travels east into the region of more constant rainfall but there are overall similarities with the western block, in physiognomy, structure and composition that reflect general habitat influences common to the whole mountain region, overriding the climatic gradient from west to east. These influences – factors such as soil structure and depth, soil moisture, altitude, aspect and slope – will determine the kind of climax vegetation that develops. On the foothills and lower slopes the climax may be composed principally of tall proteoid shrubs; on the upper slopes a shorter ericoid form is usually prominent, while restioids often dominate exposed ridges, peaks and specialized habitats.

As one ascends a mountain slope there are few distinct boundaries between these 'zones'; changes in structure, plant form and floristic composition, except where determined by fire, are nearly always gradual. Nevertheless, at the present state of our knowledge it seems possible to do no more than subdivide Mountain Fynbos broadly in terms of these rather nebulous zones.

2.1.1.1 *The proteoid zone.* Proteoid shrubs are usually about 1.5–2.5 m high but occasionally reach the stature of a small tree. Their shape ranges from rounded to laxly or narrowly erect. They have large (5–15 cm), flat, leathery, isobilateral leaves, narrowly to broadly ovate or oblong, with a hairy, waxy or glaucous surface. The proteoid form, though conspicuous where it occurs, is not as widespread or constant a feature as the ericoid.

The proteoid zone, where this form predominates, comprises the plains, lower slopes and plateaus. On poorly insolated slopes in the west with a rainfall of about 500–1000 mm per year this zone extends up to an altitude of 900–1000 m. In the east, on cool slopes with a higher rainfall, the upper limit is lower; on hot northerly slopes it is higher. The soil is either coarse-grained, shallow sand, white or grey, derived from Table Mountain Sandstone, or a deeper, more fertile, brown or reddish sandy loam where the substratum is granite or shale. In the best developed parts, fynbos of the proteoid zone has a more complex structure than any other type. In such places there are commonly three layers, an upper, usually discontinuous layer of proteoid shrubs, a dense middle layer of ericoids, lower proteoids with smaller leaves and some larger tufted restioids such as *Tetraria bromoides*, and a ground layer of small shrubs, herbs, many geophytes and low restioids, often matted and wiry. The upper layer is about three metres tall in the western parts but can be as much as six metres in favourable localities, especially further east. *Protea neriifolia* is the commonest plant in this layer, replaced in places in the southwest by *P. lepidocarpodendron*, in the north by *P. laurifolia* and in the east by many other species such as *P. mundii, P. eximia, P. lacticolor, P. longiflora* and *Leucadendron eucalyptifolium*. In the driest parts, as in the northern and inland ranges, the whole community may be quite open in character, whereas in the wettest (south and east) the upper layer is very dense and the other layers poorly defined.

Fig. 4. *Mimetes fimbriifolius* (the dark scattered bushes) forming the upper layer, 2–3 m tall, in a savanna-like community with a closed lower layer of restioids, graminoids and ericoids. Grootkop above Simonstown, c. 300 m.

A widespread community of the three-layered type is Waboomveld (Taylor 1963) commonly found on bouldery fans or screes where *Protea arborea* (Waboom) with its distinctive broad, grey, glaucous leaves and whitish stem forms an open, savanna-like upper layer of 2–3 m (see Figs. 6, 13 and Table 1). Below this is a rather straggly middle layer of coarse restioids (e.g. *Restio gaudichaudianus*) and caespitose shrubs with grasses, especially *Themeda triandra*, often forming an important component (Werger et al. 1972). *Protea arborea*, unlike most *Proteas*, is protected from fire damage by its thick, corky bark. *Leucospermum conocarpodendron* and *Mimetes fimbriifolius*, both similar to *P. arborea* in form and fire resistance, replace *P. arborea* as the upper layer of a similar savanna-like community on mountain slopes near the coast (Fig. 4).

Table 1. Fynbos communities, from Werger, Kruger & Taylor 1972.

Communities	A	B	C	D
number of relevés	11	7	10	3
Diospyros glabra	5	2	1	1
Rhus angustifolia	4	.	.	1
Cassytha ciliolata	3	1	1	.
Pteridium aquilinum	3	1	.	.
Podalyria myrtillifolia	3	1	.	.
Euphorbia genistoides	3	.	.	.
Helichrysum zeyheri	3	.	.	.
Rhus tomentosa	2	.	.	.
Psoralea rotundifolia	.	5	.	.
Helichrysum teretifolium	1	5	.	.
Corymbium scabrum	1	4	1	.
Brunia nodiflora	.	4	1	.
Danthonia lanata	.	4	1	.
Osteospermum tomentosum	.	4	1	.
Tetraria burmannii	.	4	.	.
Psoralea fruticans	1	3	.	.
Adenandra serpyllacea	1	3	.	.
Elytropappus glandulosus	.	3	1	.
Hypodiscus willdenowia	.	2	.	.
Restio sieberi	.	.	5	3
Anthospermum ciliare	1	1	5	1
Thamnochortus gracilis	1	.	4	1
Staberoha cernua	1	1	3	.
Prismatocarpus diffusus	1	1	3	1
Tetraria capillacea	1	.	4	.
Tetraria fasciata	.	.	3	.
Hypodiscus aristatus	.	.	4	.
Willdenowia sulcata	.	.	3	1
Blaeria dumosa	.	.	3	1
Clutia polygonoides	.	.	3	1
Pentameris macrocalycina	.	.	2	1
Widdringtonia cupressoides	.	1	2	.
Erica coccinea	.	.	2	.
Psoralea aculeata	.	.	2	.
Elegia racemosa	.	.	2	.
Cliffortia atrata	.	.	2	.
Coleonema juniperinum	.	1	2	.
Sympieza articulata	.	.	2	.
Lobelia coronopifolia	.	.	2	.
Nebelia paleacea	.	.	2	.
Scyphogyne muscosa	.	.	1	.
Cliffortia polygonifolia	.	1	2	3
Euryops abrotanifolius	1	.	.	2
Restio perplexus	.	.	.	3
Asparagus compactus	.	.	.	3
Venidium semipapposum	.	.	.	2
Tetraria cuspidata	4	5	5	2
Cymbopogon marginatus	5	5	3	3
Aristea thyrsiflora	5	5	2	3
Erica hispidula	3	5	5	.
Clutia alaternoides	5	5	3	3
Stoebe plumosa	5	3	2	3

Table 1. (contd.)

Communities	A	B	C	D
number of relevés	11	7	10	3
Danthonia stricta	5	4	2	2
Restio triticeus	2	5	3	3
Pentaschistus colorata	1	4	5	2
Bobartia indica	3	5	3	1
Leucadendron adscendens	4	5	4	.
Tetraria bromoides	3	5	4	.
Cliffortia ruscifolia	3	4	5	1
Erica globulifera	5	4	2	.
Aristea capitata	2	4	3	1
Eremia totta	3	3	3	.
Erica nudiflora	2	5	3	.
Anthospermum aethiopicum	5	5	.	.
Montinia caryophyllacea	5	5	1	.
Restio gaudichaudianus	5	3	1	1
Asparagus thunbergianus	5	4	.	.
Diosma hirsuta	4	5	1	.
Protea neriifolia	4	3	1	.
Ficinia filiformis	3	5	1	.
Lichtensteinia lacera	3	5	.	1
Rhus rosmarinifolia	5	3	.	.
Protea repens	2	3	2	.
Berkheya armata	3	4	.	.
Pentaschistis curvifolia	2	4	.	1
Protea arborea	5	1	2	2
Watsonia pyramidata	4	1	.	3
Themeda triandra	3	1	.	3
Psoralea obliqua	4	1	.	2
Ursinia filiformis	3	1	.	2
Restio filiformis	.	5	4	.
Tetraria ustulata	1	4	3	.
Metalasia muricata	.	3	5	.
Pentaschistis steudelii	.	4	3	1
Penaea mucronata	1	3	3	1
Ursinia crithmoides	.	4	2	2
Corymbium glabrum	1	3	3	1
Protea acaulis	3	4	2	1
Erica plukeneti	2	4	3	2
Haplocarpha lanata	2	4	2	2
Maytenus oleoides	4	3	1	1
Cliffortia cuneata	3	3	2	1
Centella glabrata	2	3	2	1
Corymbium villosum	3	2	1	1
Elegia juncea	1	3	2	2
Osmites hirsuta	1	2	2	2
Agathelpis dubia	1	2	2	2
Stoebe spiralis	1	2	2	1
Scabiosa columbaria	2	3	.	1
Caesia eckloniana	3	3	.	.
Gnidia inconspicua	2	3	.	.
Erica articularis	1	3	1	.
Phylica spicata	2	3	.	.
Hypodiscus albo-aristatus	.	3	2	1
Leucadendron spissifolium	1	2	2	.
Ficinia bracteata	2	1	1	1
Ficinia deusta	1	3	1	.

Table 1. (contd.)

Communities number of relevés	A 11	B 7	C 10	D 3
Ehrharta ramosa	1	.	2	3
Thamnochortus fruticosus	2	.	1	.
Schizaea pectinata	1	2	2	.
Gerbera crocea	2	.	1	.
Agathosma juniperifolia	1	3	1	.
Restio cuspidatus	1	2	1	.
Ehrharta bulbosa	2	1	1	.
Cannamois virgata	2	.	1	1
Stoebe cinerea	1	3	.	.
Erica calycina	.	2	1	1
Helichrysum odoratissimum	1	3	.	1
Muraltia heisteria	1	2	1	.
Pellaea viridis	1	.	.	2
Muraltia alopecuroides	.	.	2	.
Helichrysum cymosum	2	1	.	.
Senecio pinifolius	.	2	.	1
Linum thunbergii	1	2	.	.
Salvia africana	2	.	.	.

A: Protea arborea community of lower slopes with bouldery soils.
B: Brunia nodiflora community of lower, relatively mesic slopes.
C: Thamnochortus gracilis community of higher altitudes.
D: Restio perplexus community of steep slopes at high altitudes.

Fig. 5. Dense, mainly ericoid community, up to 2 m tall, on very steep south-facing slopes of the coastal mountains. The taller, rounded bushes towards the left foreground are *Berzelia dregeana*, an endemic species. Platberg, Kogelberg, c. 900 m.

Seed-regenerating tall proteoids are not such a constant feature of the proteoid zone because they are killed by fire. Frequent burning therefore reduces this vegetation to a two-layered or even a single-layered form. Indeed, on account of the relatively high incidence of fire on the foothills and lower slopes, such reduced communities are considerably more common than the mature form.

Variations of this vegetation include communities where the upper layer may be sparse and scattered, as in the dry *Protea laurifolia* community of the north (Fig. 3) or where it is absent for reasons other than fire. On skeletal soils on some northerly aspects, for instance, the mature community is two-layered, the upper dominated by restioids; and on cool moist, steep south-facing slopes of the coastal mountains there may be no layering in a very dense, mainly ericoid community up to 2 m in height (Fig. 5 and Table 1). On the warmer, drier and gentler lower slopes of the southwestern coastal mountains, where the soil is a coarse, quartzy, white sand, there is an open ericoid–restioid mixture only about 50 cm high, in which conspicuous emergent clumps of the tough-edged, glossy-leaved *Tetraria thermalis* are regularly spaced at intervals of two or three metres (Fig. 11). Taylor (1963) named this distinctive and unusual community 'Bergpalmietveld', after the common name of the *Tetraria* (see also Taylor 1969a, Boucher 1972, Kruger 1974).

Small trees are sometimes found in the proteoid zone. *Maytenus oleoides* and *Heeria argentea*, for example, sparsely dot the rocky hillsides of the western block from the Hottentots Holland Mountains to the Cedarberg. Where surface rock is extensive, as on screes, these species, joined by hardier forest trees like *Podocarpus elongatus, Olea africana, Maytenus acuminata* and *Olinia ventosa*, may form closed scrub or forest averaging 7–8 m in height (cf. Werger et al. 1972).

Widdringtonia cupressoides, an erect, cypress-like shrub common in fynbos of the lower slopes, coppices from the base after fire. It generally occurs in the middle layer of medium-aged proteoid vegetation, but where this has been protected from fire for about 20 years, *W. cupressoides* grows into and then overtops the upper layer, finally appearing in clumps as an emergent cupressoid small tree 4–5 m high (Fig. 6). Round the edges of forest in sheltered, fire-free kloofs, as at Boosmansbos in the Langeberg, this species can form pure, closed stands up to 7 or 8 m high.

Above the proteoid zone the shrubs are smaller and the structure simpler. The larger proteoids become scarce and confined to sheltered places; at higher altitudes they disappear. At this stage the ericoid-restioid zone is reached.

2.1.1.2 *The ericoid–restioid zone.* Ericoid shrubs with their narrow, often rolled, stiff leaves are commonly rounded in shape due to their widely divaricate branching. In this zone they vary in average height from about 20 cm to 1.5 m. Many of them contain aromatic oils or resins. The vegetation is seldom bright green except after a fire. Usually it is yellowish-green or dun-coloured with seasonal splashes of yellow, white, pink and mauve from *Ericas* in flower.

A similar small-leaved shrub form common in the family Penaeaceae has flat, hard, leathery leaves usually 1–2 cm long and up to 1 cm wide. Typically these leaves are broadly ovate in shape and often imbricate, acuminate and cordate; in other species they are narrowly ovate and truncate, in yet others broadly lanceolate or elliptic or rhombic with sharply acuminate apex and cuneate base. Variations of these 'penaeoid' shapes are quite frequent among the ericoid-leaved

Fig. 6. The vegetation beyond the diagonal line across the left foreground is tall proteoid closed fynbos protected from fire for 22 years, an unusually long period. The proteoids are mainly *Protea neriifolia*, 2–3 m tall, and the clumps of dark spire-like small trees emerging from this layer are *Widdringtonia cupressoides*, 4–5 m tall. The light grey vegetation in the right-hand middle distance is *Protea arborea* (Waboomveld), about 2–3 m tall, growing on a boulder fan at the base of a side gulley or 'kloof'. The lower restioid–ericoid vegetation in the left foreground, regularly burnt at 4-year intervals, is devoid of tall proteoids. Jonkershoek State Forest near Stellenbosch, c. 400–600 m.

genera and families such as Rhamnaceae (some *Phylica* spp. like *P. buxifolia* and *P. oleaefolia*), Rutaceae (some *Agathosma* spp.), Asteraceae, Fabaceae and others.

The ericoid–restioid zone is found on the upper slopes, ridges, plateaus and summits of the mountains. The soil is usually a shallow, coarse, acid, porous sand derived from the folded Table Mountain Sandstone strata that have built the Cape mountains; occasionally, inland, it is from the Witteberg Series. In the Table Mountain Series are one or two shalebands – narrow shelves forming level or slightly dipping collars round the upper slopes – which are prominent features of the mountains in the north and west. The soil of the shale-bands is less coarse but still shallow, and drainage is often impeded in the lower part of the band.

The rainfall increases and temperatures decrease with increasing altitude. No figures for summits are available but from extrapolation of data from the Jonkershoek Forest Influences Research Station it is estimated that the high peaks above Jonkershoek and Wemmershoek in the southwest probably receive over 5000 mm of precipitation per year! While most of this falls during winter in the western parts, it has been shown that clouds formed by the summer southeast winds deposit an appreciable amount of moisture on the vegetation, not only on coastal mountains but even on some far inland ranges such as the Swartberg (Marloth 1904, 1907b). Snow falls in winter but seldom persists for longer than a few days except on the southern slopes of the highest peaks where it may lie occasionally until December (Marloth 1902). Winds can be extremely strong and may persist, in summer especially, for a week or ten days.

This rigorous climate has stamped its mark on the vegetation. Bushes like *Cliffortia ruscifolia* that are erect in the valleys assume a prostrate form on the upper slopes. Tall shrubs are either absent or scattered; they seldom make a discernible layer. They are usually up to 1.50 m high, and in form ericoid or penaeoid rather than proteoid (e.g. *Aspalathus, Psoralea, Cyclopia, Berzelia* and some *Leucadendrons*). Grass species, conspicuous mainly after fires (Fig. 7), may be quite numerous but only in the east do they contribute significantly to the cover. Geophytes are less common than on the lower slopes. Annuals are very rare.

Trees are absent except in two limited localities. In the Cedarberg, *Widdringtonia cedarbergensis*, gnarled, stunted and heavily branched, persists in the refuge of rocky scarps and screes (Fig. 19). The species may in the past have formed a closed forest on the upper slopes and plateaus but despite the policy of protection introduced early this century, mere relics remain, thanks to previous felling and burning (vide Hubbard 1937). A taller species, *W. schwarzii*, precariously maintains its small range in the Kouga mountains to the east.

The general character of the vegetation is thus similar to the simpler communities of lower levels, with no more than two layers, in which ericoids and restioids are prominent, either separately or in mixture (see Table 1).

Fig. 7. The summit ridge of an inland eastern range. Grasses are conspicuous on this veld burnt 1–2 yrs. previously. Swartberg, c. 1600 m.

Sometimes there may be only one layer. On steep, cool slopes of the southwestern mountains, for example, a dense shrub community, only 25–50 cm tall, is dominated largely by Ericaceae, particularly *Erica hispidula* (Kruger 1974). Similar communities, with different group-dominants, are common throughout the southern coastal ranges (Fig. 5). In contrast with these are the open restioid communities of drier inland ranges containing a mixture of coarse, rigid, tufted species of *Cannamois*, *Hypodiscus* and many others, separated by bare soil (Fig. 1).

Also in the drier northern mountains such as the Skurweberg and Cedarberg, between rocky ridges and boulder-slopes at the lower limit of this zone, there are level sandy corridors with a low restioid vegetation scarcely 30 cm tall. Many communities are represented here, each determined by a narrow range of soil depth and moisture and, unlike fynbos in general, each dominated by one or two species (Figs. 8 and 9). Similar communities are found on higher plateaus in these mountains, as on Wolfberg at an altitude of c. 1550 m where, on deeper, well-drained sand, a sparse ericoid upper layer overtops the restioids in places (Taylor 1976).

Fig. 8. Short restioid and graminoid communities about 30 cm tall on sandy flats at the base of high peaks in the northern Cedarberg. The scattered tall proteoid shrubs are *Protea arborea* (Waboom) up to 3 m tall. Boontjieskloof, c. 1150 m.

The table-top summits of the northern Cedarberg also bear a vegetation similar to the foregoing except that the communities, occurring on very shallow humic soil over bedrock, are strongly differentiated structurally and each is clearly dominated by a single species. Along the edge of the bare rock surface is a narrow band of *Restio curviramis* forming dense, low, rounded, cushiony tufts 20 cm high, in the openings between which grow tiny annuals and succulents. Further back from the edge of the rock is a broader belt of *Cannamois* cf. *nitida* growing in dense erect tufts about 1 m high (Fig. 10). The same zonation of restioid communities, though less clearly differentiated and composed of *Hypolaena crinalis* and *Chondropetalum mucronatum*, can be seen on the flat summit of Table Mountain above Cape Town.

Most Cape mountain summits are not flat but conical, sharply pointed or ridge-like. On the highest of these mountains, a perceptible change in growth form may

Fig. 9. Sandy corridors with low restioid vegetation scarcely 30 cm tall between the rock spires of the Skurweberg. Koue Bokkeveld, c. 1200 m.

occur above an altitude of about 2700 m. Only about 20 summits exceed this height, most of them in the southwestern and inland ranges. At the turn of the century Marloth climbed five of these high peaks and listed 72 plants that he found growing on them. The list shows that, although the flora still consists of the typical

Fig. 10. A table-top summit of the northern Cedarberg showing distinct zonation of restioid communities on bedrock (see text). Skerpioensberg, 1600 m.

Cape element, the growth forms of many species approximate those of some high Alpine plants. The cushion plant, *Psammotropha frigida*, for instance, is very like the Alpine *Saxifraga bryoides*; the decumbent shrublets *Erica tumida* and *Acmadenia teretifolia* spread over rocks and boulders somewhat after the style of *Juniperus nana*, and *Aspalathus pedicellata* possesses horizontal stems lying flat on the ground, resembling the typical 'Spalierstrauch' growth form of *Salix herbacea*. These features show the rudiments of an alpine-type flora, but since few Cape peaks reach the altitude of permanent winter snow, 'the conditions for an entirely Alpine flora, well distinguished from that of the valleys, are not present' (Marloth 1902).

2.1.1.3 *Hygrophilous fynbos*. The communities described above all occur on substrata that become dry for at least part of the year. In Mountain Fynbos there are also communities of permanently wet or moist sites – river and stream banks, seepage and drainage lines, marshes and swamps, pans and moist flats. The range of habitats is enormous and the communities are legion, being determined not only by their position in the soil moisture gradient but also by the macroclimate of their particular locality as well as by phytogeographical factors. Nevertheless, the permanence of plentiful soil moisture is a common unifying feature, and it is true in the Cape, as elsewhere, that many hygrophilous plants are widely distributed, even in this region of highly diverse physiography. On the other hand, it is also true that most waterside plants of Capensis are strictly confined to Capensis: even *Prionium serratum*, probably the most typical and widespread plant of Capensis streams and rivers, is scarcely ever encountered outside the region of the Cape flora. (Among ubiquist waterside plants that do occur outside Capensis are *Cyperus textilis, Juncus lomatophyllus, Laurembergia repens, Cliffortia strobilifera* and *C. linearifolia*.)

The vegetation of these habitats will collectively be referred to as Hygrophilous Fynbos, the term 'hygrophilous' seeming more appropriate to the wide range of variation than 'wet' (cf. Phillips 1931: 'Hygrophilous Macchia', Adamson 1938a: 'Wet Sclerophyll Bush'). Were it not for fire, the nature of these habitats would permit the development, by succession, of a vegetation other than fynbos, since the foremost feature of Capensis ecosystems – a dry summer soil – is here wanting. Indeed, on sites where fire cannot reach, such succession has taken place, as can be seen in the stream bank forest of protected kloofs (Fig. 11). This forest is an extension (perhaps a relic) of the temperate evergreen forest that is beyond the scope of this chapter (see Chapter 11); only the earlier, fynbos stages of succession will be dealt with here.

An important stage in this succession is the *Brabeium stellatifolium* community found along the banks of larger streams with bouldery or sandy beds in broad valleys of the western and northern mountains at altitudes under about 600 m. This dense scrub up to 5 m high fringes the lower, less steep parts of streams such as the Eerste River at Jonkershoek near Stellenbosch (Werger et al. 1972), the Elands River in Du Toit's Kloof (Fig. 12) and similar streams northwards to the Jan Dissels River in the Cedarberg. Over this wide range its composition is somewhat broader than the community described for Jonkershoek by Werger et al. (1972). In the upper layer the most constant character species are *Brabeium stellatifolium, Metrosideros angustifolia* and *Salix capensis*; less constant are *Brachylaena neriifolia, Psoralea pinnata, Rhus angustifolia* and *Erica caffra*.

Fig. 11. A stream bank forest developing in a kloof protected from fire in the Langeberg. The dark scattered shrubs on the steep foreground slope are *Tetraria thermalis*. Eleven o'clock Peak, c. 1000 m.

Podalyria calyptrata, *Freylinia lanceolata* and *Halleria elliptica* are common in the southwest. There is an open lower layer of shrubs (e.g. *Diospyros glabra*, *Myrica serrata* and sometimes *Rhus tomentosa* and *Myrsine africana*) up to 1.5 m high, in which restioids like *Leptocarpus paniculatus*, *Restio subverticillatus* and *Elegia capensis*, together with grasses such as *Pennisetum macrourum* and

Fig. 12. Riparian scrub on the Eland's River in the foreground. *Brabeium stellatifolium*, though present, is not conspicuous here. Beyond, the river is choked with a thicket of the introduced *Acacia mearnsii*, and *Pinus pinaster* are scattered on the slopes above. Du Toit's Kloof Mountains, c. 350 m.

Pentameris thuarii, occur in dense but intermittent patches. In the wetter parts close to the stream, ferns like *Blechnum capense, B. punctulatum, Todea barbara* and *Pteridium aquilinum* form a dense layer of about 1 m. Grey, stout, sword-leaved tufts of *Prionium serratum* of about the same height are very characteristic of the stream edges, whether sandy or rocky. In sandy places this plant gradually colonizes the stream bed itself, forming dense, spikey mats that catch flood debris and impede water flow. *Wachendorfia thyrsiflora*, one of the plants growing socially in these mats, is conspicuous when its tall spikes of golden-yellow flowers appear in spring. None of the constituents of this community, except *Erica caffra* and the Restionaceae, are of typical fynbos form. Most of the upper layer shrubs have narrowly ovate to lanceolate leathery, stiff leaves, the lower shrubs smaller, softer leaves, sometimes glossy or hairy. The subsequent taller successional stage contains trees like *Maytenus acuminata, M. heterophylla, Hartogia schinoides, Cassine* spp., *Podocarpus elongatus* and *Olinia ventosa* which pave the way for stream bank forest of *Ilex mitis, Cunonia capensis, Halleria lucida, Podocarpus latifolius, Rapanea melanophloeos, Kiggelaria africana* and many others (Duthie 1929, Thunberg & Kotze 1940, Taylor 1955, Heyns 1957, van der Merwe 1966, Werger et al. 1972).

In the east, variations of the *Brabeium stellatifolium* community are considerable. *Laurophyllus capensis*, endemic to the southern Cape, becomes important on wet mountain slopes. *Empleurum serrulatum, Leucadendron salicifolium* and *L. eucalyptifolium* replace *Podalyria calyptrata* as major pioneers of upper streams. Eastward from the Houw Hoek Mountains *Noltea africana* appears in riverbank scrub of the lower slopes and plains. The importance of *Brabeium stellatifolium* and *Metrosideros angustifolia* diminish eastward until, on reaching the temperate evergreen forest region east of the Gouritz River, both species cease to occur. In this region, most stream banks are flanked by a wet type of forest, even in the mountains where *Virgilia oroboides* is an important pioneer in the intermediate stages from fynbos (Phillips 1931). *Gleichenia polypodioides* occurs on wet sites throughout Capensis but is most prominent in this forest region where it forms impenetrable thickets, especially in young pine plantations.

Along perennial streams of lower slopes of the dry mountains inland of the forest region at altitudes below 600 m, there is an open or sparse upper layer chiefly of *Salix capensis*, beneath which grow dense patches of *Cliffortia strobilifera* (ubiquitous along streams in Capensis) and tall Restionaceae like *Cannamois virgata* and *Elegia capensis*, 1–2.5 m high. Annual streams have a similar composition without the upper layer of *Salix* (Campbell 1975).

Other vegetation of moist and wet habitats is of more typical fynbos form and structure but the growth is denser and less xerophytic. There is an absence of hard, leathery leaves and a prevalence of soft ones, usually ericoids. The true proteoid leaf is rare, but species of *Leucadendron*, *Cliffortia* and others with long, narrow, soft, hairy leaves are fairly frequent. Very few of the species common in other fynbos communities are present, and plants with a social habit are more frequent, forming communities sometimes markedly dominated by one or two species.

Such vegetation is typical of drainage lines and seepages on lower slopes and deep sandy plateaus. A *Berzelia–Osmitopsis* community is found in these localities in the southwest, with variations north and eastward. In the southernmost part of the Cape Peninsula (Taylor 1969), *Berzelia abrotanoides* dominates the seepages, mixed with sparse *Osmitopsis asteriscoides*, a lean, erect composite, both plants about 1.8 m high (see Table 2). The canopy is dense; only a few Cyperaceae, soft shrubs and herbs occur in an open middle layer, and rosettes of *Drosera* dot the ground. Occasionally, *Berzelia abrotanoides* occurs in pure dense stands. On one 50 m² plot the only other species were *Cliffortia subsetacea*, *Leucadendron laureolum* (intruding from an adjoining community), *Merxmuellera cincta*, *Drosera curviscapa* and *Penaea mucronata*, all occasional. Nearby, other typical moist-habitat plants occurred, among them *Chironia decumbens*, *Restio ambiguus*, *Tetraria flexuosa* and *Utricularia capensis*.

Northward the upper layer is dominated by *Berzelia lanuginosa* with or without *Osmitopsis asteriscoides*, and subsidiary species vary with locality. Along the southwestern coast belt *Mimetes hirtus*, *Psoralea aphylla* and sometimes *P. pinnata* are present. These are replaced on mountain slopes immediately inland, as at Jonkershoek (Werger et al. 1972), by *Leucadendron salicifolium*. Here the community is taller and richer, with an undergrowth c. 0.5 m high, a middle layer of sedges, restioids and shrubs 1–2 m high, and an open tree layer 3–5 m high. In the lower and middle layers *Carpha glomerata*, *Elegia capensis*, *Restio graminifolius*, *Cliffortia graminea*, *Tetraria punctoria*, *Elegia thyrsifera*, *Leptocarpus paniculatus* and sometimes *Restio compressus* and *Carpacoce spermacocea* occur.

Towards the east, 'flushes' in the heath area of the foothills and lower slopes appear to be less tall (Muir 1929). Commonly these habitats are dominated by *Juncus lomatophyllus* or *Laurembergia repens* and contain also *Carpha bracteosa*, *Cyperus tenellus*, *Fuirena hirta*, *Ficinia indica*, *Utricularia capensis*,

U. ecklonii, Centella asiatica, Drosera cuneifolia and *Pulicaria capensis*.

The *Berzelia–Pseudobaeckia* tall fynbos of rocky streams in the Kogelberg (Boucher 1972) occurs also at seepage zones and pediment marshes in adjacent areas (Kruger 1974). *Berzelia lanuginosa, Brunia albiflora, B. alopecuroides, Leucadendron salicifolium, L. xanthoconus, Pseudobaeckia africana, Restio dispar, R. purpurascens* and many others form a dense community 1.5 to 3.0 m high, generally dominated by one or two species, the precise composition varying with locality and habitat. In stream beds the boulders are often covered by a thick mat of *Scirpus digitatus* which is very strongly rooted to withstand the fast-flowing streams. This form of Hygrophilous Fynbos is highly characteristic of the wet Table Mountain Sandstone sites on the area east of the Hottentots–Holland divide, as far as the Swartberg at Caledon.

Distinct from the tall, dense, predominantly ericoid vegetation of seepages and streams are the restioid tussock communities found on marshy flats where water is often more stagnant, or on wet, shallow-soiled mountain slopes. On the plateau of the southern Cape Peninsula, for example, *Elegia parviflora* occurs in almost pure stands, 30 cm high, in shallow depressions or 'pans' on bedrock, while *Elegia cuspidata*, 90 cm, is conspicuous in a more complex community on deep, poorly drained sandy plateaus (see Table 2, and Taylor 1969, plate 21). On coastal mountain slopes like the Kogelberg (Boucher 1972) and the Groenlandberg in the southern Hottentots–Holland range (Kruger 1974) where rainfall may exceed 3000 mm a year, *Restio ambiguus* forms a dense restioid mat, and *Chondropetalum mucronatum*, similar in form to *Elegia cuspidata* though stouter and taller (c. 2 m), is dominant in the more complex community. Communities of the 'pan' type occur more often in the western block of mountains where high-level plateaus are a feature of the topography, as on the Dwarsberg above Jonkershoek and northward via the Bains Kloof and Cold Bokkeveld ranges to the Cedarberg. The restionaceous communities of the table-top summits of the Cedarberg, already described, closely resemble the pan vegetation, but are distinguished by the sharply seasonal water supply of the summits. The 'slope' type of restioid tussock is more common in the eastern ranges where the mountains typically have ridge-like summits and well-defined slopes and where high-level plateaus are rare.

Conspicuous arcuate tufts of *Cannamois virgata* form stands exceeding 3 m high in poorly drained rocky situations throughout the Capensis mountain systems. Cyperaceae such as *Tetraria flexuosa* (40–50 cm) often dominate local flushes and marshy plateaus. All restioid marsh communities – there are many more besides those mentioned here – are local in occurrence, are usually dominated by a single species, and show a markedly even canopy and simple structure (Figs 10 and 13).

2.1.2 Arid Fynbos

Along the inland margin of Capensis there occurs a type of fynbos that Adamson (1938a) called Dry Sclerophyll Bush, and Acocks (1953) Arid Fynbos. Since within Mountain Fynbos itself there are many habitats bearing vegetation that could be described as dry, the term Arid Fynbos for this separate and distinct type is preferred. It occurs as a narrow belt along the inland lower slopes of the innermost Cape folded ranges. Aspect is north-easterly in the Cedarberg–Swartruggens area and northerly in the Swartberg–Baviaanskloof. Included in this type

Table 2. Hygrophilous fynbos communities, from Taylor 1969a.

Community number of relevés	A 4	B 5	C 8
Restio compressus	4	.	.
Osmitopsis asteriscoides	4	.	.
Berzelia lanuginosa	4	1	.
Danthonia cincta	3	.	.
Psoralea pinnata	3	.	.
Psoralea aphylla	3	.	.
Elegia cuspidata	1	5	5
Serruria glomerata	.	4	4
Ursinia tenuifolia	.	4	2
Prismatocarpus sessilis	.	5	.
Rafnia crassifolia	.	4	.
Bobartia indica	.	.	3
Restio dodii	3	3	5
Berzelia abrotanoides	4	4	5
Chondropetalum nudum	1	4	4
Cliffortia subsetacea	1	5	5
Erica parviflora	2	5	5
Erica corifolia	2	5	5
Pentaschistis curvifolia	2	5	5
Restio bifurcus	2	5	5
Restio quinquefarius	1	5	5
Scyphogyne muscosa	1	4	4
Tetraria fasciata	.	5	5
Tetraria flexuosa	1	2	5
Staberoha distachya	.	2	1
Erica bruniades	2	3	4
Erica capensis	3	3	2

A: Restio compressus community of seepage steps;
B: Elegia cuspidata community, prismatocarpus subcommunity of recently burnt, seasonally moist sites;
C: Elegia cuspidata community, Bobartia indica subcommunity of unburnt, marshy flats.

are minor ridges and uplands further inland, from the Bonteberg near Karoopoort eastward via the Witteberg to the Swanepoelspoortberge near Willowmore (Marloth 1923), as well as outlying mountains in Namaqualand of which the Kamiesberg is the best developed (Adamson 1938b). Though records for the mountains are lacking, rainfall is probably at the lower limit for fynbos (i.e. 250 mm). On the lower slopes of the Witteberg it is reported to be only about 125–150 mm (Compton 1931). In Namaqualand, Arid Fynbos occurs on granite, elsewhere on Witteberg quartzite or Table Mountain Sandstone. Altitude may vary from 500–1000 m or even higher on the slopes furthest inland where the winter rains and summer south-east clouds rarely penetrate.

The clear-cut geographical range and specialized habitat of Arid Fynbos justify its separation from Mountain Fynbos. The low rainfall is the chief determinant of its most distinctive physiognomic feature – the open character of the vegetation

Fig. 13. A restioid marsh community of *Elegia capensis* 1 m tall, on the sandy flats of the northern Cedarberg. The boulder slopes beyond are a typical habitat of *Protea arborea*, the Waboom. Boontjieskloof, c. 1150 m.

(Fig. 14). Whereas in even the driest type of Mountain Fynbos total crown cover generally exceeds 70 per cent, that of Arid Fynbos is seldom over 50 per cent (Taylor unpubl. data). The structure is simpler with less distinct layering. In other features its general physiognomy resembles the simpler Mountain Fynbos of the proteoid zone, with a preponderance of ericoid forms and few proteoids (Fig. 15). Restioids are only locally conspicuous. The northern portions are characterized by communities in which bushes with small flat leaves predominate (Adamson 1938a). In the eastern parts the taller succulents, especially *Aloe ferox*, become conspicuous (Campbell 1975). Geophytes and annuals are common in the extensive openings.

Arid Fynbos has been little studied in detail. Adamson's (1938b) account of the Kamiesberg vegetation includes some communities of this type. On the upper, north-facing slopes of this range *Passerina glomerata* is very abundant and other ericoids like *Cliffortia ruscifolia, Cullumia rigida, Anthospermum tricostatum* and *Diosma hirsuta* are common, while Restionaceae occur in separate tufts, not forming a complete ground layer. On the eastern ridges additional species like *Indigofera spinescens, Lobostemon glaucophyllus, Dodonaea viscosa, Pelargonium scabrum* and *Chrysanthemoides monilifera* appear, with a ground layer again of isolated tufts, mainly of *Restio sieberi*, a widespread high-altitude species.

Near Wuppertal, the boundary between Arid Fynbos and Karoo types is clearcut, coinciding with the contact between the Table Mountain and Bokkeveld series at the base of a fault valley (Fig. 14). Relevés taken by the writer revealed this Arid Fynbos to be an open community with only 33 to 45 per cent total cover. A sparse upper layer contains *Protea laurifolia* and the stout, erect, rhizomatous

Fig. 14. The hamlet of Wuppertal lies on the contact of two geological series of the Cape System. The Table Mountain Sandstone on the right bears semi-open Arid Fynbos (33% total cover; predominant layer mainly ericoids, 1–1.5 m), while on the left a low, open succulent Karoo occurs on the Witteberg series. Wuppertal, c. 650 m.

Fig. 15. Arid fynbos with a preponderance of ericoid forms at the northeastern edge of the Cedarberg. Near Wuppertal, c. 800 m.

Cannamois dregei (1.5 to 2.0 m) as the most conspicuous component, with *Protea glabra* emergent to 2.5 m in the driest parts. There is an indistinct middle layer (0.5 to 1.5 m) of chiefly ericoids like *Phylica rigidifolia* and *P. pulchella*, a *Passerina* sp., *Diosma hirsuta*, *Anthospermum aethiopicum*, *Eriocephalus umbellulatus*, one important *Restio*, *R. ocreatus*, and the widespread dry-site proteoid, *Leucadendron pubescens*. The very open ground layer consists of small succulents, ericoids and restioids. Marloth's (1923) species lists of Arid Fynbos from the Witteberg near Matjiesfontein eastward to the Blydeberg (Anysberg) near Willowmore, show rather few species in common with the Wuppertal area. From recent distribution studies in the Ericaceae there is some evidence of a transition zone between Arid Fynbos constituents of the west and east, in the vicinity of Seven Week's Poort to Prince Albert (E. G. H. Oliver, pers. comm.). East of this transition, Campbell (1975) reports extensive areas of 'dry shrubland' on xeric valley slopes below c. 800 m. *Dodonaea viscosa* is the characteristic species of this shrubland on the north slopes of the Kouga mountains. *Elytropappus rhinocerotis* is often co-dominant, or *Themeda triandra* where grazing pressure is low. On more rocky sites *Elytropappus* gives way to *Eriocephalus* spp., with woody shrubs, e.g. *Tarchonanthus*, *Rhus*, *Euclea* and *Cussonia*, and *Aloe ferox* associated. On the drier slopes *Portulacaria afra* becomes dominant and a number of small succulents are found. A single restiad indicates this shrubland's link with fynbos.

Marloth (1907a, 1915, 1923) considered that the summer cloud of the southeast wind provides an important source of moisture even for Arid Fynbos, but this seems based on scant evidence, at least for sites at comparatively low altitudes far inland. Marloth (1923) also pointed out that the Arid Fynbos patches occurring inland of the high mountain ranges are 'islands' surrounded by a Karoo vegetation of different floristic composition. The boundary between the two vegetations is abrupt (see also Levyns 1950). The Arid Fynbos is confined to the Witteberg quartzites, the Karoo vegetation to the Dwyka conglomerates. Marloth considered this due to the fact that the fynbos plants, though strongly adapted to withstand seasonal drought, as he had already discussed in a previous paper (Marloth 1892), cannot persist through the very dry summers of the inland fynbos fringe unless their roots penetrate deeply into rock fissures that store moisture. The quartzites have such fissures, the Dwyka does not.

The fact that each Arid Fynbos 'island' is separated from its nearest neighbour by many kilometers of Karoo vegetation of quite different composition, prompts the view that Arid Fynbos of such situations is a relic vegetation. Marloth (1929) quoted geological and biological evidence to show that the climate of the west coast belt was, at periods during the Tertiary, considerably wetter and warmer than at present. At such times fynbos would have covered the inland plains as well as the mountains. With the onset of aridity the mountain habitats provided the only retreat for the fynbos element. The probability of such a migration, or of a series of advances and retreats, is supported by Acocks (1953) who cites common genera in the non-succulent Karoo flora like *Chrysocoma*, *Hermannia*, *Euryops*, *Pteronia*, *Eriocephalus*, *Selago*, *Walafrida* and *Lightfootia* that are also well represented in the Arid Fynbos. This, together with the transitions sometimes found between Arid Fynbos and Karoo vegetation is seen as evidence for the past intermingling of these floras that must have occurred during periods of fluctuating rainfall. Even today the Arid Fynbos is an unstable type that tends to develop, 'under conditions of grazing mismanagement, a generic composition much like that of the Karroid False Fynbos that invades the grassveld of the mountains of the Upper Plateau, in which such typical families as Proteaceae, Rutaceae and Ericaceae are poorly represented or not at all' (Acocks 1953, p. 153).

2.1.3 Coastal Fynbos

Unlike the mountains which have provided a refuge for Capensis vegetation since its beginnings, the land between the mountains and the present shore line has had a much shorter history of plant colonization. These coastal plains have several times been modified by changes in sea level (Breuil 1945, Krige 1927, quoted by Levyns 1952). During the mid-Tertiary, when marine limestones were laid down from the vicinity of present-day Bredasdorp to Riversdale, the coastline lay much nearer the mountains than it does today. In Pleistocene times much of the continental shelf was exposed by shoreline retreat. Later in the Quaternary era the sea reached levels 130 to 170 metres higher than at present, inundating all low-lying ground on the west and south coasts. At this time the spine of the Cape Peninsula and the isolated mountains of the west coast like the Paardeberg, Paarlberg, Kasteelberg and Piketberg would have been islands in a sea whose shores washed the foothills of the major mountain chains. The warm Moçambique Current would have flowed between the present Cape Peninsula and the mainland and then curved northward up the west coast creating a climate warmer and moister than that of today. Remains of marine animals, now found only in the warmer waters of the east coast, have been reported from many places in the west, and rivers in the northwest exhibit features of rejuvenation impressed on mature streams, indicating warmer and wetter conditions than those existing at present (Haughton 1927, 1928, quoted in Marloth 1929, Smith 1950, quoted in Levyns 1952).

With the emergence of the Cape Flats and other coastal areas in fairly recent times, the Moçambique Current was deflected southward from False Bay, and the west coast became drier than the east. This change is reflected in the floras, the Malmesbury flats bearing a markedly different one from that of similar recent flats in the Bredasdorp area. Today, therefore, two main subdivisions of Coastal Fynbos are discernible, one on marine sands of the west coast from the Cape Flats northward to the Elands River, the other on limestone eastward from Danger Point to near Mossel Bay. The difference is more floristic than physiognomic. Both the western and southern variations are typical fynbos with restioid, ericoid and some proteoid forms, but both have many more grass species and annuals than Mountain Fynbos.

Being further from orographic influences, the coasts have a lower rainfall than the mountains, in the range of 250 to 500 mm with the lower limit on the west coast where sea mists are however fairly frequent (see Chapter 2). Because of the oceanic influence, temperature fluctuations are not extreme and frosts are absent.

The general appearance of Coastal Fynbos on sand of the west coast is predominantly ericoid and rather open, the bushes being somewhat rounded and up to 1 m high (Fig. 16). On limestone in the south the ericoids occur beneath an upper layer of proteoids that are sparse to dense depending on fire history, and exceed 1 m in height. Species common to both west and south coasts include the ericoids *Metalasia muricata*, *Anthospermum aethiopicum*, *Eriocephalus*

Fig. 16. Coastal Fynbos on the dune systems of the Cape Flats. The vegetation is predominantly ericoid and rather open, the bushes being somewhat rounded and up to 1 m tall. The clump of darker, larger bushes near the centre is the intrusive pest-plant, *Acacia cyclops*, which, if unchecked, will form extensive impenetrable thickets. Near Swartklip, Cape Flats, c. 35 m.

racemosus and *E. umbellulatus*, the grasses *Cynodon dactylon* and *Ehrharta calycina* (*E. villosa* occurring on open dunes, more frequent in the west), and other species like *Crassula ramosa, Cynanchum africanum, Nylandtia spinosa, Pelargonium triste* and the annual *Nemesia versicolor*. A conspicuous restioid found throughout Coastal Fynbos is *Thamnochortus erectus* forming coarse tufts up to 1 m high. In some situations it is strongly dominant on loose dune sand inland of the littoral zone where the Cape Dune Mole Rat (*Bathyergus suillus*) regularly disturbs the surface (Fig. 17). *Thamnochortus spicigerus* occupies similar habitats but is less common.

Confined or almost confined to the south coast belt are the conspicuous 1- to 2-metre proteoids *Protea obtusifolia, P. susannae, Leucadendron muirii* and *L. coniferum*, ericoids such as *Clutia ericoides, Erica spectabilis, Eroeda imbricata, Helichrysum teretifolium, Lightfootia calcarea, Limonium scabrum, Phylica selaginoides* and *Simocheilus* spp. and restioids including *Chondropetalum microcarpum, Ficinia truncata, Restio eleocharis, R. cuspidatus, Scirpus membranaceus, Tetraria cuspidata, Thamnochortus paniculatus* and *T. insignis* (locally in Riversdale Division). Many grasses also have a markedly southern distribution, for example *Ehrharta erecta, Festuca scabra, Koeleria cristata, Pentaschistis patuliflora* and *Themeda triandra*, but a number of the above-mentioned species are not confined to the coast belt. Other plants characteristic of south coast fynbos are *Chascanum cernuum, Hermannia trifoliata, Pelargonium betulinum, Senecio arnicaeflorus* and *Zygophyllum fulvum*. Table 3 illustrates the floristic composition of a fynbos community on dunes and sandy flats at the south coast.

Fig. 17. Coastal Fynbos: *Thamnochortus erectus* forming coarse tufts up to 1 m tall, dominant on loose sand where the Cape Dune Mole Rat disturbs the surface. Nachtwacht, Bredasdorp Distr., c. 20 m.

West coast fynbos, though less rich, also contains characteristic species scarcely if ever found in the south. Among the proteoids are *Leucospermum hypophyllocarpodendron* subsp. *canaliculatum*, *L. rodolentum* and *Protea scolymocephala*, ericoids are represented by *Cliffortia juniperina*, *Leyssera gnaphaloides*, *Limonium longifolium* and *Phylica cephalantha*, while characteristic restioids are *Willdenowia arescens* and *W. striata*. The grasses *Pentaschistis patula* and *P. triseta* are almost confined to the west coast and there are several distinctive geophytes, among them *Albuca major*, *Antholyza ringens*, *Caesia contorta*, *Homeria miniata*, *Oxalis hirta*, and annuals, for example *Dimorphotheca pluvialis*, *Ursinia cakilefolia* and *Crassula* (=*Vauanthes*) *dichotoma* (Acocks unpubl. data).

2.1.4 Fire in Fynbos

'Terra de Fume' was the name that Vasco da Gama gave to the Cape of Good Hope when he sailed round the coast nearly 400 years ago, because of the columns of smoke seen from the sea (Wicht 1945); and even today the frequent recurrence of veld fires in the dry season remains one of the striking features of this Cape (Kruger 1974). Under suitable conditions, natural fires can be caused by lightning strikes and friction from rolling stones. Wicht (1945) quoted authentic cases of fires caused by lightning and by a falling rock in the Jonkershoek mountains near Stellenbosch. Immediately after the earthquake of September, 1969, many fires started simultaneously where rockfalls occurred; I saw evidence of this myself in the Cedarberg.

Table 3. Coastal fynbos, from Van der Merwe 1974.

Community number of relevés	A 9	B 10
Protea obtusifolia	5	4
Anthospermum ciliare	4	5
Leucadendron adscendens	4	2
Ficinia elongata	2	4
Oxalis depressa	2	4
Phylica selaginoides	3	3
Astephanus neglectus	4	2
Festuca scabra	3	3
Acmadenia assimilis	3	2
Aspalathus calcaria	2	3
Leucadendron muirii	3	2
Helichrysum teretifolium	3	2
Crassula fastigiata	2	3
Heliophila arenaria	2	3
Eroeda imbricata	2	3
Phylica parviflora	3	1
Chironia baccifera	2	1
Stoebe muirii	2	2
Lightfootia calcaria	2	1
Myrica quercifolia	2	1
Colpoon compressum	2	1
Helichrysum paniculatum	2	2
Euchaetis bolusiae	2	1
Acrostemon vernicosus	3	1
Centella dentata	1	2
Diosma hirsuta	2	1
Erica coccinea	2	1
Felicia sp.	2	1
Ficinia brevifolia	4	.
Rhus mucronata	3	.
Asparagus capensis	2	.
Anthospermum prostratum	2	.
Elytropappus rhinocerotis	2	.
Berkheya coriacea	2	5
Erica spectabilis	2	4
Ficinia praemorsa	.	5
Metalasia gnaphaloides	1	3
Rhus lucida	.	3
Peucedanum sieberianum	.	2
Agathosma serpyllacea	1	2
Helichrysum niveum	.	2

A: Protea obtusifolia community, Ficinia brevifolia subcomm. of sandy flats;
B: Protea obtusifolia community, Berkheya coriacea subcomm. of dunes.

In the precipitous mountains of the southwestern Cape, fires could have started by these means long before early man discovered how to make and use fire. The latter event, which in southern Africa is associated with the Chelles–Acheul and perhaps the Fauresmith industries of over 53,000 years ago (West 1965), would nevertheless have increased the frequency of fire. The Cape fynbos, in which fire has played an important role for so long, can therefore be regarded as a fire-type vegetation, that is, one in which component species are adapted to survive the recurrent fires of a climate '... characterized by coincidence of low temperatures with high moisture, and, conversely, high temperatures with low moisture ...' (Sweeney 1967).

Although stringent laws to control wild-fires were passed as early as 1687 (Botha 1924), veldburning continued unabated, and it was only in the beginning of the present century that botanists began to write with misgivings on the subject. Bolus & Wolley-Dod (1904) were perhaps the first to suggest that the practice of veldburning might be 'more harmful than beneficial from the economic point of view, and that from a botanical point of view it tends to the destruction of species and the consequent greater deformity, not necessarily to the greater usefulness, of the vegetation'. Similar views were repeated and enlarged upon by, inter alia, Michell (1922), Marloth (1924), Pillans (1924), Compton (1926, 1929) and Sim (1943). It was claimed, for instance, that fires diminished summer streamflow and accentuated winter floods, reduced the power of the vegetation to capture moisture from the southeast cloud, caused desiccation, impoverishment and erosion of the soil and created 'sandy wastes'. Other writers, for example Phillips (1936), Hubbard (1937, Adamson (1952), Heyns (1957) and Van der Merwe (1966) noted that fire had reduced the former extent of forests. Except for the systematic observations of Levyns (1924, 1929a, 1935a) and Adamson (1935), both of whom described post-burn regeneration and postulated succession after a fire, no exact work was done during this period, however.

In an attempt to assess the limited and contradictory evidence available at the time, Wicht (1945) observed that burning under control, in a chosen season and in the absence of sustained pasturing, would appear to be less harmful to fynbos than the intensely hot wild-fires that swept through huge areas at the driest time of year. He stressed the urgent need for research to determine the precise effect of fire on vegetation and the ecosystem. Though support for Wicht's views was prompt (e.g. Wroughton 1948), the research, except for his own well-planned but uncompleted statistical experiment (Wicht 1948) and some scant but significant hydrological observations (e.g. Rycroft 1947, Banks 1964), was long delayed.

Only fairly recently, following intensive development in agriculture, forestry and botany, has experimental research gained impetus. Studies on the effect of fire, or its exclusion, on vegetation and water supplies have been undertaken (Wicht & Banks 1963, Martin 1966, Van der Merwe 1966, Plathe & Van der Zel 1969, Van der Zel & Kruger 1975). Following earlier research (e.g. Jordaan 1949, 1965), the effect of burning at different seasons is being investigated (Kruger 1972a, b), as is the use of fire in managing Cape vegetation (le Roux 1966, Trollope 1973, reviewed by Wicht & Kruger 1973), a subject which has already received attention elsewhere in Africa (West 1965).

2.1.4.1 *Fire survival.* Wicht (1945) recognized four ways in which fynbos species can survive fire: (1) geophytes that regenerate from underground storage organs, (2) sprouters that regrow from rootstocks, (3) plants with thick bark that protects dormant stem buds from fire, and (4) woody shrubs that are killed by fire and regenerate from seed.

In his study of the regeneration of species after a fire at Swartboskloof near Stellenbosch, Van der Merwe (1966) termed these means of fire survival, 'fire life forms'. For comparison with normal life forms he erected ten fire life form classes using as criterion the highest position of the buds from which a plant forms shoots after a fire. Wicht's four fire survival mechanisms of fynbos species are represented in the following fire life forms: (1) fire geophytes, (2) fire hemicryptophytes, (3) fire chamaephytes and nanophanerophytes and (4) fire therophytes. The latter class includes normal therophytes as well as woody plants surviving only in the form of seed. According to Van der Merwe, the aerial perennating buds from which post-fire coppice originates are always closer to the soil surface than the overwintering buds of the normal life forms.

Fire geophytes are probably the most successful of these fire life forms because fire scarcely affects soil temperatures at depths of 2 or 3 cm below the surface. Marloth (1908) ascribed the prevalence of certain Cape geophytes to the frequent occurrence of fires. Most geophytes flower profusely for the first season after a burn, but as the vegetation ages, they flower less frequently and less abundantly – indeed, the fire lilies (*Cyrtanthus* spp.) only flower profusely immediately after a fire and seldom bloom again until the next burn – but there is no evidence, as yet, that geophytes die out if fynbos is unburnt for periods of thirty years or more.

In Coastal Fynbos, annuals appear temporarily in great numbers after a fire, but in Mountain Fynbos they play a minor role compared to their importance in Mediterranean countries. Walter (1968) ascribed this difference to the nutrient-poor soils of Cape mountains.

Martin (1966) found that the rapid transit of fire through fynbos in the Grahamstown area, together with the insulatory properties of the soil, resulted in only a momentary rise in surface temperature to above 500°C at any one spot, while temperatures at even 1 cm depth often remained below 50°C. Under these circumstances, sprouting species, both fire hemicryptophytes and fire chamaephytes, can usually survive an average fire because their perennating organs are close to the soil surface.

The remarkable recovery of fynbos after a burn (Taylor 1973) is due largely to the rapid growth of sprouting species. In his study of mountain fynbos, Kruger (1972b) found that both canopy and basal cover had, within 25 to 30 months of burning, reached 70 to 90 per cent of the levels measured before treatment. Sprouting species comprised 90 to 98 per cent of this basal cover and 80 to 95 per cent of canopy cover. Graminoid and restioid species (fire hemicryptophytes) were the quickest to regenerate and became temporarily dominant, but later there was a relative increase in the cover of sprouting herbs and shrubs (fire chamaephytes). Taylor (1972b) measured a maximum shoot length of 100 cm, and an average length of 30 cm, in a sprouting *Asparagus* sp. three months after a midsummer burn in Coastal Fynbos on the Cape Flats – but much of this growth probably occurred within the first two or three weeks. At three months after fire, only one species out of 18 that were regenerating in four 0.025 ha transects was not a root-sprouting species. Van der Merwe (1966) reported that 67 per cent of the species at Swartboskloof regenerated vegetatively and 33 per cent regenerated only from seed. These data confirm that, at least in earlier successional stages, the majority of fynbos plants reproduce vegetatively. Such species, in general, comprise the bulk of the basal cover and help to reduce erosion during the period immediately after a burn, not only by quickly covering the ground but also because their basal tufts are seldom destroyed by fire and remain in position to anchor the surface soil (Kruger 1972b).

Woody shrubs that can survive fire only in the form of seeds are the most vulnerable. In fynbos, the 'youth period' of these plants, from establishment to the age at which viable seeds are formed, is about five or six years on average, so that fires recurring at shorter intervals will gradually exterminate them. Since many species of Proteaceae and Ericaceae belong to this group, it is important to ensure that burning intervals in a controlled burning programme exceed the maximum youth period for a particular community, if the full complement of fynbos species is to be perpetuated.

Martin (1966) found two types of seed-regenerating shrubs on the Grahamstown Nature Reserve. The first regenerates from seeds present in the soil. Such plants, e.g. *Anthospermum aethiopicum* and *Selago corymbosa*, have an early period of importance in the regenerating fynbos but give way to species of the second type that regenerate entirely from light seed borne into the area after the fire. *Erica demissa* and *E. chamissonis* which belong to this group assume dominance at about 8 years after fire and retain their competitive advantage through their production of a dense cover and extensive shallow root systems. Kruger (1972) noted the dominance of seed-regenerating shrubs like *Erica hispidula* and *Leucadendron xanthoconum* and observed that, in the mountains near Grabouw, these and other shrubs increase in dominance with the age of the community.

We see, then, a rhythmic pattern in the succession after fire, in which each of the four fire regeneration forms plays a role of temporary ascendancy, ending with the dominance, in some communities, of the fire therophytes. These latter, especially the taller, longer-lived members of the Proteaceae, may persist as dominants for at least 25 years, as *Protea neriifolia* has done at Jonkershoek, and in the absence of fire may only give way to the next stage on the senescence of the *Proteas*. From the few patches of old fynbos available for observation there are indications that in the mountain foothills and kloofs in the vicinity of Stellenbosch, with an annual rainfall of c. 800 to 1800 mm per annum, this senescence may only begin to occur at an age in excess of thirty years. What the next stage may be is largely conjectural, again because of the death of older fynbos stands. In a catchment at Jonkershoek protected for 35 years, *Protea neriifolia* seedlings are coming up in openings where the old proteas have died, suggesting a possible 'cyclic succession'; but there is also some slight colonization by woody shrubs and small trees derived from the forest flora.

Since fynbos species are so effectively adapted to survive fire, the flora appears to have evolved over a long period in equilibrium with the average fire interval (Martin 1966). This fire-type vegetation will maintain itself as a relatively stable community only if the succession is periodically interrupted before the senescent stage of the larger fire therophytes is reached. In practice, this interruption can only be effectively accomplished by burning the vegetation at an age between the maximum youth period of the most retarded, and the senescence of the most precocious fire therophyte, which is usually at an age of 8 to 15 years in the southwestern Cape. To maintain vigorous fynbos in all its attractive diversity of forms and species, such a cycle of catastrophic interruptions must be perpetuated. This is the object of controlled burning.

2.1.4.2 *Season of burn.* Wild-fires occur mainly, though not always, in summer, but intentional burns can be artificially induced at virtually any time of year except the rainiest periods in high rainfall areas.

The time of year at which burning occurs, especially when frequently repeated at the same season, can have an important cumulative effect on the composition of the vegetation. If it is to be an effective tool in vegetation management, controlled burning should be carried out at a season when it will least harm the composition and regeneration of a plant community.

Thirty years ago, though concrete evidence was lacking, there were conflicting opinions on the effect of burning at different times of year, such as that of which Wicht (1945) wrote as follows: 'The theoretical argument that a spring burn will be less harmful, because the veld will not be so dry, and the fire will, therefore, be less intense, while moisture remaining in the soil will permit almost immediate regrowth, may be countered by the contention that in spring the plants are actively growing and in flower, and reserve materials are low, while in summer and autumn many plants are dormant and reserve materials, in the form of starch and oils, are available for regrowth.'

Since that time, little research has been done on the important question of burning season. Jordaan's (1949) embryological study, supported later by field observations (Jordaan 1965) indicated that late summer, January to March, was the safest time to burn *Protea repens* and *P. pulchra.* He pointed out that similar studies on other species would help to provide an answer to the question of an optimal burning season for the vegetation as a whole. Van der Merwe (1966) could find no evidence of extermination of species in any of the seven communities at Swartboskloof following a February burn of the fifteen-year-old vegetation there.

The experimental investigation of the effects of burning season and fire interval, recently initiated (Kruger 1972a, b), is continuing. Tentative results so far indicate (1) Proteaceous seedlings germinate immediately after a September or November burn but, if burnt in March, most germination is delayed until spring; (2) graminoid and restioid species appear to grow chiefly in spring and early summer; and (3) geophytes, except *Cyrtanthus*, appear to flower normally in their appropriate seasons, whether the fire occurs in spring or autumn, but autumn burns strongly stimulate the flowering of *Watsonia pyramidata.*

Complex questions such as the effect on vegetation of burning season and fire interval, weather before and after a burn, fire intensity and its relation to amount of litter, wind strength etc., are being fully investigated. Consideration of the evidence now accumulating has led, in the meantime, to the adoption of a policy of prescribed burning in mountain catchments.

2.2 *Strandveld*

The term Strandveld has been used by previous authors for a wide variety of concepts that are often the equivalent of a landscape type and can include more than one vegetation unit. The Strandveld of Muir (1929) and Jordaan (1946), for example, include communities treated here as Coastal Fynbos. The Strandveld described in this chapter comprises chiefly the broad-sclerophyll woody scrub of dunes near the coast, but because of the complex interplay of fynbos and forest elements in the succession from pioneer littoral vegetation to scrub, especially along the southern coast belt, the dune succession sometimes includes an ericoid, fynbos-like scrub and sometimes proceeds from pioneer vegetation to a typical Coastal Fynbos that forms a fairly stable sub-climax.

Since the south coast receives intermittent showers throughout the year and the west coast is subject to mists and fog that compensate somewhat for the lack of summer rain (see Chapter 2), the coastal climate permits the development of a stunted scrub derived for the most part from the temperate forest flora, to which elements from the Karoo flora are added on the drier west coast. Because the coast is a fairly uniform habitat and the littoral strip forms an easy, uninterrupted pathway for plant migration, we cannot speak of two separate blocks of Strandveld as we can of Coastal Fynbos or Lowland Renosterveld. There are, nevertheless, considerable differences between the extremes of south and west coast plant communities due to increasing aridity, especially northward up the west coast.

2.2.1 Pioneer littoral dune vegetation

The strand vegetation along the whole of the south coast is fairly uniform. On the edge of the beach where it has been moistened by sea spray, *Agropyrum distichum* is a common pioneer grass. It has greatly increased in abundance since it was first used for driftsand reclamation about forty years ago. Other pioneers of the crest or seaward slope of the littoral dune include another grass, *Ehrharta villosa*, and low, sprawling succulents or semi-succulents like *Senecio elegans* (which, though normally an annual, has a biennial succulent form near the coast), *Chenolea diffusa*, *Arctotheca populifolia*, *Hebenstreitia cordata*, *Silene crassifolia*, *Cnidium suffruticosum*, *Tetragonia decumbens*, *Carpobrotus acinaciformis* and other Mesembryanthemae. *Scaevola thunbergii*, with broad, flat, succulent leaves, occurs only east of Cape Aghulhas; it colonizes the littoral just above high water mark, building up the sand into low mounds (see Phillips 1931, Figs. 5 & 6, and Chapter 13, Fig. 7 for excellent illustrations of this species and the community it forms). The spiny *Eragrostis cyperoides* occurs chiefly on the west coast but also sporadically along the south coast, for example in the Knysna region (Phillips 1931).

In the northern part of the west coast the pioneer community shows signs of increasing aridity in its simpler composition and in the presence of more succulents, among them *Didelta carnosa* var. *tomentosa*.

Occurring on an unstable substratum and subject to strong onshore winds, the littoral plant covering is easily disturbed and destroyed. Because it is palatable and nutritious to stock it has, moreover, been heavily grazed and frequently burnt since the early days of European settlement. The natural fragility of the ecosystem together with its maltreatment over three centuries has resulted first in a thinning of the plant cover and then in its inundation by drifting sand. Since Keet's (1936) report, the reclamation of driftsands on the southern Cape coast has been undertaken, with considerable success, by the Department of Forestry (King 1939, Gohl 1944, Walsh 1968). The Department has recently acquired dune areas on the west coast for the same purpose.

2.2.2 Successional littoral dune communities

Where sand drift is partly stabilized, shrubs like *Metalasia muricata*, *Myrica cordifolia*, *Passerina vulgaris*, *P. ericoides* and other *Passerina* species, *Stoebe plumosa*, *Sutherlandia frutescens*, *Chironia baccifera*, *Chrysanthemoides monilifera* and *Colpoon compressum* are typically found. Though they can exist as a dwarf, wind-sheared scrub on the top of the dunes, these shrubs reach their maximum height of one to three metres in the lee of the dunes where some, especially *Myrica cordifolia*, form dense clumps or thickets. In different combinations, these and other shrubby elements form a series of transitory stages in the succession from pioneer vegetation to Coast Scrub. Some of these stages, especially where dominated by the ericoid *Metalasia muricata* and *Passerina* species, may be regarded as a type of pseudo Coastal Fynbos, and if *Restio eleocharis* and *Erica* species enter, the community becomes a sub-stable Coastal Fynbos where fire precludes further development. If protected from fire, the succession proceeds to Coast Scrub.

2.2.3 Coast scrub

This community, embracing the Littoral Scrub of Phillips (1931), the Dune Valley Scrub of Muir (1929), the *Sideroxylon* Scrub of Taylor (1969a) and the Dune Scrub of Boucher (1972), is the usual climax of Strandveld where it escapes fire for long periods. It develops best in the slacks, troughs or 'valleys' between the dunes and towards the foot of steep, south-facing dune slopes where soil moisture conditions are most favourable. Woody species common to both south and west coasts are *Euclea racemosa*, *Maytenus heterophylla*, *Pterocelastrus tricuspidatus*, *Rhus glauca* and *R. laevigata* (=*R. mucronata*).

South Coast Scrub dominated by *Sideroxylon inerme* and *Pterocelastrus tricuspidatus* contains characteristically *Carissa bispinosa*, *Diospyros dichrophylla* and *Olea exasperata* among other woody shrubs (compare also Chapter 13). The highest development of South Coast Scrub is found in the small forest patches dominated by *Celtis africana*, *Olinia ventosa* and *Apodytes dimidiata* in sheltered, well-watered depressions in the Stanford area near Danger Point described by Taylor (1961). These forests differ considerably from the remnant forest patches found in kloofs and on screes of the southern slopes of mountain ranges in the southwestern Cape, which will be dealt with in Chapter 11, but show a resemblance to the coastal forests of the Knysna region further east, described by Phillips (1931).

West Coast Scrub is poorer in species but more widespread in occurrence, extending inland beyond the sand dunes to the limestone plains where it merges gradually into Coastal Fynbos. *Diospyros austro-africana*, *D. glabra*, *Euclea tomentosa*, *Rhus dissecta* and *Solanum guineense* are characteristic of West Coast Scrub. *Pterocelastrus tricuspidatus* is important in the succession of both east and west, starting as the nucleus of 'bush clumps' which later coalesce to form patches of scrub. The morphological variation of this species is remarkable. In the Knysna forests it is found 'as a large tree, 60 ft to 80 ft in height by 6 ft or 7 ft g.b.h. ... throughout the forests but is more frequent in the drier than in the moister forests' (Laughton 1937). Along the coast it occurs frequently as a bushy shrub sometimes only 1–2 m high, from east of Port Elizabeth westward to the Cape Peninsula and thence northward to the vicinity of Lamberts Bay.

Due to the greater aridity of the west coast, the scrub there is generally shorter (1.25 m), more thorny (*Putterlickia pyracantha*, *Lycium* and many *Asparagus* spp. being common) and more succulent (*Cotyledon paniculata*, *Crassula* and *Euphorbia* spp.) and it exhibits an interesting drought-deciduous feature in the abundant

Zygophyllum morgsana which, in this locality, loses its leaves completely in summer (Boucher & Jarman 1977). This 'Dense Strandveld Scrub' of Alcocks' (1953) Veld Type 34 seems more closely related both physiognomically and floristically to the Sundays River Scrub, a variety of the Valley Bushveld, than to the South Coast Scrub.

A different type of West Coast Scrub is found on the deeper weathered granite soils of the Langebaan Peninsula. The canopy, more open than that of Dense Scrub, is formed of leathery-leaved short evergreen trees, chiefly *Maurocenia frangularia, Pterocelastrus tricuspidatus* and *Olea africana*, 1.20 m to 4.0 m tall (Boucher & Jarman 1977). Yet another scrub type of the west coast, dominated by *Maytenus lucida*, is described by Boucher & Jarman (1977) as the *Maytenus–Kedrostis* community.

2.2.4 Marsh and pan communities

The vegetation of moist depressions near the coast, where the salt content of the soil is high, is a low scrub or sward (30–45 cm) in which *Scirpus nodosus, Juncus kraussii, Plantago carnosa, Cnidium suffruticosum* and the grass *Sporobolus virginicus* are the chief components. *Apium graveolens, Chironia decumbens, Conyza pinnatifida* and *Samolus valerandi* are confined to the wettest parts, while *Chondropetalum nudum, C. tectorum, Merxmuellera cincta, Orphium frutescens* and *Samolus porosus* are shared with other marshy communities inland. *Helichrysum orbiculare* is locally dominant in marshy areas with less salt in the soil (Taylor 1969a, 1972b).

Marsh communities with belts of *Typha latifolia* subsp. *capensis* and *Phragmites australis*, dominant either singly or together, are found right round the coast, as at Langebaan (Boucher & Jarman 1977), Groenvlei (Martin 1960) and Knysna (Phillips 1931), and halophilous marshes, meadows or pans with *Salicornia* and *Arthrocnemum* spp. are found in similar situations, also in the Riversdale (Muir 1929) and Port Elizabeth (Schonland 1919) districts.

2.2.5 Literature and further research

Local areas of both Coastal Fynbos and Strandveld have been described in works by Talbot (1947) and Boucher & Jarman (1977) for the west coast, Adamson (1934) for Robben Island, Taylor (1969a) for the southern Cape Peninsula, Taylor (1972b) for the Cape Flats, Boucher (1972) for the Cape Hangklip area, Taylor (1961) for the Stanford area, Jordaan (1946) for the Bredasdorp and Caledon districts, Muir (1929) for the Riversdale area, Martin (1960) for Groenvlei and vicinity, Phillips (1931) for the Knysna region, Schonland (1919) for Port Elizabeth and Dyer (1937) for Bathurst.

Both Coastal Fynbos and Strandveld are valuable but vulnerable as grazing veld, yet their areas are being rapidly reduced by township and industrial development, and especially by the encroachment of alien weed vegetation. A botanical survey for the whole of the coastal lowlands to determine priorities for development, for conservation and veld management is overdue but is shortly to be put in hand.

2.3 Coastal Renosterveld

Between the Strandveld and Coastal Fynbos of the coastal plain and the Mountain Fynbos of the inland ranges, an elevated platform bears a vegetation that has little resemblance to fynbos: the Coastal Renosterveld, dominated by *Elytropappus rhinocerotis*, the renosterbos (Fig. 18). The same shrub dominates another transitional area, inland of the mountain ranges between fynbos and Karoo. This Mountain Renosterveld, more akin to Karoo in habitat and floristics, will not be dealt with here (see Chapter 9).

The platform that bears Coastal Renosterveld is built of shales that weather to fine-grained soils, heavier and more fertile than the sands of the mountains or the coast. Like Coastal Fynbos, Coastal Renosterveld occurs in two separate blocks. The western block, the so-called 'Swartland', is a gently undulating landscape

Fig. 18. Coastal Renosterveld: a well developed sample with numerous ericoids besides *Elytropappus*, some coarse tufted grasses, and *Aloe ferox* (foreground). Near Riversdale, c. 150 m.

which has been so extensively ploughed for wheat-growing that very little trace of the natural vegetation remains, and the vernacular name is no longer appropriate (Talbot 1947). This block, predominantly on shales of the Malmesbury and Klipheuwel formations, is less than 300 m above sea level and receives most of its 250–500 mm of rain in the six winter months, April to September. The southern block, built largely of Bokkeveld shales, is more incised, steeply rolling country with greater altitudinal variation and a somewhat higher rainfall that becomes more evenly distributed through the year as one travels east.

Ranging widely in different habitats, *Elytropappus rhinocerotis* has at least three distinct ecotypes (Levyns 1929b, p. 166). The one most commonly found in

Coastal Renosterveld is a densely-branched, waxy grey shrub up to about a metre high, with minute, closely adpressed, cupressoid leaves, giving a drab and uniform appearance to the landscape (Fig. 18). Contrary to first impressions created by the overwhelming dominance of renosterbos, this vegetation is rich in species. Acocks (unpubl. data) has listed a total of 938 species for the western block and 1320 for the southern, compared with 1338 and 1054 for the western and southern blocks of Coastal Fynbos respectively. In the southern block, 39 species have been recorded in a 64 m^2 quadrat (Taylor 1970), which is comparable to the general level of diversity in fynbos. Families typical of the Cape Flora, especially the restioid and proteoid elements, are, however, conspicuously absent in Coastal Renosterveld, while tropical affinities are evident in, for example, the Acanthaceae (*Barleria, Blepharis*) and in the grasses, succulents and trees.

From the days of early botanical travellers like Sparrman (1785, quoted in Muir 1929), Coastal Renosterveld has been recognized as a disturbance community replacing, probably, a type of grassy shrubland that had been frequently burnt by the early settlers and overgrazed by their domestic stock. Since practically no undisturbed vegetation exists today, its possible composition can only be reconstructed from scattered remnants. In the southern block, though there were reports of *Themeda* grassland between Swellendam and Mossel Bay (Bateman 1961), much of the vegetation appears to have been a dense and thorny scrub dominated by *Olea africana* and *Sideroxylon inerme*, with other species of coast scrub like *Carissa bispinosa, Diospyros dichrophylla* and *Pterocelastrus tricuspidatus* occurring in between. The lower, drier valleys would have had a semi-succulent scrub in which *Acacia karroo, Aloe arborescens* and *A. ferox* were prominent, while at higher, moister altitudes the scrub appears to have been transitional to the Langeberg forests, judging from the presence today of dry-forest trees like *Cassine* spp., *Rapanea melanophloeos, Maytenus acuminata* and *Heteromorpha arborescens*. The west coast scrub was probably more succulent and related to the Strandveld Scrub (Acocks 1953).

Grass species are still numerous in the Coastal Renosterveld though their cover is usually low. Of the species occurring with a presence of over 40 per cent in both the western and the southern blocks, about one-sixth are grasses; of these, nearly 50 per cent are introduced annuals emanating chiefly from the Mediterranean regions of Europe. This is an interesting parallel with the Soft Chaparral or Coastal Sage (*Salvia mellifera*) of California, the North American analogue of Renosterveld, which has also been heavily invaded by European annual grasses (R. H. Whittaker, pers. comm.). Many of the indigenous perennial grasses frequent in the southern block but rare in the west, e.g. *Digitaria eriantha, Hyparrhenia hirta, Panicum stapfianum* and *Sporobolus capensis*, have migrated down the east coast belt from centres in the warmer summer-rain regions of South Africa. In the western block these species are replaced by common non-tropical forms like *Lasiochloa echinata* and *Pentaschistis patula*, both annuals.

Other annuals are also more numerous in the west, both in species and in individuals, e.g. *Cotula turbinata, Dimorphotheca pluvialis, Dischisma ciliatum* and *Nemesia barbata*. Geophytes are frequent throughout but there are more species in the south, with *Watsonia aletroides* and *Tritonia crocata* characteristic (Acocks unpubl. data).

Associated with *Elytropappus rhinocerotis*, narrow-leaved asteraceous shrubs of similar form are subdominant in limited areas. *Relhania genistaefolia* and *R.*

cuneata are practically confined to the south coast block, *Chrysocoma tenuifolia*, *Pteronia incana* and *Eriocephalus umbellulatus* are frequent in both blocks, while *Relhania ericoides* and *Leyssera gnaphaloides* are more common in the west. It seems likely that, before disturbance, these shrubs were dominants or codominants of distinct communities that are now obscured by the proliferation of *Elytropappus*. Similarly, where degraded mountain fynbos of lower slopes has been invaded by *Elytropappus rhinocerotis*, only a few remnant *Corymbiums* or *Leucadendrons* bear testimony to the original vegetation. Examples of such vegetation were to be found on the Stellenbosch flats (Duthie 1929) and today still occur on the slopes of Helshoogte above them, in the mountain foothills of the Caledon and Bredasdorp districts (Jordaan 1946), in the Bontebok Park near Swellendam (Grobler 1967) and on the Witteberg Series in the Riversdale district (Muir 1929).

There is no foundation at all for the fanciful old belief that the renosterbos is an introduced plant. *Elytropappus* is a genus endemic to Capensis, in the drier parts of which *E. rhinocerotis* forms stable natural communities (Adamson 1938a).

In an attempt to understand, explain and control the aggressiveness of this species, there has been considerable research not only on its taxonomy (Levyns 1935c), its biology (Levyns 1927, 1935b, 1956) and autecology (Scott & Van Breda 1937) but also on its behaviour when subjected to burning.

Levyns studied the effect of experimental burning and clearing in two climatically different areas, one near Stellenbosch in the west (Levyns 1929a) and the other in a drier region near Riversdale about 230 km to the east (Levyns 1935a). The annual rainfall of the Stellenbosch area is rather more than 625 mm, most of it falling in winter, whereas at Riversdale the rainfall of about 430 mm is fairly evenly distributed throughout the year and the district is subject to periodic hot, dry winds that have a deleterious effect on the vegetation. The Renosterveld in the Riversdale district is therefore much more open than the mesophytic type at Stellenbosch, and the bushes are lower in stature, usually less than a metre high. The uniform shading of the ground by renosterbos at Stellenbosch is consequently not present at Riversdale. The renosterbos needs sunlight for its early growth (Levyns 1927, 1935b), and well-established seedlings cease to develop even if moderately shaded. After the burning treatment at Stellenbosch, seedlings of renosterbos germinated in great numbers. Seedlings of this species are extremely sensitive to drought (Levyns 1927), and the damp winter months of the Stellenbosch climate evidently favoured their establishment and growth, since renosterbos regained its dominance without any intermediate successional stages. At Riversdale with its lower and more erratic rainfall, conditions were less favourable to the establishment of renosterbos and, after burning, the vegetation underwent a series of successional stages before Renosterveld was once more established. After clearing, however, the results were similar at both stations: grasses were temporarily abundant, but the successional stages noted at Riversdale were not apparent.

It seems clear, therefore, that in the wetter western part of its range where the dense cover of renosterbos hinders its own regeneration, Renosterveld is a secondary community that can only be perpetuated by recurrent fires: here, the original community may have been dominated by other ericoid asteraceous shrubs, for example *Relhania ericoides* in the 'Grey Bush' of Duthie (1929). In the drier eastern parts the Renosterveld can be a stable, mixed community; here its spread and dominance are probably due to destruction of the more advanced successional stages, usually by overgrazing. In these eastern parts it is desirable to restore the stable community by providing a lengthy rest period without grazing, particularly because the grasses, important in this community, are mostly 'sweet', that is, palatable to stock in the dry season (Muir 1929, Jordaan 1946).

3 Conservation and utilization

3.1 *Conservation*

The Cape flora is in danger, and the need for its preservation because of its uniqueness, its scientific value and aesthetic appeal has been repeatedly stressed. Recommendations have included the development of local reserves, State acquisition of land for large nature reserves, the incorporation of private property into 'national park lands' and the perpetuation of wild plant species by growing them in botanical gardens (e.g. Kanthack 1908, Compton 1932, Wicht 1945, Adamson 1953, Rycroft 1955/56, Taylor 1957, 1962, Control of Alien Vegetation Committee 1962, Chater 1970). Some of these authors recognized the need for safeguarding the water supplies of Cape mountain catchments, and the Report of the Interdepartmental Committee on the Conservation of Mountain Catchments in South Africa (1961) recommended that land be acquired on a large scale for this purpose. In 1962 a sub-committee resolved that a series of experiments be established in Cape mountains to determine the effects of veldburning on the hydrology and vegetation of these catchments.

The Department of Forestry, by far the largest single controller of mountain land under natural vegetation in the Republic, now owns over 650,000 ha of unafforested mountain land in the western and southern Cape, much of which constitutes important catchment areas. Research in catchment hydrology (reviewed by Wicht 1947, 1965) and later in ecology, including the influence of vegetation on water supplies (e.g. Rycroft 1955, Banks 1961, Plathe & Van der Zel 1969, Wicht 1971b, Van der Zel & Kruger 1973) has led the Department to adopt a policy of controlled or prescribed burning as the principal management tool on these mountain lands (le Roux 1966, Wicht 1971a, 1971c, reviewed by Wicht & Kruger 1973).

A related activity of the Department of Forestry from the 1930's is the reclamation of coastlands where overgrazing with overburning destroyed the natural vegetation and led to the spread of driftsands (see Sect. 2.2.1).

Concurrently with these developments, the National Parks Board, the Cape Provincial Department of Nature Conservation and local Divisional Councils and municipalities have acquired and managed land chiefly for the conservation of the faunal component of ecosystems. The reserves administered by these authorities in the Capensis Veld Types total 24,700 ha or a mere 3.6 per cent of the total reserved land of 686,000 ha, the remainder being non-afforested land in State Forest Reserves (see Table 4 and Chapter 41).

The International Biological Programme of the 1960's, which created an awareness of the need to preserve natural ecosystems as open-air laboratories and storehouses of biological material of potential benefit to mankind (Du Toit 1968), also stimulated the conservation of fynbos ecosystems.

Table 4. Conserved areas in Capensis (extracted from Edwards 1974)

Veld Type		Area reserved km² ($\times 100$ = ha)			
		Prov. & Nat.	For. Dept.	Total	% of Veld Type
34 Strandveld		3.90	60.10	64.00	0.96
46 Coastal Renosterveld		49.16	76.40	125.56	0.90
47 Coastal Fynbos		54.22	128.10	182.32	2.10
69 Macchia	Mtn. & Arid Fynbos	139.09	2587.20	2726.29	15.30
70 False Macchia		0.57	3761.00	3761.57	21.10
		246.94	6612.80	6859.74	

The concept of outdoor recreation as an activity to be planned, fostered and managed, and as one that is not incompatible with nature conservation, started with the formation of the South African Nature Union and the reports by Opperman (1957) and Brockman (1961). The Department of Forestry was prompt in recognizing the recreation potential of its vast estate (De Wet 1961) and gradually the principle of multiple use – for open-air recreation, for wilderness trails, for angling and ultimately perhaps even for hunting, besides the major aim of water conservation – took hold (Taylor 1962, Jooste & Venter 1965, Möhr 1966, De Villiers 1969, Wicht 1969, Ackerman 1972, Taljaard 1973, Bigalke 1974, Van Zyl 1974).

To make provision for these manifold and diverse needs, and to assess the conservation status of ecosystems in South Africa, stocktaking of conserved areas at the national level has been done by Van der Merwe (1962), Codd (1968), Rycroft (1968) and Edwards (1974). The table above shows that conservation in Strandveld, Coastal Renosterveld and Coastal Fynbos is far from adequate. It remains for the State to embark on a national plan for acquiring specific sites, if possible of large extent, for the conservation of ecosystems and rare species that are not yet represented in existing reserves.

Apart from State organizations responsible for nature conservation in the Cape, three advisory bodies with representatives from many conservation-minded societies, the Council for the Habitat, the Co-ordinating Council for Nature Conservation in the Cape, and the Society for the Protection of the Environment, keep a watching brief on developments that threaten rare species and ecosystems, and alert the authorities to problems and dangers in conservation, environmental degradation and pollution.

3.1.1 Shrinking ecosystems and vanishing species

The Working Group for Rare and Endangered Plant Species of the National Programme for Environmental Sciences, established in 1973, has examined the herbarium collections for the area south of the Orange River and west of a line near Port Elizabeth (26°E). Besides nearly the whole of Capensis, this includes the western portion of the Karoo–Namib Region. In that area, 1700 candidate species for extinct, threatened or endangered status were found. A large proportion of these species were from Acocks' (1953) fynbos Veld Types 47, 69 and 70. The present indications are that some 500 species in this area may indeed be endangered, and about 60 probably extinct. Most of the latter appear to be in heavily used areas along the southern and southwestern Cape coasts. The original area of Acocks' mapped Veld Types 47, 69 and 70 has been calculated as between 4.4 (Edwards 1974) and 4.6 million hectares. Using satellite images coupled with a general knowledge of the region, the upper limit of the present area for these Veld Types appears to be about 1.8 million hectares (A. V. Hall pers. comm., 19th August 1975). These three veld types have therefore contracted by some 60 per cent through intensive land use for farming, forestry and other activities; and Coastal Renosterveld appears to have been reduced to only about 9 per cent of its former extent, almost fully replaced by wheat farming!

These most alarming figures emphasize the urgency for conserving the remnants of Capensis ecosystems and species.

3.1.2 Pest-plants

From the time of the earliest records in 1834, a number of introduced woody trees and shrubs have spread, largely by natural means, over an area of some 430,000 ha in the southwestern and southern Cape. Several of the species form extensive thickets that exclude other taxa. In the lowland veld types, Strandveld and Coastal Fynbos, the Australian *Acacia cyclops*, *A. saligna* and *Leptospermum laevigatum* have invaded approximately 300,000 ha of land in degrees of cover varying from scattered to dense. Mountain fynbos has been invaded principally by *Hakea sericea* and *Pinus pinaster*, but also by *Hakea gibbosa*, *H. suaveolens* and *Pinus halepensis*, to an extent of some 130,000 ha (Taylor 1969b, c). *Acacia longifolia*, *A. mearnsii*, *A. melanoxylon* and, more recently, *Albizia distachya*, *Sesbania punicea*, *Lantana camara*, *Homalanthus populifolius* and *Pittosporum undulatum* are encroaching on lower slopes, in kloofs and on stream banks, and new candidate weeds continue to be recorded (Taylor 1975).

There is no doubt that these invaders are an important factor in causing the rarity and extinction of species in Capensis vegetation, and that the pest-plant invasion is now the greatest single threat to the intact structure of plant communities in Cape mountains.

About R100,000 is being spent by public authorities each year in removing the infestations, mainly by manual eradication (cf. Taylor 1974). Perhaps as much again is being spent by private individuals. Reports suggest that this has local success in curbing marginal infestations and in clearing smaller nature reserves. The very large areas elsewhere cannot be effectively controlled by cutting (Hall 1974). Chemical means have been tried without great practical success so far (Schütte 1953, Jooste 1965, 1966/67), but need further testing with new and more effective herbicides. Biological control of *Hakea*, using seed-eating insects from its home range in Australia, has been studied since 1962 (Neser & Annecke 1973). A similar long-term study of *Acacia* species is under way, following preliminary investigations into their history of introduction, present distribution, biology and autecology (Roux 1961, Middlemiss 1963, Roux & Middlemiss 1963, Jones 1963, Roux & Warren 1963, Roux & Marais 1964).

Speculations on possible reasons for the phenomenal spread of these plants have included the theory that they may be better adapted than the indigenous flora to the trace element deficient Cape soils (Schütte 1960); that the Cape flora is out of harmony with a drying climate, creating vacant niches that can be more successfully filled by the Australian invaders (Cone 1973); and that in the Capensis fauna there are no suitable insects capable of preying upon the invading plants, to take the place of natural predators in their home range.

3.2 *Utilization*

Compared to forest, savanna and grassland, the Cape flora, for all its beauty and diversity, has yielded few useful plants.

Modern man would find it hard to survive for more than a day or two on the food of this veld. In Strandveld, *Carpobrotus edulis* and *C. acinaciformis* fruits are palatable but hardly sustaining. *Nylandtia spinosa* berries are thirst quenching but bitter, and the young inflorescences of some *Trachyandra* species can make fine flavouring for a stew. Most of the bulbs, corms, tubers and rhizomes that helped to feed the Strandlopers, Bushmen and Hottentots are scorned today.

The veld plants that early settlers used have long since been superseded by manufactured goods. Pioneers were happy to sleep on mattresses made from dried flower-heads of *Helichrysum vestitum.* They bound their bundles with strips of *Passerina* bark, cured hides with tanning from the bark of Waboom (*Protea arborea*), used its wood for the brake-blocks of their wagons and as fuel for their fires. In the north they made their floors and ceilings – and their coffins – of the aromatic and durable wood of the Clanwilliam Cedar (*Widdringtonia cedarbergensis*, Fig. 19), fenced their fields with its poles and thatched their roofs with *Thamnochortus* reeds. Aloe juice and buchu brandy were in the panaceae for their ills; Rooi and Heuning teas quenched their thirst. Of all these, modern man needs only the thatch, the Buchu and the tea.

The aromatic fynbos flora contains some valuable medicinal herbs, among them Buchu. Leaves and twigs of the Round-leaf Buchu (*Agathosma betulina*) are gathered in the wild from the high slopes of the northern mountains, and the plant is cultivated on the lower slopes, usually on land unsuitable for other crops. Oval-leaf Buchu (*A. crenulata*), of less commercial value, comes from the central and eastern mountains (Compton & Mathews 1921, Werner 1949, Blommaert 1972). The dried leaves of buchu, exported to Britain, Europe and the U.S.A., may bring in a revenue of R300,000 in a good year. On distillation, the dried leaves yield various oils, including diosphenol, which are used mainly in the flavour and pharmaceutical industries (Blommaert & Bartel 1976).

Besides the two Buchu species, there are 132 other indigenous species of *Agathosma* and 170 indigenous species of other rutaceous genera that contain volatile oils (Blommaert 1972). The systematic chemical exploration of this family could thus yield many more species of commercial value.

'Rooi Tea', for which there is a considerable local market, is cultivated in the foothills of the northern mountains from *Aspalathus linearis* subsp. *linearis*. This species has a wide distribution in the southwestern Cape but the wild forms from which the original material for cultivation was gathered are reported to derive mainly from the Pakhuis Mountains in the northern Cedarberg (Dahlgren 1968). Different mountain species of *Aspalathus*, *Cyclopia* and *Rafnia* yield other 'teas' of varying quality.

Capensis vegetation has provided grazing for domestic animals for many centuries. Since the time of Christ at least, Hottentots grazed their fat-tailed sheep (Schweitzer & Scott 1973) and cattle on the lowland area and burnt the veld to

Fig. 19. Clanwilliam Cedar, *Widdringtonia cedarbergensis* Bassonsklip, Cedarberg, c. 1300 m.

provide young pasturage. This practice was emulated by the early Europeans but the veld soon deteriorated under the more intensive grazing systems of the permanent settlers in contrast to the nomadic habits of the indigenous tribes. The introduction of goats made matters worse. As a result, the extensive 'sweet' grasslands of the coastal platform have been almost completely replaced by Coastal Renosterveld, and the Coastal Fynbos has been heavily invaded by alien *Acacia* species (see Sect. 3.1.2). Present grazing is, therefore, largely limited to the 'sour' grasses of the mountain foothills and lower slopes which provide seasonal pasturage in spring and summer for herds that are returned to the Karoo during winter.

The most recent and perhaps the most promising use of fynbos is as a source of cut flowers for the export market. Long before the days of air freightage, the remarkable lasting qualities of the Chincherinchee (*Ornithogalum thyrsoides*) made it the only suitable candidate for sea transport to Britain and Europe. Today, many species of Proteaceae, some picked in the veld but more and more being selectively cultivated in 'orchards', are exported by air to Europe. Apart from their bizarre beauty and their excellent lasting qualities, selected varieties have the advantage of coming into flower in the northern winter when there is a dearth of cut flowers in Europe. The exported Proteaceae are being increasingly supplemented by 'Cape greens', species of *Erica, Brunia, Leucadendron, Phaenocoma* and others that are gaining wide popularity overseas. During the months of October, November and December 1975, fresh wild flowers to the value of nearly R500,000 were exported from Cape Town airport alone (S. Afr. Protea Producers & Exporters Association, newsletter no. 10, February 1976).

Dried Cape flowers of many species including Restionaceae, available at all times of the year, are also becoming increasingly popular on world markets.

References

Ackerman, D. P. 1972. The proclamation of wilderness areas by the Department of Forestry. S. Afr. For. J. 82:19–21.

Acocks, J. P. H. 1953. Veld types of South Africa. Mem. Bot. Surv. S. Afr. 28:1–192.

Adamson, R. S. 1929. The vegetation of the South-Western Region. In: The botanical features of the South-Western Cape Province, pp. 15–32. Speciality Press, Cape Town and Wynberg.

Adamson, R. S. 1934. The vegetation and flora of Robben Island. Trans. Roy. Soc. S. Afr. 22:279–296.

Adamson, R. S. 1935. The plant communities of Table Mountain. III. A six years' study of regeneration after burning. J. Ecol. 23:44–55.

Adamson, R. S. 1938a. The vegetation of South Africa. British Empire Vegetation Committee, London.

Adamson, R. S. 1938b. Notes on the vegetation of the Kamiesberg. Mem. Bot. Surv. S. Afr. 18:1–25.

Adamson, R. S. 1952. The flora of the Cape Province. Cape Dept. of Nature Conservation Report 9:29–33.

Adamson, R. S. 1953. Can we preserve the Cape flora? J. Bot. Soc. S. Afr. 39:11–12.

Adamson, R. S. 1958. The Cape as an ancient African flora. Pres. Add. Sect. K (Botany), Brit. Assoc. Glasgow. The Advancement of Science, 15:118–127.

Adamson, R. S. & Salter, T. M. 1950. Flora of the Cape Peninsula. Juta, Cape Town and Johannesburg.

Aubréville, A. 1975. Essais sur l'origine et l'histoire des flores tropicales Africaines. Application de la théorie des origines polytipiques des angiospermes tropicales. Adansonia, ser. 2, 15:31–56.

Banks, C. H. 1961. The hydrological effects of riparian and adjoining vegetation. For. in S. Afr. 1:31–45.

Banks, C. H. 1964. Further notes on the effect of autumnal veldburning on stormflow in the Abdolskloof catchment, Jonkershoek. For. in S. Afr. 4:79–84.

Bateman, J. A. 1961. The mammals occurring in the Bredasdorp and Swellendam Districts, C. P., since European settlement. Koedoe 4:78–100.

Beard, J. S. 1959. The origin of African Proteaceae. J. S. Afr. Bot. 25:231–235.

Bews, J. W. 1916. An account of the chief types of vegetation in South Africa, with notes on the plant succession. J. Ecol. 4:129–159.

Bews, J. W. 1917. The plant ecology of the Drakensberg Range. Ann. Natal Mus. 3:511–565.

Bews, J. W. 1925. Plant forms and their evolution in South Africa. Longmans, Green & Co., London.

Bigalke, R. C. 1974. Wild life on forest land: problems and prospects. S. Afr. For. J. 89:16–20.
Blommaert, K. L. J. 1972. Cultivation of Buchu. Fruit & Fruit Technology Res. Inst., Bulletin No. 74.
Blommaert, K. L. J. & Bartel, E. 1976. Chemotaxonomic aspects of the buchu species Agathosma betulina Pillans and A. crenulata Pillans from local plantings. J. S. Afr. Bot. 42:121–126.
Bolus, H. 1886. Sketch of the flora of South Africa. In the 'Official Handbook of the Cape of Good Hope', Cape Town.
Bolus, H. 1905. Sketch of the floral regions of South Africa. In: Science in South Africa, pp. 199–240. Maskew Miller, Cape Town.
Bolus, H. & Wolley-Dod, A. H. 1904. A list of the flowering plants and ferns of the Cape Peninsula, with notes on some of the critical species. Trans. S. Afr. Phil. Soc. 14:207–373.
Botha, C. G. 1924. Note on early veldburning in the Cape Colony. S. Afr. J. Sci. 21:351–352.
Boucher, C. 1972. The vegetation of the Cape Hangklip area. Unpubl. M.Sc. thesis, University of Cape Town.
Boucher, C. & Jarman, M. L. 1977. The vegetation of the Langeaan area. Trans. Roy. Soc. S. Afr. 42:241–272.
Boughey, A. S. 1957. The origin of the African flora. Oxford Univ. Press, London.
Breuil, H. 1945. The old Palaeolithic Age in relation to quaternary sea-levels along the southern coast of Africa. S. Afr. J. Sci. 41:361–374.
Brockman, C. F. 1961. Outdoor recreation in relation to nature conservation in South Africa. Report to S. Afr. Nature Union.
Campbell, B. (ed.). 1975. A preliminary report on the Sapree River catchment in the Kouga Mountains, Southern Cape. Rondebosch: University of Cape Town.
Chater, S. W. 1970. Nuweberg as a floral reserve. J. Bot. Soc. S. Afr. 56:24–26.
Codd, L. E. 1968. The conservation status of ecosystems in South Africa. S. Afr. J. Sci. 64:446–448.
Comins, D. M. 1962. The vegetation of the district of East London and Kingwilliamstown, Cape Province. Mem. Bot. Surv. S. Afr. 33:1–32.
Compton, R. H. 1926. Veld-burning and veld deterioration. S. Afr. J. Nat. Hist. 6:5–19.
Compton, R. H. 1929. The results of veld-burning. Dept. of Publ. Educ., Cape of Good Hope.
Compton, R. H. 1931. The flora of the Whitehill District. Trans. Roy. Soc. S. Afr. 19:269–329.
Compton, R. H. 1932. Local nature reserves. J. Bot. Soc. S. Afr. 18:10–16.
Compton, R. H. & Mathews, J. W. 1921. The cultivation of Buchu. National Botanic Gardens, Kirstenbosch. Economic Bulletin no. 1. 8pp.
Cone, G. B. 1973. The floras of the south-western part of South Africa. Tuatara 20:160–164.
Control of Alien Vegetation Committee, 1962. The case for National Park Lands. Publ. by Bot. Soc. S. Afr.
Dahlgren, R. 1968. Revision of the genus Aspalathus. II. The species with ericoid and pinoid leaflets. 7. Subgenus Nortieria. With remarks on Rooibos Tea cultivation. Bot. Not. 121:165–208.
De Villiers, P. C. 1969. Die gebruik van die boseiendom vir buitelugontspanning. Lecture to S. Afr. Inst. of For. Symposium, Stellenbosch. 15pp.
De Wet, D. R. 1961. Our forests and the public. Cape Dept. Nature Cons. Report 18:23–31.
Drege, J. F. & Meyer, E. 1843. Zwei Pflanzengeographische Documente nebst einer Einleitung von Dr. E. Meyer. Besondere Beigabe zur 'Flora', 2.
Drude, O. 1887. Atlas der Pflanzenverbreitung. In: Berghaus, Physik. Atlas, Karte 6. Gotha.
Duthie, A. V. 1929. Vegetation and flora of the Stellenbosch Flats. Ann. of the Univ. of Stellenbosch, 7:1–59.
Du Toit, A. L. 1966. The geology of South Africa. Third ed. Oliver & Boyd, Edinburgh and London.
Du Toit, C. A. 1968. The present-day challenge to biologists. S. Afr. J. Sci. 64:3–12.
Dyer, R. A. 1937. The vegetation of the divisions of Albany and Bathurst. Mem. Bot. Surv. S. Afr. 17:1–138.
Dyer, R. A. 1966. Impressions on the subject of the age and origin of the Cape Flora. S. Afr. J. Sci. 62:187–190.
Edwards, D. 1967. A plant ecological survey of the Tugela River Basin. Mem. Bot. Surv. S. Afr. 36:1–285.
Edwards, D. 1974. Survey to determine the adequacy of existing conserved areas in relation to vegetation types. Koedoe 17:2–37.

Engler, A. 1882. Versuch einer Entwicklungsgeschichte der extratropischen Florengebiete der suedlichen Hemisphere. Engelmann, Leipzig.

Engler, A. 1903. Über die Frühlingsflora des Tafelberges bei Kapstadt, nebst Bemerkungen über die Flora Süd-Afrikas, us.w. Notizblatt des Königl. bot. Gart., II. Leipzig.

Gohl, C. R. 1944. Driftsand reclamation and coast stabilization in the south-western districts of the Cape Province. J. S. Afr. For. Assoc. 12:4–18.

Goldblatt, P. 1972. Iridaceae. In: Lectures on the Cape Flora at the University of Cape Town's Public Summer School, Jan.–Feb. 1972. Univ. of Cape Town.

Good, R. 1947. The geography of the flowering plants. 1st edition. Longmans, Green and Co., London.

Good, R. 1964. The geography of the flowering plants. 3rd edition. Longmans, London.

Good, R. 1974. The geography of the flowering plants. 4th edition. Longmans, London.

Grisebach, A. 1872. Die Vegetation der Erde. 2 vols. Engelmann, Leipzig.

Grobler, P. J. & Marais, J. 1967. Die plantegroei van die Nasionale Bontebokpark, Swellendam. Koedoe 10:132–148.

Hall, A. V. 1961. Distribution studies of alien trees and shrubs in the Cape Peninsula. J. S. Afr. Bot. 27:101–110.

Hall, A. V. 1974. Research group on invasive trees and shrubs in Cape vegetation. Unpublished circular, 14th June 1974.

Hamilton, G. N. G. & Cooke, H. B. S. 1965. Geology for South African students. 5th ed. Central News Agency, South Africa.

Haughton, S. H. 1927. Notes on the river-system of south-west Gordonia. Trans. Roy. Soc. S. Afr. 14:225–232.

Haughton, S. H. 1928. In: P. Wagner & H. Merensky, The diamond deposits on the coast of Little Namaqualand. Trans. Geolog. Soc. S. Afr. 31:35–41.

Haughton, S. H. 1969. Geological History of Southern Africa. Geological Society of South Africa.

Heyns, A. J. 1957. Flora, fenologie en regenerasie van 'n inheemse woudgemeenskap naby Stellenbosch. J. S. Afr. Bot. 23:111–119.

Hubbard, C. S. 1937. Observations on the distribution and rate of growth of Clanwilliam Cedar, Widdringtonia juniperoides Endl. S. Afr. J. Sci. 33:572–586.

Hutchinson, J., 1946. A botanist in Southern Africa. Gawthorn, London.

Jones, R. M. 1963. Preliminary studies of the germination of seed of Acacia cyclops and Acacia cyanophylla. S. Afr. J. Sci. 59:296–298.

Jooste, J. v. d. W. 1965. Experiments in the control of Acacia mearnsii De Wild. S. Afr. J. Agric. Sci. 8:1165–1166.

Jooste, J. v. d. W., 1966/67. Die bestryding van Hakea. Tydksr. Natuurwet. 6/7:315–318.

Jooste, J. & Venter, H. 1965. Opelugontspanningspotensiaal van sekere Staatseiendomme in Suid Afrika. (Quoted in De Villiers, 1969.)

Jordaan, P. G. 1946. Plantegroei van die distrikte Bredasdorp en Caledon. Tydskr. Wetensk. en Kuns, nuwe reeks 6:47–58.

Jordaan, P. G. 1949. Aantekeninge oor die voortplanting en brandperiodes vir Protea mellifera Thunb. J. S. Afr. Bot. 15:121–125.

Jordaan, P. G. 1965. Die invloed van 'n winterbrand op die voortplanting van vier soorte van die Proteaceae. Tydskr. Natuurwet. 5:27–31.

Kanthack, F. E. 1908. Destruction of mountain vegetation. Agr. J. Cape of Good Hope. 33:194–204.

Keet, J. D. M. 1936. Report on driftsands in South Africa. Forest Dept. Bull. 172.

Killick, D. J. B. 1963. An account of the plant ecology of the Cathedral Peak area of the Natal Drakensberg. Mem. Bot. Surv. S. Afr. 34:1–178.

King, N. L. 1939. Reclamation of the Port Elizabeth driftsands. J. S. Afr. For. Assoc. 2:5–10.

King, L. C. 1962. The morphology of the earth. Oliver & Boyd, Edinburgh and London.

King, L. C. 1963. South African scenery, a textbook of geomorphology. 3rd edition. Oliver & Boyd, Edinburgh and London.

Krige, A. V. 1927. An examination of the Tertiary and Quaternary changes in sea-level in South Africa. Ann. Univ. Stell. 5, Sect. A. no. 1. 81 pp.

Kruger, F. J. 1972a. The effect of early summer controlled burns at four and twelve year intervals on the vegetation at Zachariashoek. Progress Report, Jonkershoek Forest Research Station.

Kruger, F. J. 1972b. Jakkalsrivier catchment experiment: investigation of the effects of spring and autumn burns on vegetation. Progress Report, Project 116/25, Jonkershoek Forest Research Station, June, 1972.

Kruger, F. J. 1974. The physiography and plant communities of the Jakkalsrivier catchment. Unpubl. M.Sc. thesis, University of Stellenbosch.

Laughton, F. S. 1937. The sylviculture of the indigenous forests of the Union of South Africa with special reference to the forests of the Knysna Region. Dept. of Agriculture and Forestry, Science Bulletin no. 157, Forestry Series no. 7.

Le Roux, H. H. 1966. Veldbestuur in die wateropvanggebiede van die Winterreënstreek van Suidwes-Kaapland. For. in S. Afr. 6:1–32.

Levyns, M. R. 1924. Some observations on the effect of bush fires on the vegetation of the Cape Peninsula. S. Afr. J. Sci. 21:346–347.

Levyns, M. R. 1927. A preliminary note on the Rhenoster Bush (Elytropappus rhinocerotis) and the germination of its seed. Trans. Roy. Soc. S. Afr. 14:383–388.

Levyns, M. R. 1929a. Veld-burning experiments at Ida's Valley, Stellenbosch. Trans. Roy. Soc. S. Afr. 17:61–92.

Levyns, M. R. 1929b. The problem of the Rhenoster Bush. S. Afr. J. Sci. 26:166–169.

Levyns, M. R. 1935a. Veld-burning experiments at Oakdale, Riversdale. Trans. Roy. Soc. S. Afr. 23:231–243.

Levyns, M. R. 1935b. Germination in some South African seeds. J. S. Afr. Bot. 1:89–103.

Levyns, M. R. 1935c. A revision of Elytropappus Cass. J. S. Afr. Bot. 1:161–170.

Levyns, M. R. 1938. Some evidence bearing on the past history of the Cape Flora. Trans. Roy. Soc. S. Afr. 26:401–424.

Levyns, M. R. 1950. The relations of the Cape & Karoo floras near Ladismith, Cape. Trans. Roy. Soc. S. Afr. 32:235–246.

Levyns, M. R. 1952. Clues to the past in the Cape Flora of today. S. Afr. J. Sci. 49:155–164.

Levyns, M. R. 1954. The genus Muraltia. J. S. Afr. Bot. Suppl. Vol. 2.

Levyns, M. R. 1956. Notes on the biology and distribution of the Rhenoster Bush. S. Afr. J. Sci. 52:141–143.

Levyns, M. R. 1958. The phytogeography of members of Proteaceae in Africa. J. S. Afr. Bot. 24:1–9.

Levyns, M. R. 1962. Possible Antarctic elements in the South African flora. S. Afr. J. Sci. 58:237–241.

Levyns, M. R. 1963. The origins of the flora of South Africa. Lantern 13, 1:17–21.

Levyns, M. R. 1964. Migrations and origin of the Cape Flora. Trans. Roy. Soc. S. Afr. 37:85–107.

Levyns, M. R. 1972. The Rhenosterbush. Veld and Flora 2:7–9.

Marloth, R. 1892. Some adaptations of South African plants to the climate. Trans. S. Afr. Phil. Soc. 6:31–38.

Marloth, R. 1902. Notes on the occurrence of Alpine types in the vegetation of the higher peaks of the south-western districts of Cape Colony. Trans. S. Afr. Phil. Soc. 11:161–168.

Marloth, R. 1903. The historical development of the geographical botany of Southern Africa. Rep. S. Afr. Assoc. Adv. Sci. 1903:251–257.

Marloth, R. 1904. Results of experiments on Table Mountain for ascertaining the amount of moisture deposited from the south-east clouds. Trans. S. Afr. Phil. Soc. 14:403–408.

Marloth, R. 1906. The phytogeographical subdivisions of South Africa. Rep. Brit. Assoc. Adv. Sci. 1905:589–590.

Marloth, R. 1907a. On some aspects in the vegetation of South Africa which are due to the prevailing winds. Rep. S. Afr. Ass. Adv. Sci., 1905 & 1906:215–218.

Marloth, R. 1907b. Results of further experiments on Table Mountain for ascertaining the amount of moisture deposited from the south-east clouds. Trans. S. Afr. Phil. Soc. 16:97–105.

Marloth, R. 1908. Das Kapland. Gustav Fischer, Jena.

Marloth, R. 1915. The effects of drought and some other causes on the distribution of plants in the Cape region. S. Afr. J. Sci. 12:383–390.

Marloth, R., 1923. Observations on the Cape flora: its distribution on the line of contact between the south-western districts and the Karoo. S. Afr. J. Nat. Hist. 4:335–344.

Marloth, R. 1924. Notes on the question of veldburning. S. Afr. J. Sci. 21:342–345.

Marloth, R. 1929. Remarks on the realm of the Cape flora. S. Afr. J. Sci. 26:154–159.

Martin, A. R. H. 1960. The ecology of Groenvlei, a South African fen. I. The primary communities. J. Ecol. 48:55–71. II. The secondary communities. J. Ecol. 48:307–329.

Martin, A. R. H. 1965. Plant ecology of the Grahamstown Nature Reserve. I. Primary communities

and plant succession. J. S. Afr. Bot. 31:1–54.
Martin, A. R. H. 1966. Plant ecology of the Grahamstown Nature Reserve. II. Some effects of burning. J. S. Afr. Bot. 32:1–40.
Michell, M. R. 1922. Some observations on the effects of a bush fire on the vegetation of Signal Hill. Trans. Roy. Soc. S. Afr. 10:213–232.
Middlemiss, E. 1963. The distribution of Acacia cyclops in the Cape Peninsula area by birds and other animals. S. Afr. J. Sci. 59:419–420.
Möhr, M. v. N. 1966. Buitelugontspanning op bosreservate. For. in S. Afr. 7:11–16.
Muir, J. 1929. The vegetation of the Riversdale area, Cape Province. Mem. Bot. Surv. S. Afr. 13:1–86.
Neser, S. & Annecke, D. P. 1973. Biological control of weeds in South Africa. Dept. Agr. Tech. Serv. Entom. Mem. 28:1–27.
Nordenstam, B. 1974. The flora of the Brandberg. Dinteria 11:1–67.
Opperman, R. W. J. 1957. Our vanishing heritage. Vigor 2:1–6.
Ornduff, R. 1974. An introduction to California plant life. Univ. of California Press, Berkely, Los Angeles.
Phillips, J. F. V. 1931. Forest-succession and ecology in the Knysna region. Mem. Bot. Surv. S. Afr. 14:1–327.
Phillips, J. F. V. 1936. Fire in vegetation: a bad master, a good servant, and a national problem. J. S. Afr. Bot. 2:35–45.
Pillans, N. S. 1924. Destruction of indigenous vegetation by burning in the Cape Peninsula. S. Afr. J. Sci. 21:348–350.
Plumstead, E. P. 1961. Ancient plants and drifting continents. S. Afr. J. Sci. 57:173–181.
Plumstead, E. P. 1969. Three thousand million years of plant life in Africa. Alex L. du Toit Memorial Lectures No. 11. Geol. Soc. S. Afr., Annexure to Vol. 72.
Plumstead, E. P. 1974. Trees of the distant and more recent past in South Africa. S. Afr. For. J. 88:1–5.
Pole Evans, I. B. 1920. The veld: its resources and dangers. S. Afr. J. Sci. 17:1–34.
Pole Evans, I. B. 1936. A vegetation map of South Africa. Mem. Bot. Surv. S. Afr. 15:1–23.
Raven, P. H. 1973. The evolution of Mediterranean floras. In: Di Castri, F. & Mooney, H. A. (eds.), Mediterranean Type Ecosystems. Chapman & Hall, London.
Rehman, A. 1880. Geobotanische verhältnisse von Süd-Afrika. Denkschr. d. Akad. d. Wiss. in Krakau, 5.
Rennie, J. V. L. 1935. On the flora of a high mountain in South-West Africa. Trans. Roy. Soc. S. Afr. 23:259–263.
Report of the Interdepartmental Committee. 1961. Conservation of mountain catchments in South Africa. Dept. Agr. Tech. Serv. Pretoria.
Riley, D. & Young, A. 1966. World vegetation. University Press, Cambridge.
Roberts, B. R. 1961. Preliminary notes on the vegetation of Thaba 'Nchu. J. S. Afr. Bot. 27:241–251.
Roberts, B. R. 1966. Observations on the temperate affinities of the vegetation of Hangklip mountain near Queenstown, C. P. J. S. Afr. Bot. 32:243–260.
Robyns, W. 1959. Some problems in the tropical afromontane flora. Adv. of Sci. 15:323–328.
Rourke, J. P. 1972. Taxonomic studies on Leucospermum R. Br. J. S. Afr. Bot. Suppl. vol. 8.
Roux, E. R. 1961. History of the introduction of the Australian Acacias on the Cape Flats. S. Afr. J. Sci. 57:99–102.
Roux, E. R. & Marais, C. C. G. 1964. Rhizobial nitrogen fixation in some South African Acacias. S. Afr. J. Sci. 60:203–204.
Roux, E. R. & Middlemiss, E. 1963. The occurrence and distribution of Acacia cyanophylla and A. cyclops in the Cape Province. S. Afr. J. Sci. 59:286–294.
Roux, E. R. & Warren, J. L. 1963. Symbiotic nitrogen fixation in Acacia cyclops A. Cunn. S. Afr. J. Sci. 59:294–295.
Rycroft, H. B. 1947. A note on the immediate effects of veld-burning on stormflow in a Jonkershoek stream catchment. J. S. Afr. For. Assoc. 15:80–88.
Rycroft, H. B. 1955. The effect of riparian vegetation on water-loss from an irrigation furrow at Jonkershoek. J. S. Afr. For. Assoc. 26:2–9.
Rycroft, H. B. 1955/56. Saving our flora. J. Bot. Soc. S. Afr. 41/42:13–15.
Rycroft, H. B. 1968. Cape Province. In: Hedberg, I. & O. (eds.), Conservation of vegetation in Africa south of the Sahara. Acta Phytogeogr. Suec. 54:235–239.

Schimper, A. F. W. 1898. Pflanzen-geographie auf Physiologischer Grundlage. Fischer, Jena.
Schimper, A. F. W. 1903. Plant geography on a physiological basis. Clarendon Press, Oxford.
Schönland, S. 1919. Phanerogamic flora of the divisions of Uitenhage and Port Elizabeth. Mem. Bot. Surv. S. Afr. 1:1–118.
Schütte, K. H. 1953. Hakea eradication by means of new herbicides. J. S. Afr. For. Assoc. 23:30–36.
Schütte, K. H. 1960. Trace element deficiencies in Cape vegetation. J. S. Afr. Bot. 26:45–49.
Schweitzer, F. R. & Scott, K. J. 1973. Early occurrence of domestic sheep in sub-Saharan Africa. Nature 241 (5391):547.
Scott, J. D. & Van Breda, N. G. 1937. Preliminary studies on the root system of the Rhenosterbos (Elytropappus rhinocerotis) on the Worcester Veld Reserve. S. Afr. J. Sci. 33:560–569.
Sim, J. T. R. 1943. Mountain fires. Farming in S. Afr. 18:283–286.
Smith, J. L. B. 1950. Pomadasys operculare. Ann. Mag. Nat. Hist. 3:778–785.
Sparrman, A. 1785. Voyage to the Cape of Good Hope, vol. I. pp. 250–254. London.
Stewart, C. M. 1904. A note on the quantities given in Dr. Marloth's paper 'on the moisture deposited from the south-east clouds.' Trans. S. Afr. Phil. Soc. 14:413–417.
Story, R. 1952. Botanical survey of the Keiskammahoek District. Mem. Bot. Surv. S. Afr. 27:1–184.
Sweeney, J. R. 1967. Ecology of some 'fire type' vegetation in northern California. Proc. Tall Timbers Fire Ecol. Conf. 7:111–125.
Talbot, W. J. 1947. Swartland and Sandveld. Geoffrey Cumberlege, Oxford Univ. Press, Cape Town.
Taljaard, E. P. S. 1973. Opelug-ontspanning op boseiendomme. S. Afr. For. J. 86:34–37.
Taylor, H. C. 1955. Forest types and floral composition of Grootvadersbos. J. S. Afr. For. Assoc. 26:33–46.
Taylor, H. C. 1957. Upsetting nature's balance. Cape Dept. Nature Cons. Report 14:61–63.
Taylor, H. C. 1961. Ecological account of a remnant coastal forest near Stanford, Cape Province. J. S. Afr. Bot. 27:153–165.
Taylor, H. C. 1962. Some thoughts on mountain reserves. Cape Dept. Nature Cons. Report 19:31–37.
Taylor, H. C. 1963. A bird's-eye view of Cape mountain vegetation. J. Bot. Soc. S. Afr. 49:17–19.
Taylor, H. C. 1969a. A vegetation survey of the Cape of Good Hope Nature Reserve. Unpubl. M.Sc. thesis, University of Cape Town.
Taylor, H. C. 1969b. Pest-plants and nature conservation in the Winter Rainfall Region. J. Bot. Soc. S. Afr. 55:32–38.
Taylor, H. C. 1969c. Pesplante en natuurbewaring. For. in S. Afr. 10:41–46.
Taylor, H. C. 1970. I.B.P. Check Sheet for the Riversdale Nature Reserve. Unpubl. Report.
Taylor, H. C. 1972a. Fynbos. Veld and Flora 2:68–75.
Taylor, H. C. 1972b. Notes on the vegetation of the Cape Flats. Bothalia 10:637–646.
Taylor, H. C. 1973. Fire in fynbos. Veld and Flora 3:18–19.
Taylor, H. C. 1974. Combat and control of Hakea. Fruit and Fruit Technology Research Inst., Information Bull. 264:1–3.
Taylor, H. C. 1975. Weeds in South Western Cape vegetation. S. Afr. For. J. 93:32–36.
Taylor, H. C. 1976. Notes on the vegetation and flora of the Cedarberg. Veld & Flora 62:28–30.
Thiselton-Dyer, W. 1909. Geographical distribution of plants. Seward's Darwin and modern science. (ex Bews, 1925).
Thunberg & Kotze, J. J. 1940. Some Langeberg forests. J. S. Afr. For. Assoc. 5:32–39.
Trollope, W. S. W. 1973. Fire as a method of controlling macchia (fynbos) vegetation on the Amatole mountains of the Eastern Cape. Proc. Grassld. Soc. Sth. Afr. 8:35–41.
Van der Merwe, C. 1974. 'n Plantegroei opname van die De Hoop-Natuurreservaat. Unpubl. Rep. Cape Prov. Admin., Stellenbosch.
Van der Merwe, N. J. 1962. The position of nature conservation in South Africa. Koedoe 5:1–122.
Van der Merwe, P. 1966. Die flora van Swartbosskloof, Stellenbosch en die herstel van die soorte na 'n brand. Annale Univ. Stell. Vol. 41, ser. A, no. 14.
Van der Zel, D. W. & Kruger, F. J. 1975. Results of the multiple catchment experiments at the Jonkershoek Research Station, South Africa. 2. Influence of protection of fynbos on stream discharge in Langrivier. For. in S. Afr. 16:13–18.
Van der Zel, D. W. & Plathe, D. J. R. 1969. 'n Veldbrand eksperiment op meervoudige opvanggebiede in Jakkalsrivier, Lebanon. For. in S. Afr. 10:63–71.

Van Vuuren, D. R. J. 1973. Die oorsprong en verwantskappe van die Suid-Afrikaanse flora. Publ. v.d. Univ. v.d. Noorde, Reeks C, nr. 25.
Van Zinderen Bakker, E. M. 1967. Upper Pleistocene and Holocene stratigraphy and ecology on the basis of vegetation changes in Sub-Saharan Africa. In: Bishop, W. W. & Clark, J. D. (eds), Background to evolution in Africa. Univ. of Chicago Press.
Van Zyl, P. H. S. 1974. State forest wilderness areas. J. Mountain Club of S. Afr. 77:5–15.
Walsh, B. N. 1968. Some notes on the incidence and control of driftsands along the Caledon, Bredasdorp and Riversdale coastline of South Africa. Dept. of Forestry Bull. 44.
Walter, H. 1968. Die Vegetation der Erde in öko-physiologischer Betrachtung. II. Die gemässigten und arktischen Zonen. Fischer, Jena.
Warming, E. 1909. Oecology of plants. Clarendon Press, Oxford.
Weimarck, H. 1941. Phytogeographical groups, centres and intervals within the Cape Flora. Lund Univ. Årsskrift Avd. 2. Bd. 37 Nr. 5.
Wellington, J. H. 1955. Southern Africa, a geographical study. Vol. I: Physical geography. University Press, Cambridge.
Werger, M. J. A., Kruger, F. J., & Taylor, H. C. 1972. A phytosociological study of the Cape fynbos and other vegetation at Jonkershoek, Stellenbosch. Bothalia 10:599–614.
Werner, H. F. 1949. The cultivation of Buchu. J. Bot. Soc. S. Afr. 35:13–14.
West, O. 1965. Fire in vegetation and its use in pasture management. Commonw. Bureau of Pastures and Field Crops.
Wicht, C. L. 1945. Preservation of the vegetation of the South Western Cape. Special publ. of the Roy. Soc. S. Afr.
Wicht, C. L. 1947. Hydrological Research in South African forestry. Brit. Emp. For. Conf. in S. Afr.
Wicht, C. L. 1948. A statistically designed experiment to test the effects of veldburning on a sclerophyll scrub community. I. Preliminary account. Trans. Roy. Soc. S. Afr. 21:479—501.
Wicht, C. L. 1965. Forest hydrological research in the South African Republic. In: Sopper, W. E. & Lull, H. W. (eds.), Forest Hydrology. Intern. Symp. Forest Hydrol. Pergamon Press, Oxford.
Wicht, C. L. 1969. Forestry and human environment in Southern Africa. S. Afr. For. J. 71:5–11.
Wicht, C. L. 1971a. The task of forestry in the mountains of the Western and Southern Cape Province. Proc. Grassld. Soc. Sth. Afr. 6:20–27.
Wicht, C. L. 1971b. The influence of vegetation in South African mountain catchments on water supplies. S. Afr. J. Sci.: 67:201–209.
Wicht, C. L. 1971c. The management of mountain catchments by forestry. S. Afr. For. J. 77:6–12.
Wicht, C. L. & Banks, C. H. 1963. Die invloed van beheerde brand en bebossing op die wateropbrengs van bergopvanggebiede in die Wintereënstreek. Tydskr. Aardryksk. 2:23–29.
Wicht, C. L. & Kruger, F. J. 1973. Die ontwikkeling van bergveldbestuur in Suid-Afrika. S. Afr. For. J. 86:1–17.
Wicht, M. L. 1971. Creeping invasion of the 'green cancers'. Afr. Wild Life 25:11–14.
Wroughton, F. H. 1948. To burn or not to burn. J. S. Afr. For. Assoc. 16:76–78.

9 The Karoo–Namib Region

M. J. A. Werger*

1. Introduction 233
2. Phytogeography 234
2.1 Boundaries and overlaps with other phytochoria 234
2.2 Floristic characterization and subdivision 236
2.3 Arid disjunctions 240
2.4 Origin of the Karoo–Namib flora 241
3. Vegetation 243
3.1 Structure and physiognomy 243
3.2 Zoogenic changes in vegetation structure and physiognomy 246
3.3 Vegetation of the Namib Domain 248
3.4 Vegetation of the Namaland Domain with exception of the southern Kalahari 259
3.5 Vegetation of the southern Kalahari 265
3.6 Vegetation of the Western Cape Domain 277
3.7 Vegetation of the Karoo Domain 286

References 295

---

* I am very grateful to Dr. O. A. Leistner, Pretoria, for his comments on the manuscript. Photographs are by the author unless otherwise stated.

The Karoo–Namib Region

1. Introduction

The extensive desert and semi-desert areas in the southwestern part of southern Africa, south and west of the Zambezian Domain, house a rich and distinct flora and are classified as the Karoo–Namib Region. North of the coastal mountain ranges parallel to the Indian Ocean, these dry lands cover an area wider than 1000 km from the Atlantic coast to the centre of the plateau at about 26°E. Northwards this arid area gradually becomes narrower, following more or less the escarpment in the northern half of South West Africa until it tapers off on the Atlantic coast about 150 km south of Lobito in Angola (Chapter 7, Fig. 12). In the Kalahari sand area on the high plateau of southeastern South West Africa, southwestern Botswana and the northern Cape Province, the boundary with the adjacent Sudano–Zambezian Region is not clear cut (see Section 2.1). The two floras interdigitate here to a certain extent, but the 250 mm isohyet coincides satisfactorily with a chorological boundary and is here taken as the borderline. Further south on the loamier substrates of the Karoo and South African Highveld the chorological borderline coincides more closely with the 400 mm isohyet. Towards the Atlantic coast the annual precipitation rapidly decreases to less than 100 mm, while off the escarpment, in a narrow coastal belt, mist and fog constitute an important source of moisture (Walter 1936, 1973, Stengel 1971). Parts of the Namib received less than 10 mm of rain in fifty per cent of the years during which records have been kept (Van Zinderen Bakker 1975), but experienced mist on more than 120 days per year on average. At Gobabeb the precipitation derived from fog was 31 mm per year and at Swakopmund 130 mm per year (see Chapter 2). This precipitation, however, is rich in salts (120 kg/ha/yr at Swakopmund; Goudie 1972). Pre-dawn dewfall too is an important source of moisture in the driest parts of the Karoo–Namib Region. Apart from being important to plant life, it is also a main agent in the formation of desert varnish in these areas (cf. Bauman 1976). Linked with a decrease in annual precipitation is a decrease in reliability of the precipitation: whereas the variability in annual rainfall at the boundary between the Karoo–Namib and Sudano–Zambezian Regions lies between 25 and 40 per cent of the average annual rainfall it reaches values of over 80 per cent at the coast near Walvis Bay. Such a variability in rainfall is of equal importance to plant life as the amount of precipitation. Plants are adapted to these unreliable water conditions by differential germination, seed longevity, seed diversification and complex germination mechanisms which are triggered by a precise combination of environmental factors or a sequence of events (cf. Ihlenfeldt 1971, Noy-Meir 1973). Within the Karoo–Namib Region the seasonality of the rainfall also differs greatly: the area north and east of the line Willowmore–Prince Albert–Fraserburg–Pofadder–Aus–Swakopmund receives more than 60 per cent of its annual precipitation in the period from October to March and has to be regarded as summer rainfall area; west of this line follows a narrow belt of about 100 to 150 km wide or less which can be designated as uniform rainfall area; the area between the line Sutherland–Calvinia–Gamoep–Vioolsdrif–Spencer Bay and the Atlantic coast receives less than 40 per cent of its annual precipitation in the period October to March and has to be regarded as winter rainfall area (Wellington 1955, Weather Bureau 1957, and Chapter 2).

This pattern of precipitation is reflected in flora and vegetation patterns of the area.

2. Phytogeography

2.1 *Boundaries and overlaps with other phytochoria*

In the first phytogeographical subdivisions of southern Africa (e.g. Grisebach 1872), the Karoo–Namib Region was not distinguished as a distinct phytochorion but since Bolus (1875) the 'Karoo' was recognized as a phytogeographical unit distinct from the 'Kalahari' and the 'Sudan' (see Chapter 7). Later authors dealing with southern African phytogeography were mainly concerned with defining the boundary between Capensis and the Karoo–Namib Region. In 1947 Lebrun published a phytogeographical map of Africa which has, with several changes and amendments, formed the basis of our present conception of the pattern (see Chapter 7). Lebrun (1947) was still doubtful on the hierarchical status of the Karoo–Namib area, and included it very reluctantly as a separate domain in his Sudano–Zambezian Region. At least since Monod's (1957) review of African chorology, the Karoo–Namib area is generally regarded as a region of its own within the African part of the Palaeotropic Kingdom.

In the coastal lowlands of Angola about 150 km south of Lobito, the boundary between the Karoo–Namib and Sudano–Zambezian Regions is not very sharp. The sandy coastal strip on which the two floras meet is dry and allows some mixing of floras in its communities. Further south the boundary follows more or less the escarpment, though locally, particularly along the wider river valleys the Sudano–Zambezian element deeply penetrates the Karoo–Namib Region. In southernmost Angola and in Kaokoland a species like *Colophospermum mopane* nearly reaches the Atlantic coast along drainage lines.

From west of Windhoek the boundary runs southeast onto the plateau to include the southern Kalahari. Monod (1957) and White (1965, 1970 in Chapman & White 1970) drew the boundary between the Karoo–Namib and Sudano–Zambezian Regions in the Kalahari area just north of the Orange River to coincide largely with the southern distribution limit of *Acacia haematoxylon*. In this way the sandy southern Kalahari was included in the Sudano–Zambezian Region. White (1965) based his decision on a study of the distribution of tree species of the Zambezian Domain, and pointed out that he viewed the Kalahari as somewhat transitional between the two regions. Recently, White (1976) rejected the hierarchical phytochorological classification of Africa and devised a new system in which the sandy southern Kalahari is also not included in the Karoo–Namib Region but in a wide Kalahari–Highveld regional transition zone (see Chapters 7 and 11, Fig. 2).

Appreciating the somewhat transitional character of the flora of this vast sandy area Volk (1966a), Leistner & Werger (1973) and Werger (1973a) found that in the total flora of the southern Kalahari, as delineated by Leistner (1967), the Karoo–Namib element certainly predominates over the Zambezian or Sudano–Zambezian elements. These authors therefore included the southern Kalahari in the Karoo–Namib Region. Using a procedure suggested by

Tolmachev (1971), Werger (1973a) provided further support for this view. He compared the flora of the southern Kalahari with floras of various areas well within the Karoo–Namib and the Sudano–Zambezian Regions. He found that the southern Kalahari flora shows much greater similarity to floras of the Karoo–Namib Region than to floras of the Sudano–Zambezian Region.

From the southern Kalahari as defined by Leistner (1967) the border between the Karoo–Namib and Sudano–Zambezian Regions continues southeastwards to a point east of Kuruman (Coetzee & Werger 1975) and then southwards along the minor escarpment between the Cape Middleveld and the South African Highveld where it crosses the Orange River near Luckhoff, just upstream of the Orange–Vaal confluence (Werger 1973b). Although the chorological border is quite clear in this area there is a considerable transgression of the Karoo–Namib element into the grassland flora of the Sudano–Zambezian Highveld. This is due to the severe mismanagement, particularly overgrazing, of the vegetation for more than a century. Karoo–Namib species prove to be more competitive than the Sudano–Zambezian ones under these circumstances and this has resulted in the replacement of Sudano–Zambezian grasslands by Karoo–Namib dwarf shrub vegetation on a fairly large scale. Elsewhere, over extensive areas, the Karoo–Namib dwarf shrubs have intruded the Sudano–Zambezian grasslands, so that a vegetation with a mixed flora has resulted (cf. Acocks 1953, 1964, Werger 1973b). The Karoo–Namib species have intruded into the South African Highveld over large distances and some have advanced deep into the Orange Free State and Lesotho. Other species have not yet progressed that far but the process is continuing at a fast rate. Satellite imagery enabled Jarman & Bosch (1973) to conclude that karroid vegetation had intruded the grasslands in a northeasterly direction over distances varying from 43 to 70 km since 1953.

In the eastern Cape Province the boundaries of the Karoo–Namib Region against the other phytochoria are very difficult to define. Local relief is strongly varied and the floras of five phytochoria which meet here (Capensis, Karoo–Namib, Sudano–Zambezian, Afromontane and Indian Ocean Coastal Belt) result in a mosaic of communities each with different chorological affinities or in communities composed of a chorologically mixed flora.

In the south and extreme southwest the Karoo–Namib Region is clearly segregated from Capensis and since Bolus (1875), this boundary always has been recognized as such. It is not a simple line, however. Several mountain ridges run parallel to the coast in the southern and southwestern Cape Province, causing large differences in the amount of precipitation received at various sites. Depending on their height above the surroundings, the direction of their slopes and their distance from the coast, the mountain ridges, koppies and other elevated areas receive a fair to large amount of rainfall, particularly on their southern slopes and tops. They are separated by narrow to very wide arid valleys and depressions. The Karoo–Namib flora occupies these arid parts whereas the wetter slopes, tops and ridges carry 'renosterbosveld' or 'fynbos' communities constituted by Cape flora (see Chapter 8) (Compton 1929a). Levyns (1950) has described in detail such a situation from the vicinity of Ladismith in the Little Karoo where exclaves of Cape vegetation and flora cap nearly all koppies. According to Levyns (1950, 1964) four koppies, rising about a hundred metres above the surrounding plains which carry a Karoo–Namib flora, are situated in a wet to dry rainfall gradient. The first and wettest koppie carries fynbos on its summit and southern slopes, and renosterbosveld, dominated by *Elytropappus rhinocerotis*, on its northern slope; on the second koppie fynbos is found on the top, renosterbosveld on the southern slope, and succulent karroid bush on the northern slope; on the third koppie renosterbosveld covers the top and the southern slope, whereas succulent karroid bush covers the northern slope; and the fourth and driest koppie carries succulent karroid bush throughout, although with a more luxuriant growth on the summit and southern slope (see Chapter 8, Fig. 14).

In the coastal lowlands of the southwestern Cape the boundary between Capensis and Karoo–Namib Region is not always as clear as in the mountains and some mixing of floras occurs here in the strandveld and coastal renosterbosveld (see Chapter 8).

2.2   *Floristic characterization and subdivision*

The Karoo–Namib flora is characterized by the strong development of the families Asteraceae, Poaceae (particularly the tribe Stipeae), Aizoaceae, Mesembryanthemaceae, Liliaceae, and Scrophulariaceae (Werger 1973a). Amongst the grasses the 'whitish' coloured tenuous to robust Stipeae are prominent and rich in species. Woody dwarf shrubs preponder over wide areas and occur particularly in two growth forms: dwarf shrubs with small, ericoid or finely dissected, sometimes strongly rolled leaves with hairs or a strongly xeromorphic structure; and succulents, both stem and leaf succulents. The former growth form is especially widespread among the Asteraceae, while succulents occur in numerous families, though most commonly in Mesembryanthemaceae, Euphorbiaceae, Crassulaceae, Asclepiadaceae and Liliaceae. Trees are not common except on levees along the larger rivers. They mostly belong to the Mimosaceae, Ebenaceae and Anacardiaceae.

Many genera are endemic to the Karoo–Namib Region, such as *Leucophrys, Kaokochloa, Calicorema, Hypertelis, Plinthus,* most Mesembryanthemaceae, *Grielum, Adenolobus, Ceraria, Sisyndite, Augea, Nymania, Ectadium, Microloma, Welwitschia,* and many more (Volk 1964, 1966). A very large number of species is endemic to the Karoo–Namib Region. These comprise both widespread Karoo–Namib species as well as less widespread ones and local endemics.

Since Monod (1957) the Karoo–Namib Region has mostly been regarded as consisting of three domains: the Karoo Domain, the Namaqualand Domain, and the Namib Domain. Later authors, like Troupin (1966) and Aubréville (1975), redrew the boundaries between these three domains. Troupin's map cannot be regarded as an improvement compared with that by Monod, however. The subdivisions need adjustments as some boundaries cut across floristically and ecologically homogeneous areas while some clear chorological boundaries are not recognized. Domains should be clearly characterized both chorologically and ecologically. The new subdivision of the Karoo–Namib Region proposed below (Chapter 7, Fig. 12) is based on this premise.

(a) Namib Domain

The Namib Domain comprises the extremely arid, sandy, rocky or calcareous coastal strip from about 150 km south of Lobito to about 30 km south of

Alexander Bay at the mouth of the Orange River. This coastal strip is situated between the Atlantic Ocean and the escarpment, thus varying in height from sea-level to locally about 1000 m. Virtually the entire area receives less than 100 mm of rain per year. These gradients in altitude and consequently in precipitation, which correlate with other environmental gradients and generally run parallel to the coast, implicate a biological gradient. Generally speaking, however, this coastal strip is sufficiently homogeneous to be regarded as a phytochorological unit. The small area south of the Orange River traditionally is not included in the Namib. In that area the flora and habitat are similar to that north of the Orange River, however, and there is no reason why it should be excluded from the Namib. Zoogeographically too this area is similar to the area further north, and can be regarded as an extension of the Pro-Namib (Namib fringe), according to Coetzee (1969). There are a considerable number of endemics or near-endemics in the Namib Domain, e.g. *Acanthosicyos horrida, Arthraerua leubnitziae, Sarcocaulon mossamedense, Euphorbia virosa* subsp. *arenicola, Stipagrostis subacaulis, S. gonatostachys, S. namibensis, S. garubensis, S. lutescens, S. sabulicola, Trianthema hereroensis, Adenolobus pechuelii* and *Welwitschia mirabilis* (cf. De Winter 1965, Giess 1971).

(b) Namaland Domain

The Namaland Domain covers the narrow escarpment belt inland of the Namib Domain, but southwards it broadens gradually, and south of Windhoek it comprises extensive areas on the plateau, including the southern Kalahari as delineated by Leistner (1967). The southern border runs mainly a few km south of the Orange River from the vicinity of Vioolsdrif via the Pofadder area to the vicinity of Upington. The substrate differs markedly within the Namaland-Domain: litholithic and sandy loam soils on the escarpment and the southwestern part of the plateau, as against deep sandy (and to a small extent calcareous and silty) soils in the southern Kalahari. This brings about a clear contrast in flora and vegetation and it is relevant to separate the southern Kalahari as a distinct sub-domain from the remainder of the Namaland Domain. The Namaland Domain contains many endemics or near-endemics, e.g. *Panicum arbusculum, Stachys burchelliana, Schotia afra* var. *angustifolia, Berkheya chamaepeuce, Barleria lichtensteiniana, Curroria decidua, Euphorbia gariepina, E. gregaria, E. avasmontana, Leucosphaera bainesii, Zygophyllum dregeanum* and *Z. microcarpum* (cf. Volk 1964, 1966, De Winter 1965, Merxmüller 1966–1972, Werger 1973b, Werger & Coetzee 1977).

(c) Western Cape Domain

The Western Cape Domain comprises the coastal strip and the escarpment mountains in South Africa, south of the Namib Domain. From the Richtersveld it extends northwards into South West Africa as far north as the vicinity of Aus and occurs in a wedge of rugged, mountainous country between the Namib and Namaland Domains. In the south the Little Karoo, lying between the Cape mountain ridges, the Langeberg and the Swartberg ranges, is included. The area thus included in the Western Cape Domain receives its rainfall during winter and, for a small part, throughout the year. The summer rainfall area is excluded from the domain. There is some chorological variation within the Western Cape Domain: the Little Karoo has a number of endemic species, and so has the Van Rhynsdorp area and the Richtersveld, but the differences and endemisms are not of such a magnitude that distinction of separate domains or subdomains would be justified. Along the western slopes of the mountains in the Richtersveld–Aus area, which run parallel to the coast, fog and air humidity from the Atlantic cause some interesting distribution patterns. There are apparently quite distinct and stable zones of humidity on these mountains, and as a result some species are distributed in belts, a phenomenon particularly noticeable in the Crassulaceae (Tölken, pers. comm.). The Western Cape Domain is very rich in succulent species, mainly belonging to Mesembryanthemaceae, Crassulaceae and Euphorbiaceae, but also Asclepiadaceae, Portulacaceae, Asteraceae, Geraniaceae, Zygophyllaceae, Liliaceae, Chenopodiaceae, etc. Many species are endemics or near-endemics, e.g. many Mesembryanthemaceae and Crassulaceae, *Euryops annuus*, *E. dregeanus*, *E. namibensis*, *E. namaquensis*, *Leucoptera nodosa*, *Ehrharta barbinodis*, *Aristida dasydesmis* and *Stipagrostis zeyheri* subsp. *macropus* (cf. De Winter 1965, Nordenstam 1969, 1976, Acocks 1971, Herre 1971).

(d) Karoo Domain

The Karoo Domain comprises the summer rainfall area east of the Western Cape Domain and south of the Namaland Domain. In the centre of the South African plateau it borders on the Zambezian Domain of the Sudano–Zambezian Region. In the Karoo Domain the Asteraceae are represented by a great number of species most of which are strongly xeromorphic dwarf shrubs. Endemics or near-endemics are numerous and include *Nestlera humilis*, *Pteronia glauca*, *Thesium hystrix*, *Zygophyllum gilfillani*, *Euphorbia aequoris*, and many others (cf. Werger 1973b).

Many species are not restricted to one domain but are more widespread; some are common to two or three domains, others occur over virtually the entire Karoo–Namib Region, and again others are not restricted to the region but penetrate into the drier areas of the adjacent Sudano–Zambezian Region. Overlap with Capensis is not very strong, however. There is also a large group of species which is mainly distributed in the central plateau area, where the Karoo–Namib and Sudano–Zambezian Regions meet. Of all these groups many examples can be found in Volk (1964, 1966), Acocks (1971), Werger (1973b) and Werger & Coetzee (1977).

Within phytochoria there are usually definite areas of relatively small size showing concentration of endemics or of closely-related species (cf. Weimarck 1941, Croizat 1952, 1968, Volk 1966, Nordenstam 1969, Burtt 1971, Exell & Gonçalves 1973). Within the four Karoo–Namib domains such centres can also be distinguished.

Volk (1966) briefly mentioned the Namib centre, defining it as the narrow coastal belt, and pointing out that too little was still known to give a definite opinion on its phytogeographical status. This 'centre' is obviously identical to what is here distinguished as Namib Domain.

Volk (1966, cf. 1964) also distinguished a Kaokoland centre. This is an escarpment area extending from the Brandberg northwards into southern Angola.

Nordenstam (1974) also recognized this centre. It is characterized by several endemic species like *Kaokochloa nigrirostris, Asthenatherum mossamedense, Sesamothamnus guerichii, S. benguellensis, Acacia robynsiana, A. montis-usti, Balanites welwitschii,* several species of *Commiphora, Gossypium triphyllum,* several species of *Petalidium, Microloma hereroense, Engleria africana, Pachypodium lealii,* and several more (cf. Volk 1964, 1966, Giess 1971, Tinley 1971, Nordenstam 1974). It is interesting to note that some mountains or small mountain ranges within the Kaokoland centre have their own endemics. Nordenstam (1974) presented such a list for the Brandberg.

Within the Namaland Domain another centre can be distinguished. The hot and rugged mountainous area of the lower Orange River between Vioolsdrif and Augrabies and the mountains in the Warmbad–Karasburg area, constitute the Karas centre. It is characterized by several endemics or near-endemics, including *Monechma spartioides, Antherothamnus pearsonii, Sutera ramosissima, Commiphora gracilifrondosa, Ozoroa namaensis, Rhus populifolia, Schotia afra* var. *angustifolia, Crassula densa, Maerua gilgii, Pellaea deltoidea* and *Hermbstaedtia glauca.*

The Gordonia centre, tentatively distinguished by Volk (1966), and chacterized by species like *Acacia haematoxylon* and *Stipagrostis amabilis,* is not regarded as a mere centre. It is here distinguished as the southern Kalahari Subdomain of the Namaland Domain.

In the Western Cape Domain three centres have been recognized: the Little Karoo centre, the Van Rhynsdorp centre, and the Gariep centre.

The Little Karoo centre is situated between mountains capped by Cape fynbos and renosterbosveld, and is geographically a virtual exclave. As in the entire Western Cape Domain, succulent species are well represented (cf. Compton 1929b), particularly members of the Mesembryanthemaceae, Crassulaceae and Asclepiadaceae (tribe Stapelieae). Endemic or nearly endemic genera of the Mesembryanthemaceae include *Gibbaeum, Antegibbaeum, Cerochlamys* and *Zeuktophyllum,* while *Glottiphyllum* has a marked centre within this area (Nordenstam 1965, 1969, Herre 1971, Leistner 1977).

The Van Rhynsdorp centre is located in the lowlands near Van Rhynsdorp and comprises sandy plains strewn with quartzite pebbles such as the Knersvlakte, and rocky outcrops. Endemic species found here include *Euryops namaquensis, Leucoptera oppositifolia, L. subcarnosa* and some species of *Senecio (Kleinia), Othonna, Babiana, Zygophyllum* and *Pelargonium.* Also some genera of the Mesembryanthemaceae are endemic or near-endemic (e.g. *Argyroderma, Dactylopsis, Oophytum*) (Nordenstam 1965, 1969, 1976, Herre 1971).

Particularly important is the Gariep centre, recognized by Nordenstam (1969), which largely coincides with Volk's (1966) Lüderitzland centre. It comprises the mountainous area of the Richtersveld and the escarpment mountains extending further north as far as the vicinity of Aus. Several genera of Mesembryanthemaceae are endemic or nearly so, such as *Arenifera, Dracophilus* and *Juttadinteria,* while others, like *Conophytum, Lithops* and *Drosanthemum* have endemic species here. Species of *Aloe (A. pillansii, A. ramosissima, A. pearsonii), Pachypodium (P. namaquanum), Acacia (A. redacta), Adenoglossa, Crassula, Euphorbia, Pelargonium* and *Pteronia* are endemic too in the Gariep centre (Volk 1966, Nordenstam 1969, 1976, Edwards & Werger 1972).

Within the presently defined Karoo Domain, Nordenstam (1969) recognized a

Western Upper Karoo centre situated in the mountainous areas of the Nieuweveld, Roggeveld and Hantam Mountains. It is a marked centre of the genus *Euryops*. It is not typical of the Karoo Domain, however, since it has strong affinities with Capensis due to its high location in the mountains. The same area was already distinguished as part of the North–West Cape floral centre by Weimarck (1941).

2.3 *Arid disjunctions*

Early in the present century it was noticed that there is a remarkable disjunct distribution pattern in many taxa of the African dry areas. This pattern exists both in plant and animal taxa. The most common disjunct pattern is that of a taxon occurring in the Karoo–Namib Region and again (or a closely related taxon) in the arid areas of northeastern Africa (Somali, Kenya, Ethiopia). Sometimes the northern distribution area is larger and includes part of the Arabian peninsula and areas deeper into Asia or a vast part of the Sahara or Sahel. Engler (1921) was the first author to draw attention to this remarkable pattern, and since then the topic has been dealt with by numerous authors (Range 1932, Lebrun 1947, Moreau 1952, Balinsky 1962, Volk 1964, 1966, De Winter 1965, 1966, 1971, Winterbottom 1967, Verdcourt 1969, Burtt 1971, Lebrun 1971, 1976, Monod 1971, Knapp 1973, Werger 1973a, b, c). Detailed surveys of the plant taxa involved are presented by Monod (1971) and De Winter (1971). De Winter distinguished six categories of disjunctions:

(1) families limited to the southern and northern arid areas of Africa and in some cases also present in Asia, e.g. Neuradaceae, Salvadoraceae, Wellstediaceae;
(2) genera limited to the southern and northern arid areas, e.g. *Echidnopsis, Citrullus, Kissenia, Stipagrostis*;
(3) similar as the previous category, but also more widespread, e.g. *Aizoon, Moringa, Forskohlea, Fagonia, Rogeria*;
(4) species limited to the southern and northern arid areas of Africa and in some cases also occurring in Asia or Europe, e.g. *Corbichonia decumbens, Geigeria acaulis, Suaeda fruticosa, Kissenia spathulata, Cypholepis yemenica, Enneapogon desvauxii, Eragrostis trichophora, Fingerhuthia africana, Stipagrostis obtusa, Triraphis pumilio, Mesembryanthemum nodiflorum, Commicarpus squarrosus, Rogeria adenophylla, Oligomeris linifolia, Zygophyllum simplex*;
(5) species with vicariant subspecies or varieties in the southern and northern arid areas, e.g. *Cleome angustifolia, Stipagrostis ciliata, S. hirtigluma*;
(6) vicariant, closely related species, one confined to the northern, the other to the southern arid areas, e.g. *Huernia insigniflora* vs. *H. somalica, Stapelia revoluta* vs. *S. prognatha, Heliotropium hereroense* vs. *H. rariflorum, Stipagrostis uniplumis* vs. *S. papposa* and *S. foexiana, Tricholaena capensis* vs. *T. teneriffae, Wellstedia dinteri* vs. *W. socotrana*.

Apart from these disjunctions there are also a number of taxa which show a contiguous distribution but are particularly strongly represented in both arid areas.

Engler (1921) mentioned the possibility of long-distance dispersal to explain these disjunctions, but many more details have become known since. Well-developed antitelechoric mechanisms have been observed in numerous species of the arid regions, in some cases even in taxa with a widely disjunct distribution (cf. Stopp 1958, Ihlenfeldt 1971). A more likely explanation has been generally adopted by botanists and zoologists (Balinsky 1962, Volk 1964, De Winter 1966, 1971, Winterbottom 1967, Verdcourt 1969, Burtt 1971, Monod 1971, Werger 1973b, c), who interpreted the present-day disjunctions as a relict situation of formerly continuous distribution areas. In the past the two arid areas were connected on one or more occasions by an arid corridor stretching across Africa from Somali via Kenya, Tanzania, and Zambia, to Botswana, South West Africa, and South Africa. Also Cooke's (1962) hypothetical vegetation map of southern Africa at 50 to 60 per cent of the present rainfall suggested such a possible arid corridor. Balinsky (1962) showed that the northeastern and southwestern arid regions are at present still connected by an 'arid' tract in which the precipitation is less than 10 mm per month on average during at least three consecutive months per year (Fig. 1). The course of this 'arid' tract coincides fully with that of the suggested arid corridor (see also Chapters 2 and 6).

Opinions on how long ago the arid area was continuous seem to differ; Volk (1964) and Burtt (1971) seem to think of a more remote period of time (Tertiary, older than the tropical African flora) than other authors, who speculate on comparatively recent changes in precipitation. Evidence for such fluctuations in the rainfall pattern over Africa is provided by palaeoecological and other studies (cf. Van Zinderen Bakker 1964, 1969, 1975, Van Zinderen Bakker & Coetzee 1972, Rust & Wieneke 1973, Cooke 1975, see Chapter 6).

2.4 *Origin of the Karoo–Namib flora*

It is generally believed that arid conditions have persisted in southwestern Africa for a very long time. This is concluded mainly from the large number of endemic species in the arid areas, the numerous and advanced adaptations of taxa (both plant and animal) to arid conditions, and the antiquity of relicts like *Welwitschia mirabilis*. This does not necessarily imply that the arid conditions were geographically stable; there is evidence that the arid zones have sometimes shifted in latitude (cf. Goudie 1972, Cooke 1975, Van Zinderen Bakker 1975, and Chapters 6 and 26). According to Raven & Axelrod (1974, see Chapter 5) arid and semi-arid conditions occurred in Africa on a limited scale at least since Cretaceous times when Africa and South America were still joined in West Gondwanaland. Some 85 million years ago Africa became separated from South America and the southern Atlantic became wide enough for a cold water current to develop. This resulted in the development of the Namib desert and the dry hinterlands which began during the Oligocene. Subsequently, Antarctic and northern hemisphere glaciations repeatedly brought about changes in the atmospheric and oceanic circulation and caused the Benguela Current to shift further northwards. This involved a latitudinal shift of the Namib desert and dry hinterlands, so that arid and semi-arid conditions prevailed as far north and inland as northeast Angola and parts of Zaïre (Van Zinderen Bakker 1975, see Chapter 6). It also involved a northward shift of the southern boundaries of the dry areas and a northward extension of the Cape flora (cf. Levyns 1964, compare Chapter 8). Adamson (1960) had previously also postulated a former more extensive distribution of the temperate flora. He had explained the reduction of this area not by means of shifting climatic belts, but by an increase in aridity and in the size of the

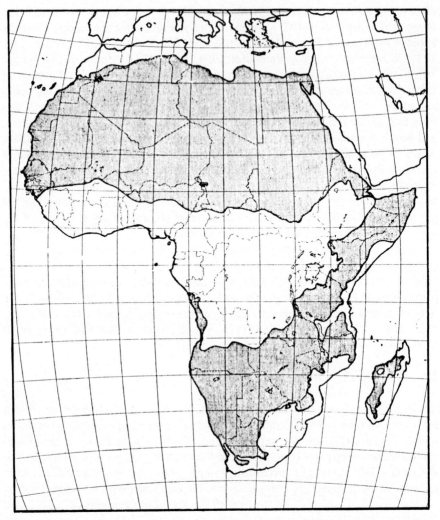

Fig. 1. Arid corridor. In the hatched area the rainfall is less than 10 mm per month in at least three consecutive months (after Balinsky 1962).

distribution area of the dry flora.

Bews (1925) regarded the Karoo flora to be a product of specialization of the tropical flora bordering on it to the north and east, in response to arid conditions. Also Levyns (1964), though recognizing the links between the Karoo and fynbos floras, considered the Karoo flora to have a northern origin. She suggested two separate migration waves and regarded the Karoo flora as younger than the fynbos flora.

Acocks (1953), followed by Axelrod & Raven, Chapter 5, considered the Karoo flora to be derived both from the Cape flora and the Tropical flora, but with a more dominant influence of the Cape flora. He thus also regarded the Karoo flora to be younger than the Cape and Tropical floras. Particularly the similarity in growth form of the dwarf shrubs dominant in the Karoo and the fynbos, as well as the affinities between the two floras are stressed by Acocks. There is, according to him, a very clear transition from Fynbos and Arid Fynbos through Mountain Renosterbosveld and Western Mountain Karoo to the typical karoo vegetation, both in species composition and growth form. However, one should consider that possibly not only evolutionary but rather (palaeo)ecological factors may be relevant in explaining this transition in species composition, and that the way in which a veld type is defined may also be relevant in this respect.

3. Vegetation

3.1 *Structure and physiognomy*

In the arid Karoo–Namib Region the vegetation shows a remarkably high degree of structural and physiognomic diversity and includes some most peculiar formations. Patches of evergreen and deciduous forests and woodland are confined to the banks of perennial streams or occur along the major watercourses. Thickets and open shrublands, or bushlands, or thorn shrubs, sometimes including low trees, develop locally in overgrazed areas, or in wide, shallow depressions and washes in which seasonal run-off concentrates. These are predominantly deciduous. Extensive patches of open, deciduous, low shrubland can be dominated by the bignoniaceous species *Rhigozum trichotomum* or *Catophractes alexandri*. Evergreen shrublands with an understorey of dwarfshrubs and grasses cover many slopes and rocky sites in the Karoo Domain. Evergreen succulent shrublands and thickets are particularly important in the southern part of the Karoo–Namib Region fringing Capensis whereas thorny and non-thorny evergreen succulent shrublands and thickets dominated by species of *Euphorbia*, are found on some rocky slopes, particularly in the Namaland Domain. Open thorn savanna vegetation occurs particularly in the southern Kalahari while open savanna with succulent, liliaceous trees is locally common, especially in the Namaland Domain. Rugged areas, as they occur along the escarpment and on extensive rocky outcrops, carry open shrubland vegetation with a varying proportion of succulents, deciduous shrubs and grasses.

Open, steppe-like grasslands are common on sandy areas, such as the southern Kalahari, the Namib and parts of Bushmanland, but they also cover some pan floors. Sandy parts of Bushmanland locally carry an open grassland formation consisting of shrubby and spiny, caespitose grasses, mixed, after good rains, with ephemeral succulents and other herbs. In the Namib fringes, the so-called 'Vornamib' or 'Pro-Namib', low shrubs of *Commiphora* and other genera can occur widely spaced in the grasslands (Fig. 5). Open, ephemeral or annual grasslands can develop in the Namib Domain after good rains, whereas dense grasslands seasonally cover the silty beds of the major dry rivers, notably the Nossob and Auob Rivers in the southern Kalahari. Reed swamps and semi-aquatic and aquatic communities sporadically cover very limited areas in pools, along streams and at the coast.

Dwarf scrub is very typical for the Karoo–Namib Region. It is nearly always evergreen, and can be both succulent or narrow-leaved. Dwarf scrub formations of the Karoo–Namib Region range from fairly closed on relatively moist sites to extremely open formations in the desert and severely overgrazed areas. The narrow-leaved dwarf scrub formations are typical of the Karoo Domain, where they cover the wide plains between the koppies and mountain ridges, while on the more broken sites and on slopes with stony soils larger shrubs and grass tufts can be common and frequently lead to shrublands (Fig. 31). The grass component in these dwarf scrub formations varies greatly depending on soil type and degree of overgrazing. Succulent dwarf scrub formations are particularly characteristic of the Western Cape Domain. They include formations with and without emergent shrubs or occasional low trees. The emergents are often, though not always, succulent too. In dwarf scrub formations of mountain slopes emergents are more common than in those of the plains. Grasses are rare in the succulent dwarf scrub formations, but temporary formations of ephemeral succulents and other herbs can develop following rains. Adamson (1938a) and Levyns (1962) associated the increase of succulence towards the western and southern continental margins as compared with the less succulent central parts of the Karoo–Namib Region with a decrease in likelihood of severe frost during winter in the same directions. It is doubtful, however, whether Knapp's (1973) suggestion that succulents are particularly abundant in Africa in areas with a relatively high air humidity holds as a general rule for southern Africa.

True desert formations are restricted to the Namib Domain (Figs. 10 and 11), though entirely or nearly bare areas also are found on some pan floors in the other domains. The true desert formations of the Namib range from bare areas absolutely devoid of vegetation, to areas with bare rocks or shifting sands and widely interspaced grass tufts, dwarf shrubs or shrubs. Near the coast the mostly succulent dwarf shrubs collect wind-blown sand between their branches so that they are partially covered; but due to rapid growth of the branches these shrubs are seldom entirely covered by the sand. Thus, hummocks up to 2 m high develop consisting of dwarf shrub and sand, with their surface covered entirely by the tips of the branches. A similar development can be observed in the spiny, mainly leafless shrub *Acanthosicyos horrida* which is especially common at the mouth of the Kuiseb River. In the Namib desert lichen formations are found which consist mainly of unattached lichens. In some areas, such as in the vicinity of the Swakop River, *Welwitschia mirabilis* occurs in peculiar formations in which it completely dominates the vegetation. Further northwards this species, which may be regarded as a dwarf tree, occurs locally in the grassland of the Namib fringes and does not determine the vegetation physiognomically.

The restricted possibilities for plant life in the Karoo–Namib Region have induced a large variety of growth and life forms, designed to withstand the severity of the climate and the habitat and to survive the unfavourable periods.

Adaptation to the dry climate is particularly expressed in various forms of succulence, but also in other forms of xerophytism like narrow or ericoid leaves, finely dissected leaves, hairy surfaces, sclerophylly, leaflessness for prolonged periods of the year, the development of desiccation-tolerant foliage, but also in ephemerality and in the development of a dense cushion form and a relatively minor development of plant tissue above-ground. Adaptation to the wind-blown and shifting sands is particularly expressed in the cushion form of dwarf shrubs, a caespitose shrubby form in grasses, and the development of a life form transitional between hemicryptophytes and geophytes, which either possesses underground storage organs or a marked ability to form suckers. A trailing growth form is common among herbaceous and slightly woody plants on sandy sites.

Among the species with succulent growth form there are:
(1) plants which are mainly stem succulents, such as *Pachypodium namaquanum*,

Euphorbia gregaria, E. avasmontana, E. virosa, Sarcocaulon spp., *Trichocaulon* spp., *Huernia* spp., *Stapelia* spp., *Hoodia gordonii, Commiphora* spp., *Moringa ovalifolia, Psilocaulon absimile*;
(2) plants which are mainly leaf succulents, such as *Lithops* spp., *Conophytum* spp., *Fenestraria* spp., *Drosanthemum* spp., *Zygophyllum* spp., *Salsola* spp., *Aloe claviflora, A. hereroensis, Sansevieria aethiopica, Anacampseros* spp. These include both plants with an erect, woody stem, and rosette plants;
(3) plants which are both stem and leaf succulents, such as *Aloe pillansii, A. dichotoma, A. ramosissima, Cotyledon* spp., *Crassula* spp., *Ceraria namaquensis, C. fruticulosa, Portulacaria afra, Talinum caffrum.*

Many succulents, particularly those usually growing under or near other taller species, have short and shallow root systems. Some of them possess deciduous, slender rootlets in the upper soil layers. These rootlets drop off in the dry season (cf. Cannon 1924). Several others have short but somewhat succulent roots (cf. Marloth 1908).

Among these succulents *Pachypodium namaquanum* (Fig. 29), *Aloe pillansii* (Fig. 28), *A. ramosissima* and *A. dichotoma* (Fig. 17) possess a relatively thick stem with a 'rosette' of leaves on the top (pachycaul).

It should be noticed that there are various types of succulents. In the Karoo–Namib Region there are at least two clear types: a xerophytic form in which the plants are somewhat woody and possess a strong, waxy cuticula, and a form in which these characters are less developed. Succulents of the first type include the species of *Aloe* and *Euphorbia*. These plants are more abundant in the arid parts of the Karoo–Namib Region where there is no influence of fog.

Non-succulent xeromorphic species, with small leaves of various types include many of the asteraceous dwarf shrubs, such as *Pentzia* spp., *Nestlera* spp., *Eriocephalus* spp., *Pteronia* spp., *Osteospermum* spp., *Felicia muricata, Chrysocoma tenuifolia, Pterothrix spinescens, Phymaspermum parvifolium,* but also *Sutera* spp., *Hermannia* spp., *Aizoon* spp., *Plinthus* spp., *Lightfootia albens, Nenax microphylla, Walafrida* spp., and many others. Evergreen sclerophyllous species include various species of *Euclea* and *Maytenus, Rhus erosa, Rhus undulata, Montinia caryophyllacea, Olea africana* and *Welwitschia mirabilis,* whereas mainly leafless, non-succulent or hardly succulent species include *Sisyndite spartea, Hermbstaedtia glauca, Thesium hystrix, Acanthosicyos horrida, Monechma spartioides* and *Cadaba aphylla.* All of these usually have deep root systems, variously with or without shallow lateral roots. Some of the narrow-leaved dwarf shrub species show an anomalous, eccentric type of secondary growth resulting in the formation of periderm bridges and ultimately leading to the longitudinal splitting of the main stem into a number of separate plants (cf. Theron et al. 1968). Desiccation-tolerant foliage is found in several grasses such as *Enneapogon desvauxii, Oropetium capense, Eragrostis nindensis, Sporobolus lampranthus,* in several ferns and in other species such as *Lindernia intrepida* (Gaff 1971, Gaff & Ellis 1974, Walter 1973).

Herbaceous ephemerals as well as succulent ephemerals are common; some examples of the former are *Cleome angustifolia, Euphorbia glanduligera, Semonvillea* spp., *Arctotis leiocarpa, Ursinia* sp., *Manulea schaeferi,* whereas the latter include *Trianthema triquetra, Mesembryanthemum magniflorum, M. cryptanthum* and *Pherolobus maughanii.*

The caespitose shrubby form is shown by the grasses *Stipagrostis namaquensis, S. amabilis* (Fig. 19), and in a different way also by *S. brevifolia* (Fig. 18).

True geophytes are common and include many species of the Liliaceae, Amaryllidaceae and Iridaceae, as well as some other families. They vary from species with enormous bulbs, such as *Pseudogaltonia, Brunsvigia* and *Boophane*, to those with small bulbs or corms, like *Cyperus, Moraea* and *Ixia*.

Species with a life form transitional between hemicrytophytes and geophytes, as distinguished by Leistner (1967), are typical of loose sandy substrates and include *Vahlia capensis, Neuradopsis austro-africana, Cassia mimosoides, Jatropha erythropoda, Adenia repanda, Adenium oleifolium, Cynanchum orangeanum, Harpagophytum procumbens* and others. Some species, like *Pachypodium succulentum*, are intermediate between geophytes and chamaephytes.

A comparative study of life forms in the southern Kalahari has been carried out by Leistner (1967) and is shown in Table 1.

Table 1. Life form spectra of different habitats and ecological species groups of the southern Kalahari (after Leistner 1967).

	No. of spp.	S	M	N	Ch	H			G		T			
						H1		H2	G1	G2	T1	T2	T3	T4
						H1.1	H1.2							
Southern Kalahari (total flora)	444	2.2	2.7	8.3	12.2	7.2	3.2	10.1	14.0	7.4	19.9	7.9	2.7	2.2
All species growing on pan soils	58	1.7	0	5.2	10.3	13.8	12.0	3.5	3.5	13.8	20.6	3.5	3.5	8.6
All species growing on river soils	163	1.9	4.3	9.2	11.6	8.0	5.5	11.0	5.5	8.6	22.1	4.9	2.5	4.9
Species confined to fine soils	41	0	0	0	2.4	17.1	19.5	12.2	2.4	4.9	19.5	4.9	4.9	12.2
Species confined to calcrete outcrops	26	7.6	0	11.6	23.0	11.6	0	27.0	0	0	11.6	0	7.6	0
All species growing on dune crests	114	0	8.7	16.7	7.9	7.0	3.5	2.6	22.9	4.4	16.7	5.3	2.6	1.7
Species growing only on fairly compact to very loose sand	49	0	4.1	18.4	8.2	10.2	2.0	2.0	32.8	2.0	12.2	4.1	2.0	2.0

S	: succulents		H1.1	: non-rhizomatous caespitose H
M	: microphanerophytes		H1.2	: rhizomatous caespitose H
N	: nanophanerophytes		H2	: non-caespitose H
Ch	: chamaephytes		T1	: summer therophytes
G	: geophytes (cryptophytes)		T2	: winter therophytes
G1	: intermediate to H		T3	: all year therophytes
G2	: typical geophytes		T4	: potential biennials
H	: hemicryptophytes			

3.2 Zoogenic changes in vegetation structure and physiognomy

As in most arid and semi-arid vegetation, autogenic succession is not common in the dry vegetation types of the Karoo–Namib Region, though some allogenic succession does occur (Werger & Leistner 1975, cf. Bews 1925, Noy-Meir 1973).

Marked changes in the vegetation correlated with metereological and climatic fluctuations in the southern Kalahari have been described by Werger & Leistner (1975), and zoogenic action, particularly overgrazing and trampling, can bring about dramatic changes in vegetation structure and physiognomy as well as in floristics in these arid regions. These changes are most common and radical in the marginal areas of the Karoo–Namib Region fringing the Sudano–Zambezian Region. Here the vegetation is most suitable for pasture lands, and overgrazing is most common. As in the savanna areas, overgrazing and trampling disturbs the balance between the grasses and woody plants in the vegetation (see Chapter 10) and leads to an increase of woody plants. Roux (1966) has shown that in the Upper Karoo the moisture brought by rains early in the season is mainly consumed by the shallow but intensive root systems of the grasses, while moisture from rains later in the season is taken up mainly by the deeper and extensive root-

Fig. 2. Severe overgrazing in the southern Kalahari has led to the dwarf shrub vegetation at this side of the fence, where the unpalatable *Chrysocoma polygalifolia* has replaced the grasses which are still abundant at the undisturbed other side of the fence. The dwarf shrub *Rhigozum trichotomum* is also abundant (photo O. A. Leistner).

systems of the dwarf shrubs. This difference is probably not only caused by differences in their root systems but also by an intrinsic difference in their physiological periodicity. Severe grazing results in a lessening of the amount of water used by the grasses and an increase of the amount of water available for the woody plants. In the long term areas with a relatively high grass cover can become nearly pure dwarf shrublands (Fig. 2). Locally thickets of thorn shrubs can also develop (Shaw 1875, Acocks 1953, Volk 1974, Werger 1973b, 1977a, b, Werger & Leistner 1975). Other effects of overgrazing can be denudation of the land leading to an increased run-off and erosion (e.g. Walter 1939, Phillips 1956, Roberts 1965) or a considerable change in species composition owing to a decrease of palatable species due to selective grazing and a strong increase of poisonous or unpalatable species, such as *Chrysocoma tenuifolia, C. polygalifolia, Hertia pallens, Eriocephalus* spp., *Euryops* spp., *Gnidia polycephala, Requienia sphaerosperma, Acrotome inflata* and *Schmidtia kaliharensis* (Leistner 1967, Mostert et al. 1971, Werger 1973, 1977b, Werger & Leistner 1975).

Locally also rodents change the vegetation structure. In the southern Kalahari particularly *Parotomys brantsi* digs its tunnels in fairly compact soils and gnaws the branches of shrubs, especially *Rhigozum trichotomum*, into short pieces which it carries to its burrows. Large populations of these animals may denude an extensive surface almost completely (cf. Leistner 1967).

A potentially more short-term effect on vegetation structure is caused by termites and locusts. Large densities of termite colonies, particularly *Hodotermes mossambicus* and *Trinervitermes* spp., can denude fairly large areas of all grass within a short period of time. Since the natural predators of termites have been diminished considerably by man during the last decennia, the effects of termites on the vegetation has probably increased (cf. Mostert et al. 1971, Werger 1977b).

3.3 Vegetation of the Namib Domain

In the vicinity of the Caporolo River, south of Benguela, Karoo–Namib species are already present in what is generally a dry thorn savanna of Sudano–Zambezian affinity (see Chapter 10). In a narrow coastal belt south from Cape Santa Marta the vegetation is of a definite Karoo–Namib character. The vegetation has the appearance of a semi-desert with patchily distributed succulents of the genus *Euphorbia, Cyphostemma uter, Sarcocaulon mossamedense, Aloe littoralis, Hoodia currori* and *H. parvifolia*, and shrublets such as *Salvadora persica, Petalidium tomentosum* and *Blepharis* sp. which occur on compact and rocky soils. Sandy areas occasionally carry an open grass vegetation with *Stipagrostis* spp. A structure diagram of such a *Sarcocaulon* stand is presented by Jessen (1936, plate 32). Matos & Sousa (1970) describe the vegetation near Moçâmedes in somewhat more detail. In the littoral zone on saline soils a community occurs characterized by, amongst others, *Salsola zeyheri, Chloris pubescens, Sesuvium mesembryanthemoides, S. sesuvioides, Suaeda fruticosa, Leucophrys mesocoma, Paspalum vaginatum* and *Aizoanthemum mossamedense*. Somewhat further inland, where calcrete plays a more important role, the vegetation consists of *Aizoon virgatum, Aizoanthemum mossamedense, Euphorbia bellica, Zygophyllum orbiculatum, Z. simplex, Opophytum (Mesembryanthemum) dactylinum, Leucophrys mesocoma, Indigofera daleoides, I. alternans, Orthanthera stricta* and some other species. Nearer Moçâmedes, on locally more gravelly and rocky sites, *Euphorbia* sp., *Stipagrostis prodigiosa* and *S. hirtigluma* become important. Extensive patches with a sparse population of the succulent *Sarcocaulon mossamedense* also occur. During most of the year no other plants are visible here, but following the first rains here and there individuals of *Polygala mossamedense, Sesuvium sesuvioides, Petalidium engleranum* and *Hibiscus micranthus* appear.

Stands of *Welwitschia mirabilis* occur particularly in slight depressions, sometimes with *Maerua welwitschii, Salvadora persica* and *Acacia tortilis* subsp. *heteracantha* as companion species. Grasses like *Stipagrostis prodigiosa, S. subacaulis, S. hirtigluma, Eragrostis porosa, Enneapogon cenchroides*, and some other herbs and dwarf shrubs mentioned above may also be prominent in these stands. Elsewhere vast sandy areas carry this type of vegetation but with a prominence of *Stipagrostis hochstetterana* var. *secalina*, *Aristida hordeacea* and *Danthoniopsis dinteri*. Further east, more than 40 km away from the coast, this type of vegetation passes into the shrubby vegetation which is more typical of the Namaland Domain.

Along the larger riverbeds *Acacia albida* is important together with *Tamarix usneoides*, and locally *Ficus* sp. and *Hyphaene benguellensis*. Other species of *Acacia*, and *Tamarix angolensis* may also be found. Smaller riverbed plants include *Pluchea dioscoridis*, *Sesuvium sesuvioides*, *Commicarpus africanus*, *Atriplex halimus*, *Cordia gharaf*, *Epaltes gariepina*, and locally *Phragmites mauritianus*.

True desert with shifting sand dunes occurs in the Iona National Park between Porto Alexandre and the Cunene River. Here large areas are completely devoid of vegetation, while elsewhere *Acanthosicyos horrida* encroaches upon the sand. On more compact substrates *Zygophyllum orbiculatum*, *Z. simplex*, *Galenia africana*, *Sesuvium portulacastrum*, *Vogelia africana*, *Pterodiscus aurantiacus*, *Stipagrostis subacaulis*, and other species can be found. The grasses *Odyssea paucinervis* and *Sporobolus spicatus* inhabit compact substrates (Engler 1910, Jessen 1936, Gossweiler & Mendonça 1939, Airy Shaw 1947, Barbosa 1970, Diniz 1973, Huntley 1973).

In northern South West Africa the coastal stretch of desert consists largely of dunes and is very sparsely vegetated. Several species, such as *Hermannia* sp., *Indigofera cunenensis*, *Merremia multisecta*, *Acanthosicyos horrida*, and even *Odyssea paucinervis*, act as sand collectors and emerge through small dunes. *Eragrostis cyperoides* and *Stipagrostis ramulosa* are the two most common grasses. Elsewhere, *Salsola nollothensis* occurs on the dunes or a community of the shrubby *Ectadium virgatum* var. *rotundifolium* with *Merremia multisecta*, *Citrullus ecirrhosus* and sometimes *Welwitschia mirabilis* can be found. Rocky substrates harbour succulents, including *Lithops ruschiorum*, *Sarcocaulon mossamedense* and *Othonna lasiocarpa*, whereas dense cushion-like growths of *Salsola* sp., *Zygophyllum clavatum* and *Z. stapffii* occupy some dry river beds. Small waterholes or wet places can be densely covered by *Odyssea paucinervis* or a community of *Phragmites australis*, *Typha latifolia* var. *capensis*, *Scirpus dioicus*, *S. littoralis* and *Juncellus laevigatus*. *Tamarix usneoides* and *Salvadora persica* occur sporadically along river beds. Towards the Namib fringes *Balanites welwitschii*, *Colophospermum mopane* and *Maerua schinzii* become more common riverbed companions while on the sandy flats several species of *Stipagrostis* and *Kaokochloa nigrirostris* are locally abundant (Giess 1968, 1971).

The central Namib comprises the area between the Huab and the Kuiseb Rivers. In this part of the Namib, where sand dunes occur only in a coastal strip south of the Swakop River, a narrow vegetation belt of about 200 m wide follows the coast line. The wind-pruned, rounded form of dwarf shrubs filled up with sand between the branches is a common growth form again. Near the beach such plants include *Psilocaulon salicornioides*, *Zygophyllum clavatum*, *Salsola aphylla* and *S. nollothensis* (Fig. 3). Nearest to the sea *Arthrocnemum affine* pioneers, whereas *Zygophyllum stapffii* and *Arthraerua leubnitziae* are the only prominent plants on more gravelly flats particularly in slight depressions. At rocky sites the dwarf shrubs *Drosanthemum luederitzii*, *Ruschia* cf. *sedoides* and *Tetragonia arbusculoides*, as well as the annuals *Stipagrostis hermannii*, *S. namibensis*, *S. sub-*

Fig. 3. In the central Namib north of the Kuiseb River *Zygophyllum clavatum* and *Psilocaulon salicornioides* form small dunes (photo W. Giess).

acaulis, and some other species, occur. Lichens of many species, including *Teloschistes flavicans*, *Parmelia convoluta* and *Usnea* sp., are abundant on otherwise quite barren, quartzite or gypsum plains (Fig. 4). Succulent dwarf shrubs and annuals, such as *Hydrodea bossiana*, *Pentzia hereroensis*, *Sesuvium sesuvioides* and *Zygophyllum simplex* are locally common near salt pans, whereas some dry riverbeds may show a dense growth of *Sporobolus robustus* or more

Fig. 4. Locally near the coast lichens like *Teloschistes* reach high cover values (photo W. Giess).

Fig. 5. Grassy, calcrete hills in the Namib fringes, with scattered individuals of *Commiphora*.

open communities with *Eragrostis spinosa*. Further away from the coast several other species of dwarf shrubs, shrubs, grasses and ephemerals become characteristic. Following a good rain storm the annual grasses *Stipagrostis namibensis*, *S. subacaulis*, *S. hermannii* and *S. hirtigluma* var. *hirtigluma* become particularly abundant. Towards the Namib fringes (Pro-Namib) the deeply dissected, but sloping calcrete surface carries a rather dense grasscover of

Fig. 6. Kuiseb River with trees and shrubs of *Acacia albida*, *A. erioloba*, *Tamarix usneoides* and *Euclea pseudebenus* (photo W. Giess).

mainly perennial species of *Stipagrostis* with widely scattered shrubs, predominantly of the genus *Commiphora* (Fig. 5). Near the Kuiseb River *Welwitschia mirabilis*, which prefers washes and similar habitats, reaches its southernmost distribution boundary, while *Acanthosicyos horrida* here occurs optimally on sandy substrates. Fairly big trees of *Acacia erioloba* can be seen here and there along dry riverbeds or drainage lines, while they can form good stands of tall trees together with *A. albida* along the Kuiseb and Swakop Rivers (Fig. 6). Large stands of *Tamarix usneoides* and of the exotic *Nicotiana glauca* are also common here (Engler 1910, Cannon 1924, Giess 1962, 1968, 1969, 1971, Kers 1967, Walter 1973).

A detailed analysis of the vegetation of the central Namib is presented by Robinson (1977), who distinguished four major habitat types: salt marshes, pans, sand dunes, and a sandy plains–washes–rock outcrops-complex. Near the coast four different salt marsh communities, one comprising three subcommunities, were distinguished showing clear relationships to degree of salinity and flooding (Fig. 7). These communities frequently occur in a zonation pattern and may be regarded as successional. All communities contain few species only. Pans included clay pans with a species-poor *Platycarpha carlinoides* Community, loamy pans with a slightly richer *Calicorema capitata* Community, and pans in the Pro-Namib region with a floristically richer *Aristida adscensionis* Community. In the dune areas Robinson recognized five communities which occur in a west–east gradient and are related to amount and nature of precipitation and stability of substrate (Fig. 8). All communities are extremely poor in species and frequently stands contain only the one typical species, but further inland the number of species increases somewhat (*Stipagrostis lutescens–Limeum fenestratum* Subcommunity) though these species are mainly annuals, whereas nearer the coast the species are perennial. Out of the dune area stands of *Salsola nollothensis* are locally common on sandy sites close to the coast. Washes and rock outcrops present relatively moist habitats in the Namib, owing to the accumulation of water in the sandy washes and in cracks. Therefore, the communities of these sites have some floristic similarities with one another but also with the communities of the sandy plains. Slope (often slope just 'upstream' of the site of the community), substrate material, and particularly amount and kind of moisture, as well as soil depth are important ecological factors determining the communities. The floristic composition of the communities in these habitats is shown in Table 2, whereas their ecological and possible successional relationships are apparent from Fig. 9 (Robinson 1977).

These results are confirmed by those of the minor survey by Moisel (1975) of Welwitschia Plain. Moisel, sampling only on perennial species, found five clearly segregated communities: (1) the *Adenolobus pechuelii–Commiphora saxicola* community occurs on the rocky slopes of the marble hills of the Husab range. The community in which also *Trianthema triquetra*, *Euphorbia virosa* and *Stipagrostis* sp. are constant is confined to the shallow washes and drainage lines, and canopy covers only about 5 per cent of the surface. (2) Granite outcrops with a coarse sandy soil carry the *Sarcocaulon mossamedense* community. *Monechma arenicola*, *Zygophyllum stapffii*, *Stipagrostis* sp. and *Welwitschia mirabilis* are also common in this community which has a slightly higher canopy cover than the previous one. (3) A very open community of nearly exclusively *Welwitschia mirabilis* with a canopy cover of less than 5 per cent is found on gently sloping,

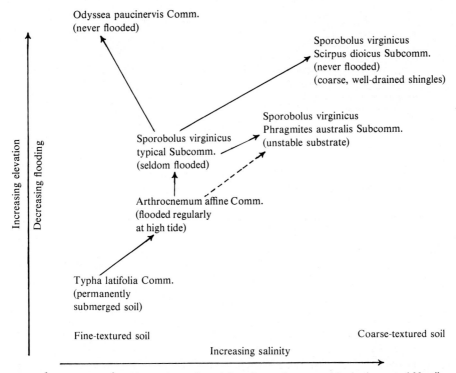

Fig. 7. Ecological relationships and zonation of the salt marsh communities in the central Namib. Broken arrow indicates the less frequently observed course of vegetation development. The *Odyssea paucinervis* community frequently grades into a community determined by *Eragrostis cyperoides* on foot slopes of dunes near the coast (adapted from Robinson 1977).

gravelly, coarse sand (Fig. 10). (4) Plains of fine, saline sand harbour the *Salsola aphylla* (? *S. tuberculata*) community, covering between 20 and 50 per cent. *Zygophyllum stapffii* is fairly common, and locally *Welwitschia mirabilis* and *Arthraerua leubnitziae* may be found. (5) Moist brack sites and salt marshes support dense stands of the *Phragmites australis–Juncus arabicus* community, sometimes with *Tamarix usneoides* or *Sporobolus virginicus*, and covering between 50 and 100 per cent.

The southern Namib is largely an area of shifting sand dunes of up to 300 m high and largely devoid of vegetation. Typical are the very sparsely scattered tufts of the perennial *Stipagrostis sabulicola* in this area (Fig. 11). This grass, as well as the shrubby *Acanthosicyos horrida*, *Trianthema hereroensis* and *Psilocaulon marlothii* locally form small dunes. Following good rains the dune valleys can show large stands of *Stipagrostis gonatostachys*. *Salsola nollothensis* is again common in this part of the coastal Namib, and on rocky substrates scattered individuals of *Othonna furcata*, *Pelargonium* sp., *Salsola* spp., *Lycium decumbens*, *Drosanthemum luederitzii*, *Jensenobotrya lossowiana* and *Osteospermum crassifolium* can locally be encountered. Sandy sites and washes harbour other species, including *Eragrostis cyperoides*, *Crotalaria schultzei*, *Mesembryanthemum hypertrophicum*, *Limonium membranaceum*, and *Lebeckia multiflora*, while some brack water pools can contain *Ruppia maritima*, *Arthrocnemum dunense* or *Sporobolus virginicus* (Engler 1910, Range 1932,

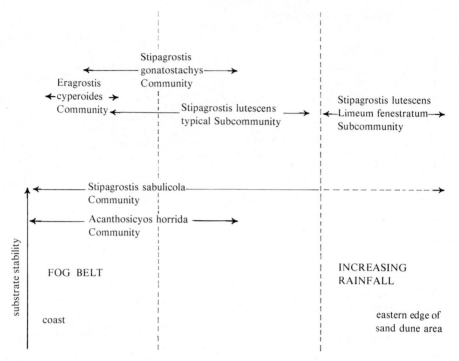

Fig. 8. Distribution of the sand communities in the central Namib along the west–east climatic gradient. The vertical dimension indicates the stability of the substrate. Arrows indicate the extent of distribution (adapted from Robinson 1977).

Giess 1971, 1974, Robinson & Giess 1974). Away from the coast these communities, though sparse and usually covering less than one per cent of the ground surface, give way to an even sparser growth of *Sarcocaulon spinosum* or of the 1.5 m tall, succulent shrub *Euphorbia gummifera* on gravelly sand plains or on lower slopes. On the latitude of Lüderitz, about 50 km inland an almost completely bare plain is reached. This plain extends for some 50 km and is occasionally interrupted by very sparsely vegetated mountainous outcrops. Fog usually does not penetrate up to here any more and, if it does so sporadically, its influence is too small and the precipitation is still too low to allow noticeable plant growth. About a hundred km from the coast the Pro-Namib starts with a low and open grass steppe of up to about five per cent total aerial cover. The main species are *Stipagrostis lanipes*, *S. obtusa*, *S. ciliata*, *Ehrharta pusilla* and *Eragrostis nindensis*. Further inland other species, including herbs, dwarf shrubs and geophytes become more common and the aerial cover of the vegetation increases to 15 or even 25 per cent. Here lies the transition to the more shrubby vegetation of the Namaland Domain (Coetzee & Werger 1975).

In the southern tip of the Namib Domain, some 30 km south of the Orange River, there is a rapid transition to the dense vegetation of succulents typical of the 'strandveld' (Acocks 1953) of the Western Cape Domain. In a narrow sandy zone the vegetation has a mixed character with *Babiana sambucina*, *Asparagus juniperoides*, *Stipagrostis subacaulis*, *S. schaeferi*, *S. ciliata*, *S. geminifolia*, *S. dregeana*, *Felicia namaquana*, *Dimorphotheca polyptera*, *Sarcocaulon multifidum*, *Zygophyllum cordifolium*, *Z. morgsana*, *Z. prismatocarpum*, *Hermannia*

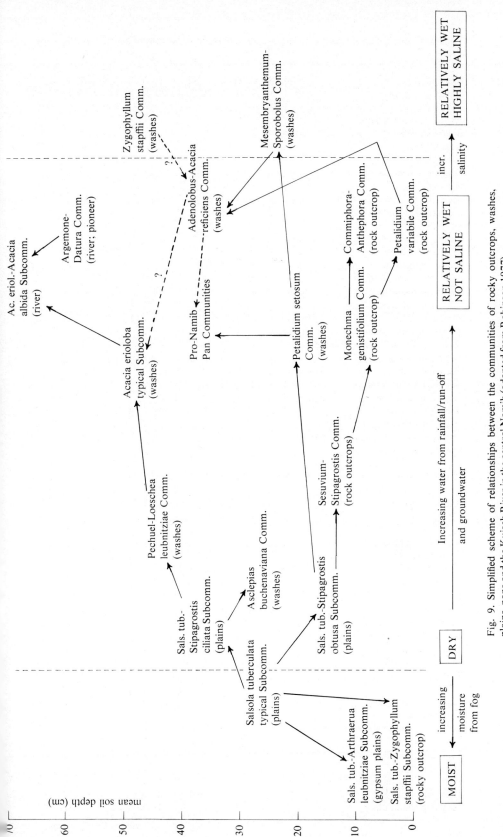

Fig. 9. Simplified scheme of relationships between the communities of rocky outcrops, washes, plains, pans and the Kuiseb River in the central Namib (adapted from Robinson 1977).

Table 2. Communities of washes, outcrops and sandy plains of the central Namib (after Robinson 1977).

Community	A	B_1	B_2	C	D	E	F	G	H	I	J	K	L	M_1	M_2	M_3	M_4	M_5
Number of relevés	10	23	32	9	30	4	28	20	13	20	23	9	11	12	19	11	25	47
Asclepias buchenaviana	5	.	.	.	.	.	.	.	.	.	.	.	.	.	.	.	.	.
Acacia erioloba	1	5	4	2	.	.	1	.	.	.	.	.	.	.	.	.	.	.
Acacia albida	.	.	4	2	.	.	.	.	.	.	.	.	.	.	.	.	.	.
Tamarix usneoides	.	.	3	2	1	3	.	.	.	.	.	.	.	.	.	.	.	.
Euclea pseudebenus	.	1	4	.	1	.	.	1	.	1	1	.	.	.	.	.	.	.
Nicotiana glauca	.	1	3	3	.	.	.	.	.	.	.	.	.	.	.	.	.	.
Salvadora persica	.	.	2	2	.	1	.	.	.	.	.	.	.	.	.	.	.	.
Chenopodium ambrosioides	.	.	2	3	.	.	.	.	.	.	.	.	.	.	.	.	.	.
Datura innoxia	.	.	2	5	.	.	.	.	.	.	.	.	.	.	.	.	.	.
Datura stramonium	.	.	2	4	.	.	.	.	.	.	.	.	.	.	.	.	.	.
Argemone ochroleuca	.	.	2	4	.	.	.	.	.	.	.	.	.	.	.	.	.	.
Ricinus communis	.	.	1	3	.	.	.	.	.	.	.	.	.	.	.	.	.	.
Eragrostis trichophora	.	.	1	3	.	.	.	.	.	.	.	.	.	.	.	.	.	.
Nidorella resedifolia	.	.	2	4	.	.	.	.	.	.	.	.	.	.	.	.	.	.
Adenolobus pechuelii	.	1	.	.	4	.	1	2	2	.	1	.	.	.	.	.	.	.
Acacia reficiens	.	.	.	.	3	.	1	.	.	.	.	.	.	.	.	.	.	.
Asthenatherum glaucum	.	1	.	.	2	.	1	.	.	.	.	.	.	.	.	.	.	.
Mesembryanthemum guerichianum	.	.	.	.	4	.	.	.	.	1	.	.	.	1	.	.	.	.
Sporobolus nebulosus	.	.	.	.	4	1	.	.	.	1	.	.	.	.	.	.	.	.
Galenia papulosa	.	.	.	.	3	.	.	.	.	.	.	.	.	.	.	.	.	.
Schmidtia kalihariensis	.	1	.	2	4	1	.	.	.	.	.	.	.	.	.	.	.	.
Petalidium setosum	.	1	.	.	1	1	5	.	1	.	.	.	2	.	1	.	1	.
Tribulus terrestris	.	1	.	.	1	.	2	.	.	.	.	1	.	1	.	.	1	1
Monechma desertorum	.	.	.	.	1	.	2	1	.	.	.	.	.	.	.	.	2	1
Trianthema triquetra subsp. parvifolia	.	.	.	.	.	.	2	.	2	.	.	.	.	.	.	.	2	1
Sesuvium sesuvioides	.	.	.	.	2	.	.	3	.	.	.	1	.	1	.	.	.	.
Euphorbia phylloclada	.	1	.	.	2	.	1	3	.	.	.	2	1	1	2	.	1	1
Cleome diandra	.	1	.	.	1	.	1	3	1	.	.	1	1	.	.	.	1	.
Monechma genistifolium	.	1	.	.	1	.	1	.	2	2	1	.	.	.	.	.	.	.
Adenia pechuelii	.	.	.	.	.	.	.	.	2	.	.	.	.	.	.	.	.	.
Sutera maxii	.	1	2	1	1	1	1	.	1	.	.	.	.	.	.	.	.	.
Aptosimum angustifolium	.	1	.	.	1	.	1	.	1	2	1	.	.	.	.	.	.	.
Monsonia umbellata	.	1	.	.	1	.	1	1	3	2	1	.	.	.	1	.	1	.
Gisekia africana	.	1	.	.	1	.	1	1	2	2	.	.	.	.	.	.	.	.
Cleome luederitziana	.	2	1	.	1	2	1	1	2	1	.	1	.	.	1	1	1	.
Petalidium variabile	.	1	.	.	.	.	1	.	1	5	.	.	.	.	.	.	.	.
Indigofera auricoma forma	.	.	.	.	.	.	.	.	1	2	1	.	.	.	.	.	.	.
Hirpicium gazaneoides	.	.	.	.	.	.	.	.	1	2	.	.	.	.	.	.	.	.
Pegolettia senegalensis	.	.	.	.	1	.	1	.	1	1	2	.	.	.	.	.	.	.
Commiphora glaucescens	.	.	.	.	.	.	.	.	1	4	.	.	.	.	.	.	.	.
Anthephora pubescens	.	.	.	.	.	.	.	.	.	4	.	.	.	.	.	.	.	.
Ruelia diversifolia	.	.	1	.	.	.	1	.	.	1	3	.	.	.	.	.	.	.
Triraphis pumilio	.	.	.	.	.	.	1	.	.	1	2	.	.	.	1	.	.	.
Rhus marlothii	.	.	.	.	.	.	1	.	.	.	2	.	.	.	.	.	.	.
Enneapogon scoparius	.	.	.	.	.	.	.	.	.	.	2	.	.	.	.	.	.	.
Codon schenkii	.	.	.	.	.	.	.	.	.	.	2	.	.	.	.	.	.	.
Helichrysum tomentulosum subsp. aromaticum	.	.	.	.	.	.	.	.	.	.	2	.	.	.	.	.	.	.
Panicum arbusculum	.	.	.	.	.	.	.	.	.	.	2	.	.	.	.	.	.	.
Senecio flavus	.	.	.	.	.	.	.	.	1	.	2	.	.	.	.	.	.	.
Euphorbia avasmontana	.	.	.	.	.	.	.	.	1	1	3	.	.	.	.	.	.	.
Commiphora virgata	.	.	.	.	.	.	.	1	.	3	3	.	.	.	.	.	.	.
Blepharis obmitrata	.	.	.	.	1	.	1	.	1	1	3	.	.	.	.	1	.	.

Community	A	B_1	B_2	C	D	E	F	G	H	I	J	K	L	M_1	M_2	M_3	M_4	M_5
Number of relevés	10	23	32	9	30	4	28	20	13	20	23	9	11	12	19	11	25	47
Monechma arenicolum	.	1	.	.	.	.	2	.	2	3	4	2	.	.	.	.	.	1
Tribulus zeyheri	.	1	.	2	.	1	1	.	2	4	2	.	.	.	.	.	1	.
Stipagrostis uniplumis var. uniplumis	.	.	.	.	.	.	1	1	1	3	3	1	.	.	1	.	3	1
Eragrostis nindensis	.	.	.	.	.	.	.	.	1	2	4	.	.	.	.	.	1	1
Curroria decidua	.	.	.	.	.	.	.	.	2	1	1	.	.	.	.	.	.	.
Solanum rigescentoides	.	.	.	.	.	.	.	1	1	1	2	.	.	.	.	.	.	.
Osteospermum microcarpum	.	1	1	.	.	1	1	1	1	1	3	.	1	1	.	.	.	.
Forskohlea candida	.	.	.	.	.	2	1	.	1	1	2	.	.	.	.	.	.	.
Trichodesma africana	.	.	.	.	.	.	1	1	.	2	1	4	.	.	.	.	.	.
Asparagus denudatus	.	.	.	.	1	.	.	1	.	1	1	3	.	.	.	1	.	.
Pechuel–Loeschea leubnitziae	1	1	2	1	1	.	.	.	.	.	.	5	1	1	1	.	.	.
Stipagrostis hochstetterana	.	3	.	.	1	1	2	.	1	1	1	3	.	1	.	.	1	1
Zygophyllum stapffii	.	1	.	.	1	.	.	1	.	.	.	2	4	.	5	2	.	.
Salsola tuberculata	3	1	.	.	1	.	1	1	1	.	.	.	.	5	4	4	3	1
Arthraerua leubnitziae	.	.	.	.	.	.	.	.	.	.	.	.	.	.	1	5	.	.
Aloe asperifolia	.	.	.	.	.	.	.	.	.	.	.	.	.	.	2	.	.	.
Dyerophytum africanum	1	1	.	.	.	.	1	1	.	.	2	.	.	.	2	.	.	.
Stipagrostis obtusa	1	1	.	.	1	1	3	4	2	2	1	1	.	1	1	.	5	2
Zygophyllum cylindrifolium	1	.	.	.	2	.	3	1	2	2	2	1	.	1	.	.	4	.
Stipagrostis ciliata	4	4	.	.	5	1	4	5	4	2	3	4	2	3	3	2	3	5
Zygophyllum simplex	1	1	1	2	3	4	3	3	4	1	3	2	2	1	2	.	4	3
Stipagrostis uniplumis var. intermedia	.	2	1	.	2	1	3	1	4	4	4	.	.	.	.	.	.	.
Enneapogon desvauxii	.	.	.	.	1	1	3	2	4	4	4	.	.	1	1	.	4	2
Tephrosia dregeana	1	1	.	.	3	.	3	2	1	2	3	2	1	1	2	.	2	1
Indigofera auricoma	.	2	1	.	1	.	3	2	2	1	2	2	.	1	1	.	1	1
Euphorbia glanduligera	.	1	.	.	1	1	2	1	3	2	3	1	.	.	.	.	2	1
Mollugo cerviana	.	1	.	1	2	.	2	1	1	1	1	1	.	1	.	.	1	1
Brachiaria glomerata	.	2	.	.	1	.	1	1	1	1	1	2	1	.	1	.	.	1
Commiphora saxicola	.	.	.	.	1	.	1	1	2	2	3	.	.	.	1	.	.	1
Geigeria alata	.	1	.	.	1	1	.	2	1	2	1	1	.	.	.	.	1	1
Hermannia modesta	.	1	1	2	1	.	2	1	2	1	1	1	.	.	.	.	1	1
Eragrostis annulata	.	1	1	1	1	3	2	.	1	1	1	1	1	.	.	.	1	1
Stipagrostis hirtigluma	1	1	.	.	1	3	2	.	2	1	1	.	.	.	.	.	1	1
Eragrostis spinosa	.	1	2	4	.	.	.	.	.	.	.	.	.	.	.	.	.	.
Crotalaria podocarpa	.	2	.	.	1	.	1	.	.	1	.	1	.	.	.	.	1	.
Cyperus marginatus	.	.	2	2	.	.	.	.	.	.	.	.	.	.	.	.	.	.
Sutera canescens	.	.	2	2	.	.	.	.	1	.	.	.	.	.	.	.	.	.
Solanum nigrum	.	.	2	2	.	.	.	.	.	.	1	.	.	.	.	.	.	.
Psoralea obtusifolia	.	.	1	2	.	4	1	.	.	.	.	.	.	.	.	.	.	.
Flaveria bidentis	.	.	1	2	.	.	.	.	.	.	.	.	.	.	.	.	.	.
Heliotropium ovalifolium	.	1	1	2	.	.	.	.	.	.	.	.	.	.	.	.	.	.
Cynodon dactylon	.	.	1	2	.	.	.	.	.	.	.	.	.	.	.	.	.	.
Senecio marlothianus	.	.	.	2	.	.	.	1	.	.	1	1	.	.	.	.	.	.
Tagetes minuta	.	.	1	3	.	.	.	.	.	.	.	.	.	.	.	.	.	.
Suaeda plumosa	.	.	1	2	.	.	.	.	.	.	.	.	.	.	.	.	.	.
Helichrysum argyrosphaera	.	.	1	2	.	.	.	.	.	.	.	.	.	.	.	.	.	.
Setaria verticillata	.	1	1	2	.	.	.	.	.	.	.	.	.	.	.	.	.	.
Lannea intybacea	.	.	.	2	.	.	.	.	.	.	.	1	.	.	.	.	.	.
Kissenia capensis	.	.	1	1	.	2	.	.	.	.	.	.	.	.	.	.	.	.
Chloris virgata	.	.	1	1	.	2	.	.	.	1	.	.	.	.	.	.	.	.
Vahlia capensis	.	.	1	1	.	2	.	.	.	.	.	.	.	.	.	.	.	.
Sporobolus consimilis	.	.	1	1	.	2	.	.	.	1	.	.	.	.	.	.	.	.
Geigeria ornativa	.	1	.	.	.	.	2	.	.	1	.	1	.	.	1	.	1	1
Kohautia lasiocarpa	.	1	.	.	1	.	2	.	1	1	.	.	.	.	.	.	2	1
Euphorbia inaequilatera	.	.	.	.	1	.	2	.	1	.	.	.	.	1	.	.	1	1

Community	A	B₁	B₂	C	D	E	F	G	H	I	J	K	L	M₁	M₂	M₃	M₄	M₅
Number of relevés	10	23	32	9	30	4	28	20	13	20	23	9	11	12	19	11	25	47
Amaranthus thunbergii	.	1	.	.	1	.	.	1	.	1	2	.	.	.	.	.	1	.
Sarcocaulon mossamedense	.	.	.	.	1	.	.	1	1	.	2	.	.	.	.	.	.	.
Helichrysum roseonivium	.	.	.	.	.	.	.	.	.	.	2	.	.	.	.	.	.	.
Aristida parvula	.	.	.	.	.	.	.	.	.	.	.	2	.	1	1	.	1	1
Galenia africana	1	1	1	.	1	.	.	.	.	.	.	2	1	.	2	.	.	.
Helichrysum hernaroides	.	1	.	.	1	.	1	.	1	.	.	2	.	.	.	.	.	.
Lotononis sp.	.	.	.	.	.	.	.	.	.	.	.	2	.	.	.	.	.	.
lichens	.	1	1	.	.	.	.	.	.	.	1	.	.	.	2	2	2	.

A: Asclepias buchenaviana community.
B: Acacia erioloba community with typical (1) and A. albida (2) subcommunities.
C: Argemone – Datura community.
D: Adenolobus – Acacia reficiens community.
E: Mesembryanthemum – Sporobolus community.
F: Petalidium setosum community.
G: Sesuvium – Stipagrostis obtusa community.
H: Monechma genistifolium community.
I: Petalidium variabile community.
J: Commiphora – Anthephora community.
K: Pechuel–Loeschea leubnitziae community.
L: Zygophyllum stapffii community.
M: Salsola tuberculata community with typical (1), Zygophyllum stapffii (2), Arthraerua leubnitziae (3), Stipagrostis obtusa (4) and Stipagrostis ciliata (5) subcommunities.

Fig. 10. *Welwitschia mirabilis* on the Welwitschia Plains north of the Swakop River (photo W. Giess).

Fig. 11. Dunes of the southern Namib with scattered tufts of *Stipagrostis sabulicola*.

patellicalyx, Euphorbia ephedroides, Grielum humifusum and several other species (Werger unpubl.).

3.4 Vegetation of the Namaland Domain with exception of the southern Kalahari

Although in the northernmost part of the Karoo–Namib Region the Namaland Domain cannot be clearly separated from the Namib Domain, they are already well distinct at the latitude of Moçâmedes. Here the Namaland Domain comprises the more shrubby vegetation characteristic of the broken country of the escarpment zone. Typical in this area are the low and open, shrubby communities with the odd succulent *Cyphostemma uter*, and *Othonna arborescens, Helichrysum mossamedense, Felicia mossamedensis, Dicoma foliosa, Osteospermum microcarpum*, and other species, as well as the taller shrub communities with *Acacia mellifera* subsp. *detinens, A. reficiens, Commiphora angolensis, Sterculia setigera, Grewia bicolor, Boscia albitrunca, Salvadora persica, Rhigozum virgatum, Balanites welwitschii, Sesamothamnus benguellensis, Maerua angolensis, Gossypium anomalum, Hoodia currori, Sansevieria pearsonii*, and many others. Higher up the escarpment and further inland this shrub vegetation comprises more Sudano–Zambezian species and passes into *Colophospermum mopane* and other Sudano–Zambezian savanna and woodland communities (see Chapter 10). On less rocky sites with a deeper soil, grasses like *Stipagrostis hochstetterana* var. *secalina* dominate and *Welwitschia mirabilis* still occurs. In the east this community type fringes on shrubby mopane communities. Along drainage lines and in valleys the Sudano–Zambezian element penetrates deepest into this Karoo–Namib area. In these contact areas the shrub communities of the rocky slopes contain several species of *Commiphora* and *Euphorbia*, as well as *Phaeoptilum spinosum, Catophractes alexandri, Rhigozum virgatum*, and many of the species mentioned above (Jessen 1936, Matos & Sousa 1969, Barbosa 1970, Diniz 1973, Huntley 1973).

In Kaokoland the vegetation is similar to that of southernmost Angola. Arid hills carry an open shrub formation dominated by *Commiphora* species such as *C. virgata, C. giessii, C. wildii*, and *C. kraeuseliana*, which are generally less than 2 m high. Other scattered woody plants include *Maerua schinzii, Colophospermum mopane, Boscia albitrunca, Ceraria longipedunculata, Acacia mellifera* subsp. *detinens, A. robynsiana, A. montis-usti, Moringa ovalifolia* and *Sterculia africana*. Dwarf shrubs and grasses are common. Towards the higher edge of the escarpment mopane becomes evermore plentiful together with other Sudano–Zambezian species (see Chapter 10). *Acacia albida* is still common along the seasonal rivers, and is accompanied by a mixture of Karoo–Namib and Sudano–Zambezian species: *Acacia erioloba, Ficus sycomorus, Combretum imberbe, Diospyros mespiliformis, Ziziphus mucronata, Peltophorum africanum* and *Hyphaene ventricosa* in the tree layer, and *Diospyros lycioides, Euclea pseudebenus, E. divinorum, Tamarix usneoides, Combretum hereroense, Salvadora persica*, and *Maytenus senegalensis* in the shrub layer, which sometimes occurs as thickets. Where the canopy is open, shrubs such as *Pechuel–Loeschea leubnitziae, Mundulea sericea*, and *Rhigozum brevispinosum* may be common. Wide flats of gravelly sand can be covered with dense grasslands of *Kaokochloa nigrirostris* following good rains, though heavy grazing has reduced the extent and cover of this community (Malan & Owen-Smith 1974).

Further south along the escarpment the vegetation keeps its shrubby character, (Fig. 12) and many species still occur though some are not found south of the Brandberg where they are replaced by others. *Commiphora* species are still important, and so are the succulent trees or shrubs *Aloe dichotoma, Cyphostemma currorii, C. juttae* and *C. bainesii*. Also *Euphorbia guerichiana* is a common constituent of the escarpment vegetation. Apart from many of the species listed above for Kaokoland, *Adenolobus garipensis, Phaeoptilum spinosum, Terminalia prunioides, Sesamothamnus guerichii, S. benguellensis* and several acanthaceous species are also common. Flat areas with a deeper sandy soil carry a savanna type of vegetation with many of the same woody species in a matrix of *Stipagrostis* species (Fig. 13) (Giess 1971, Nordenstam 1974).

South of Windhoek the Namaland Domain comprises vast stretches of the plateau area. The soils are mostly a sandy loam, sometimes gravelly or bouldery, sometimes with concretions or thick layers of calcrete. Stony ridges or even mountains break the flat plateau surface here and there. A very open shrub and dwarf shrub community is most typical here. *Rhigozum trichotomum* is very common, but locally *Catophractes alexandri* is frequent. Other common shrubs include *Acacia mellifera* subsp. *detinens, A. nebrownii, Phaeoptilum spinosum, Boscia foetida, B. albitrunca, Cadaba aphylla, Parkinsonia africana* and *Nymania capensis*. Dwarf shrubs include *Zygophyllum suffruticosum, Dyerophytum africanum, Aizoon schellenbergii, Thesium lineatum, Montinia caryophyllacea* and *Pteronia lucilioides*. Tufted grasses such as *Stipagrostis obtusa, S. ciliata, S. uniplumis, Eragrostis nindensis* and *Enneapogon desvauxii* are scattered between these woody plants. On rocky ridges the conspicuous liliaceous succulent tree *Aloe dichotoma* can be very abundant, and *Rhigozum obovatum, Pappea capensis, Euphorbia virosa* and *Adenolobus garipensis* are some common shrubs, while the grasses include *Panicum arbusculum, Triraphis ramosissima*,

Fig. 12. Escarpment near Karibib with *Acacia senegalensis* var. *rostrata* (foreground right), *A. reficiens*, several species of *Commiphora*, *Boscia foetida*, *Euphorbia guerichiana*, and others. The grasses are mainly *Stipagrostis* spp. (photo W. Giess).

and *Anthephora ramosa*. Near the western edge of the plateau *Zygophyllum* cf. *meyeri* and *Z. suffruticosum* are sometimes dominant in dwarf shrub communities which are 0.5 m high and have a canopy cover of 10 to 25 per cent. Over wide flat or slightly undulating areas with angular quartzitic or granitic boulders and coarse gravel covering 75 per cent of the surface, the glaucous, leafless, succulent shrub *Euphorbia gregaria* is the sole dominant (Fig. 14). These shrubs are rounded and

Fig. 13. On deeper soils between the escarpment slopes grasses (here mainly *Stipagrostis hochstetterana* var. *secalina*) dominate (photo W. Giess).

Fig. 14. Extensive, rocky plains with *Euphorbia gregaria* as the sole prominent between Aus and Ai-Ais.

about 1–2 m in diameter. The companion species of this community include many of those named above. Drainage lines support a denser woody community with low trees up to 6 m and shrubs up to 4 m tall. Patchily this community forms small thickets, in which large woody species are *Acacia erioloba, A. karroo, A. mellifera* subsp. *detinens, Ziziphus mucronata* and *Euclea pseudebenus*.

Fig. 15. Overgrazed plains in the vicinity of Keetmanshoop carry an open dwarf shrub vegetation with *Salsola tuberculata* and *Zygophyllum dregeanum* and sparse grass tufts of mainly *Stipagrostis ciliata*. The drainage line in the background supports a denser scrub with *Acacia mellifera* subsp. *detinens, A. karroo, Ziziphus mucronata*, and others.

Elsewhere on deep riverine sand the shrubby *Stipagrostis namaquensis* is prominent.

Further eastward on the plateau calcrete layers become more important, until the vast sand deposits of the southern Kalahari are reached west of the Auob River. The calcareous substrate carries an open dwarf shrub vegetation with *Salsola tuberculata* and *Zygophyllum dregeanum* dominating on the more brackish sites (Fig. 15), but also with *Aizoon schellenbergii, Petalidium linifolium, Leucosphaera bainesii, Aptosimum albomarginatum, Monechma australe* and other woody species, and with the tufted grasses listed above (Engler 1910, Range 1932, Volk & Leippert 1971, Coetzee & Werger 1975).

South of the Orange River near Pofadder in Bushmanland *Aloe dichotoma* and *Euphorbia gregaria* occur together as dominants over fairly large areas (Figs. 16 and 17). Deep sandy sites in this area, but also north of the river, support a community dominated by the spiny, dense, bushy, suffrutescent *Stipagrostis brevifolia*

Fig. 16. In the vicinity of Pofadder and Onseepkans trees of *Aloe dichotoma* and shrubs of *Euphorbia gregaria* are co-dominant over fairly large areas.

(Fig. 18) (Desert False Grassveld, Acocks 1953, cf. Adamson 1938a). In the dry season this bushy grass is often nearly the only species visible, but following early rains many geophytes and annuals sprout from the sand and present a considerable soil cover. They include *Arctotis leiocarpa, Felicia namaquana, Heliophila integrifolia, Manulea schaeferi, Hebenstreitia parviflora, Lapeirousia caudata* and *Grielum obtusifolium* (Werger unpubl.). Occasionally *Aloe dichotoma* and the shrubs *Rhigozum trichotomum* break the monotony of this grass community. An example of a *Stipagrostis brevifolia* community with several succulents is given in Table 5, community E.

A detailed study of the vegetation near the southern border of the Namaland Domain (Acocks 1953: Orange River Broken Veld as redefined in the 1975 edition) has been carried out by Werger & Coetzee (1977) in the Augrabies Falls

Fig. 17. Scenery near Onseepkans with *Aloe dichotoma* on quartzite outcrop and forming a type of 'savanna' in the distance. In the middle distance grassy vegetation with *Stipagrostis* spp., mainly *S. brevifolia*.

National Park. This area can be broadly divided into three major physiographic units: (1) rocky outcrops or shallow soil on rock, (2) sandy plains, and (3) alluvial deposits of the Orange River. The first of these units covers most of the area and largely determines the rugged semi-desert character of the landscape. Two rock

Fig. 18. *Stipagrostis brevifolia* community on deep sand, c. 70 km west of Pofadder, showing the bushy growthform. The area is rather strongly overgrazed.

types are important, namely pink gneiss and quartz-rich granulite. The latter type outcrops in a ridge of hills which support a *Commiphora gracilifrondosa* shrub community which shows a clear segregation into two subcommunities of different slope aspect. The pink gneiss carries three different communities: (1) the *Ceraria namaquensis* community, an extremely open, succulent shrub formation, of smooth domes which are devoid of soil; (2) the *Indigofera heterotricha–Zygophyllum suffruticosum* community, an open shrub community of rocky outcrops with sand pockets, rocky plains with a very shallow soil layer, and washes or shallow sandy drainage lines. Each of these three habitats supports a different subcommunity and owing to a 'foam' structure in the uppermost soil layer the vegetation shows a marked pattern of fairly dense plant growth and scalded areas (Volk & Geyger 1970, Werger & Coetzee 1977); (3) the *Antherothamnus pearsonii* shrub community of relatively protected, narrow gorges and ravines in the pink gneiss. In the contact zone of the quartz-rich granulite and the pink gneiss there is a fairly long, gentle slope with a deeper rocky soil, carrying an *Enneapogon scaber–Euphorbia gregaria* shrub savanna. Two other savanna communities at Augrabies are the *Stipagrostis hochstetterana* var. *secalina* community on deep, loose, white sand, and the *Eragrostis trichophora–Acacia mellifera* community of fairly narrow, though deep, sandy depressions in which the erosion material of the gneiss domes is accumulated. A third savanna community, the *Monechma australe–Acacia erioloba* community, occurs on somewhat more compact, yellow sand in a limited part of the area in which ground squirrels (*Xerus inauris*) are very active. On rocky dry river sides a dense shrub community of *Schotia afra* var. *angustifolia* is found (Table 3). The alluvial deposits support a number of different communities depending on substrate and periodicity of flooding. The most important of these are the *Ziziphus mucronata–Euclea pseudebenus* community, a gallery forest or woodland on silty soil, the *Ficus cordata* community on rocky sides at the bottom of the deep Orange River gorge, the *Sisyndite spartea* community on strongly weathered gneiss in drainage lines, and the *Stipagrostis namaquensis* community on wind-blown river sand.

3.5 *Vegetation of the southern Kalahari*

The southern Kalahari is so distinct geomorphologically, and therefore also floristically from the remainder of the Namaland Domain, that it can be regarded as a subdomain of its own. Dune sand, clayey or silty soils of pans and of dry riverbeds, and calcrete screes along the major dry rivers are the four major habitats of the southern Kalahari. Dune sand covers at least 90 per cent of the area. The dunes are largely stable; under conditions of overgrazing and trampling, which are easily brought about in the southern Kalahari, the vegetation is destroyed and the sand dunes begin to shift again. There are three types of sandy soils: red, pink and white. Red sand is by far the commonest. Its colour is due to a coating of ferric oxide. This coating can be removed by washing. Pink sand is largely intermediate between red and white sand, and occurs in the contact zone of the two, and where calcrete is capped by only a thin layer of sand. White sand is least common and is found in the immediate vicinity of pans and riverbeds and on calcrete outcrops. White sand generally has an appreciably higher mineral content than the other two types. The soils of the major dry riverbeds, Auob, Nossob,

Table 3. Main communities of Augrabies Falls National Park (after Werger & Coetzee 1977).
(*: the species reaches exceptionally high cover values).

Community	A	B_1	B_2	B_3	C	D	E	F	G_1	G_2	H
Number of relevés	6	6	9	13	5	5	2	3	8	8	4
Ceraria namaquensis	5	.	.	.	1	.	.	.	.	.	.
Panicum arbusculum	3	1	.	.	.	.	.	.	.	.	.
Acacia mellifera subsp. detinens	1	1	4	4	.	2	2	.	.	.	1
Indigofera heterotricha	3	5	4	2	1	.	2	2	.	.	1
Zygophyllum suffruticosum	2	.	3	4	1	.	.	1	.	.	.
Aristida congesta subsp. barbicollis	.	3	3	2	.	.	.	.	.	1	.
Stipagrostis anomala	.	2	2	3	1	.	.	.	1	.	.
Limeum dinteri	.	3	2	2	1	1	.	.	.	.	.
Blepharis mitrata	.	.	3	2	.	.	.	.	.	.	.
Sarcostemma viminale	.	.	3	2	.	1	1	.	.	.	.
Sericocoma avolans	1	.	3	2	.	.	1	.	.	.	.
Polygala leptophylla	.	2	2	2	.	.	.	1	1	.	1
Boscia foetida subsp. foetida	1	.	2	2	2	2	1	.	.	.	.
Euphorbia rhombifolia	.	3	1	2	.	.	.	1	.	.	.
Boscia albitrunca	1	2	3	1	1	1	1	.	.	.	2
Hermannia stricta	.	.	2	2	.	.	.	.	.	.	.
Rhynchosia totta	.	2	2	.	.	.	1	2	.	.	.
Barleria rigida	1	2	3	1	.	.	.	1	.	1	1
Tetragonia arbusculoides	.	.	2	1	2	1	.	.	.	.	.
Microloma incanum	.	.	2	.	.	.	.	.	.	.	1
Thesium lacinulatum	.	.	2	.	.	.	1	.	1	.	1
Zygophyllum dregeanum	.	.	2	5	.	2	.	.	.	.	.
Oropetium capense	1	1	.	4	2	.	.	.	.	.	.
Mollugo cerviana	.	.	.	2	.	1	.	.	.	.	.
Senecio longiflorus	.	.	1	2	.	.	.	.	.	.	.
Salsola tuberculata	.	.	.	2	1	.	.	.	.	.	.
Eragrostis nindensis	.	.	.	.	2	.	.	.	.	.	.
Chascanum gariepina	.	.	1	1	3	.	.	.	.	.	.
Osteospermum amplectens	.	.	.	.	4	.	.	.	.	.	.
Oxalis sp.	.	.	.	1	3	.	.	.	.	.	.
Tephrosia dregeana	1	.	.	.	3	2	.	.	.	.	.
Phyllanthus maderaspatensis	.	.	2	1	3	.	.	.	.	.	.
Enneapogon desvauxii	2	5	3	5	2	1	1	.	2	.	1
Limeum aethiopicum	2	1	3	4	3	1	.	.	.	.	1
Hermannia spinosa	3	.	4	3	4	.	2	.	.	.	1
Dyerophytum africanum	1	3	3	3	3	.	1	1	.	2	.
Eragrostis porosa	2	3	4	2	4	2	1	2	1	.	.
Aptosimum spinescens	1	.	4	3	3	.	1	.	.	.	1
Hibiscus elliottiae	1	2	3	2	2	.	.	.	1	2	.
Asparagus denudatus	1	.	3	2	3	1	2	.	.	.	1
Tribulus zeyheri	2	1	2	2	3	1	.	.	1	.	.
Rhigozum trichotomum	1	.	2	2	5	1	.	.	.	1	.
Stipagrostis hochstetterana var. secalina	.	.	2	1	.	5	.	.	.	.	1
Kohautia cynanchica	1	.	1	1	.	4	.	1	.	.	.
Ptycholobium biflorum	.	.	.	1	.	3	.	.	.	.	.
Parkinsonia africana	.	.	.	1	.	2	.	.	.	.	.
Cadaba aphylla	1	.	.	1	.	3	.	.	.	.	.
Eragrostis annulata	1	2	5	5	4	2	1	1	.	.	.
Euphorbia gregaria	2	5	1	3	5*	5	.	1	.	.	1
Indigofera argyroides	.	1	4	4	3	2	.	.	.	1	.
Lotononis platycarpa	2	3	2	2	4	2	.	.	2	1	.
Nymania capensis	2	2	2	2	4	5	.	.	1	.	1
Dicoma capensis	1	1	.	2	3	2	.	.	.	.	.
Gisekia africana	.	.	.	1	2	3	1	.	.	.	.
Cenchrus ciliaris	.	.	.	.	.	.	2	.	.	.	2

Species	A	B1	B2	B3	C	D	E	F	G1	G2	H
Eragrostis trichophora	.	.	.	.	.	.	2	.	.	.	.
Anthephora pubescens	.	3	1	.	.	.	2	1	2	.	1
Pappea capensis	.	2	3	.	.	.	2	.	.	.	1
Panicum maximum	.	.	.	.	.	.	2	.	.	.	.
Ocimum canum	.	.	.	.	.	.	2	.	.	.	.
Androcymbium sp.	.	.	.	.	.	.	2	.	.	.	.
Berkheya spinosissima var. namaensis	1	1	.	.	2	.	2	2	.	.	.
Oxalis obliquefolia	.	.	.	.	.	.	2	.	.	.	.
Chloris virgata	.	.	.	.	.	.	2	.	.	.	.
Tragus berteronianus	.	.	.	1	2	.	2	.	.	.	.
Setaria verticillata	.	.	1	.	.	1	2	.	.	.	.
Lycium austrinum	.	.	2	.	1	1	2	.	.	.	.
Rhynchelytrum repens	1	1	1	.	.	.	2	1	.	2	1
Antherothamnus pearsonii	.	1	.	.	.	.	.	3	.	.	.
Barleria lancifolia	.	1	2	.	.	.	.	1	.	.	.
Stachys burchelliana	.	.	.	.	.	.	1	2	1	.	.
Commiphora gracilifrondosa	1	1	.	.	.	.	.	.	5	4	.
Trichodesma africana	1	.	.	.	.	.	1	1	2	2	1
Abutilon pycnodon	.	.	.	.	.	.	1	2	2	2	2
Cleome angustifolia subsp. diandra	2	2	.	1	5	.	.	.	2	2	.
Euphorbia glanduligera	.	.	.	.	.	.	.	.	2	2	.
Berkheya chamaepeuce	.	.	.	.	.	.	.	.	4	2	.
Rhus populifolia	.	.	.	.	1	.	2	2	3	1	2
Hermannia minutiflora	.	3	.	.	1	.	1	3	3	.	1
Adenolobus gariepina	1	.	1	.	.	.	.	.	2	4	1
Cleome oxyphylla	.	.	.	.	.	.	.	.	1	3	.
Sisyndite spartea	.	.	.	.	.	2	.	.	1	2	1
Hibiscus engleri	.	2	2	.	.	.	.	.	.	2	.
Schotia afra var. angustifolia	.	.	.	.	.	.	.	.	.	.	4
Triraphis ramosissima	.	5	2	.	.	.	.	2	2	2	4
Stipagrostis uniplumis	5	5	5	5	5	5	2	2	2	4	4
Enneapogon scaber	5	5	5	4	5*	.	1	3	5	5	4
Monechma spartioides	3	5	5	4	5	4	1	3	5	3	2
Forskohlea candida	3	5	3	3	4	3	2	3	5	5	1
Aristida curvata	3	5	5	4	5	.	2	1	4	5	2
Schmidtia kalihariensis	4	3	2	3	3	5	.	.	3	4	.
Enneapogon cenchroides	2	3	2	3	.	.	2	2	.	.	.
Codon royeni	3	4	2	1	1	1	.	2	2	3	.
Sutera tomentosa	.	2	1	.	1	1	1	1	2	.	1
Cucumis dinteri	1	.	1	.	.	.	1	1	.	2	1
Senecio sisymbrifolius	1	1	.	.	1	1	1	.	2	.	.
Montinia caryophyllacea	.	2	1	1	.	1	.	.	.	.	1
Phaeoptilum spinosum	.	.	1	1	.	2	.	.	.	.	1
Peliostomum leucorrhizum subsp. junceum	1	2	.	.	.	.	.	2	.	.	.
Rogeria longiflora	.	.	.	.	.	1	.	1	.	2	.
Curroria decidua	1	1	1	.	.	.	.	.	.	2	.
Indigofera pungens	2	.	.	.	.	.	.	.	.	.	.
Solanum sisymbrifolium	.	.	2	.	.	.	.	.	1	.	.
Boerhavia repens	.	.	.	.	.	.	.	.	.	2	.
etc.											

A: Ceraria namaquensis community.
B: Indigofera heterotricha–Zygophyllum suffruticosum community with subcommunities of Triraphis ramosissima (1), Monechma spartioides (2), and Zygophyllum dregeanum (3).
C: Enneapogon scaber–Euphorbia gregaria community.
D: Stipagrostis hochstetterana community.
E: Eragrostis trichophora–Acacia mellifera community.
F: Antherothamnus pearsonii community.
G: Commiphora gracilifrondosa community with subcommunities of Rhus populifolia (1) and Adenolobus gariepina (2).
H: Schotia afra community.

Molopo and Kuruman, are compact sandy clays or silts. Some branches of these dry rivers are raised and blocked off from the main course by a shallow threshold of sand, particularly in the Nossob. These areas are called alluvial pans and possess a sandy loam soil. Pans are of various types: salt pans, sand pans, calcrete pans and clay pans, the latter being the most common ones. All these substrates, and also the calcrete screes, are clearly characterized by their own plant communities.

The plant ecology of the southern Kalahari has been studied in detail by Leistner (1967), Leistner & Werger (1973), and Werger & Leistner (1975). General outlines were given by Giess (1971) and Leser (1971), and to some extent by Range (1932), Weare & Yalala (1971) and Blair Rains & Yalala (1972). Leistner (1967) also presented an account of the phenology.

In the southern Kalahari essentially two seasons can be distinguished: a warm season with a maximum of precipitation (summer), and a colder, drier season (winter). These seasons are separated from one another by short transitional periods. The calendar disposition of these seasons varies from year to year depending on meteorological fluctuations, though it can be said by approximation that summer covers the period from November to March, and winter the period from June to August.

Germination is strongly seasonal in the southern Kalahari. At least 85 per cent of the species germinate in the period from October to March, while most of the remainder germinate from April to September. Few perennials are among this latter group. There is also a number of species that germinates throughout the year.

Late winter to early spring is the period of maximal vegetative growth, during which trees, shrubs and plants with underground storage organs start to sprout even before the first rains have fallen. During late May the frosts start and summer annuals die off, while most perennials start losing their leaves.

With the exception of species with deep roots or storage organs, the species in the southern Kalahari flower only after having received rains. In spring before the rain starts about 4.0 per cent of the species investigated flower, while 1.7 per cent start after the first small rains. During November and early December these groups start setting fruit. In early summer 16.5 per cent start flowering and after the first summer rains 25.1 per cent of the perennials and 15.0 per cent annuals follow suit. Only few species (1.2 per cent) flower only during the hottest months of the year (December to February). In late summer and autumn 4.6 per cent of the species flower. In autumn the percentage is 5.5, comprising many geophytes. These species continue flowering until the first frosts set in at the end of May. Species typically flowering in winter comprise 13.3 per cent and are mainly annuals. Some perennials (6.0 per cent) flower both in spring and autumn, while 6.9 per cent of the species do not show any periodicity in their flowering season (cf. Leistner 1967).

The vegetation of sandy habitats consists of an extremely open shrub or tree savanna, whereas the other substrates harbour an open dwarf shrub or grassland formation.

The characteristic association of the coarse sand of dune tops and slightly undulating sand plains is the Stipagrostietum amabilis. It is characterized by the large shrubby *Stipagrostis amabilis* and some other species. In the northeastern marginal areas of the southern Kalahari the subassociation terminalietosum of this association occurs. Here the precipitation is somewhat higher and trees and shrub species of a Sudano–Zambezian distribution type, like *Terminalia sericea* and *Albizia anthelminthica*, are thus able to grow. The usual trees and shrubs in the Stipagrostietum amabilis are *Acacia erioloba*, *A. haematoxylon* and *Boscia albitrunca* (Table 4).

On the lower slopes of the dunes and in most of the dune valleys ('streets') the association Hirpicio echini–Asthenatheretum is found (Fig. 19). The same shrubs and trees as well as *Grewia flava* build up a sparse layer of emergents, while tuft grasses, particularly *Asthenatherum glaucum*, dominate. Both in this association and in the Stipagrostietum amabilis prostrate annuals and perennials, sometimes with long, trailing runners, are abundant on the sandy surface. Most common are *Acanthosicyos naudinianus* and *Citrullus lanatus*, the latter forming the main water source for nomadic tribes during their short visits to the desolate area in days passed (cf. Story 1958).

Fig. 19. View of the dune country of the southern Kalahari between Nossob and Auob. The foreground (right) shows the Stipagrostietum amabilis of the dune tops, while the valleys support the Hirpicio–Asthenatheretum, here with rather many trees of *Acacia erioloba* and (less) *A. haematoxylon*.

The pink sand carries the *Monechma incanum–Stipagrostis ciliata* community: in deep dune valleys where calcrete comes near the surface, and in sand pans, the community is treeless and has a high abundance of *Indigofera alternans* and *Aptosimum albomarginatum*, while the pink riverine dune sand community contains many shrubs, particularly *Acacia mellifera* subsp. *detinens*. *Rhigozum trichotomum* is also common in both types of this community.

The communities discussed so far typically show a catena type of pattern conform the main northwest–southeast direction of the dunes.

On the more or less compact white sand around most pans and along the main river courses the Peliostomo–Stipagrostietum obtusae, an open dwarf shrub and

Table 4. Plant communities of the southern Kalahari (after Leistner & Werger 1973).

Community	A_1	A_2	B	C_1	C_2	D	E	F	G	H	J
Number of relevés	5	14	13	9	13	19	16	10	5	7	4
Stipagrostis amabilis	3	5	.	.	1	.	.	.	.	.	.
Eragrostis trichophora	1	4	1	.	.	.	.	.	.	.	.
Crotalaria spartioides	.	3	.	.	.	.	.	.	.	.	.
Terminalia sericea	4	.	.	.	.	.	.	.	.	.	.
Rhus tenuinervis	3	1	.	.	.	.	.	.	.	.	.
Ehretia rigida	3	.	.	.	.	.	.	.	.	.	.
Albizia anthelmintica	2	.	.	.	1	.	.	.	.	.	.
Grewia retinervis	2	.	.	.	.	.	.	.	.	.	.
Crotalaria sphaerocarpa	.	1	4	.	1	.	.	.	.	.	.
Hirpicium echinus	.	1	5	.	.	.	.	.	.	.	.
Dicoma schinzii	.	.	4	1	1	.	.	.	.	.	.
Cassia italica	.	.	3	1	.	.	.	.	.	.	.
Merremia verecunda	1	.	2	2	.	.	.	.	.	.	.
Portulaca kermesina	.	.	3	.	.	.	.	.	.	.	.
Neuradopsis austro-africana	.	.	2	.	1	.	.	.	.	.	.
Dimorphotheca polyptera	.	.	2	1	.	.	.	.	.	.	.
Cyamopsis serrata	.	.	2	.	1	.	.	.	.	.	.
Melhania burchellii	.	.	2	.	.	.	.	.	.	.	.
Asthenatherum glaucum	4	4	5	1	1	.	.	.	.	.	.
Oxygonum delagoense	5	5	4	1	2	.	.	.	.	.	.
Limeum arenicolum	5	5	3	1	1	.	.	.	.	.	.
Stipagrostis uniplumis	4	3	4	1	1	.	.	.	2	.	.
Hermannia tomentosa	3	4	4	1	.	.	.	.	.	.	.
Plinthus sericeus	2	4	2	.	1	.	.	.	.	.	.
Sericorema remotiflora	.	4	4	.	.	.	.	.	.	.	.
Aristida meridionalis	3	4	4	.	.	.	.	.	.	.	.
Chascanum pumilum	2	4	4	1	.	.	.	.	.	.	.
Citrullus lanatus	2	4	2	1	.	.	.	.	.	.	.
Indigofera flavicans	3	4	2	2	1	.	.	.	.	.	.
Acanthosicyos naudinianus	3	2	4	.	.	.	.	.	.	.	.
Sesamum triphyllum	.	2	2	1	1	.	.	.	.	.	.
Merremia tridentata	.	4	2	1	.	.	.	.	.	.	.
Indigofera daleoides	1	2	4	.	.	.	.	.	.	.	.
Hermannia burchellii	.	2	2	.	.	.	.	.	.	.	.
Cleome kalachariensis	.	2	2	1	.	.	.	.	.	.	.
Ipomoea hackeliana	.	1	2	.	.	.	.	.	.	.	.
Limeum viscosum	1	2	2	.	.	.	.	.	.	.	.
Phyllanthus omahekensis	1	1	2	.	.	.	.	.	.	.	.
Indigofera aspera	1	.	1	2	.	.	.	.	.	.	.
Monechma incanum	.	1	1	4	4	1	.	.	.	.	.
Chrysocoma polygalifolia	.	1	1	2	1	.	.	.	.	.	.
Triraphis fleckii	.	.	.	2	2	1	.	.	.	.	.
Harpagophytum procumbens	.	.	1	1	2	.	.	.	.	.	.
Peliostomum leucorrhizum	.	.	1	.	.	4	.	.	.	.	.
Comptonanthus molluginoides	.	.	.	.	.	3	.	.	.	.	.
Eragrostis brizantha	.	.	.	.	.	2	.	.	.	.	.
Camptorrhiza strumosa	.	.	.	.	1	2	.	.	.	.	.
Ledebouria undulata	.	.	.	1	.	2	.	.	.	.	.
Ophioglossum polyphyllum	.	.	.	.	.	2	.	.	.	.	.
Brachiaria glomerata	5	5	4	5	3	2	.	.	.	.	.
Eragrostis lehmanniana	5	4	5	3	2	1	.	.	.	.	.
Requienia sphaerosperma	5	5	5	4	2	.	.	.	.	.	.
Fimbristylis hispidula	4	4	5	3	3	1	.	.	.	.	.
Limeum fenestratum	5	5	3	3	3	1	.	.	.	.	.

Community Number of relevés	A_1 5	A_2 14	B 13	C_1 9	C_2 13	D 19	E 16	F 10	G 5	H 7	J 4
Limeum sulcatum	3	4	4	3	2	2	.	.	.	.	.
Heliotropium ciliatum	5	5	5	1	2	.	.	.	.	.	.
Acacia haematoxylon	1	4	4	2	1	.	.	.	.	.	.
Jatropha erythropoda	1	2	4	2	2	2	.	.	.	.	.
Cynanchum orangeanum	2	2	3	3	2	2	.	.	.	.	.
Celosia linearis	.	2	1	3	2	2	.	.	.	.	.
Aptosimum depressum	.	.	2	2	1	1	.	.	.	.	.
Acacia mellifera subsp. detinens	2	.	1	.	5	.	1	1	.	.	.
Grewia flava	3	1	3	.	3	.	.	.	.	.	.
Acacia erioloba	2	1	2	.	3	1	.	.	.	.	.
Indigofera alternans	.	.	1	4	3	3	.	.	.	.	.
Aptosimum albomarginatum	.	.	.	3	1	3	1	1	.	1	.
Limeum myosotis	.	.	1	2	2	3	.	.	.	.	.
Grielum humifusum	.	.	1	2	2	3	.	.	.	.	.
Plinthus cryptocarpus	.	.	.	.	3	.	.	.	.	.	.
Hybanthus densifolius	.	.	1	2	1	2	.	.	.	.	.
Monsonia angustifolia	.	.	1	2	1	1	.	.	.	.	.
Indigofera auricoma	.	.	.	.	.	.	5	1	.	.	.
Barleria rigida	.	.	.	.	.	.	4	2	.	.	.
Aizoon schellenbergii	.	.	.	.	.	.	5	.	.	.	.
Euphorbia glanduligera	.	.	.	.	.	.	5	.	.	.	.
Cleome angustifolia subsp. diandra	.	.	.	.	.	.	4	.	.	.	.
Fagonia sinaica var. minutistipula	.	.	.	.	.	.	3	2	.	.	.
Limeum aethiopicum	.	.	.	.	.	.	4	.	.	.	.
Zygophyllum pubescens	.	.	.	.	.	.	3	.	.	.	.
Sylitra biflora	.	.	.	.	.	.	2	.	.	.	.
Hermannia abrotanoides	.	.	.	.	.	.	2	.	.	.	.
Phyllanthus maderaspatensis	.	.	.	.	.	.	2	.	.	.	.
Enneapogon scaber	.	.	.	.	.	.	2	.	.	.	.
Tribulus cristatus	.	.	.	.	.	.	2	.	.	.	.
Enneapogon cenchroides	.	.	.	.	.	.	2	.	.	.	.
Zygophyllum tenue	.	.	.	.	.	.	1	5	.	.	1
Sporobolus lampranthus	.	.	.	.	.	.	.	4	.	.	.
Eragrostis truncata	.	.	.	.	.	.	.	4	.	.	.
Indigofera argyroides	.	.	.	.	.	.	.	4	.	.	.
Salsola rabieana	.	.	.	.	.	.	.	1	3	1	3
Hirpicium gazanioides	.	.	.	.	.	.	.	.	3	.	.
Asparagus denudatus	.	.	.	.	.	.	.	.	3	.	.
blue-green algae	.	.	.	.	.	.	.	.	3	.	.
Sporobolus coromandelianus	.	.	.	.	.	.	.	.	5	2	2
Sporobolus rangei	.	.	.	.	.	.	.	.	.	5	.
Lycium tenue	.	.	.	.	2	.	1	1	.	.	4
Rhigozum trichotomum	.	.	1	4	5	5	5	3	.	.	2
Stipagrostis obtusa	.	.	1	4	5	5	4	2	.	.	1
Dicoma capensis	.	1	1	3	4	5	2	3	.	.	.
Stipagrostis ciliata	.	1	4	5	5	1	3	1	.	.	.
Hermannia modesta	.	.	2	4	3	4	3	3	.	.	1
Kohautia lasiocarpa	.	.	2	2	2	5	2	1	.	1	1
Talinum caffrum	.	.	2	2	2	2	1	.	.	1	.
Tribulus terrestris	.	.	.	1	.	5	2	5	.	.	1
Enneapogon desvauxii	.	.	.	1	1	1	5	5	.	3	4
Tragus racemosus	.	.	.	1	.	3	3	5	.	3	3
Monechma australe	.	.	.	.	.	3	5	1	.	.	1
Trianthema triquetra subsp. parvifolia	.	1	1	.	.	4	5	5	3	4	3
Aptosimum lineare	.	.	.	.	.	1	2	5	4	.	1
Eragrostis annulata	.	.	.	.	2	1	3	4	3	1	2

271

Community Number of relevés	A₁ 5	A₂ 14	B 13	C₁ 9	C₂ 13	D 19	E 16	F 10	G 5	H 7	J 4
Geigeria ornativa	.	.	.	.	.	2	.	3	.	1	3
Oropetium capense	.	.	.	.	.	2	.	3	.	3	.
Plinthus karooicus	.	.	.	.	.	2	.	3	.	.	1
Eriocephalus pubescens	.	.	.	.	.	2	2	4	.	.	1
Gisekia africana	3	3	4	5	5	5	3	1	.	.	.
Schmidtia kalihariensis	3	2	4	5	5	4	2	1	.	1	.
Tribulus zeyheri	4	2	2	4	5	2	1	1	.	1	1
Euphorbia inaequilatera	2	2	3	2	2	4	3	2	.	.	1
Lycium cinereum	4	1	.	.	5	.	2	.	.	.	.
Lotononis platycarpa	.	.	2	3	2	2	2	3	.	.	.
Cucumis africanus	1	.	1	3	2	4	3	2	.	1	2
Mollugo cerviana	.	2	1	4	5	3	3	.	.	1	.
Asparagus suaveolens	2	.	.	3	1	2	2	.	.	.	.
Eragrostis porosa	.	.	.	1	2	3	2	.	.	1	1
Boscia albitrunca	4	1	1	.	3	.	1	.	.	.	.
Ornithogalum amboense	.	.	.	.	.	1	1	2	.	4	.
Acrotome inflata	.	1	1	3	.	2	1	.	.	.	.
Cleome paxii	.	1	1	1	2	1	.	.	.	.	.
Limeum argute-carinatum	.	.	.	.	.	1	3	.	2	.	.
Pollichia campestris	.	1	1	.	2	.	.	.	.	.	.
Lophiocarpus polystachyus	.	1	.	.	2	1	.	.	.	.	.
Cleome gynandra	.	.	1	.	2	.	.	.	.	.	.
Trachyandra laxa var. laxa	.	.	2	.	.	1	.	.	.	.	.
Pentzia globosa	.	.	.	.	.	2	1	1	.	.	.
Geigeria pectidea	.	.	.	.	.	2	1	.	.	.	1
Felicia hyssopifolia	.	.	.	2	1	1	.	.	.	.	.
Heliotropium lineare	.	.	.	.	.	2	.	.	.	.	.
Phyllanthus pentandrus	.	.	.	.	2	.	.	1	.	.	1
Amaranthus schinzianus	.	1	.	.	1	.	2	.	.	.	.
Aristida congesta	.	.	.	.	.	.	.	.	2	.	.
Aristida adscensionis	.	.	.	.	.	.	.	2	.	.	.
etc.											

A: Stipagrostietum amabilis (2) with subassociation of Terminalia sericea (1).
B: Hirpicio echini–Asthenatheretum.
C_1: Monechma incanum–Stipagrostis ciliata–Indigofera alternans community.
C_2: Monechma incanum–Stipagrostis ciliata–Acacia mellifera community.
D: Peliostomo–Stipagrostietum obtusae.
E: Aizoo–Indigoferetum auricomae.
F: Sporobolo lampranthi–Zygophylletum tenuis.
G: Sporoboletum coromandeliani.
H: Sporoboletum rangei.
J: Lycium tenue community.

grasses community, is found. It is of varying physiognomy owing to differences in the occurrence of *Rhigozum trichotomum*. For year after year the annual grass *Schmidtia kalihariensis* can form dense stands of up to 0.6 m high in this and the previous community, particularly in disturbed areas. It is possible that this species has an allelopathic effect on the germination of other species.

An open dwarf shrub community, the Aizoo-Indigoferetum auricomae, is restricted to the calcrete screes on the banks of the main rivers. Characteristic dwarf shrubs include *Aizoon schellenbergii, Barleria rigida, Zygophyllum pubescens, Ptycholobium* (=*Sylitra*) *biflorum* and *Limeum aethiopicum*. Sometimes taller shrubs of *Lycium austrinum* are present (Fig. 20).

Fig. 20. Nossob River south of Twee Rivieren. On the calcrete bank occurs the Aizoo-Indigoferetum and at the edge of the bare dry river floor the Peliostomo-Stipagrostietum obtusae. The shrub in the foreground is *Rhigozum trichotomum* and the dwarf shrubs include *Aizoon schellenbergii* (photo O. A. Leistner).

Fig. 21. Zonation of the vegetation around a pan. The centre left shows the bare surface of the pan centre, surrounded by the grass tufts that form the Sporoboletum rangei. The light coloured belt in the foreground consists of the Peliostomo-Stipagrostietum obtusae, and the dark coloured belt along the dune base in the background is the *Monechma incanum–Stipagrostis ciliata* community with much *Aptosimum albomarginatum* and *Rhigozum trichotomum*. The dune slope supports the Hirpicio-Asthenatheretum, which can be distinguished in this picture as the grey belt below the bushy zone along the dune crests with the Stipagrostietum amabilis (photo O. A. Leistner).

Another dwarf shrub community is the Sporobolo lampranthi–Zygophylletum tenuis of the alluvial pans. This soil type is intermediate in mineral contents and texture between that of the dunes and that of the dry riverbeds. On pan floors, along edges of pans, and sometimes in or along dry riverbeds small dunes of loamy sand and usually less than 1 m high are found. Such sites carry a community dominated by the dwarf shrub *Lycium tenue*. This rhizomatous species, which is also capable of suckering from exposed major roots, is probably instrumental in the formation of the small dunes.

Most pan floors consist in their centres of very compact and almost impervious clayey, highly alkaline soils with a high mineral content. No perennial species are found here; the pan floors are either bare or support the Sporoboletum coromandeliani, an association in which the ephemerals *Sporobolus coromandelianus* and *Trianthema triquetra* subsp. *parvifolia* are constant, and which is encountered only for a few weeks following rains. However, its cover values can reach up to 30 per cent, hence the community can provide useful grazing for wild animals. Where the hard pan floors are overlain by a thin layer of white, calcareous sand, as is usual along their periphery, a very open community is found, the Sporoboletum rangei, characterized by the perennial tufted grass *Sporobolus rangei*. Salt pans are usually bare, with sparse plant growth restricted to the marginal zone. *Suaeda fruticosa*, *Hypertelis salsoloides*, *Trianthema triquetra* subsp. *parvifolia* and *Zygophyllum microcarpum* are common species here. Pans often show a more or less concentric arrangement of plant communities around them. In the centre they are bare or carry the Sporoboletum coromandeliani, often followed by a zone of Sporobletum rangei; then follows a zone of Peliostomo–Stipagrostietum obtusae, surrounded by a ring of *Monechma incanum–Stipagrostis ciliata* community with *Aptosimum albomarginatum*, which gives way to the communities of the red sand (Fig. 21).

The dry riverbeds with their heavy clayey or silty soils which are rich in nutrients, characteristically carry a dense grassland community, the Panicetum colorati. This community can be overgrown locally by dwarf shrubs, e.g. *Galenia africana*, or shrubs, e.g. *Lebeckia linearifolia*. Also trees of *Acacia erioloba* and *A. haematoxylon* occur rather commonly in the dry riverbeds. Unequal penetration of water into the ground, or destruction of the grass cover in the Panicetum colorati can lead to the domination by different species at different sites in this community. Local small depressions occur in the riverbeds. These contain water for a much longer time than the riverbeds themselves, and sometimes harbour the waterplants *Marsilea macrocarpa* and *Aponogeton junceus*. *Eragrostis rotifer* usually forms dense stands around such depressions.

The ecological relationships between these Kalahari communities were presented by Leistner & Werger (1973) in the two schemes shown in Figs. 22 and 23, which are not intended, however, to suggest direct succession schemes. As pointed out before, succession in these areas is mainly allogenic, not autogenic (Werger & Leistner 1975).

Not far north of the Orange River the boundary of the continuous sand cover is reached. This boundary is usually, though not always, abrupt and the vegetation changes markedly. Small, or sometimes even larger, exclaves of wind-blown Kalahari sand do occur further southwards into the Karoo Domain, however, some even south of the Orange River. All these sites carry the characteristic Kalahari vegetation with *Stipagrostis obtusa* and *S. ciliata*, though sometimes as

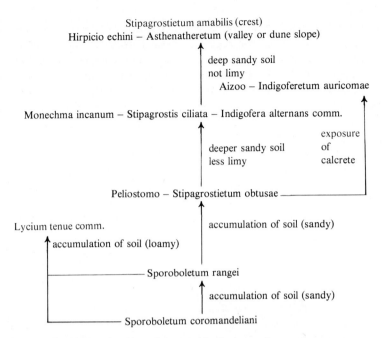

Fig. 22. Relationships of communities from pan floor to dune crest.

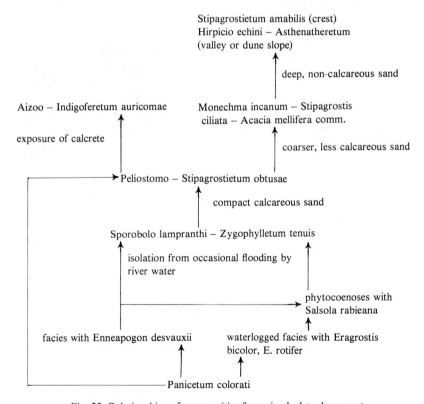

Fig. 23. Relationships of communities from riverbed to dune crest.

a result of overgrazing, in a hardly recognizable form. Werger (1973b) described two such communities from the region between Petrusville and Douglas in the Upper Orange River area: the Enneapogono desvauxii–Stipagrostietum on slightly loamy sands, and the Pentzio calcareae–Stipagrostietum on sandy soils on calcrete. Of this latter association a subassociation with *Acacia erioloba* occurs on deep sands near Douglas. Frequently stands belonging to one of these associations can be encountered which are strongly dominated by the shrub *Rhigozum trichotomum*, probably as a result of overgrazing.

Eastwards the boundary of the southern Kalahari runs in the vicinity of Kuruman but is generally not as sharp as the southern boundary. Between Kuruman and Vryburg polycorms of *Acacia erioloba* are distributed patchily on scattered deep sand accumulations (Fig. 24) on the limy and dolomitic Ghaap Plateau (Coetzee & Werger 1975). The grasses are often of Sudano–Zambezian

Fig. 24. Polycorm of *Acacia erioloba* on sand between Kuruman and Vryburg. Note the amount of top soil eroded away (photo O. A. Leistner).

affinity in these patches, like *Themeda triandra* and *Cymbopogon plurinodis*, but *Stipagrostis uniplumis* and *Schmidtia pappophoroides* can also be prominent, particularly following severe grazing. Over large areas the calcrete substrate supports a grassland, also of Zambezian affinity, and locally there are bushclump patches with a mixed, though mainly Zambezian flora comprising *Olea africana*, *Tarchonanthus* sp., *Acacia karroo*, *A. tortilis* subsp. *heteracantha*, various species of *Rhus*, *Grewia flava*, *Boscia albitrunca*, and others (Chapter 32, Fig. 32). Towards the southeastern boundary, near the Orange River, and in mountainous areas like the Langeberg and Asbestos Mts., both the Zambezian and the Karoo elements become prominent (cf. Acocks 1953).

3.6 Vegetation of the Western Cape Domain

The Western Cape Domain is entirely located in South Africa except for a small outlier which extends into South West Africa north of the Richtersveld as far as the vicinity of Aus. It comprises the highly succulent dwarf shrub vegetation along the western coast, on the western escarpment mountains which consist mainly of gneiss and quartzite, and on the western edge of the plateau, while in the south it is fringed by Cape fynbos. The boundary between the Western Cape Domain and the more centrally situated Karoo Domain lies in the uniform rainfall area in the vicinity of the winter rainfall area (see Sect. 1 and 2.2). Hence, the Western Cape Domain receives virtually all its rain during winter.

The most comprehensive description of this area is given by Acocks (1953). The following of Acocks' Veld Types belong to the Western Cape Domain: the main part of the Strandveld (VT 34), the Succulent Karoo (VT 31), the Namaqualand Broken Veld (VT 33), the Western Mountain Karoo (VT 28), the False Succulent Karoo (VT 39), the western parts of the Succulent Mountain Scrub or Spekboomveld (VT 25) and of the Karroid Broken Veld (VT 26), and marginally also the Mountain Renosterbosveld (VT 43). The Western Cape element still occurs in various degrees in the more eastern parts of VT 25 and VT 26, in the eastern parts of Valley Bushveld (VT 23) and in the Noorsveld (VT 24).

The Strandveld has strong affinities with some fynbos communities, particularly at its southern boundary where annual rainfall is 250 mm or more. Further north the proportion of succulents is greater and there is a gradual transition into Succulent Karoo. Dwarf shrubs usually initiate the formation of small dunes or sandy hummocks. Typical species include *Zygophyllum morgsana, Blackiella inflata, Salvia nivea, S. africana-lutea, Ruschia utilis, R. rupigena, Euphorbia burmannii, Grielum humifusum* and *Heliophila remotiflora* (Acocks 1953, Werger unpubl.).

The Succulent Karoo forms a long belt between the Strandveld of the coast and the mountains of the escarpment in Namaqualand. Southwards this Veld Type also occurs further inland in the Tanqua Karoo, east of the fynbos of the Cedarberg and the Cold Bokkeveld Mts., as well as locally near the southern fringes of the Western Cape Domain. The precipitation is less than 200 mm here, and the altitude is less than 600 m. The vegetation is strongly dominated by succulents or, following good rains, by annuals, mainly of the families Asteraceae and Brassicaceae, which can cover large areas in dense, colourful stands. The Namaqualand form of this Veld Type occurs both on the sand of the coastal plain and on the heavier rocky soil of the foothills of the escarpment (Fig. 25). Mists are very important here. The succulents range in height from virtually subterraneous to dwarf shrubs and shrubs of up to 2.5 m. Many species of *Ruschia, Lampranthus, Drosanthemum, Sphalmanthus, Mesembryanthemum, Psilocaulon, Conophytum, Cephalophyllum* and several other genera of the Mesembryanthemaceae are present. Other succulents in this vegetation include several species of *Euphorbia*, of which *E. mauritanica* is the most important, *Crassula, Cotyledon, Adromischus, Huernia, Stapelia, Sarcocaulon* and *Aloe*, as well as *Augea capensis, Othonna floribunda*, and others. Among the common dwarf

Fig. 25. Dry form of Succulent Karoo near Sendelingsdrift between the escarpment and the coast. The rocky hills and gravelly and sandy flats and washes carry an open succulent dwarf shrub vegetation in which Aizoaceae are prominent.

shrubs are *Salsola zeyheri, Galenia africana, Asparagus capensis, Zygophyllum retrofractum, Z. stapffii, Z. lichtensteinianum* and *Pteronia* spp. Some grasses of the genera *Stipagrostis, Chaetobromus, Ehrharta, Pentaschistis,* and *Lasiochloa* are also present (Marloth 1908, Acocks 1953, 1971). Very little detailed work has been carried out in this area. Table 5 summarizes the results obtained from line transect studies made by officers of the Dept. of Agricultural Technical Services. Only perennial species are included here and several species have not been identified to species or even genus level. The transects have been surveyed in the area from Hondeklipbaai southwards to the vicinity of the Groen Rivier. The substrate was brown-reddish to white sand. Three communities were clearly distinguished: a *Stipagrostis zeyheri* community, a *Chaetobromus dregeanus* community, and a *Tetragonia fruticosa–Stipagrostis ciliata* community. These three communities which indicate differences in the sandy substrate, are geographically separated, the *Stipagrostis zeyheri* community occurring furthest south. The three communities have many species in common.

The Succulent Karoo in the Tanqua River valley, an area surrounded by mountains and receiving less than 150 mm rain annually, is severely tramped out and eroded. The substrate consists largely of gravelly sand or of brackish alluvial silt. Where damage is less severe, the vegetation is dominated by short and stemless succulents. Species of *Ruschia, Sphalmanthus, Rhinephyllum, Drosanthemum, Hereroa, Psilocaulon, Mesembryanthemum, Galenia, Salsola, Euphorbia, Crassula, Cotyledon* and many other genera are common again, but the genera of Mesembryanthemaceae are usually represented by other species than in the

Table 5. Some communities from the northwestern Cape Province, South Africa (after data from the Dept. of Agricultural Technical Services).

Community	A	B	C	D	E
Number of relevés	2	7	4	34	20
Lightfootia thunbergiana	2	.	.	1	1
Aspalathus spinosa	2	.	.	.	.
Passerina vulgaris	2	.	.	.	.
Struthiola leptantha	2	.	.	.	.
Wiborgia sericea	2	.	.	.	.
Zygophyllum divaricatum	2	.	.	.	.
Elytropappus rhinocerotis	2	.	.	1	.
Eriocephalus ericoides	2	.	.	.	.
Muraltia rhamnoides	2	.	.	.	.
Pteronia glauca	2	.	.	.	.
Pentzia incana	2	.	.	.	.
Lebeckia sericea	2	.	.	1	.
Stipagrostis zeyheri	.	5	.	.	.
Asparagus capensis	.	4	.	.	.
Sarcocaulon sp. 1	.	3	.	.	.
Grielum humifusum	.	4	.	1	.
Hermannia multiflora	.	5	.	1	.
Mesembryanthemaceae 2	.	5	2	1	.
Asparagus stipulaceus	.	5	3	1	.
Euphorbia sp. 1	.	4	4	.	.
Chrysanthemoides monilifera	.	.	4	1	.
Chaetobromus dregeanus	.	.	3	1	.
Berkheya spinosa	.	.	3	.	.
Leontonyx sp.	.	.	3	1	.
Mesembryanthemaceae 4	.	.	3	2	.
Zygophyllum microcarpum	.	.	4	4	.
Limeum africanum	.	.	3	3	.
Sarcocaulon sp. 2	.	.	3	3	1
Tetragonia fruticosa	1	.	2	4	.
Stipagrostis ciliata	2	.	2	3	.
Asparagus asparagoides	2	.	1	4	1
Mesembryanthemaceae 3	.	.	.	3	.
Pelargonium quinatum	1	1	1	3	.
Mesembryanthemaceae 5A	.	1	.	4	.
Mesembryanthemaceae 5	.	1	.	2	.
Pteronia onobromoides	.	.	.	2	.
Didelta carnosa	.	.	.	2	.
Anthospermum aethiopicum	.	4	3	3	.
Ficinia sp.	.	4	4	2	.
Salvia lanceolata	.	4	1	2	.
Senecio aloides	.	4	4	5	.
Helichrysum hebelepis	.	4	1	3	.
Lebeckia multiflora	.	3	2	5	.
Pteronia paniculata	.	3	4	3	.
Putterlickia pyracantha	.	3	.	3	.
Osteospermum grandiflorum	.	5	4	4	.
Zygophyllum morgsana	.	4	4	5	.
Euphorbia mauritanica	.	2	2	4	.
Monochlamys albicans	.	2	.	2	.
Cephalophyllum cf primulinum	.	2	1	3	.
Pharnaceum aurantium	.	2	.	2	.
Microloma sagittatum	.	1	3	2	.
Conicosia sp.	.	1	2	2	.

279

Community	A	B	C	D	E
Number of relevés	2	7	4	34	20
Staberoha sp.	.	3	.	1	.
Limonium sp.	.	2	3	1	.
Willdenowia striata	.	2	.	2	.
Taragonia spicata	1	5	3	5	.
Hermannia trifurca	2	2	3	5	.
Ehrharta calycina	2	4	4	3	.
Hermannia disermifolia	1	4	2	1	.
Ehrharta barbinodes	2	1	2	1	.
Pteronia divaricata	2	2	1	2	.
Berkheya cuneata	1	3	.	2	.
Herrea blanda	1	2	.	2	1
Stipagrostis brevifolia	.	.	.	.	4
Stipagrostis obtusa	.	.	.	1	3
Stipagrostis uniplumis	.	.	.	.	2
Zygophyllum retrofractum	.	.	.	.	2
Psilocaulon absimile	.	.	.	.	2
Lycium sp.	.	3	1	3	3
Eriocephalus africanus	1	5	1	3	2
Galenia secunda	.	1	4	3	3
Euphorbia cherisina	1	.	4	3	1
Psilocaulon utile	.	.	3	3	2
Rhus undulata	2	1	.	2	1
Mesembryanthemaceae 11	.	.	3	1	3
Galenia africana	1	.	.	1	2
Schismus sp.	.	3	4	1	2
Monechma pseudopatulum	.	.	.	2	1
Rhus incisa	1	.	.	1	.
Wiborgia mucronata	1	.	.	1	.
Ifloga sp.	.	.	2	1	.
Diospyros ramulosa	.	.	.	2	.
Euclea tomentosa	.	.	2	1	.
Mesembryanthemaceae 1	.	2	.	1	.
Euclea racemosa	.	2	.	1	.
Chrysocoma coma-aurea	.	1	1	.	.
Euryops sp.	.	.	2	1	.
Cotyledon wallichii	.	1	1	1	.
Deverra aphylla	.	2	1	1	.
Melianthus sp.	.	2	.	1	.
Mesembryanthemaceae 6	.	1	.	2	.
Pteronia sp.	.	.	1	1	.
Hypertelis salsoloides	.	.	.	1	1
Cheiridopsis cf candidissima	.	.	2	.	1
Euphorbia sp. 2	.	2	.	1	1
Salsola zeyheri	.	.	.	1	1
Chrysocoma polygalifolia	.	.	.	1	1
etc.					

A: Elytropappus rhinocerotis community in Western Mountain Karoo.
B: Stipagrostis zeyheri community in Succulent Karoo.
C: Chaetobromus dregeanus community in Succulent Karoo.
D: Tetragonia fruticosa–Stipagrostis ciliata community in Succulent Karoo.
E: Stipagrostis brevifolia community of sandy flats in Bushmanland.

Namaqualand form of this Veld Type. Grasses are not common, but *Stipagrostis obtusa* and some other species are locally abundant. In the southwestern part, just north of the Wittebergen, transitions to Karroid Broken Veld occur, where the vegetation is an admixture of succulents and narrow-leaved dwarf shrubs (Cannon 1924, Compton 1929a, Acocks 1953). Several of the succulent genera also are important on silty flats near the southeastern edge of the Karoo–Namib Region, but the Karoo element is stronger in that flora, and there is already some mixing with Sudano–Zambezian species. This type of vegetation is therefore not directly comparable to the other two forms of Succulent Karoo.

The Namaqualand mountain ridges are covered by a scrubby succulent vegetation classified by Acocks (1953) as Namaqualand Broken Veld and, on the highest parts, Western Mountain Karoo. The mountains are rugged and consist of colourful granites, gneiss, schists and quartzites, with a shallow or deeper soil of gravelly sandy loam on the dry, level to slightly sloping plains between the ridges.

Fig. 26. Early morning mist covers the mountains of the Richtersveld, while the highest peaks in the background stick out above the mist layer.

On the higher slopes there are clear differences in the vegetation of different aspects, the vegetation of southern slopes being less succulent and more like the Western Mountain Karoo which covers the highest parts. It is likely that this decrease in succulence in the highest parts is due to a decrease of the effect of mist at those altitudes, as these higher peaks often stick out of the mist belt as 'nunataks' (Fig. 26) (Werger unpubl.). It is however also possible that the increased hazard of frost at such higher altitudes is important in this respect (cf. Adamson 1938a, Levyns 1962). Small trees and shrubs of the Namaqualand Broken Veld include *Aloe dichotoma*, *Ceraria namaquensis*, *Euclea tomentosa*,

Rhus undulata var. *undulata, Ficus ingens, Putterlickia pyracantha, Ozoroa dispar, Boscia albitrunca* and *Pappea capensis*, while among smaller shrubs and dwarf shrubs *Dodonaea viscosa* var. *angustifolia, Othonna arbuscula, Galenia africana, Euphorbia burmannii, E. dregeana, E. mauritanica*, with its underground parasite *Hydnora africana, Pteronia incana, Ruschia* spp., *Cotyledon wallichii, C. paniculata* (Fig. 27), *Didelta spinosa, Hermbstaedtia glauca, Montinia caryophyllacea, Crassula* spp., *Pelargonium* spp. and many others are common. Grasses are sparse. On southerly exposed slopes *Pteronia leptospermoides, P. undulata, P. divaricata, Rhus horrida* and *Indigofera pungens* are sometimes dominants. In the Richtersveld the flora is extraordinarily rich. Several endemics or near-endemics can be added to the lists above, such as the tree *Aloe pillansii*, the shrub *Aloe ramosissima, Pachypodium namaquanum* (Figs. 28 and 29) and the

Fig. 27. Stand of *Cotyledon paniculata* and other succulents, mainly Aizoaceae, in the Richtersveld.

dwarf shrub *Acacia redacta*. Also common in this area, where the succulents frequently have grey-blue colours similar to the quartzitic rock on which they grow, are *Psilocaulon arenosum, Commiphora namaensis, Euphorbia virosa, E. gummifera, E. mixta, E. cherisina, E. gregaria, E. hamata, E. gariepina, Ceraria fruticulosa, Sarcocaulon herrei, Cheiridopsis candidissima, Lithops marmorata, Drosanthemum* spp., *Stoeberia beetzii, Aspazoma amplectens, Nycteranthus noctiflorus, Carissa haematocarpa, Crassula brevifolia, C. namaquensis, C. columella, Othonna* spp., *Rhus populifolia, Ehrharta delicatula, Sutera fruticosa* and very many more (Acocks 1953, Herre 1965, Werger unpubl.). An impression of the structure of one of these succulent communities of the Richtersveld can be gained from the following sample (20 × 20 m), surveyed on a gently sloping

Fig. 28. *Aloe pillansii* as sole emergent in the succulent dwarf shrub vegetation of mainly Aizoaceae and Euphorbiaceae north of Stinkfontein (Eksteenfontein) in the Richtersveld (photo K. E. Werger-Klein).

Fig. 29. *Pachypodium namaquanum* with *Euphorbia gummifera*, *Zygophyllum suffruticosum* (right), *Putterlickia* sp. (shrub left) and several succulent Aizoaceae, Geraniaceae, and others, in the Richtersveld.

terrain with boulders, and gravel of quartzite and schist less than 30 cm in diameter and covering 85 per cent of the surface area; the total canopy cover of the vegetation amounts to about 15 per cent (Werger unpubl.):

	height	canopy diameter (m)	type	individuals mature	individuals seedling	cover %	growth form	taxa
shrubs	1.0	0.7–1	dense	5	5	2	stem succ.	Euphorbia dregeana
	1.0	0.7–1	dense	5	5	2	stem succ.	Euphorbia gregaria
	1.2	0.7–1	dense	1	5	<1	stem-leaf succ.	Cotyledon paniculata (stem diam. 25 cm)
dwarf shrubs	0.1–0.6	0.2–1	dense	70		3	stem succ.	Euphorbia 3 spp.
	0.1–0.6	0.2–0.5	dense	140		6	leaf succ., woody stem	Aizoaceae c. 20 spp.
	0.1–0.6	0.1–0.3	dense	many		<1	stem-leaf succ.	Crassulaceae 3 spp.
	0.1–0.6	0.05–0.3	dense	many		1	stem succ.	Geraniaceae 2 spp.
	0.1–0.6	0.2–0.3	dense	many		<1	succ. (mixed)	Asteraceae 3 spp.
	0.1–0.6	0.2–0.3	dense	many		<1	leaf succ.	Zygophyllaceae 2 spp.
field layer	0–0.1	—	open	many		2	herbs / herbs / ann. + perenn. (leaf succ.)	Asteraceae 3 spp. / Brassicaceae 1 sp. / Aizoaceae 4 spp.

This type of vegetation extends along the escarpment north of the Orange River, but the species diversity decreases gradually towards the northern boundary of the domain at about the latitude of Aus.

South of Springbok at 30° 30′ SL and 18° EL lies the Kamiesberg massif on the top of which arid fynbos occurs (see Chapter 8). In a wide zone at medium altitudes, the vegetation is less succulent than that of the surrounding areas but it contains more narrow-leaved dwarf shrubs and is classified as Mountain Renosterbosveld by Acocks (1953). This vegetation is particularly well-developed on deeper soils on which *Elytropappus rhinocerotis* is dominant. There is still a strong floristic affinity with fynbos (Adamson 1938b). An example of such a type of community is presented in Table 5, community A.

Further inland on the plateau east of the Namaqualand escarpment, the climate gets somewhat drier again and the vegetation shows a transitional character from the succulent dwarf shrub and shrub vegetation to the narrow-leaved dwarf shrub vegetation of the Karoo Domain. Acocks (1953) classifies this transitional type as False Succulent Karoo. Many species of *Ruschia* are still important, particularly *Ruschia robusta*, and so are other Mesembryanthemaceae, but eastwards species of Asteraceae gradually gain in importance.

South of Namaqualand the vegetation of the plateau belonging to the Western Cape Domain and fringing the Karoo Domain is classified as Western Mountain Karoo (Acocks 1953). This vegetation covers the undulating to rolling country, usually with a shallow, rocky soil on granite or sandstone. North and west of Calvinia succulent dwarf shrubs are still fairly common in this vegetation, but further south and east where rainfall is slightly higher, succulents are less common, and the dwarf shrubs are tall (up to 1 m high). Frequently encountered species include *Pentzia* sp., *Eriocephalus ericoides*, *Galenia africana*, *Pteronia glauca, P. glomerata, Zygophyllum gilfillani, Salsola zeyheri, Euphorbia mauritanica, Ruschia ferox, Drosanthemum lique, Asparagus capensis, Cotyledon wallichii, Pterothrix spinescens,* and others. Particularly in the northern parts *Salsola zeyheri* becomes abundant. Grasses are few, apart from *Ehrharta calycina* and *Merxmuellera stricta.* Locally, particularly in the Tanqua Karoo, this type of vegetation has been overgrazed severely and has become virtually a desert, while the dwarf shrubs have taken on a dense, thorny, tangled 'cushion' form (Acocks 1953).

Higher up the mountain ridges between Calvinia and the southern Cape mountains which are covered with fynbos, this karroid vegetation merges into Mountain Renosterbosveld already mentioned above. It is a karroid dwarf shrub type of vegetation but was much more grassy in former days, according to Acocks (1953). One of the more important grasses is and has been *Merxmuellera stricta.* Typical species include *Elytropappus rhinocerotis, Relhania squarrosa, R. genistaefolia, Eriocephalus africanus, Euryops lateriflorus, Pentzia incana, Chrysocoma tenuifolia, Walafrida saxatilis, Pteronia incana* and *Ruschia multiflora.* Thus, the Karroo element is of considerable importance in this vegetation type. There are also transitions from this Veld Type to the adjacent ones, Western Mountain Karoo and Succulent Karoo, as well as to fynbos, particularly in the southwestern part of its distribution area.

Enclosed between the high mountain ridges of the southwestern Cape lie the dry karroid areas of the Little Karoo and Robertson Karoo. Acocks (1953) classifies the vegetation of these areas mainly as Karroid Broken Veld and Spekboomveld. The Karroid Broken Veld of this area comprises a vegetation of shrubs and dwarf shrubs with a dominance of succulents and few grasses. The area covered by this vegetation generally lies between 300 and 600 m, is hilly and rocky, and has predominantly shallow, stony soils. Just as in Namaqualand many succulents are similar in colour to the substrate they grow on. Locally the surface is uniformly covered with white quartzite fragments on a clayey subsoil. Plants on such sites are pubescent and glaucous to almost white. The most important shrub or small tree of the Little Karoo is *Euclea undulata,* which is frequently accompanied by *Cotyledon paniculata* (Fig. 27), *Carissa haematocarpa, Lycium austrinum, L. arenicolum, Rhigozum obovatum, Rhus undulata* var. *undulata, Cadaba aphylla, Schotia afra* var. *afra, Euphorbia mauritanica, Nymania capensis,* and others. Common karoo dwarf shrubs include *Eriocephalus ericoides, Pentzia incana, Nestlera humilis, Blepharis capensis, Pachypodium succulentum,* and many others, while a few examples of the very many common Mesembryanthemaceae are *Sphalmanthus blandus, Aridaria noctiflora, Conophytum petraeum, Psilocaulon utile, Ruschia ferox, R. multiflora, R. stellata, Gibbaeum perviride, G. pubescens, Hereroa stanleyi, Lampranthus haworthii, Drosanthemum hispidum,* and *D. lique.* Some locally occurring grasses include *Stipagrostis ciliata, S. obtusa, Enneapogon desvauxii* and *Ehrharta calycina.* Higher on the hills and ridges this vegetation merges into Mountain Renosterbosveld and the Cape element gradually increases in importance. At some places, particularly at some geomorphologic boundaries, the karroid vegetation merges directly into fynbos. In such transitional habitats the shrubs and dwarf shrubs are generally higher, and the vegetation becomes denser. *Dodonaea viscosa* var. *angustifolia, Acacia karroo* and *Rhus lucida* are often abundant at such sites. *Acacia karroo, Rhus viminalis* and *R. lancea* are common riverine trees in this part of the Western Cape Domain (Marloth 1908, Engler 1910, Compton 1929a, Acocks 1953, Levyns 1954, cf. Adamson 1938a).

Joubert (1968) studied the vegetation of the Robertson Karoo in somewhat more detail. He distinguished eight dominance communities as follows:
(1) the *Euphorbia mauritanica* community occurs in the plains and on the lower gentle slopes. It is up to 1 m high and has a canopy cover of about 65 per cent; (2) the *Pteronia paniculata* community occurs on the same sites, is up to 0.35 m high and scores a canopy cover of about 40 per cent; (3) the *Pteronia incana* community is restricted to the lower, but steep, southfacing slopes. The community is up to 0.7 m high, dense and covers nearly 75 per cent; (4) the *Salsola glabrescens–Acacia karroo* community is riverine and best developed on the brackish silt along the minor rivers. It contains trees up to 4 m in height and reaches canopy cover values of 65 per cent; (5) the *Elytropappus rhinocerotis* community is only found on the higher parts of the hills, at first only on south-facing slopes, but with sufficient altitude on slopes of any aspect. The community is up to 0.7 m high and covers about 55 per cent; (6) the *Euclea undulata* community is a shrub community of low altitudes in valleys between the hills. It becomes up to 3 m tall and covers less than 50 per cent; (7) the *Crassula rupestris* community is restricted to the steep, north-facing slopes. The community is very open (canopy cover about 35 per cent) and up to about 0.5 m high; (8) the *Elytropappus–Willdenowia* community covers a very limited area in the low lying plains. It is poor in species, up to 0.7 m tall, and the canopy covers about 55 per cent.

In the eastern half of the Little Karoo, where rainfall has increased to about 250 mm, the steep sandstone, quartzite and shale slopes support a dense succulent scrub dominated by *Portulacaria afra* (Spekboomveld). In its western parts this Veld Type consists of nearly pure stands of this several metres high shrub, but further towards the Karoo Domain other shrub species become admixed. On steep southern slopes *Portulacaria afra* is often rare or absent and the vegetation is a more or less non-succulent scrub of Mountain Renosterbosveld or fynbos affinity. The admixture of other shrubs and dwarfshrubs in Spekboomveld consists partly of species also occurring in the Karroid Broken Veld and listed above. Along the southern boundary it is rather sharply separated from fynbos ('False Macchia'), while in the eastern parts it merges into Valley Bushveld and other communities transitional between the Indian Ocean Coastal Belt and the Karoo Domain. The *Portulacaria afra* community provides good grazing and has been destroyed over large areas due to overgrazing. Locally *Opuntia* spp. have strongly invaded the community (Acocks 1953).

3.7 *Vegetation of the Karoo Domain*

The Karoo Domain comprises the vegetation of the summer rainfall area of the dry South African plateau. Only its western and southwestern margins fall in the uniform rainfall area. The vegetation is typically an open dwarf shrub formation on wide, rolling plains and pediments. The dwarf shrubs are xeromorphic, often possess resinous glands, and have mostly narrow, ericoid or pubescent leaves. Succulents are not so common, except in the areas transitional to the Western Cape Domain, but grasses are conspicuous, particularly towards the eastern margin. On hillsides scattered larger shrubs or small trees emerge from the dwarf shrub and grass layer, and along the riverbeds a woodland or gallery forest is common. The most important Veld Types (Acocks 1953) of the Karoo Domain are Arid Karoo (VT 29), False Arid Karoo (VT 35), Central Upper Karoo (VT 27), large parts of Karroid Broken Veld (VT 26) and Central Lower Karoo (VT 30). A number of other Veld Types are transitional to this and other phytochoria.

The Arid Karoo lies in the western half of the South African plateau adjacent to the Namaland and the Western Cape Domains, and covers an enormous, extremely flat area, mostly at about 900 m altitude. The annual rainfall ranges from 50 to 200 mm and falls mostly in late summer and autumn. Several rivers in the area have no outlets, resulting in endoreic catchment areas, while the larger rivers drain into the Orange River. Because the countryside is so flat, many rivers form large to enormous brackish, silty flats or pans which are practically bare, such as Verneukpan or support an open *Salsola aphylla* community, such as Groot Vloer. Large shrubs are rare over the entire region. Along the northern edge of the Arid Karoo, the transition area to the Namaland Domain, the vegetation consists mainly of an open to fairly dense dwarf shrub and grass formation on calcareous tufa, characterized by *Salsola tuberculata* subsp. *tuberculata*, *Stipagrostis obtusa* and *S. ciliata*. Several grasses with desiccation-tolerant leaves are also common, e.g. *Enneapogon desvauxii*, *Eragrostis nindensis*, *Sporobolus lampranthus* and *Oropetium capense*, and on sandy patches sometimes the shrubby *Stipagrostis brevifolia*. The fern *Ophioglossum polyphyllum* can be abundant at places and annuals and geophytes are numerous following rains. Where the vegetation is overgrazed *Rhigozum trichotomum* and several dwarf shrub species widely distributed over large tracts of Karoo tend to invade this *Salsola* community, the most common ones being *Pentzia spinescens* and *Eriocephalus spinescens*. Most of the Arid Karoo consists of an open and often overgrazed dwarf shrub formation on stony, but silty soils over shales. Erosion occurs on a large scale. The most frequent species is *Pentzia spinescens*, with *Eriocephalus spinescens* on the most stony parts and on sand *Rhigozum trichotomum* occurring in patches owing to its usual vegetative way of reproduction. Other common species include *Salsola tuberculata*, *Galenia sarcophylla*, *Ruschia ferox*, *Zygophyllum microphyllum*, *Z. gilfillani*, *Aptosimum depressum*, *Pteronia mucronata*, *Hermannia spinosa*, *Leyssera tennella*, *Gazania lichtensteinii*, *Mesembryanthemum annuum*, *Salsola zeyheri* and the above-mentioned grasses. After rains many annuals and geophytes become abundant and the barer parts can become temporarily covered by *Tribulus* spp. Acocks (1953) regarded this open *Pentzia spinescens* veld as a degraded form of the *Salsola tuberculata* community commonly found in the north and thought to be the climax community of the whole of the Arid Karoo.

To the east and south of the Arid Karoo lies the Central Upper Karoo, at somewhat higher altitudes (over 1000 m) and receiving somewhat more precipitation (200–250 mm annually, mainly in late summer) than the Arid Karoo. It is largely a flat country of lithosolic or sandy loam soils on shale, sandstone or calcrete, interrupted by dolerite hills with very rocky soils of sandy loam. Wide silt flats occur along some of the rivers. Sheat erosion is common. The vegetation of the plains consists of dwarf shrubs and a fair amount of grasses, while the hillsides are far more grassy and also support shrubs, of which *Rhus undulata* var. *tricrenata* is the most common. Shrubby vegetation, with *Lycium* spp. and

Rhigozum trichotomum, also occurs on floodplains. Common dwarf shrubs include *Eriocephalus ericoides, E. spinescens, E. pubescens, Pentzia globosa, P. incana* and several other spp., *Plinthus karooicus, Nenax microphylla, Pteronia* spp., *Nestlera* spp., *Felicia muricata, Chrysocoma tenuifolia, Pegolettia retrofracta, Salsola* spp., *Gnidia polycephala, Moraea polystachya, Osteospermum leptolobum, Sutera* spp., and many others. Common grasses are e.g. *Eragrostis lehmanniana, E. obtusa, Aristida diffusa* var. *burkei, A. congesta, Stipagrostis obtusa, S. ciliata, Enneapogon desvauxii, Fingerhuthia africana* and *Digitaria eriantha.* On different substrates different species become dominant. Near Victoria West, for example, *Euphorbia mauritanica, Limeum aethiopicum* and *Felicia muricata* are the dominant species on dolerite outcrops, whereas on level sandstone these are *Ruschia ferox* and *Eriocephalus spinescens,* and on level calcrete *Osteospermum spinescens, Nestlera conferta, Helichrysum obtusum, Plinthus karooicus, Gnidia polycephala* and *Salsola tuberculata* (Toss 1974). On some floodplains and along some drainage lines the vegetation is very dense and grassy with a few species of *Eragrostis* and *Sporobolus* being most important. The higher mountains in this area carry on their highest reaches karroid grassland with *Merxmuellera disticha.* Sometimes, however, following continuous overgrazing, only relics of this type of vegetation remain.

According to Acocks (1953) large tracts of the Central Upper Karoo have become invaded by species of the Arid Karoo to such an extent that they are distinguished as a separate Veld Type, the False Arid Karoo. The vegetation here contains a mixture of species of both Veld Types. Acocks (1953) pointed out that the 'invasion' of Arid Karoo species in this area to a large extent just means an increase in their abundance rather than a real 'invasion'. South of the Arid Karoo in the area adjacent to the Western Mountain Karoo, a variation of False Arid Karoo occurs in which *Salsola tuberculata* is dominant and *Rhigozum trichotomum* is absent. Apart from the above-mentioned dwarf shrubs several succulents are common in this vegetation (1975 edition of Acocks 1953).

In the middle Orange River valley upstream of Upington there is some influence of the Namaland element in the flora. The vegetation of the mountain slopes in this area is more shrubby than in the flat Karoo further south. North of the Orange River lies the transition to the Namaland Domain. Here and there are sandy patches with a vegetation similar to that of the southern Kalahari, but the vegetation is mainly shrubby with a mixture of the floras of the two domains concerned and, to a lesser extent, of Sudano–Zambezian species. On quartzite or ironstone slopes, e.g. of the Langeberg and the Asbestos Mts., or on calcrete important shrubs or small trees include *Tarchonanthus* sp., *Acacia mellifera* subsp. *detinens, Rhus ciliata, R. undulata* var. *tricrenata, R. dregeana, Euclea crispa* var. *ovata, E. undulata, Rhigozum obovatum, R. trichotomum, Olea africana, Maytenus heterophylla, Putterlickia pyracantha, Grewia flava* and, particularly near the Orange River valley *Cadaba aphylla, Boscia albitrunca* and *Phaeoptilum spinosum.* Locally even *Euphorbia avasmontana* and *Aloe dichotoma* occur in abundance (Fig. 30). The mostly leafless and extraordinary spiny shrub *Lebeckia macrantha* dominates locally. Many of the grasses mentioned for the Central Upper Karoo also occur here, together with such species as e.g. *Anthephora pubescens, Themeda triandra, Heteropogon contortus, Eustachys mutica, Eragrostis curvula, Cymbopogon plurinodis,* and *Sporobolus fimbriatus.* On some hills *Euphorbia avasmontana* and *Croton gratissimus* can be found.

Fig. 30. Near Koegas *Aloe dichotoma* and *Euphorbia avasmontana* are abundant on some rocky slopes. Companion species visible in the photograph are *Zygophyllum suffruticosum*, *Acacia mellifera* subsp. *detinens* and *Rhigozum trichotomum*.

Common dwarf shrubs of the Karoo occur in this vegetation with varying abundance. Usually their importance tends to increase with overgrazing, while locally particularly *Acacia mellifera* subsp. *detinens* tends to form thickets under such circumstances. Often *Acacia tortilis* subsp. *heteracantha* and *Rhigozum trichotomum* also occur in such thickets. On sandy soils *Eriocephalus ericoides* tends to increase following overgrazing, while on sandy loam *Chrysocoma tenuifolia*, and on sand-covered calcrete *Othonna pallens*, *Gnidia polycephala* and several other dwarf shrubs play that role. On rocky hill sides and calcrete the encroaching dwarf shrubs comprise a wide range of common Karoo species.

The vegetation of these northeastern parts of the Central Upper and False Arid Karoos near the Orange River valley was studied in detail by Werger (1973b), who indicated that his findings could be extrapolated over a wide adjacent area. The karroid communities, which usually have a canopy cover of about 30 to 40 per cent of the surface area, though sometimes as low as 10 per cent, were classified into the class Pentzietea incanae (Table 6) comprising five associations: (1) The Zizipho–Rhigozetum obovati covers dolerite slopes of the easternmost parts; this association includes the subassociation with *Cheilanthes eckloniana* on the steeper slopes, and the subassociation 'inops' on the more gentle slopes. (2)

Fig. 31. Calcrete plateau with Nestlero-Pteronietum in fore- and background. The main species are the composites *Pentzia calcarea*, *Nestlera humilis*, *Pteronia sordida* and *Chrysocoma tenuifolia*; *Zygophyllum gilfillani* is also common. The top of the very gradually sloping terrain (centre) is somewhat different in geological structure and is characterized by taller shrubs of *Rhigozum trichotomum* and the occasional specimens of *Boscia albitrunca* and *Acacia tortilis* (background). Near Douglas, but south of the Orange River.

The Melhanio rehmannii–Hermannietum spinosae is found on andesitic lava slopes. (3) The Monechmatetum incani occurs along drainage lines and washes, where the soil is alkaline. Where such sites are covered with Kalahari sand a subassociation with *Pentzia calcarea* is found. (4) The Nestlero humilis–Pteronietum sordidae is encountered on calcrete (Fig. 31) or on relatively shallow, calcrete-rich, sandy soils. These latter sites are typified by the subassociation with *Stipagrostis ciliata* and are usually found just below the minor scarp in the plateau, where the incision of the Orange River starts. (5) The Eriocephalo–Eberlanzietum is restricted to nearly level sites with slightly acid, loamy sands in the easternmost parts. The associations (1) and (2) have many species in common and are combined into one alliance (Enneapogono scabri–Rhigozion obovati); the same applies for associations (3) and (4) (Zygophyllion gilfillani). Rather different is the community of calcrete pans in this area: it is dominated by the low round cushions and mats of *Eragrostis truncata*, accompanied by several species of which the rosette succulent *Titanopsis*

Table 6. Pentzietea incanae communities of the Karoo Domain (after Werger 1973b).

Community Number of relevés	A_1 12	A_2 15	B 23	C_1 13	C_2 9	D_1 16	D_2 10	E 8
Zizipho-Rhigozetum obovati species								
Ziziphus mucronata	5	4	1	.	.	.	.	.
Rhus undulata	5	4	1	1	.	.	.	.
Hibiscus pusillus	3	3	1	.	.	.	.	.
Digitaria eriantha	3	2	1	.	.	.	.	.
Pollichia campestris	3	1	1	.	.	.	.	.
Argyrolobium lanceolatum	2	2	.	.	.	.	.	.
Rhus ciliata	2	2	.	.	.	.	.	.
Acacia karroo	2	2	.	1	.	.	.	.
Themeda triandra	3	1	.	.	.	.	.	.
Polygala asbestina	1	2	1	1	.	2	.	.
Viscum rotundifolium	2	1	.	.	.	.	.	.
Asclepias fruticosa	2	1	.	.	.	.	.	.
Sutera albiflora	4	1	.	.	.	.	.	1
Cheilanthes eckloniana	5	1	.	.	.	.	.	.
Cymbopogon plurinodis	3	1	1	.	.	.	.	.
Eustachys mutica	3	1	.	.	.	.	.	.
Hibiscus marlothianus	3	1	1	1	.	.	.	.
Chascanum pinnatifidum	3	1	1	.	.	.	.	.
Melolobium microphyllum	3	1	1	.	.	1	1	.
Sutera halimifolia	3	1	.	.	.	.	.	.
Solanum coccineum	3	1	1	.	.	.	.	.
Hermannia pulchra	3	1	1	.	.	.	.	.
Aloe broomii	3	.	.	.	.	.	.	.
Anthospermum rigidum	2	.	.	.	.	.	.	.
Rhynchelytrum repens	2	1	1	.	.	.	.	.
Mariscus capensis	3	.	.	.	.	.	.	.
Pellaea calomelanos	3	.	.	.	.	.	.	.
Tarchonanthus camphoratus	.	2	1	.	.	.	.	.
Selago albida	1	2	.	.	.	.	1	.
Melhanio rehmannii-Hermannietum spinosae species								
Hermannia spinosa	1	1	5	2	.	1	1	2
Lasiocorys capensis	.	.	4	1	.	3	3	.
Melhania rehmannii	1	1	3	.	.	.	.	.
Acacia tortilis	.	1	2	.	2	2	.	.
Grewia flava	.	1	2	.	.	.	.	.
Blepharis mitrata	.	.	2	1	.	.	.	.
Barleria lichtensteiniana	.	.	2	.	.	.	.	.
Oropetium capense	.	.	2	.	.	1	.	1
Eragrostis nindensis	.	1	2	1	.	.	.	.
Aloe claviflora	.	1	2	1	.	1	.	1
Enneapogono scabri-Rhigozion obovati species								
Phyllanthus maderaspatensis	4	3	5	1	.	.	1	.
Aristida diffusa	4	3	3	1	.	1	.	.
Sporobolus fimbriatus	5	3	3	2	.	.	.	.
Pegolettia retrofracta	3	4	2	2	.	.	1	.
Enneapogon scaber	3	3	4	.	.	.	1	1
Corbichonia decumbens	2	2	4	1	.	.	.	.
Lantana rugosa	3	3	3	.	.	.	.	.
Indigofera sessilifolia	2	3	3	.	.	1	.	.
Rhigozum obovatum	4	2	2	.	.	.	.	.

Community Number of relevés	A_1 12	A_2 15	B 23	C_1 13	C_2 9	D_1 16	D_2 10	E 8
Enneapogon scoparius	5	2	2	.	.	.	.	.
Heteropogon contortus	4	2	2	1	.	.	.	.
Hermannia candidissima	3	2	2	.	.	.	.	.
Solanum supinum	3	2	2	1	.	1	.	.
Senecio longiflorus	2	1	2	1	.	1	.	1
Helichrysum lucilioides	3	2	1	.	.	1	.	.
Abutilon austro-africana	1	1	2	1	.	.	.	.
Enneapogon cenchroides	1	2	2	.	.	.	.	.
Aptosimum depressum	1	1	1	.	.	2	.	.
Monechmatetum incani species								
Monechma incanum	.	1	2	5	5	1	1	.
Pentzia lanata	1	1	1	2	2	.	.	1
Nestlero humilis-Pteronietum sordidae species								
Nestlera humilis forma	.	.	.	1	2	5	5	.
Pteronia sordida	1	.	1	.	.	4	5	.
Eragrostis truncata	.	.	.	1	.	4	3	1
Hermannia pulverata	.	.	1	1	1	4	4	.
Aptosimum albomarginatum	.	.	.	1	.	3	2	.
Lycium pilifolium	.	.	.	1	.	1	2	1
Microloma massonii	1	.	1	.	.	2	1	1
Monechma desertorum	.	.	.	1	.	2	1	.
Lessertia pauciflora	.	1	.	1	2	3	1	.
Stipagrostis ciliata	1	1	.	1	2	3	1	2
Zygophyllion gilfillani species								
Zygophyllum gilfillani	.	.	1	2	1	5	5	1
Pentzia calcarea	.	.	.	2	4	2	4	.
Stipagrostis obtusa	.	1	1	4	4	5	3	2
Pentzio incanae-Rhigozetalia trichotomi species								
Cyphocarpha angustifolia	4	3	5	3	.	1	1	1
Fingerhuthia africana	3	3	4	4	.	3	.	1
Rhigozum trichotomum	2	2	4	3	2	4	.	.
Acacia mellifera	1	1	5	3	2	4	.	.
Boscia albitrunca	3	2	4	2	.	3	.	.
Ehretia rigida	3	3	3	3	.	1	.	.
Cenchrus ciliaris	3	1	2	2	2	1	.	.
Pteronia glauca	1	1	1	1	.	2	.	.
Nestlera humilis	1	1	1	2	.	.	.	.
Eriocephalo-Eberlanzietum species								
Eberlanzia spinosa	.	1	1	2	.	1	1	3
Aptosimum spinescens	1	.	1	2	.	3	3	4
Eriocephalus spinescens	1	1	.	1	.	.	.	5
Salsola glabrescens	.	1	1	2	1	1	1	3
Hermannia comosa	.	1	.	.	2	.	.	2
Pentzietea incanae species								
Enneapogon desvauxii	3	2	3	4	4	5	5	5
Barleria rigida	3	4	5	5	2	5	4	4
Pentzia incana	1	3	4	5	4	5	5	4
Limeum aethiopicum	4	4	4	3	1	4	3	3
Aptosimum marlothii	3	3	4	3	3	4	3	2
Phaeoptilum spinosum	2	2	4	4	4	2	1	5

Community Number of relevés	A_1 12	A_2 15	B 23	C_1 13	C_2 9	D_1 16	D_2 10	E 8
Plinthus karooicus	.	2	2	4	2	3	5	2
Peliostomum leucorrhizum	2	1	3	3	2	2	2	2
Thesium hystrix	3	3	2	2	.	3	4	3
Eriocephalus pubescens	1	3	3	2	2	1	1	.
Polygala hottentotta	1	1	1	2	.	1	.	1
Intruding species of Pentzio-Chrysocomion								
Chrysocoma tenuifolia	5	4	5	4	4	3	4	4
Aristida congesta	5	4	5	4	3	2	2	5
Eragrostis lehmanniana	4	2	4	3	4	2	3	2
Gnidia polycephala	3	1	1	1	3	1	4	3
Tragus koelerioides	4	4	3	3	1	1	1	2
Eragrostis obtusa	4	1	3	2	.	1	.	4
Lycium salinicolum	2	4	4	4	3	4	4	5
Indigofera alternans	1	2	2	2	.	1	.	4
Felicia muricata	2	1	2	1	.	2	1	.
Gazania krebsiana	1	1	2	2	2	1	.	3
Companion species								
Aristida curvata	3	3	3	3	.	2	.	4
Dicoma macrocephala	2	2	3	3	3	1	.	4
Asparagus suaveolens	4	4	3	4	1	2	1	1
Tragus berteronianus	1	1	2	1	3	2	4	3
Eragrostis porosa	3	2	2	1	2	2	.	.
Geigeria filifolia	2	1	2	2	.	3	3	2
Polygala leptophylla	.	2	1	1	3	2	1	2
Chenopodium album	1	2	1	2	.	1	.	1
Asparagus laricinus	1	2	1	2	.	1	1	.
Kohautia amatymbica	1	1	1	.	.	.	.	2
Commelina africana	.	1	2	.	1	.	.	.
Cadaba aphylla	.	1	1	2	.	.	.	.
Berkheya pinnatifida	.	.	1	1	.	.	.	2
Plinthus cryptocarpus	1	1	.	.	2	1	.	.
etc.								

A: Zizipho-Rhigozetum obovati with subassociations of Cheilantes eckloniana (1) and 'inops' (2).
B: Melhanio rehmannii–Hermannietum spinosae.
C: Monechmatetum incani with subassociations 'typicum' (1) and of Pentzia calcarea (2).
D: Nestlero humilis–Pteronietum sordidae with subassociations of Stipagrostis ciliata (1) and 'typicum' (2).
E: Eriocephalo–Eberlanzietum.

schwantesii and the tiny dwarf shrub *Polygala pungens* are most prominent. Wide brackish washes support an open shrub community dominated by *Lycium prunus-spinosa* and *Salsola glabrescens* accompanied by several common Karoo species. The gallery woodlands and forests in this area belong all to one association, the Zizipho–Acacietum karroo, in which, apart from *Ziziphus mucronata* and *Acacia karroo*, *Rhus viminalis* and *Diospyros lycioides* subsp. *lycioides* are important trees. Other constant species in this association are *Setaria verticillata*, *Lycium arenicolum*, *L. hirsutum*, *Asparagus setaceus*, *Clematis brachiata* and, along the inner edges, *Nicotiana glauca*. On the outer edges of the levees and on the floodplains of the larger rivers *Salsola glabrescens* and *Lycium arenicolum* form a shrub community, while the shrubby grass *Stipagrostis namaquensis* together with *Eragrostis lehmanniana* dominates on wind-blown alluvial sand.

In the east these vegetation types border on the vegetation of the Sudano–Zambezian Region. This boundary is well defined and follows clear geomorphological boundaries in the Upper Orange River area (Werger 1973b). The vegetation east of this boundary remains karroid in appearance, though it is much grassier. Acocks (1953) and other authors impute this karroid character of the vegetation to a massive invasion of Karoo species into these originally grassy areas following severe overgrazing and mismanagement for more than a century (see Sections 2.1 and 3.2). Acocks (1953) called this vegetation False Upper Karoo (VT 36) and Werger (1973b) classified the various communities in the class Rhoetea erosae and the alliance Pentzio–Chrysocomion (see Chapter 10).

The southern stretches of the Karoo Domain are largely made up of, what Acocks called, Karroid Broken Veld in the southwest and Central Lower Karoo further east. The Karroid Broken Veld of the Great Karoo is very different from that of the Little Karoo, particularly because of the insignificance of succulents in the Great Karoo. The area consists mainly of undulating, eroded stony plains with little soil. Larger shrubs are scarce in this area and restricted to rocky slopes, while the plains carry a very open dwarf shrub vegetation of partly the same species as listed for the Arid Karoo and Central Upper Karoo, but also including some succulents (Fig. 32). This dwarf shrub flora is very rich in species. The most important ones are again *Pentzia spinescens* and *Eriocephalus spinescens* (Acocks 1953).

The Central Lower Karoo is also very similar to the Arid Karoo but is less arid. It also comprises flat, stony country, but the substrate changes from calcrete to sandstone, shale and silt. The vegetation consists of an open to dense low dwarf shrub formation with a fair amount of succulents. Typical species include *Salsola tuberculata, Pentzia incana, Eriocephalus spinescens, Eberlanzia vulnerans, Ruschia ferox, Zygophyllum microphyllum, Stipagrostis obtusa, S. ciliata*, and many more (Acocks 1953). Towards the mountain ranges of the southern Cape the Karroid Broken Veld and the Central Lower Karoo merge into Cape fynbos (False Macchia), sometimes via a rather narrow belt of clearly distinct Spekboomveld, dominated by *Portulacaria afra* (Marloth 1908, Acocks 1953).

In the eastern Cape, south of the area of the False Upper Karoo, the landscape is dissected and the vegetation changes rapidly, resulting in a complicated pattern (see Section 2.1). Here the boundary of the Karoo Domain is vague and all kinds of floristic mixtures occur. Acocks (1953) has mapped a number of Veld Types for this area. The most important of these are the Noorsveld (VT 24), the False Karroid Broken Veld (VT 37), the Karroid Merxmuellera Mountain Veld (VT 60), the False Thornveld of the Eastern Cape (VT 21), and Valley Bushveld (VT 23). The Noorsveld and part of the Valley Bushveld are raher similar. The Noorsveld occupies the undulating middle part of the Sundays River valley around Jansenville and consists of a fairly dense scrub up to 2 m high. The thorny succulent *Euphorbia coerulescens* is strongly dominant and accompanied by several other shrubs, like *Rhigozum obovatum, Maytenus polyacantha, Euclea undulata, Schotia afra, Nymania capensis, Portulacaria afra*, and others, but these are usually inconspicuous. Several common Karoo dwarf shrubs occur on the more open patches, probably mainly as a result of heavy grazing. The Valley Bushveld occurs in a large number of different forms of which the Fish River and Sundays

Fig. 32. Open dwarf shrub vegetation of the Karoo Domain on rocky plains near Beaufort West. Between the mostly composite dwarf shrubs some succulent Aizoaceae dwarf shrubs are seen in flower (photo W. J. J. Colaris).

River scrub and the Addo Bush still have a considerable number of Karoo species in the undergrowth. Physiognomically these vegetation types are scrub of a mixed succulent/non-succulent character. The constituent species are partly widespread, partly of Karoo, resp. Zambezian, resp. Indian Ocean Coastal Belt affinity. These communities are very rich in species. Fish River Scrub is the most succulent, at places being dominated entirely by *Euphorbia bothae*. Other important species are *Portulacaria afra*, *Grewia robusta*, *Rhoicissus digitata*, *Asparagus striatus*, *Crassula lycopodioides*, and many more. Addo Bush is less succulent and is possibly derived from Indian Ocean Coastal Belt forest (Alexandria forest, see Chapter 13). Sundays River Scrub contains many of the same species as Fish River Scrub, though in different quantities and with *Euphorbia ledienii* replacing *E. bothae* of the Fish River Scrub (Acocks 1953). False Thornveld and False Karroid Broken Veld are secondary vegetation types of Karoo species invading the eastern Cape grasslands and other vegetation types of Zambezian affinity under excessive grazing pressure. *Acacia karroo* and several of the more common Karroo species are the principal invaders and often occur in bush clumps. Karroid Merxmuellera Mountain Veld is mainly of subalpine affinity (see Chapter 12) and under normal conditions the Karoo element is not very well represented in it, but increases under heavy grazing pressure.

References

Acocks, J. P. H. 1953. Veld Types of South Africa. Mem. Bot. Surv. S. Afr. 28:1–192. (2nd ed., 1975. Mem. Bot. Surv. S. Afr. 40:1–128).

Acocks, J. P. H. 1964. Karoo vegetation in relation to the development of deserts. In: D. H. S. Davis (ed.), Ecological studies in southern Africa. Junk, The Hague. pp. 100–112.

Acocks, J. P. H. 1971. The distribution of certain ecologically important grasses in South Africa. Mitt. Bot. Staatssamml. München 10:149–160.

Adamson, R. S. 1938a. The vegetation of South Africa. Brit. Emp. Veg. Comm., London.

Adamson, R. S. 1938b. Notes on the vegetation of the Kamiesberg. Mem. Bot. Surv. S. Afr. 18:1–25.

Adamson, R. S. 1960. The phytogeography of Molluginaceae with reference to southern Africa. J. S. Afr. Bot. 26:17–35.

Airy Shaw, K. 1947. The vegetation of Angola. J. Ecol. 35:23–48.

Aubréville, A. 1975. Essais sur l'origine et l'histoire des flores tropicales africaines. Application de la théorie des origines polytopiques des Angiospermes tropicales. Adansonia, Sér. 2, 15:31–56.

Balinsky, B. I. 1962. Patterns of animal distribution of the African continent. Ann. Cape Prov. Mus. 2:299–310.

Barbosa, L. A. Grandvaux. 1970. Carta fitogeográfica de Angola. Inst. Inv. Cien. Angola, Luanda.

Bauman, A. J. 1976. Desert varnish and marine ferromanganese oxide nodules: congeneric phenomena. Nature 259:387–388.

Bews, J. W. 1925. Plant forms and their evolution in South Africa. Longmans, Green & Co., London.

Blair Rains, A. & Yalala, A. 1972. The Central and Southern State Lands, Botswana. Land Res. Stud. No. 11. D.O.S., Tolworth.

Bolus, H. 1875. Letter from Mr. Bolus to Dr. J. B. Hooker. J. Linn. Soc. 14:482–484.

Burtt, B. L. 1971. From the South: an African view of the flora of western Asia. In: P. H. Davis, P. C. Harper & I. C. Hedge (eds.), Plant life of South-West Asia. Bot. Soc., Edinburgh. pp. 135–154.

Cannon, W. A. 1924. General and physiological features of the vegetation of the more arid portions of southern Africa, with notes on the climatic environment. Carn. Inst. Publ. 354, Washington.

Chapman, J. D. & White, F. 1970. The evergreen forests of Malawi. Commw. For. Inst., Oxford.

Coetzee, B. J. & Werger, M. J. A. 1975. A west–east vegetation transect through Africa south of the Tropic of Capricorn. Bothalia 11:539–560.

Coetzee, C. G. 1969. The distribution of mammals in the Namib desert and adjoining inland escarpment. Scient. Pap. Namib Desert Res. Stn. 40:23–36.

Compton, R. H. 1929a. The vegetation of the Karoo. J. Bot. soc. S. Afr. 15:13–21.

Compton, R. H. 1929b. The flora of the Karoo. S. Afr. J. Sc. 26:160–165.

Cooke, H. B. S. 1962. The pleistocene environment in Southern Africa – Hypothetical vegetation in Southern Africa during the pleistocene. Ann. Cape Prov. Mus. 2:11–15.

Cooke, H. J. 1975. The palaeoclimatic significance of caves and adjacent landforms in western Ngamiland, Botswana. Geogr. J. 141:430–444.

Croizat, L. 1952. Manual of phytogeography. Junk, The Hague.

Croizat, L. 1968. Introduction raisonée à la biogéographie de l'Afrique. Mem. Soc. Brot. 20:7–451.

De Winter, B. 1965. The South African Stipeae and Aristideae (Gramineae). Bothalia 8:199–404.

De Winter, B. 1966. Remarks on the distribution of some desert plants in Africa. Palaeoecology of Africa 1:188–189.

De Winter, B. 1971. Floristic relationships between the northern and southern arid areas in Africa. Mitt. Bot. Staatssamml. München 10:424–437.

Diniz, A. Castanheira. 1973. Características mesológicas de Angola. Missao Inq. Agric. Angola, Nova Lisboa.

Edwards, D. & Werger, M. J. A. 1972. Threatened vegetation and its conservation in South Africa. In: R. Tüxen (ed.), Gefährdete Vegetation und ihre Erhaltung. Ber. Int. Symp. Rinteln 1972. Cramer, Lehre (in press).

Engler, A., 1910. Die Pflanzenwelt Afrikas. I. Allgemeiner Überblick über die Pflanzenwelt Afrikas und ihre Existenzbedingungen. Die Vegetation der Erde. IX. Engelmann, Leipzig.

Engler, A. 1921. Die Pflanzenwelt Afrikas, insbesondere seiner tropischen Gebiete. III (2). Engelmann, Leipzig.

Exell, A. W. & Gonçalves, M. L. 1973. A statistical analysis of a sample of the flora of Angola. Garcia de Orta, Sér. Bot. 1:105–128.

Gaff, D. F. 1971. Desiccation-tolerant flowering plants in southern Africa. Science 174:1033–1034.

Gaff, D. F. & Ellis, R. P. 1974. Southern African grasses with foliage that revives after dehydration. Bothalia 11:305–308.

Giess, W. 1962. Some notes on the vegetation of the Namib desert. Cimbebasia 2:3–35.

Giess, W. 1968. A short report on the vegetation of the Namib coastal area from Swakopmund to Cape Frio. Dinteria 1:13–29.
Giess, W. 1969. Welwitschia mirabilis Hook. f. Dinteria 3:3–55.
Giess, W. 1971. A preliminary vegetation map of South West Africa. Dinteria 4:1–114.
Giess, W. 1974. Zwei Fahrten zur Jensenobotrya lossowiana Herre. Dinteria 10:3–12.
Gossweiler, J. & Mendonça, F. A. 1939. Carta fitogeográfica de Angola. Gov. Geral de Angola, Lisboa.
Goudie, A. 1972. Climate, weathering, crust formation, dunes and fluvial features of the Central Namib Desert, near Gobabeb, South West Africa. Madoqua, Ser. 2, 1:15–31.
Grisebach, A. 1872. Die Vegetation der Erde nach ihrer klimatischen Anordnung. 2 Bde. Engelmann, Leipzig.
Herre, H. 1965. Eine Pflanzen-Sammelreise nach Namaqualand. Der Palmengarten 29 (12 papers).
Herre, H. 1971. The genera of the Mesembryanthemaceae. Tafelberg, Cape Town.
Huntley, B. J. 1973. An Eden called Iona. Afr. Wildlife 26:136–145.
Ihlenfeldt, H. D. 1971. Some aspects of the biology of dissemination of the Mesembryanthemaceae. In: H. Herre, The genera of the Mesembryanthemaceae. Tafelberg, Cape Town. pp. 28–34.
Jarman, N. G. & Bosch, O. 1973. The identification and mapping of extensive secondary invasive and degraded ecological types (test site D). In: O. G. Malan (ed.), To assess the value of satellite imagery in resource evaluation on a national scale. pp. 77–80. C.S.I.R., Pretoria.
Jessen, O. 1936. Reisen und Forschungen in Angola. Reimer, Berlin.
Joubert, J. G. V. 1968. Die ekologie van die weiveld van die Robertson Karoo. Ph.D. thesis Univ. Stellenbosch (Unpubl.).
Kers, L. E. 1967. The distribution of Welwitschia mirabilis Hook f. Svensk Bot, Tidskr. 61:97–125.
Knapp, R. 1973. Die Vegetation von Afrika. Fischer, Stuttgart.
Lebrun, J. 1947. La végétation de la plaine alluviale au sud du Lac Édouard. Explor. Parc Nat. Albert Fasc. 1:1–800. Parcs Nat. Congo belge, Bruxelles.
Lebrun, J.-P. 1971. Quelques phanérogames africaines à aire disjointe. Mitt. Bot. Staatssamml. München 10:438–448.
Lebrun, J.-P. 1976. Quelques aires remarquables de phanérogames africaines des zones sèches. Boissiera 24:91–105.
Leistner, O. A. 1967. The plant ecology of the southern Kalahari. Mem. Bot. Surv. S. Afr. 38:1–172.
Leistner, O. A. 1977. Southern Africa. In: R. A. Perry (ed.), I.B.P. Arid Lands Synthesis. Vol. 1. Univ. Press, Cambridge.
Leistner, O. A. & Werger, M. J. A. 1973. Southern Kalahari phytosociology. Vegetatio 28:353–399.
Leser, H. 1971. Landschaftsökologische Studien im Kalahari sandgebiet um Auob und Nossob. Steiner, Wiesbaden.
Levyns, M. R. 1950. The relations of the Cape and the Karroo floras near Ladismith, Cape. Trans. Roy. Soc. S. Afr. 32:235–246.
Levyns, M. R. 1962. Past plant migrations in South Africa. Ann. Cape Prov. Mus. 2:7–10.
Levyns, M. R. 1964. Migrations and origin of the Cape flora. Trans. Roy. Soc. S. Afr. 37:85–107.
Malan, J. S. & Owen-Smith, G. L. 1974. The ethnobotany of Kaokoland. Cimbebasia (B)2:131–178.
Marloth, R. 1908. Das Kapland, insonderheit das Reich der Kapflora, das Waldgebiet und die Karroo, pflanzengeografisch dargestellt. Wiss. Ergebn. Deutsch. Tiefsee-Exped. "Waldivia", 1898–1899. Bd. 2 T.3. Fischer, Jena.
Matos, G. Cardoso de & Sousa, J. N. Baptista de. 1968. Reserva parcial de Moçâmedes. Carta da vegetação e memória descritiva. Inst. Inv. Agron. Angola. Nova Lisboa.
Merxmüller, H. (ed.). 1966–1972. Prodromus einer Flora von Südwestafrika. Cramer, Lehre. 35 parts.
Moisel, A. 1975. A Braun–Blanquet survey of the vegetation of the Welwitschia Plain. Univ. Cape Town, unpubl. report.
Monod, T. 1957. Les grandes divisions chorologiques de l'Afrique. C.S.A./C.C.T.A. Publ. No. 24:1–150. (C.S.A./C.C.T.A., London.)
Monod, T. 1971. Remarques sur les symmétries floristiques des zones sèches nord et sud en Afrique. Mitt. Bot. Staatssamml. München 10:375–423.
Moreau, R. E. 1952. Africa since the Mesozoic: with particular reference to certain biological problems. Proc. Zool. Soc. London 121:869–913.

Mostert, J. W. C., Roberts, B. R., Heslinga, C. F. & Coetzee, P. G. F. 1971. Veldmanagement in the Orange Free State region. Dept. Agric. Tech. Serv. Bull. (Pretoria) 391:1–97.
Nordenstam, B. 1965. Synpunkter på karroo floran. Bot. Notis. 118:458–459.
Nordenstam, B. 1967. Phytogeography of the genus Euryops (Compositae). Opera Bot. 23:1–77.
Nordenstam, B. 1974. The flora of the Brandberg. Dinteria 11:1–67.
Nordenstam, B. 1976. Re-classification of Chrysanthemum L. in South Africa. Bot. Notis. 129:137–165.
Noy-Meir, I. 1973. Desert ecosystems: environment and producers. Ann. Rev. Ecol. Syst. 4:25–51.
Phillips, J. 1956. Aspects of the ecology and productivity of some of the more arid regions of southern and eastern Africa. Vegetatio 7:38–68.
Range, P. 1932. Die Flora des Namalandes I. Feddes Repert. 30:129–158.
Raven, P. H. & Axelrod, D. I. 1974. Angiosperm biogeography and past continental movement. Ann. Miss. Bot. Gard. 61:539–673.
Roberts, B. R. 1965. Applied plant ecology in land-use planning of catchment areas. S. Afr. J. Sc. 61:111–117.
Robinson, E. R. 1977. A plant ecological study of the Namib Desert Park. Unpubl. M.Sc. Thesis. Univ. of Natal, Pietermaritzburg.
Robinson, E. R. & Giess, W. 1974. Report on the plants noted in the course of a trip from Lüderitz Bay to Spencer Bay, January 10–21, 1974. Dinteria 10:13–17.
Roux, P. W. 1966. Die uitwerking van seisoenreënval en beweiding op gemengde Karooveld. Proc. Grassl. Soc. Sth. Afr. 1:103–110.
Rust, U. & Wieneke, F. 1973. Grundzüge der quartären Reliefentwicklung der Zentralen Namib, Südwestafrika. J. S.W.A. Wiss. Ges. 27:5–30.
Shaw, J. 1875. On the changes going on in the vegetation of South Africa through the introduction of the merino sheep. J. Linn. Soc. London 14:202–208.
Stengel, H. W. 1971. Die tyd staan nie stil in die Namibwoestyn nie. S. Afr. J. Sc. 67:103–108.
Stopp, K. 1958. Die verbreitungshemmenden Einrichtungen in der südafrikanischen Flora. Bot. Stud. 8:1–103.
Story, R. 1958. Some plants used by the Bushmen in obtaining food and water. Mem. Bot. Surv. S. Afr. 30:1–115.
Theron, G. K., Schweickerdt, H. G. & Van der Schijff, H. P. 1968. 'n Anatomiese studie van Plinthus karooicus Verdoorn. Tydskr. Natuurw. 1968:69–104.
Tinley, K. L. 1971. The case for saving Etosha. Afr. Wildl. suppl. 25:1–16.
Tolmachev, A. I. 1971. Über einige quantitative Wechselbeziehungen der Floren der Erde. Feddes Repert. 82:343–356.
Toss, G. M. 1974. A study of the vegetation of the Karoo near Victoria West. Univ. Cape Town. Unpubl. report.
Troupin, G. 1966. Étude phytocénologique du Parc National de l'Akagera et du Rwanda oriental. Publ. Inst. Nat. Rech. Sc. Butare, Rwanda, 2:1–293.
Van Zinderen Bakker, E. M. 1964. Pollen analysis and its contribution to the palaeoecology of the Pleistocene in southern Africa. In: D. H. S. Davis (ed.), Ecological studies in southern Africa. Junk, The Hague. pp. 24–34.
Van Zinderen Bakker, E. M. 1969. Quaternary pollen analytical studies in the southern hemisphere with special reference to the Sub-Antarcti Palaeoecology of Africa 5:175–212.
Van Zinderen Bakker, E. M. 1975. The orig and palaeoenvironment of the Namib Desert biome. J. Biogeogr. 2:65–73.
Van Zinderen Bakker, E. M. & Coetzee, J. A. 1972. A reappraisal of late-Quaternary climate evidence from tropical Africa. Palaeoecology of Africa 7:151–181.
Verdcourt, B. 1969. The arid corridor between the north-east and south-west areas of Africa. Palaeoecology of Africa 4:140–144.
Volk, O. H. 1964. Die afro-meridional-occidentale Floren-Region in SW-Afrika. In: K. Kreeb (ed.), Beiträge zur Phytologie. Ulmer, Stuttgart. pp. 1–16.
Volk, O. H. 1966. Die Florengebiete von Südwestafrika. J. S.W.A. Wiss. Ges. 20:25–58.
Volk, O. H. 1974. Gräser des Farmgebietes von Südwestafrika. S.W.A. Wiss. Ges., Windhoek.
Volk, O. H. & Geyger, E. 1970. 'Schaumböden' als Ursache der Vegetationslosigkeit in ariden Gebieten. Zeitschr. Geomorph. 14:79–95.
Volk, O. H. & Leippert, H. 1971. Vegetationsverhältnisse im Windhoeker Bergland, Südwestafrika. J. S.W.A. Wiss. Ges. 25:5–44.

Walter, H. 1936. Die ökologischen Verhältnisse in der Namib–Nebelwüste (Südwestafrika). Jb. Wiss. Bot. 84:58–222.
Walter, H. 1939. Grasland, Savanne und Busch der arideren Teile Afrikas in ihrer ökologischen Bedingtheit. Jb. Wiss. Bot. 87:750–860.
Walter, H. 1973. Die Vegetation der Erde. Vol. I. Fischer, Stuttgart.
Weare, P. R. & Yalala, A. 1971. Provisional vegetation map of Botswana. Botswana Notes and Records 3:131–148.
Weather Bureau. 1957. Climate of South Africa. Part 4: Rainfall maps. WB 22. Govt. Printer, Pretoria.
Weimarck, H. 1941. Phytogeographical groups, centres and intervals within the Cape flora. Lunds Univ. Årsskr. N.F. Adv. 2, Bd. 37. No. 5:1–143.
Wellington, J. H. 1955. Southern Africa, a geographical study. Vol. 1. Univ. Press, Cambridge.
Werger, M. J. A. 1973a. Notes on the phytogeographical affinities of the southern Kalahari. Bothalia 11:177–180.
Werger, M. J. A. 1973b. Phytosociology of the Upper Orange River Valley, South Africa. V & R, Pretoria. (diss. Nijmegen).
Werger, M. J. A. 1973c. American and African arid disjunctions. In: N.B.M. Brantjes & H. F. Linskens (eds.), Pollination and dispersal. Fac. Science, Nijmegen. pp. 117–124.
Werger, M. J. A. 1977a. Environmental destruction in southern Africa: the role of overgrazing and trampling. In: A. Miyawaki & R. Tüxen (eds.), Environmental Conservation. Int. Symp. Tokyo 1974. Maruzen & Co., Tokyo. pp. 301–305.
Werger, M. J. A. 1977b. Zoogene Anderungen in der Vegetation der südafrikanischen Trockengebiete. Int. Symp. Rinteln 1976. Cramer, Lehre (in press).
Werger, M. J. A. & Coetzee, B. J. 1977. A phytosociological and phytogeographical study of Augrabies Falls National Park, South Africa. Koedoe 20: in press.
Werger, M. J. A. & Leistner, O. A. 1975. Vegetationsdynamik in der südlichen Kalahari. In: W. Schmidt (ed.), Sukzessionsforschung. Ber. Int. Symp. Rinteln 1973. pp. 135–158. Cramer, Lehre.
White, F. 1965. The savanna woodlands of the Zambezian and Sudanian Domains. Webbia 19:651–681.
White, F. 1971. The taxonomic and ecological basis of chorology. Mitt. Bot. Staatssamml. München 10:91–112.
White, F. 1976. The vegetation map of Africa – The history of a completed project. Boissiera 24: 659–666.
Winterbottom, J. M. 1967. Climatological implications of avifaunal resemblances between south western Africa and Somaliland. Palaeoecology of Africa 2:77–79.

10 The Sudano–Zambezian Region*

M. J. A. Werger and B. J. Coetzee

Phytogeography 303
  1.   Characterization 303
  2.   Phytogeographical subdivision 303
  3.   Boundaries with other phytochoria 308
Vegetation 310
  4.   Vegetation structure and general ecology 310
  4.1   Structurally and floristically defined vegetation types 310
  4.2   The structural types 310
  4.3   Ecology of vegetation structure 312
  4.4   Biotic influences on vegetation structure 316
  4.5   Vegetation structure and fire 318
  5.   Communities 318
  5.1   Miombo and related vegetation types 318
  5.1.1 General remarks on distribution, phenology and phytosociology .. 318
  5.1.2 Berlinia–Marquesia communities (Berlinio–Marquesion) 320
  5.1.3 High rainfall miombo on mesic soils (Mesobrachystegion) and 'chipya' 324
  5.1.4 Highland rainfall miombo on drier soils (Xerobrachystegion) 326
  5.1.5 Significance of burning in Meso- and Xerobrachystegion 330
  5.1.6 High rainfall miombo on Kalahari sand and Cryptosepalum dry forest (Guibourtio–Copaiferion) 331
  5.1.7 Dwarf miombo (anharas do ongote) 339
  5.1.8 Miombo of the remaining areas 339
      a. B. boehmii–B. allenii miombo of the escarpments 340
      b. B. spiciformis–J. globiflora miombo of the plateaux 341
      c. The high rainfall areas of Moçambique and Malawi 345
      d. Miombo limits in southern Moçambique 348
  5.2   Baikiaea vegetation 349
  5.3   Mopane vegetation 352
  5.3.1 Distribution and general ecology 352
  5.3.2 The Luangwa, Shire and Zambezi valleys 355

---

* We are very much indebted to Mr Bob Drummond of the National Herbarium, Salisbury, to Mrs Estelle van Hoepen and other members of staff of the National Herbarium, Botanical Research Institute, Pretoria, and to Mrs Anja Finne of the Forest Department, Kitwe, for their generous and extensive help concerning taxonomical nomenclature. Thanks to them the species nomenclature in this chapter could be updated till December 1976. Any remaining failures are entirely our responsibility.

5.3.3	Botswana, the Rhodesian plateau, Transvaal and southern Moçambique	356
5.3.4	South West Africa and Angola	359
5.4	Other woody vegetation types and grasslands	363
5.4.1	General remarks on distribution, phytosociology, physiognomy and phenology	363
5.4.2	Shaba	365
5.4.3	The northern and central Angolan plateau	367
5.4.4	The Angolan escarpment and coastal zone	369
5.4.5	The central parts of tropical southern Africa	372
5.4.6	The lowlands of northern and central Moçambique	378
5.4.7	Upland (Temperate) Sub-humid Mountain Bushveld	381
5.4.8	Lowveld (Subtropical) Sub-humid Mountain Bushveld	389
5.4.9	Dry Mountain Bushveld	391
5.4.10	Arid Mountain Bushveld (Androstachys johnsonii)	393
5.4.11	Broad-orthophyll Plains Bushveld	396
5.4.12	Microphyllous Thorny Plains Bushveld	413
5.4.13	Spiny Arid Bushveld	421
5.4.14	Dry Cold-temperate Grassland	424
5.4.15	Moist Cold-temperate Grassland	427
5.4.16	Moist Cool-temperate Grassland	431
5.5	Azonal vegetation	437
5.5.1	Flood plain and dambo grasslands	437
5.5.2	Woody vegetation	442
5.6	The transition to the Karoo–Namib Region	446
	References	454

10 The Sudano–Zambezian Region

Phytogeography

1. Characterization

The Sudano–Zambezian Region comprises the vast stretches of woodland, savanna and grassland vegetation with occasional dry forests and thickets, and patches of edaphically controlled swampy vegetation, in a wide zone in Subsaharan Africa around the Guineo–Congolian Region. This latter region consists mainly of humid forests. About 75 per cent of southern Africa falls in the Sudano–Zambezian Region and, more precisely, into its Zambezian Domain. The remaining 25 per cent of the area are made up mainly by the dry Karoo–Namib floral Region and the more temperate floras of the Afromontane Region and of Capensis. Smaller areas are comprised by the Indian Ocean Coastal Belt, the Afro-alpine Region, and the southernmost outliers of the Guineo–Congolian Region (Chapter 7, Fig. 12).

The woodlands, savannas and grasslands of the Zambezian Domain cover virtually the entire high plateau of southern Africa, but also reach the Atlantic Ocean in central Angola and spread out over the low and wide coastal plains of Moçambique on the Indian Ocean, where they have replaced much of the forests of the Indian Ocean Coastal Belt and are presumably favoured by anthropogenic activities like burning and cutting. Comprehensive enumerations of species typical for the Zambezian Domain are given by Mullenders (1954) for the local flora of Kaniama (Shaba = Katanga), by White (1965) for the woody flora of Zambia and by Monteiro (1970) for the local flora of Bié (Angola). These works as well as Lebrun (1947), Troupin (1966), Chapman & White (1970) and Lewalle (1972) also give an enumeration of the more wide-spread Sudano–Zambezian and pluriregional species in the local floras.

2. Phytogeographical subdivision

Over large parts of the enormous area covered by the Zambezian Domain the rich flora (cf. Lebrun 1960) only gradually changes, possibly as a result of the lack of strong relief and other contrasting physiographic factors. This makes a further phytogeographical subdivision more difficult, but at the same time less necessary. Some species show to a certain extent a geographical concentration, and on this basis a subdivision of the Zambezian Domain in sectors or in centres has been attempted in some areas. In an area like the Zambezian Domain, where floristic changes are mainly gradual and a large proportion of the species have a very wide distribution area, it is probably more sensible to distinguish centres of endemism with gradual boundaries than to subdivide the area into sharply bounded sectors. However, as pointed out in Chapter 7, the centres of endemism as understood here are comparable to sectors in phytogeographical value and sectors proposed for various areas can often be transferred into geographically less sharply circumscribed centres of endemism. In this discussion both terms are used and understood as being of equal phytogeographical rank.

Monteiro (1970) proposed to divide the Angolan part of the Zambezian Domain into six sectors: 1) Along the coast from not far north of Luanda to just south of Benguela a littoral zone is recognized as the Luanda Sector. The sector is characterized by plants adapted to the local arid conditions, growing in savannas interrupted by patches of thicket. 2) Inland from this sector, interrupted by the Guineo–Congolian sector of Amboim (see Chapter 14), the Muxima Sector is distinguished. This sector is characterized by much the same woody species as the Luanda Sector and the savannas and thickets of the Muxima Sector are interrupted by gallery forests. 3) The Bié Sector consists of the higher plateau areas of central Angola from the area between Dalatando (Salazar) and Malanje in the north to Lubango (Sá da Bandeira) in the south, including a wide area around Huambo (Nova Lisboa) and Bié (Silva Porto). The vegetation consists mainly of savannas and woodlands interrupted by plateau grasslands and by gallery forests along river courses. 4) East of the Bié Sector on the south central African plateau environmental conditions remain more or less homogeneous over large distances in eastern Angola and adjacent western Zambia. This vast area of woodlands and savannas, with extensive floodplain grasslands in the Upper Zambezi area, is distinguished as the Huila–Moxico–Lunda–Malange Sector by Monteiro (1970). The sector, which continues into western Zambia has two outliers to northern Angola, according to Monteiro (1970), one around Malanje and one around Saurimo (Henrique de Carvalho). The important woody species in this sector are largely similar to those of the Bié Sector. 5) In northern Angola, around the Cuango River, Monteiro (1970) distinguished a Cuango Sector. The sector comprises a relatively low lying area with a mosaic of humid riverine and escarpment forests and dry *Marquesia* forests alternating with slightly undulating plains carrying a tall savanna vegetation. This sector is floristically clearly transitional between the Sudano–Zambezian and Guineo–Congolian Regions. 6) Along the southern border of Angola running into western Zambia and Caprivi, the woodlands and savannas are floristically sufficiently distinct to justify a classification into a separate sector, the Cubango Sector.

It is clear that the criteria used by Monteiro (1970) to delimit the sectors of the Angolan part of the Zambezian Domain include physiognomic characters of the dominant vegetation types and not only its floristics. If more weight is given to the floristic criterium, however, some of Monteiro's sectors should be united, so that the Luanda and Muxima Sectors become one, while for other sectors another delimitation seems desirable. The Bié Sector is certainly characterized by a very high degree of endemism, particularly concentrated in the high escarpment areas, and by a rich flora (Lebrun 1960, Exell & Gonçalves 1973), but as a meaningful phytogeographical unit it should be redefined so as to include the southwestern, central and northern parts of the Huila–Moxico–Lunda–Malange Sector of Monteiro. The southeastern parts of Monteiro's Huila–Moxico–Lunda–Malange Sector which carry a somewhat different flora and vegetation on the deep Kalahari sands can be separated from the western parts and form together with the major part of Monteiro's Cubango Sector and the adjacent parts of the neighbouring countries a separate chorological unit, the Barotse Sector or Centre (see p. 305).

In Shaba (Katanga), Duvigneaud (1958) distinguished three different sectors within the Zambezian Domain. In southwestern Shaba adjacent to Angola the Lunda Sector is recognized. This sector is a continuation of the Bié and parts of the Huila–Moxico–Lunda–Malange Sectors distinguished by Monteiro (1970). It is best to rename this large sector of the Zambezian Domain. Since Monteiro's combined names are too long, and Duvigneaud's name unsuitable because there is already a Lunda Sector further north in the Guineo–Congolian Region, it is proposed to call this the Central Angolan Sector (see Chapter 7, Fig. 12).

In Central Shaba around the Lualaba River downstream from Bukama, Duvigneaud distinguished a Lower Katangan Sector. This sector, in which gallery forests and 'muhulu' forests are frequent, covers an area that is strongly transitional between the Guineo–Congolian and Sudano–Zambezian Regions and has not yet been studied in detail. It is probable that this sector should be regarded as only a transitional area between two regions.

Southern and eastern Shaba fall into the Katango–Zambian (Northern Rhodesian) Sector. This sector, which includes large parts of Zambia, particularly around Lake Bangweulu, is ecologically rather varied and, consequently, includes a great variety of vegetation types. Miombo woodlands are most common, but open forests of *Marquesia* and *Cyperus papyrus* swamps also occur near the large lakes Mweru and Bangweulu. Floristically the Katango–Zambian Sector is clearly characterized and several typical woody plants are listed below. On the high plateaux in the area covered by this sector the afromontane element is relatively strongly represented (Lisowski et al. 1971) and the highlands of Marungu and Muhila in the northernmost part of this sector, are floristically so distinct, that they are included with the Afro-oriental Domain of the Sudano–Zambezian Region.

Duvigneaud (1958) has proposed a subdivision of the sectors into districts and subdistricts using ecological and physiognomic as well as floristic criteria. This subdivision will not be discussed here, because its chorological value is not readily apparent, while the ecological features of the area will be dealt with in the following pages.

Analysing the distributions of woody species occurring in Zambia, White (1965) concluded that 35 per cent of them are widely distributed in the Zambezian Domain, while an equally large percentage of the species is clearly confined to three centres of endemism, the Katangan, the Barotse and the Zambezi Centres. The Katangan Centre is the same as the Katango–Zambian Sector of Duvigneaud (1958) and comprises most of High Shaba, the Zambian Copper Belt and the Lake Bangweulu–Lake Mweru Districts of Zambia. It is characterized by *Anisophyllea pomifera, Baphia bequartii, Boscia cauliflora, Brachystegia glaberrima, B. gossweileri, B. puberula, B. taxifolia, B. wangermeeana, Bridelia duvigneaudii, Chrysophyllum bangweolense, Combretum celastroides* subsp. *laxiflorum, Meiostemon tetrandus, Craterosiphon quarrei, Cryptosepalum exfoliatum, Diospyros mweroensis, Garcinia pachyclada, Euphorbia williamsonii, E. fanshawei, Magnistipula bangweolensis, Lannea asymmetrica, Lonchocarpus nelssii* subsp. *katangensis, Marquesia macroura, Memecylon flavovirens, Monotes adenophyllus, M. angolensis, M. dasyanthus, M. discolor, M. elegans, M. magnificus, Olax obtusifolia, Oldfieldia dactylophylla, Salacia rhodesiaca, Terminalia erici-rosenii, Uapaca benguelensis, U. robynsii, U. pilosa, Uvariastrum hexaloboides* and *Viridivia suberosa*.

The Barotse Centre is defined by species confined to the Kalahari sand area of Barotseland and adjoining areas of the Upper Zambezi, the Cuando–Cubango area of southeastern Angola and adjacent Caprivi, Botswana and western

Rhodesia. It is identical with Monteiro's (1970) enlarged Cubango Sector and includes the northern Kalahari as defined by Giess (1971). The typical species of this centre is *Baikiaea plurijuga* but also restricted to it are, according to White (1965), *Acacia fleckii, Alchornea occidentalis, Baphia massaiensis* subsp. *obovata, Brachystegia bakerana, Dialium engleranum, Erythroxylum zambesiacum, Grewia schinzii, Hannoa chlorantha,* and *Lonchocarpus nelsii.*

The Zambezi Centre, and most clearly its Kariba Subcentre, is characterized by a number of species confined to the hot and dry low-lying valley of the central Zambezi. The distribution areas of some of these species also extend further in other hot valleys like those of the Luangwa and Kafue Rivers, while others occur outside the Zambezi Centre on termite mounds which provide very special conditions (see Chapter 39). The Zambezi Centre is characterized by *Acacia eriocarpa, Commiphora karibensis, Drypetes mossambicensis, Euphorbia cooperi* var. *calidicola, Ficus zambesiaca, Guibourtia conjugata, Pterocarpus brenanii, Triplochiton zambesiacus,* and many other species which optimally occur here, but are often found in several other dry and hot environments in southern Africa (cf. White 1965). It is likely that the hot valley of the middle Limpopo has to be included in this Zambezi Centre s.l.

The 35 per cent of widespread woody species of the Zambezian Domain occurring in Zambia, although not showing geographical patterns, are clearly segregated ecologically, according to White (1965). About two thirds of these species are typical of the miombo woodlands dominated by species of *Brachystegia, Julbernardia* and *Isoberlinia,* covering most of the poor soils of the plateau and the sides of the larger valleys. One third of them is typical of the thorn savannas and woodlands (munga) of the richer and heavier soil types. These two types of vegetation form complicated mosaics over large regions of southern Africa but they each have a typical floristic composition and the species of one type are never important in the other type.

Croizat (1952, 1967) sketched a Kalaharian Centre in the general area of central southern Africa. Croizat kept its description and peculiarities rather vague, however, but it seems (cf. Croizat 1967) that White's Katangan, Barotse and Zambezi Centres all fall under Croizat's Kalaharian Centre. Similarly, Croizat (1968) points out other centres in southern Africa, e.g. in Angola, explaining the patterns on the basis of vicariant speciation from postulated very ancient (Jurassic) ancestor populations.

The main floristic elements in the Rhodesian flora are discussed by Wild (1968c). Three of these elements are typically Zambezian and identify various centres. The Kalahari sand element discussed by Wild equates to the Barotse Centre. Wild mentions much the same species typical for this element as White (1965) and lists a few other species which are important in the Barotse Centre but have a more widespread distribution. Wild's (1968c) savanna element of lower altitudes on more basic soils is identical with the Zambezi Centre s.l. Wild explicitly mentions that this element is important in the hot, deep and wide river valleys of the Zambezi, Sabi, Limpopo and Shire Rivers in Zambia, Rhodesia, Moçambique and Malawi. The woodland element on acid soils at medium altitudes is the same as the common *Brachystegia spiciformis–Julbernardia globiflora* miombo vegetation of Rhodesia and wide surrounding areas in Moçambique, Malawi and eastern Zambia. This woodland area is not yet distinguished as a specific centre or sector. The area has many species in common with the Central Angolan Sector but also differs from it in several typical species. It is vicariant to the Central Angolan Sector and it is proposed to name it the Rhodesian Sector or Centre. As its typical and common species Wild (1968c) lists *Brachystegia allenii, B. glaucescens, B. utilis, Diospyros kirkii, Flacourtia indica, Monotes glaber, Strychnos spinosa, Uapaca kirkiana*, and several other widespread species. Within the area of this Rhodesian Sector the rivers are often fringed by a gallery forest, its flora being described by Wild as an element of its own and characterized by *Albizia glaberrima, Antidesma venosum, Cleisthanthus schlechteri, Cordyla africana, Diospyros mespiliformis, Kigelia africana, Lecaniodiscus fraxinifolius* and *Trichilia emetica*. These species do not indicate a separate sector or centre, but occur azonally out of their main distribution range.

The flora of northern Moçambique is still insufficiently collected. It contains a number of species typical of the Oriental Domain of the Sudano–Zambezian Region, and also possesses a number of endemics, for example *Euphorbia graniticola, E. ramulosa, E. decliviticola, Rhodognaphalon mossambicense* and *Faroa involucrata*. Although there is some evidence that the high plateaux of northern Malawi and Lake Malawi together seem to form a phytogeographical barrier between the Zambian and northern Moçambique flora (Exell & Wild 1961), our present floristical knowledge of northern Moçambique does not justify the recognition of that area as a separate chorological sector.

Attempts to subdivide the area south of the Limpopo have usually been based on physiographic and physiognomic criteria. Chorological subdivisions have been few and most of them had the Cape flora as a starting point. The Transvaal plateau represents a distinct centre characterized by several indigenous species, both woody and non-woody. Examples are *Acacia permixta, A. rehmanniana, Aloe castanea, Canthium suberosum, Cassine burkeana, Catha transvaalensis, Combretum moggii, Elephantorrhiza burkei, Encephalartos eugene-maraisii, Erythrophysa transvaalensis, Kirkia wilmsii, Protea caffra, Rhus engleri, R. zeyheri, Vitex mombassae, V. zeyheri*, and perhaps also the following species which all have a slightly wider distribution: *Acacia davyi, Aloe marlothii, Brachylaena rotundata, Canthium gilfillanii, Clerodendrum myricoides, Olea capensis* var. *enervis, Pavetta eylesii*, and *Vitex rehmannii*.

Brachystegia does not cross the Limpopo southwards and there has been some speculation about the reasons for this. Also *Julbernardia globiflora* and *Uapaca kirkiana* do not occur south of this river, while otherwise it does not seem to form a major phytogeographical barrier. Specific climatological factors, such as severity of frost south of the Limpopo, have been suggested as an explanation. On the Transvaal plateau the influence of frosts is far more severe than north of the Limpopo and certainly this forms a salient factor in the characterization of the area. Exell & Wild (1961) and Wild (1968c) also suggested, however, that in the case of *Brachystegia* the Limpopo is possibly only a temporary boundary because several *Brachystegia* species at present seem to be in the process of evolving new species at an unusually high rate. These presumably are filling new niches and are migrating southwards. The Limpopo valley would thus merely be a transitional limit reached presently and crossing it would just be a matter of time.

It is not clear to what extent the Barberton–Lydenburg Centre distinguished by Croizat (1952, 1965, 1967, 1968, 1972) and by Nordenstam (1969) as important centres of endemism in *Euphorbia, Aloe, Stapelia* and *Euryops*, is identical with the Transvaal plateau Centre or only represents a part of it. Croizat (1967) includes a number of species of the afromontane, afro-alpine and the Cape element in his characterization of the Barberton–Lydenburg Centre. These species occur on the high Transvaal Drakensberg and on the Lebombos. There are, however, also a number of endems of lower altitudes and more tropical affinity in the mountainous country around Barberton and a short distance further to the north (cf. Palmer & Pitman 1972). Nordenstam lists as typical species *Euryops transvaalensis*, *E. laxus*, *E. pedunculatus*, *E. gilfillanii* and *E. discoideus*.

The eastern Transvaal lowveld is floristically similar to the Limpopo valley.

The northwestern Transvaal and central Botswana together with the eastern parts of central South West Africa have a flora that is somewhat different from that of the Transvaal plateau. Bremekamp (1935) mentions that over 20 per cent of the flora (known at that time) is endemic to the area. Intensive collecting since 1935 undoubtedly will have lowered this percentage, but it is not known what the percentage of endemics of the central Kalahari is at present. It is likely, however, that the central Kalahari of which Bremekamp said that it forms 'a separate subprovince of the great East and South African savanna', should indeed be recognized as a distinct centre, different from the Barotse Centre and the Transvaal plateau Centre.

3. Boundaries with other phytochoria

There are a number of species with a distribution area including the highlands of Natal, Swaziland, the Transvaal plateau, the Central Kalahari and northern South West Africa. They are characteristic of the Zambezian Domain in its transitional area from typically tropical to a more temperate area with occasional frosts in winter.

Further to the south, on the South African Highveld, the influence of frosts is even more severe. Many Transvaal species do not occur here and the woody species are mainly restricted to the broken country. While the Sudano–Zambezian species remain important in the lower strata, the Afromontane element is relatively strong in the shrub and tree layers. In the southern part of the Highveld, Afromontane species dominate on the southfacing slopes resulting in an archipelago of Afromontane enclaves penetrating from the Drakensberg highlands deep into the plateau (Werger 1973a).

In the high country of Lesotho and the eastern Orange Free State the grassland and scrub vegetation of the slopes and the ravine forests cannot be assigned to the Zambezian Domain any longer but are Afromontane and Afro-alpine (see Chapters 11 and 12).

In the Eastern Cape the floristic composition changes again. Here the flora constitutes a complicated mixture of endemic, Sudano–Zambezian, Karoo–Namib, Afromontane and Cape species (cf. Dyer 1937, Story 1952, Acocks 1953). The endemic group is strong enough to justify the recognition of the Albany Centre by Croizat (1965, 1968, 1972) based on endems in *Euphorbia* and by Nordenstam (1969) based on *Euryops*. Weimarck (1941) had already distinguished this centre as the Southeastern Centre, or a part thereof, based on an analysis of the Cape flora. The Eastern Cape represents a marginal area and its inclusion within the Sudano–Zambezian Region is disputable.

On the coastal plains along the Indian Ocean in Natal and Moçambique there is no clear boundary between the Zambezian flora and the flora of the Indian Ocean Coastal Belt. Anthropogenic activities have sharply reduced the Coastal

Belt forests and thickets and these are often replaced by secondary vegetation types with a Zambezian composition. Cutting and burning destroys the litter covering the soil and with it the micro-organisms responsible for nitrogen production in the soil. Furthermore, the andropogonid grasses which are common in savannas and woodlands, seem to secrete antibiotic–antibacterial substances thus adding to the effect of fire. It seems to be this lack of mineral nitrogen produced in the soil that prevents the quick regeneration of forest vegetation in climatologically favourable areas like the Moçambiquan coastal plains (cf. De Rham 1970). However, different findings are reported by Trapnell et al. (1976) from woodland areas.

In the very north of Moçambique near the Rovuma River, the flora is also mixed with species which mainly occur in the Oriental Domain of the Sudano–Zambezian Region.

Westwards the Zambezian Domain borders on the Karoo–Namib Region. Generally this border is floristically quite clearly defined and follows major physiographic lines. In southwestern Angola, about halfway between Moçâmedes and Benguela, and in northern South West Africa, the border runs gradually further inland along the eastern edge of the escarpment. There occurs a narrow transitional belt of semi-desert, and very open facies of mopane scrub (Mendonça 1961, Volk 1964, 1966a, Barbosa 1970, Monteiro 1970, Giess 1971). In South West Africa the border generally follows the escarpment until the vicinity of Windhoek (Volk 1966a, Werger & Coetzee 1977). South of Windhoek the border between the Karoo–Namib and Sudano–Zambezian Regions is formed by the eastern frontier of the dune area of the southern Kalahari as defined by Leistner (1967) and runs further south along the minor escarpment between the Cape Middleveld and the Highveld, crossing the Orange River just upstream of the Orange–Vaal confluence near Luckhoff (Werger 1973a, b).

As early as 1935 Bremekamp noted that there is a distinct floristic boundary between the central and the southern Kalahari. In the southern Kalahari the flora is to some extent transitional (White 1976). The tree and shrub flora is still mainly Sudano–Zambezian (cf. White 1965) whereas the flora of the lower vegetation layers which are by far the most important, has stronger affinities with the Karoo–Namib Region (Werger 1973b). Thus, the situation here differs from the southern part of the Highveld where the lower strata are floristically mainly Sudano–Zambezian and the tree and shrub flora contains a strong Afromontane element. Such mixed floras are most probably a result of past changes in climate. During former wetter and slightly cooler epochs the Sudano–Zambezian flora will have extended its range in a southwestern direction over part of the Karoo–Namib Region. Similarly the Afromontane Region will also have increased its range down the Lesotho highlands during cooler periods. Subsequent warmer periods will have had the opposite effect, but fragments of the original flora have still been able to hold out, resulting in local floras of mixed affinity (cf. Werger & Leistner 1975, and Chapter 6).

The border area between the Karoo and the Sudano–Zambezian grasslands of the Highveld is floristically also transitional. Karoo–Namib species have widely intruded into the grasslands, enhanced by overgrazing by domestic livestock (Acocks 1953, Werger 1973, 1976, Jarman & Bosch 1973).

The border with the Guineo—Congolian Region in the north is rather diffuse, particularly in areas where anthropogenic influences on the vegetation are strong. Details are included in Chapter 14.

Vegetation

4. Vegetation structure and general ecology

4.1 *Structurally and floristically defined vegetation types*

Several systems for the structural–physiognomical classification of vegetation types have been proposed and literature on the various aspects involved is voluminous. A widely used system in Africa is the one presented at the Yangambi conference (C.S.A. 1956), which has been commented upon by Boughey (1957), Monod (1963), Guillaumet & Koechlin (1971), Descoings (1973), and some other authors, both on terminology and on definitions of the types. Many accounts on the vegetation of southern Africa have used as their basis the Yangambi system, or a slightly modified form of it. In the present account the vegetation types of the Zambezian Domain are not discussed according to their structure only but wherever sufficient knowledge of the types is available their floristic affinities are emphasized which is more useful (cf. Cole 1963b, Werger 1977b). Indication of the structure of the various communities is given using a widely acceptable terminology. The structural principle is not used here as the only basis for discussing the vegetation because the structure of floristically very similar and very closely related communities can vary considerably, as a result of edaphic factors, and even more so, as a result of fire and biotic and anthropogenic influences (cf. Shantz & Turner 1958). Clear examples of such changes in structure of floristically very similar or very closely related communities are provided by *Colophospermum mopane* communities, which occur as woodlands and as savannas, and both with trees and with shrubs as the dominant growth form; communities characterized by an abundance of *Pteleopsis myrtifolia* and several companion species can have the structure of thickets, woodlands or savannas, while floristically closely related dry forests and woodlands characterized by *Adansonia digitata* and several companion species occur, and various miombo and related fire-degraded stages show remarkable structural differences, although they are floristically very similar. Furthermore, structurally different types like *Cryptosepalum* forest and *Brachystegia* woodland are floristically not directly identical, but they are rather closely related, and they should certainly be discussed with reference to one another. Apart from their floristic similarities, structurally transitional stands of these vegetation types are not rare. Another major advantage of not using a structural–physiognomical classification system strictly is that it is not necessary to rigidly define the used terminology to indicate the structure of a floristically defined community. It is this rigidity that has so often made a failure of a structural–physiognomical classification system because of the unbridgeable difficulties arising when it is applied to the gradually changing features in nature.

4.2 *The structural types*

The structural–physiognomical terms used in this chapter can be described as follows (cf. C.S.A. 1956, Tinley 1966, 1975, Greenway 1973):
a. A forest is a plant community with a closed tree canopy layer of touching or interlocking crowns normally in more than one layer, and usually also with a

shrub layer and saplings, and with a discontinuous herb layer or a moss layer. Climbers and epiphytes may be present. A forest can be evergreen or deciduous or contain both evergreen and deciduous woody species.

b. A thicket is a very dense, often almost impenetrable plant community of large shrubs, sometimes including trees, and often somewhat stratified. Scattered, large, emergent trees can occur. Climbers can be abundant, while a sparse grass layer is discontinuous or even absent. Epiphytes can be present, but are usually rare. A thicket can be evergreen, but is often deciduous or of a mixed character.

c. A woodland is a stratified plant community with an open tree layer with crowns less than one crown diameter apart or touching but predominantly not overlapping. Scattered shrubs may be present and the undergrowth of grasses and forbs is important although usually not very dense. Grasses are frequently of a tufted growth form. Epiphytes and climbers can be present but are generally rare. African woodlands are mostly deciduous, but evergreen elements can occur, and then usually among the scattered shrubs.

d. A savanna is a plant community with a discontinuous layer of woody species whose individuals are spaced more than one crown diameter apart, and with a, usually dense, ground layer of grasses and forbs. Epiphytes and climbers, when present, are generally rare. Grasses can be tall and tufted (tall grass savanna) or short (short grass savanna). Woody species can be trees and/or shrubs. If only trees or only shrubs are present the plant community is called tree savanna or shrub savanna, respectively. If the woody species occur in discontinuous, island-like thickets as, for example, on termitaria separated by grassland, the formation is called bushclump savanna. If the woody species are mainly thorny the savanna is called (microphyllous) thorn savanna; if they are non-thorny and largely possess non-dissected leaves the savanna is called broad-leaved or broad-orthophyll savanna. Mixed types of savanna consisting of both thorny species and non-thorny species with non-dissected leaves are also common. If the woody elements are spaced far apart the community may be called open savanna. A savanna is usually deciduous, but savannas with evergreen trees occur also, for example, palm savannas. Degraded savannas where, owing to destructive biotic or anthropogenic agents, the normal undergrowth of perennials and annuals is totally replaced by annuals can be encountered although not commonly. Some authors prefer the term wooded grassland instead of savanna. If the woody species are low trees with short boles and shrubs the savanna is sometimes called bushland. A type of bushland is the bushveld, a term that has long been widely used in southern Africa by laymen and scientists for vegetation consisting of a mixture of small to medium sized trees, shrubs of varying height, grasses and forbs, with these elements occurring in widely varying proportions, commonly over small distances (Acocks 1953, Van Wyk 1971). Four main physiognomic types of bushveld have been distinguished:

(i) Broad-orthophyll bushveld, with broad-leaved mesophytic trees and shrubs;

(ii) Broad-sclerophyll arid bushveld, with species with smallish to decidedly broad, coriaceous leaves and other xerophytic characters such as volatile aromatic oils;

(iii) Spiny arid bushveld with woody plants being typically tangled and shrubby with small leaves and short, hard, stubby branchlets on long drooping or profusely angularly branched stems;

(iv) Microphyllous thorny bushveld with woody plants with compound pinnate leaves with rather small, somewhat sclerophyllous leaflets, and with long, straight or curved thorns, which are often modified stipules.

e. A grassland is a plant community predominantly covered by grasses. Other graminoid plants, forbs and geophytes can also be common in grasslands; dwarf shrubs may be present but are not prominent and shrubs or trees are only occasional. The grasses can be of various growth forms and can be annual or perennial. In open grasslands the individual grasses are fairly widely spaced apart, while in a closed grassland the grass individuals form a closed cover. Grasslands can also be floating, when a dense grasscover floats on the water surface.

A special term used in the following account, namely sudd, denotes dense herby, floating communities frequently entirely dominated by *Cyperus papyrus*, but sometimes consisting of other species.

The vegetation map of southern Africa (Fig. 1) could not be constructed solely on a floristic basis, mainly due to the state of our knowledge. A number of floristically quite different but structurally similar vegetation types had to be mapped in the same shading (e.g. swampy herbaceous vegetation and various grasslands; the various savanna and woodland types mapped as 'other woodland, savanna and thicket'; the various karroid vegetation types).

4.3 *Ecology of vegetation structure*

The distribution of the broad, structurally–physiognomically defined vegetation categories is mainly determined by a number of factors which often interplay or are interdependent. These factors result from the climate, soil (substrate) and landform (drainage) complex. It is by far not always one and the same factor which is of decisive significance in determining the structure of the vegetation at any particular site. Often a large number of factors are of importance, and it is impossible to say which one is of overriding significance, while elsewhere one factor clearly wields the strongest effect (Coetzee & Werger 1975). Generally, one can say that moisture as determined by climate and physiography (including rainfall, temperature, seasonality, humidity, geology, geomorphology, drainage – many of which are interdependent) plus degree of frost and perhaps soil nutrients (often correlated with several of the above-listed factors) are basically the determinants of vegetation structure. For zonal vegetation the climate is therefore predominant in determining vegetation structure, but even on a coarse scale it is impossible to explain the distribution of the various structural types of vegetation on the basis of one overriding factor complex only, as has been tried by Cole (1963a) on the basis of landform. Apart from the factors mentioned, there is another one thwarting the others and determining vegetation structure on a local scale, namely the effects of biotic and humanly induced influences, predominantly in the form of termite activity, elephant activity, overgrazing, cutting and fire (cf. Shantz & Turner 1958, De Vos 1975).

The precipitation in the entire Zambezian Domain has a clearly seasonal character (see Chapter 2) and shows roughly a gradient from north to south except for the coastal regions and the south African part of the Zambezian Domain, where isohyets tend to run more parallel to the coastline. Precipitation decreases along the north–south gradient from about 1800 mm to 250 mm annually. The seasonality of the precipitation already rules out the occurrence of tropical rain forests which cannot be sustained under a climatic regime with more

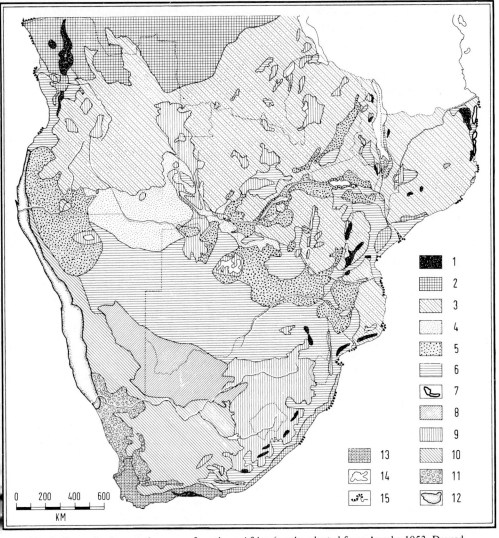

Fig. 1. Generalized vegetation map of southern Africa (partly adapted from Acocks 1953, Devred 1958, Wild & Barbosa 1967, Barbosa 1970, and Giess 1971).
1: forest (lowland and montane), 2: forest–savanna mosaic, 3: miombo and related vegetation, 4: *Baikiaea plurijuga* vegetation, 5: *Colophospermum mopane* vegetation, 6: other woodland, savanna, thicket and bushveld vegetation, 7: coastal thicket, 8: open *Acacia* savanna of the southern Kalahari, 9: grassland and (semi-aquatic) herbaceous vegetation, 10: karroid dwarf shrub and open shrub vegetation, 11: succulent dwarf shrub vegetation, 12: desert vegetation of the Namib, 13: fynbos vegetation, 14: bare pan or lake, 15: mangrove.

than three or four months with less than 100 mm of rainfall, irrespective of temperature considerations. Only locally along relatively high escarpments rainfall conditions (or frequent cloud forming) allow the occurrence of rain forests, but due to the altitude of those sites, these forests are usually Afromontane. Other forests of a much simpler structure than proper rain forests, occur patchily distributed in the high fainfall areas of Moçambique, or are edaphically controlled. The latter forests include the *Marquesia* forests on fertile, sandy soils in the Lake

Bangweulu area and the *Cryptosepalum* forests on Kalahari sands in northwestern Zambia, both under a rainfall regime of more than 1000 mm annually, as well as the *Baikiaea* forests further south on Kalahari sands receiving an annual ranfall of about 600 to 1000 mm. Gallery forests also belong to this group of edaphically controlled forests. Small patches of geomorphologically controlled forests occur here and there in ravines and below cliffs, where moisture conditions are favourable.

Major zonal structural types in the Zambezian Domain of the Sudano–Zambezian Region are moist-tropical miombo woodland, warm-temperate and dry-tropical bushveld and dry to sub-humid cool-temperate grasslands. Important also, but of lesser extent and occurring locally throughout much of the region are thickets and palm savannas.

Miombo woodlands are related to moist, frost-free to nearly frost-free tropical conditions (Fig. 2). The woodland structure changes from tall miombo in the north, where it often but by far not exclusively occurs on the flat plateaux, to a shorter miombo further south. Within the vast miombo area there are quite a number of differences in vegetation structure correlated with differences in soils and landform, as for example, the miombo on Kalahari sands in eastern Angola and western Zambia, versus the miombo on more loamy soils in much of the remainder of the area, and the miombo on rocky crests.

Moist-tropical miombo woodlands in the northern part of southern Africa are separated from warm-temperate and dry-tropical bushveld in the south by a belt of broad-sclerophyll arid *Colophospermum mopane* bushveld. The latter stretches across southern Africa in the low, hot and dry Limpopo–Shashi, Makarikari and Okavango drainage systems but occurs locally in hot and dry valleys further north too. Patches of broad-leaved, thorny, and mixed woodlands and savannas or bushveld occur locally in the miombo area. Bushveld predominates, however, on the warm-temperate low interior plateau south of the Cunene, Okavango and Limpopo Rivers where occasional light frosts occur, and on the tropical and subtropical, virtually frost-free but dry areas between the continental escarpment and the east coast. Broad-leaved woodland and savanna and broad-orthophyll mesophytic bushveld, predominates on sandy soils in moderate to high rainfall areas. Low run-off, good infiltration, good aeration and low mechanical resistance to roots growing to deep moist levels, render the moisture regime of sandy soils favourable for mesophytic woody plants (Grunow 1965).

Thorn savanna and woodland or microphyllous thorny bushveld occurs on fine-textured soils in moderately low to high rainfall areas. Such conditions prevail over extensive plains on suitable geological formations but also commonly in bottomlands and depressions elsewhere where clay and minerals accumulate. These are the fertile and physiologically drier types of soil. Arid features include bad infiltration and high run-off and although fine-textured soils hold much water, this is strongly retained against uptake by plant roots and concentrated in upper soil layers which are more vulnerable to desiccation. Root penetration to deeper levels, which are less susceptible to quick drying, is retarded mechanically and shallow-rootedness is further enforced by poor aeration with associated humic and toxic CO_2 conditions (Grunow 1965). Aridity is emphasized as a determining habitat characteristic of fine-textured soils by the fact that in arid climates, species typical of microphyllous thorny plains bushveld occur also on coarse-textured soils, as prominent constituents of arid bushveld types.

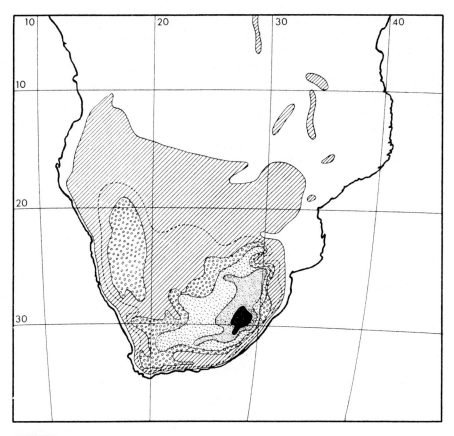

Fig. 2. Map of southern Africa showing duration of frostless period in days per year (adapted from various sources).

Spiny arid bushveld or dry woodland occurs on sandy soils in arid warm-temperate to tropical areas. Broad-sclerophyll arid bushveld or woodland, dominated by *Colophospermum mopane*, occurs optimally on particularly dry, clayey soils in arid areas such as the wide, hot and dry valleys of major rivers traversing the plateau. On sandy soils in these arid parts broad-sclerophyll arid bushveld grades into spiny arid bushveld. Low winter temperatures seem to be important in determining the southernmost boundary of the *Colophospermum mopane* vegetation. Another type of broad-sclerophyll arid bushveld, dominated by *Androstachys johnsonii*, occurs on arid mountains.

Grasslands predominate further south in the dry to sub-humid cool-temperate to cold-temperate climate with moderate to severe frost, on the high South African plateau (Fig. 2). Nowhere in the Sudano–Zambezian Region is the rainfall too low to support both a fairly dense grass cover and trees or shrubs, because woody plants occur in the lowest rainfall parts. It would, however, appear that the region lacks frost resistant woody species for zonal sandy soils. Regular fires occur throughout the region in grasslands as well as bushveld and occurred long before the advent of modern civilization. Fire damage is much more severe after frost damage so that effective frost tolerance does not necessarily mean merely surviving the frost but surviving the same in a state to still withstand field fires (Phillips 1930, Acocks 1953, 1975, Coetzee & Werger 1975).

The probably most frost-tolerant woody species of the Sudano–Zambezian Region are microphyllous thorn trees occurring locally on clayey soils of bottomlands and more extensively on upland pediplains with shallow clayey soils on the frosty high continental plateau. It seems reasonable to assume that frost tolerance evolved first in this group because:

(i) cold air in marginally frosty areas, where selection for frost tolerance would commence, accumulates in low areas where clayey soils and microphyllous plants predominate; and

(ii) similar physiological adaptations may benefit desiccation and frost tolerance, e.g. restriction of water movement through cell membranes.

Different types of grassland occur on sites with impeded drainage such as extensive floodplains, dambos and other waterlogged places (cf. Cole 1963a, Tinley 1971, Malaisse 1975). They further occur on serpentine and other toxic soils as, for example, on the Great Dyke of Rhodesia (Chapter 40) and on the gradual slopes of some watersheds (Michelmore 1939, Trapnell 1953, Vesey-Fitzgerald 1963, Cole 1963b).

The field layer in these various woodland, savanna or bushveld types consists mainly of grasses with a good admixture of forbs. Those and the grassland grasses may be conveniently categorized as either 'sour' or 'sweet'. Sour grasses typically have a high fibre content, a wiry appearance and lose their nutritive value as grazing early in the dormant season. Sweet grasses have a softer appearance owing to a lower fibre content and retain their nutritive value well into the dry, dormant season. Sour grasses are associated with high rainfall areas and poor acid, sandy soils, whereas sweet grasses are typical of dry regions and clayey, brackish soils. As the factors mentioned are continuous variables and to some extent independently variable, field layers may consist of various mixtures of sweet to sour grasses (Roberts et al. 1972).

Thickets occur widespread in patches throughout most of the warm-temperate to tropical bushveld and woodlands, often as regenerative stages after disturbance such as shifting cultivation, overgrazing or severe fire damage. They occur also as natural vegetation types, however, and are then frequently restricted to specific landforms, such as upper slopes of valley sides or to certain soils, namely loamy sands, with a clayey, impervious layer lower down (cf. Tinley 1975). Palm savannas, also occurring widespread over southern Africa, indicate impeded drainage and usually alluvial soils.

4.4 *Biotic influences on vegetation structure*

Biotic factors like termite activity, elephant activity and grazing, as well as human activities and fire can bring about great changes in vegetation structure at a specific site. The most important effect on vegetation structure produced by termites is indirect. The mounds built by them present entirely different conditions with respect to aeration, drainage and nutrient content of the soil. This obviously has its bearing on the vegetation carried by the mounds (see Chapter 39). This can result in, e.g., a mosaic pattern in woodland or bushclump savanna.

The effects of elephants on the vegetation as well as those of overgrazing, usually by domestic livestock, have been the topic of a large amount of literature which has recently been reviewed briefly by Werger (1977a). The effects of elephants on the structure of the vegetation is noticeable because of their destructive activity. Elephants frequently ringbark trees on an extensive scale, often during the dry season when they want to eat the bark. There are a few savanna and woodland species, for example *Sclerocarya caffra*, that can usually survive this treatment because even when completely ringbarked the cambium remains, but others, such as *Acacia nigrescens* with a more fibrous bark that strips off more readily, taking with it all phloem and cambium, die (Fig. 27) (Van Wyk pers. comm.). Often grass growth is enhanced when the tree cover is less dense, thus providing more fuel, and regrowth of woody species is often prevented by the fiercer fires on these sites during the dry season. In this way large patches of former woodland or savanna can be denuded of trees, particularly when over several years elephants successively increase the area in which they ringbark trees. A similar effect is produced by another elephant activity: pushing over the trees (Fig. 19). Over fairly large areas the tree layer can be destroyed in this way within a short time, thus opening up a woodland to a savanna or to a secondary grassland (cf. Laws 1970, Werger 1977a).

Overgrazing is particularly important in the savanna areas with an annual rainfall of 600 mm or less. Here there is a balance in the vegetation between the extensive and deep rooting woody plants and the intensive and shallow rooting grasses. Early in the rainy season the precipitation mainly benefits the grasses which grow quickly. In the course of the rainy season the soil moisture deepens and the water becomes more beneficial to the woody species. With overgrazing the grasses are considerably damaged and evapotranspire less water, thus leaving more water in the soil for the woody species and thus increasing their growth possibilities. Over the years this works cumulative, and the effect is a bush encroachment resulting in a densely wooded area or even a thicket. Bush encroachment also occurs sometimes in the somewhat higher rainfall areas. The most common encroachers are *Acacia mellifera* subsp. *detinens*, *A. tortilis* subsp. *heteracantha*, *A. hebeclada*, *A. karroo*, and *Dichrostachys cinerea*, but encroachment of *Terminalia sericea* or other species can also occur. At other places the effects of overgrazing are somehow different and result in denuded areas with severe erosion. Less severe overgrazing can act selectively on the species composition of a community, and in this way can change the community structure as well. Overgrazing is often associated with domestic livestock, but it can also occur when indigenous game reaches a too high population density and does not move, as for instance in some parts of game parks. The carrying capacity of the vegetation in the areas with less than 600 mm annual precipitation is rather low (Walter 1939, 1973, Walter & Volk 1954, Rattray 1960, Volk 1966b, 1974, Leser 1975, Coetzee & Werger 1975). Since domestic animals use only a limited number of species in a community it is often suggested that a mixed population of domestic animals and indigenous game with browsers complementing grazers can be larger than a population of domestic livestock only, and that it can make better use of the vegetation without leading to overgrazing effects (cf. Dasmann 1964, Werger 1977a). However, population density is not merely dependent on food supply but also on social behaviour and other ecological factors in respect of which wild and domestic animals differ, and preliminary observations in the Kruger National Park indicate that the matter is not as simple as suggested above (Joubert pers. comm.).

4.5 *Vegetation structure and fire*

Fire must be considered an important ecological factor in miombo woodlands as well as some other woodland, savanna and grassland types (Nolde 1938a, b, Aubréville 1949, Bourlière & Hadley 1960, Walter 1973). The strong seasonality in precipitation leaves the vegetation dry for several months per year, and 'dry' thunderstorms which are common at the start of the rainy season, can thus easily set the vegetation alight. In this respect fire may be regarded a natural ecological factor in the Sudano–Zambezian woodland, savanna and grassland vegetation. But man also has burnt the vegetation for centuries, and, in fact, has burnt the woodlands, savannas and grasslands during the present century often more than once a year. Therefore, most fires by far cannot be regarded as natural any longer, but fire largely has become an artificial tool in managing the vegetation cover. Burning is practised principally to remove grass litter in order to stimulate the growth of young grass tillers for cattle grazing, and also to rouse game for hunting, but increasing population pressures have led to a high frequency of burning. At the same time fire is a selecting agent in killing the most sensitive species. It also keeps the vegetation open in that it more easily kills seedlings and saplings than full grown trees and tall shrubs. Time of the year of burning is most important in this selective process because it influences the fierceness of a fire. Burning at the end of the dry season can result in fires hot enough to kill even large trees and relatively fire-resistant species with thick, corky barks (cf. Trapnell 1959). Similarly, the time of the day of burning, the period of time which has elapsed since the last burning and since the last shower, and several meteorological factors can be of importance. As the fierceness of a fire depends also directly on the amount of available fuel, grazing pressure is of importance. Not only thick barks are favourable in selection by fire; also cryptogeal germination, which is fairly wide-spread in savanna and woodland species (Jackson 1974), the shrubby hemicryptophytic growthform ('underground or geoxylic trees') developed in various families with savanna and woodland species (cf. Glover 1968, West 1971), facultative geophytism as developed in various taxonomical groups and probably originating as an adaptation to seasonal dryness (cf. Kornaś 1975), and coppicing from buds just below the soil surface, as found in many woody savanna species, are advantageous under a heavy fire regime.

5. Communities

5.1 *Miombo and related vegetation types*

5.1.1 General remarks on distribution, phenology and phytosociology

By far the most important zonal vegetation in southern Africa north of the

Cunene–Okovango and Limpopo Rivers is miombo, a woodland or sometimes a tree savanna vegetation, dominated by species of *Brachystegia, Julbernardia* and *Isoberlinia* (Fig. 1). Although some species, for example *Brachystegia spiciformis*, occur over most of this enormous area (Chapter 14, Fig. 2), the miombo vegetation is far from homogeneous. Edaphic and climatological differences over the area as well as differences in the historical–genetical development of local floras and anthropogenic influences, have resulted in a large number of different vegetation types within this broad category.

Miombo generally occurs between 800 and 1800 m on the central African plateau, though in the south of Rhodesia and in coastal Moçambique some types occur at altitudes of 675 m and less than 100 m, respectively. Below 800 m miombo is often, although not always, restricted to sandy soils. Elsewhere it occurs on a large variety of soil types, though mostly with a low humus content. Near the Chimanimani Mts. on the eastern Rhodesian escarpment miombo-like vegetation occurs as high as 2500 m. In all of the area covered by miombo rainfall is between 500 and 1800 mm per annum with about 4 to 7 months of less than 25 mm in the winter season (Fig. 1 and Chapter 2). Absolute minimum temperatures in the miombo area are generally not lower than −4°C, and occasional lower temperatures cause heavy frost damages even to *Brachystegia spiciformis* and *Julbernardia globiflora* which appear to be the most frost resistant miombo species (Aubréville 1949, 1962, Wild & Barbosa 1967, Ernst 1971, Knapp 1973, Schnell 1976–1977). Along drainage lines and low-lying, periodically flooded areas (dembos or dambos) the miombo woodland opens up into an open savanna or grassland.

Heavy metal bearing and other toxic soils occur widely distributed over the miombo area. The floristic, vegetational and ecological peculiarities of these habitats are discussed in Chapter 40.

Most of the miombo tree and shrub species shed their leaves only late in the dry season and the miombo vegetation is bare for a short period of time, usually less than three months. A few weeks to a month before the rains start the trees flush again and colour the countryside with their predominantly bright reddish new foliage. Brilliantly green, young foliage also occurs. After some days the leaves attain their normal green to greyish green colours. Most of the miombo trees and shrubs also flower in the same period before the advent of the rains. During the rainy season only few woody species flower. In most species the fruits ripen during the dry season, but species whose fruits ripen during the rainy season are not rare. In the understorey a large number of herbs also flower and set seed before the advent of the rains, and often before the appearance of leaves. These species are frequently geophytes, with large bulbs, corms or rhizomes. Grasses flower much later during the rainy season, and there is also a seasonal rhythm apparent in the appearance of the various grass species in miombo vegetation (Nolde 1938a, b, 1940, Trapnell 1959, Lawton 1963, Mildbraed & Domke 1966, Jackson 1968, 1969b, Malaisse & Malaisse-Mousset 1970, Fanshawe 1971, Knapp 1973, Ernst & Walker 1973).

General surveys of the main vegetation types in the miombo area are provided by Gossweiler & Mendonça (1939), Pedro & Barbosa (1955), Wild & Barbosa (1967), Barbosa (1970), Fanshawe (1971) and Diniz (1973). These surveys have resulted in classifications of physiognomic and dominance types of which usually only the dominant and prominent species are mentioned. The outlines presented by Knapp (1965, 1968) were intended to result in a floristic classification. This classification cannot be used here, however, since the outlines give only brief floristic descriptions of the broadly defined vegetation types without referring to specific associations and without integrating the findings of local studies. Lacking complete floristic site records, it is very difficult to integrate the results of these and physiognomic or dominance type surveys with those of detailed phytosociological studies.

The most detailed phytosociological surveys of miombo vegetation have been carried out in Shaba. These surveys resulted in a classification of the miombo communities into four alliances: the Berlinio–Marquesion, the Mesobrachystegion, the Xerobrachystegion, and the Guibourtio–Copaiferion baumianae, which together make up the order Julbernardio–Brachystegietalia spiciformis (Lebrun & Gilbert 1954, Devred 1958, Schmitz 1963, 1971). This order comprises most of the woodlands dominated by Caesalpiniaceae in Angola, Shaba, Zambia, Malawi, Moçambique and Rhodesia. Together with the Lophiretalia lanceolatae comprising similar, vicariant woodlands north of the equator in the Sudanian Domain, the Julbernardio–Brachystegietalia spiciformis are placed in the Erythrophleetea africani circumguineensia by Schmitz (1963, 1971).

5.1.2 Berlinia–Marquesia communities (Berlinio–Marquesion)

Communities of the Berlinio-Marquesion are not always typically miombo according to their physiognomy particularly when *Marquesia macroura* is dominant as in the Brachystegio–Marquesietum (Figs. 3 and 4). These communities often have the character of a semi-evergreen forest, because the tree crowns form an interlocking canopy and many species in this community only shed their leaves for a very short period while some species are evergreen. Such communities occur only over small areas, however, and most often *Marquesia macroura* is not dominant but associated with species of *Brachystegia* and other genera. The Berlinio–Marquesion communities are best developed in the periguinean girdle of woodlands and savannas in northern Angola and Shaba (see also Chapter 14), particularly in the Central Angolan Sector where annual rainfall is more than 1200 mm, but they also occur in the Katango–Zambian Sector in Shaba and the Western and Northern Provinces of Zambia. In these latter areas they are always restricted to the relatively mesic sites (Lebrun & Gilbert 1954, Duvigneaud 1958, Wild & Barbosa 1967).

Near Lubumbashi and in the area near Lake Mweru and northeast of Lake Bangweulu (Wild & Barbosa 1967, type 15) stands of the Brachystegio–Marquesietum dominated by tall trees of *Marquesia macroura* and *Brachystegia taxifolia* occur as a secondary succession on sites where the 'muhulu' or 'mateshi' vegetation (Diospyro–Entandrophragmion delevoyi; see Chapter 14) is destroyed and which are not too rocky and exposed. Under the fairly dense tree canopy there is an open layer of shrubs and small trees with *Landolphia kirkii*, *Salacia rhodesiaca*, *Alafia caudata*, *Margaritaria discoidea*, *Erythroxylum emarginatum*, *Baphia bequaertii*, *Diplorhynchus condylocarpon* and others (Trapnell 1953, Schmitz 1963, 1971, Malaisse 1973). Of the area north of Lake Bangweulu, Lawton (1963) mentions as emergent species *Entandrophragma delevoyi*, *Syzygium guineense* subsp. *afromontanum*, *Parinari curatellifolia*, *Marquesia macroura*, *Faurea saligna* and occasionally *Albizia*

Fig. 3. The evergreen forest canopy species, *Marquesia macroura*, is surrounded by a lower *Uapaca* spp. tree canopy in this advanced regeneration stage in the Lake Basin area of Mansa, Zambia (photo R. M. Lawton).

adianthifolia, while *Brachystegia taxifolia* forms a closed tree canopy. Important shrubs and climbers in this community are, in addition to those named above: *Acalypha* sp., *Bridelia duvigneaudii*, *Canthium* sp., *Clematis brachiata*, *Combretum gossweileri*, *Dichapetalum bangii*, *Leptoderris nobilis*, *Phyllanthus muelleranus*, *Monanthotaxis buchananii*, *Sapium cornutum*, *Tecomaria capensis*, *Tinnea zambesiaca* and *Uvaria angolensis*. Stands of this community are repeatedly damaged by annual or even more frequent fires and their area is

Fig. 4. A *Marquesia macroura* forest remnant with chipya vegetation in the foreground. The upper canopy is *Marquesia* and *Syzygium guineense* subsp. *afromontanum*. Luwingu, Lake Basin, Zambia (photo R. M. Lawton).

gradually reduced. Only few species in this community, such as *Parinari curatellifolia* and *Erythrophleum africanum*, are fire-resistant. Other species are all fire-sensitive in various degrees (see p. 331).

On steep slopes, particularly in Shaba, a forest community, floristically rather similar to the Brachystegio–Marquesietum but dominated by *B. microphylla*, is common.

Northeast of Lake Bangweulu a very similar community occurs which is transitional to *Brachystegia spiciformis* woodland. It is briefly described by Lawton (1963, 1972) as *Marquesia–Brachystegia* woodland. It occurs on lower slopes fringing a Mesobrachystegion woodland on gentle watersheds. Possibly it has developed after fire destruction of a 'muhulu' type of vegetation. In the *Marquesia–Brachystegia* woodland dense patches dominated by *Marquesia macroura* occur locally, but the vegetation is mostly more open and trees like *Brachystegia spiciformis, B. longifolia* and *Julbernardia globiflora* are also prominent. Other important tree species include *Faurea saligna, Parinari curatellifolia, Erythrophleum africanum, Pterocarpus angolensis, Afzelia quanzensis, Uapaca kirkiana, U. nitida*, and among the shrubs and climbers *Bridelia dubigneaudii, B. micrantha, Strophanthus welwitschii,* and species of *Ochna, Strychnos* and *Landolphia* (compare also Mansfield et al. 1975–76).

Another closely related community in this area is constituted by the *Marquesia acuminata* forests, described by Lawton (1964). These forests in which the evergreen *Marquesia acuminata* grows as trees of more than 30 m tall and makes up some 90 per cent of the upper canopy, are possibly relicts of a formerly extensive forest covering most of the area north of Lake Bangweulu. Under exploitation the forest can easily degenerate to a 'chipya' type of vegetation (see below). The community which occurs on fairly moist, and fertile, deep sandy soils, contains a few species of Guineo–Congolian and Afromontane affinities. Apart from *Marquesia acuminata, Podocarpus latifolius* is a locally abundant canopy species. Other important tall tree species include *Brachystegia spiciformis, Marquesia macroura, Daniellia alsteeniana, Faurea saligna, Syzygium guineense, Olea capensis* and *Combretum mechowianum*, and the discontinuous layer of understorey trees contains *Combretum celastroides* subsp. *laxiflorum, Brachystegia taxifolia, Aidia micrantha, Anthocleista schweinfurthii, Margaritaria discoidea, Rothmannia fischeri* and *Vitex fischeri*. The shrub layer is well-developed, consists of predominantly evergreen shrubs, and includes scandent shrubs, climbers and saplings of the tree species. *Sorindeia katangensis* is abundant, and other common species are *Ouratea welwitschii, Sapium schmitzii, Leptaulis zenkeri, Craterispermum schweinfurthii, Grewia barombiensis, Tricalysia myrtifolia, Memecylon buchananii, Enneastemon schweinfurthii, Bequaertiodendron magalismontanum, Phyllanthus muelleranus, Erythroxylum emarginatum, Vernonia bellinghamii* and *Psychotria djumaensis*, while *Combretum gossweileri, Strophanthus welwitschii, Dichapetalum bangii, Landolphia camptoloba, Uvaria angolensis, Rutidea olenotricha, Adenia* sp., *Opilia amentacea, Cissus petiolata, Hippocratea africana* and *Leptoderris nobilis* are common climbers. Epiphytic ferns, mosses and lichens are common in these forests, and mosses are also important on the forest floor. On light patches the grass *Bromuniola gossweileri* occurs, but otherwise there are very few herbs and grasses in these forests. On the forest fringes *Canthium vulgare, Brachystegia taxifolia* and other species can be abundant (Lawton 1964, Fanshawe 1971).

Similar vegetation types with *Marquesia macroura, Berlinia giorgii, Uapaca nitida, U. benguelensis, U. gossweileri, Pericopsis angolensis, Afrosersalisia cerasifera, Daniellia alsteeniana, Brachystegia spiciformis, B. longifolia, Julbernardia paniculata*, and many others, are also reported from the Cuango, Baixa de Cassange and Luanda Districts of northern Angola. Detailed phytosociological studies of this vegetation have not been made. These communities occur mainly on sandy ferralitic soils and form a mosaic with communities of the Guibourtio–Copaiferion baumianae and other miombo woodlands (Gossweiler & Mendonça 1939, Nolde 1938b, Diniz & Aguiar 1969, Barbosa 1970, Diniz 1973).

5.1.3 High rainfall miombo on mesic soils (Mesobrachystegion) and 'chipya'

The Mesobrachystegion (Table 1) comprises miombo communities on relatively mesic and fertile, loamy soils, and on large termitaria, in an area with more than 1100 mm precipitation annually (see Chapter 2). The alliance occurs over a large area mainly in Shaba and Zambia, and probably comprises numerous associations. Schmitz (1963, 1971) described two associations from the vicinity of Lubumbashi, the Combreto–Annonetum senegalensis and the Boscio–Fagaretum, each with several subassociations. Deep, mesic and rich loamy soils form the habitat of the Combreto–Annonetum senegalensis. The vegetation consists of a tree canopy of several species which can be up to 25 m tall and a well-developed layer of shrubs and low trees of 5 to 10 m in height. The herbaceous layer is dense and high, and provides fuel for fierce fires at the end of the dry season. This association also occurs in the northern part of Zambia and carries in its typical form the vernacular name 'chipya' which means 'fierce fire' (Lawton 1963, 1972, Cottrell & Loveridge 1966) but this name refers also to related degradation stages of this miombo community (see below) in which a tall and dense layer of grasses and herbs provides abundant fuel. In these degradation stages as described by Trapnell (1953) and Lawton (1963, 1972) *Brachystegia* and *Julbernardia* species are absent while the fire-tolerant miombo species are more frequent. Often 'chipya' vegetation contains fragments of Berlinio–Marquesion communities and even of 'muhulu' forest in small areas protected against fires. This indicates the successional relationships between these communities. Stands dominated by *Uapaca* spp. form an intermediate stage in this successional series from 'chipya' to Berlinio–Marquesion communities or dry everygreen 'muhulu' forest. The *Uapaca* spp. which establish themselves under the scattered specimens of *Syzygium guineense* subsp. *macrocarpum* in the 'chipya' and grow through the 'chipya' canopy, form a canopy at a height of 4–10 m. This provides a shady habitat in which the ground vegetation is suppressed. Thus, fire hazard is diminished and Berlinio–Marquesion species can become established. When the Berlinio–Marquesion community is fully regenerated, the species overtop the *Uapaca* spp. which die out (Lawton 1972) (Figs. 3 and 4). According to Lawton (1972) *Uapaca* spp. play also an important role in the regeneration of *Brachystegia–Julbernardia* woodland from severely burnt grassland (see below). Lawton (1963, 1972) lists among the species of this community *Burkea africana*, *Amblygonocarpus andongensis*, *Syzygium guineense* subsp. *afromontanum* and *Oldfieldia dactylophylla*. For the lower tree and shrub layer Lawton also names *Maprounea africana*, *Chrysophyllum bangweolense*, *Diospyros batocana*, *Friesodielsia obovata*, *Hexalobus monopetalus*, *Combretum celastroides*, *Ochna pulchra*, *Syzygium guineense* subsp. *guineense*, *Lonchocarpus capassa*, *Bridelia micrantha*, *Phyllanthus muelleranus*, *Tecomaria capensis*, *Harungana madagascariensis* and *Paropsia brazzeana*. In patches which have been exploited for wood or where the soil has been ploughed for agricultural purposes *Aframomum biauriculatum*, *Pteridium aquilinum* and *Smilax kraussiana* are particularly abundant in the herbaceous layer. Trapnell (1959), Astle (1968–69) and Mansfield et al. (1975–76) also briefly describe this vegetation from northern Zambia.

On more sandy loams a subassociation of the Combreto–Annonetum with abundant occurrence of *Brachystegia wangermeeana* is encountered, while

Table 1. Main floristic differences and similarities between Meso- (M) and Xerobrachystegion (X). Species typical for the various associations are not shown (after Schmitz 1971).

Syntaxon	M	X	Syntaxon	M	X
number of relevés	21	25	number of relevés	21	25
Mesobrachystegion species					
Setaria thermitaria	5		Anisophyllea boehmii	1	4
Costus spectabilis	5		Julbernardia paniculata	2	3
Asparagus flagellaris	4		Fadogia kassneri	1	3
Cyperus diffusus	4		Fadogia fuchsioides	2	3
Tacca leontopetaloides	4		Commelina droogmansiana	1	3
Cussonia corbisieri	4		Agathisanthemum globosum	1	3
Coccinea adoensis	4		Albizia antunesiana	1	3
Dioscorea schimperana	3		Cyphostemma obovato-oblongum	2	3
Cyphostemma hildebrandtii	3		Brachystegia spiciformis	2	2
Oxalis anthelmintica	3		Indigofera sutherlandioides	2	2
Nephrolepis undulata	3		Baphia bequaertii	2	2
Vitex andongensis	3		Pericopsis angolensis	2	2
Phyllanthus sp.	3		Pavetta schumanniana	2	2
Erythrina abyssinica	2		Urginea altissima	2	2
Combretum zeyheri	2		Pseudolachnostylis maprouneifolia	2	2
Acalypha senensis	2		Thunbergia gentianoides	2	2
Cissampelos mucronata	2		Parinari curatellifolia	1	2
			Ocimum fimbriatum	1	2
Xerobrachystegion species			Peucedanum wildemanianum	2	1
Tristachya bequaertii		5	Sterculia quinqueloba	1	1
Cryptolepis hensii		4	Vitex madiensis	1	1
Trichopteryx bequaertii		3	Vangueriopsis lanciflora	1	1
Heteropholis sulcata		3	Julbernardia globiflora		1
Thyrsia undulatifolia		3			
Mariscus pubens		3	Erythrophleetea species		
Panicum mueense		3	general		
Sacciolepis transbarbata		3	Brachiaria brizantha	3	3
Rhytachne rottboellioides		3	Cassia sangueana	1	1
Cyperus angolensis		3	Psorospermum febrifugum	1	1
Canthium crassum		3	Gardenia jovis-tonantis	1	1
Uapaca nitida		2	Swartzia madagascariensis	1	1
Melanthera albinervia		2	Afzelia quanzensis	1	1
Eragrostis racemosa		2	Faurea speciosa	1	1
Elephantopus scaber		2	Piliostigma thonningii		1
Sporobolus sanguineus		2	local mesophilous		
Erlangea trifoliata		2	Kaempferia aethiopica	3	3
Urelytrum henrardii		2	Clematopsis scabiosifolia	2	1
			Rhynchosia resinosa	1	1
Julbernardio–Brachystegietalia species			Erythrophleum africanum	2	
Ochna schweinfurthiana	2	5	local termitophilous		
Commelina africana	4	4	Ziziphus abyssinica	2	
Geophila obvallata subsp. ioides	3	4	Cayratia gracilis	2	
Asparagus abyssinicus	3	3	Cassine aethiopica	1	
Diplorhynchus condylocarpon	2	4	Haemanthus multiflorus	1	
Pterocarpus angolensis	2	4			
Chlorophytum engleri	2	4			

Brachystegia boehmii is abundant on less fertile, but heavy soils with some gravel or lateritic concretions near the surface. This habitat frequently fringes the 'dambos' and the plateaus with yellow latosols carrying a Xerobrachystegion vegetation. The same communities with *Brachystegia wangermeeana* on more sandy loams and *B. boehmii* on ironstone soils occur in adjacent northwestern Zambia in the Mwinilunga and Solwezi areas (Trapnell & Clothier 1937, Wild & Barbosa

1967, Fanshawe 1971) and in northern Zambia north of Lake Bangweulu (Astle 1968–69). Well-drained loamy, red soils in wide sink-holes on dolomite are characterized by a subassociation with *Erythrophleum africanum* strongly dominant, while over a large area in Shaba and northern Zambia on the most mesic sites, where the deep soil is often partly alluvial, a subassociation dominated by *Oxytenanthera abyssinica* is found. This bamboo is monocarpous (dies after flowering once). The vegetation then becomes more open and is invaded by *Aframomum biauriculatum* (Schmitz 1963, 1971). *Oxytenanthera abyssinica* also occurs associated with other vegetation types in northern Angola and in many other regions in Africa on mesic sites often fringing gallery forests or on old termite mounds (Pole Evans 1948, Wild & Barbosa 1967, Barbosa 1970).

The vegetation typical of the large, inhabited termitaria of northern Angola, Shaba and northern Zambia is the Boscio–Fagaretum. This association which includes three subassociations, is preliminarily placed in the Mesobrachystegion by Schmitz (1963, 1971). In Chapter 39 the termitophilous vegetation and ecology is discussed in more detail.

In northern and central Angola several of the Mesobrachystegion species are mentioned for various vegetation types (Gossweiler & Mendonça 1939, Barbosa 1970, Diniz 1973). No data exist on communities that can clearly be placed in the Mesobrachystegion, however, and it is not certain whether this alliance extends into Angola. Some authors emphasize a clear phytogeographical boundary in the vicinity of the northeastern political boundary between Angola and Shaba–gzambia, dividing the miombo area into an eastern and western part, more or less corresponding to the Katango–Zambian and the Central Angolan Sectors respectively (Exell 1957, Barbosa 1970). It is probable that the Mesobrachystegion is confined to the Katango–Zambian Sector, and that the miombo vegetation with abundant *Brachystegia wangermeeana*, *B. gossweileri*, *B. boehmii*, *Cussonia angolensis*, several species of *Combretum*, *Isoberlinia angolensis*, *Julbernardia paniculata*, and many other species, as described in types 17B and 18B from Barbosa (1970) should be classified into another related alliance. Only more detailed vegetation surveys can solve this question.

The southern edge of the central Angolan plateau, between Lubango (Sá da Bandeira) and Menongue (Serpa Pinto), is an area of mainly ferralitic sandy loams, at altitudes of 1100 to 1450 m, and carries a similar vegetation. The dominant trees include *Julbernardia paniculata*, *Brachystegia boehmii*, *B. spiciformis*, *B. gossweileri*, and locally *B. wangermeeana*. Other common species are *Swartzia madagascariensis*, several species of *Strychnos*, *Uapaca benguelensis*, *Combretum collinum*, *C. psidioides*, *Baphia bequaertii*, *Protea trichophylla*, *Clematis welwitschii*, *Pericopsis angolensis*, *Monechma scabridum*, *Triumfetta glechomoides*, and others. The herbaceous undergrowth is generally well-developed (Barbosa 1970, type 18A).

5.1.4 High rainfall miombo on drier soils (Xerobrachystegion)

Large areas in central southern Africa, mainly within the regime of the relatively moist Congo airmasses and receiving more than 1000 mm of rain per year, are covered by the communities of the Xerobrachystegion (Table 1). These communities occur on the drier and poorer, often rather shallow soils of the brown and ochre latosol type (Fig. 5). Important species in the tree layer in Xerobrachystegion communities are often the same as in the Combreto–Annonetum senegalensis, although most communities have their characteristic species of *Brachystegia*. The species composition of the shrub and herbaceous layer in the Xerobrachystegion communities is however different from that of the Mesobrachystegion communities. In particular, the herbaceous layer is less dense and shorter (Fig. 6) and as a consequence thereof, fires are less fierce in these communities, even at the end of the dry season. Schmitz (1963, 1971) described three associations of the Xerobrachystegion from Shaba. These and closely related communities also occur on a limited area in eastern Angola (Barbosa 1970, type 19) and over large stretches of northern, central, and eastern Zambia and northern and central Malawi (Wild & Barbosa 1967, types 17 and 18), interrupted by the wide valley of the Luangwa, which carries another type of

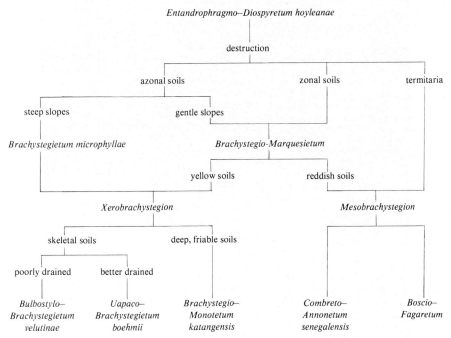

Fig. 5. Ecological and successional relationships between the various miombo communities in Shaba (after Schmitz 1971). Subassociations have not been included in this scheme.

Brachystegia vegetation on its sides and a *Colophospermum mopane* vegetation on the valley floor.

On yellow, well-drained latosols with occasionally some gravel at the surface the Brachystegio–Monotetum katangensis occurs. An important tree species in this association is *Brachystegia spiciformis* var. *latifoliata*. Trees can be up to 17 m tall. The association occurs over a very wide area in Shaba, including the edges of the high plateaux (Lisowski et al. 1971), and is most typically developed on slightly sloping terrain. On the poorest soils the herbaceous undergrowth is virtually absent and replaced by an abundant growth of *Cryptosepalum maraviense*. Locally poor or eroded soils on fairly steep slopes carry a subassociation with *Philippia pallidiflora*

Fig. 6. High rainfall miombo at Chililabombwe near the Zambia/Zaïre border, with tall trees of *Brachystegia* and *Julbernardia*, and saplings of *Uapaca kirkiana* (photo M. J. A. Werger).

subsp. *pallidiflora* with an open tree layer, in which *Uapaca pilosa* and *Monotes caloneurus* are frequent, and a well-developed shrub layer. Non-eroded, steep slopes and rocky ridges are widely covered with a subassociation in which *Brachystegia utilis* is absolutely dominant. *Brachystegia microphylla* is locally also abundant and indicates a successional stage towards the development of a *Brachystegia microphylla–Marquesia macroura* dry forest. Fringing some dambos on poorly drained, sandy soils is a subassociation with *Pericopsis angolensis* in which several species of the Mesobrachystegion communities can be abundant, while poor yellow latosols containing laterite concretions carry a *Uapaca nitida* subassociation, dominated by several species of *Uapaca* and by *Monotes katangensis*. The tree and herbaceous layers are relatively low in this subassociation although the tree layer can be dense. On sites with laterite concretions *Aloe nuttii* is sometimes abundant (Schmitz 1963, 1971).

The Uapaco–Brachystegietum boehmii is typical of the vast stretches of insufficiently drained, poor and shallow, gravelly or lateritic soils of slightly sloping terrain. During the dry season this habitat becomes parched. In Shaba the association bears a floristic resemblance to the *Uapaca nitida* subassociation of the Brachystegio–Monotetum katangensis, but in the present community *Brachystegia boehmii* is abundant, conspicuous and characteristic. There are also other floristic differences between the two communities. On the poorest and most loamy, shallow soils the association is most typically developed, and at sites where the drainage is somewhat improved owing to the sloping terrain a subassociation with *Brachystegia utilis* exists (Schmitz 1963, 1971, Malaisse & Malaisse-Mousset 1970, Malaisse 1975).

The Bulbostylo–Brachystegietum velutinae is restricted to the infertile, pale-coloured, loamy, gravelly laterites, where poor drainage causes a flooding of the surface during the wet season. This vegetation, characterized by the abundance of *Brachystegia stipulata* var. *velutina* and by *Bulbostylis mucronata* and *B. filamentosa*, is frequently interrupted at the lowest lying sites by a grassland or savanna of *Loudetia* spp., *Uapaca pilosa* and *Combretum psidioides* (Duvigneaud 1956, Schmitz 1963, 1971). On the heavily laterized soils this community is most typically developed, and on less laterized soils a subassociation with *Brachystegia utilis* is encountered. The poor drainage is partly caused by the perfect evenness of the terrain. The *Brachystegia utilis* subassociation represents a transition to the Brachystegio–Monotetum katangensis of the rocky ridges. In Shaba several of the communities of the Meso- and Xerobrachystegion mentioned above include facies dominated by *Julbernardia globiflora*. Also in Zambia, northeast and north of Lake Bangweulu this species is locally abundant in these communities (Lawton 1963, Wild & Barbosa 1967, Fanshawe 1971). This species can replace locally the *Brachystegia* species typical of the association or subassociation with the exception of *Brachystegia spiciformis* var. *latifoliata*, while the remainder of the species composition typical for the community stays unchanged. The sites invaded by *Julbernardia globiflora* are not characterized by specific soil conditions or other obvious habitat factors (Schmitz 1963, 1971). The suggested ecological and successional relationships between these various miombo communities are shown in Fig. 5.

In eastern Angola, in the vicinity of Calunda, Xerobrachystegion vegetation is found in a restricted area. No detailed information is available, but communities on well-drained as well as those on poorly drained soils seem to be represented (Barbosa 1970, Diniz 1973).

Over vast areas of central and northern Zambia Xerobrachystegion communities occur. From the available data it can be concluded that the Brachystegio–Monotetum katangensis is widespread (Fig. 6). Wild & Barbosa (1967, types 17 and 18) mention *Brachystegia floribunda*, *B. longifolia*, *B. boehmii* and *Julbernardia paniculata* as the most frequent species in this woodland which is up to 15 m tall. *Isoberlinia angolensis*, *Uapaca kirkiana* and *Erythrophleum africanum* can also be important. In northern Zambia *Monotes angolensis, Protea angolensis* and *Uapaca kirkiana* are more frequent, the latter particularly near dambos, and on lake basin soils near Lake Bangweulu *Brachystegia spiciformis* can be strongly dominant. In central Zambia scrubby woodland with *Parinari curatellifolia* and *Diplorhynchus condylocarpon* as prominent species interchange with more open stands dominated by *Julbernardia paniculata* and with *Uapaca kirkiana* in the lower storey. Locally *Brachystegia utilis* is also conspicuous. Fine textured, poorly drained soils in the Luangwa valley and in central and northern Malawi carry a low, and often heavily browsed and stunted scrub of *Brachystegia stipulata* var. *velutina*. In eastern Zambia *Brachystegia manga* and *B. spiciformis* become important; around Chipata *B. stipulata* may be added, though *Julbernardia paniculata* keeps its prominence. This type of woodland occurs also in the northern and central Malawian plateau area. On the broken areas *Julbernardia globiflora* is more conspicuous. *Brachystegia manga* is mainly found in the southern half of Malawi at altitudes between 1000 and 1300 m, while in that region *B. floribunda* reaches its optimum at higher altitudes. Particularly in the southern parts of Malawi which are more densely populated, most of these communities have been cleared for arable land, however (see Chapter 32, Fig. 4). On the higher slopes facing Lake Malawi *Brachystegia boehmii* and *B. utilis* are co-dominant while on the lower slopes *B. boehmii* is the sole dominant. In adjacent Moçambique, north of Tete, the same species are common and associated with many of the character species of the order Julbernardio–Brachystegietalia spiciformis. With decreased altitude *Julbernardia globiflora* and *Brachystegia manga* increase in number, and still lower down, below 700 m, stands dominated by *B. bussei* are frequent. At insufficiently drained sites on loamy, yellow laterites the community with *Brachystegia stipulata* var. *velutina* occurs again locally, and on more gravelly soils *Brachystegia boehmii* can dominate. In northern Moçambique around Vila Cabral, east of Lake Malawi, the *Brachystegia floribunda, B. longifolia, B. utilis, B. boehmii, Parinari curatellifolia* woodland forms a mosaic with savanna communities consisting of other species (Trapnell & Clothier 1973, Willan 1940, Brass 1953, Trapnell 1953, 1959, Fanshawe 1962a, 1971, Hursh 1962, Pike & Rimmington 1965, Wild & Barbosa 1967, Jackson 1968, Astle et al. 1969, Mansfield et al. 1975–76, Pedro & Barbosa 1955). Most if not all of these communities mentioned from Malawi, Zambia and northern Moçambique probably belong to the Xerobrachystegion.

5.1.5 Significance of burning in Meso- and Xerobrachystegion

The successional relationships of the *Brachystegia floribunda–Julbernardia paniculata* woodland communities with the 'chipya' and 'mateshi' or 'muhulu' types of vegetation (see Fig. 5) have been clearly demonstrated in the burning experiments near Ndola, Zambia (Trapnell 1959, cf. Lawton 1972, Malaisse et al. 1975). Since 1933 woodland plots have been treated consistently in one of three ways: 1) late burning, at the end of the dry season in October, involving a severe burn in hot weather; 2) early burning, at the start of the dry season, involving a very light burn; and 3) complete protection. By 1944 it was proven that late burning can destroy the canopy dominants *Brachystegia spiciformis, B. longifolia, B. floribunda, Julbernardia paniculata* and *Isoberlinia angolensis*, while early burning allows a maintained, regeneration of the woodland, and a dense dry evergreen forest of the 'muhulu' type develops under complete protection. Trapnell (1959) concluded that *Parinari curatellifolia, Erythrophleum africanum, Pterocarpus angolensis, Anisophyllea boehmii, Diplorhynchus condylocarpon, Dialiopsis africana, Uapaca nitida, Strychnos innocua, S. spinosa, Maprounea africana, Syzygium guineense, Swartzia madagascariensis, Hymenocardia acida, Vitex madiensis* and *Dombeya rotundifolia* are certainly fire-tolerant. Semi-tolerant are *Parinari polyandra, Baphia bequaertii, Pseudolachnostylis maprouneifolia, Strychnos pungens, Uapaca kirkiana, U. pilosa, Ochna schweinfurthiana, Lannea discolor, Bridelia carthartica, Hexalobus monopetalus* and *Xylopia odoratissima,* while the above-named species of *Brachystegia, Julbernardia* and *Isoberlinia* as well as *Chrysophyllum bangweolense, Garcinia huillensis, Randia* (?) *kuhniana, Bridelia duvigneaudii, Ochna afzelii, Entandrophragma delevoyi,* and some other species are fire-tender or at best hardly semi-tolerant. The tendency shown in 1944 was confirmed in 1956 (Trapnell 1959, Fanshawe 1962b) and 1976 (pers. observ. M.W.): the completely protected plots had developed to a dry evergreen 'muhulu' forest under the high rainfall regime of northern Zambia; the early burning plots had been maintained as a miombo woodland, and the late burning plots were, except for some fairly tall specimens of *Diplorhynchus condylocarpon,* virtually devoid of trees (Fig. 7). From 1973 some early and late burning plots which had shown to be in a steady state since many years, were completely protected and in 1976 these plots showed already a rapid development towards a dry evergreen forest. Lawton's (1972) presumptions on the role of *Uapaca* spp. in this succession have already been outlined in section 5.1.3.

5.1.6 High rainfall miombo on Kalahari sand and Cryptosepalum dry forest (Guibourtio–Copaiferion)

The fourth alliance of miombo woodlands, the Guibourtio–Copaiferion baumianae, is typical of the huge deposits of Kalahari sand (see Chapter 1) in central Angola, western Zambia and southwestern Shaba and the central Angolan plateau further westward with more loamy ferralitic and paraferralitic soils (roughly types 16, 17A and 24 and parts of 17B and 18, of Barbosa 1970). The annual precipitation in this area is between 700 and about 1300 mm. It comprises more or less the Central Angolan Sector. Up to the present time, the most detailed study of the communities belonging to this alliance has been carried out in central Angola in the Bié district (Monteiro 1970), although in this study only the woody species have been taken into account and the floristic analysis has been incomplete. Typical of the well-drained Pleistocene Kalahari sands under a rainfall regime of about 1000 mm per year and a dry season of about 160 days is the *Copaifera baumiana–Brachystegia spiciformis* Association (Monteiro 1970, Barbosa 1970, Diniz 1973). The community is homogeneous over large areas in central Angola and consists of a tree layer of 6 to 25 m tall. The trees have wide crowns and a light canopy foliage resulting in an aerial cover of between 30 and 90 per cent. Lower woody plants are between 0.2 and 5 m tall and usually cover between 50 and 60 per cent. Tall grasses up to 2 m are also present. Characteristic and abundant woody species include *Copaifera baumiana, Guibourtia coleosperma, Brachystegia spiciformis, B. longifolia* and *B. bakerana* (Table 2), and in the grasslayer several species of *Loudetia, Hyparrhenia, Tristachya* and *Monocym-*

Fig. 7. Miombo burning experiment at Ndola, Zambia, 43 years after initiation.
(a) Typical miombo as maintained under early burning scheme; (b) contact zone between dry evergreen forest under total protection (right) and open grassy vegetation with few trees under late burning regime (left). The contact zone has a tall herb vegetation of *Aframomum biauriculatum* not unlike chipya; (c) dry evergreen forest developed under total protection. See text for further explanation (photos M. J. A. Werger).

Table 2. Phytosociological table of Guibourtio–Copaiferion baumianae, adapted from Monteiro (1970). Only woody species are listed.

Communities number of relevés	A1 10	A2 76	B 19	C 39
Association A				
Landolphia camptoloba	4	3		
Diospyros undabunda	4	3		
Diospyros batocana	4	3		
Guibourtia coleosperma	2	2		
Dialium engleranum	3	2		
Psorospermum tenuifolium	2	1		
Monotes adenophyllus subsp. homblei	2	1		
Brachystegia bakerana	2	1		
Xylopia tomentosa	3	3	1	1
Paropsia brazzeana	4	4	1	1
Phyllocosmus lemaireanus	4	4	1	1
Subassociation of Bequaertiodendron				
Bequaertiodendron magalismontanum		2		
Association B				
Faurea speciosa		1	3	1
Protea welwitschii		1	3	1
Association C				
Brachystegia tamarindoides		1		3
Monotes loandensis		1	2	3
Syzygium guineense subsp. guineense		1	1	3
Alliance				
Copaifera baumiana	3	3	1	
Strychnos sp.	1	1	1	
Ochna pulchra	1	3	2	1
Syzygium guineense subsp. afromontanum	1	3	3	3
Cryptosepalum exfoliatum subsp. pseudotaxus	5	4	3	1
Memecylon flavovirens	2	3	2	2
Pteleopsis anisoptera	2	3	2	2
Anisophyllea gossweileri	1	2	4	3
Uapaca benguelensis		1	5	4
Order				
Uapaca kirkiana			1	2
Uapaca nitida	3	3	2	
Monotes elegans	1	1	3	1
Brachystegia longifolia	4	4	4	4
Brachystegia spiciformis	4	5	4	4
Diplorhynchus condylocarpon	2	4	2	3
Pterocarpus angolensis		1	2	3
Julbernardia paniculata	1	2	4	4
Parinari curatellifolia	2	2	2	4
Isoberlinia angolensis	1	2	1	2
Pericopsis angolensis		1	1	2
etc.				
Class				
Swartzia madagascariensis	1	2	2	2
Hymenocardia acida	2	2	2	1
Erythrophleum africanum	4	4	1	3
etc.				
Companions				
Combretum molle		1		2
Marquesia macroura	2	2	1	2
Burkea africana	1	3	3	

Table 2 (contd.)

Communities number of relevés	A1 10	A2 76	B 19	C 39
Cryptosepalum maraviense		1	1	1
Ekebergia benguelensis		1	2	1
Brachystegia glaberrima	2	1	1	1
Terminalia sericea	1	2		2
etc.				

A = Copaifera baumiana–Brachystegia spiciformis Association.
A1 = typical Subassociation.
A2 = Bequaertiodendron magalismontanum Subassociation.
B = Syzygium guineense subsp. afromontanum–Brachystegia longifolia Association.
C = Monotes loandensis–Brachystegia tamarindoides Association.

bium ceresiiforme (Rattray 1960). Apart from the typical form Monteiro (1970) distinguishes in the Bié area on mesic sites a subassociation with *Bequaertiodendron magalismontanum* which is richer in woody species. The *Copaifera baumiana–Brachystegia spiciformis* Association is distributed in a typical catena pattern on the slightly undulating sandy plains of central Angola, frequently interrupted by savannas and grasslands fringing the dambos and temporary watercourses. Types 17A and 24 of Barbosa (1970) mainly consist of the association but also in types 17B, 18A and 18B, and 24 the association occurs locally. From sites between Moxico (Luso) and Saurimo (Henrique de Carvalho) Barbosa (1970) mentions many of the species listed in Table 2.

In a wide area around Cuito-Cuanavale in northern Cubango *Brachystegia bakerana* is the most important tree together with *Guibourtia coleosperma*, *Julbernardia paniculata*, *Dialium engleranum* and *Cryptosepalum exfoliatum* subsp. *pseudotaxus*, and in the shrub and small tree layer *Copaifera baumiana*, *Xylopia odoratissima*, *Paropsia brazzeana*, and others. Lianas like *Baissea wulfhorstii*, *Bauhinia petersiana*, *Uvaria angolensis* and *Landolphia* sp. also occur. The undergrowth is very sparse here. In this broadly undulating area the lower, less drained sites are occupied by a more open woodland with *Parinari curatellifolia*, *P. capensis*, *Baphia massaiensis* subsp. *obovata*, *Copaifera baumiana*, *Trichilia quadrivalvis*, *Landolphia* and *Diospyros* spp. and several other species mentioned above as dominants. In the lowest lying, insufficiently drained parts the woodland is again replaced by savanna and grasslands dominated by *Loudetia simplex*, *Trachypogon spicatus*, and others. The association also occurs over a small area in the National Park of Bicuar further southwest towards Lubango (Teixeira 1968a).

In western Zambia the *Copaifera baumiana–Brachystegia spiciformis* Association is common on Kalahari sand (Wild & Barbosa 1967, type 19). In Barotseland common species in this community include nearly all those named above and in Table 2, as well as some others. In fringes around dambos where sand is deposited as a result of sheet flow from the surrounding higher lying area, *Terminalia sericea* is often conspicuously abundant.

In northern Barotseland *Cryptosepalum exfoliatum* subsp. *pseudotaxus*, *Brachystegia longifolia* and *Marquesia macroura* are locally common in stands which are transitional to other communities, including the *Cryptosepalum* dry forest discussed below. Transitions of the *Copaifera baumiana–Brachystegia spiciformis* Association to *Colophospermum mopane* and *Baikiaea plurijuga* communities occur in the less humid areas of southern Barotseland, southern Angola and also in western Rhodesia where Kalahari sands carry a community floristically similar to the *Copaifera baumiana–Brachystegia spiciformis* Association (Wild & Barbosa 1967, Farrell 1968b).

The *Syzygium guineense* subsp. *afromontanum–Brachystegia longifolia* Association occurs patchily distributed in the Bié district and adjacent areas on somewhat drier sites, and with a finer grained soil. Structurally this vegetation is very similar to the *Copaifera baumiana–Brachystegia spiciformis* Association though the grass layer is usually denser in this latter association (Monteiro 1970) and species with persistent and coriaceous leaves are more important. On relatively mesic sites transitions to the previous association occur. A vegetation type which probably represents a *Baikiaea plurijuga* subassociation of the *Syzygium guineense* subsp. *afromontanum–Brachystegia longifolia* Association is common in Bicuar and distributed in a regular catena pattern corresponding with the drainage lines and the soil types in the flat area (Fig. 8) (Teixeira 1968a).

On the driest sites with a medium to fine textured sandy-loam soil, but a hardened layer near the surface, mainly in the north of Bié and in adjacent districts, a vegetation type classified as *Monotes loandensis–Brachystegia tamarindoides* Association* is found. Trees are from 6 to 30 m high here and usually cover 50 to 70 per cent, although open stands with a tree canopy cover as low as 15 per cent occur. Lower woody species are again up to 5 m and cover about 50 per cent on average. The grass layer is usually fairly open and sometimes sparse. *Monotes loandensis, Uapaca kirkiana, U. benguelensis, Syzygium guineense* subsp. *guineense* and *Isoberlinia angolensis* belong to the characteristic and often dominant tree species in this association (Monteiro 1970). Barbosa's (1970) type 16 and part of type 18A largely correspond with the *Monotes loandensis–Brachystegia tamarindoides* Association here and there interrupted with stands that can be classified as *Syzygium guineense* subsp. *afromontanum–Brachystegia longifolia* Association. The most important tree species in this area are *Julbernardia paniculata, Brachystegia spiciformis, B. gossweileri, B. tamarindoides, B. puberula, B. floribunda, Isoberlinia angolensis, Pterocarpus angolensis, Pericopsis angolensis* and *Parinari curatellifolia*. Frequent low tree and shrub species include *Monotes caloneurus, Diplorhynchus condylocarpon, Anisophyllea gossweileri, Ekebergia benguelensis, Pseudolachnostylis maprouneifolia, Rothmannia englerana, Ximenia americana, Bridelia angolensis, Maytenus senegalensis, Burkea africana, Dombeya quinqueseta, Pteleopsis anisoptera, Albizia antunesiana* and *Uapaca benguelensis* (Engler 1910, Nolde 1938a, b, Barbosa 1970, Diniz 1973). As important grasses Rattray (1960) mentions *Hyparrhenia diplandra, H. filipendula, H. rudis, H. umbrosa, H. variabilis, Trachypogon spicatus*, and others. A structural diagram of this type of vegetation showing the spatial distribution of the main species of the various vegetation layers is presented by Jessen (1936, plate 1). Further north *Brachystegia boehmii* and *B. wangermeeana* become local dominants and represent communities of types 17B and 18B of Barbosa (1970) (Chapter 41, Fig. 2).

On the highest points of the central Angolan plateau, at altitudes between 1900

---

* It is still uncertain whether *B. tamarindoides* and *B. glaucescens* are conspecific or must be regarded as different species.

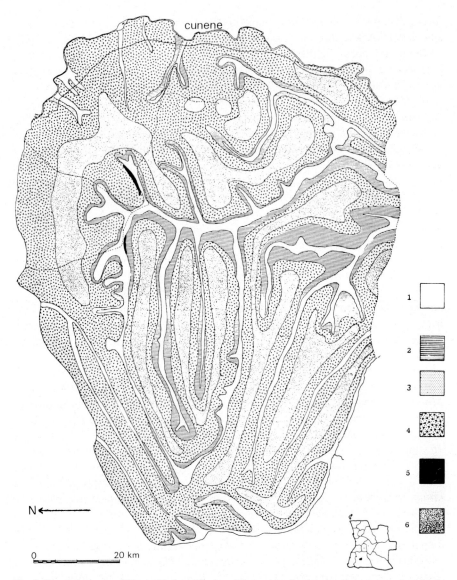

Fig. 8. Vegetation map of Bicuar National Park, southern Angola, showing the catena pattern in the very slightly undulating country (from Teixeira 1968a). 1: grasslands of *Loudetia superba* with *Parinari capensis* and *Pygmaeothamnus zeyheri*; 2: *Burkea africana* savanna with patches of *Themeda triandra* grassland; 3: *Acacia sieberana* var. *woodii* savanna; 4: *Brachystegia spiciformis–Julbernardia paniculata* woodland (very similar to Monteiro's (1970) *Syzygium guineense* subsp. *afromontanum–Brachystegia longifolia* Association); 5: *Brachystegia bakerana* woodland (very similar to *Copaifera baumiana–Br. spiciformis* Ass. of Monteiro 1970); 6: woodland and thicket of *Hippocratea parviflora–Baphia massaiensis* var. *obovata* with or without *Baikiaea plurijuga*.

and 2200 m in the Huíla district around Lubango, on quartzitic soils, the woodland is usually not more than 5 m tall with *Brachystegia spiciformis* or *B. floribunda* and *Julbernardia paniculata* as dominants. *B. tamarindoides* is less common, but several montane species of *Faurea*, *Protea*, *Syzygium*, *Cussonia*, *Ochna*, and *Parinari*, are present, and sometimes in the undergrowth species like

Stoebe cinerea, Helichrysum kraussii and *Vernonia* sp. which are more typical of the montane grassland. On dolomitic soils the *Brachystegia*s disappear, and species like *Carissa edulis, Peltophorum africanum, Securidaca longepedunculata*, and *Strychnos potatorum*, join *Julbernardia paniculata* as prominent species, whilst on dolerite *Brachystegia gossweileri* together with *Tarchonanthus camphoratus, Pteleopsis anisoptera, Commiphora mollis, Steganotaenia araliaceae*, and many others, are found. A structure diagram of this vegetation is presented by Jessen (1936, plate 12).

When cultivated fields are abandoned on the central Angolan plateau regeneration starts with a massive growth of *Rhynchelytrum repens* which is associated with several other grasses like *Digitaria milanjiana, D. longiflora, Eragrostis aspera, E. patens, Eleusine indica, Pogonarthria squarrosa*, and herbs like *Emilia coccinea* and *Vernonia* sp. Later *Pennisetum polystachion* and other grasses become co-dominants with *Rhynchelytrum repens* and several asteraceous species appear. Soon young plants of *Isoberlinia angolensis, Julbernardia paniculata, Brachystegia tamarindoides* and *B. spiciformis* become established and locally become dominant. In this stage also woody species like *Albizia adianthifolia, Vitex doniana, Monotes caloneurus, Strychnos* sp., *Diplorhynchus condylocarpon, Bridelia atroviridis, Anisophyllea gossweileri, Combretum* sp., *Lannea rubra* and *Clerodendrum* sp. can be abundant. After two to three years the ruderal grasses mentioned above are replaced by denser and taller species, notably *Hyparrhenia filipendula, H. rufa*, and locally *Cymbopogon citratus*. In the woody layer, which is still not high, the number of species increases with, for example, *Ochna afzelii, Rothmannia englerana, Mucuna stans, Combretum platypetalum* subsp. *baumii* and *Schrebera trichoclada*. Until the third to fourth year the tall grass species are dominant but from then on the trees and shrubs mentioned, augmented by *Albizia antunesiana, Indigofera hirsuta* and *Adenodolichos rhomboideus*, begin to dominate the vegetation. During the following years fire will exercise an important selective effect on the vegetation, favourable for the most fire-resistant species, like *Hymenocardia acida*, various species of *Combretum, Diplorhynchus condylocarpon, Dalbergia carringtoniana* and *Dombeya quinqueseta*, and from the fifth year onwards the usual species of *Julbernardia, Brachystegia, Isoberlinia* and *Monotes* will become strongly dominant. The undergrowth now is varied with several woody and rhizomatosous species and grasses, including *Annona stenophylla, Lannea rubra, Calanda rubricaulis, Secamone* sp., *Vitex domiana*, several species of *Hyparrhenia, Cymbopogon citratus, Panicum maximum, Pennisetum polystachion, Brachiaria jubata* and *Digitaria diagonalis* (Diniz & Aguiar 1972). It is believed that the dominant miombo trees take 100 to 150 years to mature (Lawton 1963).

Near Balovale and Mwinilunga in the Zambian–Angolan border region, at altitudes of 1100 to 1200 m, on deep Kalahari sands under a high rainfall regime, *Cryptosepalum exfoliatum* subsp. *pseudotaxus* can be the absolute dominant or even be pure in a dense tree canopy layer (Wild & Barbosa 1967, type 3, Barbosa 1970, type 4; Chapter 14, Fig. 2). This dense, evergreen *Cryptosepalum* dry forest called 'mavunda' (Figs. 9 and 10), with trees from 6 to 19 m high, is floristically closely related to the *Copaifera baumiana–Brachystegia spiciformis* Association. Apart from *Cryptosepalum*, also *Brachystegia spiciformis, B. longifolia, B. floribunda, Syzygium guineense* subsp. *afromontanum, Guibourtia coleosperma*,

Bersama abyssinica, *Erythrophleum africanum*, and *Combretum elaeagnoides* can occur in the tallest tree layer. The lower tree layer of about 6 to 12 m tall includes *Baphia bequaertii*, *Bauhinia mendoncae*, *Pteleopsis anisoptera*, *Ochna pulchra*, *Diospyros undabunda*, *D. batocana*, *Tricalysia* sp., *Bequaertiodendron magalismontanum*, *Combretum celastroides* subsp. *laxiflorum* and *Vitex mombassae*. Among the common shrubs are *Paropsia brazzeana*, *Psychotria* sp., *Pavetta assimilis*, *Sapium cornutum*, *Canthium huillense*, *C. schimperanum*, *Xylopia tomentosa*, *Erythrococca menyhartii*, *Grewia flavescens*, *Phyllanthus polyanthus* and *Copaifera baumiana*. Lianas and climbers such as *Landolphia camptoloba*, *Uvaria angolensis*, *Combretum gossweileri*, *C. paniculatum* subsp.

Fig. 9. Sandy road through well-developed dry evergreen *Cryptosepalum* forest, with vigorous growth of shrubs along the roadside. About 100 km south of Mwinilunga on Kabompo Road (photo J. P. Loveridge).

microphyllum, *Artabotrys monteiroae*, and *Byrsocarpus orientalis* can be common, and in the undergrowth *Megastachya mucronata*, *Setaria* sp., *Danthoniopsis viridis*, *Panicum heterostachyum*, *Cyperus* sp., *Phaulopsis imbricata*, *Biophytum helenae*, and many mosses are found. Epiphytic lichens are prominent. After clearing the forest nearly impenetrable regeneration stages with numerous lianas, shrubs and small trees can develop. In typically developed *Cryptosepalum* forest fire is not an important factor owing to the sparse herbaceous undergrowth (Trapnell & Clothier 1937, Fanshawe 1961, 1971, Cottrell & Loveridge 1966, Barbosa 1970, Verboom & Brunt 1970).

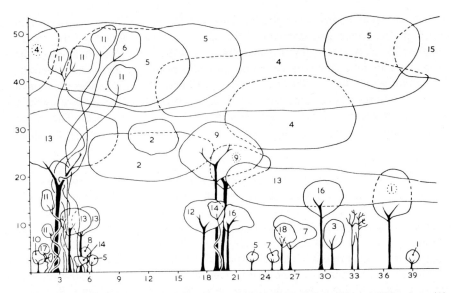

Fig. 10. Profile diagram of *Cryptosepalum* forest. The transect is 3.6 m wide and 12.8 m long (10 feet = 3.05 m). Vertical scale = 2.1 times horizontal scale. Canopies of trees not in transect but overhanging it, shown without trunks (from Cottrell & Loveridge 1966). 1: *Bauhinia mendoncae*, 2: *Baphia* sp., 3: *Brachystegia longifolia*, 4: *B. spiciformis*, 5: *Cryptosepalum exfoliatum* subsp. *pseudotaxus*, 6: *Clematis brachiata*, 7: *Copaifera baumiana*, 8: *Combretum gossweileri*, 9: *Diospyros undabunda*, 10: *Grewia flavescens*, 11: *Landolphia camptoloba*, 12: *Phyllanthus polyanthus*, 13: *Pteleopsis anisoptera*, 14: *Rytigynia umbellulata*, 15: *Syzygium guineense*, 16: *Tricalysia* sp., 17: *Uvaria angolensis*, 18: *Canthium schimperanum*.

5.1.7. Dwarf miombo (anharas de ongote)

A peculiar type of *Brachystegia* vegetation is the so-called 'anharas do ongote', a scrub formation of about 0.40 m in height, containing very few grasses and herbaceous plants, except geophytes. The shrubby species possess thick, woody rhizomes. Owing to the lack of herbaceous species, fire is not an important ecological factor in this formation. This vegetation type is wide-spread in the districts of Bié, Huambo and Huíla, growing at altitudes between 900 and 1600 m along the edges of the shallow valleys separating the typical miombo vegetation from the grassy valley bottoms, particularly in the upper reaches of the rivers. The soils are generally sandy loams with loose lateritic material above a hardened plinthite. The habitat is characterized by strong changes in the water regime. In the rainy season there is an excess of moisture for several months, followed by a sudden but long period of severe dryness of the soil (Diniz & Aguiar 1966, 1972, Barbosa 1970). *Brachystegia russelliae* (ongote) is absolutely dominant, but associated with various species of the group *Cryptosepalum maraviense* s.l. at various places, such as *C. crassiusculum* and *C. curtisiorum* and sometimes *Parinari capensis* and *Protea paludosa* (Gossweiler & Mendonça 1939, Airy Shaw 1947, Mildbraed & Domke 1966, Barbosa 1970, Diniz 1973).

5.1.8 Miombo of the remaining areas

The miombo vegetation in the lower rainfall areas not yet discussed has not been studied in detail phytosociologically. This vegetation has been classified into broad physiognomic and dominance types. The available floristic data are insufficient to allow a full phytosociological comparison and classification of the various communities but it is clear that they also belong to the order Julbernardio–Brachystegietalia spiciformis, although not to one of the alliances mentioned above. In most of this wide area of miombo vegetation the undergrowth is primarily characterized by species of *Hyparrhenia* (Rattray 1960). The major types, mainly as distinguished by Wild & Barbosa (1967), will be briefly discussed here.

a) *B. boehmii–B. allenii miombo of the escarpments.* On the escarpments and plateaux the most widespread are the *Brachystegia boehmii–B. allenii* woodland (type 29) and the *Brachystegia spiciformis–Julbernardia globiflora* woodland (type 23). The major escarpments of the plateau area in Zambia, Malawi and Rhodesia, such as the Luangwa escarpment, the northern escarpment of the Zambezi in Moçambique, both escarpments of the Zambezi in Zambia and Rhodesia, the escarpment of the Shire River in Malawi and the western escarpment of Lake Malawi, are characterized by *Brachystegia boehmii–B. allenii* woodland. This type is most typical at altitudes of about 1000 to 1350 m, with an annual precipitation of 800 to 1100 mm and on medium to fine ferralitic, sometimes stony, soils. In Moçambique this vegetation occurs also at lower altitudes. The woodland generally is up to 13 m high with *Brachystegia boehmii* and *Julbernardia globiflora* as most important trees, commonly associated with *Piliostigma thonningii, Monotes engleri, Pericopsis angolensis, Afzelia quanzensis, Diospyros kirkii, Peltophorum africanum, Kirkia acuminata* and various species of *Combretum*. Locally on well-drained alluvial sands, for example in the Luangwa valley, *Erythrophleum africanum* can be dominant. The grass cover is sparse and includes *Hyparrhenia filipendula, Hyperthelia dissoluta, Brachiaria brizantha, Pogonarthria squarrosa, Tristachya rehmannii, Rhynchelytrum nyassanum, Craspedorhachis rhodesiana* and various species of *Eragrostis* as its most prominent constituents (Trapnell & Clothier 1937, Willan 1940, Rattray 1960, 1961, Hursh 1962, Pike & Rimmington 1965, Wild & Barbosa 1967, Farrell 1968b, Jackson 1969b, Jacobsen 1968, Fanshawe 1971). Some of the communities of this woodland occurring in the Luangwa area are most probably Xerobrachystegion communities. In eastern Zambia *Brachystegia manga* and *B. stipulata* are locally important and on the Luangwa and Zambezi escarpments *B. allenii* often dominates. Northwest of Salisbury *B. boehmii* is often solidly dominant, although the usual companion trees like *Psorospermum febrifugum, Pseudolachnostylis maprouneifolia, Diospyros kirkii,* and also *Faurea saligna* and *Protea gaguedi* are present. *Becium obovatum, Aspilia pluriseta, Andropogon schirensis, Tristachya nodiglumis* and *Bothriochloa glabra* are most frequent in the undergrowth (Trapnell 1953, Jacobsen 1973). In the Tete District of Moçambique most of the above-mentioned tree species are frequent in this vegetation. They are associated with *Pterocarpus angolensis, Burkea africana, Swartzia madagascariensis,* and locally with *Brachystegia bussei, B. utilis, Acacia macrothyrsa, Terminalia sericea, Albizia antunesiana,* or others. *Oxytenanthera abyssinica* also occurs patchily as a result of local soil differences. In northern Moçambique this woodland occurs on the escarpment of the Vila Cabral plateau facing Lake Malawi at altitudes from 500 to 1000 m and in a largely similar floristic composition as in the Tete area (Pedro & Barbosa 1955, Wild & Barbosa 1967, Astle et al. 1969).

Below 1000 m on the central plateau escarpments, particularly on the Zambezi and Sabi escarpments, the *Brachystegia boehmii–B. allenii* woodland gives way to almost pure *Julbernardia globiflora* woodland (Wild & Barbosa 1967, type 30). This woodland alternates with mopane woodland (see below), or *Acacia* zones, along drainage lines. These lower zones of the escarpment are somewhat drier and warmer. *Julbernardia globiflora*, which is 7 to 13 m high in this woodland, is sometimes accompanied by *Brachystegia spiciformis* or *B. boehmii*, and lower down also by *Colophospermum mopane*, *Kirkia acuminata*, or *Sclerocarya caffra*. The grass cover is sparse and includes *Heteropogon contortus*, and species of *Eragrostis* and *Aristida*. In the Matopos Hills, with a fairly dry climate, there occurs a *Julbernardia globiflora* woodland above 1000 m. In Moçambique, along the Sabi and Zambezi escarpments and also in southern Malawi on the Shire escarpment this type of woodland occurs below 700 m, on clayey to sandy loam soils, under a rainfall regime of 800 to 900 mm. *Julbernardia globiflora* up to 12 m tall is accompanied by *Terminalia sericea*, *Dalbergia melanoxylon* and *Acacia nigrescens*, as well as the normal miombo species (Pedro & Barbosa 1955, Rattray 1960, 1961, Rattray & Wild 1961, Boughey 1961, Wild & Barbosa 1967, Farrell 1968a).

b) *B. spiciformis–J. globiflora miombo of the plateaux*. *Brachystegia spiciformis–Julbernardia globiflora* woodland becomes important in Zambia south of Kabwe (Fig. 11) and is found widespread in Rhodesia, but covers also extensive stretches of land in Moçambique. It generally occurs on varied, although always well drained soils with a rainfall of 750 to 1200 mm and most often above 1350 m altitude. On the Limpopo escarpment it descends to 675 m altitude, however, and in Moçambique it is found mainly between 600 and 1000 m, but also occurs lower down near Mocuba. It is a short type of woodland with trees from 6 to 13 m tall. Both *Brachystegia spiciformis* and *Julbernardia globiflora* can be dominant (Fig. 12), but dominance of the latter often indicates a secondary stage in this woodland. Other locally prominent tree species include *Uapaca kirkiana* (Fig. 13), *Brachystegia boehmii*, *Monotes glaber*, *Faurea saligna*, *F. speciosa*, *Combretum molle*, *Albizia antunesiana*, *Strychnos spinosa*, *S. cocculoides*, *Flacourtia indica* and *Vangueria infausta*. The grass cover is rather sparse and made up of the same prominent species as in the *Brachystegia boehmii–B. allenii* woodland. Irregularly, the miombo is interspersed with patches of *Acacia* savanna or grasslands where drainage is impeded, often with *Protea gaguedi* and *Parinari curatellifolia* at its fringes. Extensive areas with *Parinari curatellifolia* savanna on sandy soils also interrupt the miombo, and so do granitic domes with a pioneer rupicolous vegetation in which *Xerophyta retinervis* can be dominant (Engler 1910, Henkel 1931, Gilliland 1938, Rattray 1960, Rattray & Wild 1961, Boughey 1961, Barclay-Smith 1964, Wild & Barbosa 1967, Farrell 1968a).

A small number of relevés are available from this type of woodland at about 100 km northwest of Salisbury. The tree canopies cover here usually between 50 and 80 per cent, the shrub layer is open and always covers less than 50 per cent, and the herbaceous undergrowth usually covers between 50 and 80 per cent again. Three communities can be clearly distinguished: an *Ochna schweinfurthiana–Uapaca kirkiana* community (A) on granitic outcrops and

Fig. 11. *Brachystegia spiciformis–Julbernardia globiflora* woodland with tall *Hyparrhenia* sp. and *Dactyloctenium* sp. (foreground) under heavy but irregular burning regime on sandy soil near Kalomo, Zambia (photo M. J. A. Werger).

rocky granitic soils, a *Brachystegia boehmii* community (B) on lower slopes of pyroxenite or flat areas with pyroxenite outcrops, and a *Digitaria milanjiana–Brachystegia glaucescens* community (C) on the tops of pyroxenite ridges (Table 3) (Werger et al. 1978).

It has been observed that regeneration on abandoned fields of the two dominant species in this type of woodland, *Brachystegia spiciformis* and *Julbernardia globiflora*, mainly develops from root-sucker growth and not from germination of

Fig. 12. *Brachystegia spiciformis–Julbernardia globiflora* woodland on derived sandy soils 20 km west of Salisbury, Rhodesia (photo M. Leppard).

Fig. 13. *Uapaca kirkiana* dominated woodland about 150 km northwest of Salisbury, Rhodesia, on fairly shallow, granitic soil. Note the large amount of lichens on the stems (photo M. J. A. Werger).

seed (Strang 1966). In the high rainfall areas of eastern Rhodesia *Brachystegia utilis* is sometimes dominant, whilst *B. glaucescens* becomes dominant on rocky ridges (Fig. 14) with *Monodora junodii, Vangueria infausta, Diospyros usambarensis, Canthium huillense, Manilkara mochisia* and *Maerua kirkii* in an open shrubby layer on sandstones. On granitic substrate the undergrowth is much denser and contains more species (Farrell 1968a).

Fig. 14. Flushing foliage of *Brachystegia glaucescens* miombo on a rocky outcrop near Umtali, eastern Rhodesia, some weeks before the rains, gives the impression of a thin vegetation cover in reddish colours. The undergrowth has been burnt (photo M. J. A. Werger).

Table 3. *Brachystegia spiciformis–Julbernardia globiflora* woodland near Salisbury (after Werger et al. 1978) (+: with high cover values).

Community	A	B	C	Community	A	B	C
Number of relevés	4	3	4	Number of relevés	4	3	4
Differential species A:				Aspilia mossambicensis	4	1	
Ochna schweinfurthiana	4			Themeda triandra	3	2	
Bulbostylis macra	4	1		Phyllanthus sp.	3	1	
Uapaca kirkiana	3	1		Vigna pygmaea	3	2	1
Cyphostemma junceum	3	1		Clematopsis scabiosifolia	3	2	
Helichrysum nudifolium	3			Gerbera viridifolia	2	2	
Parinari curatellifolia	3			Brachiaria brizantha	2	2	
Syzygium guineense	3			Gloriosa superba	2	2	
Mariscus pubens	3			Conyza aegyptiaca	2	1	
Berkheya zeyheri	3			Conyza sumatrensis	2	1	
Digitaria flaccida	3			Acalypha alleni	2	1	
Endostemon obtusifolius	3			Indigofera rhynchocarpa	2	1	
Stomatanthes africanus	3			Gnidia kraussiana	2	1	
Pericopsis angolensis	2			Mariscus psilostachys	1	3	3
Monotes engleri	2			Annona stenophylla		3	3
Indigofera wildiana	2			Turraea nilotica		3	2
Plectranthus esculentus	2			Cyphostemma gigantophyllum		3	2
Differential species B:				Andropogon gayanus		2	4+
Brachystegia boehmii		3	1	Setaria lindenbergiana	1	2+	4+
Strychnos spinosa		3	1	Clerodendrum myricoides		2	3
Launaea rarifolia	1	3	1	Diplorhynchus condylocarpon		2	3+
Dombeya rotundifolia		2		Hymenodictyon floribundum	1	1	3
Bauhinia petersiana		2		Tarenna neurophylla		2	2
Differential species C:				Rhoicissus revoilii		2	1
Brachystegia glaucescens	3	1	4+	Adenia gummifera		2	1
Digitaria milanjiana		1	4+	Chlorophytum macrosporum		2	1
Pellaea viridis			4	Dolichos kilimandscharicus	4	3	4
Pellaea calomelanos		1	4	Lannea discolor	3	3	4
Aristida leucophaea			4	Commelina africana	3	3	4
Cleome monophylla			4	Digitaria gazensis	4+	3	2
Melanthera albinervia	1	1	4	Brachiaria serrata	3	1	4
Tapiphyllum velutinum		1	4	Cyphostemma crotalarioides	3	2	3
Albizia antunesiana	1	1	4	Heteropogon contortus	3	3	2
Bulbostylis contexta			3	Pavetta schumanniana	2	3	2
Spermacoce subvulgata	1		3	Becium obovatum	4	2	2
Diheteropogon amplectens			3	Aneilema johnstonii	2	1	4
Elephantorrhiza goertzei			3	Faurea saligna	3	1	2
Xerophyta equisetoides			3	Hypoestes verticillatus	2	2	2
Habenaria filicornis			3	Pseudolachnostylis maprouneifolia	1	3	2
Ipomoea verbascoidea		1	3	Arthropteris orientalis	2	1	2
Pterocarpus angolensis		2	2+	Microchloa kunthii	3	2	3
Loudetia simplex		1	3+	Combretum molle	2	1	1
Indigofera setiflora		1	3	Ledebouria revoluta	2	2	3
Other species:				Scleria bulbosa	2	2	3
Acalypha senensis	2	3		Cyanotis longifolia	2	2	2
Blumea alata	3	2		Commelina eckloniana	1	1	2
Hyparrhenia cymbaria	3	2		Thunbergia lancifolia	2	1	2
Julbernardia globiflora	2+	2		Cussonia arborea	2	1	1
Brachystegia spiciformis	2+	1+		Vangueria infausta	1	1	2
Temnocalyx obovatus	3	2	1	Pseudarthria hookeri	2	1	1
Sporobolus sanguineus	3	1		Cassia sangueana	1	2	1
Flacourtia indica	2	2		Asparagus laricinus	1	2	1
Rhynchelytrum setifolium	4	2		Dyschoriste fischeri	1	1	2

Table 3 (contd.)

Community	A	B	C	Community	A	B	C
Number of relevés	4	3	4	Number of relevés	4	3	4
Indigofera emarginella	1	1	2	Raphionacme longifolia	1		2
Tephrosia decora	2		2	Ozoroa reticulata	1	2	
Eriosema affine	2		2	Tripogon minimus		2	1
Mariscus alternifolius	1	2		Combretum zeyheri		1	2
Schistostephium heptalobum	2		1	Cymbopogon excavatus		1	2
Celosia trigyna		1	2	Justicia elegantula	2	1	
Zanha africana		2	1	Heteromorpha arborea	2		1
Sphenostylis marginata		2	1	Commelina ceciliae		1	3
Clematis brachiata	2	1		Crassula nodulosa	1		2
Mariscus leptophyllus	1		2	etc.			

At Inyanga in eastern Rhodesia this woodland still grows at almost 2500 m on westerly slopes, where the trees are only up to 4 m high (Boughey 1961, Wild & Barbosa 1967), at Chimanimani up to about 1700 m altitude (Crook 1956, Goodier & Phipps 1962, Phipps & Goodier 1962), and at Mt. Mlanje in southern Malawi up to about 1600 m as a stunted, lichen-hung woodland (Chapman 1962, Chapman & White 1970, Morris 1970). In Moçambique it occurs on clayey and sandy loam soils east of Beira in a wide area along the Rhodesian border and in the area between Nampula and the Malawian border. Here the trees are 8 to 17 m tall, the shrub layer sparse, and the grass layer, as usual with *Hyparrhenia* species, up to 3.5 m high. Many of the widespread miombo species are also common in this woodland while on secondary areas *Entada abyssinica, Piliostigma thonningii, Dichrostachys cinerea* subsp. *nyassana*, and some other species are particularly abundant (Pedro & Barbosa 1955, Rattray 1961, Wild & Barbosa 1967). More detailed species lists of this woodland are available from the Serra da Gorongosa area, northwest of Beira. On sandy loam or sand at altitudes between 400 and 800 m, and receiving an annual precipitation of 700 to 1200 mm, the woodland is well developed and reaches 20 to 25 m in height. Macedo (1970) recorded the woody species listed in Table 4. While *Brachystegia spiciformis* generally dominates between 600 and 800 m altitude, *B. boehmii* does so lower down. Dominance of *Julbernardia globiflora* usually indicates a secondary stage. This species regenerates much faster than the *Brachystegia* species. Several successional stages of the regenerating woodland communities are encountered and correlated with various degrees of human destruction of the original vegetation and the time elapsed since the destruction took place. Higher on the mountains, where the precipitation is also higher, mosaics of miombo woodland communities, *Parinari curatellifolia* savannas and humid montane forest communities, are found together with stands of miombo woodland that are floristically and ecologically transitional to the montane forests (Macedo 1970, Tinley 1971).

c) *The high rainfall areas of Moçambique and Malawi.* In the higher rainfall areas of the Vila Pery region and between Nampula and the Zambezi in Moçambique the *Brachystegia spiciformis–Julbernardia globiflora* woodland changes and *Brachystegia spiciformis* becomes the sole dominant (Wild & Barbosa 1967, type 21). The precipitation is between 1200 and 1800 mm in these areas and the soil is a red, compact, ferralitic clay. The trees are from 12 to 22 m tall and the woodland is generally dense. In the Vila Pery region associated species include

Table 4. Brachystegia spiciformis–Julbernardia globiflora woodland at Serra da Gorongosa (after Macedo 1970).

trees	A	B	small trees, shrubs and climbers	A	B
Brachystegia spiciformis	x		Bauhinia galpinii	x	
Brachystegia boehmii	x	x	Bauhinia petersiana		x
Julbernardia globiflora	x	x	Dombeya sp.	x	x
Burkea africana	x	x	Byrsocarpus orientalis	x	x
Pterocarpus angolensis	x	x	Psorospermum febrifugum	x	x
Pterocarpus rotundifolius subsp. polyanthus		x	Cassia petersiana	x	
Pseudolachnostylis maprouneifolia	x	x	Swartzia madagascariensis	x	x
Cussonia spicata	x		Ozoroa reticulata	x	x
Heteropyxis natalensis	x	x	Strychnos innocua	x	x
Piliostigma thonningii	x	x	Annona senegalensis	x	x
Tabernaemontana elegans	x	x	Erythrina abyssinica	x	x
Brackenridgea zanguebarica	x		Diplorhynchus condylocarpon	x	x
Pericopsis angolensis	x	x	Gardenia sp.	x	
Vitex doniana	x		Antidesma venosum	x	
Vitex sp. (payos)	x		Hymenocardia sp.	x	x
Albizia versicolor	x	x	Dichrostachys sp.	x	
Sclerocarya caffra	x		Lannea schimperi var. stolzii	x	x
Combretum molle	x	x	Oxytenanthera abyssinica	x	x
Erythrophleum africanum	x	x	Securidaca longepedunculata	x	x
Amblygonocarpus andongensis	x		Harrisonia obtusifolia		x
Millettia stuhlmannii	x	x	Holarrhena pubescens		x
Entada abyssinica	x	x	Vangueria infausta		x
Cussonia arborea	x	x	Ormocarpum sp.		x
Lonchocarpus capassa		x	Xeromphis obovata		x
Terminalia sericea		x	Combretum zeyheri		x
Bersama abyssinica		x			
Markhamia obtusifolia		x			
Diospyros mespiliformis		x			

A: *Brachystegia spiciformis–Julbernardia globiflora* woodland of sandy soils between 600 and 800 m altitude;
B: *Brachystegia boehmii–Julbernardia globiflora* woodland of sandy loam soils between circa 400 and 700 m altitude.

Pterocarpus rotundifolius subsp. *rotundifolius*, *Vitex payos*, *V. doniana*, *Dombeya burgessiae*, *D. rotundifolia*, *Cussonia spicata*, *Schrebera alata* and *Harungana madagascariensis*, with *Pteridium aquilium* and *Setaria megaphylla* in the undergrowth. In particularly well drained areas, stands dominated by *Uapaca sansibarica* and *U. kirkiana* alternate with the *Brachystegia spiciformis* dominated woodland. At protected places this woodland type may pass into evergreen forest patches, as discussed in Chapter 13. In the area between Nampula and the Zambezi the shrub layer is densely developed and this *Brachystegia spiciformis* woodland contains species of the evergreen forest including epiphytes. On sandy soils with a lateritic layer near the surface *Brachystegia utilis*, *B. allenii* or *B. boehmii* dominate locally, and on rocky slopes *B. glaucescens* is common, accompanied by succulent species of *Aloe*, *Kalanchoe* and *Crassula* (Pedro & Barbosa 1955, Wild & Barbosa 1967).

Also in northern Malawi dense patches dominated by *Brachystegia spiciformis*, and *Dracaena* sp., *Saba florida*, *Megastachya mucronata* and *Panicum trichocladum* in the understorey occur, and on the plateau around Mzimba a woodland dominated by *Brachystegia taxifolia* (Fig. 15) and *B. glaucescens* is found (Hursh 1962, Wild & Barbosa 1967, Jackson 1968).

The vast mesoplanaltic regions mainly above 500 m in northern Moçambique,

Fig. 15. *Brachystegia taxifolia* woodland near Mtanga-tanga forest, Vipya plateau, Malawi, alt. 1500 m. This species together with *Brachystegia spiciformis* and *Isoberlinia tomentosa* is typical of the woodland along the western fringe of the Vipya grasslands. The transition from woodland to *Protea* grassland with relict forest patches is usually fairly abrupt. However small patches of woodland occur scattered here and there in the grasslands (photo J. D. Chapman).

including the Amaramba, Marrupa and Macondes areas, are generally covered by a *Brachystegia utilis–B. boehmii* woodland. The annual rainfall is between 900 and 1400 mm in this area and is of the monsoon type (see Chapter 2). Orange ferralitic clays carry a woodland with *Brachystegia utilis*, *B. boehmii*, *Julbernardia globiflora*, and several other common miombo species. On red ferralitic clays the woodland is denser and contains in addition to these species more broadleaved species, while on stony, shallow soils *B. manga* is common (Pedro & Barbosa 1955, Wild & Barbosa 1967, type 28).

The lower lying river valleys in this area down to 150 m altitude are covered by about 8 m tall woodland with *Brachystegia boehmii*, *B. allenii*, *B. spiciformis*, *Julbernardia globiflora*, *Pterocarpus angolensis*, *P. rotundifolius* subsp. *polyanthus*, *Burkea africana*, *Afzelia quanzensis*, *Swartzia madagascariensis*, *Erythrophleum africanum*, *Pseudolachnostylis maprouneifolia*, *Piliostigma thonningii*, *Stereospermum kunthianum*, *Dalbergia melanoxylon*, *Sterculia quinqueloba*, *S. africana*, and several others. The woodland communities alternate with dry thicket and forest patches and with thorn savannas (Pedro & Barbosa 1955, Wild & Barbosa 1967, type 31). Still lower down on loamy soils a similar *Brachystegia* woodland and in the sandy sublittoral zone a *Brachystegia spiciformis–Berlinia orientalis* woodland alternates with *Pteleopsis myrtifolia* woodland or with *Adansonia* or *Acacia* savannas. The *Brachystegia–Berlinia* woodland also contains *Uapaca nitida*, *U. sansibarica*, *Phyllocosmus lemaireanus*, *Syzygium guineense*, *Parinari curatellifolia*, *Maprounea africana*, and other trees, locally also *Julbernardia globiflora* (Pedro & Barbosa 1955, Wild & Barbosa 1967, types 33 and 32).

South of the Rio Lúrio to about 16°S in the subplanaltic zone between 50 and 500 m a fairly dense woodland is patchily distributed on the clayey soils. The rainfall is about 1000 mm per year here, and the common miombo trees are 12 to 18 m tall and near the coast other species intrude in this vegetation. The miombo woodland forms a mosaic here with forest and thorn savanna communities. Still further southwards as far as the Zambezi the annual rainfall gradually decreases to 800 mm. In the subplanaltic zone adjacent to a vast *Brachystegia spiciformis–Julbernardia globiflora* woodland on the higher country, a mosaic of communities of this woodland and the *Brachystegia boehmii–B. allenii* woodland with thicket and thorn savanna communities is found, strongly correlated with small changes in soil type. In the sublittoral zone of this area the annual rainfall is somewhat higher and reaches 1200 mm. The woodland is up to 25 m high and consists of *Brachystegia boehmii, Julbernardia globiflora, Pseudolachnostylis maprouneifolia, Pterocarpus angolensis, Pericopsis angolensis, Hirtella zansibarica, Ochthocosmus lemaireanus*, and many other prominent species. Other savanna and riverine communities occur patchily on the wetter, poorly drained sites (Pedro & Barbosa 1955, Wild & Barbosa 1967, types 27, 24 and 26 respectively).

In wide areas north and south of the Zambezi delta tall *Brachystegia spiciformis–Julbernardia globiflora* woodland frequently containing the epiphytic fern *Platycerium alcicorne* also intersperse in a close mosaic with the dense, semi-deciduous forests of the Indian Ocean Coastal Belt (see Chapter 13).

d) *Miombo limits in southern Moçambique.* South of Beira down to the Rio Save small patches of miombo woodland occur in the sublittoral zone on sandy or calcareous soils in what is mainly a savanna with small trees of *Xeroderris stuhlmannii, Burkea africana, Julbernardia globiflora, Sclerocarya caffra, Albizia versicolor, Pseudolachnostylis maprouneifolia* and other species (Wild & Barbosa 1967, type 42).

The southernmost extensions of miombo vegetation are found in the Moçambique lowlands between the Rio Save and the Limpopo. The sandy soils of the sublittoral dunes carry a dense *Brachystegia spiciformis* woodland, with *Sclerocarya caffra, Trichilia emetica, Albizia versicolor, A. adianthifolia, Garcinia livingstonei, Pterocarpus angolensis*, and others. The undergrowth is generally very sparse, notwithstanding an annual precipitation of 900 to 1200 mm. The woodland is regarded as a secondary succession to the destroyed wet forests of the Indian Ocean coast (see Chapter 13) as indicated by several species in it which belong to those forest communities. Further inland the rainfall decreases from 1000 to 700 mm annually. Here, on sandy soils, a *Brachystegia spiciformis–Julbernardia globiflora* woodland with *Pterocarpus angolensis, Afzelia quanzensis, Garcinia livingstonei, Strychnos innocua, Dialium schlechteri*, and other species, alternates with deciduous *Balanites* forest and mopane communities on more loamy soils often in a sharply outlined mosaic. Still further inland, into the semi-arid lowland areas with 400 to 800 mm annual precipitation, the sandy soils carry a *Julbernardia globiflora* savanna with largely the same species. Interspersed in other savanna communities, thickets, mopane vegetation, and on the poorly drained sites, palm savannas occur. Also *Androstachys johnsonii* forms gregarious patches in the *Julbernardia globiflora* savanna mosaic (Pedro & Barbosa 1955, Wild & Barbosa 1967, types 20, 25 and 36, Guerreiro 1966, Tinley 1976).

5.2 Baikiaea vegetation

The Barotse Centre is characterized by deep Kalahari sands and an annual precipitation of 500 to 1000 mm strongly concentrated in the period November to April. Typical of this vast, flat area that stretches over southeastern Angola, western Zambia, the westernmost part of Rhodesia, Caprivi, northern Botswana and northeastern South West Africa, are vegetation types dominated by *Baikiaea plurijuga* and azonal, *Loudetia* grasslands on alluvium (Fig. 1). There is considerable floristic and physiognomic variability in *Baikiaea* vegetation, particularly towards the edges of the Barotse Centre. Detailed phytosociological data allowing a full comparative study with the adjacent vegetation types are however unavailable.

In the northern parts transitions to miombo woodlands occur and over a considerable area stands contain many species of the Guibourtio–Copaiferion baumianae. In the eastern, southern and western parts transitions towards the mopane vegetation and *Acacia* savannas and woodlands are common (Wild & Barbosa 1967, Blair Rains & McKay 1968, Barbosa 1970, Verboom & Brunt 1970, Fanshawe 1971, Menezes 1971, Giess 1971, Weare & Yalala 1971, Diniz 1973, Knapp 1973). It is difficult to draw the boundary between *Baikiaea* vegetation and the surrounding woodland and savanna communities, because *Baikiaea plurijuga* has been exploited rigorously as it provides a valuable timber. The species is rather sensitive to fire (Mitchell 1961a, Cumming 1962) and does not regenerate easily in frequently burnt areas, so that more fire-resistant species become dominants quickly. In well-developed *Baikiaea* communities species of *Brachystegia* and *Julbernardia* as well as *Colophospermum mopane* are totally absent, however. *Baikiaea plurijuga* then is the sole dominant, forming a fairly dense, dry semi-deciduous forest with trees up to 20 m in height. There is a dense and shrubby lower storey of *Combretum engleri, Pteleopsis anisoptera, Pterocarpus antunesii*, but also of *Guibourtia coleosperma, Dialium engleranum*, and *Strychnos* species, *Parinari curatellifolia, Ochna pulchra, Baphia massaiensis* subsp. *obovata, Diplorhynchus condylocarpon, Terminalia brachystemma, Burkea africana, Copaifera baumiana* and *Bauhinia petersiana* subsp. *serpae*. In a typical *Baikiaea* forest near Sesheke in Zambia the principal companion trees are *Acacia erioloba, Combretum collinum, Lonchocarpus nelsii, Strychnos potatorum, Entandrophragma caudatum, Markhamia obtusifolia* and *Vangueria randii*, while climbers and lianas include *Combretum elaeagnoides, C. celastroides, Dalbergia martinii, Acacia ataxacantha, Bauhinia petersiana* subsp. *serpae, Friesodielsia obovata* and *Strophanthus kombe*. Present in the shrub layer are *Acalypha chirindica, Canthium frangula, Croton scheffleri, Fagara trijuga, Grewia retinervis, Tarenna luteola* and many others (Trapnell & Clothier 1937, Martin 1940, Wild & Barbosa 1967, Verboom & Brunt 1970, Fanshawe 1971).

In the Sesheke district these *Baikiaea* forests also occur sometimes in a circular pattern with a dense outer shell of normal *Baikiaea* forest around a thicket with or without a light stocking of *Baikiaea*. Also dwarf-shell forests of *Baikiaea plurijuga*, about 1 to 1.5 m in height, are reported from here (Fanshawe & Savory 1964). These are situated along dambos surrounded by normal size *Baikiaea* forests. The woody species in the dwarf forests are of a characteristic growth form: the tap root has branched just below the surface into three to six short twisted branches and from the ends of some of the branches clumps of two to ten stool shoots grow. These shoots appear not to become older than four years. The dwarf forests occur in shallow, circular depressions of a few feet deep with the taller forest surrounding it. A dense but vaguely reticulate pattern of minor drainage channels covers the depression floor. There is a sparse grass cover in these dwarf forests. Sometimes dwarf *Baikiaea plurijuga* is replaced by dwarf *Guibourtia coleosperma* and usually there are some intruding species from the *Acacia–Terminalia* savannas. The floristic composition is apparent from Table 5. Although there is, at least periodically, abundant water and a bad soil aeration in the depressions covered by dwarf *Baikiaea* forests, the situation is, according to Fanshawe & Savory (1964), not identical with the normal dambo margins which commonly occur in this area. In dambos a phreatic or vadose watertable exists near, and periodically over, the soils surface and at the dambo margins the soil is mottled due to impeded drainage. Under the dwarf-shell forests the soil is not mottled. Fanshawe & Savory suggest that drainage is inherently free here, but that these sites receive quantities of telluric water resulting in reduced aeration of the soil which leads to the peculiar growth form. The nature of the impermeable or semi-permeable substratum under the pervious soil at the sites of the dwarf forests is not yet known.

Strongly associated with the *Baikiaea plurijuga* dominated forests but on more alluvially influenced soils on very slightly sloping terrain towards the rivers, are forest patches dominated by *Ricinodendron rautanenii* with trees up to 12 m tall. Associated tree species can include *Sclerocarya birrea*, *Baikiaea plurijuga*, *Pterocarpus angolensis*, *Burkea africana*, *Combretum psidioides*, *Dialium engleranum*, and several others. These patches are particularly frequent in the vicinities of the lower Cubango, Cuito and Cuando Rivers and their numerous tributaries in Angola, in southern Zambia and in Caprivi (Seiner 1909, Barbosa 1970, Giess 1971, Diniz 1973).

In southwestern Angola transitions both towards the *Brachystegia* woodlands and towards the drier *Colophospermum* woodland exist. In the former transitions *Brachystegia bakerana*, *Cryptosepalum exfoliatum* subsp. *pseudotaxus*, *Baissea wulfhorstii*, *Tylosema fassoglensis* and *Gymnema sylvestre* are locally conspicuous in the upper layer, while the lower layer includes *Melhania velutina*, *Steganotaenia araliacea*, *Leucas martinicensis* and *Hibiscus rhodanthus*. In the drier transitions on slightly convex, excessively drained, sandy sites *Baikiaea plurijuga* becomes much sparser, and more xerophytic species increase in importance. They include *Croton pseudopulchellus*, *Combretum aureonitens*, *C. apiculatum*, *Ptaeroxylon obliquum*, *Commiphora angolensis*, *Terminalia sericea*, *Hexalobus monopetalus*, *Grewia villosa* and, of course, *Colophospermum mopane*, while on slightly more mesic sites *Baphia massaiensis* subsp. *obovata*, *Pteleopsis anisoptera*, *Bauhinia macrantha*, *Ochna pulchra*, *Combretum zeyheri*, *Vitex mombassae*, *Hippocratea parvifolia*, *Clerodendrum uncinatum*, *Grewia suffruticosa*, *Gloriosa simplex* and *Perotis patens* are common (Teixeira 1968a, Barbosa 1970, Menezes 1971, Diniz 1973, Diniz & Aguiar 1973).

In northeastern South West Africa, northern Botswana and western Rhodesia transitional stands of *Baikiaea plurijuga* with *Colophospermum mopane* and much *Burkea africana*, but also *Erythrophleum africanum*, *Afzelia quanzensis*, *Pterocarpus angolensis*, *Boscia albitrunca*, *Acacia erioloba*, *A. fleckii* and *Terminalia sericea* can be common, the latter species being particularly abundant in a fringe around slight depressions and dambos, where sand is deposited by sheet

Table 5. Floristic composition of Baikiaea dwarf forest (after Fanshawe & Savory 1964).

Woody		Jasminum streptopus	o-f
Acacia ataxacantha	o	Lonchocarpus nelsii	o
Allophylus cataractarum	lo	Markhamia acuminata	o
Baikiaea plurijuga	a	Ochna cinnabarina	o
Baissea wulfhorstii	f	Ozoroa longipes	lr
Baphia massaiensis subsp. obovata	f-c	Pavetta schumanniana	lo
Bauhinia petersiana subsp. serpae	f	Rhus tenuinervis	o
Boscia albitrunca	o	Rhynchosia caribaea	r
Bridelia duvigneaudii	r	Schrebera trichoclada	lo
Canthium burtii	r	Strychnos innocua	lo
Canthium frangula	r-o	Tarenna luteola	o
Canthium huillense	lo	Tephrosia cephalantha	lf
Cissampelos mucronata	o	Terminalia sericea	f
Clerodendrum capitatum	o	Tinnea zambesiaca	lo
Clerodendrum myricoides	r	Tricalysia alleni	o
Cocculus hirsutus	lo	Tricalysia cacondensis	lf
Combretum celastroides	o-f	Vangueria infausta	r
Combretum elaeagnoides	f	Vitex amboniensis	o
Combretum engleri	o	*Herbaceous*	
Combretum collinum	o	Asparagus asiaticus	lo
Combretum mossambicense	o	Blepharis maderaspatensis	f
Combretum psidioides	o-f	Blainvillea latifolia	lo
Combretum zeyheri	o-f	Borreria scabra	f
Commiphora angolensis	r-o	Chlorophytum brachystachyum	r
Croton gratissimus	lo	Corallocarpus sp.	r
Croton pseudopulchellus	f-c	Cyphostemma crotalarioides	r
Dichrostachys cinerea	o	Dicerocaryum zanguebarium	o
Diospyros lycioides	lo	Justicia heterocarpa	f
Erythrococca menyhartii	o	Monechma debile	f
Euphorbia espinosa	lr	Phyllanthus leucanthus	o
Fagara trijuga	r	Plectranthus biflorus	f
Friesodielsia obovata	o	Pupalia lappacea	lo
Gardenia brachythamnus	lf	Sansevieria kirkii	lo
Grewia avellana	o	Sclerocarpus africanus	lo
Grewia bicolor	lo	Tephrosia lupinifolia	o
Grewia falcistipula	o	Thesium sp.	r
Grewia flavescens	o	Vernonia poskeana	f
Grewia retinervis	r-o	*Grasses and Sedges*	
Guibourtia coleosperma	v.lo	Cyperus diffusus subsp. sylvestris	r-o
Hippocratea parviflora	lo	Eragrostis spp.	f
Indigofera ormocarpoides	lo	Mariscus dubius	o
Ipomoea verbascoidea	o		

a = abundant   o = occasional
f = frequent   r = rare
c = common   l = local

flow (Fig. 16). With frequent fires a dense shrub layer develops with *Baphia massaiensis* subsp. *obovata*, *Bauhinia petersiana* subsp. *serpae*, *Paropsia brazzeana* and other shrubs and climbers. The grass layer is sparse when the shrubby understorey is densely developed, but when it is more open species like *Aristida meridionalis*, *A. congesta*, *Eragrostis pallens*, *Pogonarthria squarrosa*, *Brachiaria nigropedata*, *Perotis patens* and *Eragrostis lehmanniana* are found

Fig. 16. *Baikiaea plurijuga* woodland with *Guibourtia coleosperma, Pterocarpus angolensis, Terminalia sericea, Burkea africana,* etc. on deep sand in northeastern South West Africa (photo W. Giess).

(Seiner 1909, Mitchell 1961b, Wild & Barbosa 1967, Wild 1968a, Blair Rains & McKay 1968, Giess 1971, Weare & Yalala 1971, Simpson 1975). When fire damage is severe or following cultivation *Baikiaea plurijuga* can disappear completely.

In Rhodesia *Baikiaea* does not form a closed canopy, possibly because of anthropogenic influences (Fig. 17). Here, *Baikiaea* is commonly associated with *Pterocarpus angolensis, Guibourtia coleosperma* and *Ricinodendron rautanenii,* and in the lower storeys again *Paropsia brazzeana, Baphia massaiensis* subsp. *obovata* and *Dirichetia rogersii.* Common intruders are *Kirkia acuminata, Brachystegia spiciformis* and *Julbernardia globiflora.* The sparse grass layer includes *Aristida stipitata, A. pilgeri, Triraphis schinzii, Tristachya rehmannii* and species of *Digitaria* and *Eragrostis* (Henkel 1931, Martin 1940, Rattray 1961, Boughey 1961, Wild & Rattray 1961, Wild 1964, Wild & Barbosa 1967, Wild 1968b).

5.3 *Mopane vegetation*

5.3.1 Distribution and general ecology

Mopane, *Colophospermum mopane,* is indigenous to southern central Africa. Where the species occurs, it is nearly always the sole dominant of a woodland community (broad sclerophyll arid bushveld) or, in some cases, a tree savanna,

Fig. 17. *Baikiaea plurijuga* woodland in southwestern Rhodesia, showing the open tree canopy as a result of exploitation (photo B. H. Walker).

and usually makes up about 90 per cent or more of the total phytomass of the community. Mopane woodland mainly occurs in the more or less flat and wide valley bottoms of the large rivers of southern Africa, the Zambezi, Luangwa, Shire, Save, Limpopo, Okavango and Cunene, and on the adjacent wide plains at altitudes between 100 and 1200 m (Fig. 18). The annual precipitation in the distribution area of mopane is between 400 and 800 mm, with exception of the northern Luangwa valley where annually 1000 mm are measured, and the transition to the Karoo–Namib Region at the west coast, where rainfall is about

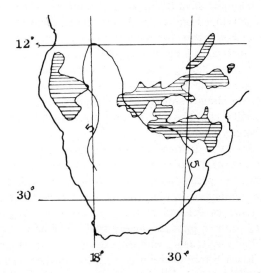

Fig. 18. Distribution map of *Colophospermum mopane*. The 5°C isotherm of the coldest month (July) is also shown (mainly after Henning & White 1974).

100 mm yearly. Henning & White (1974) state that the southern distribution boundary of *Colophospermum mopane* largely coincides with the 5°C isotherm of mean daily minimum temperature for the coldest month, July, and that low winter temperatures also explain the gap in its distribution area along the Okavango River at the border of Angola with South West Africa. Towards its southern boundary frost damage to the trees can be observed frequently (Van der Schijff 1969, Knapp 1973). Mopane woodland grows on fine grained, sandy to loamy and clayey, usually deep soils, though sometimes calcrete layers occur near the surface. Soils under mopane tend to develop a high exchangeable sodium content, which inevitably results in reduced permeability and increased susceptibility to erosion.

Mopane vegetation varies from fairly dense to open woodland or, in some cases, to tree savanna, with broad, somewhat sclerophyll leaves and with trees usually from 5 to 17 m tall. Woodlands up to 25 m tall are occasionally found in the northernmost reaches of its distribution area. A shrubby growth form up to 6 m in height is also common, and locally, particularly on heavy, impervious soils, a low scrub mopane can dominate the countryside. As a rule, the height of the individuals decreases from the more constantly warm, high rainfall areas in the northern parts of its distribution area to the periodically cooler and drier parts further south and west.

The position of the leaves on the trees is such that a large quantity of light reaches the soil surface. However, the understorey in mopane woodland is usually poorly developed showing a sparse growth of grasses, herbs and some woody species, but there is a steady rejuvenation of the tree layer. Particularly in shrubby mopane vegetation the grassy understorey is poorly developed, which may be caused by the pronounced tendency of the species to form shallow lateral roots (Ellis 1950, Henning & White 1974). The sparse grass cover results in a lack of sufficient fuel for hot fires, so that fire does not play an important role in this vegetation. Although *Colophospermum mopane* is not particularly sensitive to fire, it can be damaged. Fire damage, just as frost damage, usually results in strong coppicing from the main stem (Seagrief & Drummond 1958, Tinley 1966). Even with this poorly developed undergrowth mopane woodland represents an important range land for both game, particularly elephant, and domestic cattle, since its leaves possess a high nutritional value, particularly of proteins, and are readily browsed by the animals during all seasons. *Colophospermum mopane* stays green until far into the dry season. Several months after the start of the dry season, when the undergrowth has already entirely died down and many companion trees stand bare, mopane is still green. The trees first have yellowish leaves, later on becoming reddish brown before they are shed.

The woody species in mopane woodlands have their main flowering period from October to February, with a maximum in January. This maximum coincides with the maximum in precipitation. From March onwards flowering decreases quickly and the number of fruiting woody species increases until it reaches its peak in May. Towards the end of the dry season, in August, the number of flowering species starts to increase rapidly again. The herby undergrowth mainly flowers during the rainy season, with the exception of most geophytes and some succulents that flower before the rains come. Most of the undergrowth species are only physiologically active during the rainy season and die with decrease of rainfall in March and April (Knapp 1973).

5.3.2 The Luangwa, Shire and Zambezi valleys

In the Luangwa, Shire and Zambezi valleys, where the annual precipitation values are 600 mm or more, *Colophospermum mopane* often occurs in pure and dense stands, forming a tall woodland with trees generally from 10 to 17 m high. Here and there other tree species like *Kirkia acuminata, Sterculia africana, Adansonia digitata*, several species of *Commiphora, Combretum* and *Acacia* can be mixed with the mopane.

In the Luangwa valley, a wide, flat-bottomed trough bounded by steep, dissected escarpments that rise to 700 and 800 m above its floor, mopane woodland covers extensive areas in the alluvial zone, frequently interrupted by patches of *Acacia* savanna and *Combretum–Terminalia sericea* woodland. The floor of the trough, which rises from about 400 m at the Zambezi junction to 1000 m at its upper end, comprises a number of different soil types correlated with these various woodlands. Where a sandy sheet overlays a hard and compact, alkaline, sandy loam with a columnar structure a tall *Colophospermum mopane* woodland is found with trees up to 25 m high, interspersed with shrub mopane and a short grass cover. On the heavier textured solonetzes the mopane canopy is only up to 15 m in height. Common woody species in these woodlands include *Adansonia digitata, Combretum elaeagnoides, C. obovatum, Diospyros quiloensis, Holarrhena pubescens, Ximenia americana* and *Markhamia obtusifolia*, while in the sparse undergrowth *Eragrostis viscosa, Andropogon gayanus, Aristida adscensionis, Chloris virgata, Brachiaria eruciformis, Echinochloa colonum, Urochloa mosambicensis, Kyllinga alba*, and several other species occur. Elephants push over erratically a large number of mopane trees. This action together with the moderately hot fires result in a pollarded growth of mopane with stems up to 2 m tall (Rattray 1960, Astle et al. 1969, Lawton 1971).

In the Shire valley pure mopane woodland, up to 20 m tall, is found both on deep soils of the valley floor near Lake Malawi and on stony hills in southern Malawi. In the southern Shire valley mopane woodlands are nearly pure. Infrequently *Dalbergia melanoxylon, Pterocarpus brenanii, Acacia nigrescens, Diospyros kirkii*, and a few other species occur in the tree layer, while in the shrub layer *Euclea undulata, E. divinorum, Maytenus senegalensis, Asparagus buchananii* and *Ximenia americana* can be encountered. The commonest grasses are *Sehima nervosum* and *Digitaria milanjiana*, but *Panicum maximum, Aristida adscensionis*, and other species occur. The mopane woodlands abruptly border on *Combretum fragrans* savanna on poorly drained sites and are elsewhere interspersed with patches of *Combretum apiculatum* woodland. Further north in Malawi mopane vegetation is rarely encountered and only in fairly small patches (Pike & Rimmington 1965, Wild & Barbosa 1967, Agnew & Stubbs 1972, Hall-Martin 1972).

Medium size to tall mopane woodland with trees from 8 to 18 m high, occurs in the Zambezi valley from Caprivi to Malawi. On the deep, alluvial soils in southern Barotseland *Colophospermum mopane* woodlands adjoin the *Baikiaea* forests or woodlands and miombo woodlands particularly along drainage lines. Woodland of pure *Colophospermum mopane* is common here, but stands mixed with *Commiphora* spp. and *Terminalia sericea* are not rare. *Acacia nigrescens, Albizia amara, A. harveyi, Euphorbia candelabrum* (*E. ingens*?) and *Strychnos potatorum* can also be important in the tallest tree layer, with *Boscia mossam-*

bicensis, Combretum elaeagnoides, Balanites aegyptica and *Ximenia americana* frequently in the lower canopy layer and *Cissus quadrangularis, Fockea multiflora, Turbina shirensis* and *Asparagus africanus* as climbers. The undergrowth is sparse but includes *Enteropogon macrostachyus* and *Sporobolus panicoides* (Trapnell & Clothier 1937, Verboom & Brunt 1970).

Around Victoria Falls mixed *Colophospermum mopane–Combretum* woodland, frequently over 10 m tall, covers large areas on shallow soil over basalt. The most important woody species here are, apart from mopane, *Combretum apiculatum, C. imberbe, C. molle, C. hereroense, Acacia nigrescens, Dalbergia melanoxylon, Terminalia randii*, and several more, while the undergrowth consists as a result of severe grazing, mainly of annual species of *Aristida, Heteropogon contortus* and *Hyparrhenia* spp. (Hill 1969).

From Lake Kariba further downstream 7 to 15 m high mopane woodland covers the slopes of the escarpments. Commonly associated woody species near Kariba include *Croton gratissimus, Combretum elaeagnoides, C. apiculatum, C. celastroides, Strychnos innocua, Diospyros quiloensis, Holmskoldia tettensis, Commiphora* spp., and several others. On top of the ridges the *Combretum* species can become dominant, while on deep, sandy soils *Meiostemon tetrandus* becomes prominent. The Zambezi River is locally fringed by an evergreen gallery forest or woodland with *Kigelia africana* and *Trichilia emetica* on the levees, and on the alluvial soils of the floodplain grasslands alternate with savannas with *Acacia albida* (Fig. 69), a species that flushes and bears fruit during the dry season (cf. Pole Evans 1948, Magadza 1970, Jarman 1971).

In the Tete district the woodland is 10 to 15 m tall, consisting of almost pure mopane with some *Adansonia digitata*. The soils are fairly deep and clayey and often contain calcareous material. The rainfall measures 500 to 700 mm annually and the altitude varies between 200 and 500 m here. On more stony soils the mopane gets a more mixed character with *Commiphora, Combretum* and several of the other common species listed above for other parts of the Zambezi valley. The undergrowth is then sparse and grassy and contains species of *Andropogon, Setaria* and *Cenchrus ciliaris* (Pedro & Barbosa 1955).

5.3.3 Botswana, Rhodesian plateau, Transvaal and southern Moçambique

Near the northern edge of the Okavango delta, where the annual precipitation measures about 500 mm, there are shallow east–west valleys fringed with tree mopane and *Burkea africana*, while farther east, in and around the Moremi Wildlife Reserve and the Mababe Depression, mopane woodland with tall trees normally up to 14 m and near floodplains even over 20 m in height alternates with shrub mopane of the broad-sclerophyll arid bushveld type and with microphyllous thorny *Acacia* bushveld and broad-orthophyll *Terminalia sericea* bushveld (Fig. 19). Still further north, near Caprivi, the mopane vegetation reaches the *Baikiaea plurijuga* woodlands on the deep sandy soils (Seiner 1909, Engler 1910, Pole Evans 1948a, Tinley 1966, Wild & Barbosa 1967, Blair Rains & McKay 1968, Weare & Yalala 1971). On Kalahari sands in Rhodesia almost pure mopane woodland and savannas are common, while stands mixed with *Brachystegia boehmii* also occur.

Around the grassy, flat plains of Makarikari, where the soils consist of sand, silt and clay, *Colophospermum mopane* grows locally as trees with *Acacia nigrescens*,

Fig. 19. Tall mixed mopane woodland at Moremi, Botswana, showing elephant damage (photo H. Vahrmeijer).

Terminalia prunioides, Sclerocarya caffra, and *Combretum imberbe,* and elsewhere as shrubs with *Dichrostachys cinerea* and *Maytenus heterophylla* (Holub 1890, Seiner 1912, Weare & Yalala 1971). In the riverine woodland *Colophospermum mopane* is up to 9 m tall with *Acacia nigrescens* as the most important tree companion and *Grewia flavescens* var. *olukondae* and *G. flava* in the shrub layer. Further away from the rivers the canopy becomes reduced in height and cover. *Indigofera flavicans* is an important species in the very open ground layer here. Nearer the Makarikari salt pan a low open bushveld of mopane mixed with some species of *Acacia* and *Grewia,* 3 to 5 m high with an occasional emergent is present on the silty soil (Seagrief & Drummond 1958).

On soils derived from basalt, mainly in southeastern Rhodesia, mopane also forms extensive patches of open shrub land or bushveld. The open grassy undergrowth is mainly made up of annuals such as *Enneapogon cenchroides, Eragrostis viscosa* and *Aristida* spp. with a few scattered perennials like *Schmidtia pappophoroides* and *Cenchrus ciliaris.* Locally *Grewia bicolor* and *Combretum apiculatum* are frequent in the shrubby undergrowth on level ground of alluvial and of granitic origin respectively, and on stony, sloping country mopane is associated with many of the common (woody) species including *Sclerocarya caffra, Acacia nigrescens, Combretum apiculatum, Kirkia acuminata, Terminalia prunioides, Adansonia digitata, Adenium obesum* var. *multiflorum, Sansevieria bainesii* and species of *Grewia* and *Commiphora* (Fig. 20). This type of vegetation is rather similar to Arid Bushveld (see below). On basalt soils *Acacia* and *Commiphora* species become important, on shallow basalt-sandstone contact soils *Boscia mossambicensis* and *B. albitrunca* appear, on sandstone *Sesamothamnus lugardii,* while *Catophractes alexandri* patchily dominates on soils with calcrete nodules. The grass cover changes with soil type and grazing pressure, with *Eragrostis rigidior, E. superba, E. curvula, Brachiaria nigropedata* and *Urochloa*

Fig. 20. *Colospermum mopane* woodland near Birchenough Bridge, Rhodesia (photo M. Leppard).

pullulans on sandy soils, *Cenchrus ciliaris*, *Schmidtia pappophoroides*, *Panicum maximum*, *Pennisetum* sp., *Setaria* sp., *Ischaemum afrum* and *Digitaria milanjiana* frequent on basaltic soils, and on heavily grazed areas more annuals like *Enneapogon cenchroides*, *Aristida adscensionis*, *Dactyloctenium aegyptium*, *D. giganteum*, *Tragus berteronianus*, and others. Elsewhere, on heavy and badly drained soils, mopane is stunted and shrubby and only up to 4 m tall. Few grasses occur here, but ephemerals such as *Craterostigma plantagineum* and *Portulaca hereroensis* can be found. Stunted mopane is also typical of gypsiferous soils in Rhodesia and Botswana (Chapter 40, Fig. 6) (Engler 1910, Henkel 1931, Rattray & Wild 1955, Wild 1955, 1974, Rattray 1960, 1961, Boughey 1961, Wild & Barbosa 1967).

In eastern Botswana and the adjacent areas in Rhodesia and Transvaal, in the Limpopo valley and further south, the mopane vegetation is largely similar to that in southeastern Rhodesia and usually occurs in a shrubby growth form of up to 5 m high (Fig. 21), or as low tree mopane up to 8 m tall (Fig. 22). Annual rainfall is about 400 mm or less here. The southern boundary of mopane vegetation in the Transvaal is formed by the Soutpansberg, and east of the Transvaal Drakensberg by the Olifants River valley (Acocks 1953, Brynard 1964, Wild & Barbosa 1967, Louw 1970, Van Wyk 1971, Weare & Yalala 1971).

In the interior of southern Moçambique mopane vegetation occurs as a mixed tree and shrub savanna over an extensive area largely including the Save, Changane and Limpopo catchment areas. The annual rainfall is about 400 to 500 mm here, the altitude about 100 to 250 m and the soils are of a loamy sand type and represent lacustrine, calcareous formations. Differences in mopane dominated vegetation are associated with differences in soil type. In the Save valley mopane also occurs locally on solonetzic soils (Myre 1972). Generally mopane forms a tall woodland here, with trees up to 14 m and with shrubs. The grass layer is sparse and usually covers less than 10 per cent. The floristic composition of these vegetation types is apparent from Table 6. The *Neuracanthus*

Fig. 21. Low shrub mopane (*Colophospermum mopane*), 1.5–2 m tall, on clayey basaltic soil with ±400 mm rainfall and a tri-annual burning regime, Kruger National Park (photo P. van Wyk).

africanus–Colophospermum mopane Community (A) occurs on grey to greybrown compact loamy sands with a considerable amount of calcareous material in the upper layers. The *Sporobolus fimbriatus–Colophospermum mopane* Community (B) occurs on more acid dark grey to grey-brown compact loamy sands. The phytosociological position of these communities cannot be established yet owing to lack of comparable data. Near drainage lines the mopane woodlands are frequently interrupted by savanna and bushveld communities described below. Further southwards *Colophospermum mopane* most commonly is accompanied by *Ximenia americana*, *Salvadora angustifolia* var. *australis*, *Azima tetracantha*, *Adenium obesum* var. *multiflorum*, *Boscia albitrunca*, *Pachypodium saundersii*, *Dombeya kirkii*, *Sansevieria* sp., *Courbonia glauca*, *Albizia harveyi*, *Euphorbia* spp., and others. Along rivers and on waterlogged sites the mopane savanna is normally interrupted here by stands of *Acacia xanthophloea* and *Hyphaene natalensis*, while on the most sandy patches a *Julbernardia globiflora* savanna or a *Terminalia sericea–Guibourtia conjugata* bushveld is regularly encountered (Wild & Barbosa 1967, Tinley 1976).

5.3.4 South West Africa and Angola

Northwestern Botswana has the Okavango gap in the mopane distribution but northwestern South West Africa and adjacent southeastern Angola again contain a huge area dominated by mopane. The eastern margin of this area receives about 500 to 600 mm of rain yearly and mopane forms tall shrubs and trees, but further westward to the escarpment this amount of precipitation is quickly reduced and down the Kaokoland and Chela escarpments into the Namib low shrubs of *Colophospermum mopane* together with *Balanites welwitschii* still occur in

Fig. 22a. Early spring aspect of heavily grazed tall mopane woodland near Sibasa, Transvaal, showing *Albizia harveyi* in flower and in the undergrowth *Sansevieria* sp. and *Adenium obesum* var. *multiflorum* (left) (photo M. J. A. Werger).

Fig. 22b. Summer aspect of tall mopane woodland on loamy granitic soil and with ±400 mm rainfall in the Kruger National Park. Note the good grass cover (photo P. van Wyk).

Table 6. Mopane communities of the Save valley, Moçambique (adapted from Myre 1972).

Community	A	B	Community	A	B
Number of relevés	17	17	Number of relevés	17	17
Enneapogon scoparius	5		Mariscus dubius		3
Neuracanthus africanus	4		Mariscus macrocarpus		2
Sporobolus fimbriatus		5	Tricliceras tanacetifolium		2
			Cyphostemma sp.		2
herby species			Tragia sp.		2
Seddera suffruticosa	5	4	Urginea altissima		2
Decorsea schlechteri	5	3	Tragus berteronianus		2
Schmidtia pappophoroides	5	3	Commelina eckloniana		2
Eragrostis superba	5	3	Secamone sp.		2
Heteropogon contortus	5	3	Hibiscus micranthus		2
Ledebouria hyacinthia	4	3	Sporobolus ioclados		2
Digitaria eriantha	3	4	Enteropogon macrostachyus		2
Eragrostis rigidior	3	4	etc.		
Tephrosia purpurea	3	4			
Urochloa mosambicensis	2	5	woody species		
Aristida adscensionis subsp. guineensis	2	4	Colophospermum mopane	5	5
Ruellia sp.	2	4	Dichrostachys sp.	5	5
Elytraria acaulis	3	3	Commiphora africana	5	4
Rhynchosia totta	3	2	Dalbergia melanoxylon	5	3
Corchorus asplenifolius	3	2	Grewia bicolor	4	4
Stylochiton sp.	2	2	Cissus cornifolia	4	4
Pogonarthria squarrosa	2	2	Acacia nigrescens	4	3
Urochloa pullulans	2	2	Maerua parvifolia	3	4
Brachiaria deflexa	2	2	Grewia sp.	2	4
Talinum crispatulum	2	2	Acacia sp.	3	2
Indigofera sp.	2	2	Combretum sp.	3	2
Talinum caffrum	2	2	Maerua sp.	2	2
Bothriochloa sp.	3		Combretum hereroense	3	
Chrysopogon montanus	3		Sclerocarya caffra	3	
Tricliceras longepedunculatum	3		Ochna sp.	3	
Phyllanthus sp.	3		Ximenia americana	3	
Asparagus africanus	3		Vitex petersiana	2	
Ptycholobium plicatum	2		Combretum imberbe	2	
Corbichonia decumbens	2		Ozoroa sp.	2	
Digitaria memoralis	2		Phyllanthus maderaspatensis	2	
Stathmostelma pauciflorum	2		Lantana rugosa	2	
Eragrostis tremula	2		Phyllanthus kirkianus		3
Aristolochia petersiana	2		Diospyros usambarensis		3
Asparagus sp.	2		Cleistochlamys kirkii		2
Indigofera schimperi	2		Drypetes mossambicensis		2
Pyrenacantha kaurabassana		4	Maerua grantii		2
Evolvulus alsinoides		3	Combretum apiculatum		2
Enneapogon cenchroides		3	Courbonia glauca		2
Melhania forbesii		3	Markhamia acuminata		2
Stylochiton puberulus		3	Fagara humilis		2
Asparagus setaceus		3	etc.		

predominantly dry riverbeds under a rainfall of somewhat less than 100 mm per year and at an altitude of about 250 m (Fig. 23). Here *Colophospermum mopane* crosses the border of the Sudano–Zambezian Region into the Karoo–Namib Region. Southwards mopane is dominant until the vicinities of the Brandberg and Outjo, while northwards it occurs as far as the Caporolo River, about a hundred km south of Benguela, where it is still prominent, just as at the foot of the Serra da Chela. In Kaokoland *Colophospermum mopane* is locally associated with *Sesamothamnus benguellensis*, *S. guerichii*, several species of *Commiphora*,

Fig. 23. In the western parts of Kaokoland, S.W.A., shrubby *Commiphora* spp. occur scattered on the slopes while small mopane is restricted to the valleys and lower lying areas (photo W. Giess).

Ceraria longipedunculata, and many acanthaceous species, whilst in the more eastern parts, on fine red sands, mopane is accompanied by *Dichrostachys cinerea, Acacia mellifera* subsp. *detinens, Combretum apiculatum, Terminalia sericea, Catophractes alexandri* and a wide variety of herbs and grasses. On termitaria in this area up to 10 m large trees of mopane and *Combretum imberbe* preponderate. In the mountainous parts of Kaokoland, such as the escarpment and Baynes Mountains *Colophospermum mopane* woodland also contains several *Commiphora* spp. in abundance, together with *Adansonia digitata, Acacia recifiens, Sclerocarya birrea, Cyphostemma currori, Pachypodium lealii, Elephantorrhiza suffruticosa, Terminalia prunioides*, and sometimes *Entandrophragma spicatum* and *Euphorbia edouardoi*. An open shrub layer is composed here mainly of *Rhigozum brevispinosum, Croton gratissimus, Monechma genistifolium* and *Barleria prionitoides*, although *Myrothamnus flabellifolius* is also found. The trees can still be tall in Kaokoland, particularly along rivers, where they reach up to 15 m. Frequent grass species are here *Entoplocamia aristulata, Stipagrostis hirtigluma* and *Schmidtia kalihariensis*. On calcrete or calcareous rich soils around Etosha pan mopane forms a shrubby bushveld or a fairly dense woodland with trees up to 9 m tall, depending on soil type, soil thickness and availability of soil moisture. Other species occurring here under the mopane include *Croton gratissimus, Petalidium engleranum, Triraphis ramosissima*, and many others. Wide areas on the plateau west of Ethosa are covered by a low, shrubby bushveld approximately 1 m tall, dominated by *Colophospermum mopane* and *Catophractes alexandri* and accompanied by a mixture of Sudano–Zambezian and Karoo–Namib species in the undergrowth (Giess 1971, Joubert 1971, Malan & Owen-Smith 1974).

In Angola near Ngiva (Pereira d'Eça) at altitudes from 1000 to 1200 m and at the foot of the Chela escarpment, between 800 and 1100 m altitude, *Colophosper-*

mum mopane grows over vast areas as trees from 7 to 15 m in height, accompanied by *Terminalia prunioides, Commiphora angolensis, Combretum oxystachyum, Acacia erubescens, Balanites angolensis, Cordia ovalis, Hexalobus monopetalus, Croton* sp., *Ximenia caffra, Grewia bicolor, Euclea* sp., and in the herb layer *Schmidtia pappophoroides, Aristida rhiniochloa, A, adscensionis, Anthephora pubescens, Eragrostis annulata, E. porosa, E. superba* and *Pegolettia senegalensis*. Elsewhere 7 m tall mopane is associated with bushveld species as in the mountains of Kaokoland, or it occurs in nearly pure stands which mostly indicate insufficiently drained soils. During the rainy season these stands contain temporarily flooded ponds of varying size with aquatic vegetation. Better drained soils can carry massive stands of *Sansevieria cylindrica* under the canopy cover of the woody elements. On sites with locally more water available *Spirostachys africana* becomes subdominant or dominant and companion species include *Pteleopsis diptera, Pterocarpus lucens* subsp. *antunesii, Commiphora angolensis, C. mollis, Combretum psidioides, C. zeyheri* and various species of *Acacia*. Among the common grass species are *Eragrostis superba, E. rigidior, E. variegata, Tricholaena monachne, Chloris roxburghiana, Urochloa bolbodes, Schmidtia pappophoroides* and *Aristida rhiniochloa*. On the transition towards the *Julbernardia–Brachystegia* woodland a dry mopane bushveld with *Pterocarpus lucens* subsp. *antunesii, Croton gratissimus, Combretum hereroense* and *Securidaca longepedunculata* is normally encountered. Further south in the sandiest habitats *Colophospermum mopane* is replaced by *Baikiaea plurijuga*, while toward the Karoo–Namib Region nearer the coast on dry and often rocky soils *Catophractes alexandri, Rhigozum virgatum* and *Phaeoptilum spinosum* are frequent among up to 3 m tall mopane. Acanthaceous species and various species of *Commiphora* are common in the lower strata here, just as in adjacent South West Africa.

South of Lubango there is a large area with predominantly impermeable black clays, occasionally with a gilgai microrelief. *Colophospermum mopane* is often much smaller here and forms a microphyllous thorny bushveld. It is associated with *Acacia kirkii, A. nilotica* subsp. *subalata, A. hebeclada* subsp. *tristis, Securinega virosa, Spirostachys africana, Peltophorum africanum, Dichrostachys cinerea, Jatropha campestris, Indigofera schimperi, Melanthera marlothiana, Dichanthium papillosum*, and others. Particularly on alluvial soils *Acacia kirkii* can be abundant. The mopane bushveld is frequently interrupted by periodically inundated sites of various sizes with an open grass and herb growth (Gossweiler & Mendonça 1939, Airy Shaw 1947, Diniz & Aguiar 1969, Barbosa 1970, Menezes 1971, Diniz 1973, Huntley 1974, Malan & Owen-Smith 1974, Humbert in Schnell 1976–1977).

5.4 *Other woody vegetation types and grasslands*

5.4.1 General remarks on distribution, phytosociology, physiognomy and phenology

Most of the thorny, microphyllous and broad-leaved thickets, woodlands, savannas and bushveld types as well as most grasslands of the Zambezian Domain are floristically closely related although there exist gradual, or sometimes abrupt, differences in floristic composition of the various communities. These vegetation types, which cover vast stretches south of, and to some extent also within, the miombo area, have not been studied in detail and on a comparative basis except in southern Zaïre and in South Africa. Consequently, a useful general classification system for these vegetation types in the entire Zambezian Domain, either floristically or physiognomically, has not been worked out yet. In floristic systems many local names for communities have been used but most often the floristic definition of these communities is unclear because they are characterized by widespread species with only local differentiating value or which only locally constitute a characteristic species combination. For southern Zaïre a floristic system has been proposed, however (see below), but in South Africa outlines of parts of such a system are emerging but are incomplete and based on only a small portion of the variation in these vegetation types, while in the other southern African areas these kinds of studies are non-existent. In this type of vegetation relationships are complex and it would be premature to suggest what emphasis to place on these various affinities when grouping community types into broader classes. Thus, in the following account strong emphasis is placed on floristic affinities but alternative and better floristic arrangements may be possible, groups are not necessarily identified at the broadest possible level, nor are they necessarily of comparable rank. As the floristic composition (but frequently also the physiognomy) changes, mainly as a result of differences in climate, particularly in rainfall regime and severity and length of frosty period, in soil types and drainage, and as a consequence of fire and human activities, it was decided to use a physiognomic–ecological terminology to identify these various vegetation types in this account, except for Shaba. For the tropical areas, where the vegetation is relatively tall, the terms 'woodland' and 'savanna' will again be used, while for the usually low-branching vegetation of the area south of the Cunene and Limpopo Rivers the term 'bushveld' will be used in combination with physiognomically and climatically (see Schulze & McGee, Chapter 2) descriptive adjectives based on Thorntwaite's system.

Apart from the tropical woodland, savanna and grassland communities some thickets and scrub formations will also be mentioned which are floristically similar to the savanna communities, but have developed a deviating physiognomy owing to edaphic or anthropogenic influences, such as overgrazing or frequent burning. Besides, certain thickets are included which have developed as regeneration stages after forest destruction, particularly in the periguinean zone of northern Angola and patchily along the escarpment in Moçambique and Rhodesia.

Locally, as in northern Zambia and in the coastal zone of northern Moçambique, thickets with very restricted or no floristic relationships to the widespread Zambezian woodland, savanna and bushveld communities occur. These, as well as floristically unrelated vegetation types like semi-aquatic or saline grasslands and some riverine communities, which alternate patchily distributed or in catenas with the widespread Zambezian vegetation types, are also included here.

Seasonal aspects in woodlands, savannas, bushveld and grasslands vary with ecology, floristic composition and physiognomic type. Some general remarks can, nevertheless, be made. With the onset of the rainy season the grasses and herbs grow out fast and the shrubs and woody species flush with new leaves. Some trees already start to get new leaves shortly before the rains commence. The second half of the rainy season marks the main flowering period for the grasses, which afterwards dry out. Soon the grass cover takes on the characteristic yellowish-brown to reddish colour, but in drainage lines, dambos and other depressions, the grasses stay green for a few more weeks. The herbs also flower mainly during the second half of the dry season, and they stay green somewhat longer than the grasses. The leaves of the woody species are still green when the grasses are already dry, but they also dry out after the beginning of the dry season and are gradually dropped. There is a difference according to the various genera. *Adansonia* and *Sterculia* are usually bare first, followed by such genera as *Acacia*, *Commiphora*, *Terminalia* and *Lannea*. *Combretum* and *Burkea* usually only drop their leaves towards the end of the dry season. Some of the woody species flower during the wet season, but more species in the middle of the dry season and still many others just before or very early in the wet season. The species which flower during the dry season often have conspicuous, bright coloured flowers in white, red, yellow or purple. Most of the geophytes flower just before or early in the wet season, frequently with large flowers. Many trees that flower during the wet period have inconspicuous flowers. The first half of the dry season is also the period in which most species bear ripe fruits (Knapp 1973, Rutherford 1975). Cole & Brown (1976) relate the flushing and flowering of woody species in a rather dry (400 mm rainfall) bushveld area in Botswana to root structure: deep rooting trees and shrubs, which are able to draw on ground water, respond to rising temperatures and increasing air humidity from September onwards and flower and come into leaf before the rains come. These plants carry their leaves into the dry season and shed them after the first frosts. Trees and shrubs with lateral rooting systems remain leafless and without flowers until the soil has been well moistened by several good rainstorms.

5.4.2 Shaba

The thorny, microphyllous, and broad-leaved woodlands and savannas and some grasslands of Shaba were classified by Duvigneaud (1949) in the classes Erythrino–Acacietea campylacanthae and Combreto–Hymenocardietea, but this classification could not be maintained and Schmitz (1963) described the class Hyparrhenietea which comprises, amongst others, both classes distinguished by Duvigneaud. The Hyparrhenietea is a grouping of vegetation types which are common on more heavy, sometimes alluvial and periodically inundated or badly drained soils. Several species are widespread in the communities belonging to this class, but most conspicuous is the general importance of species of the genus *Hyparrhenia* (Schmitz 1963, 1971, cf. White 1965). On the heaviest, and often insufficiently drained soils *Acacia* species can be important too, but thorn savanna is not restricted to such sites. So far two orders belonging to the Hyparrhenietea have been described from southern Zaïre (see Table 7), although these types of vegetation extend outside this area. The Hyparrhenio–Acacietalia campylacanthae are the thorny, microphyllous savannas of the more or less rich, heavy, and moist, alluvial soils; the Themedetalia, comprise most of the broad-leaved savannas and some thorn and mixed savannas.

The Ctenio–Parinarietalia latifoliae of the class Ctenio–Loudetietea simplicis include open grasslands, sometimes with an abundance of low shrubs and more rarely with some taller shrubs, common on the poor sandy or shallow, rocky soils of the high plateaux of Shaba and in the dambos further southwards. The alliance Tephrosion hockii-manikensis represents the vegetation of the high plateaux of Shaba and will not be further discussed here. The Thesio–Gnidion comprises the

Table 7. Phytosociological classification of the savanna and grassland communities of Shaba and adjacent areas (after Schmitz 1963, 1971, and Streel 1963).

Syntaxon	Broad habitat characterization
Hyparrhenietea	
Hyparrhenio–Acacietalia campylacanthae	rich, heavy, moist alluvium
Hyparrhenion confinis	rich, deep, moist soils
Acacio–Beckeropsidetum	moist, non-eroded soils
Acacietum albido-sieberanae	nutrient-rich, well-drained levees
Acacio–Hyparrhenion cymbariae	oligo- to mesotrophic, moist soils
Acacietum campylacanthae katangense	clayey to silty (inundated) alluvium
Acacio–Hyparrhenietum cymbariae	heavy mesotrophic soils (Lufira valley)
Albizietum harveyi	oligo- to mesotrophic soils (Lufira valley)
Acacio–Pennisetetum purpurei	light alluvium
Acacio–Hyparrhenion diplandrae	
Euhyparrhenietum diplandrae	
Paspaletum auriculati	meso- to eutrophic, moist soils,
Pennisetetum angolensis	periodically flooded
Mesohyparrhenietum diplandrae	
Acacietum pilispinae	
Themedetalia	
Andropogonion schirensis	dry, poor soils (NW Shaba)
Elymandro–Sclerietum	quartzitic soils
Loudetio–Ochnetum	granitic soils
Sopubio–Andropogonetum	heavily eroded soils
Combreto–Hyparrhenion	meso- to eutrophic, temp. water deficient
Acacietum macrothyrsae	heavy, mesotrophic alluvium, light topsoil
Combreto–Hyparrhenietum collinae	heavy colluvium
Combreto–Andropogonetum gayani	heavy, meso- to eutrophic alluvium
Themedo–Acacietum hockii	heavy, rich alluvium, Mg-rich
Combreto–Terminalietum katangense	badly drained laterite and clay near dambos
Uapaco–Combretetum katangense	poor, gravelly or lateritic, impervious soils
Hyparrhion rufae	rich soils, temp. inundated, temp. very dry
Ctenio–Loudetietea simplicis	
Ctenio–Parinarietalia latifoliae	dambos or poor sandy to shallow rocky soils
Tephrosion hockii-manikensis	high plateaux of Shaba
Thesio–Gnidion	dambos
Zonotricho–Alloteropsidetum	
Oligohyparrhenion diplandrae	marshy, oligotrophic clay soils
Tristachyo–Sclerietum nyaensis	sandy clay
Nanocyperetum flavescentis	heavy clay
Oligohyparrhenietum diplandrae	moist, oligo- to mesotrophic sandy clay
Nanocyperetum proceri	temp. water deficient, sandy clay
Cryptosepalion maraviensis	rocky and lateritic soils fringing dambos
Acrocephalo–Cryptosepaletum maraviensis	
Acrocephalo–Eragrostidetalia cupricolae	metalliferous soils

vegetation types of the dambos of Shaba, Zambia, Rhodesia and part of Angola. These flat and wide, oval depressions at the upper parts of drainage lines are characterized by badly drained poor and impermeable clayey soils with sometimes a shallow toplayer of sand. The impermeable layer can be lateritic. This results in sites with alternating extremely dry and flooded periods. Vegetation types of this alliance are also found on similar soils in wide alluvial valleys and on gravelly and lateritic plateaux in Shaba. So far only the Zonotricho–Alloteropsidetum of the dambos has been described in this alliance, with *Zonotriche decora*, *Fimbristylis complanata*, *Alloteropsis homblei*, *Aeschynomene solitariiflora*, *Moraea unifoliata*, *Thesium quarrei*, *Gnidia chrysantha*, *G. hockii*, *Hyparrhenia bracteata*, *Rhytachne rottboellioides*, and several other species. The

Oligohyparrhenion diplandrae comprises the grassland communities of marshy sites with oligotrophic clayey soils and the Cryptosepalion maraviensis those of rocky and lateritic soils and fringing dambos (Streel 1963, Schmitz 1963, 1971).

The second order of the Ctenio–Loudetietea simplicis, the Acrocephalo–Eragrostidetalia cupricolae, comprising the communities of metalliferous soils, has recently been distinguished as a separate class (Ernst 1974) and is discussed in Chapter 40.

It is not yet determined how far this phytosociological classification system based on findings in Shaba is applicable to the larger Zambezian area, but it is likely that it will not be of much use outside the high rainfall area (approximately 1000 mm isohyet) (see Chapter 2).

5.4.3 The northern and central Angolan plateau

In the periguinean zone of Angola, where rainfall is usually over 1300 mm annually, thicket and savanna vegetation containing several Guineo–Congolian species is common and has been briefly described in Chapter 14. Tall, tufted grass species of *Hyparrhenia* and *Andropogon* are common there and among the woody plants *Dichrostachys cinerea, Annona arenaria, Sterculia quinqueloba, Albizia gummifera, A. adianthifolia, Piliostigma thonningii, Cochlospermum angolense, Psorospermum febrifugum, Hymenocardia acida, Diplorhynchus condylocarpon, Adansonia digitata, Terminalia sericea*, and at more compact, wetter soils *Combretum psidioides, C. zeyheri, Strychnos psidioides, S. floribunda, S. cocculoides, S. henningsii, Pteleopsis diptera, Acacia sieberana* and others can be abundant. Locally, at marshy sites the palms *Elaeis guineensis* and *Hyphaene* spp. are common. Often these formations occur in mosaics or in a catena pattern with forest patches, marshy grasslands, and gallery or semi-deciduous escarpment forests corresponding with minor variations in relief and with differences in soil types. Towards Shaba, formations form mosaics with *Brachystegia, Berlinia, Daniellia* and *Marquesia* woodlands. The herbaceous layer is usually well developed in this area under a fairly high rainfall regime, and fires at the end of the dry season can be fierce and destructive. Rhizomatous suffrutescent plants, particularly of the genera *Anisophyllea, Parinari* and *Landolphia* are locally well represented, and *Smilax kraussiana* is also frequent. In the Lunda area, and locally in Alto Zambeze at the Zambian border, patches of parkland have developed on fairly well-drained, poor, sandy soils near rivers, where in an open stratum of grasses and rhizomatous low shrubs large individuals of the palm *Borassus aethiopum* (Fig. 29) dominate the landscape.

In the Lunda area and in Moxico there are also extensive patches of deep, nutrient poor, leached sands supporting a typical vegetation type, mostly less than 1 m in height, which is locally called 'chanas da borracha'. These patches cover the highest parts of the slightly undulating sandy plains and it has been suggested that they represent former sand dunes. The 'chanas da borracha' consist of an open tufted grassland vegetation in which rhizomatous shrubs are common. Among the grasses *Loudetia arundinacea, L. simplex, Ctenium newtonii, Hyperthelia dissoluta, Hyparrhenia filipendula, Andropogon schirensis, Elyonurus argenteus* and some other species are commonest. For most of the year the grasses conceal the creeping, rhizomatous, woody plants, but the soil is so excessively drained that the vegetation remains open, and fires are never fierce.

The rhizomatous plants include *Parinari pumila, P. capensis* var. *latifolia, Anisophyllea fruticulosa, Landolphia parvifolia, L. camptoloba, L. lanceolata, Napoleona gossweileri, Morinda angolensis*, and a few others. The high proportion of white milky juice producing Apocynaceae among these is noteworthy, so that the formation is actually named 'borracha' (rubber). The local population exploited this formation intensively for rubber before the second world war (Gossweiler & Mendonça 1939, Airy Shaw 1947, Mildbraed & Domke 1966, Barbosa 1970, Diniz 1973). Similar formations on comparable sites, have been recorded in adjacent Zaïre and described by Duvigneaud (1949) as Landolphio–Trachypogonetum thollonii of the alliance Carpodinio–Trachypogonion, order Ctenio–Parinarietalia latifoliae, class Ctenio–Loudetietea simplicis.

Further south on the central Angolan plateau the rainfall decreases gradually. The area north of Quibala, where the yearly precipitation measures about 1200 to 1300 mm, has vast broad-leaved savanna areas containing *Terminalia sericea, Piliostigma thonningii, Dalbergia nitidula, Diplorhynchus condylocarpon, Cochlospermum angolense, Hymenocardia acida, Burkea africana, Pericopsis angolensis, Dombeya quinqueseta, Annona stenophylla, Combretum psidioides, C. platypetalum* subsp. *baumii, Erythrophleum africanum, Cussonia angolensis, Pseudolachnostylis maprouneifolia* and *Maprounea africana* among the trees and large shrubs. *Acacia sieberana, A. macrothyrsa, Pterocarpus rotundifolius* subsp. *rotundifolius, P. angolensis, Entada abyssinica* and *Erythrina abyssinica* are abundant on more fertile soils containing some organic material. At some places *Acacia sieberana* represents the only tree, particularly on the fertile sites at the base of inselbergs and on strips of alluvial soil. Along rivers *Pandanus* sp. sometimes gives a fringing formation. Important grass species include *Hyparrhenia* spp., *Andropogon gayanus* and *Panicum maximum*. Here and there the savanna vegetation is mixed with patches of *Brachystegia* miombo. Elsewhere, small patches of this savanna vegetation occur in vast stretches of miombo. Such patches are often encountered on ferralitic soils containing organic material mainly at the foot of inselbergs (Fig. 24) and other places of marked relief (Diniz & Aguiar 1972, Diniz et al. 1972).

The inselbergs support a floristically quite different rupicolous vegetation with *Aloe andongensis, Sarcostemma viminale, Xerophyta* sp., *Plectranthus concinnus, P. welwitschii, Selaginella dregei, Pellaea doniana*, and others, on conglomerate, but *Xerophyta stenophylla, Myrothamnus flabellifolius*, etc. on crystalline rock (Chapter 14, Fig. 4).

North and east of Huambo (Nova Lisboa) and near Lubango (Sá da Bandeira) some high plateaux rise above 1700 m in altitude. They carry a grassland formation with low broad-leaved trees and shrubs. Partially these are the same species as those of the savanna, but they also include *Protea angolensis, P. dekindtiana, P. petiolaris, Oldfieldia dactylophylla, Ochna* sp., *Ekebergia benguelensis, Syzygium guineense* subsp. *huillense, Kotschya strigosa*, and Afromontane species like *Philippia benguelensis, Myrsine africana* and *Stoebe* spp. Graminoids are of the genera *Hyparrhenia, Andropogon, Eragrostis, Loudetia, Ctenium, Fimbristylis* and *Xyris* with dominance varying with drainage efficiency of the soil. Locally also *Drosera indica* is found. Near Lubango shallow soils on these high plateaux contain several rhizomatous species of *Eriosema*. Some of these high plateau communities doubtless belong to the Tephrosion hockii-manikensis described from Shaba (Engler 1910, Gossweiler & Mendonça 1939, Barbosa

1970, Diniz 1973, Diniz et al. 1972). At lower altitudes in this area, around 1200 m, savanna communities dominated by *Burkea africana* are frequent. These communities contain also many of the more common species of the savannas north of Quibala, and they typically occur in belts between the badly-drained *Loudetia superba* grasslands of the valleys and the miombo or *Baikiaea* communities on the higher lying parts of the slightly undulating plateau surface (Fig. 8) (Teixeira 1968a, Menezes 1971).

5.4.4 The Angolan escarpment and coastal zone

High on the central Angolan escarpment broad-leaved savanna communities with tall grasses of *Hyparrhenia*, *Andropogon* and *Heteropogon* and dominant trees and shrubs of *Cochlospermum angolense*, *Piliostigma thonningii*, *Maytenus senegalensis*, *Ozoroa insignis*, *Crossopteryx febrifuga*, *Bridelia angolensis* and *Lonchocarpus pallescens* replace the central plateau savannas described above. Here and there they are intermixed with miombo communities and gallery forests or stands of *Pandanus welwitschii*. *Adansonia digitata* becomes important in some of these savanna communities and occasionally even infiltrates the *Brachystegia tamarindoides* miombo on the plateau. Very locally, cloud forests occur, mainly in the north (see Chapter 14).

Lower down on the escarpment the decrease in rainfall to about 600 mm is noticeable in the increase in xerophytism in the vegetation. The savanna communities become lower and frequently close to a thicket or scrub formation. Common woody species include *Adansonia digitata*, *Sclerocarya caffra*, *Spirostachys africana*, *Combretum zeyheri*, *C. psidoides*, *Pteleopsis anisoptera*, *Kirkia acuminata*, *Pterocarpus lucens* subsp. *antunesii*, *Aloe palmiformis*, *Trema orientalis*, *Croton angolensis*, *Commiphora mollis*, *C. angolensis* and *Sterculia quinqueloba* (Fig. 24). Some of these species, together with *Dichrostachys cinerea*, *Strychnos floribunda*, *Ziziphus abyssinica*, *Ximenia americana*, *Acacia brevispica*, and others, dominate in the thickets (compare Jessen 1936, plate 14). Further south on the escarpment savanna communities alternate with miombo and mopane communities and mixed stands occur in which *Spirostachys africana*, and several species of *Commiphora*, *Combretum* and *Strychnos* are prominent.

In the coastal lowlands of Angola in a wide area from Ambrizete to south of Lobito extensive stretches of land are covered by a savanna community with *Andropogon gayanus* var. *squamulatus*, *Hyperthelia dissoluta* and species of *Hyparrhenia*, *Panicum maximum*, *Heteropogon contortus* and *Digitaria milanjiana* in the undergrowth. Important woody emergents are *Adansonia digitata* (Fig. 25), *Sterculia setigera* and the tall succulent *Euphorbia conspicua* (=*E. candelabrum* sensu Leach 1974). Other common woody species include several of the species occurring in the escarpment communities (Fig. 24). These communities grow over extensive areas, mainly on fine-textured fersialitic soils. At the southern boundary of this savanna vegetation, between Lobito and the Rio Caporolo, *Terminalia prunioides* locally becomes dominant and under tree canopies *Sansevieria cylindrica* occurs in large populations. In the littoral area the grass layer generally remains lower and *Schizachyrium sanguineum*, species of *Combretum*, and *Strychnos spinosa* become more important. These communities are very similar in appearance to Spiny Arid Bushveld. Elsewhere in the coastal

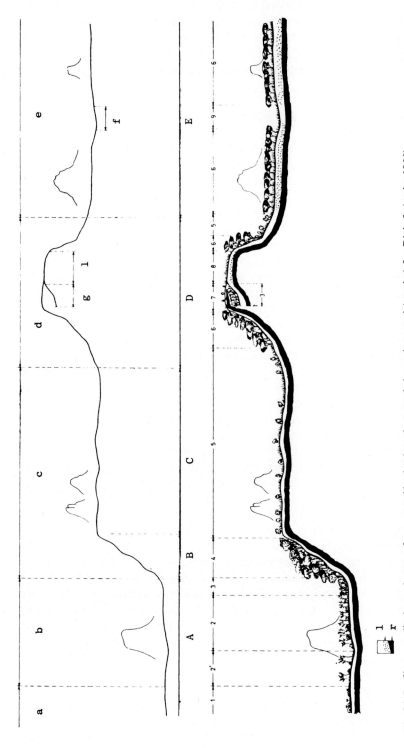

Fig. 24. Profile perpendicular to the coast from coastal lowlands to the central plateau in central Angola (after Diniz & Aguiar 1966).
A: littoral plain; B: escarpment, C: pediplain (lower level), D: mountain ridge of edge of plateau; E: pediplain; a: dry coastal soils; b: fersialitic soils; c: paraferralitic soils; d: lithosol, litholithic soils and paraferralitic soils; e: feralitic soils, f: hydromorphic soils; g: colluvial soil; l: laterite; r: rock.
1: grassland and savanna with *Setaria welwitschii, Euphorbia conspicua,* and *Dichrostachys*; 2 & 2' open and denser woodland with *Adansonia digitata*; 3: (evergreen to semi-evergreen) thicket; 4: evergreen to semi-evergreen forest; 5: savanna; 6: miombo (woodland) with *Brachystegia, Combretum,* etc.; 7: rupicolous vegetation; 8: montane grassland; 9: dambo grassland.

Fig. 25. In the dry coastal lowlands of central Angola near Ngunza (Novo Redondo) *Adansonia digitata* stays relatively low (photo K. E. Werger–Klein).

zone, where relief is stronger, dense stands with *Adansonia digitata, Sterculia setigera, Euphorbia conspicua, Commiphora angolensis, C. africana, Maerua angolensis, Balanites angolensis, Croton angolensis, Boscia urens* and *Carissa edulis* (Fig. 26) alternate with the *Schizachyrium* grasslands. Structural diagrams of these types of vegetation are shown in Jessen (1936).

Fig. 26. A dense woodland to thicket formation in which *Euphorbia conspicua* (=*E. candelabrum* s. Leach 1974) is prominent covers the upper valley sides of the Rio Cuanza at Quiçama, Angola. Extensive sudds of *Cyperus papyrus* are visible in the Rio Cuanza, and here and there a tree of *Elaeis guineensis* on stable alluvium (photo M. J. A. Werger).

371

On more calcareous soils a dense shrubby vegetation is common, with various species of *Combretum* and *Grewia*, *Croton angolensis*, *Pteleopsis diptera*, *Bauhinia tomentosa*, *Tarchonanthus camphoratus*, *Hymenostegia laxiflora* and *Ximenia americana* from which taller trees of *Guibourtia gossweileri*, *Adansonia digitata*, *Acacia welwitschii* and *Albizia versicolor* emerge, while locally on deeper and more mesic soils *Pterocarpus tinctorius*, *Lannea antiscorbutica* and *Ficus welwitschii* are also included. Coarse-textured soils support a thicket dominated by *Strychnos henningsii*, often associated with *S. floribunda*, *Combretum camporum* and a sparse undergrowth of *Kyllinga triceps*, *Sansevieria cylindrica*, *Sarcostemma viminale*, *Gloriosa superba*, *Eragrostis ciliaris*, *Blepharis maderaspatensis*, and others. Near the coast where the coarse-textured soil is sometimes affected by salt, grassy communities occur dominated by *Eragrostis superba*, accompanied by *E. prolifera*, *Digitaria milanjiana*, *Dichanthium papillosum*, *Aristida adscensionis*, *Heteropogon contortus* and *Chloris virgata*. Emergents here are *Hyphaene gossweileri*, and some of the other common littoral species. Open grasslands of *Setaria welwitschii* dominate the heavy black and dark brown clays in the coastal zone, sometimes accompanied by *Bothriochloa radicans* and *Dichanthium papillosum* at slightly better drained places in a somewhat undulating countryside. Other frequent companions include *Neuracanthus scaber*, *Maerua angolensis*, *Rhynchosia minima*, *Clitoria ternata* and *Aloe zebrina*. The grasslands are interrupted by small islands of *Dichrostachys cinerea* subsp. *forbesii* scrub on degraded patches of soil and by bush clumps of *Acacia welwitschii*, *Adansonia digitata* and *Euphorbia conspicua* on local patches of better drained or more shallow soils. In the main riverbeds such as the Rio Cuanza these savanna and thicket communities are interrupted by extensive sudds of *Cyperus papyrus* (Fig. 26), grasslands dominated by *Echinochloa crus-pavonis*, *E. pyramidalis* and *Oryza stapfii*, or by semi-evergreen or evergreen gallery forests with *Ceiba pentandra*, *Elaeis guineensis*, *Millettia thonningii*, *Albizia glaberrima* and *Adina microcephala* (Gossweiler & Mendonça 1939, Airy Shaw 1947, Rattray 1960, Teixeira et al. 1967, Teixeira 1968b, Diniz & Aguiar 1969, Diniz 1973, Huntley 1974).

Southwards from Lobito the coastal vegetation develops a more thorny character with several *Acacia* species joining the common woody plants, and gradually the Karoo–Namib element becomes stronger also (Barbosa 1970, Matos & Sousa 1970, Diniz 1973).

5.4.5 The central parts of tropical southern Africa

In the central parts of tropical southern Africa the communities alternating with the miombo, mopane or *Baikiaea* vegetation are often physiognomically rather similar to the extensive bushveld types further south. There is also a floristic affinity between some of these communities of eastern Angola, Zambia, Malawi, and parts of Rhodesia and Moçambique and the bushveld types further south, but generally the differences are large enough to justify a separate discussion.

In eastern Angola and Barotseland dense patches of low shrub entirely dominated by *Diplorhynchus condylocarpon*, sometimes mixed with some *Burkea africana* and a few other savanna species alternate with the *Loudetia* floodplain grasslands. On the higher lying areas with Kalahari sand, miombo, mopane, *Cryptosepalum exfoliatum* subsp. *pseudotaxus*, and *Baikiaea plurijuga* communities alternate with the grasslands. Further south in central Angola rhizomatous shrublands of *Landolphia parvifolia* var. *tholloni* and *Chamaeclitandra henriquesiana* as well as *Diplorhynchus* dominated shrublands form extensive stands.

In southern Angola, southern Zambia and adjacent parts of Caprivi, northern Botswana and Rhodesia the predominant plains of Kalahari sand covered by *Baikiaea plurijuga* communities and to a lesser extent by miombo and mopane vegetation are frequently interrupted by zones of heavy alluvial and alluvio-colluvial soils (Fig. 27). Here one finds grasslands as well as mixed savannas with *Acacia sieberana* var. *woodii*, *Diospyros mespiliformis* and sometimes *Acacia erioloba* along the rivers, while at other places *Acacia kirkii* is dominant and accompanied by low trees and shrubs of *Ziziphus mucronata*, *Acacia seyal*,

Fig. 27. Stand of *Acacia tortilis* on the edge of the Chobe floodplain in northern Botswana. The bark of the tree in the foreground is stripped by an elephant (photo P. N. F. Niven).

Peltophorum africanum and *Dichrostachys cinerea*. Grey-brown soils of medium fine texture carry stands of broad-sclerophyll arid or spiny arid woodlands or bushveld with *Colophospermum mopane*, *Kirkia acuminata*, *Grewia bicolor*, *Combretum* spp., *Spirostachys africana*, *Terminalia prunioides*, *T. sericea*, and others, while some higher lying areas, e.g. along the upper Cunene, support a woody community with *Diospyros mespiliformis*, *Ficus sycomorus* and *Combretum imberbe*, but dominated by the tall *Acacia albida* which is in full leaf during the dry season. On calcareous soils, along drainage lines, in Cuando–Cubango savanna communities occur with *Acacia nigrescens*, *A. erioloba*, *Kigelia africana*, *Trichilia emetica*, *Lonchocarpus capassa*, *Albizia versicolor*, *A. harveyi*, *Afzelia quanzensis*, *Diospyros mespiliformis* and locally also with *Piliostigma thonningii*, *Dalbergia melanoxylon*, *Acacia hebeclada* subsp. *tristis*,

A. erubescens, A. fleckii, A. brevispica, Combretum zeyheri, Ximenia caffra, Maytenus senegalensis, Strychnos pungens, S. spinosa, and *S. cocculoides.* In Zambia, Malawi, Rhodesia and central Moçambique a thorn savanna (called 'munga' in Zambia) is common with *Acacia xanthophloea* and *A. polyacantha* on deep greyish or black, moist soils (Fig. 28) and *A. sieberana* var. *woodii* on slightly drier soils. On alluvial soils *A. albida* or *A. tortilis* subsp. *heteracantha* can also predominate and in the Luangwa valley *Terminalia sericea, Sclerocarya caffra* or *Combretum imberbe* are prominent on slightly better drained sites in extensive *Acacia* savannas on recent alluvium, while older, more alkaline alluvials carry a mopane woodland. Thickets containing many savanna species as well as *Commiphora mollis, Euphorbia candelabrum,* and other species, also occur here. The field layer consists of tall grasses, like *Hyparrhenia filipendula* and *Andropogon gayanus.*

Fig. 28. *Acacia polyacantha* subsp. *campylacantha* savanna, here with *Combretum erythrophyllum*, 20 km east of Salisbury, Rhodesia (photo M. Leppard).

Locally the palms *Borassus aethiopum* (Fig. 29) and *Hyphaene benguellensis* subsp. *ventricosa* are common, the latter frequently in island thickets on old termitaria in Cuando–Cubango together with *Ficus sycomorus* and *Combretum imberbe*, or it dominates in low tufts on sandy, well-drained sites. *Phoenix reclinata* occurs in small stands on moist sites. Heavily grazed areas are invaded by a dense scrub in which *Acacia mellifera* subsp. *detinens, A. nilotica* subsp. *subalata, A. seyal* and *Dichrostachys cinerea* can be prominent (Trapnell & Clothier 1937, Trapnell 1953, Feely 1965, Wild & Barbosa 1967, Astle et al. 1969, Barbosa 1970, Fanshawe 1971, Menezes 1971, Diniz 1973, Diniz & Aguiar 1973).

Quite different from other woody vegetation in southern Africa, is the vegetation commonly called 'itigi'. It consists of extensive thickets in the area between Lake Mweru Wantipa and Lake Tanganyika in the very north of Zambia. They

Fig. 29. On sites with a somewhat impeded drainage the dense, thorny vegetation opens up and the palm *Borassus aethiopum* is common (Mazabuka, Zambia) (photo M. J. A. Werger).

grow on shallow, heavy and stony soils in flat areas and on skeletal soils on the gentle slopes, both areas having impeded drainages and with seasonal changes from extremely dry to very wet. These thickets are the southern extension of a thicket more common in Tanzania, and floristically strongly different from other thickets or woodlands further south. The thickets contain evergreen species on low mounds, possibly relic termitaria, while the other species are deciduous and grasses are virtually absent. Prominent species include *Bussea massaiensis*, *Baphia massaiensis*, *Boscia angustifolia*, *Burttia prunoides*, *Diospyros mweroensis*, *Combretum mweroense*, *Meiostemon tetrandus*, and many other species (Wild & Barbosa 1967, Fanshawe 1971).

Already in central Zambia *Pterocarpus angolensis* occurs in the savannas but in south Luangwa, around Petauke, and in central and southern Malawi as well as in Moçambique east of Lake Chilwa and in the upper parts of the Rio Lugenda area the *Pterocarpus* savanna is well developed. On shallow and drier soils *Pterocarpus rotundifolius* subsp. *rotundifolius* and *polyanthus* replace *P. angolensis* as the most prominent woody species. Also *Combretum zeyheri* on fertile soils, and *C. apiculatum*, *C. elaeagnoides* and *Kirkia acuminata* on stony ridges, are abundant, while locally dense thickets of *Oxytenanthera abessynica* develop. In Rhodesia this savanna is restricted to the Mazoe valley north of Salisbury. Everywhere the *Pterocarpus* community is intermixed with patches dominated by species of *Acacia* (Fig. 28), most frequently *A. polyacantha*, on heavy, black and alluvial soils, while sometimes it passes into a savanna with *Adansonia digitata*, *Cordyla africana*, *Kigelia africana*, and even *Borassus aethiopum* on poorly drained areas.

A common type of savanna or dry woodland in Malawi, particularly in the south along Lake Malawi and in the Shire valley, occurs under a rainfall regime of about 700 mm and contains the following tree species: *Adansonia digitata*, *Cordyla africana*, *Lonchocarpus capassa*, *Sterculia africana*, *S. appendiculata*, *Kirkia acuminata*, *Acacia nigrescens*, *Afzelia quanzensis*, *Combretum imberbe*, *Securidaca longepedunculata*, and others, with *Hyphaene benguellensis* subsp. *ventricosa* mainly along the drainage lines. Common grasses include *Digitaria gazensis*, *D. milanjiana*, *D. diagonalis*, *Setaria sphacelata*, *Panicum maximum*, *Hyparrhenia filipendula*, *H. rufa*, *H. nyassae*, *H. gazensis*, *H. variabilis*, *H. dichroa*, and locally *Urochloa mosambicensis*. On more sandy soils this type also contains *Terminalia sericea*, *Pseudolachnostylis maprouneifolia* and *Diplorhynchus condylocarpon*, while on more rocky sites *Tamarindus indica* and several species of *Euphorbia* are common. In low lying areas with badly drained soils this savanna community is replaced by a community with *Acacia nilotica* and *Combretum ghasalense*. After shifting cultivation the *Adansonia digitata* woodland regenerates to a thicket dominated by *Stereospermum kunthianum*, *Dichrostachys cinerea*, *Commiphora pyracanthoides*, *C. mollis*, *Schrebera trichoclada*, *Dalbergia arbutifolia*, and *Hippocratea crenata* (Fig. 30).

In northern Moçambique, between Quelimane and Nampula, woodland communities similar to those in the Shire valley occur locally in the lower lying areas in catenas with miombo vegetation. Sometimes the vegetation also forms thickets. Many of the species named for the Shire valley are present, and some other species, like *Pterocarpus polyanthus*, are also common locally. Particularly in secondary woodlands or thickets in depressions or wide fertile valleys this species is abundant, together with *Piliostigma thonningii*, *Sclerocarya caffra*, *Sterculia quinqueloba*, *Combretum ghasalense*, *Stereospermum kunthianum*, *Dalbergia melanoxylon*, *Dombeya burgessiae*, *D. rotundifolia*, *D. shupangae*, and others, but patches of *Terminalia* and *Acacia* savannas are also scattered over the area.

Very similar dry thickets, related to dry mountain bushveld, are common in the hot and dry Zambezi valley in Moçambique, Rhodesia and to a limited extent also in Zambia and often they occur higher on the escarpment than the mopane woodlands. Much the same thickets with a larger variety of *Commiphora* species and with *Combretum elaeagnoides* and *C. celastroides* are also extensive locally in the Zambezi valley, mainly on sandy soils. On the Rhodesian plateau similar thicket and dry woodland communities occur, with various species, for example

Fig. 30. A formerly disturbed dry woodland with *Adansonia digitata* at the western shore of Lake Malawi is regenerating to a thicket (photo M. J. A. Werger).

Combretum elaeagnoides, *C. apiculatum*, *Pterocarpus lucens* subsp. *antunesii*, *Pteleopsis myrtifolia*, and others, as local dominants in accordance with differences in habitat.

A floristically related and common deciduous thicket community in the Shire valley in Malawi is described by Hall-Martin (1972, 1975). Important trees are *Pterocarpus lucens* subsp. *antunesii*, *Adansonia digitata*, *Pteleopsis myrtifolia*, *Phyllanthus kirkianus*, *Newtonia hildebrandtii*, *Acacia welwitschii*, *Manilkara mochisia*, *Drypetes mossambicensis*, *Balanites pedicellaris*, *B. maughamii*, *Xanthocercis zambesiaca*, *Berchemia discolor*, *Cassia abbreviata*, *Lannea stuhlmannii*, *Cordyla africana*, *Vitex volkensii* and along drainage lines *Lecaniodiscus fraxinifolius*. Among the shrubs *Maerua edulis*, *Anisotes sessiliflorus*, *Grewia flavescens* and *Euphorbia lividiflora* are prominent while *Cissus rotundifolia*, *C. quandrangularis* and *Hippocratea crinita* are important climbers. In the undergrowth *Oplismenus hirtellus*, *Leptochloa uniflora*, *Hypoestes verticillatus*, *Peperomia pellucida* and *Sansevieria thyrsiflora* are among the common species. Locally this community has the structure of a dry deciduous forest. The soils carrying this community are of the acid sandy loam type becoming increasingly clayey with depth. Elsewhere in southern Malawi a riverine thicket is common, dominated by *Cola mossambicensis*, also containing *Diospyros senensis*, *Tamarindus indica* and several of the species of the *Pterocarpus lucens* subsp. *antunesii* community.

In the Zambezi valley of Moçambique and in the lower Shire valley some other dry savanna communities cover extensive areas between the 600 and 800 mm isohyets on clayey soils which sometimes contain calcareous material. Many of the species listed above for the dry savanna communities occur also in these communities, but on slopes with reddish, stony soils *Pterocarpus brenanii*, *Bolusanthus speciosus*, *Diplorhynchus condylocarpon*, *Terminalia sericea*, *T. stuhlmannii*, and *Combretum apiculatum* are prominent. In sheltered valleys the vegetation gets denser and sometimes forms dry forests with many of the same species and with *Millettia stuhlmannii, Albizia versicolor, Xeroderris stuhlmannii, Sclerocarya caffra*, and others.

On dark grey and blackish clay with a high water table *Acacia nigrescens* is dominant in places, and associated with *Combretum imberbe, Albizia harveyi, Dalbergia melanoxylon, Ziziphus mucronata*, and several species already mentioned. *Urochloa mosambicensis* and *Panicum maximum* are important grasses in this savanna. Frequently grasslands with many species of the Panicoideae and Andropogoneae interrupt these savanna communities, particularly the *Acacia–Hyphaene* savannas on poorly drained sites as for example in the Gorongosa National Park. Here some of the grasslands occur on saline sites, however, and contain *Sporobolus iocladus* with, at drier sites, *Salvadora persica*. On sandy patches in most of these savanna communities *Terminalia sericea* can be dominant, sometimes over considerable areas as for example, north of Lake Chilwa and at Kota-Kota on Lake Malawi (Pole Evans 1948b, Pedro & Barbosa 1955, Rattray 1960, Lemon 1964, Pike & Rimmington 1965, Wild & Barbosa 1967, Farrell 1968b, Jackson 1969a, b, c, Macedo 1970, Tinley 1971b, Agnew & Stubbs 1972, Hall-Martin 1972, 1975).

5.4.6   The lowlands of northern and central Moçambique

The wide zone of coastal lowlands in Moçambique shows a rather diverse vegetation pattern, corresponding with the variation in soil texture and soil moisture regime. Thickets and dry forests, in some cases of a unique floristic composition and in other cases rather similar to those of the dry Zambezi and Shire valleys, alternate with miombo, with *Acacia nigrescens* savannas of the Microphyllous Thorny Plains Bushveld type, with Broad-orthophyll Plains Bushveld, with Spiny Arid Bushveld, with gallery forests and with flood plain grasslands.

In the sublittoral zone of northern Moçambique patches of *Pteleopsis myrtifolia–Erythrophleum suaveolens* forest intersperse the miombo and savanna communities, and along water courses *Mimusops zeyheri* and *Pandanus* sp. can be abundant. In the area around the delta of the Zambezi the *Pteleopsis myrtifolia–Erythrophleum suaveolens* forest communities are rather dense on humid sites. The trees can grow up to 30 m tall and include locally more hygrophytic species, some of which belong phytogeographically to the Indian Ocean Coastal Belt (see Chapter 13). At places *Hirtella zanzibarica* is dominant in these forests while other important tree species include *Albizia adianthifolia* and *Anthocleista grandiflora*. Near Nampula dense dry woodlands with *Adansonia digitata, Rhodognaphalon* sp., *Sterculia appendiculata, Cordyla africana*, and others, similar to the communities found at the southern end of Lake Malawi, form a patchy mosaic together with miombo, *Acacia* savannas and low *Androstachys johnsonii* Arid Mountain Bushveld on shallow soils near inselbergs. The rocky outcrops carry a vegetation in which succulent species of *Euphorbia* and *Aloe* are prominent. These dry woodlands and the *Acacia* savannas are also not absent from the higher plateau area north of Nampula, although various miombo communities are more extensive here. In the lower valleys of the major rivers in this area the dry woodland becomes far more important, and the valleys have patches of dry forests and thickets of *Pteleopsis myrtifolia* and *Millettia stuhlmannii*.

Secondary thickets in this area often have a partially different floristic composition, with *Markhamia obtusifolia, Baphia gomesii, Dalbergia boehmii*,

Dichapetalum stuhlmannii, and other species. In the sublittoral lowlands these thickets and woodlands are still frequent, but in its northernmost parts, where the rainfall is of the monsoon type and averages just over 1000 mm annually, the dry woodlands and the *Pteleopsis myrtifolia* thickets are floristically somewhat different and contain some species of the Oriental Domain. Common constituents of the shrublayer in these thickets and woodlands include *Balanites maughamii*, *Erythrophleum suaveolens*, *Tetracera boiviniana*, *Dichrostachys cinerea*, *Dichapetalum barbosae*, *D. stuhlmannii*, *Sterculia schliebenii*, *Fernandoa magnifica*, and locally much *Oxytenanthera abyssinica*, while on sand, near the more recent dune areas, the Indian Ocean Coastal Belt species *Trachylobium verrucosum*, *Sideroxylon inerme*, *Garcinia livingstonei* and *Combretum constrictum* are common intruders. Grassy patches dominated by *Pennisetum polystachion* are also found in consolidated dune areas, and everywhere marsh communities, *Acacia* savannas or *Berlinia orientalis* communities are distributed patchily. The thickets usually occupy the calcareous grey and brown clays and clayey sands, with dense woodland in patches on more stony soils in the higher lying areas, while the *Acacia* savannas are found on compact black clays. Badly-drained clays carry the marsh communities and the better drained, sometimes stony soils the *Berlinia* and some other miombo communities.

The sandy, littoral zone of northernmost Moçambique carries another thicket community which is floristically quite different from the other Sudano–Zambezian communities. Deciduous woody species such as *Guibourtia schliebenii*, *Pseudoprosopis euryphylla*, *Baphia macrocalyx*, and many other species are frequent in these thickets which are sporadically interrupted by *Androstachys johnsonii* bushveld or *Pteleopsis myrtifolia* communities, often on wetter sites. The juvenile, sandy soils along the coast are colonized by prostrate pioneer species, binding the sand and starting the succession towards these thicket communities (Pedro & Barbosa 1955, Wild & Barbosa 1976, Tinley 1971a).

The deltas and large alluvial deposits at the mouth of the Zambezi and other major rivers, like Pungue, Buzio, Save, Limpopo and Maputo, characteristically carry extensive grasslands on the badly-drained and periodically-flooded plains interspersed with patches of savanna and fringing forests on the higher lying areas and various aquatic and semi-aquatic communities at the lowest lying sites. A scheme of such a pattern is presented in Fig. 31. The fringing forests in the more northern parts are similar to the dry forests with *Adansonia digitata*, *Sterculia appendiculata* and *Cordyla africana*, but in the southernmost parts of Moçambique species such as *Ficus sycomorus*, *Trichilia emetica*, *Antidesma venosum*, *Xanthocercis zambesiaca*, *Combretum microphyllum*, *C. imberbe*, *Ekebergia capensis*, *Kigelia africana*, *Diospyros mespiliformis*, *Berchemia discolor*, *Lonchocarpus capassa*, and various others, are more common constituents of the river fringing forests. The microphyllous savanna patches are usually dominated by *Acacia* species and are similar to those on the plateau. Depending on the soil type and drainage, the dominant species can be *A. polyacantha* subsp. *campylacantha*, *A. albida*, *A. nilotica* subsp. *kraussiana* or *A. xanthophloea*, of which the latter two are typical of the worst-drained sites. Also trees common in the fringing forests occur scattered over the savanna patches, and around smaller swampy depressions in the savannas, communities with *Pandanus* sp., *Borassus aethiopum*, *Hyphaene natalensis*, or *Phoenix reclinata* occur. On badly drained, low-lying areas moist grasslands are found.

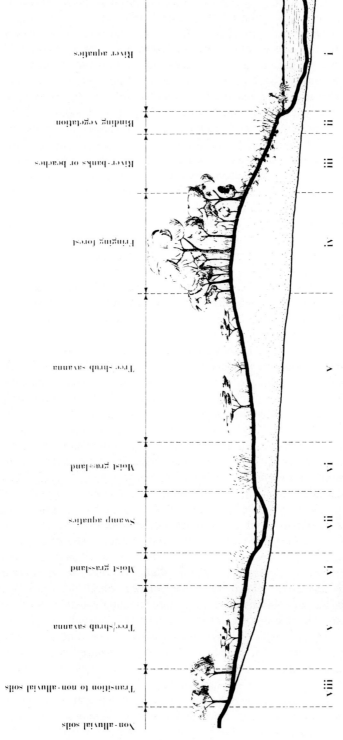

Fig. 31. Zonation of vegetation and habitat in deltas of large rivers in Moçambique (from Wild & Barbosa 1967). i: aquatic vegetation in the river; ii vegetation at the water's edge, mainly *Phragmites*, but also *Imperata cylindrica*, *Vetiveria nigritana*, *Echinochloa*, *Cyperus*, etc.; iii vegetation of river banks or beaches above normal tide level, consisting mostly of annuals in discontinuous tufts; iv: fringing forest with *Ficus* spp., *Trichilia emetica*, etc. or *Rhodognaphalon*, *Adansonia digitata* etc.; v: tree and shrub savanna with *Acacia* spp. or *Ficus* spp. and *Kigelia africana*, or *Combretum imberbe*, etc. At swampy places there is a palm savanna; vi: moist grassland with *Setaria* spp. etc. and along water's edge with *Cynodon dactylon*, *Oryza*, etc.; vii: aquatic vegetation of swamps and lagoons with *Eichhornia* etc.; viii: transition to non-alluvial soils with *Combretum imberbe*, *Lonchocarpus capassa*, *Xeroderris stuhlmannii*, etc.

In the transitional zones from alluvial deposits to non-alluvial soils the common dry woodland with *Adansonia digitata* described above is found again, in the south also containing *Sclerocarya caffra, Combretum imberbe, Lonchocarpus capassa, Albizia versicolor* and various other woody species. The undergrowth is usually fairly open and varies, of course, with the cover of the woody layer in these communities. It includes grasses like *Panicum maximum, P. deustum, Beckeropsis uniseta, Andropogon gayanus, Hyparrhenia filipendula, H. variabilis, H. lepida, Heteropogon contortus*, and some other species. South of the Save this woodland covers extensive sublittoral areas with reddish, calcareous soils having an upper layer of sandy loam over a hard horizon. Further inland a dry broad-orthophyll bushveld community dominated by *Guibourtia conjugata, Strychnos innocua* and *Terminalia sericea* occurs locally on gentle slopes fringing the alluvial soil types, and also a similar bushveld dominated by *Combretum* species, mostly *C. apiculatum* or *C. zeyheri*, and often accompanied by several other dry woodland species. These bushveld types also extend into Rhodesia, and in Moçambique they replace the miombo communities which gradually decrease in extent from the Zambezi southwards (see below).

Elsewhere in the sublittoral zone of Moçambique, off the alluvial deposits, extensive areas with badly-drained, but often sandy soils carry a *Borassus aethiopum* and *Hyphaene natalensis* palm savanna. Locally also *Kigelia africana, Pseudolachnostylis maprouneifolia, Diplorhynchus condylocarpon, Annona senegalensis* or *Euphorbia ingens* are common woody emergents in such communities. Myre (1972) described such a community from the Save valley under the name *Andropogon schirensis–Vigna parviflora* Association (see Table 8). On somewhat better-drained soils *Parinari curatellifolia, Uapaca nitida, Hymenocardia acida* and *Piliostigma thonningii* can be present too.

5.4.7 Upland (Temperate) Sub-humid Mountain Bushveld

Bushveld of the cool to warm-temperate, sub-humid mountains on the low interior plateau of South Africa includes four major types: (1) cool-temperate evergreen sclerophyll, *Protea*-dominated Bushveld, which is closely interrelated with moist cool-temperate Grassland and occurs in cool habitats on leached soils, transitional to grassland habitat; (2) cool-temperate, mixed broad-orthophyll and evergreen sclerophyll, *Landolphia capensis–Bequaertiodendron magalismontanum* shrub Bushveld, on outcrops, including summit outcrops in bushveld regions and outcrops in upland grasslands; (3) Broad-orthophyll *Faurea saligna, Burkea africana* and *Diplorhynchus condylocarpon* Bushveld, occurring on coarse-textured soils in less leached, warmer temperate areas than grasslands and *Protea*-dominated Bushveld; and (4) microphyllous thorny *Eustachys mutica–Acacia caffra* Bushveld on fine-textured soils, also in somewhat warmer less leached temperate habitats.

Field layer grasses in these bushveld types are of the sour kind typical of the leached soils of mountains and high rainfall areas.

A. *Protea-dominated Bushveld.* This bushveld type, which is the one most closely related to grassland, is dominated by evergreen sclerophyllous *Protea* trees and shrubs (Fig. 32). As discussed in the section on Moist Cool-temperate Grassland, floristic affinities of the various *Protea*-dominated Bushveld communities with

Table 8. Andropogon schirensis–Vigna parviflora Association (after Myre 1972).

Number of relevés	10	Number of relevés	10
Differential species			
Vigna parviflora	4	Urginea altissima	2
Hypoxis sp.	3	Eragrostis superba	2
Indigofera inhambanensis	3	Asparagus setaceus	2
Alysicarpus vaginalis	2	Stylochiton natalense	2
Undergrowth species		Setaria sp.	2
Andropogon schirensis	5	Panicum infestum	2
Elyonurus argenteus	5	Hyparrhenia filipendula	2
Fimbristylis hispidula	5	Stylosanthus mucronata	2
Digitaria eriantha	5	Cassia absus	2
Merremia tridentata	4	Cyperus sp.	2
Aristida congesta	4	Murdannia simplex	2
Panicum maximum	4	Setaria sphacelata	2
Cassia mimosoides	4	etc.	
Tephrosia longipes	3	*Woody emergents*	
Alloteropsis cimicina	3	Hyphaene natalensis	5
Pogonarthria squarrosa	3	Cissus cornifolia	4
Agathisanthemum bojeri	3	Dichrostachys cinerea	3
Perotis patens	3	Pseudolachnostylis maprouneifolia	3
Schizachyrium sanguineum	3	Acacia clavigera	3
Pyrenacantha kaurabassana	3	Antidesma venosum	3
Tricliceras longepedunculatum	3	Parinari sp.	2
Eragrostis atrovirens	3	Annona senegalensis	2
Heteropogon contortus	3	Cryptolepis obtusaefolia	2
Hyperthelia dissoluta	3	Garcinia livingstonei	2
Craspedorhachis africana	2	Rothmannia sp.	2
Wahlenbergia sp.	2	Acacia sp.	2
Cyperus haspan	2	Strychnos innocua	2
Abrus precatorius	2	Diospyros usambarensis	2
Rhynchosia totta	2	etc.	

grassland communities, cut across the grassland–bushveld physiognomic distinction. The ecological relationships of *Protea*-dominated Bushveld is best shown in areas where it occurs with grassland, broad-orthophyll and microphyllous bushveld on the same mountainside. Except for shrubby, mixed broad-orthophyll and evergreen sclerophyll bushveld amongst summit outcrops, grassland typically occupies the upper, coolest, most leached zone, followed by *Protea*-dominated Bushveld lower down. Still lower and on warmer, less leached slopes, follows either of the two warmer temperate bushveld types, depending on soil texture. A community that is intermediate between *Protea*-dominated Bushveld and *Eustachys mutica–Acacia caffra* Bushveld was described from a moist summit basin with sandy clay–loam soil (Coetzee 1975). These are two main types of Upland Sub-humid Mountain Bushveld and *Protea caffra* and *Acacia caffra*, which are each very typical of the two respective types, here occur together.

In the warm-temperate bushveld region of the low interior plateau of South Africa, *Protea*-dominated Bushveld is virtually restricted to cool moist south-facing slopes, occurring also on summits that are high enough for *Protea*-dominated Bushveld but not quite high enough for Moist Cool-temperate Grassland. Further south and east on the cool-temperate part of the high interior

Fig. 32. *Protea caffra* bushveld in the foreground and on the horizon and *Acacia caffra* bushveld on the valley sides in southern Transvaal west of Pretoria (photo B. J. Coetzee).

plateau, where Moist Cool-temperate Grassland predominates, *Protea*-dominated Bushveld occurs commonly on south-facing slopes as well as on high altitude north-facing slopes. In the latter instances the cool high altitude effectively counteracts hot aspects. On the high plateau the upland *Protea*-dominated Bushveld slopes are probably less subjected to frost than the surrounding grassland plains (Louw 1951, Edwards 1967, Theron 1973, 1975, Coetzee 1974, 1975, Bredenkamp 1975).

In accordance with this *Protea*-dominated Bushveld being marginal Bushveld and transitional to grassland, the distribution of component communities largely coincides with the distribution of Upland Sub-humid Mountain Bushveld and Moist Cool-temperate Grassland. On the eastern escarpment *Protea*-dominated communities extend above Moist Cool-temperate Grassland into high areas with Afro-alpine species. Different species of *Protea* are usually dominant in *Protea*-dominated Bushveld of different geographic regions. *Protea caffra* is usually dominant in *Protea*-dominated Bushveld of the mountains on the low interior plateau and in outliers of the same bushveld occurring in the northern belt of Moist Cool-temperate Grassland. *Protea welwitschii* shrubland also occurs in these areas and so does *Protea rouppelliae*, though sparsely and largely restricted to very high points. The latter species occurs commonly on the high eastern escarpment, in the eastern belt of Moist Cool-temperate Grassland. *Protea rhodantha* is a very common dominant in *Protea*-dominated Bushveld of the northern part of the eastern escarpment belt. *Protea gaguedi* communities are also largely restricted to the northern part of the eastern belt. Additional *Protea* species for the southern part of the eastern belt include *P. multibracteata, P. subvestita, P. simplex* and *P. dracomontana* (Story 1952, Comins 1962, Edwards 1967, Van der Schijff & Schoonraad 1971, Coetzee 1974, 1975, Bredenkamp 1975, Theron 1975).

In the southern part of the eastern belt, *Protea rouppelliae* Bushveld extends to higher altitudes than *Protea multibracteata* Bushveld. Where the two types occur together, *P. rouppelliae* occurs on the shallower soils where Afro-alpine species are not uncommon. Such species are rare in *Protea multibracteata* Bushveld, where the grass stratum is more typically Moist Cool-temperate Grassland. *Protea simplex* Shrubland occurs associated with *Protea multibracteata* Bushveld and the two types of community seem to be the equivalent to the southern part of the eastern Moist Cool-temperate Grassland belt of *Protea caffra* Bushveld and *Protea welwitschii* Shrubland, which occur in the northern belt and on the mountains of the northern low interior plateau. *Protea subvestita* communities and *Protea dracomontana* Shrubland, like *Protea rouppelliae* Bushveld, have distinct tendencies for the development of Afro-alpine affinities. *Protea* communities of the southern part of the eastern belt, perhaps in other parts as well, may have been declining in their extent as a result of increased human settlement and veld burning (Story 1952, Comins 1962, Edwards 1967, compare Chapters 11 and 12).

Minor variations within *Protea caffra*-dominated Bushveld are like the parallel grassland communities, largely related to presence or absence of outcrop, stoniness, soil texture, soil depth and aspect (Coetzee 1974, 1975, Bredenkamp 1975).

B. *Landolphia capensis–Bequaertiodendron magalismontanum* Shrub Bushveld. Outcrops in the northern and northeastern belts of Moist Cool-temperate Grassland and in Upland Sub-humid Mountain Bushveld in South Africa have a very distinct shrubby bushveld vegetation. Plants grow in cracks, fissures and litholitic soil pockets amongst the rocks, and the broken character of the substrate provides sheltered niches for mesophytic plants (Fig. 35). Trees may occur but shrubs are usually very prominent and many woody species that occur widely as trees are often shrubby or stunted in this vegetation.

Among the most typical and widespread of the characteristic species for these habitats are the evergreen sclerophyllous shrub *Bequaertiodendron magalismontanum* as well as the shrubs *Landolphia capensis* and *Tapiphyllum parvifolium*. *Bequaertiodendron magalismontanum* is typical also of Warm-temperate Orthophyll Bush, found azonally in ravines and along streams, where it occurs as a tall tree. Other such shrubs occurring in *Landolphia capensis– Bequaertiodendron magalismontanum* Shrub Bushveld, particularly in the mesic variations and in azonal Cool- and Warm-temperate Broad-orthophyll Bush, include *Mimusops zeyheri, Rothmannia capensis, Pittosporum viridiflorum, Brachylaena rotundata, Fagara capensis, Myrsine africana, Nuxia congesta, Diospyros lycioides* subsp. *guerkei, Clutia pulchella, Faurea saligna* and *Canthium gilfillanii* thus showing a strong Afromontane element (compare Chapter 11). More xeric variations of this shrubland typically include a variety of woody plants characteristic also of *Diheteropogon amplectens–Burkea africana* Upland Sub-humid Mountain Bushveld, e.g. *Combretum zeyheri, C. molle, Ochna pulchra, Canthium huillense, Osyris lanceolata, Euclea crispa, Vangueria in-*

fausta, Mundulea sericea and *Diplorhynchus condylocarpon*. The shrub *Elephantorrhiza burkei* is also quite typical in parts of this vegetation.

Grasses are typically those characteristic of Moist Cool-temperate Grasslands and Upland Sub-humid Mountain Bushveld, in general including *Diheteropogon amplectens*, *Brachiaria serrata*, *Rhynchelytrum setifolium*, *Trachypogon spicatus*, *Loudetia simplex*, *Andropogon schirensis* var. *angustifolius*, *Schizachyrium sanguineum* and *Themeda triandra*. The community may also include grasses with a more restricted distribution in these grasslands and bushveld communities, e.g. *Tristachya biseriata* or *Setaria lindenbergiana*. Grass composition and cover varies with amount of soil between outcrops, with moisture-determining factors such as hot and cool aspects, and with altitude. The community usually has species in common with rocky grassland communities, including *Selaginella dregei* and others mentioned in the section on Moist Cool-temperate Grasslands (Theron 1973, Coetzee 1974, 1975, Bredenkamp 1975).

A counterpart of this community, also shrubby and rocky and physiognomically very similar, occurs at higher altitudes along the eastern escarpment, particularly in the south. This community is closely related to Afromontane forest and belongs to Afromontane vegetation.

C. *Faurea saligna, Burkea africana* and *Diplorhynchus condylocarpon* Mountain Bushveld. This vegetation is much more restricted to the major Upland Sub-humid Mountain Bushveld zone of the low interior plateau and occurs on warmer temperate slopes than *Protea*-dominated Bushveld or *Landolphia capensis–Bequaertiodendron magalismontanum* Shrub Bushveld. The vegetation is characteristically broad-orthophyll and occurs on relatively coarse-textured soils. Variation in species composition is correlated with a gradient from mesic to xeric and the position of a stand of vegetation on the gradient is determined largely by soil depth, slope and aspect (Acocks 1953, 1975, Theron 1973, Coetzee 1975, Coetzee et al. 1976).

Quite a large number of tree and shrub species may be present in this vegetation in varying proportions. *Combretum molle* and *Dombeya rotundifolia* are among those with the widest ecological amplitude within this broad vegetation type. *Faurea saligna, Burkea africana* and *Diplorhynchus condylocarpon* are good indicators of useful reference points in the maze of variation.

Faurea saligna is typical of the extreme cool-mesic variations, which occur typically on south- and east-facing slopes (Fig. 33). *Heteropyxis natalensis* is also characteristic of such cool, very mesic variations. Additional woody species that may be prominent include the widespread *Combretum molle* and *Dombeya rotundifolia* and less extreme mesic species, notably *Burkea africana* and *Ozoroa paniculosa*, as well as *Mundulea sericea* and *Acacia caffra*. Stands dominated by *Faurea saligna* are marginally warm-temperate bushveld and the habitat is transitional to that of somewhat cooler more leached *Protea caffra*-dominated Bushveld. *Faurea saligna* Bushveld is also related to *Acacia caffra* Bushveld with *Acacia caffra* occurring very prominently in some stands of *Faurea saligna*. Prominent grasses in these most mesic variations include *Setaria perennis*, *Tristachya biseriata*, *Rhynchelytrum setifolium*, *Andropogon schirensis* and others, indicating distinct relationships with Moist Cool-temperate Grassland as well (Theron 1973, Coetzee 1975).

Burkea africana is characteristically prominent in less extreme mesic situations

Fig. 33. *Faurea saligna* Mountain Bushveld on south-facing shale slope near Zeerust, Transvaal (photo M. J. A. Werger).

Fig. 34. *Burkea africana* (left and centre) Bushveld on sandy soil at the foot of a quartzite slope near Zeerust, Transvaal. The broad-leaved tree (right) is *Combretum zeyheri* (photo M. J. A. Werger).

(Fig. 34), together with species such as *Ochna pulchra, Ozoroa paniculosa, Strychnos cocculoides* and *Terminalia sericea* and the usual, widespread *Combretum molle* and *Dombeya rotundifolia*. Such communities occur typically on relatively sandy mountain-associated alluvial plains, on high sandy bushveld plateaux and on slopes of various aspects that are neither too xeric nor too mesic for *Burkea* Bushveld. On shallow soils near summits, in habitats transitional to grassland, *Burkea africana* and *Ochna pulchra* occur as stunted coppicing shrubs, in places forming dense stands almost to the exclusion of grasses. On deep-sandy, cool upland plateaux, *Burkea* trees, with or without *Ochna pulchra* trees and shrubs, are typically the only prominent woody plants. *Burkea africana* Bushveld communities of slopes and alluvial plains are not quite as mesic and are richer in woody species. *Burkea* communities are notorious for the characteristic presence of poisonous plants. These include *Dichapetalum cymosum* and *Pygmaeothamnus zeyheri*, which have underground woody stem systems with only twig tips and leaves protruding (geoxylic), and *Fadogia monticola*. *Dichapetalum cymosum*, in particular, is responsible for considerable cattle losses early in the growing season when these plants are among the first in the field layer with fresh green growth. Similar critical periods may arise with drought spells in summer when shallow rooted plants have wilted but the deep rooted *Dichapetalum cymosum* is still fresh. Prominent grasses in *Burkea* Bushveld include *Setaria perennis*, which is usually overwhelmingly dominant in deep-sandy, cool upland plateau communities, *Tristachya biseriata*, which is a typical dominant on stony slopes, *Trachypogon spicatus, Andropogon schirensis, Loudetia simplex, Diheteropogon amplectens, Brachiaria serrata, Rhynchelytrum setifolium, Elyonurus argenteus, Themeda triandra, Schizachyrium sanguineum*, and others – usually a mixture of several of these species. Examples, from the Magaliesberg, of a high sandy plateau community completely dominated by *Burkea africana* trees, and a floristically richer and less mesic *Burkea africana* community on slopes, are described in detail by Theron (1973) and Coetzee (1974, 1975).

Although conveniently treated in the section on Broad-orthophyll Plains Bushveld, the sub-humid plains bushveld types on deep, poor sandy soils of aeolian origin are virtually inseparable in woody species composition from *Burkea africana* Mountain Bushveld. Even more arid deep-sandy plains bushveld communities have distinct affinities with *Burkea africana* and also *Diplorhynchus condylocarpon* Mountain Bushveld. The plains bushveld communities on deep sand are, however, in their typical expression, also quite distinct from Mountain Bushveld in their field layers.

The xeric group of warm-temperate Mountain Bushveld communities is characterized by the prominence of *Diplorhynchus condylocarpon*. In very mountainous regions of the low interior plateau the xeric group of communities occurs typically, although not exclusively, on steep, hot, north-facing slopes with shallow, stony soils. Outliers on low stony hills of the warm bushveld plains are not aspect-bound. Typical prominent species for the xeric communities, in addition to *Diplorhynchus condylocarpon*, include *Combretum apiculatum, C. zeyheri, Strychnos madagascariensis* and *Pterocarpus rotundifolius*, all of which extend to sub-humid, semi-arid plains bushveld and semi-arid Mountain Bushveld, as well as *Lannea discolor, Pseudolachnostylis maprouneifolia* and *Elephantorrhiza burkei*, which like *Diplorhynchus condylocarpon*, are more restricted to Mountain

Bushveld and related deep-sandy plains bushveld. Hot talus or talus-like slopes represent the extreme xeric end of the gradient of Warm-temperate Mountain Bushveld. In addition to the typically xeric species mentioned, these slopes characteristically include species such as *Croton gratissimus* and *Pouzolzia hypoleuca*, which have leaves with reflecting silver or white ventral surfaces (Van Vuuren 1961), *Commiphora marlothi, C. schimperi, C. neglecta, C. pyracanthoides, Urera tenax, Sarcostemma viminale* and *Euphorbia tirucalli* which are semi-succulent or succulent, and *Boscia albitrunca* and *Ochna holstii*, which are sclerophyll. Many of these woody species that differentiate the xeric variations of Warm-temperate Mountain Bushveld indicate strong affinities with more tropical bushveld types and so do grasses such as *Enneapogon scoparius, E. pretoriensis* and *Rhynchelytrum villosum*, which are prominent in some of the xeric variations. Other prominent grasses in *Diplorhynchus condylocarpon* Bushveld include *Loudetia simplex* and *Rhynchelytrum setifolium*, which are more restricted to Sub-humid Mountain Bushveld and Moist Cool-temperate Grassland, as well as *Themeda triandra* and *Heteropogon contortus* (Van der Schijff 1964, 1971, Theron 1973, Van Wyk 1973, Van Rooyen 1976, Coetzee et al. 1976).

D. *Eustachys mutica–Acacia caffra Mountain Bushveld.* Microphyllous thorny *Eustachys mutica–Acacia caffra* Mountain Bushveld grows typically in areas of water, nutrient and clay accumulation, usually on concave mountain slopes mainly in the Transvaal (Figs. 32 and 35). Soils are sandy clay-loam and clay-loam and/or pH and conductivity are high compared to the surrounding areas with broad-orthophyll bushveld. Van Vuuren (1961), Theron (1973) and Coetzee (1974, 1975) described a number of communities belonging to this broad type. Typical differential species include the trees *Acacia caffra, A. karroo, Ziziphus mucronata* and *Rhus leptodictya*, the grasses *Eustachys mutica, Heteropogon contortus* and *Eragrostis curvula,* and the forbs *Thunbergia artriplicifolia, Clematis oweniae, Sida dregei* and *Teucrium capense. Setaria perennis* which is often dominant, seems to be strongly associated with the presence of *Acacia karroo.*

An outlier belonging to this community type occurs in relatively warm sheltered dolomite valleys in the predominantly Moist Cool-temperate Grassland area. Soil texture in these valleys is sandier than usual for this broad type and the soils are also distinctively base-rich. This outlier, particularly the xeric variation of steep north-facing slopes, is distinguished by the grasses, *Chrysopogon montanus, Enneapogon scoparius* and *Anthephora pubescens* (Coetzee 1974).

The main gradient of variation of this vegetation in the predominantly bushveld area is from cool mesic to dry xeric. Typical species for the hot xeric end of the gradient such as on hot north-, northwest- and west-facing slopes and on low, hot alluvial plains, include the trees *Lannea discolor, Combretum zeyheri* and *C. molle* and the shrubs *Psidia punctata, Pouzolzia hypoleuca* and *Croton gratissimus* subsp. *gratissimus*. Characteristic species for cool mesic variations such as on steep south-facing slopes and a high cool moist plateau, include the trees *Rhus eckloniana* and *R. pyroides*, the shrubs *Artemisia afra* and *Maytenus heterophylla*, the grass *Setaria lindenbergiana*, which is usually dominant in the field layers of extreme mesic variations, and forbs such as *Vernonia natalensis* and *Mohria caffrorum. Faurea saligna* may also be present in mesic variations.

Fig. 35. *Acacia caffra*-dominated Upland Sub-humid Mountain Bushveld in low areas in the valley, where water, nutrients and clay accumulate. *Landolphia capensis–Bequaertiodendron magalismontanum* Shrub Bushveld occurs amongst outcrops in the right foreground and Moist Cool-temperate Grassland occurs on the upper slopes in the background on the right. *Protea caffra*-dominated Bushveld, which is floristically closely related to the grassland, occurs in the upper part of the valley in the background. Southern Transvaal (photo B. J. Coetzee).

An outlier that belongs to the mesic end of this gradient occurs on moist, nutrient rich clay loam soils on diabase intrusions in the predominantly Moist Cool-temperate Grassland area (Van Vuuren 1961, Van Vuuren & Van der Schijff 1970, Coetzee 1974, 1975).

5.4.8 Lowveld (Subtropical) Sub-humid Mountain Bushveld

The low country east of the Transvaal escarp is relatively frost free – a factor to which the Lowveld owes much of its distinctive character. Lowveld Sub-humid Mountain Bushveld, more specifically, occupies the rolling foothills of the escarp and a zone east of the Lebombo Mountain range towards the coast. These are the higher rainfall parts of the Lowveld, with the rainfall approximating or exceeding 700 mm per annum. In addition the granitic landscape and soils are very well drained and consequently the soils are highly leached. The soils of the Transvaal section in the north are predominantly sandy latosols with varying amounts of rock and lithosols, whereas those in the Swaziland region further south are deep red ferralitic or ferrisolic loams and clay-loams and shallower soils, often transitional between lithosols and fersialitic soils. Typical for these conditions, the natural woody vegetation for the northern region is predominantly broad-orthophyll, with a field layer dominated by sour grasses. Judging from apparent remnants and soil texture the Swaziland section would seem to have originally been largely microphyllous thorny bushveld (Acocks 1953, 1975, Compton 1966,

Van der Schijff 1958, 1971, Van der Schijff & Schoonraad 1971, Van Wyk 1971, 1973, Buitendach 1973).

Lowveld Sub-humid Mountain Bushveld has been a most favourable region for human settlement, winter vegetable and tropical crop production and a *Pinus* and *Eucalyptus* forestry industry. In arable areas very little natural vegetation remains. However, there are parts, particularly in the South African sector, with rugged terrain unsuited to any form of landuse but conservation, where natural vegetation endures.

The tropical Lowveld character as well as the mesic character of Lowveld Sub-humid Mountain Bushveld are emphasized by a number of characteristic species typical of sub-humid Lowveld but which elsewhere in the drier Lowveld are restricted to drainage lines. These species, shared by Lowveld Sub-humid Mountain Bushveld and azonal riparian Tropical and Subtropical Broad-orthophyll Bush, include *Syzygium guineense*, *S. cordatum*, *Trichilia emetica*, *Albizia versicolor*, *Ficus sycomorus*, *Kigelia africana*, *Diospyros mespiliformis*, *Rauvolfia caffra*, *Bridelia micrantha* and *Schotia brachypetala* (Buitendach 1973, Van der Schijff 1958, Van Wyk 1973, compare also Chapter 13).

Other woody species that emphasize the leached Sub-humid Mountain character extend further into the less tropical and more temperate zone, spanning the Lowveld versus Upland discontinuity. These species are common to Lowveld and Upland Sub-humid Mountain Bushveld and Lowveld deep poor sand, i.e. 'Sandveld', communities. Several of these are typical of mesic mountain slopes, e.g., *Heteropyxis natalensis*, *Faurea saligna*, *F. speciosa*, *Fagara capensis* and *Acacia caffra* (Van der Schijff & Schoonraad 1971, Van Wyk 1973). There are, however, others such as *Lannea discolor*, *Dombeya rotundifolia* and *Mundulea sericea*, which are also typical of leached, mountainous, but somewhat less emphatically mesic sites. Quite a number of species shared by Upland and Lowveld Sub-humid Mountain Bushveld, have wide distributions and contribute strongly also to the character of plains bushveld communities. These include *Sclerocarya caffra*, *Terminalia sericea*, *Combretum zeyheri*, *C. apiculatum*, *Pterocarpus rotundifolius*, *Strychnos madagascariensis* and *Dichrostachys cinerea*. Prominent species in the grass stratum also include those typical of leached mountain soils in general, occurring in Lowveld and Upland Mountain Bushveld. Examples are *Elyonurus argenteus*, *Diheteropogon amplectens*, *Loudetia simplex*, *Schizachyrium sanguineum* and *Tristachya hispida* which also occur commonly in Moist Cool-temperate Grassland. Another sour grass, *Hyperthelia dissoluta*, is tall and robust and because of this is particularly prominent and characteristic. More palatable grasses include *Digitaria* spp., *Setaria flabellata* and *Heteropogon contortus* (Van der Schijff 1958, Van Wyk 1971, 1973).

Sandveld communities in general, including the Lowveld variations, have exclusive affinities with Upland Sub-humid Mountain Bushveld (in species such as *Burkea africana* and *Diplorhynchus condylocarpon*). The Lowveld Sandveld, more particularly, has exclusive affinities with Lowveld Sub-humid Mountain Bushveld, further emphasizing that tropical Lowveld conditions are basic to the nature of this vegetation. Such affinities include *Parinari curatellifolia* subsp. *mobola*, *Albizia versicolor*, *Piliostigma thonningii*, *Pterocarpus angolensis*, *Bauhinia galpinii* and *Acacia davyii*. Other typically Lowveld species, characteristic of Lowveld Sub-humid Mountain Bushveld include *Albizia adianthifolia, Acacia sieberana* var. *woodii, Annona chrysophylla, A. senegalensis, Combretum collinum, Sterculia murex, Pterolobium exosum, Cussonia natalensis, Ficus sondersi , F. smutsii*, and, particularly on moist slopes *Acacia ataxacantha, Antidesma venosum, Trema orientalis, Catha edulis* and *Greya sutherlandii* (Van der Schijff 1958, 1971, Van der Schijff & Schoonraad 1971, Van Wyk 1971, 1973).

Lowveld Sub-humid Mountain Bushveld is mapped by Acocks (1953, 1975) as Lowveld Sour Bushveld.

5.4.9 Dry Mountain Bushveld

Dry Mountain Bushveld is mainly found on mountains and hills in the semi-arid to arid parts of the Limpopo River Lowveld and Lowveld east of the escarpment (Fig. 38), of southeastern Rhodesia, as well as in the Grootfontein area of South West Africa (Fig. 36). This vegetation, with the exception of variations amongst

Fig. 36. In the mountainous area around Otavi, central-northern South West Africa, *Kirkia acuminata* with its characteristic rounded crowns is a prominent tree of the bushveld together with *Commiphora* species, such as *C. mollis, C. crenato-serrata, C. glaucescens, Gyrocarpus americana, Croton* spp., *Terminalia prunioides*, etc. *Cissus nymphaeifolia* is visible in the right hand foreground (photo W. Giess).

rugged, bouldery outcrops, does not differ much from the bushveld of the surrounding plains (Codd 1951, Acocks 1953, 1975, Louw 1970, Giess 1971, Van der Schijff & Schoonraad 1971, Van Wyk 1973, Coetzee & Nel 1977, Gertenbach 1977).

Mountains and hills in sub-humid areas with rainfall exceeding approximately 700 mm per annum have strongly leached soils that carry the distinct floras of Upland and Lowveld Sub-humid Mountain Bushveld and species typical of these vegetation types occur on the dry plains on very poor, sandy soils. Plains Bushveld species form part of the characteristic combination of the warm and dry variations of Sub-humid Mountain Bushveld, such as Lowveld Sub-humid Mountain Bushveld or the xeric group of warm-temperate Upland Sub-humid Mountain Bushveld communities characterized by the presence of *Diplorhynchus condylocarpon*. The tendency for plains species to occur on and dominate hill slopes culminates in Dry Mountain Bushveld of semi-arid to arid areas. Here the soils are less leached and more drought-prone than in Sub-humid Mountain Bushveld and factors such as hot aspects and nutrient rich geological substrate further favour the occurrence of Dry Mountain Bushveld.

Thus Dry Mountain Bushveld on the whole is not an entirely separate floristic entity but the component communities are strongly interrelated with Broad-orthophyll and Microphyllous Thorny Plains Bushveld of sub-humid plains and with Arid Bushveld of semi-arid to arid plains. *Kirkia acuminata*, a tropical species, in the southern part of its range typical of stony and broken rocky habitats in arid and semi-arid areas, is a quite characteristic species of Dry Mountain Bushveld. Here its distribution is typically the drier parts of the rhyolitic Lebombo Mountain Range and Soutpansberg in the eastern and northeastern Lowveld; the xeric slopes of the Soutpansberg and Blaauwberg Mountains in the northern Lowveld; the broken sandstone hills in the hot dry area near the Limpopo River at Pontdrif and the broken hilly area of the Messina geological formation in the hot and dry Messina environment, both areas in the northern Lowveld; hot slopes of outliers of the Waterberg Mountain complex in the western and northwestern Lowveld; and scattered granitic and other hills all over the semi-arid and arid parts of the eastern, northern and western Lowveld (Louw 1970, Van der Schijff 1971, Van der Schijff & Schoonraad 1971, Van Wyk 1973). Louw (1970) shows *Kirkia acuminata* in the northern Lowveld largely restricted to upland areas and with optimum constancy in broken stony and rocky terrain.

Other tropical species that are characteristic of arid and semi-arid areas and particularly of Dry Mountain Bushveld, include *Adansonia digitata*, *Acacia erubescens*, *A. senegal* var. *leiorachis*, *Commiphora tenuipetiolata*, *Bridelia mollis* and *Brachylaena huillensis*. A number of species are typical of mountains and differential of Arid Mountain Bushveld but are less markedly dry-tropical in their distributions, also occurring in hot xeric phases of warm-temperate Sub-humid Mountain Bushveld, e.g., *Urera tenax*, *Pouzolzia hypoleuca*, *Euphorbia cooperi* and *Elephantorrhiza burkei*. *Ficus soldanella*, which also occurs in Sub-humid Mountain Bushveld, has a wide distribution in very rocky habitats, including Dry Mountain Bushveld. *Euphorbia confinalis* is confined to Dry Mountain Bushveld of the eastern Lowveld. *Ficus smutsii*, *Hymenodictyon parvifolium*, *Galpinia transvaalica*, *Albizia brevifolia*, *Sterculia rogersii*, *Ptaeroxylon obliquum* and *Strychnos decussata* are typical far-southern African tropical species occurring in Dry Mountain Bushveld and southwards in the low country between the eastern escarpment and Indian Ocean (Fig. 37). *F. smutsii* and *H. parvifolium* have not been recorded farther south than the northeastern Transvaal Dry Mountain Bushveld, *G. transvaalica* and *A. brevifolia* are known to occur southwards to Swaziland, *Sterculia rogersii* to northern Zululand, *Ptaeroxylon obliquum* to the eastern Cape Province and *Strychnos decussata* to the eastern and central Cape Province. *Euphorbia tirucalli* and *Afzelia quanzensis*, which have somewhat similar distributions in southern Africa, occurring in Dry Mountain Bushveld and stretching southwards to the eastern Cape, have much wider tropical distributions further north.

Fig. 37. Dry Mountain Bushveld in Natal, Mvoti River, with *Euphorbia ingens*, *E. tirucalli* and *Acacia tortilis* (photo E. J. Moll).

A number of tropical species in southern Africa are restricted to Dry Mountain Bushveld of the Limpopo River Lowveld, particularly the northern slopes of the Soutpansberg Mountain range. Notable among these are *Commiphora marlothii* and *C. edulis*. Another, *Entandophragma caudatum*, occurs in Dry Mountain Bushveld of the Lebombo Mountain range in the eastern Lowveld, and *Gyrocarpus americanus* also in this type of bushveld around Grootfontein, South West Africa (Giess 1971, Van der Schijff 1971, Van Wyk 1973).

The various differential species mentioned usually contribute little to the bulk of the vegetation in Dry Mountain Bushveld. This is made up by species that are typically dominant also on sub-humid to arid plains, notably *Pterocarpus rotundifolius*, *Combretum apiculatum*, which is characteristic of very shallow coarse-textured soils, and *Acacia nigrescens*, which is dominant on slopes with deeper, fine-textured soils. The more arid variations of Dry Mountain Bushveld characteristically also include considerable quantities of typical Arid Bushveld species, such as *Colophospermum mopane*, *Terminalia prunioides*, *Ximenia americana*, *Boscia albitrunca*, *Commiphora glandulosa* and *Acacia tortilis* (Louw 1970, Van der Schijff 1971, Van Wyk 1973, Coetzee & Nel 1977, Gertenbach 1977).

5.4.10 Arid Mountain Bushveld (Broad-sclerophyll *Androstachys johnsonii* Bushveld)

Unique, simple-structured woodland and shrubland dominated by *Androstachys johnsonii* occurs in scattered patches in dry tropical areas. The main areas of distribution include: (1) the Dry Mountain Bushveld region on the Lebombo Mountains in the eastern Transvaal, Moçambique and Swaziland; (2) alternating with

Julbernardia savanna in southeastern Moçambique in the semi-arid to arid area bordering on the sub-humid coastal zone; (3) the Arid Bushveld of the Limpopo and Save River basins of southern Rhodesia (Nuanedzi District) and the far northeastern corner of the Transvaal and Moçambique; and (4) the narrow coastal zone of Dry Deciduous Thicket (*Guibourtia schliebenii, Pseudoprosopis*) in northern Moçambique (Pedro & Barbosa 1955, Wild 1955, Rattray 1961, Wild & Barbosa 1967, Farrell 1968a, Van Wyk 1973, Van Rooyen 1976, Coetzee & Nel 1977, Gertenbach 1977).

In the Kruger National Park, *Androstachys* woodland occurs patchily on the apparently drier northern section of the Lebombo Mountain Range and on the dry northeastern section of the Soutpansberg Mountains. It is in these regions that Arid Bushveld occurs on the plains also. In accordance with its presumably xerophytic distribution, *Androstachys johnsonii* has several xeromorphic features, e.g. coriaceous leaves that are glossy above and white villose underneath (cf. Van Vuuren 1961, Van Wyk 1973), with sides that curl over and under during drought periods, white villose petioles and young twigs, and flowers that seldom appear before early summer rains. Should the rains be very late, flowers may not appear at all that season. Aromatic substances are present and may also be related to drought-tolerance (Van Wyk 1973).

Within the dry areas mentioned, *Androstachys* woodland is sharply delimited from adjoining Mountain Bushveld by very narrow boundaries. Examples of such narrow boundaries are those between *Androstachys* woodland and *Combretum apiculatum*-dominated Arid Mountain Bushveld. Such boundaries may occur without very prominent associated physiographic boundaries. The marked exclusion of individuals of other woody species from *Androstachys* woodland, with supreme dominance by *A. johnsonii*, suggests strong competitive exclusion of other vegetation types from potentially suitable habitats, along narrow boundaries (Figs. 38 and 39). Soil differences, e.g. differences that affect moisture regime, cannot be ruled out, pending the results of current investigations (Coetzee & Nel 1977).

In a well-developed, extensive stand of *Androstachys* woodland on the Lebombo Mountains, Coetzee & Nel (1977) distinguished four variations:

(a) *Woodland proper*: Dense stands – approximately 1400 individuals per hectare – of 4–8 m tall trees occur on steep slopes with relatively deep talus soils. Dominant composition varies from almost pure *Androstachys johnsonii* to a mixture of 6 m tall *A. johnsonii* (500 individuals/hectare), 4–5 m tall *A. johnsonii* (500 ind./ha.) and 6 m tall *Euphorbia confinalis* (400 ind./ha.). Canopies start from 2 m upwards and interlock. The open level under 2 m is shady and contains very few woody plants. Trees and shrubs other than those mentioned are sparsely scattered and include *Hymenodictyon parvifolium, Strychnos decussata, Entandophragma caudatum, Manilkara mochisia, Boscia albitrunca* and *Pouzolzia hypoleuca*. Lianas are common, e.g. *Combretum mossambicense*, as a scandent shrub, *Cissus rotundifolius, Dioscorea sylvatica* and *Asparagus falcatus*. Epiphytic beard-like lichens of an *Usnea* sp. are common, hanging from the branches of *A. johnsonii*. Lichens also largely cover stones and the occasional smooth-barked tree trunks such as *Strychnos decussata* and *Hymenodictyon parvifolium*. It has been noticed that dense fog is quite common in these parts of the Lebombo Range (J. Steyn, pers. comm.) and this may explain the abundant occurrence of lichens. The grasses *Panicum maximum* and *P. deustum* are dominant

Fig. 38. Dry Mountain Bushveld in a valley through the Lebombo Mountains, with dark patches of *Androstachys johnsonii* Arid Mountain Bushveld on slopes, Kruger National Park (photo P. van Wyk).

Fig. 39. Dense *Androstachys johnsonii* vegetation, belonging to Arid Mountain Bushveld, on the Lebombo Mountain Range, Transvaal (photo P. van Wyk).

in the lower stratum, with an admixture of non-graminous plants including *Sansevieria deserti*, *S. grandis*, the small fern *Actinopteris australis* and the euphorbiaceous succulent *Monadenium lugardae*.

(b) *Boulder-outcrop variation*: An example of the variation amongst boulder-outcrops is provided from a 30 m × 30 m plot with *Androstachys johnsonii* dominant, *Euphorbia confinalis* sub-dominant and also including *Albizia brevifolia*, *Ficus soldanella*, *Hymenodictyon parvifolium*, *Boscia albitrunca* and *Holmskioldia tettensis*. *Panicum maximum* is the dominant grass.

(c) *Mesophytic shrubland*: On summit areas with shallow soils but no sheet or other outcrop, 3–5 m tall *Androstachys johnsonii* shrubs form virtually pure stands. Density in three stands described varied from 1000–1700 shrubs/ha, with mainly 4–5 m tall shrubs (800–1400 ind./ha) and additional 3 m tall shrubs absent or up to 300 ind./ha. The dominant grass is *Pseudobrachiaria deflexa* with *Panicum maximum* increasing where the shrubland grades into woodland near steeper slopes. Lichens are common on stones.

(d) *Xerophytic shrubland*: Stunted 2–3 m tall, somewhat more open, xerophytic shrubland occurs on summit areas with sheet outcrop. *Androstachys johnsonii* is virtually the only shrub. Shrub density is somewhat lower than in mesophytic shrubland, namely 650 individuals/ha. Of these, 525 ind./ha were 2 m tall and only 125 ind./ha were 3 m tall. The fern *Selaginella dregei* and the sedge *Cyperus rupestris* which are both desiccation-tolerant, are common dominants in the field layer, with the grass *Pseudobrachiaria deflexa* increasing as the vegetation grades into the mesophytic shrubland of less rocky soils.

5.4.11 Broad-orthophyll Plains Bushveld

The family Combretaceae features very prominently in Broad-orthophyll Plains Bushveld, which occurs on coarse-textured soils of warm-temperate to tropical plains with 500–700 mm annual rainfall and in somewhat lower rainfall areas with deep sandy soils. *Combretum apiculatum*, *C. zeyheri* and *Terminalia sericea*, which have overlapping ecological ranges but quite different optima, are among the most prominent and distinctive woody species in this bushveld. In the exceptional communities where none of these combretaceous species are actually dominant, co-dominant or sub-dominant, these and other members of the family Combretaceae are nevertheless quite common.

The main areas of distribution of Broad-orthophyll Plains Bushveld are as follows:

(a) On sand ridges in the Botswana part of an extensive area of Quaternary and Tertiary sand deposits, such vegetation finds its main northwestern distribution. Here it forms one major component of a mosaic mapped as Northern Kalahari by Weare & Yalala (1971). The other component, which is an arid type of microphyllous vegetation, occurs on the lower areas and plains between the ridges (cf. Cole & Brown 1976). The climate is semi-arid warmer temperate. Wild & Barbosa (1967) include this Botswana mosaic in *Terminalia sericea* Deciduous Tree Savanna (Medium and Low Altitude), mapped as such also for Matabeleland in Rhodesia and the Rhodesian and Moçambique Malvernia area bordering on the northeastern corner of the Transvaal. Far northwestern outliers are recorded, for South West Africa on the deep sand of the Omeverume, Klein Waterberg and Omboroko Plateaus by Rutherford (1972, 1975) and for southern Zambia on sandy soils and fringing sandy dambos by White (1962, see Wild & Barbosa 1967). Coetzee & Werger (1975) noted from the Hotazel area in the northern Cape Province near the Botswana border that on sand dune crests *Terminalia sericea* replaces the drier Kalahari-type vegetation at approximately the 300 mm annual rainfall isohyet.

(b) The northern expanse in Matabeleland, Rhodesia, occurs on granite or gneissic sands at 1200–1500 m altitude with 500–600 mm annual rainfall. The climate is semi-arid warm-temperate. This expanse is separated from the Botswana and South African distribution areas by Arid Bushveld of the low dry Limpopo–Shashi drainage system (Lang 1952, Wild & Barbosa 1967).

(c) A northwestern South African zone, fringing the western and northern slopes of the Waterberg Mountains, borders on the Arid Plains Bushveld of the low dry Limpopo River valley, across the valley from the Botswana expanse. The soils are neutral red-sandy loams and are underlain by sediments of the Karoo and Waterberg Systems (Acocks 1953, 1975, Dept. Mines 1970, MacVicar 1973).

(d) Broad-orthophyll Plains Bushveld occurs extensively on sandy soils of the Bushveld Basin plains, i.e. on the low sub-humid warm-temperate interior South African plateau. These plains are encircled by mountains, namely the Magaliesberg and Pilansberg in the southwest, the Waterberg and Strydpoort Mountains in the west and north, the Sekhukhune Mountains in the east, and the sharp rise to the high interior plateau in the south. The low relief in the northwestern part of the basin, known as the Springbok Flats, is underlain by sediments and basic igneous basalt of the Karoo System. Here the Broad-orthophyll Bushveld occurs only on sandy sediments and on aeolian sands, which fringe and separate two large expanses of Microphyllous Thorny Bushveld occurring on the red and black montmorillonite clay of the basalt. The southern and eastern Bushveld Basin is largely undulating granite country. Here fine-textured limy soils with Microphyllous Thorny Plains Bushveld occur in bottomland sites and Broad-orthophyll Plains Bushveld occurs on weakly developed sandy and gravelly upland sites without lime (Galpin 1926, Acocks 1953, 1975, Grunow 1965, Dept. Mines 1970, Van der Schijff 1971, Coetzee et al. 1976).

(e) Another southeastern zone of Broad-orthophyll Plains Bushveld occurs in the low country east of the continental escarpment. Here the vegetation type also occurs on granite, restricted to the moderately low rainfall (500–700 mm) parts of the undulating granite landscape, such as between the Crocodile River and the Timbavati tributary of the Olifants River in the Kruger National Park. The climate is semi-arid sub-tropical to tropical. As on the granite of the Bushveld Basin on the interior plateau, Broad-orthophyll Plains Bushveld is confined to the sandy upland sites. Microphyllous Thorny Bushveld occurs on the fine-textured calcareous soils of bottomland sites. Local areas of deep sandy soils on sandstone in this region also carry Broad-orthophyll Plains Bushveld. The adjoining granite country towards the escarpment foothills, where the rainfall is more than 700 mm, belongs to Lowveld Sub-humid Mountain Bushveld, and the areas with weakly developed soils on granite and with a rainfall lower than 500 mm, north of the Timbavati River belong to Arid Bushveld types (Van der Schijff 1958, Van Wyk 1973, Hirst 1975, Coetzee & Nel 1977, Gertenbach 1977).

(f) Islands of Broad-orthophyll Plains Bushveld occur on deep sandy soils, aeolian or on sandstone, within the predominantly Arid Bushveld areas of the low dry country around and between the Limpopo and Save Rivers, where the climate is arid tropical. Examples are provided by Van der Schijff (1958, 1964, 1971) and Van Wyk (1973) from the northern and northeastern Transvaal and by Wild & Barbosa (1967) from the Malvernia area in Rhodesia and Moçambique. Northeastern outliers also occur around dambos.

Principally then, Broad-orthophyll Plains Bushveld occurs in a broad warm-temperate semi-arid, sandy belt across the continent, bordering on moister, more tropical woodland areas in the north, and drier as well as moister, cooler-temperate areas in the south. Discontinuities between the main distribution areas are at the arid tropical valley of the Limpopo River and its northwestern tributary Shashi River system.

The belt of Broad-orthophyll Plains Bushveld may be roughly divided into a northwestern, Botswana, area of *Terminalia sericea*-dominated vegetation on Kalahari sand and a southeastern, South African, zone of *Combretum*-dominated vegetation on upper granite undulations. This geographic and geological division is not absolute and *Terminalia sericea* is locally dominant in special habitats of the southeast. Such *T. sericea* habitats include remnants of Kalahari sand not stripped off by erosion, deep and poor sand underlain by sediments of the Karoo System, very leached sandy soils fringing dambos and ecologically similar grassland zones, and very leached sandy soils of high rainfall granite areas. Thus the main variation within Broad-orthophyll Plains Bushveld is related to a gradient from deep, poor-sandy soils to shallower soils that are somewhat richer in minerals. *Terminalia sericea* occupies the deep poor, sandy end of the gradient with the poor status of the soil due to geological origin (Kalahari sand, Karoo sandstone) or extreme leaching, e.g. resulting from local drainage conditions (fringing dambos) or high rainfall (Lowveld Sub-humid Mountain Bushveld). The other end of the gradient, i.e. shallow sandy soils on granite in moderate to low rainfall areas is typified by *Combretum apiculatum*. Intermediate situations, e.g. deep sand derived in situ from granite in moderate rainfall areas, are commonly typified by the dominance of *Combretum zeyheri*, which nevertheless has a wide tolerance on the gradient and occurs commonly with *Terminalia sericea* as well as *Combretum apiculatum*.

Various floristic relationships with Sub-humid and Arid Mountain Bushveld and Arid Plains Bushveld communities suggest that the main gradient of variation within Broad-orthophyll Plains Bushveld is also one of soil moisture. *Terminalia sericea*-type communities on deep sand are related to Sub-humid Mountain Bushveld whereas the *Combretum*-type communities, particularly the typical *C. apiculatum*-type, are more strongly related to and grade into Arid Mountain Bushveld, Arid Plains communities and the drier variations of mopane veld. The mesic affinities of the *Terminalia sericea* communities may be ascribed to the favourable moisture properties of sand, particularly deep below the surface. Run-off is low and infiltration good and water moves rapidly downward beyond the reach of shallow-rooted plants, also beyond the upper zone where it would otherwise be rapidly lost through evaporation at the surface. Quick root penetration to moist, deep levels is favoured by low mechanical resistance and good aeration. Accordingly the mesic relationships with Sub-humid Mountain Bushveld are in the occurrence of deep-rooted, woody species while the grass and forb layer may be quite unrelated to Sub-humid Mountain Bushveld (Grunow 1965).

(a) *Terminalia Sandveld.*
Terminalia sericea communities of temperate areas such as the Botswana dunes and

sandy plains, the South West African outliers, or deep-sandy areas in the Bushveld Basin of the Transvaal, are strongly related to moderately mesic variations of Upland Sub-humid Mountain Bushveld. This relationship includes the common dominance, co-dominance or sub-dominance of *Burkea africana* and *Ochna pulchra*. More tropical *Terminalia sericea* communities (Fig. 40), such as those of Moçambique, northeastern Transvaal and south and southeast Rhodesia, are also related characteristically to xeric phases on Upland Sub-humid Mountain Bushveld with species such as *Diplorhynchus condylocarpon* and *Pseudolachnostylis maprouneifolia* in addition to *Burkea africana*. Lowveld Sub-humid Mountain Bushveld species that occur in *Terminalia sericea* communities of the plains, particularly the relatively moist tropical variations such as in the northeastern Transvaal,

Fig. 40. *Terminalia sericea* bushveld near Wankie, southwestern Rhodesia (photo B. H. Walker).

include (Van Wyk 1973 and pers. comm.) *Pterocarpus angolensis, Piliostigma thonningii, Antidesma venosum, Annona senegalensis, Schrebera argyrotricha, Parinari curatellifolia* subsp. *mobola, Combretum collinum* subsp. *gazense, Erythrina lysistemon* and *Pavetta schumanniana*. Woody species shared by Upland and Lowveld Sub-humid Mountain Bushveld, and occurring typically also in *Terminalia sericea* communities of the plains include *Combretum molle, Dombeya rotundifolia* and *Lannea discolor*. Others such as *Conbretum zeyheri, Sclerocarya caffra* and *Strychnos madagascariensis* are typical of Lowveld Sub-humid Mountain Bushveld, warm dry variations of Upland Sub-humid Mountain Bushveld and Broadorthophyll Plains Bushveld in general, including *Terminalia sericea* communities on deep sand, particularly in semi-arid and tropical regions.

These species, mentioned in discussing the relationships with Sub-humid Mountain Bushveld, include a large proportion of the more widespread woody species of *Terminalia sericea* plains communities. The grass stratum, on the other hand, is not substantially related to Sub-humid Mountain Bushveld. *Eragrostis pallens*, the most exclusive yet quite constant species of *Terminalia* communities is dominant in arid or heavily grazed stands. Dominant species in less disturbed stands, particularly in relatively mesic areas, include *Digitaria* spp. and *Brachiaria nigropedata*, which have somewhat wider distributions in Broad-orthophyll Plains Bushveld, and occur also typically in *Combretum zeyheri* communities on deep sand. *Schmidtia pappophoroides* is a prominent grass, especially in relatively dry sandy areas. *Panicum maximum*, a very widespread warm-temperate to tropical bushveld species, is common and is usually confined to under trees and shrubs. Other prominent grasses include *Perotis patens* and *Pogonarthria squarrosa*, which are widespread on the coarse-textured soils in Broad-orthophyll Plains Bushveld.

(i) Temperate, Sub-humid. Examples of temperate sub-humid *Terminalia* Sandveld are provided by Coetzee et al. (1976) from sandy soils of the Hutton Form in the Nylsvley Nature Reserve on the Springbok Flats of the Bushveld Basin (Fig. 41), and by Rutherford (1972, 1975) from the isolated plateau outliers in South West Africa (Fig. 42). The dominant trees in *Terminalia* Sandveld on Nylsvley are *Burkea africana*, *Terminalia sericea* and *Combretum molle*, and the dominant shrubs are *Ochna pulchra* and *Grewia flavescens*, the latter species occurring as thickets on typical low mounds. On the central summit plateau at Omeverume in South West Africa the situation is the same except that *Combretum psidioides* subsp. *dinteri* replaces *C. molle*, and *C. collinum* instead of *Grewia flavescens* forms thickets on low mounds (Fig. 42b). The dominant grasses at Nylsvley are *Eragrostis pallens* and a *Digitaria* sp. *Brachiaria nigropedata* is constantly present but in small quantities. At Omeverume, which is naturally protected from cattle grazing by an encircling cliff barrier, the dominant grasses are *Digitaria polevansii*, *Brachiaria nigropedata* and others, whereas on an adjoining well-grazed plateau in South West Africa *Eragrostis pallens* is dominant as at Nylsvley, strongly suggesting that a high cover of *E. pallens* is secondary. Other woody species common to Nylsvley and Omeverume are *Dombeya rotundifolia*, *Ozoroa paniculosa* and *Securidaca longepedunculata*. Very similar communities occur on sandy temperate plains elsewhere on the Springbok Flats as well as at Nylstroom on the southeastern edge of the Waterberg and in the Waterberg area on the sandy Vaalwater enclaves of the northwestern Transvaal plains. Acocks (1975) calls these *Terminalia* Veld Proper and records similar communities for the areas underlain by sandy Karoo sediments on the northwestern Transvaal plains immediately west of the Waterberg and bordering on Arid Bushveld of the low Limpopo River region. These northwestern variations notably have much *Sclerocarya caffra* and are called *Sclerocarya-Burkea* Veld and *Burkea* Veld, which together with *Terminalia* Veld Proper belong to Mixed *Terminalia–Dichapetalum* Veld. The latter is one of the two major subdivisions of Acock's (1975) Mixed Bushveld type, which in turn corresponds with Broad-orthophyll Plains Bushveld. *Terminalia* Veld Proper was recorded as the Sandveld Association by Galpin (1926).

These sub-humid temperate examples are floristically the most closely related *Terminalia* plains communities to Upland Sub-humid Mountain Bushveld but are still well separated from the latter, particularly in the field layer. At Nylsvley, on sandy soils of the Clovelly Form in low mesic areas, variations with field layers

Fig. 41a. Summer aspect of temperate, sub-humid *Terminalia* Sandveld, belonging to Broad-orthophyll Plains Bushveld, in the Nylsvley Nature Reserve near Naboomspruit, Transvaal (photo B. J. Coetzee).

Fig. 41b. Dry season aspect of *Terminalia sericea* bushveld at Nylsvley, Transvaal (photo M. J. A. Werger).

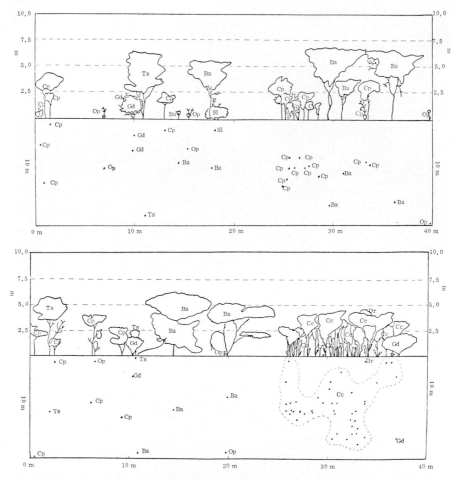

Fig. 42. Vegetation profiles of undisturbed *Terminalia sericea* sandveld community at Omeverume, South West Africa (from Rutherford 1975). Ba: *Burkea africana*; Cc: *Combretum collinum*; Cp: *C. psidioides*; Dr: *Dombeya rotundifolia*; Gd: *Grewia deserticola*; Op: *Ochna pulchra*; Sl: *Securidaca longepedunculata*; Ts: *Terminalia sericea*. Profile b shows a grouping of *Combretum collinum* on a low mound.

grading into that of Upland Sub-humid Mountain Bushveld, occur. Dominant grasses in these mesic variations include *Setaria perennis* and *Elyonurus argenteus* on the distrophic medium sand Mosdale Series and *Trachypogon spicatus* on better drained and better aerated, mesotrophic coarse sand of the Sibakwe Series. Further affinities with mesic mountain bushveld in such habitats include the presence of scattered *Faurea saligna* trees on the Mosdale Series. Sandy communities without *Terminalia sericea* and dominated by *Burkea africana* and *Faurea saligna*, and usually also including *Ochna pulchra* or *Protea caffra*, occur in the Bushveld Basin, e.g. on the upland area between Cullinan and Denilton and at the northern foot of the Magaliesberg near Rustenburg. Such communities are outliers of Upland Sub-humid Mountain Bushveld transitional to *Terminalia*-type plains communities, and examples of similar *Burkea–Faurea* Bushveld occur also in Upland Sub-humid Mountains Bushveld of the Waterberg Mountains.

Also from the Nylsvley Nature Reserve is the first of three examples discussed under Broad-orthophyll Plains Bushveld that are relevant to the problem of bush encroachment by *Dichrostachys cinerea*. These are examples of how *D. cinerea* subsp. *africana* seems to find its optimum habitat on soils that are: (1) deep and sandy like those of *Terminalia* Sandveld, but (2) more trophic than soils of typical *Terminalia* Sandveld, tending in this respect towards soils of typical Microphyllous Thorny Plains Bushveld of fine-textured soils. Although *D. cinerea* subsp. *africana* seems to have a remarkably wide tolerance, it is on this sandy part of the transitional nutrient gradient between Broad-orthophyll Plains Bushveld and Microphyllous Thorny Plains Bushveld that the species typically seems to attain its greatest height and alarming abundance as a shrub. *Dicrostachys cinerea* subsp. *africana* is microphyllous and thorny and of the Leguminosae, differing superficially from the typical *Acacia* spp. of Microphyllous Thorny Plains Bushveld only in that the thorns of *D. cinerea* are slightly modified branchlets and not stipules. The Nylsvley example is not particularly dramatic but *D. cinerea* subsp. *africana* is characteristically the dominant shrub in variations of *Terminalia* Sandveld on eutrophic deep-sandy soils situated at the bottom of sandy slopes, where nutrients moving to lowlands begin to accumulate. Such variations are transitional to Microphyllous Thorny Bushveld and include typical species of the latter, namely *Acacia karroo, A. nilotica, A. tortilis, Ehretia rigida* and *Euclea undulata*. Also present is *Acacia caffra*, characteristic of fine-textured soils on mountain slopes and indicative therefore of conditions intermediate between leached mountain soils and nutrient rich lowlands on plains.

(ii) Temperate, Semi-arid. Temperate semi-arid *Terminalia* Sandveld comprises the vast dunes and sandy plains of Botswana and parts of South West Africa and the sandy areas of the Limpopo River Lowveld in the northwestern and northern Transvaal. Typical general *Terminalia* Sandveld species such as *Terminalia sericea, Burkea africana, Peltophorum africanum* and *Combretum zeyheri* are among the most prominent woody species. *Grewia flava*, which is typically abundant in arid sandy bushveld regions including Spiny Arid Bushveld, is a very common shrub. Other common woody species include *Bauhinia macrantha, Boscia albitrunca, Croton* spp., *Acacia tortilis, A. erioloba*, and some other *Acacia* species which are also indicative of arid affinities. *Eragrostis pallens*, which is typical of *Terminalia* Sandveld, increasing in abundance with aridity and retrogressive succession, is a very prominent grass. Other grasses listed for these areas include, e.g. *Stipagrostis uniplumis, Anthephora pubescens, Perotis patens, Schmidtia pappophoroides, Aristida meridionalis* and other *Aristida* spp. and Acocks (1975, quoting Irvine 1941) suggests that *Digitaria* and *Panicum* spp. might formerly have been more abundant (Passarge 1904, Seiner 1912, Acocks 1953, 1975, Rattray 1960, Tinley 1966, Wild & Barbosa 1967, Van der Schijff 1971, Giess 1971, Weare & Yalala 1971, Leser 1972, Cole & Brown 1976).

(iii) Subtropical, Moderately dry. Included here are local *Terminalia sericea* zones on sandstone, aeolian sand or fringing dambo-related vegetation on granite, in the southern part of the eastern Transvaal Lowveld. In somewhat moister areas such as those towards the escarpment in the eastern Transvaal (Fig. 43) and in the higher rainfall parts of Rhodesia, this vegetation grades into Lowveld Sub-humid Mountain Bushveld (Henkel 1931, Gilliland 1938, Lang 1952, Goldthorpe 1957,

Fig. 43. Undulating granite landscape with moderately high rainfall (700–800 mm), very leached sandy soils and abundant *Terminalia sericea* trees (greyish), in the eastern Transvaal Lowveld near Pretoriuskop, Kruger National Park. The vegetation belongs to Broad-orthophyll Plains Bushveld, grading into Lowveld Sub-humid Mountain Bushveld (photo P. van Wyk).

Rattray & Wild 1955, 1961, Boughey 1961, Mitchell 1961b, Wild & Barbosa 1967, Farrell 1968a, b, Jacobson 1973).

In moderately dissected undulating granite terrain the following sequence from upland to lowland includes a variation of *Terminalia* Sandveld: (a) *Combretum zeyheri* occurs on summits with deep loose sand of the reddish B-horizon Hutton Form or the yellowish B-horizon Clovelly Form. (b) The *Terminalia sericea* zone is typical of the leached, grey Ferndale belt, where the topography changes from convex to concave and drainage water moves to the surface. (c) Below the *Terminalia* fringe a dambo-like grassland zone of impeded drainage and seasonal water-logging occurs. (d) A zone of Microphyllous Thorny Plains Bushveld on moderate to weak-structural clayey soils may occur in bottomlands below the grassland zone.

An example from aeolian sand at Pumbe on the Lebombo Mountains is provided by Coetzee & Nel (1977). In the stand described *Terminalia sericea* is the dominant woody species, followed by *Combretum molle*. Other woody species include *C. zeyheri*, *Sclerocarya caffra*, *Pseudolachnostylis maprouneifolia*, *Lannea stuhlmannii*, *Cassia abbreviata*, *Strychnos madagascariensis*, *Cissus loniceri folius* and *Maytenus heterophylla*. The dominant grasses are *Panicum maximum* and *Digitaria pentzii* and others include *Aristida argentea*, *Perotis patens*, *Pogonarthria squarrosa*, *Schmidtia pappophoroides*, *Tragus berteronianus* and *Heteropogon contortus*, a very typical combination of widespread *Terminalia* Sandveld species. The largely temperate *Burkea africana* is notably absent and so are many of the tropical species occurring in the northeastern corner of the Transvaal and the Malvernia area of Rhodesia and Moçambique.

Also from Coetzee & Nel (1977) is an example from a sandy enclave underlain by sandstone, in the area between the Sabie and Elephants Rivers in the Kruger National Park. *Terminalia sericea* is by far the dominant woody plant, followed by *Ormocarpum trichocarpum*. The few other woody plants in the relevé include *Dichrostachys cinerea*. The dominant grasses are *Digitaria pentzii* and *Brachiaria nigropedata*, followed by *Perotis patens*, *Pogonarthria squarrosa* and *Panicum maximum*, in that order.

The second example pointing to the strong affinities of *Dichrostachys cinerea* subsp. *africana* for soils that are sandy but more trophic than in *Terminalia sericea* Sandveld, is provided from this area by Coetzee & Nel (1977). A large sandy plain east of Kumane in the Kruger National Park is completely dominated by a fairly dense stand of 2–3 m medium-tall shrubs, mainly of *Dichrostachys cinerea* subsp. *africana*. Very indistinct structure is present in the soil as well as free carbonates. Nearby, northwest of Kumane on a very gradual slope, the following sequence of plant communities were recorded from upland to bottomland: (a) The sandy summit area has a number of indistinct ridges and troughs, with *Terminalia sericea*-dominated dense tree bushveld on the rises and an open to semi-dense shrub bushveld composed of *Dichrostachys cinerea* and *Combretum zeyheri* as well as scattered *C. apiculatum* in the slight depressions. (b) Somewhat further down the gentle slope, immediately following the *Terminalia* zone and still on sand though presumably more trophic, dense stands of tall *Dichrostachys cinerea* subsp. *africana* shrubs alternate with shrub bushveld consisting of a mixture of *D. cinerea* and *Combretum* spp. (c) Lower down the slope the *Dichrostachys* shrub bushveld grades into a very dense bushveld of *D. cinerea* and tall to very tall *Albizia petersiana* shrubs; and (d) hence the vegetation changes to Microphyllous Thorny Plains Bushveld of *Acacia welwitschii* as the clay comes closer to the surface and the soil becomes strongly structural. (e) The final type in the sequence is dense *Spirostachys africana* bushveld on very brackish soil along a drainage line. The sequence clearly shows an optimum position for *Dichrostachys cinerea* subsp. *africana* on a habitat gradient of soil texture, sand depth and nutrients, i.e. immediately alongside *Terminalia sericea* Sandveld, which occupies the nutrient-poor deep-sandy extreme (Coetzee & Nel 1977).

A related type of vegetation occurs in the southernmost parts of the sublittoral zone of Moçambique, south of the Limpopo, and in adjacent Natal, where sandy soils support a community with *Albizia adianthifolia*, *A. versicolor*, *Afzelia quanzensis*, *Ficus burtt-davyi*, *Sclerocarya caffra*, *Terminalia sericea*, *Ziziphus mucronata*, *Dialium schlechteri*, *Garcinia livingstonei*, *Strychnos spinosa*, *Syzygium cordatum*, *Lannea discolor*, and many other species. Sometimes the vegetation locally opens to a grassy savanna dominated by *Hyperthelia dissoluta*, or *Urelytrum squarrosum*. Myre (1964, 1971) found that the *Andropogono gayani–Corchoretum junodii* is the most common community on sandy soils in this sublittoral zone. The association is commonly dominated by *Andropogon gayanus* var. *squamulatus* in the grass layer, while *Strychnos innocua* often is the most abundant woody species. The association was classified by Myre in the

Table 9. Andropogono gayani–Corchoretum junodii (after Myre 1971).

Number of relevés	6	Number of relevés	6
Differential species		Salacia kraussii	2
Andropogon gayanus	5	Elyonurus argenteus	2
Alloteropsis cimicina	5	Hermbstaedtia elegans	2
Panicum cf. coloratum	3	Panicum deustum	2
Cassia petersiana	3	etc.	
Alliance, order and class		*Other herby species*	
Corchorus junodii	5	Perotis patens	5
Pogonarthria squarrosa	5	Dicerocaryum zanguebaricum	4
Bulbostylis zeyheri	5	Evolvulus alsinoides	3
Eustachys mutica	5	Panicum maximum	3
Merremia tridentata	5	Tricholaena monachne	2
Agathisanthemum bojeri	5	Abrus precatorius	2
Ledebouria sp.	5	Sacciolepis curvata	2
Commelina africana	5	etc.	
Urelytrum squarrosum	4	*Woody species*	
Aristida graciliflora	4	Strychnos innocua	5
Ipomoea pes-tigridis	4	Terminalia sericea	5
Kohautia virgata	4	Combretum hereroense	5
Aristida congesta	3	Strychnos spinosa	4
Rhynchosia totta	3	Sclerocarya caffra	4
Tephrosia longipes	3	Dichrostachys cinerea subsp. cinerea	4
Crotalaria monteiroi	3	Conopharyngia elegans	3
Stylosanthus mucronata	3	Brachylaena discolor	2
Sporobolus pyramidalis	3	Vangueria infausta	2
Kyllinga alba	2	Annona senegalensis	2
Helichrysum kraussii	2	etc.	

Andropogonion austromossambicensis, an alliance of the order Themedetalia triandrae, originally described from Zaïre (Table 9). In the littoral zone the most common grassland or savanna on sandy soils was described by Myre (1964, 1971) as the Themedo–Salacietum kraussianae together with its subassociation parinarietosum. In this association the most abundant grasses are *Themeda triandra*, *Urelytrum squarrosum* and *Trachypogon spicatus*, while *Dichrostachys cinerea* subsp. *cinerea* and *Strychnos spinosa* are the most common woody species, with in the parinarietosum also the dwarf shrub *Parinari latifolia*. The association was classified in another alliance of the Themedetalia, the Themedion triandrae littoreo-mossambicense. Most of the woodland communities on sand have been destroyed, however, owing to the heavy population pressure, and have been replaced by 'orchards' in which mainly *Anacardium occidentale* and *Mangifera indica* grow, but which also contain some indigenous useful species, most often those with edible fruits, such as *Sclerocarya caffra*, *Trichilia emetica*, *Garcinia livingstonei*, *Strychnos spinosa*, *Syzygium cordatum*, *Vangueria infausta*, etc. North of the Limpopo these 'orchards' also occur, but on a more limited scale. These communites on sand alternate with swampy areas on black, peaty soils supporting a *Raphia* or *Pandanus* swamp forest, or are crossed by belts of ancient dunes which carry a dry forest with *Afzelia quanzensis*, *Sideroxylon inerme*, *Balanites maughamii*, *Pteleopsis myrtifolia* and *Erythrophleum suaveolens* (see Chapter 13).

The *Terminalia* communities at the Victoria Falls National Park are comparable. Hill (1969) mentions a *Burkea africana–Terminalia sericea* woodland on fairly deep sand fringing miombo. An open bushveld with *Terminalia sericea, Combretum imberbe, C. hereroense* and *Ziziphus mucronata*, as common trees, and *Hyparrhenia filipendula, H. dichroa, Heteropogon contortus, Cymbopogon excavatus, Cynodon dactylon, Bothriochloa insculpta,* and *Imperata cylindrica* as common grasses covers drainage lines where a shallow sandy soil overlies basalt.

(iv) Subtropical, Sub-humid. *Parinari curatellifolia* savanna (Fig. 44; Chapter 14, Fig. 2) is typical of the Rhodesian plateau and to a more limited extent of central Moçambique. The vegetation differs structurally from other broad-orthophyll types in that it is usually more open, the tree crowns denser and the boles longer (Henkel 1931, Michelmore 1939, Rattray & Wild 1961, Wild & Barbosa 1967, White 1976b).

Parinari curatellifolia, which is practically evergreen, is the dominant tree in a very characteristic association common on the plateau of central Rhodesia on sandy soils with a high water table under an annual precipitation of about 800 to

Fig. 44. *Parinari curatellifolia* savanna near Enkeldoorn, Rhodesia (photo M. Leppard).

1100 mm, although higher in Moçambique. Such sites are restricted to areas with a subdued relief creating hydrological conditions over extensive areas which approximate those existing in the fringing areas of dambos, where *Parinari curatellifolia* is also common. If the watertable becomes too high even for *Parinari* an edaphic grassland develops, and these two vegetation types freely intergrade in central Rhodesia. *P. curatellifolia* is often the only tree species in these savannas but temperate Upland (Temperate) Sub-humid Mountain Bushveld species such as *Faurea speciosa, F. saligna, Protea gaguedi* and *P. angolensis* may be common as well as subtropical Lowveld (Subtropical) Sub-humid Mountain Bushveld species such as *Syzygium guineense* and, in the Moçambique variant, *Piliostigma thonningii* and *Pterocarpus angolensis*. Another widespread Mountain Bushveld species occurring also in the Moçambique examples is *Combretum molle*. Common grasses include *Cymbopogon validus*, *Beckeropsis uniseta*, and species of *Hyparrhenia*, *Setaria* and *Pennisetum*, while in Moçambique *Pteridium aquilinum* locally dominates the undergrowth.

The temperate affinities of this subtropical sub-humid sandveld are with the relatively cool section of a gradient within temperate bushveld, characterized by, e.g. *Protea* and *Faurea* species. The more tropical sub-humid sandveld discussed in the next section (v) also has temperate affinities but with the warm section of the gradient within temperate bushveld, characterized by, e.g. *Diplorhynchus condylocarpon* and *Pseudolachnostylis maprouneifolia*.

(v) Tropical, Sub-humid to Arid. Moist tropical sandveld (Fig. 40) such as at Punda Milia in the Kruger National Park and in parts of Matabeleland in Rhodesia, differs considerably from arid tropical sandveld such as on the red-sandy plateau at Wambia in the far northeastern corner of the Transvaal and adjoining Moçambique. The moist tropical sandveld is related much more strongly than arid tropical sandveld to the temperate *Terminalia sericea* Sandveld. This latter relationship involves species characteristic also of Upland Sub-humid Mountain Bushveld such as *Burkea africana*, *Ochna pulchra*, *Diplorhynchus condylocarpon* and *Pseudolachnostylis maprouneifolia*. A similar type of distinctive ecological relationship exists between moist tropical sandveld and Lowveld Sub-humid Mountain Bushveld. The species involved are listed in the first paragraph on *Terminalia sericea* Sandveld.

Despite the differences, moist and arid tropical sandveld have many tropical sandveld species in common and it is to some extent on the basis of these relationships that the arid tropical sandveld communities are included in the broad concept of *Terminalia sericea* Sandveld. Examples of tropical sandveld species shared by the moist sandveld of Punda Milia and the arid sandveld of Wambia are listed by Van der Schijff (1958, 1964, 1971), Van Wyk (1973), and Van Rooyen (1976). Among these are *Guibourtia conjugata*, *Pteleopsis myrtifolia*, *Xeroderris stuhlmannii* and *Monodora junodii*.

The Punda Milia sandveld is described by Van der Schijff (1957) on the basis of three stand samples. The vegetation is mainly shrubby and composed of numerous woody species, among which *Combretum zeyheri*, *Strychnos madagascariensis*, *Alchornea schlechteri*, *Monodora junodii*, *Burkea africana*, *Diplorhynchus condylocarpon*, *Dalbergia melanoxylon*, *Pseudolachnostylis maprouneifolia* and *Combretum collinum* subsp. *gazense*. *Digitaria* sp. and *Panicum maximum* are by far the most common grasses. Other grasses include *Perotis patens*, *Schmidtia pappophoroides*, *Diheteropogon amplectens*, *Pogonarthria squarrosa*, *Cenchrus ciliaris*, *Eragrostis pallens*, *Aristida meridionalis* and *Urochloa rhodesiensis*.

In the sandy area around Wambia two distinct types of sandveld occur:
Xeric type – On the flat and level plateau of deep, red, presumably aeolian sand with pH ranging between 4.5 and 5, the vegetation is homogeneous with simple and uniform structure. Three to five metre tall shrubs (mostly 4 m tall) form a dense woody stratum dominated by *Baphia massaiensis* subsp. *obovata* and *Guibourtia conjugata* (Fig. 45). Other common woody species include *Monodora junodii*,

Pteleopsis myrtifolia, Xylia torreana, Hugonia orientalis, Hymenocardia ulmoides and *Combretum celastroides*. The grass cover is sparse and dominated by *Eragrostis pallens* (cf. Van der Schijff 1957, Van Rooyen 1976).

Mesic type — On the gentle slopes from the pleateau downwards the sand is more greyish to yellowish. The vegetation is less dense and the structure more diverse, including trees and shrubs of various sizes and a large variety of species. *Terminalia sericea, Combretum zeyheri, C. apiculatum* and *Dalbergia melanoxylon*, e.g., while largely absent from the red-sandy plateau, are common on this lower, gently sloping, more mesic terrain. Dominant grasses include *Digitaria* sp. and *Panicum maximum*.

Fig. 45. Xeric shrubby sandveld on the flat and level plateau of deep red sand, in the Wambia area, Kruger National Park. The 3–4 m tall homogeneous shrub layer is dominated by *Baphia massaiensis* subsp. *obovata* and *Guibourtia conjugata* (photo B. J. Coetzee).

In contrast with the xeric type of the red-sandy plateau, this mesic type of the greyish to yellowish sand more closely resembles the moist tropical sandveld of Punda Milia.

(b) *Combretum veld.*

Combretum-dominated Broad-orthophyll Plains Bushveld is typical of upland sites in undulating granite terrain of 500–700 mm rainfall areas. The main distribution areas of *Combretum* veld, where these conditions are met, are:

(i) in the eastern Transvaal Lowveld, on Archaic granite of the moderately semi-arid tropical area between the Crocodile and Timbavati Rivers; and locally in adjacent Moçambique;

(ii) in the Central Transvaal sub-humid, warm-temperate Bushveld Basin, on granite of the Bushveld Igneous Complex, and in a comparable situation on the high plateau of central South West Africa and Botswana.

In the western Transvaal Derdepoort–Dwaalboom–Rooibokkraal area of

Archaic granite where the climate is classified as semi-arid, warm-temperate, towards the arid Limpopo region of metamorphosis and granitization, and in the area north of the Timbavati River in the eastern Transvaal Lowveld, *Combretum apiculatum* Broad-orthophyll Plains Bushveld grades into Arid Bushveld types.

The most common type of *Combretum* veld occurs on very shallow sandy to gravelly soil and is dominated by *Combretum apiculatum*. Such vegetation accounts for one of the major variations of Acocks' (1975) Mixed Bushveld Type and is part also of his Lowveld and Arid Lowveld Types.

(i) Eastern Transvaal Lowveld. In the eastern Transvaal Lowveld, shallow sandy soil communities on granite dominated by *Combretum apiculatum* (Fig. 46), are characteristic of rapidly eroding landscapes with a high density of drainage lines and short undulations. In such strongly dissected areas *C. apiculatum* communities of uplands grade gradually into bottomland communities typified by *Acacia nigrescens*, *A. gerrardii*, *Combretum hereroense* and other woody species. Common trees of the *Combretum apiculatum* upland communities also include *Sclerocarya caffra*, *Peltophorum africanum*, *Ziziphus mucronata*, *Acacia exuvialis* and others. *A. exuvialis* is particularly abundant on the extremely shallow and gravelly upland soils towards major rivers. In this latter type of habitat several *Grewia* species tend to form a dense shrubby layer, probably at least partly owing to heavy grazing and trampling by large herbivores, which concentrate along major water courses during the dry season. Common grasses in Lowveld *Combretum apiculatum* communities include *Digitaria pentzii*, *Schmidtia pappophoroides*, *Pogonarthria squarrosa*, *Heteropogon contortus*, *Aristida congesta*, *Panicum maximum* (under trees and shrubs), *Perotis patens*, *Eragrostis rigidior* and *Trichoneura grandiglumis* (Van der Schijff 1958, Van Wyk 1973, Hirst 1975, Coetzee & Nel 1977). A comparable sequence of communities with *Combretum apiculatum* veld on the shallow, stony soils is reported from the Save valley in Moçambique (Myre 1972).

Combretum zeyheri-dominated communities also occur in upland positions in the eastern Transvaal granitic landscape. These communities are typical of deep sandy granitic soil, found in moderately dissected landscapes that erode less rapidly than the typical *C. apiculatum* veld landscapes. A considerable amount of internal drainage water, absorbed by the deep sandy soils of *C. zeyheri* summits, is forced to the surface by an impenetrable layer of strong-structural clay along a fringe around the summit areas. The summit *C. zeyheri* veld grades into *Terminalia* veld of the extremely leached sands along the fringe. In higher rainfall areas extremely leached sands and *Terminalia sericea* veld occupy the entire summit area. Further down the slope, where the drainage-impeding clay is very near the surface, a grassland zone sharply separates Broad-orthophyll Plains Bushveld of the summit from Microphyllous Thorny Plains Bushveld of bottomlands. This sharp distinction is in contrast with the gradual transition to bottomland types, found in strongly dissected areas with shallow soil and *C. apiculatum* summits. Very common shrubs in *C. zeyheri* veld include *Strychnos madagascariensis* and *Pterocarpus rotundifolius*. Both these species may form quite dense shrub layers. Common grasses are those listed for *Combretum apiculatum* veld as well as *Brachiaria nigropedata* and occasionally *Setaria flabellata* (Coetzee & Nel 1977, Gertenbach 1977).

(ii) Central Transvaal Bushveld Basin. Examples of *Combretum apiculatum* veld in the Bushveld Basin on the interior South African plateau, are provided by Grunow (1965) and by Coetzee et al. (1976) from the Soutpan Experimental Farm and Nylsvley, respectively. At Nylsvley this vegetation occurs on litholitic soils and on sandy soils of the Mispah Series, underlain by weathering felsite. At Soutpan *Combretum apiculatum* veld occurs on gravelly sandy soil on granite and ironpan. Theron (1973) at the Loskopdam Nature Reserve, found a highly significant correlation between the occurrence of *Combretum apiculatum* and high iron content of the soil. Galpin (1926) noted that the difference on the Springbok Flats between the sandy habitats of *C. apiculatum* veld on the one hand and *Terminalia sericea* Sandveld on the other is that the former occurs on red soils and the latter on grey soils.

Fig. 46. Eastern Transvaal Lowveld *Combretum* veld near Skukuza, Kruger National Park (photo P. van Wyk).

Sub-humid, temperate *Combretum apiculatum* veld of the Bushveld Basin prominently and typically share woody plants and grasses with Sub-humid Mountain Bushveld, e.g., *Acacia caffra, Dombeya rotundifolia, Ozoroa paniculosa, Rhus leptodictya* and *Burkea africana*, and the grasses *Setaria perennis, Trachypogon spicatus, Schizachyrium sanguineum, Elyonurus argenteus, Diheteropogon amplectens, Eragrostis racemosa, Brachiaria serrata* and *Digitaria monodactyla*. Such grasses are usually dominant. In the tropical Lowveld *Themeda triandra* occurs on clayey soils and is very atypical of sandy soils and of *Combretum apiculatum* veld. In the grasslands of the high interior, in Upland Sub-humid Mountain Bushveld and in *Combretum apiculatum* veld of the Bushveld Basin – i.e., in temperate climates – *Themeda triandra* is, however, quite typical and very common on sandy soils and usually also among the dominants. Other common species of *Combretum apiculatum* veld of the Bushveld Basin include *Combretum zeyheri, Vitex rehmannii, Dichrostachys cinerea, Terminalia sericea* and *Peltophorum africanum*, as well as grasses that are characteristic also of the moderately semi-arid tropical Lowveld *Combretum apiculatum* veld, e.g. *Brachiaria nigropedata, Digitaria pentzii, Eragrostis superba, Perotis patens, Heteropogon contortus* and *Aristida congesta* subsp. *barbicollis.*

The third example, showing the strong affinities of *Dichrostachys cinerea* for deep sandy soils that are somewhat more trophic than the sandy soils of typical

Fig. 47. East of Windhoek, S.W.A., *Acacia hereroensis* is a conspicuous bushveld species, and is accompanied by *Combretum apiculatum, Tarchonanthus camphoratus*, and others (photo W. Giess).

Broad-orthophyll Plains Bushveld, is provided from Soutpan (Grunow 1965). Here, *Dichrostachys cinerea* occurs optimally on habitats intermediate between acid residual sandy soil of *Combretum apiculatum* veld and alkaline transported soils with high clay and salt content of *Acacia mellifera* veld. *Dichrostachys cinerea* has a high constancy in communities on deep soil with sandy to sandy loam topsoils and sandy loam lower horizons and attains its maximum development in dry, excessively drained variations of such soils.

On the high plateau of central South West Africa in the vicinity of Windhoek similar *Combretum apiculatum* veld occurs, though the vegetation is rather open and several *Acacia* species are common (Fig. 47). Giess (1971) mentions as common woody species, e.g. *Combretum apiculatum, Tarchonanthus camphoratus, Dombeya rotundifolia, Rhus marlothii, Acacia hereroensis, A. erubescens*, and *A. hebeclada* subsp. *hebeclada*, while the grass layer contains many of the species named above.

Also in western Botswana a similar *Combretum apiculatum* veld locally covers quartzitic ridges often overlain with a sandy soil. *Grewia flava*, several *Acacia* species and *Catophractes alexandri* can be frequent companion species, and the latter dominates here and in adjacent semi-arid areas, when calcrete occurs near the surface (cf. Cole & Brown 1976).

5.4.12 Microphyllous Thorny Plains Bushveld

Acacia-dominated Microphyllous Thorny Plains Bushveld occurs on fine-textured soils in sub-humid to moderately semi-arid areas. The fine-textured soils are relatively dry and fertile. Arid features include bad infiltration and high run-off and although fine-textured soils hold much water, this is strongly retained against uptake by plant roots and concentrated in upper soil layers which are highly vulnerable to drying. Root penetration to deeper levels that are less susceptible to quick drying, is retarded mechanically and shallow-rootedness is further enforced by poor aeration with associated humic and poisonous CO_2 conditions (Grunow 1965). Aridity is emphasized as a determining habitat characteristic of fine-textured soils by the fact that in arid climates, typical species of Microphyllous Thorny Plains Bushveld occur also on coarse-textured soils, as prominent constituents of Arid Bushveld types.

The largest expanses of pure Microphyllous Thorny Plains Bushveld occur on heavy black and red montmorrillonite clays derived from basalt of the Drakensberg Stage, Stormberg Series, which belongs to the Karoo System. These clayey basaltic landscapes are very flat and level and include:
(a) the major part of the sub-humid, warm-temperate Springbok Flats on the low interior plateau Bushveld Basin; and
(b) the moderately semi-arid tropical/subtropical southern Lebombo Flats in the Swaziland and eastern Transvaal Lowveld south of the Olifants River. The basaltic Lebombo Flats north of this river are drier, with annual rainfall below 500 mm, and support Broad-sclerophyll Arid Bushveld dominated by *Colophospermum mopane*.

Extensive areas continuously dominated by Microphyllous Thorny Plains Bushveld also occur on flat loamy plains in the Bushveld Basin. In undulating granitic terrain of the Bushveld Basin and eastern Transvaal Lowveld, Microphyllous Thorny Plains Bushveld occurs in lowland sites, forming an extensive mosaic with Broad-orthophyll Plains Bushveld of upland sites. Several less extensive types of Microphyllous Thorny Plains Bushveld occur on clayey soils of clay-rich geological substrates other than basalt, such as dolerite, norite and shales, or in clayey bottomlands along drainage lines. Islands of Microphyllous Thorny Plains Bushveld occur on the grassland plains of the high interior. In Rhodesia and southern Moçambique very similar communities occur, but northwards the communities gradually change and contain more tropical species (Lang 1952, Goldthorp 1957, Rattray & Wild 1961, Wild & Barbosa 1967).

(i) *Acacia nigrescens Tropical Plains Thornveld*. *Acacia nigrescens* occurs in low tropical areas such as in Rhodesia, the eastern Transvaal and Swaziland Lowveld, and in southern Moçambique. Everywhere it is a typical and constant species of basalt plains and of bottomland sites in undulating granite terrain (Fig. 48). Similar and transitional *Acacia nigrescens* communities occur in the northern and northwestern Transvaal and eastern Botswana, bordering on Arid Bushveld

Fig. 48. *Acacia nigrescens* Tropical Plains Thornveld during the wet season with a good grass cover of *Panicum maximum*, *Urochloa mosambicensis*, *Digitaria pentzii* and other species. (Kruger National Park, Transvaal; photo M. J. A. Werger).

of the low Limpopo and Olifants River valleys. The species also contributes notably to the bulk of the woody vegetation in some variations of Dry Mountain Bushveld. *Dichrostachys cinerea* subsp. *africana* is very common over the range of *Acacia nigrescens* Plains communities and is most abundant in variations on strongly coloured red soils, indicative of good aeration and drainage. These two soil conditions plus at least a fair amount of soil nutrients as discussed on various pages under 'Broad-orthophyll Plains Bushveld', appear to be the common denominators of typical optimal *Dichrostachys cinerea* subsp. *africana* habitats. Drastic measures may be required to render aeration unsuitable for *D. cinerea* on sandy soils. However, on clayey soils, such as in *Acacia nigrescens* Bushveld, above average rainfall and careful management of soil-moisture preserving conditions, such as a tall dense grass layer, might be enough to impede aeration temporarily to the extent that it inhibits *Dichrostachys cinerea* root systems. Coetzee & Nel (1977) consider this a hypothesis worth investigating for management of *D. cinerea* bush encroachment.

A distinctive combination of dominant grasses includes *Themeda triandra*, *Panicum coloratum*, *Digitaria pentzii*, *Panicum maximum* and *Urochloa mosambicensis*. The unpalatable grass *Bothriochloa insculpta* is present and may be dominant in degraded forms of the field layer, and *Aristida congesta* subsp. *barbicollis* is also common in low seral stages (Van der Schijff 1958, Van Wyk 1973, Coetzee & Nel 1977, Gertenbach 1977).

Sclerocarya caffra is characteristically absent and *Cordia gharaf* is positively differential in a stunted, shrubby, *Acacia nigrescens*-dominated form of *Acacia nigrescens* Bushveld. Where soils in the latter type of bushveld are strongly vertic, the afore-mentioned combination of grasses is replaced by mainly *Setaria woodii*.

Communities with *Acacia nigrescens* and *Setaria woodii* are transitional to azonal, periodically inundated *Setaria woodii* grassland of vertisols, which sometimes have gilgai micro-relief. Related grasslands on vertic soils with gilgai have also been described from the temperate Bushveld Basin, i.e. linear gilgai from Rustenburg by Verster et al. (1973) and round gilgai by Coetzee et al. (1976) from the Nyslvley Nature Reserve near Naboomspruit. In the latter two areas distinct combinations of species occur, respectively on rises and in low areas, e.g., *Aristida bipartita* on rises and *Setaria woodii* in low areas. Other typical grasses of such *Setaria woodii* grasslands and related bushveld include *Sorghum versicolor* and *Ischaemum glaucostachyum*.

Acacia nigrescens Plains Bushveld with tall *Acacia nigrescens* and *Sclerocarya caffra* trees is widespread in the eastern Transvaal and Swaziland Lowveld, Natal (Figs. 48, 49, 50 and 51), southern Moçambique and Rhodesia. In the Transvaal, Swaziland and Rhodesia *Ormocarpum trichocarpum* is a quite common shrub

Fig. 49. *Acacia nigrescens–Sclerocarya caffra* Tropical Plains Thornveld on clayey basaltic soils, in the Kruger National Park (photo B. J. Coetzee).

throughout. On relatively light clay soils, such as in low positions on granite terrain or under related conditions, *Acacia gerrardii* is prominent and distinctive. *Dalbergia melanoxylon*, probably an indicator of mesic variations, is also present here as well as in a heavier basaltic soil variation without *Acacia gerrardii* and characterized by *Albizia harveyii* and *Lonchocarpus capassa*. Seemingly drier variations are characterized by *Maerua parvifolia* (Coetzee et al. 1975).

In southern Moçambique south of the Save the *Acacia nigrescens* vegetation occurs interspersed in *Julbernardia globiflora* woodland, *Adansonia digitata* dry bushveld, mopane communities, *Androstachys johnsonii* bushveld, and *Hyphaene natalensis* palm savannas, each on sites with different soil types and drainage.

Fig. 50. Tall tree savanna of predominantly *Sclerocarya caffra* and *Acacia nigrescens* on the eastern foothills of the Lebombo Mountains (photo E. J. Moll).

Along the Rio Changane these communities are also interspersed with halophytic communities on strongly saline soils, while low-lying but less saline sites carry a savanna of *Acacia nigrescens*, *A. senegal*, *A. nilotica* subsp. *kraussiana* with *Combretum imberbe*, *C. hereroense*, *C. ghasalense* and *Ormocarpum trichocarpum* in the tree layer and *Urochloa mosambicensis*, *Setaria holstii*, *Panicum coloratum*, and *Bothriochloa insculpta* among the grasses. This type of savanna is similar to the valley bottom communities north of the Save. In the Maputo district of southern Moçambique these types of communities have been studied in detail by Myre (1961, 1964, 1971), who distinguished the savanna association Themedo–Turbinetum oblongatae on dark clayey soils derived from basalt of the Lebombo Mountains with the subassociation albizietosum on more sandy soils (Table 10, Fig. 50). The association occurs on the more or less level areas in the border region of Moçambique, South Africa and Swaziland. From blackish clays he described the *Setaria holstii–Ischaemum afrum* Association in which, apart from the two name-giving species, *Sehima galpinii* is also abundant and the most common woody emergents are *Dichrostachys glomerata* subsp. *glomerata* and various species of *Acacia*, including *A. nigrescens*.

Further north in Moçambique similar *Acacia nigrescens* vegetation occurs in a sublittoral belt between the Zambezi and Save Rivers, mainly along the Buzi River, but also further north at Gorongosa. *A. nigrescens* is here often accompanied by *A. karroo*, *A. nilotica* subsp. *kraussiana* and *Combretum imberbe*, but on the valley slopes *A. nigrescens* is again accompanied by *Sclerocarya caffra*, *Terminalia sericea*, *Peltophorum africanum*, *Mundulea sericea*, *Bolusanthus speciosus*, *Ormocarpum trichocarpum* and various other species, and on lighter loams they include *Burkea africana*, *Pterocarpus rotundifolius* subsp. *rotundifolius*, *Cassia abbreviata*, *Pseudolachnostylis maprouneifolia*, and some other dry woodland species. The important species in the grass layer change from *Heteropogon contortus* and various species of *Aristida* on the valley bottoms, to *Themeda triandra* on the skeletal or heavy black soils of the slopes, and to *An-*

Table 10. Themedo–Turbinetum oblongatae with subassociations typicum (A) and albizietosum (B) (after Myre 1961, 1971)

Community	A	B	Community	A	B
Number of relevés	14	4	Number of relevés	14	4
Character species of assoc.			Sida chrysantha	5	
Turbina oblongata	5	4	Panicum coloratum	4	
Lasiosiphon capitatus	5	1	Ipomoea fragilis	4	
Jatropha sp.	5	2	Cenchrus ciliaris	4	
Differential species B			Merremia palmata	4	
Albizia anthelmintica		4	Rhynchosia totta	4	
Undergrowth species			Cassia mimosoides	4	
Urochloa mosambicensis	5	4	Lantana salvifolia	4	
Aristida barbicollis	5	2	Merremia tridentata	3	
Hibiscus pusillus	5	3	Abutilon austro-africanum	3	
Eragrostis superba	5	3	Teramnus labialis	3	
Vernonia oligocephala	5	4	Rhynchosia albissima	2	
Themeda triandra	5	4	Rhynchelytrum repens	2	
Bothriochloa insculpta	5	2	Panicum deustum		2
Heteropogon contortus	5	3	etc.		
Panicum maximum	5	4	*Woody emergents*		
Solanum panduraeforme	5	4	Acacia nigrescens	5	1
Phyllanthus maderaspatensis	5	3	Dichrostachys cinerea subsp. cinerea	4	4
Eustachys mutica	4	3	Lonchocarpus capassa	4	1
Cymbopogon excavatus	4	3	Acacia sp.	4	2
Digitaria eriantha	4	2	Ziziphus mucronata	4	2
Talinum caffrum	4	2	Sclerocarya caffra	4	2
Stylochiton natalense	3	3	Ormocarpum trichocarpum	4	1
Thesium gracile	3	3	Gossypium africanum	4	2
Digitaria argyrograpta	3	3	Maytenus sp.	3	3
Tragia sp.	2	3	Bolusanthus speciosus	1	2
Setaria sphacelata	3	1	Combretum imberbe	4	
Asparagus africanus	3	3	Ozoroa obovata	2	
Commelina africana	3	2	Euclea divinorum	2	
Abutilon guineense	2	2	Cassia petersiana	2	
Hyparrhenia filipendula	2	1	Acacia nilotica		2

dropogon and *Hyparrhenia* species on the lighter loams. In the Save (Sabi) valley of southeastern Rhodesia this vegetation is also common, though rocky slopes support Dry Mountain Bushveld (Pedro & Barbosa 1955, Myre 1961, 1964, 1971, 1972, Wild & Barosa 1967, Farrell 1968a, Macedo 1970, Tinley 1971a, b). Near drainage lines *Acacia nigrescens* is often co-dominant with *Spirostachys africana* and accompanied by *Combretum imberbe*, *Lonchocarpus capassa*, and several other trees. This type of vegetation occurs also in the middle Zambezi and the Okavango–Linyanti area (Simpson 1975, Tinley 1966, Hill 1969).

(ii) *Acacia tortilis–Acacia karroo Sub-humid Temperate Plains Thornveld.* *Acacia tortilis*, *A. karroo* and *A. nilotica* are a typical dominant combination in plains thornveld of the sub-humid temperate interior plateaux, occurring on heavy norite clay, similar basalt clays of the Sprinbok Flats and on loamy soils elsewhere in the Bushveld Basin (Fig. 52) and locally in southeastern Rhodesia. Common woody associates in all these areas include *Ziziphus mucronata*, *Rhus pyroides* and *Grewia flava*. Common grasses include *Themeda triandra*,

Fig. 51. Simple structured (probably fire-induced) *Acacia nigrescens* Tropical Plains Thornveld on basalt plains in the Hlane Wildlife Sanctuary, Swaziland (photo P. van Wyk).

Bothriochloa insculpta, Panicum coloratum and *Elyonurus argenteus.* Additional typical grasses for black, vertic clays include *Setaria woodii, Sehima galpinii, Ischaemum afrum, Sorghum versicolor, Aristida bipartita, Eragrostis chloromelas, Fingerhuthia africana* and *Enneapogon scoparius,* and for oxidized red clays *Heteropogon contortus, Aristida canescens* and *Cymbopogon plurinodis.*

Fig. 52. *Acacia tortilis–Acacia karroo* Sub-humid Temperate Plains Thornveld near Roedtan, Transvaal (photo B. J. Coetzee).

As usual, *Dichrostachys cinerea* var. *africana* is most prominent on well aerated soils, including loams and red basaltic clay. In eight reconnaissance relevés from *Acacia tortilis–A. karroo* Thornveld, those from loamy soils are differentiated from those on more clayey soils also by *Rhus leptodictya* and others. Included under the loamy-type of Thornveld are dense bushclumps associated with termite mounds, including those underlain by acid igneous rocks as well as those underlain by basic igneous rocks. A considerable number of species further differentiate termitaria thickets. Prominent among these are the woody species *Pappea capensis* and *Cassine transvaalensis* (Galpin 1926, Acocks 1975, Coetzee et al. 1976).

This basalt clay vegetation of the interior of the continent above the escarpment comprises Acocks' (1975) Springbok Flats Thornveld, which is divided into Red Turfveld and Black Turfveld. Norite clay vegetation falls under Acocks' Other Turf Thornveld. Despite erratic rainfall, which renders these clayey plains marginally suitable for crop cultivation, the soils are particularly fertile and render such good yields in favourable years that very little sub-natural vegetation has not succumbed to cultivation.

Closely related outliers, particularly of the loamy types of *Acacia tortilis–A. karroo* Thornveld, occur as isolated small enclaves in predominantly grassland areas of the higher interior. Such outliers may occur as bushclumps on termitaria, e.g. in grasslands on the Waterberg in the Transvaal where such bushclumps include *Acacia tortilis*, *A. karroo*, *A. robusta*, *Ziziphus mucronata*, *Rhus leptodictya*, *R. pyroides* and *Dichrostachys cinerea*. On the Jack Scott Nature Reserve in Highveld grassland, a similar bushclumb occurs isolated in the centre of a narrow impenetrable ring of *Acacia hebeclada*. More commonly, such outliers occur as stands of thornveld in clayey lowland sites (Louw 1951, Coetzee 1974).

(iii) *Acacia mellifera* Dry Warm-Temperate Plains Thornveld. *Acacia mellifera* subsp. *detinens* typically occurs largely in arid cold-temperate regions, particularly towards the Karoo–Namib Region, in the Kalahari where it commonly occurs with *Acacia erioloba*, and in Arid Bushveld types of the Sudano–Zambezian Region. *Acacia erioloba* is particularly common here on the more sandy soils (Fig. 53) and forms stands which are transitional to Broad-orthophyll Plains Bushveld (Giess 1971, Cole & Brown 1976).

In warmer-temperate arid and sub-humid areas, such as the Bushveld Basin and northern Botswana, *A. mellifera* is dominant on droughty, saline, clayey soils of lowland sites forming a mosaic with Broad-orthophyll Plains Bushveld communities of Upland sites. In the Transvaal Bushveld Basin *A. mellifera* subsp. *detinens* communities are the common type of bottomland vegetation in undulating granite landscapes (Passarge 1904, Seiner 1912, Weare & Yalala 1971, Leser 1972, 1975, Coetzee & Werger 1975).

In the dry warm-temperate area between Okahandja and Outjo in central South West Africa, in an extensive area of thornveld called Thornbush Savanna, *A. mellifera* subsp. *detinens* is an important species and is inclined to increase unduly with mismanagement. Other common woody species here include *A. reficiens* and *A. hebeclada* (Giess 1971). The tendency for *Acacia mellifera* subsp. *detinens* to encroach as dense bush presents a problem also in Kalahari-type vegetation (Acocks 1953, 1975, Palmer & Pitman 1972). Near Windhoek a related thornveld is important, again forming a mosaic with Broad-orthopyll Bushveld on better drained upland

419

Fig. 53. Tall *Acacia erioloba* savanna of the central Kalahari in the border region of Botswana and South West Africa (photo W. Giess).

sites. This vegetation contains a distinct Karoo–Namib element, however (Volk & Leippert 1971).

A detailed example of an *Acacia mellifera* subsp. *detinens* community is provided by Grunow (1965) from Soutpan in the Bushveld Basin. The relevant soils are fine-textured, alkaline, saline and particularly xeric. Constant woody species include *Acacia mellifera* subsp. *detinens*, *A. tortilis*, *A. gillettiae*, *Carissa bispinosa*, *Dichrostachys cinerea* and *Ehretia rigida*. Constant grasses are *Sporobolus smutsii*, *Eragrostis atherstonei*, *Panicum maximum* and *Aristida congesta* subsp. *barbicollis*. Other common grasses include *Panicum coloratum*, *Eragrostis obtusa*, *Themeda triandra*, *Digitaria pentzii* and *Urochloa mosambicensis*.

(iv) *Acacia erubescens Sandy Thornveld*. *Acacia erubescens* may occur mixed with other thorn trees but is also dominant in some stands of Thornveld on stony and sandy soils. Good examples of the stony type of *A. erubescens* thornveld occur on the Lebombo Mountains in the eastern Transvaal Lowveld near Nwanedzi. Large stands of the sandy type occur north of Soutpan in the Bushveld Basin of the Transvaal. In the western and northwestern Transvaal, in the central parts of South West Africa and in Botswana *A. fleckii* occurs mixed with *A. erubescens* in such sandy habitats (Giess 1971, Palmer & Pitman 1972, Van Wyk pers. comm., Cole & Brown 1976).

(v) *Brackish Lowland Thornveld Communities*. *Euclea divinorum*, a tall evergreen sclerophyll shrub, is a very common and strongly differential species for a range of distinctly related brackish lowland bushveld communities. Other species in a typical combination include the woody plants *Pappea capensis* and *Maytenus tenuispina*, the grasses *Dactyloctenium aegyptium*, *Sporobolus nitens*,

Fig. 54. *Acacia welwitschii* brackish Thornveld on shale, near Skukuza, Kruger National Park (photo P. van Wyk).

S. iocladus and *Chloris virgata*, and the forbs *Ocimum* spp., *Cyathula* spp., *Blepharis integrifolia*, *B. transvaalensis* and *Abutilon austro-africanum*. The grass *Enteropogon macrostachyus* and the forbs *Achyranthes* spp. and *Pupalia lapacea* are typical under trees in this community group (Coetzee & Nel 1977, Coetzee et al. 1976, Gertenbach 1977).

The thorntrees that occur with such combinations differentiate between various communities belonging to this group, e.g.: *Acacia tortilis* brackish Thornveld; *Acacia grandicornuta* brackish Thornveld occurring, e.g., extensively in the Sabie River valley near Skukuza in the eastern Transvaal Lowveld; *Acacia welwitschii* brackish Thornveld occurring, e.g., in the eastern Transvaal Lowveld on a narrow band of Karoo shale separating the basalt and granite regions (Fig. 54); *Acacia senegal* var. *rostrata* brackish Thornveld, noted, e.g., in the eastern Transvaal Lowveld and along the Limpopo River in the northern Transvaal; *Acacia tenuispina* brackish Thornveld, e.g., on clayey alluvium along the Limpopo River in the northwestern Transvaal (Bosch 1971).

5.4.13 Spiny Arid Bushveld

Spiny Arid Bushveld occurs in the arid temperate valley of the upper Limpopo River and in parts of the dry northern and eastern Transvaal Lowveld as well as in similar habitats in South West Africa, Rhodesia and Botswana. In the western, northern and eastern Transvaal it occurs geographically between arid *Colophospermum mopane* Bushveld on the one hand and semi-arid to sub-humid Broad-orthophyll and Microphyllous Thorny Plains Bushveld on the other (Acocks 1953, 1975, Van der Schijff 1971). Climatic differences only partially explain the vegetation boundaries. The sharp boundaries of *Colophospermum mopane*-dominated vegetation, the lack of consistent corresponding abiotic boundaries and indications that *C. mopane* is aggressive and competitive – such as its supreme dominance where it occurs and the characteristic presence of pioneer individuals on denuded patches – suggest that this vegetation is invading the Spiny Arid Bushveld (Gertenbach 1977).

Microphyllous, thorny species that in sub-humid areas are largely restricted to physiologically dry clayey soils, occur also on sandy soils in the climatically dry Spiny Arid Bushveld. Typical examples are *Acacia tortilis*, which is very common and often dominant or co-dominant, *A. mellifera* subsp. *detinens* and *A. luederitzii*. The microphyllous, thorny *A. erubescens*, which occurs widely on sandy soils, is also typical (Acocks 1975).

A very characteristic xeromorphic growth form in Spiny Arid Bushveld is that of a shrubby plant with small leaves and stubby or short spiny branchlets on long branches that are conspicuous owing to the small leaves, giving the plants a tangled spiny appearance (Fig. 55). Examples are *Rhigozum brevispinosum*, *Lycium* cf. *austrinum*, *Terminalia prunioides*, *Boscia albitrunca*, *B. foetida* subsp.

Fig. 55. Spiny Arid *Commiphora* Bushveld west of Birchenough Bridge, Rhodesia (photo M. Leppard).

foetida and subsp. *rehmannii*, *Sterculia rogersii*, *Commiphora pyracanthoides* and *C.* cf. *merkeri*. Other typical species of Spiny Arid Bushveld include *Sesamothamnus lugardii*, which has the afore-mentioned spiny features as well as swollen, succulent, bottle-shaped main stems; *Adansonia digitata*, a giant tree version of the swollen, bottle-shaped succulent growth form (Fig. 56); *Grewia flava* and *Dichrostachys cinerea*, which are among the most constant woody species in Spiny Arid Bushveld; *Ziziphus mucronata*, which is also quite common; and *Catophractes alexandri*, typically of temperate Arid Bushveld, particularly when calcrete occurs near the surface.

Another distinctive xeromorphic feature of Spiny Arid Bushveld is that tree species which grow considerably taller in sub-humid regions are here characteristically stunted. Examples are *Terminalia sericea*, *Combretum apiculatum* and *Acacia nigrescens*. The latter two species are also typically the ones that are among the dominants in Dry Mountain Bushveld. *Terminalia sericea*, *Rhigozum* sp., *Grewia flava* and *Acacia tortilis* are the main dominants on deep, fine grey-brown sand underlain by Karoo System sediments between the Matlabas and Mogol Rivers in the western Transvaal and the Limpopo River valley. This variation is described by Acocks (1975), following Irvine (1941), as Dwarf *Terminalia sericea–Rhigozum* sp. Veld – a variation of Acocks' Arid Sweet Bushveld, which is the equivalent of Spiny Arid Bushveld. Among the grasses for this variation are those typical of dry deep sand, including *Eragrostis pallens*, *Schmidtia pappophoroides* and *Brachiaria nigropedata*.

A variation called *Grewia flava* Veld (Acocks 1975) and in which *Grewia flava*, *Acacia luederitzii*, *A. erubescens*, *A. mellifera* subsp. *detinens*, *A. tortilis* and *Boscia albitrunca* are among the most prominent woody plants, occurs mainly on

Fig. 56. Spiny Arid Bushveld on shallow rocky soil in Vendaland, northern Transvaal, with *Adansonia digitata*, *Ficus smutsii*, *Acacia tortilis*, *Commiphora* sp., etc. (photo M. J. A. Werger).

shallow reddish soil over limestone in the Limpopo Zone of metamorphosis and granitization. Common grasses include *Schmidtia pappophoroides*, *Eragrostis* sp., *Digitaria eriantha*, *Heteropogon contortus*, *Panicum coloratum*, *Aristida congesta* subsp. *congesta* and subsp. *barbicollis* and *Enneapogon scoparius*. Stands dominated by low *Commiphora pyracanthoides*, with stunted *Acacia tortilis*, *Combretum apiculatum*, *Grewia flava* and *Terminalia sericea* occur where the limestone is at or near the surface (Acocks 1975). In such areas the grasses include *Urochloa* sp., *Sporobolus nitens*, *Panicum coloratum*, *Anthephora pubescens*, *Cenchrus ciliaris* and *Enneapogon scoparius*. On somewhat deeper red sandy loam soil of the Archaic Granite area between the Pietersburg Plateau, Soutpansberg, Drakensberg and Magabeneberg a somewhat less stunted variation of

Grewia flava Veld occurs (Acocks 1953, 1975). Related vegetation types also occupy the Olifants and Steelpoort River valleys in the low northeastern interior Transvaal plateau, and the hilly country near Ghanzi in Botswana and near Tsumeb in South West Africa (Passarge 1904, Seiner 1912, Giess 1971, Weare & Yalala 1971, Cole & Brown 1976).

Terminalia prunioides is a common constituent of shallow gravelly soils with lime, e.g. in *Terminalia/Commiphora/Acacia nigrescens* Veld on the shallow basaltic soils towards the Olifants River in the eastern Transvaal Lowveld (Van Wyk 1973), slopes in the low, dry Olifants and Steelpoort River valleys (Acocks 1975) as well as in the northern Transvaal Limpopo River Lowveld (Van der Schijff 1971). In somewhat higher rainfall areas the species occurs in physiologically dry, brackish areas (Van Wyk 1973).

5.4.14 Dry Cold-temperate Grassland (*Eragrostis superba–Eragrostis chloromelas* Grassland)

Most of the arable land on the dry cold-temperate plains of the high interior plateau is under cultivation (predominantly maize) and uncultivated areas have deteriorated as a result of overgrazing, trampling, continued light selective grazing and unsuccessful attempts at cultivation. In uncultivated areas *Themeda triandra* and *Eragrostis chloromelas* are the typical dominants although retrogressive succession has in places progressed to the stage where short-lived grasses and dwarf shrubs of the more arid Karoo–Namib Region (Chapter 9) are assuming dominance (Acocks 1953, 1975, Mostert 1958, Morris 1973, 1976, Scheepers 1975, Werger 1973a, 1977a).

The main body of this vegetation occurs at 1200–1500 m altitude on the flat to gently undulating plains of the high central interior plateau. The southern part of the area is underlain by Ecca and Beaufort Series sandstones, mudstones and shales of the Karoo System with remnants of Tertiary to Recent aeolian sand (Dept. Mines 1970). The vegetation and its catenary relationship with the environment is typically that of the Kroonstad 'Key Area' surveyed by Scheepers (1975) (Fig. 57). Here the vegetation–environment complex is largely related to four main landforms. These are, from upland to bottomland: (i) depositional aeolian sandy plains and erosional-structural plateaux; (ii) erosional-structural minor rocky escarpments and ridges; (iii) stripped clay pediplains and peneplains; and (iv) alluvial valley flats of the main river valleys. In the northern part of the area, typical Dry Cold-temperate Grassland occurs on the largely featureless plains underlain by Ventersdorp lava and by Dwyka series tillite and shale of the Karoo System (Dept. Mines 1970, Morris 1973, 1976).

The climate in Köppen's System is BSkw, i.e. Arid (Steppe), dry hot but with a low mean annual temperature and a dry winter season. Winters are severely frosty (Fig. 2) and the area falls in Thornthwaite's Cold-temperate Thermal Region. The Thornthwaite Moisture Region is Dry Sub-humid and the mean annual rainfall is 450–600 mm and erratic (Schulze & McGee, Chapter 2).

As an entity, Dry Cold-temperate Grassland differs from related Moist Cold-temperate Grassland by the prominence in the former of the following and other species:
(i) *Digitaria argyrograpta*, *Eragrostis superba* and *Aristida congesta*, which are prominent on a variety of soil types;

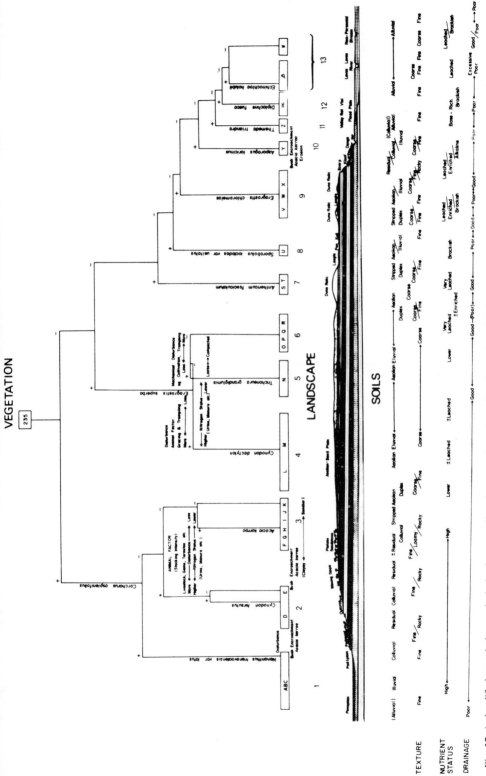

Fig. 57. A simplified association-analysis dendrogram aligned over a schematic profile of a section of the landscape in the Kroonstad area (South African Highveld) indicates the correlated occurrence of soil types and vegetation types (after Scheepers 1975).

(ii) *Panicum coloratum, Sporobolus ioclados* var. *usitatus* and *Eragrostis obtusa*, which are commonly prominent on fine-textured soils;
(iii) *Setaria woodii, Fingerhuthia sesleriiformis* and *Pennisetum sphacelatum*, typically on dark, periodically wet to inundated clays of the Rensburg Form, which are calcareous to moderately saline and alkaline;
(iv) *Sporobolus ludwigii* on soils such as the Estcourt Form, with high salinity and alkalinity levels; and
(v) *Eragrostis lehmanniana* on sandy soils.

Based on such species distributions the main Dry Cold Grassland communities may be broadly delineated as those occupying respectively: sandy soils, characterized by *Eragrostis lehmanniana*; droughty clayey soils, typical of the Arcadia Form, characterized by *Panicum coloratum*; periodically wet dark vertic clay, typical of the Rensburg Form, characterized by *Setaria woodii*; and sandy topsoil, highly saline and alkaline, typical of the Estcourt Form, characteristically with several *Sporobolus* species.

Sandy soil Dry Cold-temperate Grassland has strong affinities with sandy soil Moist Cold-temperate Grassland in that *Heteropogon contortus, Aristida diffusa, Trichoneura grandiglumis, Tristachya hispida, Eragrostis gummiflua* and *Elyonurus argenteus* occur optimally on sandy soils in both areas. Over a wide range of habitats the two areas share a characteristic combination of prominent species that distinguish them from Moist Cool-temperate Grasslands. Such characteristics are the general prominence of *Themeda triandra, Eragrostis chloromelas, Cymbopogon plurinodis, Microchloa caffra, Eragrostis plana, Setaria flabellata* and the absence or complete lack of prominence of many characteristic species of Moist Cool-temperate Grasslands. An intermediate community, clearly related to the two lower rainfall Cold-temperate Grasslands as well as the higher rainfall Cool-temperate Grassland, was described by Coetzee (1974) from a rainfall area of approximately 700 mm. This is Dolomite Grassland occurring on sandy dolomite-derived soil, which is the drier type of sandy soil in that area. The relationship of that community with high rainfall, moderately frosty Moist Cool-temperate Grassland is indicated by the prominence of *Trachypogon spicatus, Diheteropogon amplectens, Rhynchelytrum setifolium, Schizachyrium sanguineum* and *Bewsia biflora* (Table 2 in Coetzee 1974). Others such as *Loudetia simplex, Andropogon schirensis* and *Panicum natalense* are, however, lacking and *Eragrostis chloromelas* (given as *E. curvula*), typical of the severely frosty lower rainfall Cold-temperate Grasslands, is highly prominent.

Dry Cold-temperate Grassland, particularly on fine-textured soils, has affinities with vegetation of Dry Sub-humid and Semi-arid Moisture Region Bushveld areas in the occurrence of *Cymbopogon plurinodis, Setaria flabellata, Panicum coloratum* and *Setaria woodii*.

Themeda triandra should be the sole dominant of most of the Dry Cold-temperate Grassland. The droughty calcareous dark clays, typically of the Gelykvlakte Series (Arcadia Form), occurring on poorly drained but effectively dry peneplains, adjacent pediplains and lower pediment slopes, should provide valuable sweet grazing but have been overgrazed. The climax vegetation has as a result yielded successively to dominance by *Panicum coloratum, Eragrostis chloromelas, Sporobolus ioclados* var. *usitatus*, short lived grasses including *Aristida* spp., *Chloris virgata* and *Tragus racemosus* and finally Karoo–Namib dwarf shrubs such as *Pentzia globosa* and *Chrysocoma tenuifolia*. Dominance by

Sporobolus ioclados represents the last stage before the vegetation commences to break down to a stage from which recovery is unlikely, but at and above this critical stage the vegetation has a remarkable recovery potential (Scheepers 1975). Similar sequences of degradation occur on upper pediment slopes, summits and plateaux with fine-textured soils and on the heavily grazed and trampled soils on Dwyka tillite and shale (Morris 1973, 1976, Scheepers 1975).

With continued understocking and accompanying selective grazing, *Elyonurus argenteus* gains prominence with *Eragrostis chloromelas*. Such is the case on some vertic dark clay soils (Arcadia Form) but particularly on loose deep sandy soils of the upland plateaux, where the base status of the soil is low, grazing is less sweet and a tendency to understock occurs (Scheepers 1975). The typical prominence of *Elyonurus argenteus* on deep soils of the extensive plains underlain by Ventersdorp lava is also associated with light grazing (Morris 1973, 1976).

Shallow non-arable sand, underlain by impenetrable clay or rock, is typically heavily stocked, severely overgrazed and trampled. Dung and urine deposition and possibly termite activity has led to enhanced nitrogen status of the soil with concurrent prominence of *Cynodon dactylon*. A similar preference is displayed by *Cynodon hirsutus* but with optimum distribution on somewhat more clayey soil (Scheepers 1975).

High salinity and alkaline bottomlands have highly palatable and nutritious fodder but a low productivity and a tendency to overstock these habitats exist. Sheep particularly eradicate sweet grasses in these areas, inviting the invasion of dwarf shrubs such as species of *Pentzia, Salsola, Chrysocoma, Felicia* and *Lycium* (Scheepers 1975).

With this varied interaction of soil and biotic factors, quite a variety of minor communities occur (see Fig. 57).

Typical Dry Cold-temperate Grassland comprises zonal vegetation in Acocks' (1953) Pan Turfveld of the western Orange Free State and Dry and Transitional Cymbopogon–Themeda Veld (Fig. 58), which forms the largest part of this vegetation. Relict outliers of the latter occur at similar altitude and rainfall and under the same Thornthwaite Moisture Regimes, south of Lesotho between the highlands and the coast. This southern area has been largely invaded by Karoo vegetation (Acocks 1953). The Northern Variation of the Dry Cymbopogon–Themeda Veld merges to the west into semi-arid Kalahari vegetation e.g. on aeolian sand over limestone where *Stipagrostis uniplumis* is dominant or very prominent (Morris 1973, 1976). Towards higher rainfall areas and in moist bottomlands the vegetation grades into Moist Cold-temperate Grassland.

5.4.15 Moist Cold-temperate Grassland (*Tristachya hispida–Eragrostis chloromelas* Grassland)

Moist Cold-temperate Grassland occurs east of the Dry Cold-temperate Grassland at a higher altitude and with a higher rainfall, bordering on the high eastern escarpment where it merges with Moist Cool-temperate Grassland and Afro-alpine vegetation. The most extensive soil types of the region are cultivated for wheat and maize production and natural pastures are largely restricted to non-arable land. On black clay, *Themeda triandra* is overwhelmingly dominant. Sandy soil pastures have mainly deteriorated from a *Themeda triandra*-dominated stage to one where *Elyonurus argenteus, Aristida junciformis* and

Fig. 58. Dry Cold-temperate *Cymbopogon–Themeda* Grassland in the Orange Free State. Woody species occur on the slopes of the kopje in the background (photo M. J. A. Werger).

Eragrostis chloromelas are dominants (West 1951, Acocks 1953, 1975, Roberts 1966, Scheepers 1975).

The area in which this vegetation occurs lies between 1500 m and 1800 m altitude. The northern plains, which are underlain by rocks of the Ecca Series (Karoo System), have black montmorillonite clay soils. In the southern part, which is underlain mainly by sandstones of the Beaufort and Stormberg Series, the soils are mainly acid and neutral yellow and grey sands and loams (Dept. Mines 1970, MacVicar 1973, Scheepers 1975). The main landforms in the south are determined largely by geology (Scheepers 1975) (Fig. 59). At lower altitudes sedimentation of the Beaufort Series has resulted in an extensive gently undulating landscape of coalescing pediplains and the Molteno Beds of the Stormberg Series form a structural plateau at higher altitudes.

In the Cold-temperate region winters are severely frosty (Fig. 2) in the east and the humidity region is Moist Sub-humid. Rainfall increases from approximately 600 mm in the west to over 700 mm in the east (Schulze & McGeee, Chapter 2, Weather Bureau 1957) (Fig. 60).

Prominent species in this Moist Cold-temperate Grassland include *Themeda triandra, Eragrostis chloromelas, Cymbopogon plurinodis, Setaria flabellata, Eragrostis plana, Heteropogon contortus, Microchloa caffra* and *Elyonurus argenteus*, which is a typical species combination for both Dry and Moist Cold-temperate Grassland. Other prominent species which are more distinctive of this Moist Cold-temperate grassland, are *Aristida junciformis, Tristachya hispida, Brachiaria serrata, Eragrostis capensis, Digitaria tricholaenoides* and *Eragrostis racemosa*. Some of the latter group of more mesophytic species, notably *Brachiaria serrata* and *Eragrostis racemosa*, are examples of floristic affinities between Moist Cold-temperate and Moist Cool-temperate Grasslands as well as with Upland and Lowveld Sub-humid Mountain Bushveld.

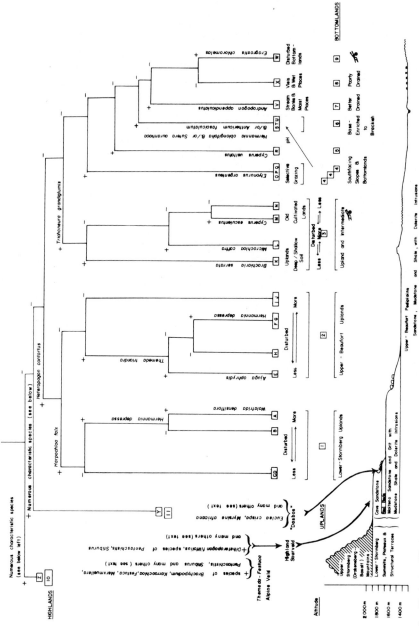

Fig. 59. A simplified association-analysis dendrogram aligned over a schematic profile of the landscape near Bethlehem, eastern Orange Free State, indicating the geology, shows the relationships between vegetation types and substrate (after Scheepers 1975).

Fig. 60. Moist Cold-temperate Grassland covers the flat surfaces in the eastern Orange Free State while the slopes carry an open shrub vegetation with *Rhus* spp. (Rhoo–Aloetum ferocis and Rhamno–Rhoetum) (photo M. J. A. Werger).

Harpochloa falx and *Digitaria monodactyla* are typical prominent grasses of the distinctively leached sandy upland soils of the Molteno Beds structural plateau, i.e. the Lower-Stormberg Uplands. The altitude is typically 1650–1800 m but may be lower with compensating mesic factors such as cool aspects and deep, leached sandy soils. *Themeda triandra* is dominant on well-preserved sites but with continued light selective grazing *Elyonurus argenteus* may replace *Themeda triandra* as dominant. *Aristida junciformis* becomes dominant with *Elyonurus argenteus* declining to sub-dominance under continuous severe selective grazing by sheep (Scheepers 1975).

On sandy and loamy soils at lower altitudes, where temperatures are higher, rainfall lower, soil moisture lower and soils somewhat less leached, typically at 1500–1650 m altitude, the grassland is less sour than the *Harpochloa falx* upland type. The vegetation of these predominantly Beaufort Uplands varies with aspect and soil type from relatively xeric to relatively mesic. Typical degradation of pasture is due to selective grazing with *Elyonurus argenteus*, *Eragrostis chloromelas* and *Aristida junciformis* replacing *Themeda triandra* as dominants (Scheepers 1975).

On previously cultivated lands which have been left fallow the first perennial successional stage is typified by the following prominent species: *Cynodon dactylon, C. hirsutus, Aristida congesta, Eragrostis chloromelas, Pogonarthria squarrosa, Trichoneura grandiglumis, Eragrostis curvula* and many forbs. At a later stage of secondary succession *Eragrostis chloromelas* becomes dominant and sub-dominant grasses may include *E. plana, Cynodon hirsutus* and *Trichoneura grandiglumis.*

The southerly Moist Cold-temperate Grassland communities, occuring on the sandy soils of the Stormberg and Beaufort Series, coincide largely with Acocks' (1975) Cymbopogon–Themeda Veld, Highland Sourveld to Cymbopogon–Themeda Veld Transition and Southern Tall Grassveld. The latter occurs east of the former two that form the main block of sandy soil Moist Cold-temperate Grassland and is separated from them by a high zone of Moist Cool-temperate Grassland on the eastern escarp. *Themeda–Hyparrhenia* Grassland, typical of the eastern outlier and described by Edwards (1967), forms a belt of vegetation immediately above the karroid vegetation in hot, dry valleys of rivers draining from the eastern escarp into the Indian Ocean. Edwards (1967) lists for this vegetation a species combination with clear Moist Cold-temperate Grassland characteristics. Similar vegetation, also fringing eastern karroid valley vegetation, is described by Story (1952) and Comins (1962) from south of the Lesotho highlands, between the highlands escarp and the coast, as the Transition zone between Sweetveld and Sourveld. This far southern outlier belongs to Acocks' (1975) Dohne Sourveld, a variation of Highland Sourveld without winter snow and occurring at relatively low, warm altitudes.

The northernmost Cold-temperate Grassland community occurring on black montmorillorite clay underlain by the Ecca Series is described and mapped by Acocks (1953, 1975) as Themeda Veld or Turf Highveld. *Themeda triandra* is typically completely dominant. Other characteristic Moist Cold-temperate Grassland features include the prominence of *Heteropogon contortus, Eragrostis racemosa, Tristachya hispida, Elyonurus argenteus, Brachiaria serrata, Cymbopogon plurinodis, Eragrostis chloromelas, E. plana* and *E. capensis* (Acocks 1953, 1975).

5.4.16 Moist Cool-temperate Grassland (*Trachypogon spicatus–Diheteropogon amplectens* Grassland)

The most sour of the three grassland types occurs mainly in two broad belts, one marking the northern extreme and one the eastern extreme of the extensive grassland area of the high interior. Quite a number of very prominent species distinguish Moist Cool-temperate Grassland from the two cold-temperate types. Notable examples of such species include *Trachypogon spicatus, Diheteropogon amplectens, Loudetia simplex* and several others, many of which are typical also of Upland Sub-humid Mountain Bushveld.

Geology in this vegetation type is varied and rocks of the Witwatersrand, Ventersdorp, Transvaal and Karoo Systems are spanned. Soils are typically sour to neutral, red and yellow latosols with varying amounts of rock and lithosols (Dept. Mines 1970, MacVicar 1973).

The northern and eastern belts of Moist Cool-temperate Grassland occur on the moderately frosty northern and eastern edges of the upper interior plateau, bordering on Bushveld where only occasional light frost occurs (Fig. 2). The

Fig. 61. Moist Cool-temperate Grassland (northern belt) in the southern Transvaal near Krugersdorp (photo B. J. Coetzee).

northern belt (Fig. 61) borders on Sudano–Zambezian Bushveld of the lower interior plateau and outliers of Moist Cool-temperate Grassland occur on the mountains of the lower interior plateau as well as in low, frosty areas with leached soils and impeded drainage. The eastern belt, which is high on the eastern continental escarpment, borders on Bushveld of the escarp foothills, i.e. Lowveld Sub-humid Mountain Bushveld of the Sudano–Zambezian Region in the north and karroid Valley Bushveld (Chapter 9) in the South. Afro-alpine vegetation separates the southern part of the eastern belt from the more westerly Cold-temperate Grasslands on which Moist Cool-temperate Grassland elsewhere adjoins. Humidity regions include Humid with 800–1000 mm annual rainfall in the east and Moist Sub-humid with 700–800 mm in the north (Schulze & McGee, Chapter 2). Most of the Moist Cool-temperate Grassland on the eastern escarpment and Soutpansberg have been replaced by *Pinus* and *Eucalyptus* plantations (Van der Schijff 1971).

A typical combination of very prominent species that differentiate this grassland type from the two types of Cold-temperate Grassland, but which are quite common also in Sub-humid Mountain Bushveld, particularly the Uplandtype, includes *Trachypogon spicatus*, *Rhynchelytrum setifolium*, *Loudetia simplex*, *Andropogon schirensis*, *Aristida aequiglumis*, *Panicum natalense*, *Bewsia biflora*, *Urelytrum squarrosum* and *Setaria perennis*. Very constant prominent grasses with somewhat wider distributions include *Diheteropogon amplectens*, which seems to have a greater tolerance for conditions such as Semi-arid Mountain Bushveld, and *Eragrostis racemosa* and *Brachiaria serrata*, which are listed as prominent species of Moist Cold-temperate Grassland also. *Themeda triandra* and *Elyonurus argenteus*, common prominent grasses in both types of Cold-temperate Grassland, are also very constantly prominent here (Louw 1951, Acocks 1953, 1975, Killick 1958, 1963, Compton 1966, Van der Schijff 1971, Du

Plessis 1972, Morris 1973, 1976, Theron 1973, Van Wyk 1973 and pers. comm., Coetzee 1974, 1975, Coetzee & Werger 1975, Bredenkamp 1975, Coetzee et al. 1976).

Although Moist Cool-temperate Grassland communities are conveniently separated as a physiognomic–floristic unit, they are in their purely floristic affinities tightly interwoven with some Upland Sub-humid Mountain Bushveld communities. As discussed earlier in greater detail, Upland Sub-humid Mountain Bushveld communities may be arranged along a floristic and physiognomic gradient associated with an abiotic gradient from moist, cool climate and strongly leached soils to drier, warmer climate with less leached soils. Moist Cool-temperate Grassland is a continuation of the cool, mesic end of this gradient but the physiognomic and floristic responses to the main environmental gradient are not equally abrupt nor do they coincide precisely. On small sections of the gradient, with many closely related communities, other habitat factors may dominate floristic affinities, causing them to cut across physiognomic boundaries. These small sections of the main vegetation–environment gradient do not necessarily represent small areas. An example of the dovetailing floristic relationships between the grassland and bushveld sections is provided from a Nature Reserve in the predominantly bushveld area of the low interior plateau (Coetzee 1975). The reserve is situated on a mountain and includes grasslands as well as Upland Sub-humid Mountain Bushveld. The grasslands, which are situated above the bushveld on slopes, represent outliers of the more extensive regional Moist Cool-temperate Grassland. The grassland community on litholitic soils is more closely related to Deciduous Broad-orthophyll and Evergreen Bushveld on litholitic soils than to grassland on deep well-developed soils, which in turn is more closely related to Evergreen Bushveld and Evergreen Shrubland on well-developed soils. The grassland community on very shallow litholitic soils with sheet outcrop is most strongly related to bushveld among outcrops. These three grassland types are nevertheless quite strongly related at a slightly broader level and belong to typical Moist Cool-temperate Grassland.

The close affinities between Moist Cool-temperate Grasslands and Upland Sub-humid Mountain Bushveld communities are also apparent from examples provided by Edwards (1967), Theron (1973), Coetzee (1974) and Bredenkamp (1975). In his classification, Coetzee (1974) grouped together grasslands and bushveld belonging to the same geological formation. He did so with dolomite vegetation, where soils are sandy but relatively base-rich, chert vegetation, where soils are sandy and very poor, and with shale vegetation where soils are clay loams with a somewhat higher nutrient status than chert soils. Bredenkamp (1975) emphasized the strong floristic relationship between grasslands and bushveld on coarse sandy soils of the Witwatersrand geological system, on the one hand, and between grasslands and bushveld on the Ventersdorp System on the other. Edwards (1967) classified the grass stratum of *Protea multibracteata*-dominated Bushveld with a faciation of Moist Cool-temperate Grassland. Theron (1973) pointed out the close floristic similarity of grassland outliers at Loskopdam with the field layers of deciduous and evergreen bushveld communities.

In ad-hoc classifications from different parts of the Moist Cool-temperate Grassland region, the main communities have been variously associated with factors such as geology, soil depth, stoniness, degree of outcrop, soil texture, geomorphology, altitude and aspect. Apart from the typical Moist Cool-temperate

Grassland features mentioned, the few areas described in some detail are rather divergent in floristic composition and furthermore variously related to bushveld communities that have not been placed in a broad floristic context. Therefore, rather than attempt to provide a generally applicable finer classification for the region, the main features of some ad-hoc classifications may be highlighted.

An example of variation in grassland outliers on the mountains in the predominantly bushveld area of the low interior plateau is provided by Coetzee (1975) from the Rustenburg Nature Reserve. The grassland soils are predominantly quartzite-derived sandy loam and sandy clay-loam underlain by Magaliesberg quartzite of the Transvaal System. The three main Moist Cool-temperate Grassland communities occurring are:

A. The community on shallow litholitic black and dark reddish-brown soils amongst sheet outcrop is characterized by several species with xeromorphic features, including:
(i) desiccation-tolerant plants such as the shrublet *Myrothamnus flabellifolius*, the fern *Selaginella dregei*, the sedges *Coleochloa setifera* and *Cyperus rupestris* (presumably desiccation-tolerant), the grasses *Oropetium capense*, *Eragrostis stapfii* (appearance suggests desiccation-tolerance), *E. nindensis* and *Microchloa caffra* and *Xerophyta viscosa* (cf. Gaff 1971, Gaff & Ellis 1974); and
(ii) a number of succulents such as *Frithia pulchra*, a rare endemic which is buried in gravel with only the lensed tips of the leaves at the surface and resembling the surrounding quartzite gravel, *Khadia acutipetala*, *Euphorbia schinzii*, *Adromischus umbraticola*, *Aloe peglerae* which is another rare endemic, a semi-succulent shrub *Lopholaena coriifolia* and the shrublet *Rhus magalismontana*.

B. The litholitic but non-rocky grasslands are characterized by the grass *Tristachya biseriata*, which is usually also entirely dominant and by a number of forbs which are very characteristic of such habitats, notably *Rhynchosia monophylla*, *Sphenostylis angustifolius* and *Cryptolepis oblongifolia*. A similar *Tristachya biseriata*-dominated community was described by Theron (1973) from Loskopdam.

C. Typical species on non-stony, well-developed soils include the grasses *Digitaria brazzae* and *Tristachya rehmannii* and several forbs, among them *Elephantorrhiza elephantina*, *Ipomoea ommanneyi*, *Crabbea hirsuta* and *Crassula transvaalensis*.

An outlier community is also described by Coetzee et al. (1976) from the Nylsvley Nature Reserve situated in a marginally frosty area dominated by bushveld of the sub-humid plains on the low interior plateau. The community, which is closely related to a Broad-orthophyll Plains Bushveld community occurs on gently sloping felsite terrain, on the frostier lower slopes with leached soils, impeded vertical drainage and underlain by a hard impenetrable flintite bank. Typical Moist Cool-temperate species include *Trachypogon spicatus* and *Setaria perennis* which have characteristically high cover values in a mesic variation. The presence of *Eragrostis racemosa* and *Aristida junciformis* are further features of the mesic variation. *Elyonurus argenteus* is distinctively prominent in a xeric variation, which is further characterized by the presence of *Brachiaria serrata*, *Loudetia flavida*, *Aristida diffusa*, *Trichoneura grandiglumis* and *Eragrostis nindensis*. Other species that are typical of Moist Cool-temperate Grassland include *Diheteropogon amplectens, Urelytrum squarrosum,* and *Schizachyrium sanguineum,* but grasses such as *Rhynchelytrum villosum, Perotis patens* and *Digitaria* cf. *pentzii* indicate relationships with Plains Bushveld and perhaps drier types of grassland. Grunow (1965) described a similar type of grassland outlier from the bushveld plains on the low interior plateau.

Moist Cool-temperate Grasslands have further been described in some detail from a private nature reserve in the central portion of the northern band of this vegetation type (Coetzee 1974). This reserve lies on the Transvaal System and spans chert and dolomite of the Dolomite Series and shale and quartzite of the Pretoria Series. The three main grassland communities here are each associated with different geological formations:

A. Dolomite-derived soils are relatively more base-rich and therefore also physiologically drier than the chert and shale-derived soils in the same area. The dolomite grasslands, which are also well-exposed to cold southerly weather, are strongly related to the somewhat drier Cold-temperate Grasslands further south. Distinctive dolomite grassland species include the grasses *Eragrostis chloromelas, E. capensis, Trichoneura grandiglumis, Triraphis andropogonoides, Aristida congesta* subsp. *congesta, Fingerhuthia sesleriiformis, Eustachys mutica, Microchloa caffra* as well as several forbs. *Elyonurus argenteus* gains prominence and becomes the sole dominant in overgrazed areas.

B. Grassland on the poor sandy soils derived from chert are characterized by the grasses *Monocymbium ceresiiforme* (particularly constant on rocky stony areas), *Digitaria brazzae* (particularly constant on non-rocky or non-stony areas), *Sporobolus eylesii,* the geoxylic shrub *Pygmaeothamnus zeyheri, Crassula transvaalensis,* and others. Some species, associated with sandy soils but not restricted to very poor sandy soils, cause floristic relationships between chert and dolomite grasslands. Such species include the grasses *Sporobolus pectinatus, Digitaria monodactyla* and *D. tricholaenoides.* Others, associated with soils that are poor in bases but less sensitive to soil texture, are responsible for affinities between chert and shale grasslands. Such species include *Loudetia simplex, Urelytrum squarrosum* and *Panicum natalense,* which are typical for Moist Cool-temperate Grasslands in general.

C. Typical species for secondary grasslands on previously cultivated lands that have been left fallow include *Eragrostis gummiflua* on sandy soils and *E. curvula, Cynodon dactylon, Hyparrhenia hirta, Sporobolus africanus* and several forbs including *Walafrida densiflora, Solanum* spp., *Wahlenbergia caledonica,* and others.

Another area representing the central portion of the northern belt of Moist Cool-temperate Grassland, in this instance on the Witwatersrand and Ventersdorp Geological Systems, is the Suikerbosrand Nature Reserve described by Bredenkamp (1975). Here the main grassland types are associated with geology, rockiness and topographic position. The soils derived from quartzite of the Witwatersrand System are relatively sandy with low pH, and variable but often low conductivity as compared with the soils of the Ventersdorp System, except for very shallow rocky soils of the Ventersdorp System, which are also sandy and acid with low conductivity. Very common grasses on both geological systems include *Trachypogon spicatus, Eragrostis racemosa, Brachiaria serrata, Heteropogon contortus, Themeda triandra* and *Elyonurus argenteus.* The main sub-divisions are as follows:

A. Characteristic species of the acid nutrient-poor sandy soils, i.e. upland and lowland Witwatersrand quartzite, as well as very shallow rocky parts of the Ventersdorp system where soils are similarly acid, poor and sandy, include the grasses *Aristida junciformis, Digitaria monodactyla, Monocymbium ceresiiforme, Rhynchelytrum setifolium, Loudetia simplex, Tristachya rehmannii, Diheteropogon filifolius* and *Digitaria tricholaenoides*. The deep sandy soils of upland plateau and lowland plains have a number of species in common with upland grasslands of the Ventersdorp System, such as *Diheteropogon amplectens* and *Tristachya hispida. Eragrostis capensis*, which is quite typical of the Ventersdorp System area, also occurs on the lowlands of the Witwatersrand System where the soils have a slightly finer texture than on the quartzite uplands.

B. Differential species that are typical for the upland sandy clay–loam soils of the Ventersdorp System include the grasses *Digitaria diagonalis, Cymbopogon excavatus* and *Setaria nigrirostris*, as well as *Berkheya setifera, Rhus discolor, Indigofera zeyheri, Vernonia natalensis*, and others.

C. Lowland communities of the Ventersdorp System, where the soils are sandy clay–loams to clay-loams, are typified by, e.g., *Setaria flabellata, Andropogon apendiculatus, Microchloa caffra* and *Sida dregei*.

D. Typical species for grasslands on very rocky areas, such as sheet outcrops, are *Selaginella dregei* and *Cyperus rupestris* (as in the Rustenburg Nature Reserve) as well as *Ursinea nana. Rhus magalismontana* and *Lopholaena coriifolia*, which are differential for quartzite sheet outcrop in the Rustenburg Nature Reserve, are here also typical for grasslands on quartzite sheet outcrop areas. At Suikerbosrand *Aristida transvaalensis* and *Sporobolus pectinatus* are also typical of this habitat. *Sutera caerulea* and *Crassula setulosa* are restricted to rocky areas of the Ventersdorp System.

E. Common grasses on fallow, previously cultivated areas include *Eragrostis curvula, Hyparrhenia hirta, Cynodon dactylon, Aristida congesta, A. scabrivalvis* and *Setaria nigrirostris*. A typical constant forb in early successional stages is *Walafrida densiflora*.

At the western extreme of the northern belt of Moist Cool-temperate Grassland the more typical communities of this type of grassland occur on chert gravel in the dolomite area. Typical species for this vegetation include *Trachypogon spicatus, Eragrostis racemosa* and *Diheteropogon amplectens*. In areas with an overburden of sand, presumably of aeolian origin, and less typical Moist Cool-temperate Grassland vegetation, *Elyonurus argenteus* is very prominent (Morris 1973, 1976).

Apart from the northern section of the eastern belt which belongs to Acocks' (1975) Northeastern Sandy Highveld, most of the Moist Cool-temperate Grassland areas have been classified by Acocks (1975) as False Grassland Types, i.e. where formerly marginal bushveld has been converted into grassland largely, according to Acocks, as a result of excessive burning. Not only does this grassland type border on bushveld but the close floristic affinities with the near-grassland end of a bushveld gradient have been discussed already. Furthermore, the tendency towards woody vegetation in the absence of fire has been demonstrated by succession studies at Frankenwald in the northern belt of Moist Cool-temperate Grassland (Roux 1969). However, equally frequent burning on the less frosty, low interior plateau does not produce grasslands and as pointed out in previous pages, frost may be an important factor in conjunction with fire in maintaining a grassland vegetation. Whereas frost in the adjoining bushveld areas is lighter than in Moist Cool-Temperate Grassland, frost in the adjoining, undisputedly climax (Moist Cold-temperate) Grassland is considerably more severe than in the contentious (Moist Cool-temperate) Grassland. Frost, provided it is moderate, may be compensated for by a light burning regime to convert marginal grassland into marginal bushveld although all parts of Moist Cool-temperate Grassland would not be equally readily convertible into bushveld.

Other important factors, also discussed in an earlier section, are moisture regime and nutrient status of the soil, including amount of leaching. Grasslands in the Rustenburg Nature Reserve are situated above marginal Evergreen Bushveld, on soils that are distinctly and extremely leached (Coetzee 1975). The sharp boundary between *Protea caffra* bushveld and Grassland, a quite common phenomenon, does not seem to be explicable in terms of fire and frost only. Furthermore, azonal bushveld in severely frosty areas such as Dry Cold-temperate Grassland is commonly associated with nitrogen-enriched fine-textured and lowland soils (Morris 1973, 1976, Scheepers 1975). In these latter instances the species involved may simply be very frost-tolerant, a trait which, as discussed earlier on, may conceivably have evolved first in nutrient rich, clayey and lowland habitats.

Tinley's (pers. comm.) contention that impeded drainage causes grassland is a quite feasible explanation for many grasslands in predominantly Plains Bushveld areas, but it is an unlikely explanation for many stands of typical Moist Cool-temperate Grassland on very well-drained, deep, loose sandy soils of the Hutton and Clovelly Forms (Morris 1973, 1976, Coetzee 1975, Scheepers 1975).

Acocks' False Grassveld Types included under Moist Cool-temperate Grassland are: (i) Bankenveld grasslands, which form the northern belt; and (ii) Piet Retief Sourveld, the Northern Tall Grassveld and Natal Sour Sandveld, which form the northern section of the eastern belt. The southern part of the eastern belt includes typical Highland Sourveld grasslands, which according to Acocks (1953, 1975) have replaced forest.

5.5 *Azonal vegetation*

5.5.1 Flood plain and dambo grasslands

In eastern Angola and in Barotseland vast stretches of flat land are drained and periodically flooded by the Upper Zambezi. Rainfall is about 800 to 1250 mm annually here. These regularly flooded areas and those where during several months each year the water table is near the surface carry short tufted grasslands with varying floristic composition in accordance with depth of the water table and duration of flooding. Most abundant on the hydromorphic soils is *Loudetia simplex*, accompanied by species of *Tristachya*, *Aristida*, *Eragrostis*, and *Ctenium newtonii*. These communities can be classified in the class Ctenio-Loudetietea simplicis and probably as Thesio-Gnidion. When the soil contains more humus Cyperaceae are prominent. On non-flooded sites rhizomatous shrubs like *Parinari pumila* are present, while locally palms like *Raphia* sp. or *Borassus aethiopum* have established themselves.

In Cuando–Cubango in southern Angola similar *Loudetia* grasslands cover large areas, here and there interrupted by thickets on old termitaria (Fig. 62).

Palms include *Phoenix reclinata* and *Hyphaene benguellensis* var. *ventricosa* in this area (Barbosa 1970, Diniz 1973).

On the Kafue floodplains of central Zambia and the floodplains in northeastern Zambia, the grasslands are mostly dominated by species of *Hyparrhenia* although *Loudetia simplex–Scleria hirtella* grasslands, sometimes with scattered trees of *Protea welwitschii*, occur. At Kafue *Hyparrhenia rufa*, *Jardinia angolensis*, *Chloris gayana* and *Setaria mombassana* are particularly important on calcareous clay. On prolonged inundated sites, in channels and lagoons, communities with *Echinochloa stagnina* and *E. pyramidalis*, or with *Vossia cuspidata*, *Vetiveria nigritana*, *Leersia hexandra*, *Oryza longistaminata* and *O. barthii* prevail while on seepages *Imperata cylindrica* often dominates. Also, more per-

Fig. 62. *Hyphaene* palms dominate in a village on the grassy Chobe floodplain in Caprivi. The land in the foreground is cultivated; in the background bushclumps mark termitaria (photo P. N. F. Niven).

manent swamps occur locally with *Phragmites mauritianus*, *Typha domingensis* and several cyperaceous plants, sometimes carrying true sudds of *Cyperus papyrus*. Similar communities cover large areas around Lake Chilwa and in the lower Shire valley in Malawi. Extensive papyrus sudds and grasslands dominated by *Echinochloa crus-pavonis*, *E. pyramidalis* and *Oryza barthii* are common too in the large rivers of northern Angola, e.g. the Rio Cuanza.

The Lake Bangweulu and the Chambeshi floodplains in northeastern Zambia carry an intricate complex of swampy grassland and waterplant communities correlated with waterdepth, duration of flooding, mobility of the water, and landform. In open water communities with *Nymphaea caerulea*, *Utricularia stellaris*, *Trapa natans* and *Pistia stratiotes* are found (see Chapter 33), while on river fringes *Vossia cuspidata* and *Echinochloa stagnina* dominate. Deep water floodplains carry communities with *Scirpus corymbosus, Thalia welwitschii, Eleocharis plantaginea*, and a number of other species, or with *Oryza barthii, Leersia hexandra, Aeschynomene cristata*, and others. On shallow water floodplains this latter community also occurs in a slightly different species composition, but here also *Cyperus digitata* sedge marshes, and *Acroceras macrum* meadows occur with *Panicum glabrescens, Echinochloa pyramidalis, Entolasia imbricata, Paspalum commersonii* and other species. On somewhat raised areas *Hyparrhenia gazensis* often dominates accompanied by several other species of *Hyparrhenia, Paspalum commersonii, Digitaria scalarum, Loudetia simplex, Setaria sphacelata, Themeda triandra*, and others, while levees carry a *Phoenix reclinata* community and *Hyparrhenia nyassae* and some other species are prominent near the margin of the floodplain. *Loudetia simplex* grasslands cover extensive areas on the terraces in the Chambeshi area, which also are flooded in the rainy season. Around Lake Bangweulu sudds with *Cyperus papyrus* and *Ludwigia stolonifera* and swamps of *Typha latifolia* subsp. *capensis* are very extensive, and they also occur at the other swampy places of northeastern Zambia. In central Zambia the grasslands are often fringed by broad-leaved savanna or bushveld, or on lower lying, wet areas by microphyllous, thorny savannas or thickets. In northeastern Zambia these communities are often lacking and the grasslands are fringed by miombo or chipya vegetation (Trapnell & Clothier 1937, Trapnell 1953, Rattray 1960, Rattray & Wild 1961, Vesey-Fitzgerald 1963, Feely 1965, Wild & Barbosa 1967, Astle 1965, 1968–69, Astle et al. 1969, Jackson 1969c, Verboom & Brunt 1970, Fanshawe 1971, Sheppe & Osborne 1971, Verboom 1975, Howard-Williams 1975).

In Moçambique the deltas and large alluvial areas at the mouth of the Zambezi and other large rivers further south, have a mosaic of extensive grasslands on badly-drained and periodically flooded soils, alternated by woody vegetation (Fig. 31). Important are *Setaria holstii, S. mombassana, S. sphacelata* and *Ischaemum afrum*, but also *I. arcuatum, Bothriochloa insculpta, Urochloa mosambicensis, Panicum coloratum*, and other species. In the Zambezi delta extensive patches with *Pennisetum purpureum* are found. Edaphic grasslands dominated by *Hyparrhenia rufa* on sandy soils, and *Themeda triandra* and *Ischaemum afrum* on clayey soils, occur over limited areas in Malawi and Moçambique (Wild & Barbosa 1967).

Similar grasslands along drainage lines occur further towards the interior plateau of southern Africa, e.g. in the Victoria Falls National Park from where Hill (1969) described a *Hyparrhenia filipendula–Ischaemum afrum* community.

Also in the Transvaal Lowveld *Ischaemum afrum* is prominent in this type of edaphic grasslands (vleis), together with *Sporobolus consimilis*, and associated with other *Sporobolus* species and *Setaria woodii* (Pienaar 1963).

Somewhat different in species composition are the floodplain grasslands in Caprivi and further south (Fig. 63). In Caprivi they contain predominantly *Miscanthidium teretifolium, Cynodon dactylon, Imperata cylindrica, Nicolasia costrata, Rhamphicara tubulosa* and *Sopubia simplex* (Vahrmeijer pers. comm.), while the extensive fresh water floodplains and swamps of the Okavango–Linyanti system for a major part are covered by a grassland dominated by *Cymbopogon excavatus*, accompanied by *Panicum repens, Andropogon eucomus* and some other species with patches of *Cynodon dactylon* and *Sporobolus spicatus*, and a *Cyperus papyrus* sudd on the permanently wet places. Other communities with

Fig. 63. An intricate mosaic of aquatic vegetation, floodplain grassland and remnants of levees, frequently in a typical horseshoe pattern, covers an extensive area in Caprivi. The floodplains contain *Miscanthidium teretifolium, Cynodon dactylon, Imperata cylindrica*, etc. and the levees support *Diospyros mespiliformis, Garcinia livingstonei, Lonchocarpus capassa, Croton megalobotrys, Ficus sycomorus, Kigelia africana*, etc. (photo H. Vahrmeijer).

Echinochloa and *Vossia* also occur, as well as *Imperata cylindrica* grasslands. Gallery forests occur along the flood channels.

At Ngami *Panicum repens* is also very common, and accompanied by *Cynodon dactylon* and various weeds, mainly as a result of severe overgrazing of the grassland. At Mababe the grassland consists mainly of *Cenchrus ciliaris, Chloris gayana, Digitaria* spp. and *Cynodon dactylon*. Makarikari and Etosha are more saline. The pan floor is bare and surrounded by an open grass community of *Sporobolus spicatus* and *Odyssea paucinervis*. The latter species, together with *Cynodon dactylon*, forms a belt around this community, particularly on the slopes of low surrounding dunes (Fig. 64). In a wider area around the pan the same species occurs in extensive grasslands together with *Aristida meridionalis*, *Heteropogon contortus*, *Cymbopogon plurinodis*, and *Sporobolus spicatus*. Swampy areas with *Phragmites australis* and drier patches with *Acacia* occur also (Seiner 1909, 1912, Pole Evans 1948a, Seagrief & Drummond 1958, Rattray 1960, Tinley 1966, Wild & Barbosa 1967, Wild 1968a, Blaire Rains & McKay 1968, Weare & Yalala 1971, Simpson 1975).

Fig. 64. The Etosha pan is a bare salt desert but at its margin halophytes like the grasses *Odyssea paucinervis* and *Sporobolus spicatus* and the dwarf shrub *Petalidium engleranum* (all in foreground) are important. *Acacia nebrownii* is also prominent (photo W. Giess).

Also in southern Moçambique saline grasslands occur locally, particularly along the Rio Changane (Pedro & Barbosa 1955, Wild & Barbosa 1967, Myre 1961, 1964, 1971).

Dambos in central and eastern Angola, Shaba, Zambia, Malawi and parts of Rhodesia and Moçambique support *Loudetia* grasslands belonging to the Ctenio–Loudetietea. Usually these grasslands are fringed by a broad-leaved savanna or bushveld. Where a humic top soil overlies the nutrient poor sand in some dambos, Cyperaceae are more important and communities are formed by the sedges *Rhynchospora rugosa*, *Mariscus deciduus*, *Acriulus greigiifolius*, *Fuirena umbellata*, together with *Lycopodium carolinianum* var. *tuberosum*, *Xyris* sp., *Drosera affinis*, *Thelypteris confluens* and various orchids of the genera *Satyrium*, *Disa* and *Eulophia*, but there can be a marked seasonal change in visible floristic composition (Kornaś 1975, Malaisse 1975).

441

Montane grasslands with *Loudetia simplex*, *Exotheca abyssinica*, and *Monocymbium ceresiiforme* or with *Andropogon schirensis*, *Elyonurus argenteus*, *Pteridium aquilinum* and many other species are more common on the highlands of Malawi, particularly the Nyika and Vipya plateaux and Mt. Mlanje and they also occur in the highlands of the Rhodesian–Moçambiquan border area. They are discussed in Chapter 11, while the grasslands on serpentine are covered in Chapter 40.

5.5.2 Woody vegetation

In the previous pages notes on woody vegetation of wet, waterlogged or riparian sites frequently have been made already, so that a short summary of the main features will suffice here.

In the periguinean part of southern Africa, where the Zambezian Domain fringes on or is transitional to the Guineo–Congolian Region, gallery and swamp forests with strong Guineo–Congolian affinities penetrate deep into the Zambezian Domain. These forests are described in Chapter 14.

In the coastal zone of northern Angola *Elaeis guineensis* and *Ceiba pentandra* are common species on riverine and swampy sites, often accompanied by *Tamarindus indica*, *Lonchocarpus sericeus*, *Mangifera indica*, and others. Further inland *Pandanus* spp. can build dense stands on swampy, riparian sites. In the miombo area of Angola, Zambia, Malawi and Moçambique various types of swamp forests occur. Fanshawe (1971) distinguishes between estuarine swamp forest, on mounds between flood channels, flooded all year round and characterized by *Ficus congensis*, *Mitragyna stipulosa*, *Raphia farinifera* and *Xylopia aethiopica*; seepage swamp forest, also on mounds between flood channels but with the water table at or just above ground level all year round and characterized by *Bequaertiodendron magalismontanum*, *Ilex mitis*, and the species of the estuarine swamp forest; and seasonal swamp forest, on alluvial flats, flooded during the rainy season but with the water table near ground level for the rest of the year, and characterized by many riparian species, e.g. *Syzygium cordatum*, *Cleistanthus milleri*, *Ficus capensis*, *Homalium africanum*, *Ilex mitis*, *Rauvolfia caffra*, and others. Seasonal swamp forest is similar to and grades into riparian forest. These swamp forests only occur in the high rainfall area of 1000 mm or more annually. Swamp forest is an evergreen formation, with trees of the above-named species up to 27 m high and a second discontinuous evergreen canopy between 9–18 m high, characterized by *Garcinia smeathmannii*, *Gardenia imperialis*, *Aporrhiza nitida* and *Phoenix reclinata*. A dense evergreen shrub layer is also common in these forests.

Riparian forest, or sometimes woodland, is widespread over the miombo area (Fig. 65). Well-developed it is an evergreen forest of some 20 m high and characterized by *Diospyros mespiliformis*, *Khaya nyasica*, *Parinari excelsa*, *Syzygium cordatum*, *S. guineense*, *Adina microcephala*, *Berchemia discolor*, *Bridelia micrantha*, *Cleistanthus milleri*, *Ilex mitis*, *Manilkara obovata*, *Trichilia emetica*, *Kigelia africana*, *Ficus sycomorus*, *F. capensis*, *Ekebergia capensis*, *Mimusops zeyheri*, *Xanthocercis zambesiaca*, *Acacia xanthophloea*, *A. albida*, and several other species. In a lower, discontinuous evergreen canopy layer *Bequaertiodendron magalismontanum*, *Gardenia imperialis*, *Diospyros lycioides* and *Phoenix reclinata* can be common. Climbers and evergreen shrubs are also

Fig. 65. Dense riparian vegetation along Rio Cuvo, Angola (photo M. J. A. Werger).

abundant and in the submontane areas the tree fern *Alsophila dregei* is frequent. From north to south in the miombo area the riparian forest changes from broad and tall evergreen forests to a narrow and frequently open semi-deciduous gallery strip. The evergreen species and species of Guineo–Congolian affinity are commonest in the high rainfall area along perennial streams whereas further southward Zambezian species become dominant and the formation fringes both perennial and seasonal streams. Several of these Zambezian riparian species reach far south and still fringe the rivers of the non-frosty Transvaal Lowveld (Fig. 66), southern Moçambique and northern Natal. The riparian vegetation often occurs in catenas or complex mosaic patterns in the lowlands along the east coast, as has been described in Section 5.4.6. Also, the well-developed spray forests around the large waterfalls in the Zambezian Domain, such as that at Duque de Bragança and the 'rain forest' at Victoria Falls, are predominantly composed of the above-listed riparian species.

When the soils are very compact and clayey the riparian vegetation is different: it usually is a more open woodland or savanna, and it is often dominated by *Acacia* species, particularly *A. polyacantha*. Fairly extensive woodlands on waterlogged soils dominated by *Acacia* species are also common (Fig. 67), the most conspicuous being those of *A. xanthophloea* (Section 5.4.5). Locally palm savannas, e.g. with *Borassus aethiopum*, dominate temporarily inundated depressions (Fig. 29).

In the area dominated by *Baikiaea* riverine and low-lying habitats are characterized by *Acacia* species, including *A. erioloba*, but *Colophospermum mopane* occurs locally also on such sites.

In the mopane area riparian forests, woodlands and savannas occur in which *Acacia albida*, *A. xanthophloea*, *Lonchocarpus capassa*, *Combretum imberbe*,

Fig. 66. Tropical riparian vegetation along the Sabie River, Kruger National Park, with abundant *Adina microcephala* (photo P. van Wyk).

Croton megalobotrys, *Ficus* spp., *Hyphaene* and *Phoenix* palms, and several other species are common (Fig. 68). These communities are transitional to mopane vegetation proper and vary in floristical composition with geographical distribution. Section 5.3 provides more detail on these communities. In the wide Zambezi and Luangwa valleys *Acacia albida* and *Trichilia emetica* form extensive woodlands or savannas, but riparian forest patches occur also (Fig. 69).

Fig. 67. *Acacia xanthophloea* woodland on a marshy site at Ndumu, Zululand (photo E. J. Moll).

Fig. 68. Zebra and buffalo in a riparian depression with *Hyphaene natalensis*, *Lonchocarpus capassa* and *Kigelia africana* in the Kruger National Park, Transvaal (photo M. J. A. Werger).

In the dry sandy southwestern fringes of the Zambezian Domain in the interior of South West Africa and in Botswana several *Acacia* species (*A. erioloba, A. albida, A. karroo, A. tortilis* subsp. *heteracantha*) fringe the drainage lines and where frosts are less severe *Combretum imberbe* and *Lonchocarpus capassa* are common (Gossweiler & Mendonça 1939, Pole Evans 1948a, Van der Schijff 1957, Boughey 1961, Myre 1964, Wild 1964, Feely 1965, Mildbraed &

Fig. 69. Zambezi floodplain in the vicinity of Mana Pools looking northwestwards into Zambia. Common trees on the floodplain include *Acacia albida* (light coloured), *Trichilia emetica* (darkest coloured) and *Kigelia africana*. Common grasses include *Vetiveria nigritana*, *Setaria sphacelata* and *Panicum maximum*. In the background the mopane woodland off the alluvium forms a sharp boundary with the floodplain (photo B. H. Walker).

Domke 1966, Wild & Barbosa 1967, Hill 1969, Barbosa 1970, Chapman & White 1970, Monteiro 1970, Bosch 1971, Fanshawe 1971, Giess 1971, Van Wyk 1973, Simpson 1975, Cole & Brown 1976).

In the light to moderately frosty parts of southern Africa, particularly in the temperate bushveld area of the low continental plateau in eastern Botswana and in the Transvaal between the Limpopo valley and the South African Highveld, as well as in the Moist Cool-temperate Grassland area, that is transitional to the high plateau of the Orange Free State, the riparian vegetation is mostly a shrubby, dense and rather low, temperate broad-orthophyll gallery forest or bush. It is floristically fairly rich, and has some of its species in common with more tropical areas, while others are shared with the more frosty parts of southern Africa. General species include the trees *Acacia caffra, Kiggelaria africana, Apodytes dimidiata, Olea capensis, Olinia emarginata* and *O. cymosa*; and the shrubs or small trees *Acokanthera oppositifolia, Brachylaena rotundata, Cassine burkeana, Cassinopsis ilicifolia, Diospyros lycioides* subsp. *guerkei* and subsp. *lycioides, D. whyteana, Grewia occidentalis, Nuxia congesta, Myrica serrata, Myrsine africana, Osyris lanceolata, Rhamnus prinoides, Halleria lucida, Hellinus integrifolius, Buddleia salviifolia*, and several others. *Combretum erythrophyllum, Acacia ataxacantha, Olea africana*, and a few other species are also common throughout this gallery strip, but they also occur in somewhat more tropical parts in riverine formations. Apart from these species the river fringes in the relatively warmest parts of this light to moderately frosty area generally support the following trees which are shared with more tropical areas further north and east: *Mimusops zeyheri, Urera tenax, Rauvolfia caffra, Pittosporum viridiflorum, Ilex mitis, Trema orientalis, Berchemia zeyheri* and *Acacia robusta*. Typical of, and abundant in the cooler parts are, particularly, *Buddleia saligna, Celtis africana* and *Calodendrum capense* (West 1951, Van Vuuren & Van der Schijff 1970, Van der Schijff 1971, Theron 1973, Coetzee 1974, 1975).

On the severely frosty, high South African plateau, drainage lines are usually fringed by grass and sedge communities and rivulets commonly have occasional exotics on their banks, such as a lone *Salix babylonica* or small stands of *Populus* spp. or Australian *Acacia* spp. Along larger streams such exotics may form fairly extensive, continuous stands. The indigenous vegetation here is a cold-temperate thornbush, or a low, thorny gallery forest along the large rivers. This formation is poor in species and its woody species have wide distribution ranges. Notable examples are *Acacia karroo, Ziziphus mucronata, Rhus pyroides, Maytenus heterophylla, Diospyros lycioides* subsp. *lycioides* and *Celtis africana*. Werger (1973a) classified this gallery forest as two associations, the Rhoo–Diospyretum (Fig. 70) and the Zizipho–Acacietum karroo, both belonging to the alliance Diospyrion lycioidis (Mostert 1958, Werger 1973a, Scheepers 1975, Morris 1976).

5.6 *The transition to the Karoo–Namib Region*

In the southwestern parts of the South African Highveld in the Orange Free State and the Cape Province, the original grassland is widely intruded by Karoo–Namib species, often to such an extent that the original grassland has been replaced by an open dwarf shrub vegetation in the plains where the soil is deep, and a slightly more grassy open dwarf shrub and shrub vegetation on the

Fig. 70. Open riverine stand of Rhoo–Diospyretum with much *Acacia karroo* along the Orange River between Bethulie and Aliwal North (photo M. J. A. Werger).

boulder-strewn slopes. This transformation from grassland to semi-desert is generally attributed to the introduction and overstocking of domestic livestock, mainly sheep, following the colonization of the area about one and a half centuries ago. That the original vegetation was grassland can be seen from the diaries and travel accounts of the early-day travellers, to which Acocks (1953) has drawn attention and which have been reviewed by Werger (1973a). It is possible that several of the karroid dwarf shrub species originally occurred already in part of the grassland area now covered by dwarf shrub vegetation, but that they were not prominent in the abundant grass growth, so that these species were not actually intruding the grassland area but merely gained absolute dominance over the grasses following mismanagement. This seems also to be indicated by the existence of a distinct group of species, including some of the intruders, which have their distribution area restricted to a fairly narrow zone at both sides of the boundary between the Sudano–Zambezian and the Karoo–Namib Regions (Werger 1973a). The intrusion of Karoo–Namib species into the Highveld vegetation is still continuing at a fairly fast rate. Jarman & Bosch (1973), using satellite imagery, recently measured an advance of karroid dwarf shrub vegetation into former grassland of 43 to 70 km in an eastern and northeastern direction over a period of 20 years.

The vegetation of this transitional area has been studied in detail by Werger (1973a). He distinguished two main groups of communities: the Pentzio–Chrysocomion communities on deep sandy loam soils of the pediplains, and the Rhoetea erosae communities of the rocky and skeletal soils of the slopes and outcrops.

The Pentzio–Chrysocomion communities are open dwarf shrub communities with a low ground cover on deep and often solonetzic soils showing a severe degree of erosion. One association, the Hermannio coccocarpae–Nestleretum confertae was distinguished by a number of subassociations indicating slight soil differences or various stages of degeneration. Dominant species in the association are virtually always the composite dwarf shrubs *Chrysocoma tenuifolia* and *Pentzia globosa*, with the grass *Eragrostis lehmanniana*, while the prostrate grass *Tragus koelerioides* is usually also prominent. In its typical form (the *Eragrostis curvula* subassociation) the association still contains a fair number of Sudano–Zambezian species, but the association is always floristically rather poor (Table 11) (Fig. 71).

The class Rhoetea erosae consists of two orders: the Grewio–Rhoetalia erosae and the Rhoetalia ciliato-erosae. The former comprises the associations Rhamno–Rhoetum, Rhoo–Aloetum ferocis and Blepharido–Rhoetum. These are fairly open shrubby communities also containing low trees, dwarf shrubs and grasses, occurring on the rocky slopes of dolerite, sandstone and shales in the southern parts of the Highveld. The Rhamno–Rhoetum occurs on the coolest and most

Table 11. Hermannio coccocarpae–Nestleretum confertae (after Werger 1973a).

Community Number of relevés	A1 13	A2 33	A3 18	A4 11	Community Number of relevés	A1 13	A2 33	A3 18	A4 11
Hermannio coccocarpae–					Pentzia globosa	4	5	5	4
Nestleretum confertae species					Walafrida saxatilis	4	5	5	4
Cynodon hirsutus	.	2	3	2	Aristida congesta	4	4	4	4
Hermannia coccocarpa	1	2	3	1	Eragrostis obtusa	2	4	2	2
Osteospermum scariosum	1	2	1	.	Lycium salinicolum	4	3	3	.
Convolvulus boedeckerianus	.	1	2	1	Gnidia polycephala	2	2	3	2
Lessertia pauciflora	2	1	1	1	Indigofera alternans	3	2	2	2
Nestlera conferta	.	1	2	1	Gazania krebsiana	2	1	3	1
Schizoglossum capense	2	1	1	.	Felicia muricata	.	2	2	1
Aptosimum marlothii	4	.	1	.	Pterothrix spinescens	1	2	2	3
Enneapogon desvauxii	4	.	.	.	etc.				
Eriocephalus spinescens	2	1	1	1					
Eragrostis curvula	.	4	.	1	Companion species				
Cyperus usitatus	.	4	.	.	Aristida diffusa	1	2	2	1
Themeda triandra	.	2	1	1	Nenax microphylla	1	2	2	1
Helichrysum dregeanum	.	2	1	1	Limeum aethiopicum	2	2	1	.
Solanum supinum	.	2	.	.	Aptosimum depressum	1	1	2	1
Oropetium capense	1	.	2	.	Hibiscus marlothianus	2	1	2	1
					Trichodiadema pomeridianum	.	2	1	.
Pentzio-Chrysocomion species					Sutera atropurpurea	1	2	1	.
Chrysocoma tenuifolia	5	5	5	5	Salsola glabrescens	2	1	1	.
Tragus koelerioides	5	5	5	3	Melolobium microphyllum	.	1	2	1
Eragrostis lehmanniana	5	5	5	5	etc.				

1: Subassociation with Aptosimum marlothii.
2: Subassociation with Eragrostis curvula.
3: Subassociation with Oropetium capense.
4: Variant with high cover of Eragrostis lehmanniana.

Fig. 71. Pediplains in the southwestern and southern Highveld with deep sandy loam soils carry the Hermannio–Nestleretum with abundant *Chrysocoma tenuifolia* and often with low termite mounds. On the hillside in the right hand background Stachyo-Rhoetum. Near Goedemoed (photo M. J. A. Werger).

Fig. 72. Rhoo–Aloetum ferocis on bouldery slope of Molteno sandstone just east of Aliwal North showing *Aloe ferox*, *Rhus erosa*, *Rh. undulata* and *Diospyros lycioides* subsp. *lycioides* (photo M. J. A. Werger).

449

Fig. 73. Osteospermetum leptolobi on very shallow, gravelly soil over slightly sloping sandstone in foreground and Stachyo–Rhoetum on steeper, bouldery slopes, partly of dolerite, in background. Near Norvalspont (photo M. J. A. Werger).

Fig. 74. Dolerite ridge supporting Mayteno–Oleetum with much *Olea africana* traverses the open vegetation of grasses and dwarf shrubs on shallow soil over mudstone in Tussen die Riviere Game Farm near Bethulie, O.F.S. (photo M. J. A. Werger).

Table 12. Rhoetalia ciliato-erosae communities (after Werger 1973a).

Community Number of relevés	A1 32	A2 8	B1 11	B2 31	B3 5	C 16	D1 10	D2 14	E 9
Osteospermetum leptolobi species									
Osteospermum leptolobum	2	3	1	1	.	.	.	.	.
Eriocephalus spinescens	2	3	1	1	2	1	.	.	.
Pentzia sphaerocephala	2	.	2	1	.	1	.	.	1
Euphorbia clavarioides	2	.	.	.	.	2	.	.	.
Phymaspermum parvifolium	1	2	.	1	.	.	.	1	.
Aptosimum marlothii	1	4	2	2	2	.	.	.	1
Enneapogon desvauxii	.	4	.	1	1	.	.	.	2
Stachyo-Rhoetum species									
Stachys rugosa var. linearis	1	.	3	3	3	.	3	3	4
Fingerhuthia africana	1	1	2	3	2	.	.	.	1
Phyllanthus maderaspatensis	.	.	2	2	3	2	.	.	2
Indigofera sessilifolia	1	1	2	2	1	.	.	1	.
Polygala ephedroides	1	.	3	.	.	1	.	1	.
Polygala uncinata	1	.	2	1	.	1	.	1	.
Hermannia candidissima	.	1	.	3	3	.	1	.	2
Polygala leptophylla	1	3	.	3	.	1	.	1	.
Talinum caffrum	.	1	.	3	1	.	1	1	1
Ehretia rigida	.	.	.	2	3	1	.	1	2
Rhigozum obovatum	.	.	.	2	2	1	.	.	1
Pachypodium succulentum	.	.	.	2	.	.	1	.	1
Ziziphus mucronata	.	.	.	1	4	.	.	.	1
Barleria rigida	.	.	.	1	3	.	.	.	.
Hermannia pulchra	.	.	.	2	1	.	1	.	1
Eragrostis chloromelas	.	.	.	1	.	1	.	.	2
Salvia namaensis	.	.	.	.	5	.	.	.	3
Nanantho vittati-Rhoetum species									
Brachiaria serrata	.	.	.	.	.	3	.	2	.
Nananthus vittatus	1	.	.	1	.	2	.	1	.
Haworthia tesselata	.	.	.	.	.	3	.	.	.
Dicoma macrocephala	1	.	.	1	.	2	1	.	1
Hyparrhenia hirta	.	.	.	1	.	2	.	.	.
Asclepias fruticosa	.	.	1	.	.	2	.	.	.
Hibisco marlothiani–Rhoion erosae species									
Pegolettia retrofracta	2	2	3	3	4	2	1	2	2
Chascanum pinnatifidum	2	1	3	2	2	4	1	2	1
Hibiscus marlothianus	4	3	3	1	.	1	1	2	.
Aptosimum depressum	3	4	4	1	.	1	.	1	.
Melolobium microphyllum	2	.	2	2	3	3	1	1	3
Helichrysum zeyheri	1	2	2	2	2	2	.	.	1
Helichrysum lucilioides	1	5	2	2	3	.	.	.	1
Blepharis villosa	2	4	3	1	.	.	.	1	.
Eriocephalus pubescens	1	2	1	2	3	1	.	1	1
Thesium spartioides	1	2	3	1	.	1	.	1	.
Trichodiadema pomeridianum	2	2	1	1	.	1	.	.	.
Anacampseros lanigera	1	.	2	1	.	.	.	.	.
Mayteno polyacanthae–Oleetum africanae species									
Maytenus polyacantha	.	.	.	1	.	.	2	3	1
Asparagus laricinus	.	.	1	1	3	1	2	3	2
Pelargonium aridum	.	.	.	.	.	1	2	2	.
Senecio hieracioides	.	.	.	.	.	.	1	2	.
Celtis africana	.	.	.	.	.	1	3	2	.
Mohria caffrorum	.	.	.	1	.	.	2	2	2
Chamarea capensis	.	.	1	.	.	.	1	3	.

Table 12 (contd.)

Community	A1	A2	B1	B2	B3	C	D1	D2	E
Number of relevés	32	8	11	31	5	16	10	14	9
Setario lindenbergianae–Buddleietum salignae species									
Setaria lindenbergiana	.	.	.	.	.	.	.	.	5
Buddleia saligna	.	.	.	.	.	.	.	1	4
Asparagus striatus	.	.	1	1	.	1	1	1	4
Osyris lanceolata	.	.	.	1	.	.	1	.	3
Solanum retroflexum	.	.	.	1	.	.	.	.	2
Cussonia paniculata	.	.	.	.	.	.	.	.	2
Rhoetalia ciliato–erosae species									
Limeum aethiopicum	2	5	3	4	4	4	1	3	.
Sutera albiflora	2	4	4	3	2	3	1	3	2
Rhus ciliata	1	1	5	3	2	5	2	3	3
Nenax microphylla	4	2	3	1	.	2	1	2	.
Lotononis laxa	2	.	5	2	.	3	1	4	.
Sutera halimifolia	1	1	4	2	1	3	2	3	2
Euclea coriacea	.	.	2	1	.	3	3	2	.
Aloe broomii	.	.	1	2	.	3	1	2	1
Rhoetea erosae species									
Aristida diffusa	5	4	5	5	3	5	4	5	3
Heteropogon contortus	2	.	5	4	2	5	3	4	3
Rhus undulata	.	.	2	5	3	4	5	4	4
Themeda triandra	2	.	4	3	2	4	3	4	5
Eragrostis curvula	1	.	2	2	2	5	4	5	2
Eustachys mutica	.	.	3	4	2	4	4	4	4
Sporobolus fimbriatus	1	.	2	4	5	3	3	3	4
Cymbopogon plurinodis	1	.	4	3	3	4	4	3	2
Enneapogon scoparius	2	2	3	3	1	4	3	2	4
Diospyros austro-africana	1	.	4	3	.	3	2	4	3
Rhus erosa	.	.	3	2	.	4	5	5	3
Hibiscus pusillus	1	2	2	4	2	3	1	2	3
Lightfootia albens	1	1	4	2	1	3	2	4	1
Digitaria eriantha	1	.	2	4	4	2	3	2	4
Diospyros lycioides	1	.	2	2	1	4	5	5	2
Olea africana	1	.	2	2	1	3	5	4	.
Helichrysum dregeanum	1	.	3	1	.	4	3	3	.
Solanum supinum	1	.	1	3	3	3	2	2	1
Dianthus basuticus	2	.	3	2	2	2	2	2	1
Selago albida	1	.	.	3	.	1	2	2	2
Argyrolobium lanceolatum	.	.	1	2	3	3	3	3	1
Hermannia cuneifolia	1	.	1	1	.	2	2	3	.
Cheilanthes hirta	1	.	2	1	1	.	3	3	4
Tarchonanthus camphoratus	.	.	1	1	3	.	2	3	4
Euclea crispa	.	.	.	1	1	1	1	2	4
Anthospermum rigidum	1	.	2	1	2	2	2	1	.
Solanum coccineum	.	.	.	1	1	1	2	1	5
Cheilanthes eckloniana	.	.	.	2	.	2	.	2	3
Adromischus rupicola	.	.	.	2	.	2	.	2	1
Cotyledon decussata-orbiculata compl.	1	.	.	2	3	2	1	2	.
Lantana rugosa	.	.	1	2	3	2	2	1	2
Pellaea calomelanos	.	.	1	1	1	2	1	1	2
Rhynchelytrum repens	.	.	1	1	1	2	.	1	.
Pentzio–Chrysocomion species									
Chrysocoma tenuifolia	5	5	5	5	5	5	5	5	4
Tragus koelerioides	5	5	5	5	4	5	4	5	4
Walafrida saxatilis	5	1	4	2	.	4	3	5	2
Aristida congesta	4	2	3	4	5	5	3	3	4
Pentzia globosa	4	4	2	3	.	4	3	5	1
Eragrostis lehmanniana	4	3	4	3	3	3	3	3	2
Gnidia polycephala	4	3	3	3	.	3	1	3	.
Lycium salinicolum	2	2	1	3	3	3	2	3	2

Community	A1	A2	B1	B2	B3	C	D1	D2	E
Number of relevés	32	8	11	31	5	16	10	14	9
Eragrostis obtusa	2	2	2	3	1	1	1	2	3
Gazania krebsiana	2	4	1	2	1	2	1	2	.
Indigofera alternans	2	4	3	1	1	2	1	2	.
Pterothrix spinescens	3	2	2	1	.	2	1	3	.
Felicia muricata	1	.	1	1	3	2	2	2	2
Hermannia coccocarpa	2	1	1	1	.	1	1	3	.
Mariscus capensis	1	.	.	2	.	2	.	2	.
Hermannia linearifolia	2	.	1	1	.	1	1	1	.
Osteospermum scariosum	1	2	.	.	.	.	.	.	.
Companion species									
Asparagus suaveolens	1	2	4	5	5	5	5	5	4
Viscum rotundifolium	.	.	2	2	1	2	4	1	.
Pollichia campestris	.	.	1	2	1	2	1	2	2
Aristida curvata	1	1	.	1	4	1	.	1	4
Acacia karroo	.	.	1	2	2	.	.	1	1
Sutera atropurpurea	2	.	.	1	.	2	1	1	.
Felicia filifolia	.	.	1	1	.	1	2	1	1
Oropetium capense	1	1	.	.	.	2	.	.	.
Elyonurus argenteus	.	.	.	.	.	1	2	.	.
Blepharis integrifolia	.	.	.	.	.	2	.	.	.
Crassula nodulosa	.	.	2	1	.	1	.	.	.
Geigeria filifolia	1	1	2	.	.	1	.	.	.
Melica decumbens	.	.	.	1	.	1	.	2	.
Pteronia glauca	1	.	.	1	1	1	.	.	.
Salsola glabrescens	1	2	.	.	.	.	.	.	.
etc.									

A: Osteospermetum leptolobi with typical (1) and Aptosimum marlothii (2) subassociations.
B: Stachyo–Rhoetum with Polygala spp. (1) and Hermannia candidissima (2) subassociations and Salvia namaensis variant (3).
C: Nanantho vittati–Rhoetum.
D: Mayteno polyacanthae–Oleetum africanae with typical (1) and Chamarea capensis (2) variant.
E: Setario lindenbergianae–Buddleietum salignae.

mesic sites and contains a distinct Afromontane element. Important species here include the shrubs or small trees *Rhamnus prinoides, Myrsine africana, Rhus divaricata, R. dentata, R. erosa, R. undulata, Olea africana, Diospyros lycioides, D. austro-africana, Grewia occidentalis, Maytenus heterophylla*, etc. and the grasses *Cymbopogon validus, Themeda triandra, Eragrostis curvula, Aristida diffusa* and *Heteropogon contortus*. The Rhoo–Aloetum ferocis (Fig. 72) occurs on warmer, mainly north-facing slopes in the same area. In this open shrub community many of the same species named above occur, but characteristic and abundant are the succulent tree *Aloe ferox* and the shrublet *Asparagus virgatus*. Also prominent are somewhat 'drier' grasses, such as *Eustachys mutica* and *Enneapogon scoparius*. The Blepharido–Rhoetum, a similar shrubby community but with several characteristic undergrowth species, occurs on again somewhat less mesic slopes and forms the transition to the Rhoetalia ciliato-erosae communities. In this order five associations are distinguished all of which contain several of the above-mentioned species, but they each also have their own group of species differentiating the community. Usually these communities are more open than the Grewio–Rhoetalia communities though occasionally fairly dense stands, dominated by *Olea africana*, occur. The communities are: the Osteospermetum leptolobi of gently sloping sandstone and mudstone outcrops

(Fig. 73); the Stachyo–Rhoetum of steep and hot dolerite slopes; the Nanantho vittati–Rhoetum of steep and hot sandstone slopes; the Mayteno polyacanthae–Oleetum africanae of steep, south-facing and thus cooler dolerite slopes (Fig. 74); and the Setario lindenbergianae–Buddleietum salignae occurring over a limited area on relatively cool dolerite slopes just under the hill tops where large boulders are frequent and influence the soil moisture regime and the microclimate (Table 12). All these Rhoetea erosae communities, and in particular the Rhoetalia ciliato-erosae communities, contain a fair amount of Pentzio–Chrysocomion species, not seldomly reaching high cover values.

References

Acocks, J. P. H. 1953. Veld Types of South Africa. Mem. Bot. Surv. S. Afr. 28:1–192.
Acocks, J. P. H. 1975. Veld Types of South Africa. 2nd ed. Mem. Bot. Surv. S. Afr. 40:1–128.
Agnew, S. & Stubbs, M. (ed.) 1972. Malawi in maps. Africana Publ. Corp., New York.
Airy Shaw, H. K. 1947. The vegetation of Angola. J. Ecol. 35:23–48.
Astle, W. L. 1965. The grass cover of the Chambeshi Flats, Northern Province, Zambia. Kirkia 5:37–50.
Astle, W. L. 1968–69. The vegetation and soils of Chishinga ranch, Luapula Province, Zambia. Kirkia 7:73–102.
Astle, W. L., Webster, R. & Lawrance, C. J. 1969. Land classification for management planning in the Luangwa valley of Zambia. J. appl. Ecol. 6:143–169.
Aubréville, A. 1949. Climats, forêts et désertification de l'Afrique tropicale. Soc. Ed. Géogr. Mar. Colon, Paris.
Aubréville, A. 1962. Extension géographique. C.S.A./C.C.T.A. Publ. 52:89–92.
Barbosa, L. A. Grandvaux 1970. Carta fitogeográfica de Angola. Inst. Inv. Cien. Angola, Luanda.
Barclay-Smith, R. W. 1964. A report on the ecology and vegetation of the Great Dyke within the Horseshoe Intensive Conservation Area. Kirkia 4:25–34.
Blair Rains, A. & McKay, A. D. 1968. The Northern State Lands, Botswana. Land Res. Stud. no. 5. D.O.S., Tolworth.
Blair Rains, A. & Yalala, A. M. 1972. The Central and Southern State Lands, Botswana. Land Res. Stud. no. 11. D.O.S. Tolworth.
Bosch, O. J. 1971. 'n Ekologiese studie van die plantegroei van 'n gedeelte van die Laer Krokodilriviervallei noordwes van Thabazimbi. Unpubl. M.Sc. Thesis, Univ. Potchefstroom.
Boughey, A. S. 1957. The vegetation types of the federation. Proc. Trans. Rhod. Sci. Ass. 45:73–91.
Boughey, A. S. 1961. The vegetation types of Southern Rhodesia. Proc. Trans. Rhod. Sci. Ass. 49:54–98.
Bourlière, F. & Hadley, M. 1970. The ecology of tropical savannas. Ann. Rev. Ecol. Syst. 1:125–152.
Brass, L. J. 1953. Vegetation of Nyasaland. Mem. N.Y. Bot. Gard. 8:161–190.
Bredenkamp, G. I. 1975. 'n Plantsosiologiese studie van die Suikerbosrandnatuurreservaat. Unpubl. M.Sc. thesis, Univ. Pretoria.
Bremekamp, C. E. B. 1935. The origin of the flora of the central Kalahari. Ann. Transvaal Mus. 16:443–455.
Brynard, A. M. 1964. The influence of veld burning on the vegetation and game of the Kruger National Park. In: D. H. S. Davis (ed.), Ecological studies in Southern Africa, pp. 371–393. Junk, The Hague.
Buitendach, E. 1973. The vegetation of the Lowveld. J. Bot. Soc. S. Afr. 59:45–50.
Chapman, J. D. 1962. The vegetation of the Mlanje mountains, Nyasaland. Govt. Pr., Zomba.
Chapman, J. D. & White, F. 1970. The evergreen forests of Malawi. Commonw. For. Inst., Oxford.
Codd, L. E. W. 1951. Trees and shrubs of the Kruger National Park. Mem. Bot. Surv. S. Afr. 26:1–192.
Coetzee, B. J. 1974. A phytosociological classification of the vegetation of the Jack Scott Nature Reserve. Bothalia 11:329–347.
Coetzee, B. J. 1975. A phytosociological classification of the Rustenburg Nature Reserve. Bothalia 11:561–580.

Coetzee, B. J., Gertenbach, W. P. D. & Nel, P. J. 1975. Korttermijnveranderings in die plantegroei van die Sentrale Distrik. Int. Report, Kruger National Park, South Africa (also Koedoe 20: in press).
Coetzee, B. J. & Nel, P. J. 1977. Preliminary report project No. BII/i/62/X. National Parks Board, S.A.
Coetzee, B. J., Van der Meulen, F., Zwanziger, S., Gonsalves, P. & Weisser, P. 1976. A phytosociological classification of the Nylsvley Nature Reserve. Bothalia 12:137–160.
Coetzee, B. J. & Werger, M. J. A. 1975. A west–east vegetation transect through Africa south of the Tropic of Capricorn. Bothalia 11:539–560.
Cole, M. M. 1963a. Vegetation and geomorphology in Northern Rhodesia: an aspect of the distribution of the savanna of Central Africa. Geogr. J. 129:290–310.
Cole, M. M. 1963b. Vegetation nomenclature and classification with particular reference to the savannas. S. Afr. Geogr. J. 45:3–14.
Cole, M. M. & Brown, R. C. 1976. The vegetation of the Ghanzi area of western Botswana. J. Biogeogr. 3:169–196.
Comins, D. M. 1962. The vegetation of the districts of East London and King William's Town, Cape Province. Mem. Bot. Surv. S. Afr. 33:1–32.
Compton, R. H. 1966. An annotated check list of the flora of Swaziland. J. S. Afr. Bot. Supp. 4:1–191.
Cottrell, C. B. & Loveridge, J. P. 1966. Observations on the Cryptosepalum forest of the Mwinilunga District of Zambia. Proc. Trans. Rhod. Sci. Ass. 51:79–120.
Croizat, L. 1952. Manual of phytogeography. Junk, The Hague.
Croizat, L. 1965, 1967, 1972. An introduction to the subgeneric classification of 'Euphorbia' L., with stress on the South African and Malagasy species. I, II, III. Webbia 20:573–706; 22:83–202; 27:1–221.
Croizat, L. 1968. Introduction raisonnée à la biogéographie de l'Afrique. Mem. Soc. Brot. 20:7–451.
Crook, A. O. 1956. A preliminary vegetation map of the Melsetter Intensive Conservation Area, Southern Rhodesia. Rhod. Agric. J. 53:3–25.
C.S.A. 1956. Phytogeography. C.S.A./C.C.T.A. Publ. 22:1–35.
Cumming, D. G. 1962. Fire protection in the Rhodesian teak forests of Northern Rhodesia. C.S.A./C.C.T.A. Publ. 52:77–79.
Dasmann, R. F. 1964. African game ranching. Pergamon, London.
Dept. Mines, 1970. Geological Map of South Africa, 1:1.000.000. Govt. Printer, Pretoria.
De Rham, P. 1970. L'azote dans quelques forêts, savanes et terrains de cultures d'Afrique tropicale humide (Côte-d'Ivoire). Veröff. Geobot. Inst. ETH, Stift. Rübel, 45:1–124.
Descoings, B. 1973. Les formations herbeuses africaines et les définitions de Yangambi considérées sous l'angle de la structure de la végétation. Adansonia, Sér. 2, 13:391–421.
De Vos, A. 1975. Africa, the devastated continent? Junk, The Hague.
Devred, R. 1958. La végétation forestière du Congo belge et du Ruanda-Urundi. Bull. Soc. Roy. For. Belg. 65:409–468.
Diniz, A. Castanheira. 1973. Características mesológicas de Angola. Missão Inq. Afric. Angola, Nova Lisboa.
Diniz, A. Castanheira & Aguiar, F. Q. de Barros. 1966. Geomorfologia, solos e ruralismo da região central angolana. Inst. Inv. Agron. Angola, Nova Lisboa.
Diniz, A. Castanheira & Aguiar, F. Q. de Barros. 1969. Regiões naturais de Angola. Inst. Inv. Agron. Angola, Sér. Cien. 7:1–6.
Diniz, A. Castanheira & Aguiar, F. Q. de Barros. 1972. Os solos e a vegetação do planalto ocidental da Cela. Inst. Inv. Agron. Angola, Sér. Cien. 26:1–24.
Diniz, A. Castanheira & Aguiar, F. Q. de Barros. 1973. Recursos em terras com aptidão para o regadio na bacia do Cubango. Inst. Inv. Agron. Angola, Sér. Téc. 33:1–28.
Diniz, A. Castanheira, Aguiar, F. Q. de Barros, Raimundo, A. R. Fonseca & Vilhena, M. Leite. 1972. Zonagem agro-ecológia de Angola. II. Memória dos trabalhos de 1971. 2 Vols. Proj. 21/66. Inst. Inv. Agron. Angola, Nova Lisboa.
Du Plessis, J. C. 1973. 'n Floristies-ekologiese studie van die plaas Doornkop in die distrik Middelburg, Transvaal. Unpubl. M.Sc. thesis, Univ. Pretoria.
Duvigneaud, P. 1949. Les savanes du Bas-Congo. Lejeunia 10:1–192.
Duvigneaud, P. 1958. La végétation du Katanga et de ses sols métallifères. Bull. Soc. Roy. Bot. Belg. 90:127–286.

Dyer, R. A. 1937. The vegetation of the divisions of Albany and Bathurst. Mem. Bot. Surv. S. Afr. 17:1–138.
Edwards, D. 1967. A plant ecology survey of the Tugela River Basin, Natal. Mem. Bot. Surv. S. Afr. 36:1–285.
Ellis, B. S. 1950. A guide to some Rhodesian soils. II. A note on mopani soils. Rhod. Agric. J. 47:49–61.
Engler, A. 1910. Die Pflanzenwelt Afrikas. I. Allgemeiner Überblick über die Pflanzenwelt Afrikas und ihre Existensbedingungen. Die Vegetation der Erde. IX. Engelmann, Leipzig.
Ernst, W. H. O. 1971. Zur Ökologie der Miombo-Wälder. Flora 160:317–331.
Ernst, W. H. O. 1974. Schwermetallvegetation der Erde. Fischer, Stuttgart.
Ernst, W. & Walker, B. H. 1973. Studies on the hydrature of trees in miombo woodland in South Central Africa. J. Ecol. 61:667–673.
Exell, A. W. 1957. La végétation de l'Afrique tropicale australe. Bull. Soc. Roy. Bot. Belg. 89:101–106.
Exell, A. W. & Gonçalves, M. L. 1973. A statistical analysis of a sample of the flora of Angola. Garcia de Orta, Sér. Bot. 1:105–128.
Exell, A. W. & Wild, H. 1961. A statistical analysis of a sample of the Flora Zambesiaca. Kirkia 2:108–130.
Fanshawe, D. B. 1961. Evergreen forest relics in Northern Rhodesia. Kirkia 1:20–24.
Fanshawe, D. B. 1962a. Floristic composition of miombo woodland. C.S.A./C.C.T.A. Publ. 52:55–57.
Fanshawe, D. B. 1962b. Burning experiments in miombo woodland. C.S.A./C.C.T.A. Publ. 52:63–64.
Fanshawe, D. B. 1971. The vegetation of Zambia. Forest Res. Bull. No. 7. Govt. Pr., Lusaka.
Fanshawe, D. B. & Savory, B. M. 1964. Baikiaea plurijuga dwarf-shell forests. Kirkia 4:185–190.
Farrell, J. A. K. 1968a. Preliminary notes on the vegetation of the lower Sabi–Lundi basin, Rhodesia. Kirkia 6:223–248.
Farrell, J. A. K. 1968b. Preliminary notes on the vegetation of Southern Gokwe District, Rhodesia. Kirkia 6:249–258.
Feely, J. M. 1965. Observations on Acacia albida in the Luangwa Valley. The Puku 3:67–70.
Gaff, D. F. 1971. Desiccation-tolerant flowering plants in Southern Africa. Science 174:1033–1034.
Gaff, D. F. & Ellis, R. P. 1974. Southern African grasses with foliage that revives after dehydration. Bothalia 11:305–308.
Galpin, E. E. 1926. Botanical survey of the Springbok Flats. Mem. Bot. Surv. S. Afr. 12:1–100.
Gertenbach, W. P. D. 1977. Preliminary report project No. BII/i/61–62/X. National Parks Board, S.A.
Giess, W. 1971. Eine vorläufige Vegetationskarte von Südwestafrika. Dinteria 4:1–114.
Gilliland, H. B. 1938. The vegetation of Rhodesian Manicaland. J. S. Afr. Bot. 4:73–101.
Glover, P. E. 1968. The role of fire and other influences on the savannah habitat, with suggestions for further research. E. Afr. Wildl. J. 6:131–137.
Goldthorp, G. D. 1957. The ecology and land use of the Belingwe Shabani Intensive Conservation Area. Rhod. Agric. J. 54:402–437.
Goodier, L. & Phipps, J. B. 1962. A vegetation map of the Chimanimani National Park. Kirkia 3:2–7.
Gossweiler, J. & Mendonça, F. A. 1939. Carta Fitogeográfica de Angola. Gov. Geral de Angola, Lisboa.
Greenway, P. J. 1973. A classification of the vegetation of East Africa. Kirkia 9:1–68.
Grunow, J. O. 1965. Objective classification of plant communities: A synecological study in the sour-mixed bushveld of Transvaal. Unpubl. D.Sc. thesis, Univ. Pretoria.
Guerreiro, M. Gomes. 1966. A floresta africana e os factores bióticos. Inst. Inv. Cien. Angola, Luanda.
Guillaumet, J. L. & Koechlin, J. 1971. Contribution à la définition des types de végétation dans les régions tropicales (exemple de Madagascar). Candollea 26:263–277.
Hall-Martin, A. J. 1972. Aspects of the plant ecology of the Lengwe National Park, Malawi. M.Sc. thesis. Univ. Pretoria. Unpubl.
Hall-Martin, A. J. 1975. Classification and ordination of forest and thicket vegetation of the Lengwe National Park, Malawi. Kirkia 10:131–184.
Henkel, J. S. 1931. Types of vegetation in Southern Rhodesia. Proc. Rhod. Sci. Ass. 30:1–23.

Henning, A. C. & White, R. E. 1974. A study of the growth and distribution of Colophospermum mopane (Kirk ex. Benth.) Kirk ex. J. Leon: the interaction of nitrogen, phosphorus and soil moisture stress. Proc. Grassld. Soc. Sth. Afr. 9:53–60.

Hill, J. C. R. 1969. Vegetation survey of Victoria Falls National Park. Dept. Conserv. Extension. Unpubl. Rep., Salisbury.

Hirst, S. M. 1975. Ungulate-habitat relationships in a South African woodland/savanna ecosystem. Wildlife Monogr. 44:1–60.

Holub, E. 1890. Von der Capstadt ins Land der Maschukulumbe. Hölder, Wien.

Howard-Williams, C. 1975. Seasonal and spatial changes in the composition of aquatic and semiaquatic vegetation of Lake Chilwa, Malawi. Vegetatio 30:33–39.

Huntley, B. J. 1974. Vegetation and flora conservation in Angola. Serv. Veter. Unpubl. Rep. no. 22, Luanda.

Hursh, C. R. 1962. Composition of the tropical dry forests of Nyasaland. C.S.A./C.C.T.A. Publ. 52:49–53.

Irvine, L. O. F. 1941. The major veldtypes of the northern Transvaal. Quinquennial Rep. Past. Res. Sta. Warmbad, Transvaal.

Jackson, G. 1968. The vegetation of Malawi. II. The Brachystegia woodlands. Soc. Malawi J. 21(2):11–19.

Jackson, G. 1969a. The grasslands of Malawi. I. Soc. Malawi J. 22(1):7–17.

Jackson, G. 1969b. The grasslands of Malawi. II. Soc. Malawi J. 22(1):18–25

Jackson, G. 1969c. The grasslands of Malawi. III. Soc. Malawi J. 22(2):73–81.

Jackson, G. 1974. Cryptogeal germination and other seedling adaptations to the burning of vegetation in savanna regions: the origin of the pyrophytic habit. New Phytol. 73:771–780.

Jacobsen, W. B. G. 1968. The influence of the copper content of the soil on the vegetation at Silverside North, Mangula Area. Kirkia 6:259–278.

Jacobsen, W. G. B. 1973. A checklist and discussion of the flora of a portion of the Lomagundi District, Rhodesia. Kirkia 9:139–207.

Jarman, N. G. & Bosch, O. 1973. The identification and mapping of extensive secondary invasive and degraded ecological types (test side D). In: O. G. Malan (ed.), To assess the value of satellite imagery in resource evaluation on a national scale, pp. 77–80. C.S.I.R., Pretoria.

Jarman, P. J. 1971. Diets of large mammals in the woodlands around Lake Kariba, Rhodesia. Oecologia 8:157–178.

Jessen, O. 1936. Reisen und Forschungen in Angola. Reimer, Berlin.

Joubert, E. 1971. The physiographic, edaphic and vegetative characteristics found in the western Etosha National Park. Madoqua 1 (4):5–32.

Killick, D. J. B. 1958. An account of the plant ecology of the Table Mountain area of Pietermaritzburg, Natal. Mem. Bot. Surv. S. Afr. 32:1–133.

Killick, D. J. B. 1963. An account of the plant ecology of the Cathedral Peak area of the Natal Drakensberg. Mem. Bot. Surv. S. Afr. 34:1–178.

Knapp, R. 1965. Pflanzengesellschaften und Vegetations-Einheiten von Ceylon und Teilen von Ost- und Central-Afrika. Geobot. Mitt. (Giessen) 33:1–31.

Knapp, R. 1968. Höhere Vegations-Einheiten von Äthiopien, Somalia, Natal, Transvaal, Kapland und einigen Nachbargebieten. Geobot. Mitt. (Giessen) 56:1–36.

Knapp, R. 1973. Die Vegetation von Afrika. Fischer, Stuttgart.

Kornaś, J. 1975. Tuber production and fire-resistance in Lycopodium carolinianum L. in Zambia. Acta Soc. Bot. Pol. 44:653–663.

Lang, P. O. 1952. The vegetation of the Insiza and Shangani Intensive Conservation Areas. Rhod. Agric. J. 49:346–351.

Laws, R. M. 1970. Elephants as agents of habitat and landscape change in East Africa. Oikos 21:1–15.

Lawton, R. M. 1963. Palaeoecological and ecological studies in the Northern Province of Northern Rhodesia. Kirkia 3:46–77.

Lawton, R. M. 1964. The ecology of the Marquesia acuminata (Gilg) R.E.Fr. evergreen forests and the related chipya vegetation types of North-Eastern Rhodesia. J. Ecol. 52:467–479.

Lawton, R. M. 1971. Destruction or utilization of wildlife habitat? In: E. Duffey & A. S. Watt (ed.), The scientific management of animal and plant communities for conservation, pp. 333–336. Blackwell, Oxford.

Lawton, R. M. 1972. An ecological study of miombo and chipya woodland with particular reference to Zambia. Unpubl. D.Phil. thesis, Univ. Oxford.

Leach, L. C. 1974. Euphorbiae succulentae Angolenses: IV. Garcia de Orta, Sér. Bot. 2:31–54.
Lebrun, J. 1947. La végétation de la plaine alluviale au sud du Lac Édouard. Exploration du Parc National Albert. Fasc. 1. pp. 1–800. Inst. Parcs Nat. Congo belge, Bruxelles.
Lebrun, J. 1960. Sur la richesse de la flore de divers territoires africains. Bull. Acad. Roy. Sc. Outremer 4:669–690.
Lebrun, J. & Gilbert, G. 1954. Une classification écologique des forêts du Congo. Publ. I.N.E.A.C., Sér. Sc. 63:1–90.
Leistner, O. A. 1967. The plant ecology of the Southern Kalahari. Mem. Bot. Surv. S. Afr. 38:1–172.
Leistner, O. A. & Werger, M. J. A. 1973. Southern Kalahari phytosociology. Vegetatio 28:353–399.
Lemon, P. C. 1964. Natural communities of the Malawi National Park (Nyika Plateau). Govt. Pr., Zomba.
Leser, H. 1972. Geoökologische Verhältnisse der Pflanzengesellschaften in den Savannen des Sandveldes um den Schwarzen Nossob und um Epukiro. Dinteria 6:1–41.
Leser, H. 1975. Weidewirtschaft und Regenbau im Sandveld. Geogr. Rundschau 27:108–122.
Lewalle, J. 1972. Les étages de végétation du Burundi occidental. Bull. Jard. Bot. Nat. Belg. 42:1–247.
Lisowski, S., Malaisse, F. & Symoens, J. J. 1971. Une flore des hauts plateaux du Katanga. Mitt. Bot. Staatssamml. München 10:51–56.
Louw, A. J. 1970. 'n Ekologiese studie von mopanie-veld noord van Soutpansberg. Unpubl. D.Sc. thesis, Univ. Pretoria.
Louw, W. J. 1951. An ecological account of the vegetation of the Potchefstroom area. Mem. Bot. Surv. S. Afr. 24:1–105.
Macedo, J. de Aguiar. 1970. Carta da vegetação da Serra da Gorongosa. Comm. Inst. Inv. Agron. Moçambique 50:1–75.
MacVicar, N. (ed.). 1973. Soil Map, Republic of South Africa. 1:2 500 000 (an interim compilation). Dept. Agric. Tech. Serv., Pretoria. Unpubl.
Magadza, C. H. D. 1970. A preliminary survey of the vegetation of the shore of Lake Kariba. Kirkia 7:253–268.
Malaisse, F. 1973. Contribution à l'étude de l'écosystème forêt claire (miombo). Note 8: Le projet miombo. Ann. Fac. Sci. Abidjan, Sér. Ecol. 6:227–250.
Malaisse, F. 1975. Carte de la végétation du bassin de la Luanza. Hydrobiol. Surv. Lake Bangweulu Luapula River Basin 18(2):1–41.
Malaisse, F., Freson, R., Goffinet, G. & Malaisse-Mousset, M. 1975. Litter fall and litter breakdown in Miombo. In: F. B. Golley & E. Medina (eds.), Tropical ecological systems. Ecol. Stud. Vol. 11, pp. 137–152. Springer, New York.
Malaisse, F. & Malaisse-Mousset, M. 1970. Contribution à l'étude de l'écosystème forêt claire (miombo): phénologie de la défoliation. Bull. Soc. Roy. Bot. Belg. 103:115–124.
Malan, J. S. & Owen-Smith, G. L. 1974. The ethnobotany of Kaokoland. Cimbebasia (B) 2:131–178.
Mansfield, J. E., Bennett, J. G., King, R. B., Lang, D. M. & Lawton, R. M. 1975–76. Land resources of the Northern and Luapula Provinces, Zambia – a reconnaissance assessment. Land Res. Stud. No. 19. D.O.S., Tolworth.
Martin, J. D. 1940. The Baikiaea forests of Northern Rhodesia. Emp. For. J. 19:8–18.
Matos, G. Cardoso de & Sousa, J. N. Baptista de. 1970. Reserva parcial de Moçâmedes. Carta da vegetação e memória descritiva. Inst. Inv. Agron. Angola, Nova Lisboa.
Mendonça, F. A. 1961. Indices fitocorológicos dà vegetação de Angola. Garcia de Orta 9:479–483.
Menezes, O. J. Azancot. 1971. Estudo fito-ecológico da região do Mucope e carta da vegetação. Bol. Inst. Inv. Cien. Ang. 8(2):7–54.
Michelmore, A. P. G. 1939. Observations on tropical African grasslands. J. Ecol. 27:282–312.
Mildbraed, J. & Domke, W. 1966. Grundzüge der Vegetation des tropischen Kontinental-Afrika. Willdenowia Beih. 2:1–253.
Mitchell, B. L. 1961a. Ecological aspects of game control measures in African wilderness and forested areas. Kirkia 1:120–129.
Mitchell, B. L. 1961b. Some notes on the vegetation of a portion of Wankie National Park. Kirkia 2:200–209.
Monod, T. 1963. Après Yangambi (1956): notes de phytogéographie africaine. Bull. I.F.A.N. 25 (A,2):594–619.

Monteiro, R. F. Romero. 1970. Estudo da flora e da vegetação das florestas abertas do planalto do Bié. Inst. Inv. Cien. Angola, Luanda.
Morris, B. 1970. The nature and origin of Brachystegia woodland. Commonw. For. Rev. 49:155–158.
Morris, J. W. 1973. Automatic classification and Ecological profiles of South-Western Transvaal Highveld Grassland. Unpubl. Ph.D. thesis, Univ. Natal, Pietermaritzburg.
Morris, J. W. 1976. Quantitative classification of the Highveld grassland of Lichtenburg, Southwestern Transvaal. Bothalia: In press.
Mostert, J. W. C. 1958. Studies of the vegetation of parts of the Bloemfontein and Brandfort Districts. Mem. Bot. Surv. S. Afr. 31:1–226.
Mullenders, W. 1954. La végétation de Kaniama, Entre-Lubishi-Lubilash, Congo belge. Publ. I.N.E.A.C., Sér. Sc. 61:1–500.
Myre, M. 1961. A grassland type of the south of Mozambique Province. C.R. IV. Réunion Plén. A.E.T.F.A.T. Junta Inv. Ultramar, Lisboa, pp. 337–361.
Myre, M. 1964. A vegetação do extremo sul da provincia de Moçambique. Estudos, Ensaios e Documentos 110:1–145.
Myre, M. 1971. As pastagens da região do Maputo. Mém. Inst. Inv. Agron. Moçambique 3:1–181.
Myre, M. 1972. Reconhecimento pascícola ao vale do Save. Comm. Inst. Inv. Agron. Moçambique 75:1–172.
Nolde, I. 1938a. Botanische Studie über das Hochland von Quela in Angola. Feddes Repert. Beih. 15:35–54.
Nolde, I. 1938b. Probeflächen verschiedener Savannenformationen im Hochland von Quela in Angola. Notizbl. Bot. Gart. Mus. Berlin-Dahlem 14:298–311.
Nolde, I. 1940. Beobachtungen über Laubfall und Lauberneuerung an tropischen Bäumen im Hochland von Quela (Angola). Engler Bot. Jahrb. 71: 233–248.
Nordenstam, B. 1969. Phytogeography of the genus Euryops (Compositae). Opera Botanica 23:1–77.
Palmer, E. & Pitman, N. 1972. Trees of southern Africa. 3 Vols. Balkema, Cape Town.
Passarge, S. 1904. Die Kalahari. Reimer, Berlin.
Pedro, J. Gomes & Barbosa, L. A. Grandvaux. 1955. A vegetação. In: Esboço do reconhecimento ecológico-agricola de Moçambique. Vol. 2:67–224. Centro Inv. Cien. Algodeira. Mém. e Trab. No. 23. Impr. Nac. de Moçambique, Lourenço Marques.
Phillips, J. F. V. 1930. Fire: its influence on biotic communities and physical factors in South and East Africa. S. Afr. J. Sci. 27:352–367.
Phipps, J. B. & Goodier, R. 1962. A preliminary account of the plant ecology of the Chimanimani Mountains. J. Ecol. 50:291–319.
Pienaar, U. de V. 1963. The large mammals of the Kruger National Park – their distribution and present day status. Koedoe 6:1–37.
Pike, J. G. & Rimmington, G. T. 1965. Malawi, a geographical study. Oxford Univ. Press, London.
Pole Evans, I. B. 1948a. A reconnaissance trip through the eastern portion of the Bechuanaland Protectorate and an expedition to Ngamiland. Mem. Bot. Surv. S. Afr. 21:1–203.
Pole Evans, I. B. 1948b. Roadside observations on the vegetation of East and Central Africa. Mem. Bot. Surv. S. Afr. 22:1–305.
Rattray, J. M. 1960. The grass cover of Africa. F.A.O. Agricultural Studies No. 49. F.A.O., Rome.
Rattray, J. M. 1961. Vegetation types of Southern Rhodesia. Kirkia 2:68–93.
Rattray, J. M. & Wild, H. 1955. Report on the vegetation of the alluvial basin of the Sabi valley and adjacent areas. Rhod. Agric. J. 52:484–501.
Rattray, J. M. & Wild, H. 1961. Vegetation map of the Federation of Rhodesia and Nyasaland. Kirkia 2:94–104.
Roberts, B. R. 1966. The ecology of Thaba 'Nchu. Unpubl. Ph.D. thesis, Univ. Natal, Pietermaritzburg.
Roberts, B. R., Opperman, D. P. J. & Van Rensburg, W. L. J. 1972. Introductory veld ecology and utilization. Univ. O.F.S., Bloemfontein.
Roux, E. 1969. Grass. Oxford Univ. Press, Cape Town.
Rutherford, M. C. 1972. Notes on the flora and vegetation of the Omuverume Plateau-Mountain, Waterberg, South West Africa. Dinteria 8:3–55.
Rutherford, M. C. 1975. Aspects of ecosystem function in a woodland savanna in South West Africa. Unpubl. Ph.D. thesis Univ. Stellenbosch.

Scheepers, J. C. 1975. The plant ecology of the Kroonstad and Bethlehem areas of the Highveld Agricultural Region. Unpubl. D.Sc. thesis, Univ. Pretoria.
Schmitz, A. 1963. Aperçu sur les groupements végétaux du Katanga. Bull. Soc. Roy. Bot. Belg. 96:233–447.
Schmitz, A. 1971. La végétation de la Plaine de Lubumbashi (Haut-Katanga). Publ. I.N.E.A.C., Sér. Sc. 113:1–390.
Schnell, R. 1976–1977. Introduction à la phytogéographie des pays tropicaux. Vols. 3–4. La flore et végétation de l'Afrique tropicale. Gauthier-Villars, Paris.
Seagrief, S. C. & Drummond, R. B. 1958. Some investigations on the vegetation of the north-eastern part of Makarikari Salt Pan, Bechuanaland. Proc. Trans. Rhod. Sci. Ass. 46:103–133.
Seiner, F. 1909. Ergebnisse einer Bereisung zwischen Okawango und Sambesi (Caprivi-Zipfel) in den Jahren 1905 und 1906. Mitt. Deutsch. Schutzgeb. 22:1–112.
Seiner, F. 1912. Pflanzengeografische Beobachtungen in der Mittel-Kalahari. Engler Bot. Jahrb. 46:1–50.
Shantz, H. L. & Turner, B. L. 1958. Photographic documentation of vegetational changes in Africa over a third of a century. Univ. Ariz.
Sheppe, W. & Osborne, T. 1971. Patterns of use of a flood plain by Zambian mammals. Ecol. Monogr. 41:179–205.
Simpson, C. D. 1975. A detailed vegetation study on the Chobe River in northeast Botswana. Kirkia 10:185–227.
Story, R. 1952. A botanical survey of the Keiskammahoek District. Mem. Bot. Surv. S. Afr. 27:1–184.
Strang, R. M. 1966. The spread and establishment of Brachystegia spiciformis Benth. and Julbernardia globiflora (Benth.) Troupin in the Rhodesian Highveld. Commonw. For. Rev. 45:253–256.
Streel, M. 1963. La végétation tropophylle des plaines alluviales de la Lufira moyenne. F.U.L.R.E.A.C., Liège.
Teixeira, J. Brito. 1968a. Parque Nacional do Bicuar. Carta da vegetação e memória descritiva. Inst. Inv. Agron. Angola, Nova Lisboa.
Teixeira, J. Brito. 1968b. Angola. Acta Phytogeogr. Suecica 54:193–197.
Teixeira, J. Brito, Matos, G. Cardosa de & Sousa, J. N. Baptista de. 1967. Parque Nacional da Quiçama. Carta da vegetação e memória descritiva. Inst. Inv. Agron. Angola, Nova Lisboa.
Theron, G. K. 1973. 'n Ekologiese studie van die plantegroei van die Loskopdam-natuurreservaat. Unpubl. D.Sc. thesis, Univ. Pretoria.
Theron, G. K. 1975. The distribution of the summer rainfall zone Protea species in South Africa with special reference to the ecology of Protea caffra. Boissiera 24:233–244.
Tinley, K. L. 1966. An ecological reconnaissance of the Moremi Wildlife Reserve, northern Okovango swamps, Botswana. Okovango Wildl. Soc., Johannesburg.
Tinley, K. L. 1971a. Determinants of coastal conservation: dynamics and diversity of the environment as exemplified by the Moçambique coast. S.A.R.C.C.U.S., Pretoria, pp. 125–153.
Tinley, K. L. 1971b. Sketch of Gorongosa National Park. S.A.R.C.C.U.S., Pretoria. pp. 163–172.
Tinley, K. L. 1975. Habitat physiognomy, structure and relationships. In: Die Soogdiernavorsingsinstituut 1966–1975. Publ. Univ. Pretoria, N.R. 97: 69–77.
Tinley, K. L. 1976. Vegetation map. In: R. H. N. Smithers & J. L. P. L. Tello, Atlas and checklist of Moçambique mammals.
Trapnell, C. G. 1953. The soils, vegetation and agriculture of North-Eastern Rhodesia. Govt. Printer, Lusaka.
Trapnell, C. G. 1959. Ecological results of woodland burning experiments in Northern Rhodesia. J. Ecol. 47:129–168.
Trapnell, C. G. & Clothier, J. N. 1937. The soils, vegetation and agricultural systems of North Western Rhodesia. Govt. Printer, Lusaka.
Trapnell, C. G., Friend, M. T., Chamberlain, G. T. & Birch, H. T. 1976. The effects of fire and termites on a Zambian woodland soil. J. Ecol. 64:577–588.
Troupin, G. 1966. Étude phytocénologique du Parc National de l'Akagera et du Rwanda oriental. Inst. Nat. Rech. Sc. Butaré (Rwanda) Publ. 2:1–293.
Van der Schijff, H. P. 1957. 'n Ekologiese studie van die flora van die Nasionale Krugerwildtuin. Unpubl. D.Sc. thesis, Univ. Potchefstroom.
Van der Schijff, H. P. 1958. Inleidende verslag oor veldbrandnavorsing in die Nasionale Krugerwildtuin. Koedoe 1:60–93.

Van der Schijff, H. P. 1964. Die ekologie en verwantskappe van die Sandveld-flora van die Nasionale Krugerwildtuin. Koedoe 7:56–76.
Van der Schijff, H. P. 1969. The affinities of the flora of the Kruger National Park. Kirkia 7:109–120.
Van der Schijff, H. P. 1971. Die plantegroei van die drie distrikte Potgietersrus, Pietersburg en Soutpansberg in die Noordelike Transvaal. Tijdskr. Natuurwet. 11:108–144.
Van der Schijff, H. P. & Schoonraad, E. 1971. The flora of the Mariepskop complex. Bothalia 10:461–500.
Van Rooyen, N. 1976. 'n Ekologiese studie van die plantgemeenskappe in die Punda Milia-Pafurigebied van die Nasionale Kruger Wildtuin. M.Sc. project, Univ. Pretoria.
Van Vuuren, D. R. J. 1961. 'n Ekologiese studie van die plantegroei van 'n noordelike en suidelike kloof van die Magaliesberge. Unpubl. M.Sc. thesis, Univ. Pretoria.
Van Vuuren, D. R. J. & Van der Schijff, H. P. 1970. 'n Vergelijkende ekologiese studie van die plantegroei van 'n noordelike en suidelike kloof van die Magaliesberg. Tijdskr. Wet. Kuns 1970: 16–75.
Van Wyk, P. 1971. Veld burning in the Kruger National Park. Proc. Ann. Tall Timbers Fire Ecology Conf. 11:9–31.
Van Wyk, P. 1973. Trees of the Kruger National Park. Perskor, Johannesburg.
Verboom, W. C. 1975. List of plant species in the main vegetation types of the Bangweulu Basin. In: J. J. R. Grimsdell & R. H. V. Bell, Ecology of the black lechwe in the Bangweulu Basin of Zambia, pp. 139–141. Anim. Prod. Res. Rep. AR1.NCSR/TR 31, Lusaka.
Verboom, W. C. & Brunt, M. A. 1970. An ecological survey of Western Province, Zambia, with special reference to the fodder resources. 2 Vols. Land Res. Stud. No. 8. D.O.S., Tolworth.
Verster, E., De Villiers, J. M. & Scheepers, J. C. 1973. Gilgai in the Rustenburg area. Agrochemophysica 4:57–62.
Vesey-Fitzgerald, D. F. 1963. Central African grasslands. J. Ecol. 51:243–274.
Volk, O. H. 1964. Die afro-meridional-occidentale Floren-Region in Südwest-Afrika. In: K. Kreeb (ed.), Beiträge zur Phytologie, pp. 1–16. Ulmer, Stuttgart.
Volk, O. H. 1966a. Die Florengebiete von Südwestafrika. J.S.W.A. Wissensch. Ges. 20:25–58.
Volk, O. H. 1966b. Einfluss von Mensch und Tier auf die natürliche Vegetation im tropischen Südwest-Afrika. In: K. Buchwald, W. Lendholt & K. Meyer (eds.), Beiträge zur Landespflege, 2, pp. 108–131. Ulmer, Stuttgart.
Volk, O. H. 1974. Gräser des Farmgebietes von Südwestafrika. Vorstand S.W.A. Wiss. Ges., Windhoek.
Volk, O. H. & Leippert, H. 1971. Vegetationsverhältnisse im Windhoeker Bergland, Südwestafrika. J. S.W.A. Wiss. Ges. 25:5–44.
Walter, H. 1939. Grasland, Savanne und Busch der arideren Teile Afrikas in ihrer ökologischen Bedingtheit. Jb. Wiss. Bot. 87:750–860.
Walter, H. 1962 (1973). Die Vegetation der Erde in ökophysiologischer Betrachtung. Bd. I. 1st, 3rd ed. Fischer, Stuttgart.
Walter, H. & Volk, O. H. 1954. Die Grundlagen der Weidewirtschaft in Südwestafrika. Ulmer, Stuttgart.
Weare, P. R. & Yalala, A. 1971. Provisional vegetation map of Botswana. Botswana Notes and Records 3:131–148.
Weather Bureau. 1957. Climate of South Africa. 4. Rainfall Maps. Govt. Printer & Weather Bureau, Pretoria.
Weimarck, H. 1941. Phytogeographical groups, centres and intervals with the Cape flora. Lunds Univ. Årsskr. N.F. Avd. 2. 37(5):1–144.
Werger, M. J. A. 1973a. Phytosociology of the Upper Orange River valley, South Africa. Diss. Nijmegen. V & R., Pretoria.
Werger, M. J. A. 1973b. Notes on the phytogeographical affinities of the southern Kalahari. Bothalia 11:177–180.
Werger, M. J. A. 1976. Zoogene Änderungen in der Vegetation der südafrikanischen Trockengebiete. In: R. Tüxen (ed.), Vegetation und Fauna. Ber. Int. Symp. Rinteln 1976. Cramer, Lehre.
Werger, M. J. A. 1977a. Effects of game and domestic livestock on vegetation in East and southern Africa. In: W. Krause (ed.), Handbook of vegetation science. Vol. 13, pp. 149–159. Junk, The Hague.
Werger, M. J. A. 1977b. Applicability of Zürich-Montpellier methods in tropical and subtropical range vegetation in Africa. In: W. Krause (ed.), Handbook of vegetation science. Vol. 13, pp. 125–145. Junk, The Hague.

Werger, M. J. A. & Coetzee, B. J. 1977. A phytosociological and phytogeographical study of Augrabies Falls National Park, South Africa. Koedoe 19: in press.

Werger, M. J. A. & Leistner, O. A. 1975. Vegetationsdynamik in der südlichen Kalahari. In: W. Schmidt (ed.), Sukzessionsforschung. Ber. Int. Symp. Rinteln 1973, pp. 135–158. Cramer, Lehre.

Werger, M. J. A., Wild, H. & Drummond, R. B. 1978. Vegetation structure and substrate of the northern part of the Great Dyke, Rhodesia. Vegetatio: in press.

West, O. 1951. The vegetation of Weenen County, Natal. Mem. Bot. Surv. S. Afr. 23:1–183.

West, O. 1971. Fire, man and wildlife as interacting factors limiting the development of climax vegetation in Rhodesia. Proc. Ann. Tall Timbers Fire Ecol. Conf. 11:121–145.

White, F. 1962. Forest flora of Northern Rhodesia. Oxford Univ. Press, London.

White, F. 1965. The savanna woodlands of the Zambezian and Sudanian Domains. Webbia 19:651–681.

White, F. 1976a. The vegetation map of Africa: the history of a completed project. Boissiera 24:659–666.

White, F. 1976b. The taxonomy, ecology and chorology of African Chrysobalanaceae (excluding Acioa). Bull. Jard. Bot. Nat. Belg. 46:265–350.

Wild, H. 1955. Observations on the vegetation of the Sabi–Lundi junction area. Rhod. Agric. J. 52:533–546.

Wild, H. 1964. A guide to the flora of the Victoria Falls. In: B. M. Fagan (ed.), The Victoria Falls, pp. 141–181. Comm. Pres. Nat. Hist. Monum. Rel., Livingstone.

Wild, H. 1968a. Bechuanaland protectorate. Acta Phytogeogr. Suec. 54:198–202.

Wild, H. 1968b. Rhodesia. Acta Phytogeogr. Suec. 54:202–207.

Wild, H. 1968c. Phytogeography of South Central Africa. Kirkia 6:197–222.

Wild, H. 1974. The natural vegetation of gypsum bearing soils in south central Africa. Kirkia 9:279–292.

Wild, H. & Barbosa, L. A. Grandvaux. 1967 (printed, 1968 publ.). Vegetation map of the Flora Zambesiaca area. Flora Zamb. Suppl. 1–71. Collins, Salisbury.

Willan, R. G. M. 1940. Notes on the vegetation of northern Nyasaland. Emp. For. J. 19:48–61.

11 The Afromontane Region

F. White*

1. Introduction 465
2. Floristics and plant geography 466
2.1 Endemism in the Afromontane flora 467
2.2 Generic geographical elements 474
2.3 Non-endemic species 475
2.4 Distant satellite populations of Afromontane near-endemic species .. 477
2.5 Intervals in the Afromontane flora 480
3. Zonation and the main vegetation types 480
3.1 Zonation in East Africa 482
3.2 Zonation and vegetation in Malawi, Rhodesia and Moçambique 483
3.3 Zonation and vegetation in South Africa 498
3.4 Afromontane forests in the Cape region 506
4. Summary and conclusions 507

References 510

---

* I am grateful to David Mabberley, who made many useful comments on an early draft, and drew my attention to some publications I might otherwise have missed, and to Ib Friis who freely placed his knowledge of the Ethiopian Afromontane flora at my disposal.

11 The Afromontane Region

1. Introduction

The vegetation of the highest mountains in tropical Africa is so different from that of the surrounding 'lowlands'* that it has attracted the attention of travellers and scientists since the earliest days of botanical exploration. Even to the layman, it is familiar because of the unusual growth-forms of its giant lobelias and senecios. In recent years the flora and vegetation of these high peaks have been regarded by specialists (Hauman 1955, Hedberg 1965) as sufficiently distinct to justify the recognition of a separate phytogeographical 'Afroalpine' Region.

It is not, however, so generally known that the floristically much richer vegetation of the lower slopes of the highest mountains and the upper slopes of lesser mountains, with a few provisos to be made below, is also totally different from the surrounding lowland vegetation. It has even stronger claims that the Afroalpine Region to chorological recognition at the rank of region. Consequently, White (1965) tentatively proposed the recognition of an archipelago-like Afromontane Region, but he was undecided whether the Natal Drakensberg should belong to it. In a later publication (White in Chapman & White 1970) it was concluded that the Afromontane Region extended at least as far as the Knysna forests in the Cape Province of South Africa. Aubréville, much earlier (1949), had already commented on the floristic similary between the Knysna forests and those of the mountains of tropical Africa.

Some phytogeographers (Lebrun 1947, Monod 1957) regard the flora of the African mountains, both Afroalpine and Afromontane, as representing no more than a series of upland facies of the surrounding lowland phytochoria. This conclusion, apparently, was not based on detailed analysis. The evidence presented in the present review suggests that it cannot be maintained.

The principal objective of this review is to describe Afromontane vegetation as it occurs in southern Africa in relation to that of the Afromontane Region as a whole, and to examine the Afromontane Region in relation to other phytochoria, both adjacent and distant.

The literature on the Afromontane flora and vegetation is voluminous but it is very incomplete and very widely scattered. No comprehensive review has previously been attempted and most publications deal with restricted areas or specialized topics.

Zonation of the Afromontane and Afroalpine vegetation on the high mountains of East Africa has been described by Hedberg (1951) and some aspects are dealt with by Lind & Morrison (1974). For West Africa, Morton (1972) has briefly commented on some general features of the Afromontane flora and made suggestions concerning its history. Chapman & White (1970) have described the montane forests of Malawi in considerable detail, and other montane vegetation more cursorily. They also attempted to place the Afromontane vegetation of Malawi in a wider ecological and chorological context. The main features of distribution of Afromontane vegetation are shown in Figs. 1 and 2, on the

---

* This term is used purely in a relative sense. 'Lowland' vegetation can occur as high as 2000 m.

A.E.T.F.A.T. 'Vegetation Map of Africa' (Keay 1959) and in more detail on the new UNESCO/AETFAT 'Vegetation Map of Africa' (White in press, A).

For southern Africa north of the Limpopo River, there is much useful information in the descriptive memoir which accompanies the 'Vegetation map of the Flora Zambesiaca area' (Wild & Barbosa 1968), and for South Africa in the 'Veld types of South Africa' by Acocks (1953), but the classifications employed in these two works are different and comparison is difficult. Wild (1968) has also briefly discussed the Afromontane species in relation to the phytogeography of South Central Africa.

The 'islands' comprising the Afromontane archipelago are very widely distributed on the African mainland. They extend from the Lome Mountains and the Tingi Hills (11°W) in Sierra Leone in the west to the Ahl Mescat Mountains (49°E) in Somalia in the east, and from the Red Sea Hills (17°N) in the Sudan Republic in the north to the Cape Peninsula (34°S) in the south. The most westerly and easterly occurrences are 7250 km apart. A similar distance separates the northernmost and southernmost.

To facilitate comparison the islands of the Afromontane archipelago can be grouped into seven regional mountain systems (Fig. 1). In each system the individual mountains are separated by smaller intervals than the interval separating that system from its nearest neighbour. It must be emphasized, however, that this arrangement is dictated purely by practical convenience. For certain purposes a somewhat different arrangement might be preferred.

The spatial relationships between the Afromontane Region and the lowland phytochoria are shown in Fig. 2. The latter are those proposed by White (1976a). Their boundaries are shown on the UNESCO/AETFAT 'Vegetation Map of Africa' (White in press, A).

In the present review very little is said of the history of the Afromontane flora. This is one of the main topics of a separate publication (White in press, B).

2. Floristics and plant geography

The Afromontane Region is an archipelago-like regional centre of endemism with most of its 4000 or more species endemic to it. It shows only moderate generic endemism, however, and has no more than three endemic families (see p. 469). Its phylogenetic relationships to other phytochoria are complex.

The Afroalpine Region, chorologically, is only feebly distinct from the Afromontane. Its total flora is small in number, and few species are not shared with the Afromontane Region, of which it is best regarded as a floristically impoverished version. It has no endemic genera.

For most purposes of chorological analysis and comparison it seems preferable to consider the Afroalpine and Afromontane Regions together, an opinion which has been voiced by others, e.g. by Weimarck (1941), who regarded the Afroalpine zone as an extreme variant of the Afromontane, and by Troupin (1966). Nevertheless, the Afroalpine Region has an extensive literature (e.g. Hauman 1933, 1955, Hedberg 1957–1975, Van Zinderen Bakker & Werger 1974) and is so different from most of the Afromontane Region in its ecology that for some purposes, ecological rather than phytogeographic, and in the interests of continuity, it could perhaps be recognized as an archipelago-like 'region of extreme floristic impoverishment' (White 1976a). White elsewhere (in press, B) has dis-

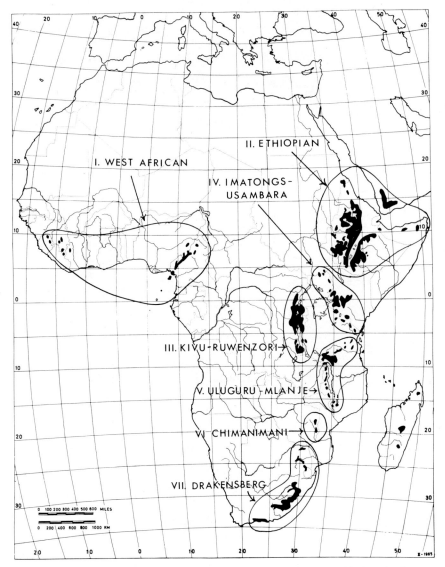

Fig. 1. Map. showing distribution of the islands of the Afromontane archipelago in the seven regional mountain systems. The distribution of montane vegetation in Madagascar is also shown.

cussed the relationships between the Afromontane and Afroalpine Regions in some detail and has made a chorological analysis of the two combined. In the present work, however, the Afroalpine vegetation of southern Africa receives separate treatment (Chapter 12).

2.1 *Endemism in the Afromontane flora*

2.1.1 General

If one excludes from consideration the transition zones between the Afromontane Region and surrounding phytochoria, and also those lowland species which are

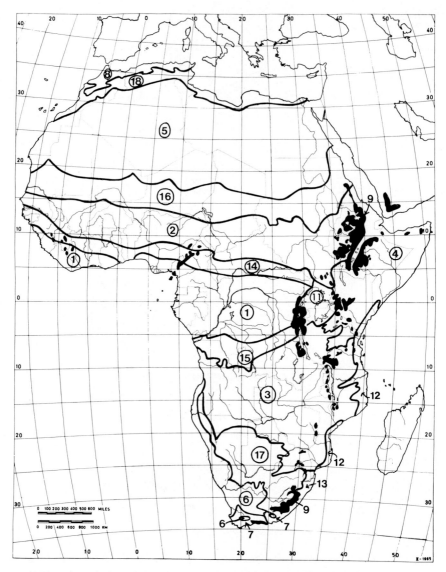

Fig. 2. Map showing the occurrence of the Afromontane and Afroalpine Regions in relation to 'lowland' phytochoria. The Afroalpine Region, which is confined to the summits of the highest mountains, is too small to show separately.

Regional centres of endemism: 1. Guineo–Congolian; 2. Sudanian; 3. Zambezian; 4. Somalia–Masai; 5. Saharan; 6. Karoo–Namib; 7. Cape; 8. Mediterranean.

Archipelago-like centre of endemism: 9. Afromontane.

Archipelago-like region of extreme-floristic impoverishment: 10. Afroalpine (not shown).

Regional mosaics: 11. Lake Victoria Basin; 12. Zanzibar–Inhambane; 13. Tongaland–Pondoland.

Regional transition zones: 14. Guineo–Congolian Sudanian; 15. Congo–Zambezia; 16. Sahel; 17. Kalahari–Highveld; 18. Sub-Mediterranean.

only of sporadic occurrence towards the lower limits of the Afromontane Region (marginal intruders, see below), and if one includes those Afromontane species which have small distant satellite populations elsewhere, then the great majority of species occurring in the Afromontane Region are either endemic to it, or, if they occur elsewhere, then, at least in Africa, they are absent from the surrounding lowlands. At present, it is impossible to estimate the total flora of the Afromontane Region with any degree of accuracy. A preliminary survey based on my own unpublished work and recently published taxonomic and ecological studies, which, however, cover only certain families and certain regions, has revealed no fewer than 1550 endemic species. The total number is almost certainly more than twice that figure.

In contrast to the very high degree of specific endemism, familiar and generic endemism are poorly developed. At most there are three endemic families. The Barbeyaceae is represented by a single tree species in Ethiopia and Arabia. The unigeneric Oliniaceae is essentially Afromontane, although some of its species are slightly 'transgressive' and also occur in small distant satellite populations (see Sect. 2.4). The unispecific *Curtisia* is usually placed in Cornaceae, though some workers give it family rank. Generic endemism, too, is not pronounced. Among larger woody plants, only 22 genera, approximately one fifth of the total, are endemic or almost so. Among herbs and smaller woody plants, generic endemism is even less pronounced. Of the few genera which are centred on the mountains of tropical Africa the following may be mentioned: *Ardisiandra* (Primulaceae), *Cincinnobotrys* (Melastomataceae) and *Stapfiella* (Turneraceae). Genera which have most species in South Africa and are essentially Afromontane include *Alepidea, *Bowkeria*, *Buchenroedera*, *Crinipes, *Kniphofia, *Macowania and *Rendlia. Those marked with an asterisk are more or less exclusively Afromontane north of the Limpopo River.

Of the Afromotane endemic species a large number are widely distributed. This is certainly true of the trees, which, in general, have been more comprehensively studied than the smaller plants. Of the 152 tree species included in the analysis discussed below, 96 species (63 per cent) occur in two or more of the seven isolated regional mountain systems into which the Afromontane Region can be divided (Fig. 1). Only 56 species (37 per cent) are confined to a single system and of these less than half are confined to a single mountain.

The behaviour of herbaceous and smaller woody plants is more complicated. There are undoubtedly a large number of local or restricted endemic species, but, in the present state of taxonomic knowledge, it is easy to over-estimate their number. Future work is more likely to lead to a decrease rather than an increase in their apparent number. It appears that some of the mountain systems and some mountains within systems are much richer in endemic species than others. There appears to be little correlation between the contemporary degree of geographical isolation and the richness of the endemic flora.

Of the seven regional mountain systems, the West African is the most isolated from the others and some of the mountains within it are more isolated from each other than are most mountains in other systems. Nevertheless, Morton (1972) found that 53 per cent of the montane flora of West Africa (which is largely composed of herbs and small woody plants) consists of widespread species, most of which also occur on the mountains of East and Central Africa. Very few of the West African endemics are confined to a single mountain. Lebrun (1958) lists 148 orophytes which are common to the mountains of eastern Zaïre and those of West Africa.

Specific endemism on individual mountains in tropical Africa, in general, appears to be low, equally for trees and smaller plants. In an analysis of the Afromontane tree flora of Malawi, Chapman & White (1970) found that not a single species (with the doubtful exception of *Dasylepis burttdavyi*, which was subsequently found to be wrongly classified) is confined to a single mountain, nor even to Malawi. In an interesting study of the Chimanimani Mountains in Rhodesia, Wild (1964) records 41 endemic species (4.6 per cent) out of an estimated total flora of 859 species (Goodier & Phipps 1961) occurring above 1220 m. There are no trees in the list. For Mt. Mlanje, Wild also lists 30 species believed to be endemic to it, which is only a small proportion of its total flora. Endemism on the relatively low-lying Western Usambara, Uluguru and Nguru Mountains in Tanzania is somewhat higher than elsewhere, though not as high as generally supposed. Some endemic species recorded by Polhill (1968) are either of uncertain taxonomic status or are now known to occur elsewhere. Nevertheless this group of mountains harbours several endemic species of *Streptocarpus* (Hilliard & Burtt 1971), *Impatiens* (Grey-Wilson pers. comm.) and other genera. Lewalle (1975) draws attention to some local endemics in Burundi.

In contrast to the situation in tropical Africa, it would appear that the great majority of smaller plants (but not of trees) which occur in the Drakensberg system do not extend to other systems, though they may be widely distributed within it. This is borne out by the monographic studies of Hilliard & Burtt (1971) on *Streptocarpus*, Nordenstam (1968, 1969) on *Euryops*, Lewis et al. (1972) on *Gladiolus*, Weimarck on *Aristea* (1940) and *Cliffortia* (1934), Codd (1969) on *Kniphofia* and Goldblatt on *Moraea* (1973). It also seems to be true of *Alepidea, Berkheya, Buchenroedera, Clutia, Erica, Helichrysum* and *Zaluzianskya* and many others.

Earlier workers have assumed that because some of the African mountains appear to be as isolated as remote oceanic islands they might be expected to have equally remarkable endemic floras. Although this assumption is rarely made explicit it is suggested by the practice of many of the older taxonomists who described new species on scanty material from each mountain in turn. Even Fries & Fries (1948), who made an outstanding study of the entire Afromontane and Afroalpine floras of Mt. Kenya and Mt. Aberdare, concluded that many species were endemic and that the proportion increases with increasing altitude. They also believed that closely-related vicarious species replace each other on different mountains.

Subsequent work, chiefly that of Hedberg on the Afroalpine flora and of many monographers on the Afromontane flora, has caused the conclusions of Fries & Fries to be modified. Hedberg (1957), on the basis of a tenfold increase in available material, was forced to reduce to synonymy a considerable proportion of the vicarious species enumerated by Fries & Fries. For some groups this process has been carried further, e.g. by Mabberley for *Senecio* and *Lobelia* (1973, 1974). Hedberg (1969) has also shown that for large parts of the Afromontane flora geographical isolation must have broken down or been considerably reduced during certain phases of the Pleistocene. The fallacy of regarding mountains on continents as comparable in their isolation to oceanic islands is discussed by White (1971).

According to Carlquist (1974) 'equatorial highlands', which he defines as occurring above 1000 m, have much in common with islands, but his statement that the genera occurring on the 'Submontane Plateau' that surrounds the East African mountains are 'temperate with few exceptions' is not supported by the facts. A recurrent theme in a more comprehensive study of the Afromontane Region (White in press, B) is that a high proportion of the species occurring on all the African mountains has been recruited from the tropical lowlands. This applies with even greater force to the mountains of tropical East Africa than to the mountains of subtropical and warm-temperate South Africa. The flora of the East-African 'Submontane Plateau' is almost exclusively of tropical affinity.

2.1.2 Patterns of endemism shown by genera and species of Afromontane trees

All of the 20 endemic genera are monotypic or consist of a few very closely related species. Most genera are widely distributed. Only six are confined to a single system. One genus, *Barbeya*, is confined to the Ethiopian system, another, *Lebrunia* to the Kivu-Ruwenzori system, and a third, *Platypterocarpus*, to the Western Usambara Mountains in the Imatongs-Usambara system. Three genera, *Gonioma*, *Platylophus* and *Seemannaralia*, are confined to the Drakensberg system. The greatest concentration of endemic genera is on the chain of East African mountains extending from the Imatongs to Mt. Mlanje. Here more than two thirds occur. There is a slight diminution towards the south, and a more rapid fall off to the north and west. The latter is a reflection of the general poverty of the Afromontane flora of the West African and Ethiopian systems.

The distributions of 152 Afromontane endemic tree species have been carefully studied and will be analysed in detail elsewhere (White in press, B). Table 1 summarizes the distributions of the 91 species which are known to occur in southern Africa. The sample referred to is based on trees which usually, at least in some parts of their range, exceed a height of 8 m. Those species belonging to difficult taxonomic groups such as *Dombeya* and certain genera of Rutaceae, which still need careful study throughout the Afromontane Region, have been excluded. It is doubtful whether their inclusion would have raised the number of the sample to more than 200. It is also unlikely that the general conclusions derived from the analysis would have shown any significant differences, though clearly some local detail would not be the same. For the purposes of this comparison, Afromontane species which either only slightly transgress the Afromontane Region by occurring as marginal intruders in adjacent phytochoria, or, elsewhere, only occur in small, widely scattered distant satellite populations (see Sect. 2.4), are regarded as endemics. Similarly those species which are strictly Afromontane in Africa but also occur in distant phytochoria are included; the latter are few in number. In Table 1, their extra-Afromontane range is broadly indicated in the last column.

The majority of Afromontane endemic tree species, as defined above, are widely distributed, with the corollary that local endemism is relatively slight. For purposes of comparison three types of distribution are recognized, viz:

(1) Afromontane endemic wides. These occur in at least 4 of the 7 regional mountain systems. They include 58 species or 38 per cent of the Afromontane endemic tree flora. There are 8 Omni-afromontane wides, and 11 wides also occur in Madagascar or have close relatives there.

Table 1. The geographical distributions of 91 species of Afromontane trees which occur in southern Africa.

I. West African system. II. Ethiopian system. III. Kivu-Ruwenzori system. IV. Imatongs-Usambara system. V. Uluguru-Mlanje system. VI. Chimanimani system. VII. Drakensburg system. As. = also occurring in Asia (excluding the Yemen). Com. = also occurring in Comores. Mad. = also occurring in Madagascar. Masc. = also occurring in the Mascarenes.

	I.	II.	III.	IV.	V.	VI.	VII.	
Acacia abyssinica		x	x	x	x	x		
Afrocrania volkensii			x	x	x	x		
Agauria salicifolia	x	x	x	x	x			Mad. Masc.
Albizia schimperana		x		x	x	x		
Aningeria adolfi-friedericii		x	x	x	x			
Anthocleista grandiflora			x	x	x	x	x	Com.
Aphloia theiformis			x	x	x	x	x	Com. Mad.
Apodytes dimidiata	x	x	x	x	x	x	x	As.
Arundinaria alpina	x	x	x	x	x			
Aulacocalyx diervilloides			x	x	x	x		
Bersama swynnertonii						x		
Bridelia brideliifolia		x	x	x				
Calodendrum capense				x	x	x	x	
Canthium mundianum						x		
C. ventosum						x		
Casearia battiscombei	x		x	x	x			
Cassine crocea						x		
C. peragua						x		
Cassipourea gummiflua	x		x	x	x	x	x	Mad.
Catha edulis		x	x	x	x	x	x	
Chrysophyllum gorungosanum			x	x	x	x		
Cola greenwayi			x	x	x	x		
Combretum kraussii						x		
Cryptocarya latifolia						x		
C. liebertiana			x	x	x	x		
C. myrtifolia						x		
Cunonia capensis						x		
Curtisia dendata						x	x	
Cussonia spicata			x	x	x	x	x	Com.
Cylicomorpha parviflora				x	x			
Diospyros whyteana					x	x	x	
Entandrophragma excelsum			x	x	x			
Fagara capensis						x	x	
F. davyi						x	x	
Fagaropsis angolensis		x	x	x	x	x		
Ficalhoa laurifolia			x	x	x			
Galiniera coffeoides		x	x	x	x			
Garcinia gerrardii s.l.			x	x	x	x	x	
G. kingaensis			x	x	x	x		
Gnidia glauca	x	x		x	x			
Gonioma kamassi						x		
Hagenia abyssinica		x	x	x	x			
Halleria lucida		x	x	x	x	x	x	
Hypericum revolutum	x	x	x	x	x	x	x	Com. Mad. Masc.
Ilex mitis	x	x	x	x	x	x	x	Mad.
Juniperus procera		x	x	x	x			
Kiggelaria africana				x	x	x	x	
Lepidotrichilia volkensii		x	x	x	x			
Leucosidea sericea						x	x	

472

	I.	II.	III.	IV.	V.	VI.	VII.	
Linociera foveolata						x		
L. peglerae						x		
Macaranga kilimandscharica		x	x	x	x			
M. mellifera					x	x		
Maesa lanceolata	x	x	x	x	x	x	x	Mad.
Maytenus peduncularis						x		
Mitragyna rubrostipulata			x	x	x	x		
Myrianthus holstii				x	x	x	x	
Myrica salicifolia s.l.	x	x		x	x	x	x	x
Neoboutonia macrocalyx				x	x	x		
Nuxia congesta	x	x		x	x	x	x	x
N. floribunda				x	x	x	x	x
Ochna holstii				x	x	x	x	x
Ocotea bullata s.l.			x	x	x	x	x	x
O. usambarensis				x	x	x		
Olinia (genus)			x	x	x	x	x	x
Oreobambos buchwaldii					x	x	x	
Platylophus trifoliatus							x	
Podocarpus falcatus s.l.			x	x	x	x		x
P. henkelii s.l.					x	x		x
P. latifolius s.l.	x			x	x	x	x	x
Prunus africana	x	x	x	x	x	x	x	Mad.
Pterocelastrus (genus)					x	x		x
Rapanea melanophloeos s.l.	x	x	x	x	x	x		x
Rhus chirindensis						x	x	
Ritchiea albersii	x	x	x	x	x			
Schefflera abyssinica	x	x	x	x	x			
S. goetzenii				x	x	x	x	
S. myriantha			x	x	x	x		Com. Mad.
S. umbellifera					x	x		x
Scolopia mundii						x		
Seemannaralia gerrardii						x		
Strombosia scheffleri	x			x	x	x	x	
Suregada procera			x	x	x	x	x	?
Syzygium masukuense					x	x		
Ternstroemia polypetala					x			
Trichocladus ellipticus			x	x	x	x	x	x
Turraea holstii			x			x	x	
T. robusta			x	x	x			
Warburgia salutaris				x	x		x	x
Widdringtonia nodiflora						x	x	x
Xymalos monospora	x			x	x	x	x	x

(2) Relatively restricted endemic species. These occur in 2 or 3 of the regional mountain systems. There are 38 species or 25 per cent of the Afromontane endemic tree flora. Although relatively restricted, most species in this category have quite extensive ranges. The range of *Diospyros whyteana* for example is 3200 km.

(3) Local endemic species are confined to a single system. There are 56 spp. or 37 per cent of the Afromontane endemic tree flora. Even local endemic species can be widely distributed. For instance most of the local endemics occurring in the Drakensberg system have quite extensive ranges as is shown by the following examples: *Combretum kraussii* (950 km), *Cunonia capensis* (1700 km) and

Scolopia mundii (1800 km). By contrast, most species in this category occurring in tropical Africa have much more restricted distributions and are confined to a single massif or a group of nearby massifs.

The mountains of East Africa (including the Kivu-Ruwenzori system) have the richest flora. Each of the three East African systems has more than 50 per cent of the endemic tree complement. Floristic impoverishment is fairly rapid to the south, and even more so to the north and the west. The woody Afromontane floras of Ethiopia and West Africa are poor in species.

Although the Drakensberg system has relatively few species it has by far the highest percentage of local endemics, but they still represent scarcely more than a quarter of its total Afromontane tree flora. The great majority of the remaining endemic Afromontane tree species occurring there are very widespread on the African mountains. In this category are many of the most important constituents of the Afromontane forests of South Africa, including *Apodytes dimidiata, Halleria lucida, Ilex mitis, Kiggelaria africana, Nuxia congesta, N. floribunda, Ocotea bullata* (including *O. keniensis*), *Podocarpus falcatus* s.l., *P. latifolius* s.l., *Prunus africana, Rapanea melanophloeos* s.l. and *Xymalos monospora*. Indeed, this assemblage of species could almost be used to define the Afromontane Region. No single species occurs throughout, but the assemblage is represented in virtually every 'island' of Afromontane vegetation, usually by several species. All except four (*Kiggelaria, Nuxia floribunda, Podocarpus latifolius* and *Xymalos*) extend as far north as Ethiopia, and all except five (*Halleria, Kiggelaria, Nuxia floribunda, Ocotea* and *Podocarpus falcatus*) are present in West Africa. In West Africa, *Podocarpus*, which is perhaps the most characteristic genus of Afromontane forests, occurs no further west than the highlands of Cameroun and southeast Nigeria. *Nuxia congesta*, however, extends as far west as the Fouta Djalon in Guinée, and *Ilex mitis* reaches the mountains of Liberia and Sierra Leone. Only five species, *Curtisia dentata, Fagara capensis, F. davyi, Leucosidea sericea*, and *Rhus chirindensis*, extend no further north than Rhodesia. The Drakensberg system has fewer Afromontane tree species than the other systems except the West African and Ethiopian. For smaller plants the situation is reversed, and, although it is not possible to produce exact figures, there is no doubt that the Drakensberg system is by far the richest.

2.2 *Generic geographical elements*

The 152 Afromontane endemic tree species mentioned above belong to 102 genera of which 20 are Afromontane endemics. The majority of non-endemic tree genera belong to the lowland tropics. 51 genera (62 per cent, or 68 per cent if subcosmopolitan genera are excluded) are better represented in the tropical lowlands of Africa than on the mountains. Some of the remaining tropical genera, e.g. *Cryptocarya, Rapanea* and *Schefflera*, although almost entirely montane in Africa, occur in the lowlands elsewhere in the tropics. Others such as *Ternstroemia* and *Cylicomorpha* are predominantly montane in Africa but occur very locally in the African lowlands. Very few genera are of northern hemispheric (e.g. *Agauria, Juniperus, Pistacia*) or southern hemispheric (*Gnidia, Podocarpus, Widdringtonia*) affinity.

For smaller plants the situation is reversed, and genera of tropical affinity are in a minority, though it is not yet possible to make a complete analysis.

2.3 Non-endemic species

Disregarding transition zones, a number of lowland species occur in the lower part of the Afromontane Region and some Afromontane species penetrate a short distance into adjacent phytochoria. These can be referred to as 'marginal intruders', but:

(1) Only a very small proportion of the species of lowland phytochoria behave in this way.

(2) The great majority of intruders into the Afromontane Region are rare and localized there. They contribute little to its phytomass, and in no way undermine the usefulness of the concept of an Afromontane Region.

Information on the Afromontane marginal intruders in Malawi is given by Chapman & White (1970). They are not discussed in general any further here. In addition to the intruders it is useful to recognize other main groups of linking elements in the Afromontane flora.

2.3.1 Forest pioneer connecting species

Some pioneer tree species of Afromontane forests, e.g. *Albizia gummifera* subsp. *gummifera*, *Anthocleista grandiflora*, *Macaranga kilimandscharica* and *Maesa lanceolata*, are Afromontane endemics or near-endemics and are absent from or very rare in the lowlands. A few others, e.g. *Bersama abyssinica*, *Croton macrostachyus* and *Polyscias fulva* are ecological and chorological transgressors (see below), which are more abundant in Afromontane forest than in lowland forest. Several other pioneer species, however, e.g. *Bridelia micrantha*, *Clausena anisata*, *Harungana madagascariensis*, are very widely distributed in lowland forest in tropical Africa, and are more abundant there. In the Afromontane Region some of them appear to be no more than marginal intruders.

2.3.2 Connecting species of secondary grassland

Secondary, fire-maintained grassland in the forest belt is composed predominantly of species which are also widespread in the lowlands. There is some evidence that, before the destruction of forest, some of them occurred as marginal intruders and were confined to edaphically specialized sites which acted as foci from which they have subsequently spread. This topic is discussed in Section 3.2.5. The relative importance of spread from Afromontane foci and of invasion from the lowlands is, at present, completely unknown. Fire and other forms of disturbance also permit the downward spread of species from the Ericaceous belt, as described by Mabberley (1975) for the Cherangani Mts. in Kenya.

2.3.3 Ecological and chorological transgressors

An 'ecological and chorological transgressor' is a species which occurs in two (or more) major phytochoria and also occurs in at least two major vegetation types. It should be equally abundant in both phytochoria, or, if more abundant in one, it should still be prominent in the other. Ecological and chorological transgressors are plants which have successfully extended their geographical and ecological ranges without having completed the processes of speciation. Some examples from the Afromontane flora of Malawi are given by White (in Chapman & White 1970), who has subsequently (1976b), described the distribution, ecology and taxonomic relationships of one of them, *Parinari excelsa*, in considerable detail.

At present 37 transgressor tree species are known or 18 per cent of the Afromontane tree flora. The distributions of the 26 species which occur in southern Africa, and some other information about them are summarized in Table 2. Further work will almost certainly produce some modifications to this list, since only a few species have been thoroughly studied both in the field and in the herbarium. In the Drakensberg system the transition zone between the Afromontane

Table 2. The geographical distributions of chorological and ecological transgressor tree species occurring in the Afromontane Region in southern Africa.

I. West African system. II. Ethiopian system. III. Kivu – Ruwenzori system. IV. Imatongs – Usambara system. V. Uluguru – Mlanje system. VI. Chimanimani system. VII. Drakensberg system. VIII. Distribution elsewhere: Bushl. = also occurring in bushland and thicket; E Afr = also occurring in lowland forest in East Africa; GC = also occurring in Guineo-Congolian lowland rain forest; Plur. = pluriregional transgressor; Mad. = also occurring in Madagascar; Masc. = also occurring in the Mascarenes: As. = also occurring in Asia.

\+ = species represented in the Afromontane Region by distinct subspecies. : = species absent over a large part of its range from the surrounding lowlands.

	I.	II.	III.	IV.	V.	VI.	VII.	VIII.
+Albizia gummifera	x	x	x	x	x	x		GC/E Afr/Mad.
+Bersama abyssinica	x	x	x	x	x	x		Plur.
Brachylaena huillensis				x				Plur.
:Caloncoba welwitschii			x	x	x			GC
Cassine aethiopica	x	x	x	x	x	x	x	Plur./Mad.
C. papillosa						x	x	Bushl.
:Cassipourea congoensis s.l.	x	x	x	x	x	x	x	GC
:Celtis africana	x	x	x	x	x	x	x	Plur.
+Craibia brevicaudata				x	x	x		Plur.
:Croton macrostachyus	x	x	x	x	x			GC
+Diospyros abyssinica		x	x	x	x			Plur.
:Ekebergia capensis	x	x	x	x	x	x	x	Plur.
+Eugenia capensis s.l.				x	x	x	x	Plur.
+Faurea saligna s.l.		x		x	x	x	x	Plur.
:Olea africana		x	x	x	x	x	x	Plur./Masc./As.
+Olea capensis s.l.	x	x	x	x	x	x	x	Plur.
:Parinari excelsa	x		x	x	x			GC
+Pittosporum viridiflorum s.l.	x	x	x	x	x	x	x	Plur./Mad./As.
:Polyscias fulva	x	x	x	x	x	x		GC
Ptaeroxylon obliquum				x			x	Bushl.
:Rawsonia lucida		x	x	x	x	x	x	E. Afr.
Schrebera alata s.l.		x	x	x	x	x	x	Plur.
Scolopia zeyheri	x		?	x	x	x	x	Bushl.
+Syzygium cordatum				x	x	?		Plur.
+S. guineense s.l.	x	x	x	x	x	x	x	Plur.
:Trichilia dregeana	x	x	?	x	x	x		Plur.
Trimeria grandiflora				x	x	x	x	Bushl.

Region and adjacent phytochoria is complex and it is not always easy to decide whether a species is truly Afromontane or a transgressor. At least a third of the transgressors are represented in the Afromontane Region by upland subspecies, though the relationship is not always simple. For instance, *Diospyros abyssinica* is represented on the mountains of Malawi by a distinct subspecies, but elsewhere there is no taxonomic difference between the upland and lowland populations. Of the remainder, seven species occur as transgressors in only one of the seven systems, and a further eleven, over a large part of their range, are absent from the surrounding lowlands.

2.3.4 Species absent from the surrounding lowlands but occurring in geographically remote phytochoria

J. D. Hooker, in a series of articles (1862, 1864, 1874, 1885), was the first to discuss in any detail the occurrence on the mountains of tropical Africa of species which are also found in Europe, the Canaries, Mauritius, Madagascar and elsewhere, or are replaced by closely related species there. This subject is also dealt with, to varying degrees, by Engler (1892, 1904a, b), Hedberg (1961a), Liben (1962), and Dejardin et al. (1973). These species are important to an understanding of the history of the Afromontane flora and have been discussed at some length elsewhere (White in press, B), but they are few in number. It is unlikely that as many as 10 per cent of Afromontane species also occur in distant phytochoria and the figure may well be much lower. The extra-Afromontane distributions of tree species included in Tables 1 and 2 are briefly indicated. Only 17 or 9 per cent of the 189 species included in the pan-Afromontane sample on which Tables 1 and 2 are based come into this category. The proportion of smaller plants showing similar distributions appears to be less, except for bryophytes and pteridophytes.

2.4 *Distant satellite populations of Afromontane near-endemic species*

The difference between this category and that of Afromontane species which occur as marginal intruders in adjacent phytochoria is one of degree: a distant satellite population is one which is separated from the nearest 'island' of Afromontane vegetation by an interval of 50 km. Some satellites may have resulted from recent invasion but most appear to be relictual. The latter are usually confined to specially favourable habitats or those from which the prevalent surrounding vegetation is excluded or enfeebled because of locally unfavourable conditions.

Distant satellite populations are of two kinds. Some are sporadic. Others form east–west extensions of the Afromontane region which may have formerly been migratory tracks, though under present conditions the latter may appear to be culs-de-sac (Wild 1956), rather than connecting links between distant Afromontane 'islands'. They will be described only briefly here since a more complete account including a discussion of their historical significance will be published elsewhere (White in press, B).

Fig. 4 shows some of the more interesting sporadic populations and three east–west extensions in southern Africa. In the Magaliesberg extension several Afromontane tree species occur in forest in sheltered kloofs. They include *Calodendrum capense, Diospyros whyteana, Halleria lucida, Ilex mitis,*

Fig. 3. Forest patch at 2500 m on Mt. Moco, Angola, including the Afromontane species *Podocarpus latifolius, Pittosporum viridiflorum* s.l. and *Ilex mitis*. *Protea* and *Philippia* are present in the surrounding grassland (photo B. J. Huntley).

Kiggelaria africana, Leucosidea sericea, Nuxia congesta, Olinia and *Pterocelastrus* (Acocks 1953, Coetzee 1974, 1975).

Wild (1956), who first drew attention to the Limpopo escarpment extension, mentions among trees *Albizia gummifera, Anthocleista grandiflora, Calodendrum capense, Diospyros whyteana* and *Ochna holstii*. Several others have been added since.

Afromontane species are recorded from the wetter parts of Zambia by White

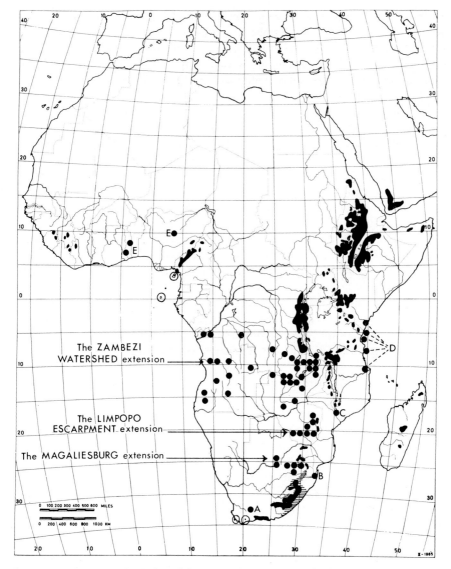

Fig. 4. Map showing the distribution of distant satellite populations of Afromontane species in relation to the Afromontane archipelago. Hollow circles represent islands of Afromontane vegetation that are too small to show otherwise.
A. Satellite of *Ilex mitis* at Ladismith in the Little Karoo. B. Satellites of *Podocarpus falcatus* in Tongaland and southern Moçambique. C. Satellite of *Calodendrum capense* in southern Malawi. D. Afromontane satellites in the Zanzibar–Inhambane Region. E. Afromontane satellites in West Africa. Horizontal lines indicate a transition zone between the Afromontane Region and other phytochoria. Elsewhere the transition zone is too narrow to show on the map.

(1962), Morton (1972) and Kornaś (1974a, b). For Shaba, Malaisse (1967) reports *Podocarpus latifolius* from the Biano, Kibara and Kundelungu plateaux. Other Afromontane species are recorded by Malaisse, and by Lisowski et al. (1970). As one proceeds westwards there is a diminution in the number of Afromontane nearendemics, but several, including *Apodytes dimidiata*, *Ficalhoa laurifolia*, *Ilex*

479

mitis, *Maesa lanceolata*, *Myrica conifera* and *Podocarpus latifolius*, reach Angola. This constitutes the Zambezi watershed extension.

In Angola well-developed Afromontane vegetation is apparently absent. Afromontane species, it seems, are always mixed with other floristic elements. They are, however, reasonably plentiful along the upper slopes of the Serra da Chela (Barbosa 1970). *Podocarpus latifolius*, associated with *Apodytes* and *Hymenodictyon floribundum*, occurs as a small tree 5–6 m tall in rocky places between 1800 and 2500 m. These populations are surrounded by stunted miombo woodland of *Brachystegia spiciformis*, *B. floribunda* and *Julbernardia paniculata* (see Chapter 10). Other Afromontane near-endemics and transgressor species occur lower down on the escarpment of the Serra de Chela between 1000 and 1500 m in forest in ravines. They include *Ficalhoa laurifolia*, *Maesa lanceolata*, *Newtonia buchananii* and *Parinari excelsa*. *Podocarpus latifolius* also occurs as a tree up to 20 m tall on Mt. Moco at 2500 m (Fig. 3) but whether this is a predominantly Afromontane community is unknown. A few Afromontane species, e.g. *Anthonotha noldeae* and *Fagaropsis angolensis*, which occur in Angola, are separated by a wide interval from their main occurrences on the mountains much further east.

Fig. 4 collectively shows the distant satellite occurrences of 29 Afromontane tree species including *Agauria salicifolia*, *Cassipourea gummiflua*, *Drypetes gerrardii*, *Podocarpus latifolius*, *Prunus africana* and *Trichocladus ellipticus*, and of ten species of shrubs and climbers and eight herbaceous species.

2.5 *Intervals in the Afromontane flora*

Many Afromontane species show remarkably wide disjunctions superimposed on the narrower disjunctions which are a consequence of the archipelago-like nature of the Afromontane Region itself.

White (in Chapman & White 1970) discussed the Malawi interval in relation to *Juniperus procera*, *Ocotea kenyensis*, *Podocarpus ensiculus/henkelii*, *P. falcatus/gracilior* and *Ptaeroxylon obliquum*. These are all forest trees, but the Malawi interval or an enlarged Malawi interval is also shown by many smaller plants including *Pentaschistis*, *Euryops* (Nordenstam 1969), *Stoebe* (Weimarck 1941), *Chrysanthemoides monilifera* (Norlindh 1946), *Lepidium africanum* subsp. *africanum* (Jonsell 1975) *Thesium triflorum* (Brummitt 1976), and *Zaluzianskya* (Hedberg 1970b). *Macowania* (Hilliard & Burtt 1976) shows both an enlarged Malawi interval between the mountains of the Transvaal and Ethiopia, and a shorter disjunction of 300 km within the Drakensberg mountain system.

Several Afromontane forest species show wide disjunctions within South Africa, e.g. *Faurea macnaughtonii*, *Prunus africana*, *Schefflera umbellifera* and *Trichocladus ellipticus* (Phillips 1931).

3. Zonation and the main vegetation types

On nearly all high mountains the vegetation diminishes in stature from the lower slopes to the summit. Tall forest gradually gives way to scrub forest followed by bushland and thicket and ultimately to dwarf shrubland. On most of the higher tropical mountains the shrubland is overtopped by arborescent senecios and lobelias, which in their pachycaul construction, however, differ greatly from more familiar trees. On well-drained soils woody plants predominate throughout. Except at the highest altitudes, grasses and Cyperaceae normally only become physiognomically dominant where the soil is shallow or the drainage impeded or where the natural vegetation has been destroyed by fire.

The Afromontane Region shows a wide and gradual change in the physiognomy of its climax vegetation and in the floristic composition of the latter which means that, even where there is a regular altitudinal replacement of vegetation types, the recognition of distinct zones is often arbitrary. Along any particular altitudinal transect regular altitudinal replacement can usually be seen, but the patterns of replacement on different mountains, or on different sides of the same mountain, are usually so different that generalized schemes, even for relatively restricted regions, are well nigh impossible to attain.

In Africa patterns of climatic change on mountains are of considerable complexity. Collectively the Afromontane archipelago occurs in, or is contiguous with, nearly all the major phytochoria occurring in Africa south of the Sahara (Fig. 2). Hence it is to be expected that the zonation including the nature of the transition zones will vary greatly. Where the climate of the surrounding lowlands is considerably hotter or drier than that of the mountains, the transition zones are narrow. This is the usual state of affairs. Reasonably extensive transitional conditions occur only in the Lake Victoria basin and in South Africa. It is only there that the relationships between the Afromontane flora and lowland floras are at all complex. Elsewhere, Afromontane species which descend into adjacent phytochoria do so only as marginal intruders or occur as distant satellite populations.

Each massif has its own peculiar features. Even on a single mountain the altitudinal range of individual species may differ greatly on different sides. This is especially true of those mountains (the great majority), which experience high rainfall on one side and low rainfall on the other.

It has long been known that the upper altitudinal limit of montane species and communities varies with the size of the mountain. On small isolated mountains the upper limit is lower than on higher massifs ('Massenerhebungseffekt'). The lower altitudinal limits may also be related to the size of the mountain. Species which occur on high mountains often descend to altitudes that are lower than the summits of nearby mountains from which they are absent (Van Steenis 1961, 1972). Various explanations of these phenomena have been offered. The 'Massenerhebungseffekt' is traditionally explained, at least in Europe, by the occurrence of higher temperatures at corresponding altitudes on the higher massifs. For tropical mountains Richards (1952) suggests that exposure to wind is at least partly responsible, whereas Grubb (1971) believes that the amount of fog cover is the operative factor, but the incidence of fog varies greatly from place to place in the Afromontane Region. Wood (1971) suggests that the occurrence of montane species on high mountains at lower altitudes than the summits of smaller mountains from which they are absent is due to historic reasons. During an earlier warmer phase they became extinct on the smaller mountains but could find refuge nearer the summits of taller mountains.

A comprehensive account of the zonation of vegetation on the East African mountains has been published by Hedberg (1951) and this must be the starting point for studies of zonation elsewhere in Africa. Hedberg recognizes three major zones which he refers to as 'belts' – the Montane Forest Belt, the Ericaceous Belt and the Afroalpine Belt. Hedberg distinguishes between a belt, which 'is an altitudinal region which can be traced on all (or most) mountains of sufficient height in a definite part of the world', and a zone which is 'a more or less local altitudinal region' within a belt. Belt is a more generalized concept and reflects climatic rather than edaphic conditions. Since in the English language, however, the words belt and zone are virtually synonymous this terminology is confusing. For 'zone' sensu Hedberg the term 'horizon', used in exactly this sense by du Rietz as long ago as 1930, is available. As Hedberg points out, there are two ways of studying altitudinal zonation. One may either investigate distribution of species and draw the zonal limits where the limits of individual species more or less coincide, or one may draw the limits according to the plant communities. Hedberg based his study on vegetation, not on altitude per se, and attached most weight to the main communities rather than to those of lower rank.

Zonation is more clearly defined on the very high mountains in East Africa that were studied by Hedberg than on the mountains in southern Africa which are lower. As is shown in the following pages the zonation on the latter is complex, though ecologically significant. On the high East African mountains the floras of the Forest belt and the Ericaceous belt are reasonably distinct. In southern Africa they are intricately intermingled.

3.1 *Zonation in East Africa*

On some East African mountains, e.g. Mt. Kenya, the forest belt consists of three horizons, true montane forest, the bamboo (*Arundinaria*) horizon and the *Hagenia – Hypericum* scrub-forest horizon. On other mountains, e.g. Ruwenzori and Elgon, the *Hagenia – Hypericum* horizon is absent, and on Kilimanjaro, both it and the bamboo horizon are lacking. In south tropical Africa *Hagenia* is only known from the Nyika Plateau where it does not form a distinct horizon. *Arundinaria* in southern Africa has no more than a sporadic distribution.

In East Africa the Ericaceous belt is occupied by bushland, shrubland or thicket* dominated by species of *Philippia*, with species of *Erica* playing an important role in the lower part, in which scattered stunted trees of *Rapanea* and *Hypericum revolutum* may also occur. The height of the vegetation ranges from 0.5–8 m. It varies from very open, especially when subjected to frequent fires, to dense and impenetrable. In East Africa the lower limit of this belt lies between 2600–3400 m and the upper varies from 3550–4100 m.

Although Hedberg (1951) suggests that 'it is tempting to compare the Ericaceous belt of the East African mountains with the "subalpine zone" described by Van Steenis (1935) from Malesia', which often consists largely of Ericaceae and occurs at comparable altitudes, he himself does not use the term and it is not currently used in East Africa. Analogous vegetation in South Africa, however, is frequently referred to as 'subalpine' (e.g. by Killick 1963, Edwards 1967, and Chapter 12).

The Afroalpine belt is rather diverse on different mountains and it is not easy to give any common vegetational features except the position above the upper

---

* Using the terminology of the new UNESCO/AETFAT 'Vegetation Map of Africa' (White in press, A).

limit of the Ericaceous belt. Giant senecios and lobelias as well as shrubby species of *Alchemilla* and *Helichrysum* are nearly always present, though their relative abundance varies greatly. In south tropical Africa the Afroalpine belt does not occur, but in South Africa the highest parts of the Drakensberg have been recognized as an Austo-afroalpine belt by Coetzee (1967) and Van Zinderen Bakker & Werger (1974) and is described in Chapter 12. On the high mountains of East Africa these three belts are reasonably distinct, but they are separated by transition zones, which are sometimes somewhat complex. The Ericaceous belt usually ascends higher on the ridges than in the valleys, as on Kilimanjaro where cold air collects in the bottom of broad U-shaped depressions. Elsewhere, as on Mt. Kenya and Meru, the Ericaceous belt ascends higher in the valleys, especially where these are narrow and V-shaped and provide protection from wind. In southern Africa, except locally, the pattern of zonation is more complex than in East Africa.

When the Afroalpine Region is kept separate from the Afromontane it is customary to include the Ericaceous belt with the latter (e.g. Hedberg 1957, 1965, Chapman & White 1970, compare Chapter 12). At least superficially, the vegetation and floras of the Forest and Ericaceous belts look quite different. This is only true, however, for the highest mountains. Elsewhere the two floras intermingle. For example, in the western Usambara Mountains in Tanzania, climatic gradients are steep, and *Ocotea usambarensis* occurs, both as an emergent tree up to 50 m tall in rainforest in the Montane forest horizon, and also as a stunted tree 10–15 m tall in *Philippia* thicket on the crests of high ridges. The nearby Uluguru Mts. are not tall enough to support an Ericaceous belt. Communities dominated by *Philippia*, with an entourage of other species from the Ericaceous belt, occur within the forest belt on steep slopes where the soil is too shallow to support forest. A multitude of comparable cases could be mentioned.

3.2 *Zonation and vegetation in Malawi, Rhodesia and Moçambique*

Information on zonation in Malawi is given by Chapman & White (1970), on the Mlanje Mts. by Chapman (1962), on the Chimanimani Mts. in Rhodesia by Phipps & Goodier (1962), and on Gorongosa Mt. in Moçambique by Macedo (1970).

Zonation in south tropical Africa is much more complicated than in East Africa and a generalized classification is probably unattainable. Most of the mountains lie within the forest belt. Several forest types can be recognized but their correlation with altitude is very imperfect. Only Mt. Mlanje and the Chimanimani Mts. are high enough to support an Ericaceous belt. On Mt. Mlanje the latter is fragmentary and is far from typical. Most of its upper slopes consist of bare rock which supports little but lichens, tufted Cyperaceae and a few scattered shrubs. Communities in which *Philippia* is common are found only in deep, sheltered clefts and gullies.

The crests of the much lower Mafingi Mts. in Malawi and the Chimanimani Mts. in Rhodesia are rocky and are very exposed. Forest is only found lower down in sheltered ravines. The crests themselves support shrubland of various types in which *Philippia* is often abundant. *Philippia* shrubland is very widespread on the upper slopes of the Chimanimani Mts. at about 2400 m.

The mountains of Malawi, Rhodesia and Moçambique are surrounded by a sea of miombo woodland. Most of the surviving patches of forest on these massifs are indisputably Afromontane in character, but those slopes which face the direction of rainbearing winds, and hence experience higher and better distributed rainfall, often have relict patches of lowland or transitional rain forest if their lower slopes are low enough. On their drier slopes miombo gives way directly to Afromontane forest or scrub forest without the intervention of lowland or transitional rain forest.

3.2.1 Transitional rain forest

Patches of forest in which species of lowland and Afromontane rain forest intermingle at all plentifully are rare. In Malawi (Chapman & White 1970) this type occurs on Lisau Saddle in the Shire Highlands and Machemba Hill near Mt. Mlanje. At about 1370 m the forests are dominated by the lowland species, *Khaya nyasica*, which is, however, associated with the following Afromontane species: *Chrysophyllum gorungosanum, Cola greenwayi, Drypetes gerrardii, Myrianthus holstii, Prunus africana, Suregada procera* and *Xymalos monospora*. It would appear that such transitional forest was formerly of very restricted extent.

Within the Eastern borders of Rhodesia small patches of forest occur which are transitional between Afromontane and lowland forest. The best known is the Chirinda forest (Swynnerton et al. 1911, Banks 1976, Goldsmith 1976), which is situated between 1076 and 1250 m, 48 km south of Chipinga. The forest is confined to soils derived from dolerite. The surrounding shale and quartzite soils support miombo woodland. The majority of species, including the tallest trees, are lowland forest species but several, including *Chrysophyllum gorungosanum* which is one of the commonest trees in the forest, are Afromontane; other Afromontane species include *Casearia battiscombei, Cola greenwayi, Cryptocarya liebertiana, Drypetes gerrardii, Halleria lucida, Myrianthus holstii, Prunus africana, Strombosia scheffleri* and *Xymalos monospora*.

3.2.2 The transition from miombo to Afromontane forest

The miombo and Afromontane floras, except for a few species like *Philippia benguelensis* discussed below, do not intermingle. In a relatively narrow transition zone patches of miombo woodland and montane forest interdigitate or occur in mosaic, the precise distribution being determined by edaphic conditions. On Mt. Mlanje (Chapman 1962) miombo, dominated by *Brachystegia spiciformis*, occurs on steep spurs radiating from the main plateau, whilst in the valleys evergreen Afromontane forest descends from above. The situation is similar in the Chimanimani Mountains (Phipps & Goodier 1962), where Afromontane forest occurs in sheltered valleys in the vicinity of ground water, but where the soil is well-drained. The soil under miombo is well-drained but dries out in the dry season.

Towards its upper altitudinal limit, which is usually between 1600 and 2100 m, miombo is very different in character from its typical lowland facies. It is a stunted, floristically impoverished, woodland not more than 6 m tall, usually dominated by *Brachystegia spiciformis*, more rarely by *B. taxifolia, B. glaucescens*, or *Uapaca kirkiana*. The trees are often festooned with *Usnea*; epiphytic orchids may also occur. The ground flora consists of a sparse cover of grasses which do not burn fiercely. This probably explains the presence of ferns, e.g. *Pellaea* and *Arthropteris orientalis*, and fire-sensitive species of *Philippia* (see also Chapter 10).

P. benguelensis is a characteristic member of the Ericaceous belt in south tropical Africa. It seems to have a wide temperature-tolerance and can occur throughout the forest belt wherever it is not excluded by competition or fire. It is absent from undisturbed forest except on rocky outcrops but is abundant as a forest precursor. In the Chimanimani Mountains *P. benguelensis* is codominant with *Brachystegia spiciformis* on very stony soils (Phipps & Goodier 1962).

3.2.3 Zonation within the forest belt

In Malawi the overall situation is too complex to allow a simple altitudinal classification, although zonation within the forest belt is apparent along any individual altitudinal transect. On the basis of floristic composition, structure and physiognomy, White (in Chapman & White 1970), distinguished two main forest types in Malawi, namely submontane rain forest* and montane forest* sensu stricto.

In general, Afromontane rain forest occurs at lower altitudes than undifferentiated montane forest but the relationship is far from exact. Submontane rain forest was so named to emphasize the fact that many of its endemic species have close relatives in the lowlands, in contrast to the species of Afromontane forest sensu stricto, which have no relatives in the lowlands at least in tropical Africa.

In Malawi, Afromontane rain forest occurs between 1370 and 2290 m (Figs. 5, 6 and 7). In structure it is similar to Guineo–Congolian lowland rainforest and many of its endemic species have their closest relatives there. Its most characteristic species include *Aningeria adolfi-friedericii*, *Entandrophragma excelsum*, *Ocotea usambarensis*, *Chrysophyllum gorungosanum*, *Cola greenwayi*, *Cylicomorpha parviflora*, *Drypetes gerrardii*, *Mitragyna rubrostipulata*, *Strombosia scheffleri* and *Myrianthus holstii*. In Malawi these species are absent from Afromontane forest sensu stricto.

Undifferentiated Afromontane forest occurs between 1675 and 2590 m. It is structurally less complex than rain forest, and, except when dominated by *Juniperus procera* or *Widdringtonia nodiflora*, both of which are largely dependent on fire for their regeneration, is rarely more than 20 m tall. Its most characteristic species include *Ilex mitis*, *Juniperus procera*, *Widdringtonia nodiflora*, *Agauria salicifolia*, *Cussonia spicata*, *Dombeya erythroleuca*, *Hagenia abyssinica*, *Kiggelaria africana*, *Nuxia congesta*, *Olinia usambarensis*, *Philippia benguelensis*, *Pittosporum viridiflorum*, *Rapanea*, *Myrica salicifolia*, *Nuxia floribunda*, *Pterocelastrus* and *Trichocladus ellipticus*.

Some species, e.g. *Podocarpus latifolius* (*milanjianus*), *Prunus* (*Pygeum*) *africana*, *Aphloia myrtiflora*, *Apodytes dimidiata*, *Xymalos monospora* and *Diospyros whyteana* occur more or less abundantly in both types.

In Malawi undifferentiated montane forest characteristically occurs above Afromontane rain forest as on the high plateaus of the Nyika and Mt. Mlanje. Several other mountains such as the Misuku Hills, Nchisi Mt., and Cholo Mt.,

---

* Designated Afromontane rain forest and undifferentiated Afromontane forest in the UNESCO/AETFAT classification (White in press, A).

Fig. 5. Afromontane rain forest at Nchisi Mt., Malawi. The photograph shows an *Aningeria adolfi-friedericii* on the ridge top, 45 m high, 12 m to the first branch and 7.5 m in girth measured over the buttresses at 2.1 m from the ground. The shrub with the right-angled branching to the left of the picture is a deciduous species of *Rytigynia* (photo J. D. Chapman).

however, are not high enough for this altitudinal replacement to take place, and only Afromontane rain forest is well-developed there, though several montane forest species may occur at the forest edge (Fig. 8). *Agauria salicifolia, Cussonia spicata, Dombeya erythroleuca, Myrica salicifolia* and *Philippia benguelensis*, for instance, occur thus in the Misuku Hills. The summit of Dedza Mt. supports undifferentiated montane forest, but any Afromontane rain forest which formerly occurred on its wetter slopes has been completely destroyed by cultivation and fire. So

Fig. 6. Afromontane rain forest, Wilindi, Misuku Hills, Malawi, alt. 1800 m. The large trees with rather massive branches in the centre of the picture are *Entandrophragma excelsum*. The short tussocky grass in the foreground is mainly *Andropogon schirensis* with some *Exotheca abyssinica* (Photo J. D. Chapman).

far as the foregoing massifs are concerned a simple altitudinal relationship between Afromontane rain forest and undifferentiated montane forest, though not necessarily for the constituent species of the latter, is reasonably clear-cut. On the Vipya Plateau, however, there is lateral rather than altitudinal replacement, since Afromontane rain forest occurs on the wetter side up to the crest and undifferentiated montane forest oc-

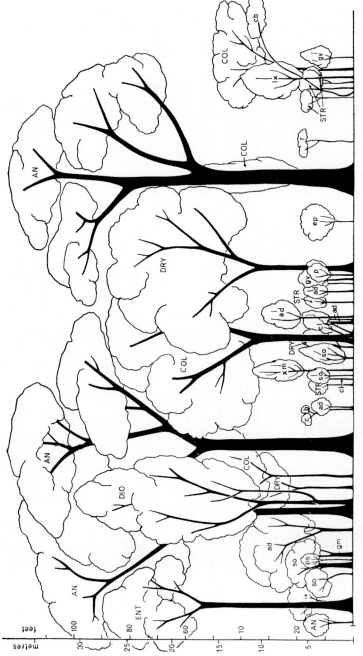

Fig. 7. Profile diagram of *Aningeria, Entandrophragma* Afromontane rain forest, Mugesse Forest, Misuku Hills (From Chapman & White 1970). ad: *Aulacocalyx diervilloides,* AN: *Aningeria adolfi-friedericii,* cb: *Craibia brevicaudata,* cl: *Coffea ligustroides,* COL: *Cola greenwayi,* DIO: *Diospyros abyssinica,* DRY: *Drypetes gerrardii,* ENT: *Entandrophragma excelsum,* ep: *Erythrococca polyandra,* gm: *Garcinia mlanjiensis,* gv: *Garcinia volkensii,* ix: *Ixora* sp., p: *Pavetta* sp., ps: *Psychotria* sp., r: Rubiaceae, so: *Sclerochiton obtusisepalus,* STR: *Strombosia scheffleri,* xm: *Xymalos monospora.*

Fig. 8. *Nuxia congesta* at the northeast of Nchisi Mt. forest. The stems and branches of both trees are covered with moss and epiphytic ferns and on the big trees the strangling epiphyte *Schefflera abyssinica* has secured a hold. Here at the windward edge of the forest the transition from grassland to trees is very abrupt, marked only by a narrow belt of *Hyparrhenia cymbaria* and *Pteridium aquilinum* (photo J. D. Chapman).

curs on the drier side. This kind of lateral relationship between Afromontane rain and undifferentiated montane forest is widespread in parts of East Africa. On Mt. Kenya, for example, the drier slopes are covered with *Juniperus* forest, in which several Malawi Afromontane species, such as *Hagenia, Nuxia, Olinia* and *Trichocladus*, are abundant. These latter are absent from the *Ocotea* Afromontane rain forests on the moister slopes (White unpubl.). In Malawi, with the exception of the Vipya plateau, the lower rain shadow slopes of the larger massifs support miombo woodland rather than forest.

One of the paradoxes of African vegetation is the fact that, although most Afromontane rain forest species are of tropical lowland affinity, in their own distributions they are often strictly confined to the mountains, whereas many species of undifferentiated Afromontane forest, which have no close relatives in the lowland tropics, descend from the mountains to the surrounding plateaux. Afromontane rain forest species have relatively restricted ranges of tolerance for temperature and humidity and are confined to the wetter parts of the mountains below altitudes at which frequent severe frosts occur. Within the greater part of this zone many, but not all, undifferentiated montane forest species can occur as pioneers at the forest edge or retain a footing on rocky ridges. In the absence of fire they are soon shaded out by more vigorous Afromontane rain forest species, except on shallow soils.

In those parts of the Afromontane Region, like Malawi and Rhodesia, where the dry season is both long and severe, fire originating from natural causes has probably been partly responsible for the intermingling of the two Afromontane forest floras. Phipps & Goodier (1962) believe that fire is a natural factor in the Chimanimani Mountains. Chapman & White (1970) review the situation in Malawi and conclude that, although the evidence for the occurrence of natural fires is conclusive, man-made fires during the last thousand years or so have had an enormously greater influence on Afromontane vegetation.

3.2.4   The transition to scrub forest and the Ericaceous belt

The forest patches that survive on the high plateau of the Nyika between 2290 and 2600 m (Fig. 9) are rather stunted. It is unusual to find trees much over 15 m in height, and most of the forests are scrub forest with an average height of 7.5–9 m. There is only one stratum of trees. The undergrowth tends to be rather open since large mammals visit the forest for shelter and browse. The most abundant species include *Afrocrania volkensii, Cassipourea congoensis, Ilex mitis, Podocarpus latifolius, Prunus africana, *Agauria salicifolia, Aphloia theiformis, Apodytes dimidiata, *Cussonia spicata, *Dombeya erythroleuca, *Hagenia abyssinica, *Kiggelaria africana, *Myrica salicifolia, *Nuxia congesta, Olinia usambarensis, Rapanea melanophloeos, Xymalos monospora, Diospyros whyteana* and **Philippia benguelensis*. Species marked with an asterisk frequently occur inside scrub forest at this altitude, but only occur at the edges of forest at lower altitudes.

Even shorter scrub forest occurs on the summits of rocky windswept hills on the Vipya Plateau at about 1980 m. There is a single dense tree storey about 6 m tall. The branches of the trees are gnarled and tightly interlaced, thickly covered with moss and lichens, and have an abundance of epiphytic ferns. The trees include *Cassipourea congoensis, Cussonia spicata, Erythroxylum emarginatum, Rawsonia lucida, Rothmannia fischeri* and *Teclea nobilis*. In physiognomy, though not in floristic composition, this is indistinguishable from the elfin forest described by Pócs (1974, 1976) from the Uluguru Mts. in Tanzania. Although this type must have recently evolved from Afromontane forest, physiognomically it is

Fig. 9. The rolling downland country of the High Nyika, Malawi. Photographed from the Nganda track at 2350 m altitude. In the small patches of stunted forest fringing the boggy hollows where streams rise the most conspicuous tree is *Hagenia abyssinica* with rounded or umbrella-shaped crown and red-brown, papery bark. Typical associates include *Myrica salicifolia*, *Agauria salicifolia*, *Rapanea melanophloeos* and *Afrocrania volkensii*. In the open hollows *Lobelia mildbraedii* is often conspicuous in the dry season and *Blechnum tabulare* is frequently abundant. The short tussocky grass of poor grounded cover in the foreground consists mainly of *Exotheca abyssinica* and *Loudetia simplex* with suffruticose *Eupatorium africanum*, *Humularia descampsii* and less frequent *Protea heckmanniana* (photo J. D. Chapman).

not a forest. In the terminology of White (in press, A) it is a type of thicket.

On Dedza Mt. the surviving forest, occurring on the plateau surface at about 1980 m, is no more than 15 m tall and floristically is somewhat similar to that of the Nyika (Fig. 10). As one ascends the rocky residual peaks the trees become increasingly stunted until towards the summit the impression is more of a dense wind-clipped shrubbery 3–5 m tall rather than of a forest. The trunks and branches of the trees are thickly padded with mosses and the crowns cluttered with lichens, epiphytic ferns and orchids. Among the rocks at the summit this low 'elfin forest' is replaced by a thicket of *Philippia benguelensis* with *Aloe arborescens*, *Cussonia spicata*, *Heteromorpha trifoliata*, *Hypericum revolutum* and *Myrsine africana*.

On Mt. Mlanje most of the montane forests are dominated by *Widdringtonia nodiflora* (Figs. 11 and 12). In Malawi it reaches its maximum size of 43 m. In other parts of its range it is exceptional to find trees over 15 m in height, and 8 m is nearer the average. On Mt. Mlanje *Widdringtonia* forests occur extensively between 1525 and 2135 m on the main plateau surface and the upper parts of the deeply incised gorges radiating from it along the western and northern slopes.

Above about 2135 m true forest does not occur and is replaced by a variation of the Ericaceous belt. The normal *Widdringtonia* is replaced by a multiple-stemmed, dwarf shrubby form, which with increasing altitude becomes more

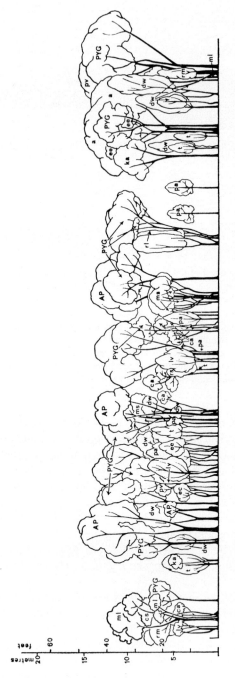

Fig. 10. Profile diagram of undifferentiated montane forest, Dedza Mt. (from Chapman & White 1970). a: *Allophylus? abyssinicus*, AP: *Apodytes dimidiata*, ca: *Clausena anisata*, cs: *Cussonia spicata*, dw: *Diospyros whyteana*, ec: *Ekebergia capensis*, ee: *Erythroxylum emarginatum*, ka: *Kiggelaria africana*, lv: *Lepidotrichilia volkensii*, ml: *Maesa lanceolata*, ms: *Maytenus senegalensis*, pa: *Peddiea africana*, pv: *Pittosporum viridiflorum*, PYG: (*Pygeum africanum*=) *Prunus africana*, rl: *Rawsonia lucida*, rm: *Rapanea melanophloeos*, rs: *Rapanea* sp.

Fig. 11. *Widdringtonia nodiflora* forest on the Chambe saddle, Mt. Mlanje, alt. 1950 m. The lichen-draped crowns of *Widdringtonia* emerge from a mixed stand of broad-leaved trees which form the actual canopy of the forest (photo J. D. Chapman).

shrubby. It persists in gullies and among rocks to at least 2440 m. Here it forms dense lichen-hung thickets rarely more than 4.5 m tall. Its associates include *Agauria salicifolia, Aphloia, Schefflera umbellifera, Myrica salicifolia* s.l., *Philippia benguelensis* and *Podocarpus latifolius*. *Philippia benguelensis* ascends higher than *Widdringtonia* and sometimes forms dense thickets with *Kotschya scaberrima*.

The *Widdringtonia* elfin thicket and *Philippia* thicket on the upper slopes of Mt. Mlanje occupy only a small fraction of the total surface. The great exposed

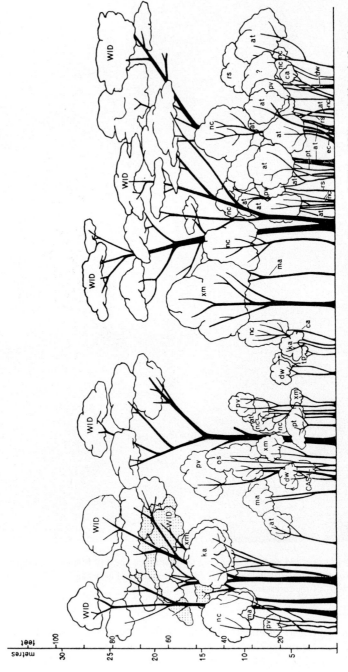

Fig. 12. Profile diagram of *Widdringtonia* montane forest, Chambe Plateau, Mt. Mlanje (from Chapman & White 1970). at: *Aphloia theiformis*, ca: *Clausena anisata*, CAS: *Cassipourea congoensis*, dw: *Diospyros whyteana*, ec: *Ekebergia capensis*, ka: *Kiggelaria africana*, ma: *Maytenus acuminata*, nc: *Nuxia congesta*, os: *Oxyanthus speciosus*, pt: *Pterocelastrus tricuspidatus*, pv: *Pittosporum viridiflorum*, rs: *Rapanea* sp., WID: *Widdringtonia nodiflora*, xm: *Xymalos monospora*.

mass of grey rock which rises 915 m above the main plateau surface and constitutes the several peaks of Mlanje, supports only scattered chasmophytes. The latter include the shrubby species *Aloe mawii, Blaeria kivuensis, Erica johnstoniana, E. whyteana, Ericinella* sp., *Muraltia flananganii, Phylica tropica* and *Xerophyta splendens,* and the grasses *Costularia nyikensis, Merxmuellera davyi* and *Eragrostis volkensii,* together with the sedge *Coleochloa setifera.*

The flora of the upper slopes of Mt. Mlanje is distinctive and includes many endemic species (Wild 1964). Zonation, however, is obscured because the hard crystalline rocks of which Mlanje is composed are very resistant to weathering. Extensive rocky outcrops occur at all levels within (and sometimes below) the Afromontane Region. The distribution of the chasmophytes they support is often determined more by the nature of the substrata than by obvious climatic factors related to altitude. According to Chapman (1962) most of the species that occur on the rocky peaks also occur in the zones below, wherever suitable rocky habitats occur.

Even the dwarf form of *Widdringtonia* can occur in suitable rocky habitats below the ordinary form. *Coleochloa setifera,* one of the most characteristic species of the peaks, is found as low as 760 m where it occurs among huge boulders in *Brachystegia* woodland at the foot of the mountains. The Afromontane forest species, *Prunus africana,* is also present in this fire-protected habitat.

In East Africa, by contrast, this kind of complication is of no more than local significance because most of the high mountains there are composed of soft volcanic rocks, or, when they are made of ancient crystalline rocks, they are more deeply weathered because of the more humid climate.

3.2.5 Afromontane grassland

The most widespread physiognomic vegetation type on the African mountains today is grassland with many floristic variants. Although some Afromontane grasslands, at first sight, look permanent, as if they have been there since time immemorial, it is now clear that most of them owe their origin and maintenance to fire. It is possible that some fire-climax grasslands are the result of natural fires, but there is now no reasonable doubt that most of them have originated or have been greatly extended relatively recently by man's destructive activity.

At least one hundred species of Gramineae are restricted to the African mountains. A few are confined to the Afroalpine Region but the majority have their distributions centred on the Afromontane Region. Many of these endemic species belong to the following genera which are widespread in temperate regions, especially in the northern hemisphere, but are absent from the tropical lowlands: *Agrostis, Aira, Anthoxanthum, Brachypodium, Bromus, Colpodium, Deschampsia, Festuca, Helictotrichon, Koeleria,* and *Poa.* Three genera, *Merxmuellera, Ehrharta* and *Pentaschistis,* which have endemic species on the African mountains, otherwise occur exclusively in the southern hemisphere, with their greatest concentration of species in the Cape. *Merxmuellera* and *Pentaschistis* are well represented in the Afromontane Region of South Africa with four and eight species respectively, but their numbers diminish rapidly to the north. *Merxmuellera* has only two species in Rhodesia and reaches its northern limit on Mt. Mlanje. *Pentaschistis* is represented by a single species, *P. natalensis,* in Rhodesia. There is then an interval in its range until the East African mountains are reached, where half-a-dozen species occur. *Ehrharta* is represented on the mountains of tropical Africa by a single species, *E. erecta*, which extends from South Africa to Ethiopia. A few other Afromontane endemic grass species belong to genera, e.g. *Pseudobromus* and *Streblochaete*, which are endemic to the African mountains but are closely related to temperate genera. Only a minority of species, including *Andropogon amethystinus* and *Eragrostis volkensii*, belong to genera which are widespread in the tropical lowlands.

The great majority of these Afromontane endemic species either occur inside forest, bamboo, or Ericaceous belt communities as normal constituents, or more frequently they occupy habitats such as rocky slopes or bogs, from which larger woody plants are excluded or are prevented from forming a closed canopy by unfavourable edaphic conditions. Some endemic Afromontane grass species are intolerant of fire, whereas others can occur in fire-climax communities. It is quite clear, however, from the general ecology of fire-tolerant endemic species, that their presence on the East African mountains is not dependent on fire.

On the mountains of tropical Africa the grasslands of the Ericaceous and Afroalpine belts are quite different from those of the Forest belt in their chorological relationships. In the former, endemic species predominate and any fire-climax grasslands that occur there are formed almost exclusively by them. In the latter the fire-maintained grasslands are dominated by species which are mostly also widespread in the lowlands and Afromontane endemics are mostly rare or absent.

On the upper slopes of the higher East African mountains the only extensive grassland is tussock grassland dominated by *Festuca pilgeri*, often in association with *Andropogon amethystinus, Koeleria gracilis* and species of *Pentaschistis* and *Agrostis* (Hedberg 1951, 1957, 1964). This grassland, usually mixed with shrubs or giant senecios and lobelias, occupies a wide range of natural habitats in the Ericaceous and Afroalpine belts, where its occurrence is clearly not always dependent on fire. Large areas of fire-induced secondary grassland, however, occur on Mt. Elgon, Mt. Kenya and Kilimanjaro.

In Malawi the Afroalpine belt does not occur, and the Ericaceous belt is occupied mostly by steep slopes of bare rock which do not provide suitable conditions for tussock grassland of the type described above. The same or related species of *Andropogon, Festuca, Koeleria* and *Agrostis*, however, do occur but they do not form extensive communities. In fire-maintained grassland they are only of local significance. *Andropogon amethystinus* occurs on rocky stream banks, *Koeleria gracilis* among rocks and in wet depressions, and *Agrostis* in boggy hollows. Only *Festuca costata* and *F. abyssinica* are found in secondary grassland, the former chiefly in rocky places (Chapman & White 1970) and the latter chiefly in peaty hollows.

On Mt. Mlanje in the lower part of the Ericaceous belt, where the slope is relatively gentle, and extending for about 300 m above the plateau surface, the spaces between the rocks are occupied by the two Afromontane endemic grasses, *Merxmuellera* (*Danthonia*) *davyi* and *Eragrostis volkensii*, amongst which appear a wide variety of low shrubby plants, including *Erica whyteana, E. johnstoniana, Blaeria kivuensis* and *Phylica tropica*. These two grasses are killed by fire. They were formerly absent from the secondary grasslands on the plateau of Mt. Mlanje, at the time when the latter were deliberately burnt by the Forest Department every year early in the dry season as a fire-protection measure. A subsequent change in policy led to the complete protection of the grasslands from fire. This was followed by their extensive invasion by *Merxmuellera* (*Danthonia*) and *Eragrostis* (Chapman 1962).

It is now beyond dispute that extensive areas of Afromontane forest have been destroyed in relatively recent times by fires started by man. It is also equally certain that natural fires started by lightning or by landslides do, from time to time, occur. The relative importance of natural and anthropogenic fire is, however, still debated. It is quite likely that their relative importance in shaping the present landscape has varied from place to place.

According to Vesey-Fitzgerald (1963) the montane grassland in the Ufipa and Mbeya highlands in Tanzania is always associated with relict patches of evergreen forest and appears to be a secondary formation replacing forest. None of the common grass species is restricted to this formation. They have been 'borrowed' from other habitats. Jackson (1969) suggests that much of the montane grassland in Malawi occurs on the site of former montane forest. He describes change in floristic composition and vigour which appear to be associated with the length of time that has elapsed since the destruction of the original forest. Chapman & White (1970) review the situation in Malawi in some detail and produce historical and circumstantial evidence for the wholesale replacement of forest by secondary fire-maintained grassland.

Jackson (1969) briefly describes four types of grassland, each less luxuriant than the last, which appear to represent four stages in the degradation and impoverishment of the original forest soil: (1) At the edges of forest relics and patches of secondary scrub, *Themeda triandra, Hyparrhenia cymbaria* and *Setaria longiseta* form a dense sward 0.9–1.2 m tall (Fig. 8). (2) Bordering the last type and representing an intermediate stage of degradation of forest sites is *Themeda triandra, Exotheca abyssinica, Setaria longiseta* grassland, 1–1.6 m tall but of low ground cover. (3) On more exposed slopes and ridges, there are wide stretches of short grassland dominated by *Andropogon schirensis* associated with *Brachiaria brizantha* and *Elyonurus argenteus*. (4) On even poorer sites the grassland is dominated by *Loudetia simplex* and *Monocymbium ceresiiforme*. Many of the above-mentioned species are widespread in the lowlands and could have been recruited from there. Jackson, however, shows that some of them also occur on shallow soils on the mountains, and could have spread from there to form extensive secondary montane grasslands. On Mt. Zomba in rocky terrain where the soils are shallow one finds mats of *Stereochlaena cameronii*. Less shallow soils are dominated by *Monocymbium ceresiiforme*, and somewhat deeper soils support patches of *Themeda-Exotheca* grassland. Other species may have spread into fire-climax grassland from montane edaphic grassland.

Virtual proof that Afromontane grassland has been recently derived from forest is provided by the occurrence of the datable remains of forest trees far from the forest edge. In one of the surviving patches of forest on the South Vipya, *Ocotea usambarensis* is one of the largest and most characteristic trees. Charred stumps have been found up to 2 km from the edge of existing forest in short *Exotheca abyssinica* – *Eragrostis racemosa* – *Elyonurus argenteus* grassland (Chapman & White 1970). Specimens collected from the outer shells of two such trees occurring half a kilometre from the forest edge have been shown to have died approximately 370 ($\pm$90) and 340 ($\pm$100) years ago. Confirmatory evidence comes from the pedologist Webster (in Chapman & White 1970), who shows that the soils under grassland on the Nyika Plateau have truncated profiles and almost certainly were formed under forest conditions. On the basis of pedological evidence alone Webster concludes that much of the Nyika Plateau was under forest perhaps no more than 1000 years ago.

In the Chimanimani Mts. in Rhodesia grassland covers the greatest area of any of the Afromontane formations (Phipps & Goodier 1962). Hydromorphic grassland occurs on seasonally waterlogged soils, and well-drained grassland elsewhere. According to Phipps & Goodier the latter is maintained by fire. They also believe that fire is not only a natural factor but has occurred sufficiently frequently and over a sufficiently long period of time to account for the occurrence of grassland on well-drained soils, which otherwise might be expected to support woody vegetation. As evidence, they cite the occurrence in the Chimanimani Mts. of no less than 17 grass species which 'are entirely restricted to montane grasslands in Rhodesia'. This explanation may very well be correct, but confirmatory studies are needed. On present evidence it is equally likely that many of these species are more characteristic of edaphic grassland and chasmophytic communities. Only one of their 17 species, *Panicum ecklonii*, is recorded by Phipps & Goodier as being characteristic of fire-maintained grassland. Three others, *Eragrostis caniflora*, *E. desolata* and *Pentaschistis natalensis* are recorded only on stony soils at higher altitudes where the effect of cloud is more marked and fires are rarer. Of the others, *Festuca costata* occurs in hydromorphic grassland. Elsewhere, *Koeleria capensis* ('*gracilis*') and *Merxmuellera* (*Danthonia*) *davyi* are chasmophytes and the latter is said to be fire-sensitive (see p. 496).

Where historic evidence for the origin of secondary grassland is lacking, 'experimental' evidence, either deliberately planned or accidentally obtained, may take its place. In south tropical Africa no planned experiments have been undertaken, but there are many accidental experiments, in which montane grassland has been protected from fire to safeguard natural stands of forest, or plantations of forest trees. Examples are described by Chapman & White (1970) from the Misuku Hills, Nchisi Mt. and Cholo Mt. One of the most extensive of such 'experiments' concerns the Mlanje Plateau (Chapman 1962). For many years, the entire plateau was completely protected from fire. The original grassland was sparse and short, no more than 50 cm tall, and dominated by *Loudetia simplex* associated with *Exotheca abyssinica*, *Themeda triandra*, *Andropogon schirensis* and *Digitaria diagonalis*. After a few years of fire-protection the grasslands were extensively invaded by the fire-sensitive species, *Merxmuellera* (*Danthonia*) *davyi* and *Eragrostis volkensii*. Extensive thickets of the shrubby forest precursor, *Kotschya scaberrima*, were formed and saplings of the secondary forest tree *Polyscias ferruginea* were not uncommon.

3.3 *Zonation and vegetation in South Africa*

Zonation within the Afromontane Region in South Africa is much more complicated than in Rhodesia and Malawi. This is partly because the individual islands of Afromontane vegetation in South Africa are surrounded by a great diversity of lowland vegetation types, and partly because the climate on the two sides of the main massif, the southern Drakensberg, is very different.

3.3.1 Relationships to adjacent phytochoria

These are complex and are summarized in Fig. 13. There are three relatively broad transition zones.

The Afromontane/Highveld transition occurs on the western side of the Drakensberg between 1525 and 1830–2150 m. It comprises the greater part of Acock's vegetation types 56, 57 and 58. The vegetation is predominantly grassland. Scrub forest is confined to sheltered kloofs and rocky slopes. Floristic details of the latter are given by Acocks (1953), Roberts (1961, 1969), Werger (1973) and Van Zinderen Bakker Jr. (1973). The Afromontane endemic tree flora, including *Podocarpus*, is almost absent from this transition zone. Apart from transgressor species it is represented by *Diospyros whyteana*, *Halleria lucida*, *Ilex mitis*, *Kiggelaria africana*, *Leucosidea sericea* and *Olinia*.

The Afromontane/Zambezian transition which comprises Acock's types 63–66, is now mostly secondary grassland. The climax was probably bushland with scrub forest in sheltered kloofs. Afromontane tree species include *Apodytes dimidiata*, *Cussonia spicata*, *Halleria lucida*, *Leucosidea sericea*, *Olinia*, *Pterocelastrus*, and *Scolopia mundii*.

The Afromontane/Tongaland–Pondoland transition is the only part of Africa where Afromontane and lowland species intermingle in any number over a relatively extensive area. Out of 523 species of larger woody plants recorded from the Tongaland–Pondoland Region 46 are Afromontane species. They are represented in nearly all the main forest types at least by a few marginal intruders. Even in Tongaland near the northern limit the Tongaland–Pondoland Region more than a dozen species, including *Podocarpus gracilior*, occur, though they contribute relatively little to the phytomass. The coastal forests and forests of the Zululand thornveld of Acocks (1953) are essentially lowland in character; the Alexandria forests are slightly more lowland than Afromontane; the Pondoland Coastal Plateau Sourveld and the eastern Province Thornveld are slightly more Afromontane than lowland and the upper part of the 'Ngongoni veld is essentially Afromontane, whereas the lower part is slightly more lowland than Afromontane. Afromontane species which occur in the canopy of Tongaland–Pondoland forest include *Calodendrum capense; Combretum kraussii*, *Fagara davyi*, *Kiggelaria africana*, *Nuxia congesta*, *Podocarpus falcatus*, *P. latifolius*, *Scolopia mundii* and *Xymalos monospora* (see also Chapter 13). The deep river valleys incised into the Drakensberg escarpment are too hot and too dry to support Afromontane vegetation. The transition from valley bushland to the Afromontane vegetation on the spurs above is often very abrupt, and may take place over a distance of a few hundred metres. This leads to the paradoxical situation that transitional vegetation may be better developed, not on the slopes immediately below the Afromontane Region, but some distance away, where intermediate conditions more extensively occur, as on the crests of lower mountain ranges such as the Lebombo Mountains or even in coastal forest itself.

3.3.2 Zonation and vegetation

The literature on Afromontane vegetation in South Africa is extensive. It is mostly of a descriptive and floristic nature. By far the most detailed study to date is that of Killick (1963) who deals with the Cathedral Peak area of the Natal

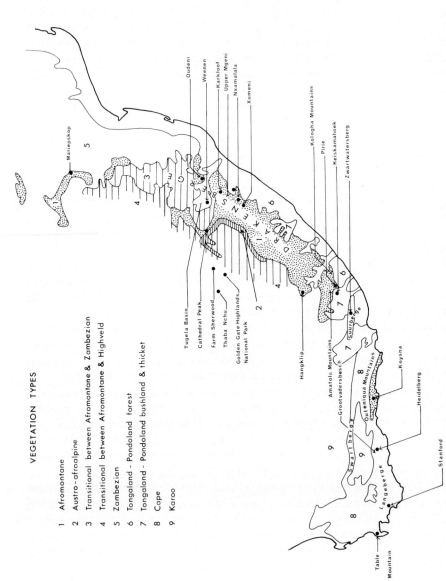

Fig. 13. Map showing occurrence of Afromontane Region in South Africa in relation to 'lowland' phytochoria and vegetation types.

Drakensberg. Edwards (1967) describes the vegetation of the headwaters of the Tugela River and its tributaries in relation to that of the Tugela basin as a whole.

Killick recognizes three altitudinal belts which coincide with three major physiographic features and are broadly equivalent to the three belts recognized by Hedberg on the East African mountains (compare Chapter 12). They are:

(1) The Montane Belt (1280–1830 m) occurs in the deeply incised river valleys which have cut back into the terrace known as the Little Berg which lies below the main Drakensberg escarpment. Its climax community is *Podocarpus latifolius* forest. It corresponds to the Montane Forest Belt of Hedberg.

(2) The 'Subalpine' Belt (1830–2865 m) occupies the lower slopes of the main escarpment and the terrace of the Little Berg which consists of finger-like spurs which extend away from the escarpment more or less at right angles. The climax vegetation is *Passerina – Philippia – Widdringtonia* bushland. It corresponds to Hedberg's Ericaceous Belt, and is discussed in detail in Chapter 12.

(3) The Alpine Belt (2865–3353 m) occupies the summit area of the Drakensberg. It corresponds, though not very closely, to Hedberg's Afroalpine Belt, and is discussed in detail in Chapter 12.

Killick points out that in the Natal Drakensberg the belts are not as clear-cut as on many mountains, but for convenience his system will be followed here, though using the terminology of Hedberg. When more is known of the ecology of this region it might be better to describe the main features of the mosaic as it exists on the ground, rather than to attempt to represent it in terms of a generalized zonation.

3.3.3 The Afromontane forest belt in South Africa

(i) *Forest*. Acocks (1953) provides a generalized account (veld types 8, North-Eastern Mountain Sourveld; 44, Highland Sourveld and Dohne Sourveld, and 45, Natal Mist Belt 'Ngongoni veld). The wide separation in his classification of number 8, which appears among tropical types, from numbers 44 and 45, which are designed 'temperate' detracts from their essential similarity. In floristic composition veld type 8 is Afromontane with a few marginal intruders from the lowlands, e.g. *Protorhus longifolia*.

The forests of Mariepskop mountain in the Transvaal Drakensberg have been described by Van der Schijff (1963), and Van der Schijff & Schoonraad (1971). Those of the Natal Drakensberg have been described by Edwards (1967), Killick (1963), Moll (1968, 1972), Moll & Haigh (1966), Rycroft (1944), Taylor (1962) and West (1951), and those of the Eastern Cape Province by Comins (1962), Dyer (1937) and Story (1952). The Knysna forests were the subject of a detailed study by Phillips (1931). Shorter accounts have also been published by Laughton (1937) and Von Breitenbach (1972). Other smaller islands of Afromontane forest occurring in the Cape have been described by Adamson (1927), Muir (1929), Taylor (1953, 1961), Werger et al. (1972) and Campbell & Moll (1977).

The Afromontane tree flora in South Africa is remarkably uniform. For purposes of comparison the islands of Afromontane vegetation have been grouped into five local systems, namely, Transvaal, Natal–Transkei, Cape Province east of the Knysna forests, the Knysna forests themselves, and the smaller enclaves of Afromontane forest in the Cape Region west of Knysna.

The analysis is based on 51 species which are mentioned in the ecological literature and of which the taxonomy, ecology and distribution patterns are sufficiently well known. Only a few taxonomically critical species were excluded from consideration. The distribution in South Africa of these 51 species, of which 45 are Afromontane endemics or near-endemics and six are ecological and chorological transgressors, is summarized in Table 3. There is a gradual diminution in number from north to south. The small enclaves of Afromontane forest in the Cape are floristically somewhat impoverished, but, even so, they contain exactly half of the entire complement. Each of the five systems shares not less than 70 per cent of its tree flora with the two adjacent systems, or the one adjacent system in the case of systems 1 and 5. The two most distant systems which are 1400 km apart have more than 50 per cent of their combined tree floras in common. Relatively few Afromontane tree species have restricted distributions in South Africa. Only *Platylophus trifoliatus* is confined to the two southernmost systems. Seven species are restricted in South Africa to one or both of the two northernmost systems. Some, such as *Combretum kraussii*, are endemic, others such as *Cryptocarya liebertiana* are widespread tropical species which extend southwards to the Transvaal or Natal but no further.

The Afromontane forests in South Africa as well as those elsewhere vary greatly in luxuriance. At one extreme are forests which are almost as tall as typical Guineo–Congolian lowland rain forest or the Afromontane rain forest mentioned on p. 485. In the Mistbelt forests in Natal the main canopy is at 18–25 m and emergents may reach a height of 37 m. In the tallest of the Knysna forests the main canopy is at 20–30 m and emergents occur up to 40 m in height. Lianes and epiphytes however are much rarer in these Afromontane forests than in true rain forest elsewhere in Africa. At the other extreme are forests which are floristically similar to the more luxuriant types but are of lower stature. Scrub forest only 5–7 m tall is widely distributed in the Ericaceous belt (see p. 490). *Widdringtonia nodiflora* does not appear in Table 3, since, in South Africa, in contrast to south tropical Africa, it rarely occurs as a forest species. Scrub forest and forest edges are its normal habitat. Van der Schijff & Schoonraad (1971), however, mention the occurrence of almost pure stands up to 15 m tall on Mariepskop Mt. In 1967, they were destroyed by a fire which was probably caused by lightning.

(ii) *Bushland*. Within the forest belt, bushland, which is usually 3–5 m tall, occurs wherever the site is too rocky or too unstable to allow the development of forest. Many bushland species also behave as forest precursors when secondary grassland is protected from fire. In the Natal Drakensberg there are three main sites: boulder beds, outcrops of Cave Sandstone and basalt outcrops.

The beds of the rivers are filled with large boulders up to 15 m across, though mostly smaller. They have fallen from the vertical cliffs above and are undergoing further weathering and transportation to the sea. Among them the following species frequently grow: *Buddleia salviifolia, Euclea crispa, Greyia sutherlandii, Leucosidea sericea, Olinia, Podocarpus latifolia* and *Widdringtonia nodiflora* (White unpub.).

The Cave Sandstone cliffs, which may be up to 150 m high, and the horizontal or sloping pavements at their summits, provide a wide variety of habitats for

Table 3. Distribution of Afromontane tree species in South Africa.

1. In Afromontane forest outliers in the Cape Floristic Region.
2. In the Knysna enclave.
3. In the Eastern Cape Province.
4. In the Natal Drakensberg.
5. In the Transvaal Drakensberg and the Soutpansberg.
tr = occurring in transitional Afromontane vegetation.
Tr = ecological and chorological transgressor.

	1	2	3	4	5
Apodytes dimidiata	x	x	x	x	x
Calodendrum capense	x	x	x	x	x
Canthium mundianum	x	x	x	x	x
C. ventosum	x	x	x	x	x
Cassine crocea	x	x	x	tr	
C. peragua		x	x	x	
Cassipourea congoensis (Tr)			x	x	x
Catha edulis			x	x	x
Celtis africana (Tr)	x	x	x	x	x
Cunonia capensis	x	x	x	x	x
Combretum kraussii				x	x
Cryptocarya latifolia				x	
C. liebertiana					x
Curtisia dentata	x	x	x	x	x
Cussonia spicata			x	x	x
Diospyros whyteana	x	x	x	x	x
Ekebergia capensis (Tr)		x	x	x	x
Fagara capensis			x		x
F. davyi		x	x	x	x
Gonioma kamassi		x	tr	tr	
Halleria lucida	x	x	x	x	x
Homalium dentatum			x	x	x
Ilex mitis	x	x	x	x	x
Kiggelaria africana	x	x	x	x	x
Leucosidea sericea			x	x	x
Linociera foveolata	x	x	x	x	
L. peglerae			x	x	
Maesa lanceolata					x
Maytenus peduncularis		x	x	x	x
Nuxia floribunda	x	x	x	x	x
Ochna arborea (Tr)		x	x	x	x
Ocotea bullata	x	x	x	x	x
Olea capensis (Tr)	x	x	x	x	x
Olinia	x	x	x	x	x
Pittosporum viridiflorum	x	x	x	x	x
Platylophus trifoliatus	x	x			
Podocarpus falcatus	x	x	x	x	x
P. henkelii			x	x	
P. latifolius	x	x	x	x	x
Prunus africana		x	x	x	x
Ptaeroxylon obliquum (Tr)			x	x	x
Pterocelastrus	x	x	x	x	x
Rapanea	x	x	x	x	x
Rawsonia lucida				x	x
Rhus chirindensis	x	x	x	x	x
Schefflera umbellifera			tr	tr	x
Scolopia mundii	x	x	x	x	x

	1	2	3	4	5
Seemannaralia gerrardii			x	x	x
Syzygium guineense subsp. gerrardii			tr	x	x
Trichilia dregeana (Tr)				x	x
Xymalos monospora			x	x	x
Total no. of spp.	25	32	44	46	43
No. of spp. shared with neighbouring systems	25	24	31	37	40
% of tree flora shared with neighbouring systems	100	75.0	70.5	78.7	93.0

chasmophytes. The more characteristic large woody plants include *Aloe arborescens, Bowkeria verticillata, Cliffortia linearis, Cussonia paniculata, Diospyros austro-africana* subsp. *rubiflora, Dovyalis zeyheri, Encephalartos ghellinckii, Erica drakensbergensis, E. westii, Euclea crispa*, a prostrate variant of *Ficus ingens* which clings closely to the rock faces, *Greyia sutherlandii, Myrica pilulifera, Passerina montana, Phylica* sp., *Protea multibracteata, P. roupelliae, Rapanea melanophloeos, Rhus lucida, Trimeria grandifolia* and *Widdringtonia nodiflora* (Killick 1963; White unpubl.). *Protea roupelliae* can establish itself in the narrowest of rock crevices.

The vegetation on the basalt cliffs is much the same as that of the Cave Sandstone cliffs but *Greyia sutherlandii* and *Cussonia paniculata* are much more abundant there.

(iii) *Grassland*. The greater part of the Montane Forest belt is occupied by grassland, chiefly *Themeda triandra* grassland, but locally *Hyparrhenia* grassland and *Miscanthidium – Cymbopogon* grassland. Small trees of *Protea multibracteata* and *P. roupelliae* are scattered through the grasslands. Their density varies greatly and is inversely proportioned to the frequency and severity of fire (Fig. 14).

The *Themeda triandra* grassland is a relatively stable community, which is prevented from successional development by recurrent grass fires (Killick 1963). Killick points out that lightening-induced fires are frequent in the Drakensberg and suggests that, because of this, grassland has been the predominant community in this zone ever since the climate included a dry season. It is extremely difficult to establish the extent to which fire-climax grasslands owe their existence to natural causes. The evidence from Malawi (p. 496) suggests that, in the absence of historical evidence, it is easy to underestimate the extent to which Afromontane forest has been destroyed by man-made fires in recent times.

Hyparrhenia grassland is not very extensive because of frequent burning. When *Themeda* grassland is protected from fire, species of *Hyparrhenia*, including *H. hirta, H. dregeana, H. temba* and *H. aucta*, invade.

Grassland dominated by *Miscanthidium capense* and *Cymbopogon validus*, both robust grasses, the former up to 2.5 m tall, is found in moist areas generally like streambanks, gullies and forest margins.

3.3.4 The Ericaceous belt in South Africa

According to Killick (1963 and Chapter 12) Afromontane forest is almost absent

Fig. 14. Small trees of *Protea roupelliae* scattered through montane *Themeda* grassland in the Cathedral Peak area of the Drakensberg. In foreground *Buddleia salviifolia* (photo M. J. A. Werger)

from this belt. Its upper limit, he says, occurs at 1860 m and the climax community is *Passerina – Philippia – Widdringtonia* bushland, which he refers to as fynbos. Because of recurrent grass fires fynbos is limited in extent. The most extensive of the surviving stands are in the upper part of the belt on steep valley and escarpment slopes at the head of the main rivers. It is, however, apparent from the distribution of Afromontane forest tree species and forest precursors that short Afromontane forest and types of scrub forest transitional to fynbos must have been widespread in the lower half of this belt, and that their destruction by fire has been relatively recent. Grassland of various types occupies most of this belt today. Exposed ridges above 2440 m support dwarf shrubland.

(i) *Forest and Scrub forest.* Short *Podocarpus* forest ascends in kloofs as high as 2075 m (Bainbridge, pers. comm., White unpubl.). In the Indumeni Valley between 2000 and 2300 m the distribution of the following Afromontane tree species, and the size of some of them, indicates that scrub forest, not fynbos, is the climax at least on relatively sheltered lower slopes: *Bowkeria verticillata, Buddleia salviifolia, Cussonia paniculata, Euclea crispa, Halleria lucida, Leucosidea sericea, Olinia emarginata, Podocarpus latifolius, Pterocelastrus* and *Rhamnus prinoides.*

Small patches of forest also survive on rather exposed slopes between 1830 and 2000 m in the High Moor Forest Reserve 70 km S.E. of the Cathedral Peak area. They are mostly about 5 m tall, though in rocky places emergent *Podocarpus latifolius* may be up to 11 m tall. The principal components are *Buddleia salviifolia, Cliffortia linearifolia, Diospyros whyteana, Euclea crispa, Halleria lucida, Heteromorpha trifoliata, Leucosidea sericea, Maytenus acuminata, M. undata, Myrsine africana, Podocarpus latifolius, Rapanea melanophloeos, Rhamnus prinoides* and *Scolopia mundii* (see Chapter 12).

The other major communities occurring in this Ericaceous belt and higher will be described in Chapter 12 under the heading 'Subalpine belt'.

3.4 *Afromontane forests in the Cape region*

The hot dry summers of the southwestern Cape preclude the occurrence of forest other than riparian scrub forest and scrub forest on screes (Werger et al. 1972 and Chapter 8) except in those few places where the rainfall is well-distributed throughout the year, or is locally augmented, and the severity of the dry season is ameliorated, by frequent cloud formation. Forest is also favoured by water-retaining soils. The best-known examples are the Knysna forests.

In the past these forests have sometimes been regarded as belonging to the Cape Region. This was the view of Grisebach (1872), and Bolus (1886), but more often they have been excluded (e.g. by Engler 1882, Schimper 1898, and Marloth 1908). It is customary for South African botanists to refer to most of the forest species as 'tropical' in contrast to the temperate Cape flora (e.g. Bews 1925, Phillips 1931, Acocks 1953). Previously no detailed assessment of their chorological relationship has been made, but some workers, e.g. Aubréville (1949) have drawn attention to the floristic similarity between the Knysna forests and those of the mountains of tropical Africa, as has Van Zinderen Bakker Jr. (1973). Several authors, e.g. Marloth (1908), Bews (1925) and Phillips (1931). have commented on the decrease in forest species as the west is approached but they did not distinguish between Afromontane species and those of tropical lowland forest affinity.

Forest is the only type of Afromontane vegetation found in the Cape. It characteristically occurs on water-retaining soils in sheltered places on the lower, especially southern, slopes of mountains. In southern Africa increasing latitude compensates for altitude and Afromontane forest locally descends to within about a hundred metres of sea-level. Enclaves of Afromontane forest within the Cape have been studied at Knysna by Marloth (1908), Phillips (1931), Laughton (1937) and von Breitenbach (1972), in the Riversdale area by Muir (1929), at Grootvadersbosch, near Heidelberg by Taylor (1953), near Stanford by Taylor (1961) and on Table Mt. by Adamson (1927) and Campbell & Moll (1977). Most of these forests are, in their floristic composition, overwhelmingly Afromontane. This is clearly shown by Table 3. They do, however, grade into scrub forest, bushland and thicket. In these formations the Afromontane character is diluted by the occurrence, according to their type, of several Cape species, or transgressor species. The latter are either widespread in tropical Africa or, elsewhere, are characteristically Tongaland–Pondoland in their distribution.

The Knysna forests vary greatly in stature and floristic composition, in relation to slope, aspect, altitude and soil moisture. The most luxuriant are comparable to Afromontane rainforest in structure. At lower altitudes with increasing temperature and increasing drought they grade into scrubforest, bushland and thicket of Tongaland–Pondoland affinity, in which Afromontane species are mixed with Tongaland–Pondoland near-endemics such as *Diospyros dichrophylla* and *Maerua caffra*, and various bushland 'wides', such as *Azima tetracantha*, *Euclea racemosa, Grewia occidentalis, Sideroxylon inerme* and *Rhus* spp. (see Chapter 13). At higher altitudes with decreasing temperature and increasing humidity forest height again falls off rapidly and tall forest gives way to scrub forest and thicket in which several Cape fynbos species such as *Berzelia intermedia, Diospyros glabra, Leucadendron eucalyptifolium* and *Protea cynaroides* are prominent, as well as the forest-edge Cape endemic, *Virgilia oroboides*.

The forests near Stanford described by Taylor occur between 90 and 245 m. The mean annual rainfall is believed to be about 550 mm, which is rather marginal for forest. Where sandy limestone overlies the prevalent Cape Sandstone the soils are coarse, shallow and not very moist. The climax vegetation is scrub forest 6–9 m tall, dominated by the bushland transgressor species *Euclea racemosa* and *Sideroxylon inerme*. The Afromontane flora is represented by a few Afromontane near-endemics, e.g. *Linociera foveolata*, and Afromontane transgressors, e.g. *Olea africana* and *O. capensis*. Taller forest, though only 9–14 m tall, occurs in a well drained depression, which is protected from unfavourable northwesterly winds yet receives high summer humidity from southeasterly mists. The soil is deeper and moister than that of scrub forest and the taller forest is predominantly Afromontane in composition at least so far as taller woody plants are concerned, though *Podocarpus* is absent, and *Euclea racemosa* and *Sideroxylon inerme* also occur.

4. Summary and conclusions

The floras occurring on the summits and upper slopes of the African mountains south of the Sahara and north of the Cape are so different from the surrounding 'lowlands' that they cannot be regarded as forming a series of upland facies of the latter.

The Afromontane flora is more complex in its origin than that of any other African phytochorion, but, it resembles them sufficiently, in its internal cohesiveness and degree of difference from adjacent phytochoria, to fully justify the recognition of an archipelago-like Afromontane Region.

For convenience of comparison and analysis the latter can be divided into the following seven regional mountain systems, of which the last three occur in southern African–West African, Ethiopian, Kivu-Ruwenzori, Imatongs–Usambara, Uluguru–Mlanje, Chimanimani, Drakensberg.

The total flora of the Afromontane Region probably exceeds 4000 species, of which more than 3000 are endemic or almost so. The Afromontane flora includes, at most, three endemic families. About 20 per cent of its tree genera are endemic and probably a somewhat smaller percentage of smaller plants. The endemic tree genera are all monotypic or oligotypic.

On any particular mountain there is usually a very wide range of vegetation types with correspondingly big changes in floristic composition. Despite this, and despite the wide intervals between adjacent mountains, and the enormous latitudinal and longitudinal extent of the Afromontane archipelago, its flora shows a remarkable continuity and uniformity. On any particular mountain extreme types of Afromontane vegetation may have few species in common but all types are intimately connected by complex series of intermediates. The differences between extreme types on a single mountain are usually greater than the differences between the Afromontane assemblage as a whole on that mountain and the assemblage found on nearby, or indeed on distant mountains.

The majority of genera and species are widely distributed. An assemblage of 12 Afromontane tree species could almost be used, alone, to define the Afromontane Region. No single species occurs throughout, but the assemblage is represented on virtually every 'island' of Afromontane vegetation, usually by several species.

The East African mountains have the richest and most diversified tree flora. Floristic impoverishment is fairly rapid towards the south and even more so to the north and west. Although the Drakensberg system is relatively poor in tree species, it has the highest percentage of local endemics, but they still represent scarcely more than a quarter of its total Afromontane tree flora. For smaller plants the Drakensberg system seems to have a richer flora than the other systems.

Very local endemic species, confined to a single mountain or to small parts of larger massifs such as the Drakensberg, appear to be few in number, equally for trees and smaller plants, and possibly do not exceed 5 per cent. Their distribution is very uneven and there appears to be little correlation between the contemporary degree of geographical isolation and the richness of the endemic flora.

Because the Afromontane Region is like an archipelago, all of its species, other than local endemics, show disjunctions. But superimposed on this are many wide or very wide disjunctions within the archipelago not merely between adjacent islands. It appears that these irregular distributions are due to complicated historical rather than ecological causes. At present this subject is poorly understood.

Throughout the Afromontane Region the lowermost vegetation type is forest, beneath which one would expect to find a transition zone connecting the Afromontane and lowland phytochoria. Nearly everywhere the transition zone has been destroyed by fire or cultivation, and its nature must be reconstructed from surviving relics and circumstantial evidence. The transition zones are usually narrow. Extensive transition zones occur only in the Lake Victoria basin and in

South Africa. It is only in South Africa where latitude compensates for altitude, and Afromontane forest descends almost to sea-level, that the relationships between the Afromontane flora and the lowland flora are at all complex. Elsewhere, Afromontane species which occur in lowland phytochoria do so only as marginal intruders or occur as distant satellite populations, or they are ecological and chorological transgressors. The greater part of lowland Africa is virtually devoid of Afromontane species.

Marginal intruders are either Afromontane species which penetrate a short distance into adjacent lowland phytochoria, or lowland species which occur in the lower part of the Afromontane Region. They are usually rare and localized and contribute little to the phytomass of the communities into which they intrude. There is evidence, however, that at least some of the species of secondary montane grassland, which are also widespread in the lowlands, formerly occurred in the Afromontane Region as marginal intruders in the forest zone, where they were confined to extremely shallow soil. Their explosive expansion has followed the destruction of forest by man.

Most satellite populations appear to be relictual and occupy specialized sites. Afromontane species which have distant satellite populations in another lowland African phytochorion are always very rare there and rarely occur in its most characteristic vegetation types. They may have much to contribute to our understanding of the history of the Afromontane flora. Some satellite populations occur sporadically, but most form part of east–west extensions of the Afromontane Region which might formerly have represented well-defined migratory tracks, though under present conditions the latter may appear as culs-de-sac. The Magaliesberg extension, the Limpopo escarpment extension and the Zambezi watershed extension are briefly described. The Zambezi watershed extension might be part of a former migratory track which linked the mountains of East and West Africa.

Ecological and chorological transgressors occur and are prominent in two or more major phytochoria and also occur in at least two major vegetation types. They thus differ from linking and pluriregional elements occurring in azonal vegetation types and from marginal intruders and near-endemics with small distant satellite population. 18 per cent of the Afromontane tree flora are transgressors. Several are represented by distinct Afromontane subspecies. The percentage of herbaceous transgressors is probably lower.

Some Afromontane species which are absent from surrounding lowland phytochoria occur also in geographically remote phytochoria, in the Cape, Madagascar, the Mediterranean, and elsewhere. Less than 10 per cent of the Afromontane flora belongs to this category.

On nearly all African mountains the vegetation diminishes in stature from the lower slopes to the summit, but this regularity is so often modified by local features of aspect, exposure, incidence of frost and depth of soil, and by overall patterns of climate dependent on the size and configuration of the mountain in relation to distance from the sea or other sources of moisture, that generalized schemes of zonation, even for relatively restricted regions, are impossible to devize. The three broad zones recognized by Hedberg for the high mountains of East Africa can usually be recognized (if the mountain is high enough), but the altitude at which zonal replacement occurs varies greatly even on different slopes of the same mountain. In the Cape region Afromontane vegetation, which is represented there only by forest, is no longer associated with the crests and summits of mountains but is confined to the lower slopes.

The most extensive vegetation type existing today in the Afromontane Region is fire-maintained grassland, consisting predominantly of species which are also abundant in the lowlands. There is incontrovertible evidence for the occurrence of natural fires caused by lightning or landslides on the African mountains. Their incidence probably varies greatly in different places. There is, however, no substantial body of evidence to support the hypothesis that fire-climax grassland due to natural fires ever occurred at all extensively. On the other hand there is much to indicate that woody vegetation, especially forest, formerly occurred on sites now occupied by grassland and that the destruction has taken place during the last 1000 years or so.

References

Acocks, J. P. H. 1953. Veld types of South Africa. Mem. Bot. Surv. S. Afr. 28:1–192 2. ed. (1975), op. cit. 40:1–128.
Adamson, R. S. 1927. The plant communities of Table Mountain: preliminary account. J. Ecol. 15:278–309.
Aubréville, A. 1949. Climats, forêts et désertification de l'Afrique tropicale. Soc. Ed. Géogr. Mar. Colon, Paris.
Banks, P. F. 1976. Chirinda forest. Rhod. Sci. News 10:39–40.
Barbosa, L. A. Grandvaux. 1970. Carta fitogeográfica de Angola. I.I.C.A., Luanda.
Bews, J. W. 1925. Plant forms and their evolution in South Africa. Longmans, Green, London.
Bolus, H. 1886. Sketch of the flora of South Africa. In Official handbook, Cape of Good Hope, Cape Town.
Brummitt, R. K. 1976. Thesium triflorum (Santalaceae) in the 'Flora Zambesiaca' area. Kew Bull. 31:176.
Campbell, B. M. & Moll, E. J. 1977. The forest communities of Table Mountain, South Africa. Vegetatio 34:105–115.
Carlquist, S. 1974. Island biology. Columbia University Press, New York & London.
Chapman, J. D. 1962. The vegetation of the Mlanje Mountains, Nyasaland. Govt. Printer, Zomba.
Chapman, J. D. & White, F. 1970. The evergreen forests of Malawi. Comm. For. Inst., Oxford.
Codd, L. E. 1969. The South African species of Kniphofia. Bothalia 9:363–513.
Coetzee, B. J. 1974. A phytosociological classification of the Jack Scott Nature Reserve. Bothalia 11:329–347.
Coetzee, B. J. 1975. A phytosociological classification of the Rustenburg Nature Reserve. Bothalia 11:561–580.
Coetzee, J. A. 1967. Pollen analytical studies in East and Southern Africa. Palaeoecology of Africa 3:1–146. Balkema, Cape Town.
Comins, D. M. 1962. The vegetation of the Districts of East London and King William's Town, Cape Province. Mem. Bot. Surv. S. Afr. 33:1–32.
Dejardin, J., Guillaumet, L. & Mangenot, G. 1973. Contribution à la connaissance de l'élément non endémique de la flore malgache (Végétaux vasculaires). Candollea 28:325–391.
Du Rietz, G. E. 1930. Classification and nomenclature of vegetation. Svensk. Bot. Tidskr. 24:489–503.
Dyer, R. A. 1937. The vegetation of the Divisions of Albany and Bathurst. Mem. Bot. Surv. S. Afr. 17:1–138.
Edwards, D. 1967. A plant ecological survey of the Tugela River basin. Mem. Bot. Surv. S. Afr. 36:1–285.
Engler, A. 1882. Versuch einer Entwicklungsgeschichte der Pflanzenwelt, 2, ed. 2. Engelmann, Leipzig.
Engler, A. 1892. Über die Hochgebirgs-flora des tropischen Afrika. Abhandl. Preuss. Akad. Wiss., Berlin.

Engler, A. 1904a. Plants of the north temperate zone in their transition to the high mountains of tropical Africa. Ann. Bot. 18:523–540.

Engler, A. 1904b. Über das Verhalten einiger polymorpher Pflanzentypen der nördlich gemässigten Zone bei ihrem Übergang in die afrikanischen Hochgebirge; p. 552–568 in Festschrift zu P. Ascherson's siezigstem Geburtstage, Berlin.

Fries, R. E. & Fries, Th. C. E. 1948. Phytogeographical researches on Mt. Kenya and Mt. Aberdare, British East Africa. Kungl. Svenska Vetenskap. Handl. ser. 3, 25 (5):1–83.

Goldblatt, P. 1973. Contribution to the knowledge of Moraea (Iridaceae) in the Summer Rainfall Region of South Africa. Ann. Missouri Bot. Gard. 60:204–259.

Goldsmith, B. 1976. The trees of Chirinda forest. Rhod. Sci. News 10:41–50.

Goodier, R. & Phipps, J. B. 1961. A revised check-list of the vascular plants of the Chimanimani Mountains. Kirkia 1:44–66.

Grisebach, A. R. H. 1872. Die Vegetation der Erde. Leipzig.

Grubb, P. J. 1971. Interpretation of the 'Massenerhebung' effect on tropical mountains. Nature 229:44–45.

Hauman, L. 1933. Esquisse de la végétation des hautes altitudes sur le Ruwenzori. Acad. Roy. Belg. Bull. Classe Sci., sér. 5, 19:602–616; 702–717; 900–917.

Hauman, L. 1955. La 'Région Afroalpine' en phytogéographie Centro-africaine. Webbia 11:466–469.

Hedberg, O. 1951. Vegetation belts of the East-African mountains. Svensk Bot. Tidskr. 45:140–202.

Hedberg, O. 1957. Afroalpine vascular plants. Symb. Bot. Ups. 15(1):1–411.

Hedberg, O. 1961a. Monograph of the genus Canarina L. (Campanulaceae). Svensk. Bot. Tidskr. 55:17–62.

Hedberg, O. 1961b. The phytogeographical position of the Afroalpine flora. Rec. Adv. Bot. 1:914–919.

Hedberg, O. 1964. Features of Afroalpine plant ecology. Acta Phytogeogr. Suecica 49:1–144.

Hedberg, O. 1965. Afroalpine flora elements. Webbia 19:519–529.

Hedberg, O. 1969. Evolution and speciation in a tropical high mountain flora. Biol. J. Linn. Soc. 1:135–148.

Hedberg, O. 1970a. Evolution of the Afroalpine flora. Biotropica 2:16–23.

Hedberg, O. 1970b. The genus Zaluzianskya F. W. Schmidt (Scrophulariaceae) found in tropical East Africa. Bot. Not. 123:512–518.

Hedberg, O. 1975. Studies of adaptation and speciation in the afroalpine flora of Ethiopa. Boissiera 24a:71–74.

Hilliard, O. M. & Burtt, B. L. 1971. Streptocarpus: an African plant study. University of Natal Press, Pietermartizburg.

Hilliard, O. M. & Burtt, B. L. 1976. Macowania in Notes on some plants of Southern Africa, chiefly from Natal. V. Notes Roy. Bot. Gard. Edinburgh 34:260–276.

Hooker, J. D. 1862. On the vegetation of Clarence Peak, Fernando Po....J. Linn. Soc. Bot. 6:1–23.

Hooker, J. D. 1864. On the plants of the temperate regions of the Cameroons Mountains and islands in the Bight of Benin. J. Linn. Soc. Bot. 7:171–240.

Hooker, J. D. 1874. On the subalpine vegetation of Kilima Njaro, E. Africa. J. Linn. Soc. Bot. 14:141–146.

Hooker, J. D. 1885. Preface to list of the plants collected by Mr. Thomson F.R.G.S. on the mountains of Eastern Equatorial Africa by D. Oliver in J. Linn. Soc. Bot. 21:392–406.

Jackson, G. 1969. The grasslands of Malawi. Soc. Malawi J. 22:7–17, 18–25, 73–81.

Jonsell, B. 1975. Lepidium L. (Cruciferae) in tropical Africa: a morphological, taxonomic and phytogeographical study. Bot. Not. 128:20–46.

Keay, R. W. J. (ed.) 1959. Vegetation map (1 : 10,000,000) of Africa south of the tropic of Cancer. University Press, Oxford.

Killick, D. J. B. 1963. An account of the plant ecology of the Cathedral Peak area of the Natal Drakensberg. Mem. Bot. Surv. S. Afr. 34:1–178.

Kornaś, J. 1974a. The Pteridophyta new to Zambia. Bull. Acad. Pol. Sci. sér. sci. biol. Cl. II. 22:713–718.

Kornaś, J. 1974b. The Pteridophyta of the Kundalila Falls, Zambia. Acta Soc. Bot. Polon. 43:479–483.

Laughton, F. S. 1937. The silviculture of the indigenous forests of the Union of South Africa with special reference to the forests of the Kynsna region. S. Afr. Dept. Agric. For., Sci. Bull. 157:1–168.
Lebrun, J. 1947. La végétation de la plaine alluviale au sud du lac Édouard. Inst. Parcs Nat. Congo Belge, Expl. Parc. Nat. Alb. Miss. Lebrun (1937–38). 1, Bruxelles.
Lebrun, J. 1958. Les orophytes africains. Comm. 6a sess. Conf. Inst. Afr. Occid. 3, Bot.:121–131.
Lewalle, J. 1975. Endémisme dans une haute vallée du Burundi. Boissiera 24a:85–89.
Lewis, G. J., Obermeyer, A. A. & Barnard, T. T. 1972. A revision of the South African species of Gladiolus. J.S. Afr. Bot. Suppl. 10:1–316.
Liben, L. 1962. Nature et origine du peuplement végétal (Spermatophytes) des contrées montagneuses du Congo oriental. Mém. Acad. Roy. Belg. bl. Sci. Coll. 4°, sér. 2, 15(3):1–195.
Lind, E. M. & Morrison, M. E. S. 1974. East African vegetation. Longman, London.
Lisowski, S., Malaisse, F. & Symoens, J. J. 1970. Plantes rares ou nouvelles pour la flore du Katanga. Bol. Soc. Brot., sér. 2, 44:225–244.
Mabberley, D. J. 1973. Evolution in the giant Groundsels. Kew Bull. 28:61–96.
Mabberley, D. J. 1974. The pachycaul Lobelias of Africa and St. Helena. Kew Bull. 29:535–584.
Mabberley, D. J. 1975. Notes on the vegetation of the Cherangani Hills, N.W. Kenya. J. E. Afr. Nat. Hist. Soc. 150:1–11.
Macedo, J. de Aquiar. 1970. Carta de vegetação da Serra de Gorongosa. Inst. Investig. Agron. Moçambique 50:1–75.
Malaisse, F. 1967. A propos d'une nouvelle station de Podocarpus milanjianus Rendle au Katanga. Publ. Univ. Offic. du Congo à Lubumbashi, 16:79–81.
Marloth, R. 1908. Das Kapland. Fischer, Jena.
Moll, E. J. 1968. A plant ecological reconnaissance of the Upper Mgeni catchment. J. S. Afr. Bot. 34:401–420.
Moll, E. J. 1972. The current status of Mistbelt mixed Podocarpus forest in Natal. Bothalia 10:595–598.
Moll, E. J. & Haigh, H. 1966. A report on the Xumeni forest, Natal. For. S. Afr. 7:99–108.
Monod, T. 1957. Les grandes division chorologiques de l'Afrique, CCTA/CSA, Publ. No. 24:1–147, London.
Morton, J. K. 1972. Phytogeography of the West African mountains. In: Valentine, D. H. (ed.), Taxonomy, Phytogeography and Evolution. pp. 221–236. Academic Press, London, New York.
Muir, J. 1929. The vegetation of the Riversdale area. Mem. Bot. Surv. S. Afr. 13:1–82.
Nordenstam, B. 1968. The genus Euryops. Part I. Taxonomy. Op. Bot. 20:1–409.
Nordenstam, B. 1969. Phytogeography of the genus Euryops (Compositae) a contribution to the phytogeography of Southern Africa. Op. Bot. 23:1–77.
Nordlindh, T. 1946. Studies in the Calenduleae. II. Phytogeography and interpretation. Bot. Not. 1946:471–506.
Phillips, J. F. V. 1931. Forest succession and ecology in the Knysna region. Mem. Bot. Surv. S. Afr. 14:1–327.
Phipps, J. B. & Goodier, R. 1962. A preliminary account of the plant ecology of the Chimanimani Mountains. J. Ecol. 50:291–319.
Pócs, T. 1974. Bioclimatic studies in the Uluguru Mountains. Act. Bot. Acad. Sci. Hungaricae. 20:115–135.
Pócs, T. 1976. Vegetation mapping in the Uluguru Mountains (Tanzania, East Africa). Boissiera 24b:477–498.
Polhill, R. M. 1968. Conservation of vegetation in Africa south of the Sahara: Tanzania. Acta Phytogeogr. Suecica 54:166–178.
Richards, P. W. 1952. The tropical rain forest. Cambridge University Press.
Roberts, B. R. 1961. Preliminary notes on the vegetation of Thaba' Nchu. J. S. Afr. Bot. 27:241–251.
Roberts, B. R. 1969. The vegetation of the Golden Gate Highlands National Park. Koedoe 12:15–28.
Rycroft, H. B. 1944. The Karkloof forest, Natal. J. S. Afr. For. Ass. 11:14–25.
Schimper, A. F. W. 1898. Pflanzengeographie auf physiologischer Grundlage. Fischer, Jena.
Story, R. 1952. A botanical survey of the Keiskamahoek District. Mem. Bot. Surv. S. Afr. 27:1–181.

Swynnerton, C. F. M., Rendle, A. B., Baker, E. G., Moore, S. & Gepp, A. 1911. A contribution to our knowledge of the flora of Gazaland. J. Linn. Soc. Bot. 40:1–245.
Taylor, H. C. 1953. Forest types and floral composition of Grootvadersbosch. J.S. Afr. For. Ass. 23:33–46.
Taylor, H. C. 1961. Ecological account of remnant coastal forest near Stanford, Cape Province. J.S. Afr. Bot. 27:153–165.
Taylor, H. C. 1962. A report on the Nxamalala forest. For. S. Afr. 2:29–51.
Troupin, G. 1966. Étude phytocénologique du Parc National de l'Akagera et du Rwanda Oriental. Inst. Nat. Rech. Sci. Butare, Rep. Rwandaise Publ. No. 2, Tervuren.
Van der Schijff, H. P. 1963. A preliminary account of the vegetation of the Mariepskop complex. Fauna & Flora Transv. 14:42–53.
Van der Schijff, H. P. & Schoonraad, E. 1971. The flora of the Mariepskop complex. Bothalia 10:461–500.
Van Steenis, C. G. G. J. 1935. On the origin of the Malaysian mountain flora. II. Altitudinal zones, general considerations and renewed statement of the problem. Bull. Jard. Bot. Buitenzorg, sér. 3, 13 (3).
Van Steenis, C. G. G. J. 1961. An attempt towards an explanation of the effect of mountain elevation. Proc. Koninkl. Nederl. Akad. Wet. Amsterdam, sér. C, 64:435–442.
Van Steenis, C. G. G. J. 1972. The mountain flora of Java. Brill, Leiden.
Van Zinderen Bakker, E. M. Jr. 1973. Ecological investigations of forest communities in the eastern Orange Free State and the adjacent Natal Drakensberg. Vegetatio 28:299–334.
Van Zinderen Bakker, E. M. Sr. & Werger, M. J. A. 1974. Environment, vegetation and phytogeography of the high-altitude bogs of Lesotho. Vegetatio 29:37–49.
Vesey-Fitzgerald, D. F. 1963. Central African grasslands. J. Ecol. 51: 243–273.
Von Breitenbach, F. 1972. Indigenous forests of the southern Cape. J. Bot. Soc. S. Afr. 58:17–47.
Weimarck, H. 1934. Monograph of the genus Cliffortia. Univ., Lund.
Weimarck, H. 1940. Monograph of the genus Aristea. Lunds Univ. Årsskr. N.F. Avd. 2, 36 (1) & Kungl. Fysiogr. Sällsk. Handl. N.F. 51 (1):1–141.
Weimarck, H. 1941. Phytogeographical groups, centres and intervals within the Cape flora. Lunds Univ. Årsskr. N.F. Avd. 2, 37 (5):1–143.
Werger, M. J. A. 1973. Phytosociology of the Upper Orange River valley, South Africa. V & R, Pretoria.
Werger, M. J. A., Kruger, F. J. & Taylor, H. C. 1972. A phytosociological study of the Cape fynbos and other vegetation at Jonkershoek, Stellenbosch. Bothalia 10:599–614.
West, O. 1951. The vegetation of Weenen County, Natal. Mem. Bot. Surv. S. Afr. 23:1–183.
White, F. 1962. Forest flora of Northern Rhodesia. Oxford University Press, London.
White, F. 1965. The savanna woodlands of the Zambezian and Sudanian Domains: an ecological and phytogeographical comparison. Webbia 19:651–681.
White, F. 1971. The taxonomic and ecological basis of chorology. Mitt. Bot. Staatssamml. München 10:91–112.
White, F. 1976a. The vegetation map of Africa: the history of a completed project. Boissiera 24:659–666.
White, F. 1976b. The taxonomy, ecology and chorology of African Chyrsobalanaceae (excluding Acioa). Bull. Jard. Bot. Nat. Belg. 46:265–350.
White, F. (ed.) in press A. Vegetation map of Africa. UNESCO, Paris.
White, F. in press B. Distribution and origins of the Afromontane flora.
Wild, H. 1956. The principal phytogeographical elements of the Southern Rhodesia flora. Proc. & Trans. Rhod. Sci. Ass. 44:53–62.
Wild, H. 1964. The endemic species of the Chimanimani Mountains and their significance. Kirkia 4:125–157.
Wild, H. 1968. Phytogeography in South Central Africa. Kirkia 6:197–222.
Wild, H. & Barbosa, L. A. Grandvaux, 1968. Vegetation map of the Flora Zambesiaca area 1:2,000,000. Supplement to Flora Zambesiaca. Collins, Salisbury, Rhodesia.
Wood, D. 1971. The adaptive significance of a wide altitudinal range for montane species. Trans. Bot. Soc. Edinb. 41:119–124.

12 The Afro-alpine Region

D. J. B. Killick*

1.	Terminology	517
2.	Phytogeography	518
3.	Ecology	521
3.1	Subalpine belt	521
3.1.1	Drakensberg	522
3.1.2	Lesotho	540
3.1.3	Amatole Mountains	542
3.1.4	Malawi	545
3.2	Alpine belt	545
3.2.1	Environment	546
3.2.2	The vegetation	550
References		558

* All photographs are by the author unless otherwise stated.

12 The Afro-alpine Region

1. Terminology

At the outset it is desirable to examine in their southern African context some of the key terms used in this chapter: they are afro-alpine, austro-afro-alpine, alpine and subalpine. Firstly, the term afro-alpine should be considered. Hauman (1955) was apparently the first to assign the flora of the high mountains of tropical Africa to the Afro-alpine Region. Following Hauman, Monod (1957) in his chorological map of Africa recognized an Afro-alpine Region occurring as scattered islands in the Sudano–Angolan Region (=Sudano–Zambezian Region of Lebrun 1947). Although the Sudano–Angolan Region extends into southern Africa, Monod did not indicate an Afro-alpine Region for southern Africa: the region is restricted to tropical East Africa. In 1965 Hedberg indicated the presence of an afro-alpine flora in South Africa, but this was limited to 16 afro-alpine species, which have extended southwards from the mountains of tropical East Africa. In other words, this was not acceptance of a local southern African Afro-alpine Region. In 1965 White produced a chorological map again with no Afro-alpine Region indicated for southern Africa. White's subsequent map (1971) based on the works of Lebrun (1947), Monod (1957) and White (1965) shows no Afro-alpine Region for Africa at all despite the inclusion of the region in the legend to the map. From the above, it is clear that chorologists have not recognized an Afro-alpine Region in southern Africa, although they may have tacitly accepted its presence there.

Van Zinderen Bakker & Werger (1974), in discussing the phytogeography of bog vegetation in Lesotho, used the term austro-afro-alpine to describe the flora of high mountains in southern Africa above the tree limit. In this they took a lead from Coetzee (1967), who distinguished an austro-afro-alpine ecological belt in Lesotho and the Drakensberg. It seems reasonable to include this floral area in the Afro-alpine Region. The prefix austro serves to distinguish the southern part from the more northerly situated central part of the Afro-alpine Region.

Among ecologists and palynologists the term alpine has been used continuously since 1946 to describe the vegetation of the summit/and the upper slopes of the Drakensberg in the Cape and Natal (Schelpe 1946, West 1951, Acocks 1953, Killick 1963, Edwards 1967) and the higher parts of Lesotho (Acocks, 1953, Van Zinderen Bakker 1955, Jacot Guillarmod 1971, Herbst & Roberts 1974). As already mentioned Coetzee (1967) coined the term austro-afro-alpine to distinguish the belt from that of the afro-alpine belt of East African mountains as recognized by Hauman (1933) and Hedberg (1951). Both Killick (1963) and Edwards (1967) considered it unnecessary to add any prefix to alpine. Werger (pers. comm. 1976) suggested that chorologists will have to consider calling the southern part the Austral Domain of the Afro-alpine Region and the central part the Central Domain of the Afro-alpine Region (cf. Chapter 7).

The term subalpine has been used in southern African ecology for a long time. Thode (1894) described the uppermost region of the Natal Drakensberg between 2130–3050 m, as the upper or subalpine region. Hutchinson (1946 used the term in approximately the same context. Killick (1963) and Edwards (1967), who both made intensive studies of the vegetation of the Natal Drakensberg used the term subalpine (belt) to describe the vegetation between 1830–1980 m and 2865–2895 m, which has fynbos as its climax community (cf. fynbos in the chapter on Capensis). Jacot Guillarmod (1971) described the subalpine belt in Lesotho as occurring in the zone between approximately 2290 m and 2900 m. Coetzee (1967), obviously following Hedberg (1951), rejected the idea of a subalpine belt in East or southern Africa, because of the absence of a zone of tall conifers (nadelwald) above montane forest as is classically found in the European Alps. She considered the upper level of montane forest to represent the tree line and the vegetation above to constitute the austro-afro-alpine belt. The fact that there is a clearly recognizable belt between the montane and alpine belts in the Drakensberg with a climax of its own, namely fynbos, and a distinct environment of its own, obviously less rigorous and microthermal than the alpine belt above, led Killick (1963) and others to designate it as subalpine in spite of the absence of coniferous forest. In this connection, it should be mentioned that elsewhere in the southern hemisphere, e.g. New Zealand and Java, the term subalpine is used for vegetation not characterized by coniferous forest (Wardle 1962, Van Steenis 1935). It is significant that Hedberg (1951) was tempted to compare the Ericaceous Belt of the East African mountains with the 'ericoid' subalpine zone described by Van Steenis (1935) for certain mountains of Malaysia.

In the account that follows the term Afro-alpine Region will be used in a phytogeographical sense, while the terms subalpine and alpine belts will be used in an ecological sense to describe the altitudinal 'zones' making up the (austral part of the) Afro-alpine Region.

2. Phytogeography

The flora of the Afro-alpine Region of southern Africa comprises a number of distinct phytogeographical elements. Weimarck (1941) distinguished the Cape, Afromontane and north hemispherical temperate elements. The elements of the Cape flora are found from the southwestern Cape northwards along the eastern mountains of Africa. The Drakensberg, in its broad sense and including the higher parts of Lesotho, is regarded as a distinct centre of the Cape flora separated from the southwestern Cape centre by several intervals or disjunctions and from the Inyangani subcentre in Rhodesia by the Limpopo interval. Weimarck cited a number of genera typical of the Cape centre, which also occur in the Drakensberg centre: they are *Pentaschistis, Ehrharta, Ficinia, Tetraria, Schoenoxiphium, Restio, Aristea, Hesperantha, Watsonia, Moraea, Monadenia, Disa, Protea, Heliophila, Cliffortia, Muraltia, Passerina, Erica, Stoebe, Metalasia, Osteospermum* and *Hirpicium*.

The Afro-montane element consists of those genera and species, which have their centre in the African mountains from Ethiopia to the Drakensberg and are also represented in Angola and the Cameroons. Weimarck considered that for phytogeographical purposes the Afro-alpine element of Hauman (1933) should be regarded as an extreme type of the Afro-montane element and united with it. He distinguished five main distribution types within the enlarged Afro-montane element. These were:

(1) A ubiquitous group containing species with a very large area and with

localities in West and East Africa, e.g. *Thalictrum rhynchocarpum, Swertia mannii, Lobelia rubescens, Helichrysum nudiflorum* and *Conyza gigantea*.

(2) An eastern group with species distributed in the mountains from Ethiopia to the Drakensberg, but not represented in Angola and in the Cameroons.

(3) A southern group with species limited to the Drakensberg and the southern areas of tropical Africa, e.g. *Leucosidea sericea, Myrica pilulifera, Buddleia salviifolia* and *Styppeichloa gynoglossa*.

(4) A northern group with species distributed only in Ethiopia and the northern mountains of Tropical Africa, e.g. *Hagenia abyssinica*.

(5) And, finally a number of groups with species endemic in one (or two) subcentre(s), e.g. certain *Dierama* species in the Drakensberg, *Lobelia stricklandiae* in the Inyangani subcentre and *Ardisiandra sibthorpioides* in the Cameroons.

According to Weimarck the north-hemispherical temperate (Afro–European or boreal) element consists of genera such as *Lithospermum, Cerastium, Juncus, Cardamine, Vaccinium, Aira, Ranunculus, Anemone, Clematis* and *Dianthus*, which occur in the Drakensberg centre, some extending to the Cape and some reaching Madagascar.

Van Zinderen Bakker (1962) recognized two floras, the Cape Flora and the Alpine Holarctic Flora. Elements of the latter are found mainly on the mountains along the eastern side of the continent as far south as Lesotho and a few of them spread further south into the cool temperate region: examples occurring in the Drakensberg and Lesotho are *Myosotis, Ajuga, Bromus, Hypericum, Valeriana, Epilobium* and *Crepis*, all of 'northern' origin. Van Zinderen Bakker states that according to the speciation which took place, these alpine elements must be very old and he dates the flora as of Tertiary age.

In 1961 Hedberg, writing on the afro-alpine vascular flora of tropical East Africa, stated that there were 25 afro-alpine species on the mountains of southern Africa and 16 in South Africa, but unfortunately the species are not mentioned. An examination of Hedberg's Afro-alpine Vascular Plants (1957) and Killick's Cathedral Peak (Drakensberg) check-list (1963) reveals that there are 84 genera and 18 species common to the two parts of the region. In other words, 80 per cent of the 105 afro-alpine genera are common to both parts, but only 6.5 per cent of the 278 afro-alpine species. The lowness of the species percentage is not surprising when it is realized that 81 per cent of the afro-alpine species are endemics. The species common to the two parts of the region are *Aira caryophyllea, Artemisia afra, Caucalis melanantha, Dryopteris inaequalis, Gnaphalium luteo-album, Hebenstreitia dentata, Helichrysum ordoratissimum, H. splendidum, Juncus dregeanus, Lithospermum afromontanum, Lycopodium saururus, Koeleria cristata, Parietaria debilis, Pleopeltis lanceolata, Ranunculus multifidus, Romulea campanuloides, Scabiosa columbaria* and *Silene burchellii*. A common feature of both parts is that *Helichrysum* and *Senecio* are the two largest genera.

To illustrate the affinities of the Drakensberg centre with the Cape centre and the Inyangani subcentre, Killick (1963) compared the flora of the Cathedral Peak area in the Drakensberg with that of the Cape centre as delimited by Bolus (1905) and the Chimanimani Mountains in Rhodesia, the most southerly of the tropical African mountains, using Goodier & Phipps's check-list (1961). For purposes of comparison, Weimarck's combination of afro-alpine and afro-montane elements was used. The results are given in Table 1. The figures suggest that the Drakensberg centre has closer floral affinities with the Cape centre than the Inyangani subcentre, but if the Drakensberg centre had been compared with the tropical African mountains as a whole the affinities might have been more nearly equal. It is surprising how few of the many genera common to the Drakensberg and Cape centres are characteristic Cape genera. Of the 282 angiosperm genera, which Weimarck (1941) lists as characteristic Cape genera, only 24 occur in the Drakensberg or, in other words, only 6.3 per cent of the Drakensberg genera are characteristic Cape genera. As might be expected the figures are even lower for the more northerly situated Chimanimani Mountains: 13 genera and 3.3 per cent. It is important to realize, however, that though the Cape genera are few in number, they are very important ecologically. For example, in the Drakensberg, the dominant species of Alpine Heath and Subalpine Fynbos belong very largely to characteristic Cape genera, e.g. *Erica, Passerina, Protea* and *Cliffortia*.

Table 1. Number of genera and species common to the Drakensberg and Cape centres and the Inyangani subcentre, and expressed (in brackets) as a percentage of the total number of genera and species in the Drakensberg.

	Cape		Inyangani subcentre (Chimanimani Mts)	
	Genera	Species	Genera	Species
Drakensberg centre (Cathedral Peak)	323 (77.09)	242 (16.68)	232 (55.38)	133 (14.67)

Van Zinderen Bakker & Werger (1974) distinguished three phytogeographical groups in the flora of Lesotho peat bogs:

(1) The South African group, consisting of species restricted to the alpine zone of the South African mountains, e.g. *Haplocarpha nervosa, Senecio cryptolanatus, Athrixia fontana, Cotula hispida* and *Helichrysum bellum*.

(2) The south and eastern African group comprising species occurring in southern Angola, S.W. Africa, South Africa, east central Africa and East Africa, e.g. *Lagarosiphon muscoides, Aponogeton junceus, Limosella capensis, L. longiflora*, and *Trifolium burchellianum*.

(3) The northern temperate group, e.g. *Agrostis subulifolia, Koeleria cristata, Scirpus fluitans* and *Ranunculus meyeri*.

According to Wild (1968) at some time in the past there must have been connections between the various centres and subcentres within the Cape flora giving an uninterrupted distribution and he examined the possibility that changes of climate in the Quaternary, as postulated by Van Zinderen Bakker (1964), Bond (1957) and others, were sufficiently great to have allowed the Zambezi and Limpopo intervals to have been crossed by elements of the Cape Flora. Considering the estimated drop of 5°C in temperature and a 140 per cent increase of rainfall, he concluded that these changes were not great enough for the intervals to have supported a Cape type of flora and that opportunities for joining the Cape centre to its northerly outposts should be pushed back to the Tertiary and beyond. He cited King's geomorphological map of Africa (1962), which shows that the centres and subcentres of the Cape flora, including Angola and the Cameroons, were continuous in the early Caenozoic of 60–70 million years ago. The Early Caenozoic (or Eocene) was apparently a period of smooth landscapes and consequently climatic conditions were probably relatively uniform and the Cape flora or its direct evolutionary precursor could have covered a large continuous area of the African continent up to the early Tertiary if the conditions were suitable. King stated that folding occurred during the Pliocene and Miocene, and it may be during this period that the Cape flora became fragmented and further north remained only on the mountains. Wild preferred this type of explanation to that put forward by Levyns (1952, 1964), who suggested that, as the most primitive species of some Cape genera such as *Phylica, Aristea, Muraltia* and *Stoebe* occur in central Africa, often showing great discontinuities with the remainder of the species of these genera, these patterns of distribution point to an old migration from central Africa to the Cape. (The question of the origin of the Cape flora is a vexed one and has been dealt with in Chapter 8.) Wild points out that the highest mountains of Southern Africa are part of the Gondwana Jurassic or Post Gondwana Cretaceous land surface (King 1962) and so go back in time to what may be the early period of development of the angiosperms.

The phytogeography of the high mountain regions in Southern Africa deserves further study. A prerequisite, however, is more intensive plant collecting, so that the mountain floras are completely known.

3. Ecology

3.1 *Subalpine Belt*

This belt is restricted in southern Africa almost entirely to the Drakensberg in the Cape and Natal, and to Lesotho, between altitudes of 1830–2895 m. It is also found in the eastern Cape on the Amatole Mountains between 1220–1938 m and on the Bosberg overlooking Somerset East (Story 1952) and probably on a number of other mountains in South Africa, information about which is meagre or lacking. The presence of this belt in the other territories of southern Africa, with the probable exception of Malawi, is doubted. Either the mountains in these territories are too low in relation to their latitudinal position or they occur in arid areas.

In Angola nothing resembling subalpine vegetation has yet been described. It is true, however, that *Philippia, Protea* and *Stoebe* occur in the higher parts of the central plateau (Barbosa 1970, Diniz 1973), but the mere presence of such fynbos plants is not necessarily an indicator of subalpine vegetation. Fynbos plants can occur at lower altitudes where they are frequently seral to montane or other forest and, indeed, can occur at almost sea-level. This point cannot be stressed too strongly, because fynbos plants are often a feature of mountain summits in southern African even at low altitudes.

In South West Africa the Brandberg reaches 2697 m but, situated as it is on the edge of the Namib Desert, it does not support subalpine vegetation (Nordenstam 1974), nor do the other mountains of the territory.

The Transvaal Drakensberg reaches 2286 m in Mount Anderson, but the general altitude of the range is considerably lower. The vegetation of the Mariepskop Complex described by Van der Schijff & Schoonraed (1971) appears to be montane, even that of the summit at 1920 m.

In Moçambique Pedro & Barbosa (1955) recognized two distinct subalpine zones, namely 'zona subalpestre de Manica', occurring in the Chimanimani Mountains (Mt. Binga, 2436 m) and the Gorongosa Mountains (1862 m), and 'zonas subalpestres da Zambesia' occurring in the more northerly mountains of Chiperone (2054 m), Namuli (2419 m) and Macua (2077 m). Later, Macedo (1970) distinguished three subalpine zones in the Gorongosa Mountains, viz. montane–subalpine (hygrophilous forest), montane–subalpine (rocky areas) and subalpine meadow. Subsequently Wild & Barbosa (1967) in their vegetation map of the Flora Zambesiaca area classified these subalpine areas as *Loudetia–Exotheca* Grasslands (submontane), thus reducing their altitudinal status in the light of their classification of high mountain areas in Rhodesia and Malawi. Wild (1965) states that there is no true alpine or montane flora in Rhodesia, but there is a well-marked submontane flora. Phipps & Goodier (1962), however, described the vegetation of the Chimanimani Mountains as Montane including in it Ericaceous and Proteaceous Scrub.

Chapman (1968) classified the vegetation of high plateaux and mountains in Malawi (including Mt. Mlanje, 3000 m; Nyika Plateau, 2606 m; Zomba, 2090 m; Dedza, 2198 m and the Vipya Plateau, 2290 m) as montane and submontane, but Chapman & White (1970) regarded the montane shrubland and montane shrub grassland occurring on the uppermost, more exposed, rocky slopes of Mt. Mlanje as belonging to the Ericaceous zone. It seems rather contradictory to designate these communities as montane and yet assign them to the Ericaceous zone. If this low-grazing grass-heath vegetation can be related to the Ericaceous Belt of East African mountains (altitudinally, Mt. Mlanje is sufficiently high for this to be possible), then it could be regarded as subalpine (in the South African sense), in which case it would be the only subalpine vegetation in southern Africa outside South Africa. The plant ecology of the Subalpine Belt in southern Africa will be described for the several mountain areas in which it occurs.

3.1.1 Drakensberg

The Subalpine Belt of the Drakensberg extends from Xalanga Peak in the eastern Cape to Mont aux Sources in Natal and occupies the Little Berg from its edge to just below the summit of the Drakensberg, i.e. from 1830–2865 m (Fig. 1). The vegetation of this Belt also occurs in a modified form on outlying plateaux such as Ntambamhlope near Estcourt. The account of environment and vegetation which follows is derived chiefly from Killick's (1963) researches at Cathedral Peak. More detailed information can be obtained from his publication.

3.1.1.1 *Environment*

Soils. The soils of the Little Berg, which are derived from the basaltic lavas, are deep and horizons are ill-defined. The surface soil ranges in thickness from 17–30 cm and consists of dark brown to blackish brown, granular to crumbly clay–loam and clay permeated by grass roots. Organic content as indicated by ig-

Fig. 1. The upper part of the Subalpine Belt in the Cathedral Peak area of the Natal Drakensberg. The peaks shown all lie forward of the general line of the scarp. In the foreground is the Camel, at left middle distance are the Pyramid and Column and in the background are the Mitre, Chessmen, Inner and Outer Horns, Bell and Cathedral Peak. The snow accentuates the stratification of the basalt.

nition loss, is high; it varies from 27 per cent on the steep slopes at the top of the higher catchments to nearly 50 per cent on the flatter areas at the base of the catchments (Nänni 1956). The high organic content is surprising, considering that the area is subject to regular burning.

Next is a layer varying from 45–120 cm and more in depth. It consists of granular brown clay loam to clay dense and compact when moist, but badly cracked when dry. Then follows stony loam with angular stones predominating and finally partly and slightly decomposed basalt. Soil reaction varies from pH 5.6 to 6.6. Exchangeable base and total adsorbed base values are high. With the exception of potassium, higher cation values are to be found in the subsoil than in the surface soil. The clay mineral, montmorillonite, is present, weakly between 33–70 cm and definite at all layers below this.

A feature of the grass slopes of the Subalpine Belt are terracettes, horizontally arranged crescentic scars, which are formed largely by frost action.

Climate. Average daily sunshine measured at 1860 m on the Little Berg at the Cathedral Peak Meteorological Station varies between 5.5 hours during the wet summer month of December and 8.3 hours during the dry winter month of June i.e. 30 and 82 per cent respectively of the possible sunshine. According to Nänni (1956) the main escarpment to the west and northwest reduces the available sunshine by amounts varying from 10 per cent during December to 1 per cent during June. This loss becomes greater the nearer the escarpment is reached.

Temperature records taken at 1860 m are illustrated in the Deasy chart in Fig. 2.1. The curves for mean diurnal fluctuation, B and C, show that air temperature is cool to mild. Only in the absolute values represented by curves A and D, does the temperature reach degrees of hot and cold. The highest temperature recorded is 31.2°C in November 1951 and the lowest 3.6°C in June 1953. Frosts are almost a daily occurrence in winter.

The two broken curves in Fig. 2.1, E and F, for grass minimum temperature, are of particular interest. The values for mean daily minimum E (below 10°C throughout the year) and absolute minimum F (below 0°C) are consistently lower than the corresponding values recorded in the Stevenson Screen. On the night of 27 June 1959 the difference in minimum temperature between the Stevenson Screen and grass level readings was 25.6°C. The lowest value recorded by the grass minimum thermometer was -16.5°C in July 1954. The cause of the low temperatures at grass level is the loss of heat by radiation at night. Radiation is especially active when the nights are calm and the sky is clear, that is during the winter months. This is illustrated by a comparison of curves C with E and D with F. The curves diverge most between May and September.

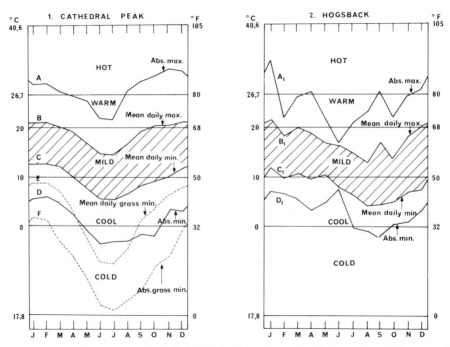

Fig. 2. Temperature charts for Cathedral Peak (air temperature, 10 years; grass temperature, 6 years) and Hogsback, Amatole Mountains (1 year).

Wind is an important ecological factor in the Drakensberg. The 'bergwinds' which blow from the west during late winter and spring often attain great velocity. These winds, though not as hot and dry as at lower altitudes (they become heated by compression as they descend), are important because they are generally accompanied by periods of low humidity and they blow at a time when soil moisture is at its minimum. Winds are also important because they are prevalent during the dry season when fire hazard is at a maximum. Once a berg fire starts and there is an attendant wind, it is very difficult to put out.

The Drakensberg derives its rain mainly from oceanic air-streams entering from east coast highs. At the beginning of summer most of the rainfall appears to be orographic. Later, the frequency of thunderstorms increases and this form of precipitation provides about 50 per cent of the total rainfall.

From Table 2 it will be seen that most of the rain falls during the summer months i.e. between October and March. The proportion is about 85 per cent. The wettest months are January, February and March and the driest June and July (compare also Chapter 2).

The average number of days per month on which rain is recorded is given by Nänni (1956) as follows: July 3; August 7; September 9; October 15; November 20; December 22; January 23; February 22; March 19; April 10; May 7 and June 3. The highest rainfall recorded in one hour at the meteorological station on the Little Berg is 81 mm in February 1954 and the highest intensity recorded for periods of 2–5 minutes is about 280 mm per hour (Nänni 1956). The greatest fall in 24 hours is 102 mm.

The altitudinal rainfall pattern in the Cathedral Peak area is typical of that of high mountains. Figure 3 illustrates schematically the variations of mean annual rainfall (calendar years) with altitude at six stations in the Cathedral Peak area, including one in Mokhotlong, Lesotho, 40 km southwest of Cathedral Peak. It will be seen that rainfall increases from 1240 mm in the Mlambonja Valley (Montane Belt) to 1418 mm, near the edge of the Little Berg, reaching a maximum of 2017 mm in the upper part of the Subalpine Belt and decreasing to 1609 mm on the summit of the Drakensberg in the Alpine Belt and 562 mm at Mokhotlong in Lesotho.

Hail as an ecological factor seems of little importance in the Drakensberg. According to Nänni (1956) hail can be expected once in every two years. The largest stones measured by the writer were 1.3 cm in diameter.

Snow can be expected any time between April and September and occurs mainly in July. Usually snow is restricted to the summit and near-summit and only occasionally reaches the lower limit of the Subalpine Belt.

Summer fog occurs mainly on the Little Berg and the summit area. This fog is common and may be continuous for two weeks at a time.

Atmospheric humidity has been recorded on the Little Berg for nearly 10 years. Relative humidities below 30 per cent are common especially in late winter. Humidities of less than 10 per cent are recorded between about 5 and 10 times annually, while 5 per cent and less is recorded occasionally (Nänni 1969). The periods of low relative humidity coincide with westerly winds.

Evaporation, measured by a Symon's evaporation tank situated at 1860 m, is highest between September and November. Mean annual evaporation is 1346 mm, which is about equal to the rainfall at the meteorological station.

Fire. Fire is an extremely important environmental factor in the Drakensberg. Fires can occur either naturally or are caused by man. An important natural cause is lightning. Lightning-induced fires occur in early spring when the grass sward is dry and inflammable. The Forestry Department at Cathedral Peak have on record several instances of grass fires being started by lightning. Killick (1963) has produced figures which show that between 6.3 and 16.9 per cent of fires recorded on forest stations in South Africa are caused by lightning. Natural fires also occur when boulders rolling down hill-slopes collide with one another or with stationary boulders producing sparks which ignite the grass sward (Nänni 1956).

It is probable that natural fires have been a factor of the environment ever since

Table 2 Mean monthly rainfall in the Cathedral Peak area (FD = Forestry Department)

	F.D. Office, Mlambonja Valley	Cathedral Peak Hotel, Mlambonja Valley	F.D. Met. Station, Little Berg	F.D., IIBr., Little Berg	F.D., IIAW., Little Berg	Organ Pipes Pass, Summit Drakensberg	Mokhotlong Lesotho
Altitude	1369 m	1469 m	1860 m	1981 m	2287 m	2927 m	2377 m
Latitude	28° 56′ S	28° 57′ S	28° 59′ S	29° 0′ S	29° 0′ S	29° 1′ S	29° 17′ S
Longitude	29° 14′ E	29° 11′ E	29° 14′ E	29° 13′ E	29° 13′ E	29° 11′ E	29° 05′ E
Period in yrs	5	17	9	10	6	3	20
	mm	mm	mm	mm	mm	mm	mm
January	184	184	197	248	284	342	99
February	279	236	289	299	392	295	81
March	170	213	212	238	284	207	68
April	52	69	57	74	76	49	35
May	35	27	35	38	52	46	21
June	9	14	13	15	24	27	5
July	9	11	13	13	13	1	10
August	27	28	39	39	45	16	14
September	70	46	67	68	93	61	20
October	88	91	108	122	164	109	55
November	143	140	157	183	265	226	76
December	174	172	231	252	325	230	78
Total	1240	1231	1418	1589	2017	1609	562

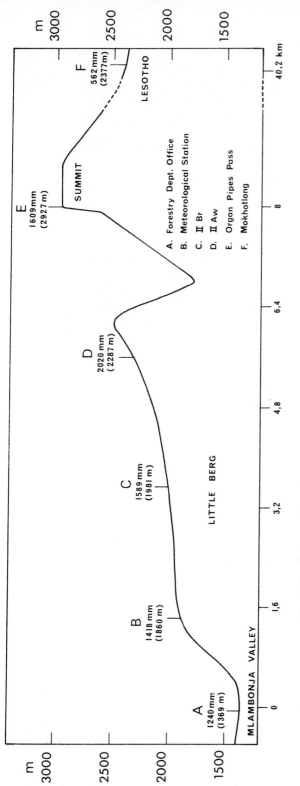

Fig. 3. Profile through Cathedral Peak area showing variation of rainfall with altitude.

the climate included a dry season. If this is the case, then it can be reasonably assumed that in the present climatic set-up grassland and not forest as contended by Acocks (1953) has been the predominating community on the slopes below the main escarpment. Bayer (1955) makes the comment that the whole behaviour of vernal aspect forbs in grassland suggests that spring burning (through lightning) is a natural factor of the climate. Thus in the Subalpine Belt grassland has the status of a fire subclimax or fire climax depending upon whether one follows the doctrines of Clements (1916) or Tansley (1935).

The mountains have long been subject to fires caused by man. It is probable that the Drakensberg Bushmen periodically set alight the grassland to produce new growth which would attract game for hunting: this was certainly the practice of Bushmen elsewhere in South Africa (Burchell 1822). There is abundant evidence (Holden 1855, Mann 1859) that the Bantu and later the European farmer fired the grassland to provide winter grazing for their cattle and sheep. Today burning of the grassland is an annual or biennial practice. The Basotho burn the grassland on the summit and occasionally the grassland on the Natal and Cape sides to attract game for hunting. The Forestry Department burns the grassland biennially in spring after the first rain of 13 mm or more. This is largely responsible for the fine sward of *Themeda triandra* on the Little Berg. In addition, the Forestry Department burns firebreaks in autumn and winter.

3.1.1.2 *Vegetation*

The vegetation of the Subalpine Belt consists mainly of bunch grassland (tussock veld according to Bews 1917), chiefly *Themeda triandra* Grassland (Fig. 4). Also present are Temperate Grasslands occurring on mesocline slopes, Tall Grassland, *Rendlia altera* Grassland and *Merxmuellera macowanii* Grassland. Woody communities include *Cliffortia linearifolia* Scrub, *Buddleia salviifolia* Scrub, *Protea* Savanna and the climax community of the Subalpine Belt, Subalpine Fynbos. In addition, there are aquatic, hygrophilous, rock outcrop and cliff communities. The suggested interrelationships of the subalpine communities are given in Fig. 5.
Aquatic and hygrophilous vegetation. Communities of aquatic and hygrophilous plants are neither abundant nor extensive in the Subalpine Belt: they occur along swiftly-flowing streams, in vleis (marshes) situated usually on level areas near the edge of the Little Berg at about 1830 m and around tarns or pans.

(a) Vlei Communities: The only regularly submerged aquatics are green filamentous algae and two mosses, *Philonotis laeviscula* and *P. afrofontana*, the latter forming small cushions.

Aquatics with their shoots slightly to almost completely emersed can be divided into dwarf and tall aquatics. The dwarf aquatics include: *Anagallis huttonii, Limosella maior, Scirpus hystrix, Bulbostylis densa, Juncus dregeanus, Eriocaulon abyssinicum, E. dregei, Pycreus rehmannianus, Xyris capensis* and *Athrixia fontana*.

The tall aquatics include *Scirpus macer, Carex cernua, Eleocharis dregeana, Rhynchospora brownii, Juncus exsertus, J. rostratus, Arundinella nepalensis, Cyrtanthus breviflorus, Gunnera perpensa* and *Epilobium hirsutum*. All these plants can form either pure or mixed stands. A combination, which is very com-

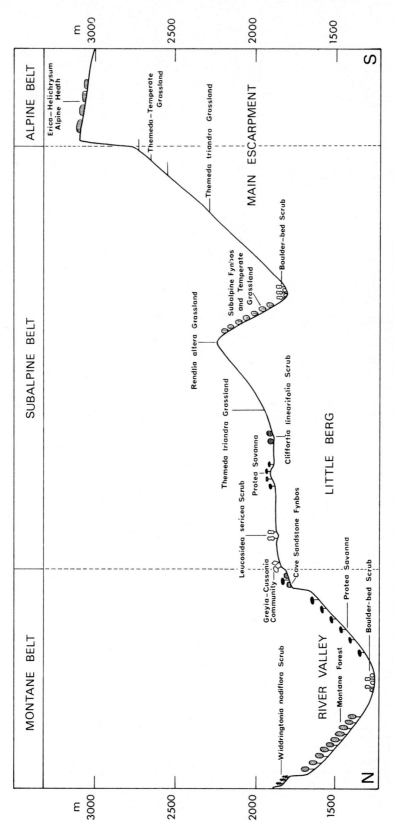

Fig. 4. Profile through the Drakensberg area showing the vegetation belts with their chief plant communities.

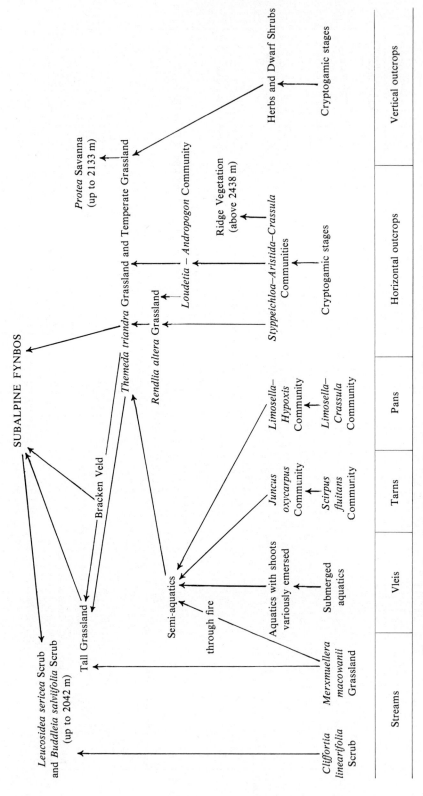

Fig. 5. Suggested interrelationships of the plant communities in the Subalpine Belt of the Drakensberg.

mon on the Little Berg, is the *Scirpus–Dryopteris–Cyrtanthus–Gunnera* Community.

Hygrophilous or semi-aquatic plants comprise a large number of species growing on moist soil. The most important species are *Pycreus oakfortensis, Rhynchospora brownii, Fuirena pubescens, Scleria welwitschii, S. woodii* and the grass *Stiburus alopecuroides*.

(b) Tarn and Pan communities: Mushroom Tarn, about 70 m in diameter, is situated in the Cathedral Peak area. Nearest the centre in water about 5–8 cm deep is *Scirpus fluitans*, then *Juncus oxycarpus*, and at the very edge on wet soil are the grasses *Agrostis huttoniae* and *Eragrostis planiculmis*.

The Ntabanyama Pan in the Giants Castle area has *Limosella capensis* and *Crassula natans* (a curious form), floating but rooted, in the middle, with *Limosella longiflora*, and *Hypoxis filiformis* at the edge.

(c) Streambank communities: *Merxmuellera macowanii* Grassland is found on approximately level streambanks between 1830–2285 m. The dominant, *M. macowanii*, is a xeromorphic grass with leaves up to 75 cm long and culms 1.2 m long and forms large tussocks some nearly 60 cm in diameter. It is very conspicuous when in flower during summer. Associated species are many, among them being: *Geranium pulchrum, G. ornithopodum* var. *album, G. ornithopodum* var. *lilacinum, Mariscus elatior, Galium wittebergense, Gunnera perpensa, Berkheya macrocephala, Alepidea amatymbica, Kniphofia* sp., *Crassula pellucida* subsp. *brachypetala, Helichrysum epapposum* var. *epapposum* and *H. fulvum*.

The hummock structure of some vleis is due to death by fire of *Merxmuellera macowanii* and subsequent revegetation of the tussock remains by hygrophilous species.

Cliffortia linearifolia Scrub occurs in deep stream gullies between 1950–2250 m. *C. linearifolia* is the dominant shrub and *Philippia evansii* is frequently subdominant. Associated shrubs include *Myrsine africana, Anthospermum aethiopicum, Diospyros austro-africana* var. *rubriflora, Athanasia punctata, Maytenus acuminata, Rhamnus prinoides, Phygelius capensis, Polemannia montana* and *Erica ebracteata*. The most characteristic herbs are: *Alepidea amatymbica, Cephalaria natalensis, Scabiosa drakensbergensis, Valeriana capensis, Helichrysum hypoleucum, Myosotis sylvatica, Printzia pyrifolia, Berkheya multijuga, Geranium pulchrum, Senecio haygarthii, Galium wittebergense, Cycnium racemosum* and *Wahlenbergia undulata*.

The tree fern, *Alsophila dregei*, occurs in this scrub at varying intervals along streams. Along the margin is *Cymbopogon validus* or *Miscanthidium capense* Grassland or the *Aristida monticola* Community.

Horizontal and Vertical Rock Outcrop Vegetation. Horizontal bare-rock areas occur along the top edge of the lowermost basalt cliffs, above and adjacent waterfalls and scattered through grassland on the Little Berg. In these areas the basalt is mainly exposed to form pavements. Vertical or nearly vertical outcrops occur on the flanks of long spurs leading to the main escarpment and on buttress slopes of the main escarpment and its outliers. These outcrops are often arranged in horizontal tiers interrupted at intervals by grassland. Occasionally the outcrops take the form of cliffs.

Table 3. Wheel and skewer point analyses of *Themeda triandra* Grassland in Catchments 3, 4 and 9 on the Little Berg (Cathedral Peak Forest Station).

Catchment 3		Catchment 4		Catchment 9	
Species	Percentage Basal Cover	Species	Percentage Basal Cover	Species	Percentage Basal Cover
Themeda triandra	11.55	Themeda triandra	12.25	Themeda triandra	12.10
Trachypogon spicatus	5.05	Trachypogon spicatus	6.15	Tristachya hispida	8.55
Tristachya hispida	4.60	Tristachya hispida	5.20	Harpochloa falx	5.20
Harpochloa falx	4.40	Heteropogon contortus	4.65	Trachypogon spicatus	3.70
Heteropogon contortus	4.10	Harpochloa falx	3.45	Heteropogon contortus	2.45
Rendlia altera	3.20	Rendlia altera	3.45	Stiburus alopecuroides	1.85
Alloteropsis semialata	2.00	Alloteropsis semialata	3.25	Andropogon appendiculatus	1.65
Elyonurus argenteus	1.80	Elyonurus argenteus	1.85	Andropogon ravus	1.35
Bulbostylis trichobasis	0.9	Andropogon ravus	0.95	Koeleria cristata	1.35
Stiburus alopecuroides	0.90	Bulbostylis trichobasis	0.85	Bulbostylis schoenoides	1.15
Andropogon ravus	0.75	Koeleria cristata	0.75	Panicum ecklonii	0.95
Other (?)	0.60	Panicum ecklonii	0.70	Rendlia altera	0.80
Koeleria cristata	0.55	Stiburus alopecuroides	0.65	Ficinia cinnamomea	0.35
Panicum natalense	0.40	Other (?)	0.65	Panicum natalense	0.20
Panicum ecklonii	0.35	Panicum natalense	0.55	Cyperus compactus	0.20
Andropogon appendiculatus	0.25	Monocymbium ceresiiforme	0.40	Oxalis obliquifolia	0.15

Bulbostylis schoenoides	0.25	Andropogon appendiculatus	0.30	Bromus speciosus	0.15
Loudetia simplex	0.15	Hypoxis acuminata	0.25	Alloteropsis semialata	0.05
Acalypha punctata	0.15	Loudetia simplex	0.20	Senecio harveyanus	0.05
Ledebouria cooperi	0.15	Acalypha punctata	0.15	Helichrysum allioides	0.05
Ficinia cinnamomea	0.10	Eragrostis racemosa	0.10		
Cymbopogon validus	0.05	Poa binata	0.10		
Eragrostis racemosa	0.05	Anthospermum hedyotideum	0.10		
Monocymbium ceresiiforme	0.05	Oxalis obliquifolia	0.10		
Alepidea capensis	0.05	Ledebouria cooperi	0.10		
Erica woodii	0.05	Scleria woodii	0.10		
Helichrysum adenocarpum	0.05	Aristea angolensis	0.05		
Helichrysum glomeratum	0.05	Ficinia cinnamomea	0.05		
Tetraria cuspidata	0.05	Lobelia flaccida var.	0.05		
Satyrium longicauda	0.05	Scleria bulbifera	0.05		
Polygala rehmannii	0.05				
Scleria bulbifera	0.05				
Scleria woodii	0.05				
Cyperus compactus var. flavissimus	0.05				
Total	42.90	Total	47.45	Total	42.30

On horizontal outcrops the first mat-builders are two low-growing mosses *Campylopus trichodes* and *Ptychomitrium cucullatifolium*. They are followed by *Pogonatum simense* and *Polytrichum commune*, considerably taller species, and then the creeping fern, *Selaginella imbricata*. Basalt pavements on the Little Berg frequently support *Styppeiochloa gynoglossa*, a xeromorphic tunic grass, *Aristida junciformis* subsp. *galpinii* and *Crassula harveyi*. These species occur separately or as co-dominants in lithophilous communities. Associated plants are numerous and include: *Eragrostis caesia, E. racemosa, E. plana, Digitaria ternata, Rhynchelytrum setifolium, Microchloa caffra, Scilla natalensis, Crassula filiformis, Notholaena eckloniana, Ledebouria cooperi, Xerophyta viscosa, Rhodohypoxis baurii, Urginea tenella, Oxalis obliquifolia, Pygmaeothamnus chamaedendrum* var. *setulosus, Psammotropha myriantha, Hermannia woodii* and *Mohria caffrorum*.

Scattered through grassland are areas supporting small stones and little soil. Frequently these areas appear as low humps. The dominants are the grasses *Loudetia simplex* and *Andropogon filifolius* with *Panicum natalense* as co-dominant. *Loudetia simplex* gives the community a grey, brown-topped colour in late January and *Andropogon filifolius* imparts a red colour in spring.

Here and there on the Little Berg are koppies of varying size, the summits of which rarely exceed 1950 m. The koppies support a mixed vegetation consisting of grassland species, *Protea multibracteata, P. roupelliae, Halleria lucida, Rubus ludwigii, Cliffortia repens* and numerous lithophytes.

On vertical outcrops the three most important herbs contributing to mat formation are the crevice plants *Scirpus falsus, Xerophyta viscosa* and *Styppeiochloa gynoglossa*. *Scirpus falsus*, about 10 cm high, readily invades moist moss mats and forms dense carpets. These are frequently invaded by *Hesperantha longituba, Athrixia fontana* and *Wahlenbergia undulata*. *Xerophyta viscosa* covers large areas of moist or dry rock surface in the Subalpine Belt; it is often accompanied by *Cymbopogon validus*. *Styppeiochloa gynoglossa* is restricted to dry cliffs below 2640 m. Other common crevice and ledge plants are xeromorphic grasses such as *Aristida junciformis* subsp. *galpinii, A. monticola, Pentaschistis oreodoxa* and the tussock grass, *Merxmuellera stereophylla*; monocotyledonous geophytes such as *Scilla natalensis, Galtonia viridiflora* and *Eucomis humilis*; and dicotyledonous herbs and small shrubs such as *Berkheya rosulata, Osteospermum thodei, Helichrysum* spp., *Crassula* spp., *Stoebe vulgaris* and many others.

Subalpine grassland. There are five major grassland types in the Subalpine Belt. They are *Rendlia altera* Grassland, *Themeda triandra* Grassland, Temperate Grassland, *Themeda triandra* – Temperate Grassland and Tall Grassland. As indicated on p. 528 these grasslands constitute a fire subclimax or fire climax.

(a) *Rendlia altera* grassland is found between 1980–2440 m on the ridges of spurs on the Little Berg. The soil is thin, black, peaty, often covered with small stones and occasionally interrupted by basalt outcrops. The grasses are mostly short and characteristic of early stages in the grassland succession. The dominant is *Rendlia altera* supported by *Sporobolus centrifugus, Eragrostis capensis, E. racemosa, Andropogon filifolius, Panicum ecklonii, Cyperus semitrifidus, Bulbostylis humilis*, and herbs such as *Rhodohypoxis baurii, Wurmbea kraussii, Moraea stricta* and *Oxalis obliquifolia*.

(b) *Themeda triandra* grassland is the most extensive in the Subalpine Belt. On xerocline slopes it extends from the top of lowermost basalt cliffs to about 2590 m and on mesocline slopes it reaches 2135 m (Fig. 6). Above 2590 m on xerocline slopes *Themeda triandra* grassland loses its identity and mixes with Temperate Grassland and ultimately *Merxmuellera–Festuca–Pentaschistis* Grassland. *Themeda triandra* itself drops out at about 2835 m.

Between 1830–2135 m, *Themeda triandra* Grassland is fairly uniform in composition and basal cover. Two catchments (3 & 4) on Cathedral Peak Forest Station of approximately 121 ha each were analysed by the Wheel Point Method of Tidmarsh & Havenga (1955) and one catchment (9) by the Skewer Point Method. The results are given in Table 3. Total basal cover is high varying between 42.30

Fig. 6. Spurs of the Little Berg projecting into Natal from the main Drakensberg escarpment between Cathkin Peak (3060 m) and Ndedema Falls. The grassland is chiefly *Themeda triandra* Grassland.

per cent in Catchment 9 to 47.45 per cent in Catchment 4. *Themeda triandra* is dominant with a basal cover of about 12 per cent. Immediately subordinate to *T. triandra* are four grasses *Trachypogon spicatus*, *Tristachya hispida*, *Harpochloa falx* and *Heteropogon contortus*, the hierarchy varying from catchment to catchment. Associated forbs are more numerous than the analyses suggest: there are about 100 species which form seasonal aspect stands of varying size.

Between 2135–2590 m *Themeda triandra* Grassland takes on new associates. *Eucomis humilis* and *Urginea macrocentra* form conspicuous communities at the foot of small cliffs. Other species occurring in this region are: *Euryops pedunculatus*, *Tetraria cuspidata*, *Kniphofia porphyrantha*, *Aloe boylei*, *Dierama igneum*, *Bupleurum mundii*, *Pimpinella caffra*, *Agapanthus campanulatus*, *Sutera breviflora* and *Heliophila rigidiuscula*.

(c) Temperate grassland can be subdivided into three types dominated by species belonging to temperate genera.

(i) *Festuca costata* Grassland. This tall, evergreen, tussock-forming grass occurs in mesic situations in the lower part of the Subalpine Belt for example streambanks and the south side of koppies, and in both mesic and xeric situations in the upper part. Its altitudinal distribution is between 1830–2865 m. *F. costata* very rarely forms a pure sward: the tussocks are separated by shorter grasses like *Themeda triandra, Koeleria cristata, Poa binata, Agrostis barbuligera, Rendlia altera, Aristida monticola, Anthoxanthum ecklonii, Andropogon appendiculatus* and *Merxmuellera stricta*. Herb associates are numerous.

(ii) *Pentaschistis tysonii* grassland. This community covers very large areas in the Subalpine Belt. It occurs between 1980–2290 m on mesocline slopes and between 2590–2865 m on xerocline slopes. Usually *Pentaschistis tysonii* is dominant in pure stands, but occasionally it mixes with *Festuca costata*. *P. tysonii* is an evergreen xeromorphic grass with tightly rolled leaves. Associates are few and include the grasses *Agrostis barbuligera, Sporobolus centrifugus, Bromus speciosus, Andropogon appendiculatus, Merxmuellera stricta* and *Rendlia altera* and herbs such as *Senecio praeteritus, S. bupleuroides, S. barbatus, Helichrysum allioides, Alepidea capensis, Aster perfoliatus, Cycnium racemosum* and *Kniphofia pauciflora*.

(iii) *Bromus speciosus* grassland. This grassland has a similar distribution to that of *Festuca costata* grassland, but is not as extensive in the Cathedral Peak; in the Giants Castle Area it is reported as being well-developed (West 1951). *Bromus speciosus* is a broad-leaved evergreen grass with leaves up to 30 cm and culms 1.2 m high. Like *Festuca costata, Bromus speciosus* does not form a close sward.

The distribution of *Festuca costata, Pentaschistis tysonii* and *Bromus speciosus* in the Drakensberg poses an ecological problem. These grasses are xeromorphic and evergreen, yet they are generally restricted to mesic sites. It has been suggested to the writer by Professor C. L. Wicht that the xerocline slopes are much drier than the mesocline slopes and support *Themeda triandra* grassland, which is able to avoid the dry winter by going into dormancy; on the mesocline slopes there is no need to go into dormancy, hence the presence of grasses like *Festuca costata, Pentaschistis tysonii* and *Bromus speciosus* which are evergreen. But why should these grasses be xeromorphic? It is very probable that during winter the mesocline slopes are liable to periods of physiological drought. The condition could be induced by: when the soil is frozen and the soil water is unavailable; when strong drying winds blow at the end of winter and in early spring when soil water content is low; when soil temperatures are low and the soil has reduced capacity to give up water to the roots.

(d) *Themeda triandra* – Temperate grassland mixes in varying proportions with the Temperate Grassland types above about 2590 m on xerocline slopes. Between Cathkin Peak and Mont aux Sources the Drakensberg faces roughly northeast, consequently the buttress slopes of the main escarpment are chiefly xeroclinical and support extensive tracts of this mixed Grassland. The buttress slopes are steep and often have only a thin covering of soil.

Between 2490–2745 m *Aristida monticola* much dwarfed in stature, assumes greater importance and in parts is even locally dominant. Also important are

Koeleria cristata, Trachypogon spicatus, Tristachya hispida, Eragrostis racemosa, Helictotrichon hirtulum, Pentaschists pilosogluma and *Ehrharta longigluma*. Associated herbs are frequent.

Above 2745 m *Merxmuellera disticha, Festuca caprina* and *Pentaschistis oreodoxa*, the dominant grasses of the Alpine Belt, enter the grassland and the result is a grassland ecotone. *Koeleria cristata, Harpochloa falx* and *Eragrostis caesia* become more prominent and are accompanied by three grasses, which seem to be exclusive to this ecotone, namely *Festuca scabra, F. killickii* and *Merxmuellera aureocephala*.

(e) Tall grassland includes three grass communities dominated by tall grasses and having very similar ecological requirements.

(i) *Miscanthidium capense* grassland. This community is found on streambanks, moist flats and in gullies below 1980 m. On the Little Berg it covers areas up to 2 ha in extent. The dominant is *Miscanthidium capense* var. *villosum* with culms 1.5–2.4 m tall. The majority of associated plants are tall herbs and shrubs, some of which are autumnal aspect plants belonging chiefly to the families Asteraceae and Lamiaceae: *Leonotis dysophylla, Rhabdosia calycina, R. grallata, Stachys albiflora, Pycnostachys reticulata, Satureia reptans, Schistostephium crataegifolium, Nidorella auriculata* subsp. *polycephala, Heteromma decurrens, Vernonia hirsuta* and *Helichrysum* spp.

(ii) *Hyparrhenia dregeana* grassland. As (i) above, but the dominant is *Hyparrhenia dregeana* and usually occupies much smaller areas.

(iii) *Cymbopogon validus* grassland. As (i) above, but the dominant is *Cymbopogon validus* and the community occupies smaller areas and occurs up to 2745 m instead of 1980 m.

Bracken Veld. Bracken Veld (*Pteridium aquilinum*) is found on moist, deep soil up to 2135 m. It invades *Themeda triandra* Grassland and the lower parts of Temperate Grassland. Closed Bracken Veld is invaded by tall herbs, for example *Tysonia africana, Anemone fanninii, Senecio isatideus, Helichrysum umbraculigerum, Kniphofia linearifolia* and *Lessertia perennans*. Bracken Veld eventually yields to fynbos.

Protea Savanna. Protea Savanna dominated by *P. multibracteata* and *P. roupelliae* is very limited in extent on the Little Berg; it is generally restricted to koppies and the stony ridges of spurs up to 2135 m, where there is some protection from fire.

Ridge vegetation. The ridges of spurs on the Little Berg above 2440 m support a mixed vegetation, which is markedly xeromorphic. The vegetation consists of dwarf shrubs, cushion and rosette plants, grasses and other herbs. The two dominant dwarf shrubs are *Passerina montana* and *Erica thodei*, supported by *Lotononis* sp. (1191), *Buchenroedera lotononoides, Polygala myrtifolia, Gnida compacta, Muraltia saxicola* and several *Erica* spp. The suffrutices are mainly grey-lanate species of *Helichrysum*, e.g. *H. argentissimum, H. odoratissimum, H. montanum, H. sutherlandii* and *H. setigerum*. Rosette plants are numerous and include chiefly composites, e.g. *Helichrysum alticolum* var. *montanum, H. confertum, H. aureum* var. *scopulosum, H. sessile, Cotula hispida* and also *Zaluzianskya pulvinata* and *Psammotropha myriantha*. Among the grasses are *Merxmuellera stereophylla, Styppeiochloa gynoglossa, Aristida junciformis* subsp. *galpinii, Themeda triandra, Pentaschistis tysonii, Merxmuellera disticha* and *Cymbopogon validus*.

Leucosidea sericea Scrub. *Leucosidea sericea* is a montane element, which occurs in sheltered situations such as streambanks and deep gullies on the Little Berg up to 2040 m. The community results from the invasion of *Cliffortia linearifolia* Scrub, fynbos, *Miscanthidium capense* Grassland and Bracken Veld by *Leucosidea sericea*. Mature *L. sericea* Scrub is dominated by *L. sericea* growing up to 6 m high with the following associates: *Buddleia salviifolia, Philippia evansii, Rhus dentata, Rhamnus prinoides, Olinia emarginata, Halleria lucida, Ilex mitis* and *Bowkeria verticillata*. Usually there is a shrub layer of *Myrsine africana* 1.2 m high. Herbs present include *Myosotis sylvatica, Carex spicato-paniculata, Alchemilla natalensis, Galium rotundifolium, Disperis fanniniae, Berkheya montana* and *Euphorbia epicyparissias*. According to Bews (1917) this scrub is resistant to grass-fires. This resistance is due to the fact that the constituent species, particularly *L. sericea*, regenerate abundantly by sprouting at the base.

Buddleia salviifolia Scrub. This is a parallel community to *Leucosidea sericea* Scrub but, as a rule, occupies drier areas and is associated with *Cymbopogon validus* rather than *Miscanthidium capense*.

Subalpine fynbos. Fynbos is the climax community of the Subalpine Belt. Because of recurrent grass fires, it is limited in extent occurring in situations providing some protection from fire. The most extensive and best developed stands of fynbos are to be found on steep valley and escarpment slopes at the head of the main rivers, for example Tseketseke, Indumeni, Mlambonja and Tugela Rivers (Fig. 7). At lower altitudes fynbos is seral to forest.

The community consists of shrubs between 0.9–3 m tall the majority of which are evergreen, though some may be deciduous. Most of the constituents have small leaves which are ericoid, elliptic or linear, variously coriaceous and glossy or grey-lanate. The vegetation is physiognomically similar to that found in the southwestern Cape. The density of the community varies considerably – from shrubs scattered in grassland to an almost impenetrable tangled mass of vegetation.

Fynbos in the Drakensberg can be either pure with one species dominant or mixed with several dominants (Fig. 8). The most important dominants are probably *Passerina filiformis* (Fig. 9), *Philippia evansii* and the conifer, *Widdringtonia nodiflora*. Pure fynbos stands can be formed by the following species: *Passerina filiformis, Philippia evansii, Widdringtonia nodiflora, Passerina montana, Erica ebracteata, Macowania conferta, Buchenroedera lotononoides, Anthospermum aethiopicum, Rhus discolor, Buddleia loricata, Protea dracomontana, P. subvestita, Syncolostemon macranthus, Calpurnia intrusa* and *Melianthus villosus*.

Shrubs which do not aggregate to any great extent include: *Senecio haygarthii, Asparagus scandens, A. microraphis, Rhus dentata, Cliffortia spathulata, Diospyros austro-africana* var. *rubiflora, Stoebe vulgaris, Euphorbia epicyparissias, Myrsine africana, Artemisia afra, Psoralea caffra, Polygala myr-*

Fig. 7. Subalpine Fynbos on escarpment slopes below the Eastern Buttress in the Tugela Valley. *Podocarpus latifolius* Forest (montane) may be seen below the Cave Sandstone cliffs at bottom left. This photograph takes in a vertical distance of 1524 m.

tifolia, *Erica straussiana, Anisodontea julii* subsp. *pannosa, Berkheya draco, Helichrysum tenax, Lasiosiphon anthylloides* and *Lotononis trisegmentata.*

Very characteristic of fynbos is the cycad *Encephalartos ghellinckii*. It is particularly abundant in the Tseketseke Valley. Subordinate to the shrubs in dense fynbos is a layer of grasses, herbs and ferns. The three most constant species are probably *Polystichum* sp. (981), *Cymbopogon validus* and *Berkheya macrocephala.*

The remaining species are: *Pellaea quadripinnata, Gleichenia umbraculifera, Festuca costata, Pentaschistis pilosogluma, Agapanthus campanulatus, Scilla natalensis, Eriospermum cooperi, Cyrtanthus erubescens, Anemone fanninii, Ranunculus baurii* (usually in wet places), *Gunnera perpensa, Alchemilla natalensis, Geranium pulchrum, Indigofera cuneifolia, I. longebarbata, Sebaea macrophylla, Diclis reptans, Helichrysum setosum, H. cooperi* and *H. umbraculigerum.*

Climbers include *Riocreuxia torulosa* var. *tomentosa, Dioscorea sylvatica* and *Clematis brachiata.*

Fig. 8. Mixed Fynbos at 2070 m in the Tseketseke Valley. The constituents are *Macowania conferta, Encephalartos ghellinckii, Widdringtonia nodiflora, Diospyros austro-africana* var. *rubriflora, Rhus dentata, Syncolostemon macranthus, Polemannia montana, Anemone fanninii, Cymbopogon validus, Berkheya macrocephala* and *Polystichum* sp. (981).

3.1.2 Lesotho

According to Jacot Guillarmod (1971) the Subalpine Belt in Lesotho lies between 2290–2900 m with variations lower or higher according to aspect.

3.1.2.1 Environment

Soils. The soils of the Subalpine Belt have been classified by Carroll & Bascomb (1967) as: 1, lithosols on lava, where the soils are shallow and stony on the higher mountain slopes, and their development is limited by steepness and a harsh climate; 2, lithosols on lava/basalt rock debris covering much of the gently undulating Upland Plateau, where much bare rock is exposed; 3, lithosols on lava/calcimorphic soils on many of the lower mountain slopes. Further details are obtainable from their publication.

Climate. Climatic data for the Subalpine Belt are meagre. Data for two stations, at Mokhotlong (2377 mm) and Oxbow (2591 mm) are relevant. Mean annual

Fig. 9. *Passerina filiformis* Fynbos in the upper Indumeni Valley. Associates present are *Helichrysum tenax, Athanasia punctata, Buddleia loricata, Senecio haygarthii* and *Berkheya draco*. The herb layer consists of *Cymbopogon validus, Pentaschistis pilosogluma* and *Berkheya macrocephala*. The outcrops in the background support *Xerophyta viscosa, Cymbopogon validus, Scilla natalensis, Buchenroedera lotononoides* and *Helichrysum sutherlandii*.

rainfall at Mokhotlong is 560 mm and Oxbow 1277 mm, most of the rain falling between October and March. Rain-shadow effects probably account for the lower rainfall of Mokhotlong.

Snow may occur at any time of the year, but especially between May and September and the higher parts are snow-covered for much of winter, particularly the southern slopes.

Hail is frequent in summer, but seldom with stones of sufficient size to damage vegetation extensively, except in highly localized areas (Jacot Guillarmod 1971).

At Mokhotlong the mean temperature during June is about 9°C with mean maxima and minima of 13.9°C and −4.1°C respectively (Bawden & Carroll 1968). In January the mean temperature is 16.6°C with mean maxima and minima of 23.9°C and 9.3°C respectively. Extreme temperatures as high as 35°C and as low as −12.5°C have been recorded. The average duration of the frost period is 177 days. At Oxbow the mean temperature during June is 1.8°C and during January 12.2°C (Herbst 1971). Absolute maximum and minimum temperatures are 21.6°C and −13.1°C respectively. It is clear that the temperature regime at the higher altitude of Oxbow is lower than that of Mokhotlong.

Average relative humidities at Mokhotlong vary from 62 per cent in September to 83 per cent in July; at Oxbow from 38 per cent in September to 66 per cent in March.

Fire. Fire is of regular occurrence in the mountain grasslands and is usually deliberate. The apparent immediate effect of fire, the provision of fresh new grass growth is the aim, but the long-term effects are either not understood or are subordinate to the immediate need for grazing (Jacot Guillarmod 1971).

3.1.2.2 *Vegetation*

The Subalpine Belt is almost entirely covered by grassland, viz. *Themeda–Festuca* Grassland (Acocks 1953). Staples & Hudson (1936) give more details about this grassland (Fig. 10). Between 2135–2750 m on northern slopes *Themeda triandra* (Seboku) is dominant with *Festuca caprina* (Letsiri) dominant above 2750 m. On these northern slopes the grasses are often replaced by *Chrysocoma tenuifolia* (Sehalahala) and to a less extent by *Felicia filiformis* as a result of overgrazing. *Chrysocoma tenuifolia* was estimated by Staples & Hudson (l.c.) as covering more than 13 per cent of the mountain area of Lesotho. On southern slopes *Themeda triandra* only predominates up to 2135 m with *Festuca caprina* Grassland above this altitude. On eastern and western slopes *Themeda triandra* Grassland attains altitudes of about 2590 m, while *Festuca caprina* Grassland tends to extend downwards to about 2515 m. According to Jacot Guillarmod (1971) with intensive grazing the mountain grasslands are rapidly changing in composition, more stoloniferous species, especially *Harpochloa falx* taking over, and also more tufted and tussocky and coarse unpalatable species such as *Aristida*. *Harpochloa falx* is found on sunny slopes and *Aristida* on the sunless slopes. Species of *Senecio* and *Helichrysum* are also becoming more prevalent and the present writer has noticed that *Catalepis gracilis* is taking over many disturbed areas such as old lands and kraal sites. Unfortunately grassland analyses by Herbst (1971) in the Tsehlanyane Valley at Oxbow do not give a true picture of the composition of Subalpine Grassland, because the analyses were carried out at altitudes between 2560–3170 m, thus including both Subalpine and Alpine Belts.

3.1.3 Amatole Mountains

These mountains lie to the north and northwest of Keiskammahoek in the eastern Cape and were included in an ecological study of the Keiskammahoek District by Story (1952). The Subalpine Belt in the Amatole Mountains lies between 1220–1938 m.

3.1.3.1 *Environment*

Soils. The two geographical formations present are the Beaufort Series and Dolerite. The soils conform very closely to the geological structure. They are residual and formed in situ. The sedimentary soils may be divided into shallow grey loams with ferruginous concretions from a few centimetres to a few metres below; then there are grey loams on clay, which are deeper soils, deficient in plant

Fig. 10. *Themeda–Festuca* Grassland in the Mokhotlong River Valley in eastern Lesotho.

foods and finally there are yellowish-brown sandy loams on sandstone. These three make up about 58 per cent of the total area. The dolerite soils comprising about 38 per cent, consist of immature black clays 30–60 cm deep containing boulders; then there are deep red clays, chocolate when virgin and red after cultivation, 2 m deep and more and finally there are black, well-developed clays forming more or less level plains.

Climate. A paper on the climate of the Keiskammahoek region by Pilson & Higgs (MS) gives the following main characteristics of the climate: 1. the 'best' rains fall in March; 2. set-in rains are probably from the south-east; 3. snow on the mountains falls three or four times in winter. It rarely extends below 1220 m; 4. the northwester is a winter wind, almost invariably hot and dry.

Story (1952) established two temperature stations in the Subalpine Belt, one on the slopes of the Hogsback at 1615 m and the other on the Wolf Plateau at 1463 m. The data from these two stations (covering one year only) are very similar, so only those from the Hogsback will be analysed in the form of a Deasy chart (Fig. 2.2). The curves for mean diurnal fluctation B_1 and C_1, show that the temperature is cool–mild. As at Cathedral Peak only the absolute values represented by curves A_1 and D_1, reach degrees of hot and cold. The absolute maximum temperature recorded was 33°C in January 1949 and the absolute minimum −2°C in September 1949. Frosts are commonplace during winter, sometimes freezing the soil.

Five rain-gauges were set up above 1400 m. The annual totals for the period 18/8/48–3/11/49 varied from 540 mm at 1420 m to 670 mm at 1620 m.

Fire. Story (1952) considered that the mountain slopes above the forests were originally under fynbos (=Story's macchia) and that at some time in South Africa's prehistory man started firing the slopes. The result was that fynbos became sparse and the fire-resistant grasslands obtained a footing and spread, until the fynbos was confined to sheltered places protected from fires. With moderate grazing and frequent fires, these grasslands remained stable. However, as the European populations and their stock increased, the grasslands were cropped short and fires became weaker and less frequent. Fynbos began migrating out of its strongholds, the patches linked, and within a comparatively short time, it had re-established itself over much of its former area. Story states that it is doubtful whether natural fires could have occurred sufficiently often to keep the vegetation at the grassland stage.

3.1.3.2 *Vegetation*

(a) *Highland sourveld.* This grassland type is very limited in extent in the Amatole Mountains. At lower altitudes it is dominated by *Themeda triandra* with typical constituents like *Harpochloa falx, Tristachya hispida, Alloteropsis semialata, Andropogon appendiculatus, Trachypogon spicatus* and many others (cf. grassland analyses at Cathedral Peak, Table 3). Forbs include *Alchemilla* spp., *Argyrolobium speciosum, Aristea schizolaena, Aspalathus* spp., *Aster bakeranus, Centella glabrata, Haplocarpha scaposa, Helichrysum* spp., *Hypoxis argentea* etc. With increase of altitude, where the soil becomes darker with a peaty texture, near the top of the Amatole Range, the following temperate grasses become more prominent: *Agrostis barbuligera, A. bergiana, Bromus speciosus, Brachypodium flexum, Festuca caprina* var. *irrasa, F. costata, Pentaschistis* sp., *Poa binata.*

Forbs include *Anthospermum aethiopicum, A. herbaceum, Athrixia phylicoides, Berkheya decurrens, Cineraria* spp., *Diascia rigescens, Eucomis* sp., *Geranium ornithopodum, Helichrysum* spp. and Restionaceae.

(b) *Outcrop Communities.* Along the escarpment edge there are outcrops of rock with a thin and scanty soil. *Aristida* species predominate accompanied by dwarf shrubs, rosette plants, Restionaceae and succulents. Characteristic plants are: *Arrowsmithia styphelioides, Aspalathus* spp., *Chrysocoma tenuifolia, Erica* spp., *Euryops dyeri, Helichrysum* spp., *Delosperma* spp., *Muraltia macroceras, Phylica galpinii, Passerina montana* and *Ursinia montana* subsp. *apiculata.*

(c) *Vlei vegetation.* This comprises *Cliffortia serpyllifolia, Pteridium aquilinum, Helichrysum fulgidum, Kniphofia* sp., *Leonotis leonurus, Nidorella auriculata, Zantedeschia aethiopica, Anthoxanthum ecklonii, Miscanthidium capense, Pennisetum macrourum, Pentaschistis* sp., *Juncus lomatophyllus, Scirpus* spp. and *Tetraria* spp.

(d) *Helichrysum argyrophyllum community.* This community occupies about 1670 ha, usually above 1220 m (Fig. 11). The community has established itself as a result of the destruction of the original grasslands. It can be controlled by burning and resting with the subsequent reappearance of grass.

(e) *Subalpine fynbos.* This is the climax community of the Subalpine Belt. According to Story (1952) fynbos is dominated by *Cliffortia paucistaminea* growing up to 2 m high and *Erica brownleeae* up to 4 m high (Fig. 11). The fynbos grows best on south-facing mountain slopes and grows in such an interlacing mass that it is practically impenetrable by man or stock. Other constituents of the fynbos are

Fig. 11. Fynbos (*Erica brownleeae* and *Cliffortia paucistaminea*) with *Helichrysum argyrophyllum* in the foreground on the Wolf Plateau in the Amatole Mountains.

Protea lacticolor, Bobartia gracilis, Rubus spp., *Pteridium aquilinum, Stoebe* spp., *Passerina* spp. and *Metalasia muricata*.

3.1.4 Malawi

As indicated on p. 522 the vegetation of the uppermost slopes of Mt. Mlanje can probably be described as subalpine. Environmental data for this region are meagre. Chapman (1962) describes the soils as black and peaty and Chapman & White (1970) state that frosts occur regularly each year.

3.1.4.1 Vegetation

The vegetation is classified by Chapman & White (1970) as montane shrubland and montane shrubby grassland consisting of *Costularia nyikensis, Merxmuellera davyi, Eragrostis volkensii* and the sedge, *Coleochloa setifera, Aloe? mawii, Xerophyta splendens, Anthospermum* spp., *Blaeria kivuensis, Crassula sarcocaulis, Diplolophium buchananii, Erica johnstoniana, E. whyteana, Ericinella* sp., *Helichrysum densiflorum, Muraltia flanaganii, Phylica tropica, Plectranthus crassus, P. sanguineus* and *Thesium whyteanum*. Chapman (1962) describes the boggy hollows as dominated by *Restio mahonii* (see Chapter 11).

3.2 Alpine Belt

As already indicated the Alpine Belt lies between about 2860 and 3484 m, the altitude of Thabana Ntlenyane, the Pretty Little Mountain, the highest point in the

Drakensberg and, indeed, in southern Africa. The fixing of the lower limit deserves further study. The Belt occupies a narrow strip along the top edge of the Drakensberg escarpment and extends downwards into Natal, the Cape and Lesotho. Parts of the escarpment lying well below 2860 m for example Qacha's Nek (2167 m) and Bushmans Nek (2438 m) obviously do not fall within the Belt. The Belt is also found on outlying peaks such as Cathedral Peak (3004 m), the Inner and Outer Horns (3018 and 3009 m respectively) and the Pyramid (2828 m), and in Lesotho on the Maluti and other mountains of sufficient height.

3.2.1 Environment

The summit plateau of the Drakensberg presents a cheerless, bleak and rather barren-looking picture to the observer. In parts the plateau forms part of the Cretaceous post-Gondwana landsurface, while the higher parts above 3075 m belong to the Gondwana Cycle of Jurassic Age (King 1972, and Chapter 1). The monotony of the landscape is broken at intervals by the presence of bogs or sponges (mokhoabo), which are a feature of the riverheads of the Orange River System on the eastern escarpment and have caused cartographers to describe the area as 'barren and boggy wastes'.

Soils. Soil studies of the Alpine Belt are few and there has been little collation of results. Carroll & Bascomb (1967) classified the soils chiefly as lithosols on lava. Van der Merwe (1941) described the soils as Mountain Black clays, which are of residual and colluvial origin and are derived entirely from basalt. Venter (1938), who studied the soils near Mont aux Sources, described the soils as varying in colour from brown through chocolate-brown to black, depending mainly on the amount of leaching and oxidation. The soils are thin and never exceed 45 cm in depth. Venter found that there were differences in the soils of southern and northern aspects: slopes with a southern aspect are blacker in colour and contain a higher percentage of moisture, K_2O, P_2O_5 and humus. Harper (1969) stated that above about 3080 m on the Injasuti and Thabana Ntlenyane there is no soil at all, but only rubble. Herbst & Roberts (1974) give the following soil properties at the summit of the Tsehlanyane Valley (c. 3050 m): pH 5.6; resistance (ohms) 3008; organic matter (%) 19.9; coarse sand (%) 17.6; fine sand 31.8; silt 15.1 and clay 15.7.

During summer the soils of the summit become very wet and water-logged, while during winter, and even during summer, according to Van Zinderen Bakker & Werger (1974), the humid soils are subjected to freezing every night and thawing every day on exposed sites (this observation deserves further study). There is much evidence of cryonival phenomena in the Alpine Belt. The most conspicuous phenomenon is the formation of needle-ice ('pipkrake', 'Kammeis'). The needle-ice heaves soil, stones and plants and produces an unstable habitat for plants. Common on the summit are what Schelpe (1946) called 'mud-patches'; these are moist, almost bare patches of soil which are subject to frost heaving. On thawing, the soil in these patches exhibits a peculiar 'raked' appearance which Troll (1944) ascribed to wind action. Other cryonival phenomena are described fully by Harper (1969). He states that the effects of solifluction are very evident in the frost-wedged boulders, which have been transported downhill and strewn below the basalt steps, and in the presence of colluvial terraces. Terracette formation is common and stone rings and polygons are frequent on the Thabana Ntlenyane massif above 3350 m. Miniature polygons have been reported by Van Zinderen Bakker (1965) for the Oxbow area at 3100 m.

A feature of bogs in Lesotho are the raised hummocks, which have a diameter of 50–70 cm and a height of 20–30 cm. These hummocks were first reported by Jacot Guillarmod (1963), and in 1965 Van Zinderen Bakker suggested that they were homologous to the rather larger and frost induced thufur of Iceland described by Thorarinsson (1951). Harper (1969 supported Van Zinderen Bakker's view. Killick (1963) described such hummocks for the Langalibalele Pass area near Giants Castle but, because of the presence of freshly-made earth hummocks and the existence of a mole-rat (*Cryptomys natalensis natalensis*) or true mole (*Chlorotalpa guillarmodii*) in the bogs, he attributed their formation to animal activity.

Climate. The climate of the Alpine Belt, as might be expected, is severe. Air temperature data kindly supplied by the Director of Hydrological and Meteorological Services, Lesotho, for the weather station at Letšeng-la-Draai at 3050 m in northeastern Lesotho have been used to produce the Deasy chart in Fig. 12. The curves for mean diurnal fluctuation, B and C, demonstrate that air temperature is cool to mild in summer, but cold in winter. The absolute temperatures, curves A and D, reach degrees of hot and frigid. The highest temperature recorded in 11 years of recording is 31°C on 29 January 1972 and the lowest, −20.4°C on 12 June 1967. Mean annual temperature is 5.7°C. The estimated mean annual frequency of days with minimum temperature below 0°C (i.e. frost days) is 183. Unfortunately there are no temperature data for ground/grass level or soil.

Rainfall data are available for three stations in eastern Lesotho. Mean annual rainfall at Sani Pass summit (2865 m) is 995.8 mm, at Organ Pipes Pass (2927 m) 1609 mm and Letšeng-la-Draai (3050 m) 713.6 mm. Annual rainfall can vary considerably from year to year: for example at Sani Pass summit the mean annual rainfall for the hyetal year 1938/39 was 1441.6 mm and for 1944/45, 439.3 mm. Most of the rain (82–88 per cent) falls between October and March during spring and summer. Droughts occur periodically: at Sani Pass summit there were three droughts of 120 days duration between 1932 and 1947: May–August 1937 (rainfall 11 mm); May–August 1941 (rainfall 20.1 mm) and June–September 1945 (rainfall 6.1 mm).

According to Shand et al. (1963) the effect of the steep mountain slopes of the Maluti Mountains is to cause the northwesterly winds, which contain only limited moisture, to drop it as orographic rain. They then continue southeastwards as dry winds producing a rain-shadow over the interior mountain region. The extremely moist southeasterly winds from the Indian Ocean produce heavy orographic rainfall on the eastern slopes of the Drakensberg mountains but, because of the 'climatic barrier', these rain-bearing winds rarely reach the interior of the mountains.

Relative humidities are high during the wet season when there is an abundance of cloud, but low during the dry season and in early spring. Humidity data are available for Letšeng-la-Draai (3050 m) for the period 1966–1976. The mean relative humidity for January is 54 per cent with the highest monthly mean being 72 per cent in January 1974. The mean humidity for August is 37 per cent, the lowest monthly mean being 18 per cent in August 1969; in September the corresponding figures are 34 and 21 per cent in September 1969. Schelpe (1946) recorded a relative humidity of 4 per cent in spring during windy weather on the summit of the Drakensberg.

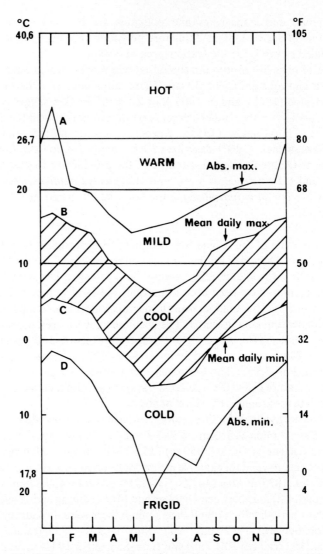

Fig. 12. Temperature chart for Letšeng-la-Draai in northeastern Lesotho.

High winds occur at any season during the year, but are most common in late winter and spring. The winds, often attain considerable velocities. They are important ecologically, because they blow at a time when both soil moisture and water-supplying power of the soil are low.

Snow falls mainly during July and can lie for periods of up to two months especially on the southern slopes. The blanket of snow protects the plants from exceedingly low temperatures and prevents the soil beneath from freezing. It also adds to the supply of soil water. Snow is also important because of its capacity for reflecting both light and heat. Evergreen alpine plants must clearly be adapted to withstand periods of high light intensities. Other ecological aspects of snow in the Drakensberg are discussed by Killick (1963).

Fire. What has been said about fire on pages 525, 528 and 542 is also applicable here.

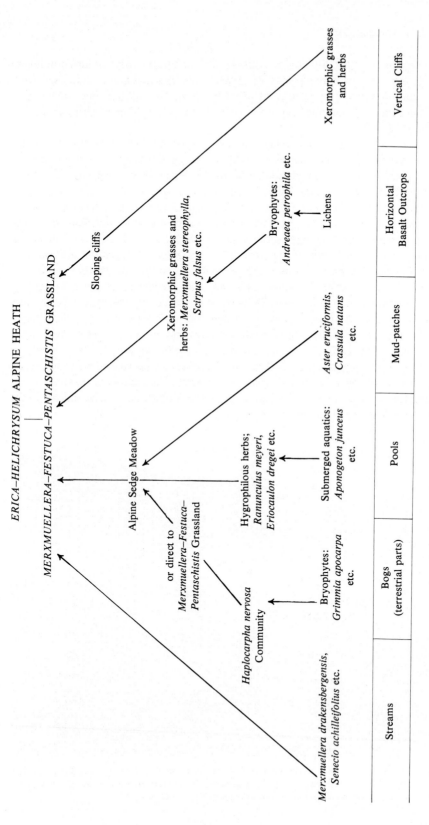

Fig. 13. Suggested interrelationships of the plant communities in the Alpine Belt of the Drakensberg and Lesotho.

3.2.2. The vegetation

The vegetation of the Alpine Belt consists of climax heath communities dominated chiefly by low, woody species of *Erica* and *Helichrysum* interspersed with grassland dominated by species of *Merxmuellera, Festuca* and *Pentaschistis*. In addition, there are aquatic, hygrophilous and lithophilous communities varying in extent (Fig. 13).

The vegetation as a whole reflects the severity of the climate: most of the plants exhibit xeromorphic features of some kind. The heath constituents are evergreen dwarf shrubs with small, ericoid, filiform or linear leaves. Species like *Erica dominans, E. glaphyra* and *Passerina montana* have hard and more or less glossy leaves, while *Helichrysum trilineatum, Eumorphia sericea* and *Athanasia thodei* have softer, grey-lanate leaves. The grasses are mostly short with filiform leaves. Cushion plants are common, some examples being *Helichrysum retortoides, H. aureum* var. *scopulosum, H. milfordiae, H. splendidum, Muraltia saxicola* and *Zaluzianskya pulvinata*. Among the perennial rosette plants are *Berkheya multijuga, Cotula hispida, Ursinia montana* and *Hirpicium armerioides* subsp. *armerioides*. Reduction in size with increase of altitude is strikingly illustrated by *Juncus exsertus*. At 1860 m it is 45 cm high, but at 2990 m it is only 5 cm high.

The vegetation of the Alpine Belt is remarkably homogeneous. During May 1961 the author walked along the summit from the Organ Pipes Pass to Mont aux Sources, a distance of about 64 km, and found very little variation in the vegetation. The main variation found was in the presence or absence of heath.

The following account of the vegetation of the Alpine Belt is based chiefly on the work of Killick (1963) at Cathedral Peak.

Fig. 14. An alpine bog with raised hummocks or thufur in the Maluti Mountains, 30 km N.W. of Mokhotlong, eastern Lesotho. The white composite is *Athrixia fontana*. Dark patches in the background are *Helichrysum trilineatum* heath (photo M. J. A. Werger).

Aquatic and Hygrophilous Vegetation. On the undulating summit area of the Drakensberg and in Lesotho there are numerous bogs and streams, which form the headwaters of the Tugela and Orange River systems. The bogs are of two types: the first, limited in area, is formed in seepage areas or flushes on mountain slopes; the second occurs in the cirque-like riverheads, where rather extensive swampy areas containing hummocks or thufur are formed (Fig. 14) – this type resembles miniature sawah areas. Both types are peat-producing. The bogs have been described by Van Zinderen Bakker (1955, 1965), Jacot Guillarmod (1962, 1963), Killick (1963) and the latter type in some detail by Van Zinderen Bakker & Werger (1974). According to Van Zinderen Bakker & Werger (l.c.) the alpine bogs of Lesotho and the Drakensberg are of late- or Post-glacial age. The oldest radiocarbon date obtained so far is 8020 ± 80 BP. The average thickness of peat layer deposited per year is only about 0.25 mm, which shows that the production of organic material is very limited in the cold alpine conditions.

The aquatic and hygrophilous vegetation may be divided into that of pools, bogs (terrestrial parts), mud-patches and streams.

(a) Pool communities: Permanent pools on the summit are rare and usually occur in bogs, but temporary ones are common in summer. Aquatics present are *Aponogeton junceus*, *Nitella dregeana*, *Limosella capensis*, *Lagarosiphon muscoides* and *Crassula inanis*. Several plants grow around the periphery of these pools, namely *Ranunculus meyeri*, *R. baurii*, *Senecio cryptolanatus*, *Utricularia* spp. (fide Jacot Guillarmod 1962), *Eriocaulon dregei*, *Colpodium* sp. nov. (Killick & Vahrmeijer 3730), *Poa binata* (dwarf form) and *Kniphofia caulescens*.

Fig. 15. Alpine Grassland with conspicuous *Euryops evansii* Community in the Tsanatalana Valley.

(b) Mud-patch communities: Scattered abundantly on the summit plateau are mud-patches described on p. 546. The mud-patches are moist throughout summer, sometimes with about 1.5 cm of water and in winter are subject to frost action. Several short plants form pure communities on these patches. They are *Aster eruciformis, Psammotropha alternifolia, Crassula vaillantii, Rhodohypoxis rubella, Limosella capensis* var., *L. longiflora* and *Juncus exsertus*. Often associated with the dominants are *Moraea* sp. (2191), *Senecio tugelensis, Ranunculus meyeri, R. multifidus* and *Felicia uliginosa*.

(c) Terrestrial bog communities: These communities show up conspicuously as shiny green carpets of semi-aquatic vegetation either in seepage areas or flushes or on thufur. Several mosses contribute initially to the formation of these carpets, the most important being *Grimmia apocarpa* and *Bryum aulacmnioides*. Others include *Grimmia drakensbergensis, Dicranella symonsii, Bartramia substricta, Mnium rostratum* and *Barbula* spp. (Van Zinderen Bakker 1955). Eventually the mosses are succeeded by a community dominated by *Haplocarpha nervosa*, a rosette herb with spathulate leaves and yellow flowers. Associates include *Scirpus fluitans, Limosella longiflora, Athrixia fontana, Agrostis huttoniae, Ericaulon dregei* and more rarely *Berkheya multijuga, Kniphofia caulescens* and *Helichrysum palustre*.

Van Zinderen Bakker & Werger (1974) classified the pool and bog communities as follows. In pools and tarns there is the Crassuletum inanis association, a community of aquatic plants. The community of very wet parts of the bogs was distinguished as the Ranunculetum meyeri association comprising two sub-associations, the aponogetonetosum subassociation in shallow stream channels with the water table at or just above the surface and the haplocarphetosum sub-association on sites with frequent sheet flow. The community of somewhat higher, drier parts is called the Senecionetum cryptolanati association, which comprises a subassociation merxmuelleretosum occurring on sites with the lowest water table. The Ranunculetum meyeri and Senecionetum cryptolanati associations are combined into the Scirpo-Limosellion longiflorae alliance.

(d) Streambank communities: The streams on the summit flow either into Lesotho or into Natal depending upon the aspect of the catchment. They are small, clear and frequently littered with grey boulders. The principal communities are formed by *Merxmuellera drakensbergensis, Kniphofia caulescens, Berkheya multijuga, Polygonum* sp., *Senecio achilleifolius, S. cryptolanatus, Juncus exsertus, Euryops montanus, Moraea alticola, Erica alopecurus, E. thodei* and the orchids *Neobolusia virginea* (on vertical banks) and *Satyrium fanniniae*. Also important is *Scirpus ficinioides*, a tall thick-stemmed sedge about 1 m high, which is frequently dominant on small islands and on moist flats adjacent streams. It is most abundant in the Langalibalele and Mont aux Sources areas, where it occupies many hectares of moist ground.

(e) Alpine sedge meadow: This community is found on moist, but firm ground, often at the edge of bogs. The constituent plants form a close continuous turf 5–10 cm high. The dominants are two sedges *Carex monotropa* and *C. killickii*. Important associates in order of abundance are *Sebaea thodeana, Geranium incanum, Schoenoxiphium filiforme, Carex glomerabilis, Trifolium burchellianum, Lobelia galpinii, Cotula paludosa, Alepidea pusilla, Scirpus diabolicus, Luzula africana, Ornithogalum paludosum, Helichrysum subglomeratum, Moraea* spp., *Eragrotis caesia, Agrostis subulifolia* and *Koeleria cristata*.

Horizontal outcrops. Horizontal basalt outcrops are fairly extensive in the Alpine Belt; they are usually found along the edge of the escarpment, adjacent streams and on the summit of high peaks.

(a) Cryptogamic communities: The pioneers are crustaceous lichens, followed by foliose species. Dendroid lichens are surprisingly rare. Mosses are the next cryptogams to appear. The chief saxicolous species according to Schelpe (1953) are *Andraea petrophila, Campylopus trichodes, Grimmia commutata* var. *brevipes, G. drakensbergensis, G. pulvinata, Ptychomitrium cucullatifolium, Brachyhymenium dicranoides* and *Anoectangium wilmsianum.* On broken outcrops are *Encalypta ciliata, Macromitrium tenue, Bartramia hampeana, Bryum argenteum* var. *lanatum, Fabbronia perciliata* and *Thuidium promontorii.*

(b) Xeromorphic herb communities: Probably the most important plant forming communities on basalt is *Merxmuellera stereophylla,* a very tufted grass closely related to and previously confused with *M. drakensbergensis,* a streambank species. *Scirpus falsus* (dwarf form) also occupies fairly extensive areas usually rather moister than *M. stereophylla.* It is particularly common on the southeast slopes of Cleft Peak. Three semi-woody plants which form cushions on rock surfaces are *Helichrysum retortoides, Euryops decumbens* and *Muraltia saxicola.* The remaining herbs, frequently occurring as chasmophytes, are *Pellaea calomelanos, Eragrostis caesia, Bulbostylis humilis, Psammotropha mucronata, P. alternifolia, Crassula schimperi* var. *lanceolata, Oxalis obliquifolia, Erica frigida, Glumicalyx montanus, Ursinia montana, Helichrysum setigerum, H. alticolum* var. *montanum, Delosperma deleeuwiae, Zaluzianskya* spp. and *Craterocapsa* spp.

Cliff vegetation. The basalt cliffs of the main escarpment up to 460 m high in parts present a variety of rock habitats – sheer faces, crevices, pockets, ledges, overhangs and moist areas near waterfalls. Owing to the inaccessibility of these cliffs, it is not possible to present an adequate picture of the cliff vegetation.

Much of the rock surface is bare, but a considerable portion is covered with alpine grassland and occasionally alpine heath. Species common on the cliffs are *Merxmuellera stereophylla, Xerophyta viscosa, Helichrysum montanum, H. aureum* var. *scopulosum* and in moist sheltered situations *Ranunculus baurii, Galtonia viridiflora* and the attractive red-flowered *Gladiolus cruentus.*

Small cliffs on the summit plateau support *Merxmuellera stereophylla, Helichrysum milfordiae, H. pagophilum, Wahlenbergia pulvillus-gigantis* and *Psammotropha alternifolia.* At their base are distinct communities formed by *Kniphofia northiae, K. ritualis, Cyrtanthus flanagani* and *Zaluzianskya longiflora,* usually mixed with alpine grasses. *H. milfordiae* often occurs on horizontal slabs of basalt at the base of small cliffs.

Overhangs contain their own characteristic flora. Usually this habitat is damp and shaded. The floors are frequently covered with the thalli of *Plagiochasma* sp., tufts of *Bryum argenteum* var. *lanatum* (Schelpe 1953) and herbaceous plants such as *Helichrysum milfordiae* which forms extensive carpets with *Crassula setulosa* var. *curta* growing up between the grey rosettes, *C. harveyi, Ranunculus meyeri, Zaluzianskya longiflora* and the fern *Woodsia montevidensis* var. *burgessiana.* The walls support the mosses *Webera depauperata* and *Fissidens latifolius* (Schelpe l.c.).

Merxmuellera–Festuca–Pentaschistis grassland. The grassland on the summit of the Drakensberg possesses an irregular physiognomy: in parts it is low, even and turf-like, while in others it is fairly tall, uneven and open. Everywhere the grassland is interrupted by mud-patches. Unfortunately the quantitative results obtained by Herbst (1971) at Oxbow in Lesotho cannot be cited here because the 56,000 points he used were distributed throughout the Tsehlanyane Valley between 2590–3050 m through a number of different plant communities occurring in both the Subalpine and Alpine Belts.

The dominants are three grasses belonging to temperate genera, namely *Merxmuellera disticha*, *Festuca caprina* and *Pentaschistis oreodoxa*. They occur in pure or mixed communities. All three grasses are xeromorphic and become dormant during the winter months. *Merxmuellera disticha* and *Festuca caprina* are very densely tufted and have filiform leaves. There seem to be two forms of *Merxmuellera disticha* on the summit, the one rather tall and lax and the other dwarf with the old leaves pronouncedly recurved. *Pentaschistis oreodoxa* has short, flat, hairy leaves which recurve with age. It seems to prefer stony areas of the summit.

Grass associates are *Koeleria cristata, Pentaschistis galpinii, P. imperfecta, P. natalensis, Harpochloa falx, Poa binata, Eragrostis caesia, Merxmuellera guillarmodiae, M. drakensbergensis*, and *Anthoxanthum ecklonii* – a meagre number of species when compared with the grasslands of lower altitudes.

Forbs are more abundant than imagined by previous workers, for example Schelpe (1946) and West (1951). Unfortunately it was not possible for the author to collect on the summit during August and September, otherwise the following list indicating time of flowering is fairly representative:

October

Moraea sp. (1854)
Basutica aberrans
Sebaea thodeana
Muraltia saxicola

Felicia rosulata
Helichrysum argentissimum
H. retortoides

November, December

Moraea alticola
M. alpina
Ornithogalum paludosum
Kniphofia caulescens
K. northiae
K. ritualis
Neobolusia virginea
Dianthus basuticus subsp.
   basuticus var. *grandiflorus*
Heliophila suavissima
Lobelia flaccida var. *hirsuta*
Alepidea galpinii

Geum capense
Lotononis galpinii
Thesium sp. (1883)
Basutica aberrans
Valeriana capensis
Euryops evansii
Senecio barbellatus
S. gramineus
Helichrysum flanaganii
H. argentissimum
Felicia rosulata
Eumorphia sericea

January–March

Dierama igneum
Moraea trifida
M. albicuspa
M. dracomontana
Disa fragans
Brownleea macroceras
Satyrium neglectum
Monadenia basutorum
Corycium nigrescens

Diascia capsularis
Craterocapsa montana
Wahlenbergia undulata
W. monotropa
Helichrysum argentissimum
H. setigerum
H. aureum
H. bellum
Senecio tugelensis

Cerastium arabidis
Psammotropha alternifolia
Crassula setulosa
Lessertia thodei
Lotononis galpinii
Alepidea thodei
Erica alopecurus
Romulea campanuloides
Alchemilla woodii
Rhodohypoxis baurii

S. cryptolanatus
Berkheya multijuga
Athrixia angustissimum
Hirpicium armerioides subsp. armerioides
Cotula hispida
Cenia macroglossa
Conium maculatum
Felicia linearis
Macowania sororis
Glumicalyx flanaganii

April, May

Oxalis obliquifolia
Scabiosa columbaria
Hirpicium armerioides subsp.
    armerioides
Ursinia montana

Helichrysum odoratissimum
H. subglomeratum
H. adenocarpum
Gymnopentzia bifurcata

July

Helichrysum odoratissimum

From the above list it will be seen that summer to early autumn is the period of maximum profusion of alpine forbs. In spring and late autumn there is very little in flower and in winter practically nothing. Schelpe (1946) states that in September, a spring month, almost the only plant in flower in quantity is *Hesperantha modesta*.

Many of the forbs form conspicuous and sometimes large socies in grassland. *Moraea alticola* is perhaps the most striking plant on the summit in summer. An iridaceous plant about 0.9 m high, with large yellow flowers, it forms dense communities in the Tsanatalana Valley. *Euryops evansii* is the tallest plant in the Alpine Belt, sometimes attaining a height of 1.2 m (Fig. 15). In winter after a heavy snowfall it is frequently the only plant visible above the surface of snow.

Fig. 16. *Erica dominans* Heath on the summit of the Drakensberg south of Mont aux Sources at c. 3048 m.

Berkheya multijuga is common in moist parts of the grassland, likewise the red and white-flowered *Kniphofia caulescens*. During January and February the orchids are a prominent feature of the flora. They include *Disa fragrans, Satyrium neglectum, Brownleea macroceras, Corycium nigrescens* and *Monadenia basutorum*. Grey communities of *Helichrysum* are common, the principal species being *H. argentissimum, H. odoratissimum* and *H. flanaganii*. Plants occurring casually without aggregating include *Dierama igneum, Gladiolus longicollis, Cerastium arabidis, Dianthus basuticus* and *Senecio tugelensis*.

In the passes leading to the summit, for example the Organ Pipes Pass, several grasses and forbs occur in *Merxmuellera–Festuca–Pentaschistis* Grassland, which are not normally found on the summit. These plants include the grasses *Aristida monticola* and *Brachypodium flexum*, the petaloid monocotyledons *Huttonaea grandiflora, Eucomis bicolor* and *E. humilis* and the dicotyledons *Bupleurum mundii, Diascia barberae, Lobelia preslii, Senecio macroalatus* and *S. bupleuroides*.

Between 2440–2745 m on the Lesotho side of the escarpment in the Cleft Peak area, i.e. adjacent to *Merxmuellera–Festuca–Pentaschistis* Grassland, *Themeda triandra* appears again, frequently in pure stands. The grass has short recurved leaves and becomes intensely red during autumn. Accompanying *Themeda triandra* are *Koeleria cristata* and *Harpochloa falx*, and in disturbed areas *Cynodon hirsutus* and *Catalepis gracilis*.

Fig. 17. Alpine vegetation on the summit of the Drakensberg looking north towards Cathedral Peak. *Erica–Helichrysum* Heath in the foreground (photo I. J. Cuthbert).

Fig. 18. Boulder-field Heath at 2926 m in the Tsanatalana Valley. The tussock grass on the streambank is *Merxmuellera drakensbergensis*.

Erica–Helichrysum alpine heath. This is the climax community on the summit of the Drakensberg. The community consists of dwarf shrubs 15–60 cm high. The term heath is used in preference to fynbos since it connotes a much shorter community. Warming (1909) defines heath as a 'treeless tract that is mainly occupied by evergreen slow-growing, small-leaved dwarf shrubs and creeping shrubs which are largely Ericaceae (ericaceous heath)'. But for fire, heath would probably occupy greater areas of the summit. Parts of the summit support no heath at all.

The dominants which cover the largest areas of the summit belong to the genera *Erica* and *Helichrysum*, hence the name *Erica–Helichrysum* Alpine Heath. Altogether there are five distinct heath communities.

(a) *Erica dominans* heath: This community is the most extensive of the heath communities. The dominant, *Erica dominans*, is a dwarf shrub 5–45 cm high with minute leathery, closely adpressed leaves. The plant has an olive-green appearance and attains full flower in October. Occurring on level portions of the summit it forms fairly dense communities invariably interspersed with alpine grasses (Fig. 16). Other constituents of this community are *Helichrysum trilineatum, Erica frigida, E. glaphyra, E. flanaganii, Chrysocoma tenuifolia, Thesium imbricatum, Cliffortia browniana, Gnidia polystachya* var. *congesta, Lotononis galpinii, Clutia nana, Euryops acraeus, E. decumbens* and *Anthospermum hispidulum*. All these shrubs are dwarf and sclerophyllous.

(b) *Erica–Helichrysum* heath: This is a common community usually found above 3200 m. The dominants are *Erica dominans* and *Helichrysum trilineatum* (Fig. 17). Casual constituents are as in the preceding community.

(c) *Erica glaphyra* heath: *Erica glaphyra* forms a pure type of heath on broken promontories at the edge of the summit plateau, a habitat which provides a certain amount of shelter and is fairly moist. This *Erica* is slightly taller than *E. dominans*, darker green in colour and the leaves are longer and patent instead of adpressed.

(d) *Helichrysum–Passerina* heath: This community appears to be limited in extent: the author has only seen it at the edge of the escarpment near Castle Buttress. The dominants are *Helichrysum trilineatum* and *Passerina montana*. The habitat is broken, hence the presence of *Merxmuellera stereophylla*, a common summit lithophyte.

(e) Boulder–field heath: Situated on the summit are fairly large areas supporting boulders varying in their density of aggregation. The habitat is stable and not to be confused with scree. The heath growing in this habitat is 0.6–1.2 m tall and sometimes quite dense. The tallness of this heath as compared with the other heath types is probably due to the protection from fire afforded by the boulders. The dominants are three composites *Athanasia thodei*, *Helichrysum trilineatum* and *Eumorphia sericea* with *Merxmuellera drakensbergensis* and other alpine grasses filling the intervening gaps (Fig. 18). Two ferns often grow in the shade of the boulders, namely *Dryopteris inaequalis* (a depauperate, high-altitude form) and *Woodsia montevidensis* var. *burgessiana*.

References

Acocks, J. P. H. 1953. Veld types of South Africa. Mem bot. Surv. S. Afr. 28:1–192.
Barbosa, L. A. Grandvaux. 1970. Carta fitogeográfica de Angola. Inst. de Investigação Científica de Angola, Luanda.
Bawden, M. G. & Carroll, D. M. 1968. The land resources of Lesotho. Land Resources Division, Directorate of Overseas Surveys, Tolworth.
Bayer, A. W. 1955. The ecology of grasslands. In: D. Meredith, The grasses and pastures of South Africa, pp. 539–550. C.N.A., Johannesburg.
Bews, J. W. 1917. The plant ecology of the Drakensberg Range. Ann Nat. Mus. 3:511–565.
Bolus, H. 1905. Sketch of the floral regions of South Africa. In: W. Flint & J. D. F. Gilchrist, Science in South Africa, pp. 198–240. Maskew Miller, Cape Town.
Bond, G., 1957. The geology of the Khami stone age. Occ. Pap. natn. Mus. Sth. Rhod. 21 A.
Burchell, W. 1822. Travels in the interior of Southern Africa. Vol. 1. Longman, London.
Carroll, D. M. & Bascombe, C. L. 1967. Notes on the soils of Lesotho. Tech. Bull. No. 1. Land Resources Division, Directorate of Overseas Surveys, Tolworth.
Chapman, J. D. 1962. The vegetation of the Mlanje Mountains, Nyasaland. Government Printer, Zomba.
Chapman, J. D. & White, F. 1970. The evergreen forests of Malawi. Commonwealth Forestry Institute, Oxford.
Clements, F. E. 1916. Plant succession. Carnegie Institution of Washington, Washington.
Coetzee, J. A. 1967. Pollen analytical studies in East and Southern Africa. In: E. M. van Zinderen Bakker, Palaeoecology of Africa, vol. 3:1–146. Balkema, Cape Town.
Diniz, A. Castanheira. 1973. Caracteristicas mesológicas de Angola. Missãio de Inquéritos Agricolas de Angola, Nova Lisboa.
Edwards, D. 1967. A plant ecology survey of the Tugela River Basin, Natal. Mem. bot. Surv. S. Afr. 36:1–285.

Goodier, R. & Phipps, J. B. 1961. A revised check-list of the vascular flora of the Chimanimani Mountains. Kirkia 1:44–66.
Harper, G. 1969. Periglacial evidence in southern Africa during the Pleistocene epoch. In: E. M. Van Zinderen Bakker, Palaeoecology of Africa, Vol. 4:71–101. Balkema, Cape Town.
Hauman, L. 1933. Esquisse de la végétation des hautes altitudes sur le Ruwenzori. Bull. Acad. Roy. Belg. Cl. Sci. sér., 5, 19:602–616, 701–717, 900–917.
Hauman, L. 1955. La "Région Afroalpine" en phytogeographie centro-africaine. Webbia 11:467–469.
Hedberg, O. 1951. Vegetation belts of the east African mountains. Svensk bot. Tidskr. 45:140–202.
Hedberg, O. 1957. Afro-alpine vascular plants. Symb. bot. upsal. 15(1):1–411.
Hedberg, O. 1965. Afro-alpine flora elements. Webbia 19: 519–529.
Herbst, S. N. 1971. 'n Ekologiese plantopname van 'n gedeelte van die Oxbow-opvanggebied, Lesotho. M.Sc. (Agric.) thesis, Univ. of Orange Free State. (Unpubl.).
Herbst, S. N. & Roberts, B. R. 1974. The alpine vegetation of the Lesotho Drakensberg: a study in quantitative floristics at Oxbow. Jl. S. Afr. Bot. 40:257–267.
Holden, W. 1855. History of the colony of Natal. Alexander Heylin, London.
Hutchinson, J. 1946. A botanist in Southern Africa. Gawthorn, London.
Jacot Guillarmod, A. 1962. The bogs and sponges of the Basutoland mountains. S. Afr. J. Sci. 58:179–182.
Jacot Guillarmod, A. 1963. Further observations on the bogs of the Basutoland mountains. S. Afr. J. Sci. 59:115–118.
Jacot Guillarmod, A. 1971. Flora of Lesotho. Cramer, Lehre.
Killick, D. J. B. 1963. An account of the plant ecology of the Cathedral Peak area of the Natal Drakensberg. Mem. bot. Surv. S. Afr. 34:1–178.
King, L. C. 1962. The morphology of the Earth. Oliver & Boyd, Edinburgh.
King, L. C. 1972. The Natal monocline: explaining the origin and scenery of Natal, South Africa. Geology Dept., Univ. of Natal, Durban.
Lebrun, J. 1947. La végétation de la plaine alluviale au sud du lac Edouard. Expl. Parc Nat. Alb. Miss. Lebrun (1937–38). 1:1–8. Inst. Parcs Nat. Congo Belge, Brussels.
Levyns, M. R. 1952. Clues to the past in the Cape flora of today. S. Afr. J. Sci. 49:155–164.
Levyns, M. R. 1964. Migrations and origins of the Cape Flora. Trans. R. Soc. S. Afr. 37:85–107.
Macedo, J. De Aguiar. 1970. Carta de vegetação da serra da Gorongosa. Communicação No. 50:1–74. Instituto de Investigação Agrónomica de Moçambique, Lourenço Marques.
Mann, R. J. 1859. The colony of Natal. Jarrold, London.
Monod, T. 1957. Les grandes divisions chorologiques de l'Afrique. CCTA/CSA, Publ. No. 24, London.
Nänni, U. W. 1956. Forest hydrological research at the Cathedral Peak Research Station. Jl. S. Afr. for. Ass. 27:1–35.
Nänni, U. W. 1969. Veld management in the Natal Drakensberg. S. Afr. For. J. 68:5–15.
Nordenstam, B. 1974. The flora of the Brandberg. Dinteria 11:3–67.
Pedro, J. Gomes & Barbosa, L. A. Grandvaux. 1955. A vegetação. In Esboço do reconhecimento ecológico-agricola de Moçambique. Vol. 2. Memórias e Trabalhos, No. 23, pp. 67–224. Centro de Investigação Científica Algodoeira, Lourenço Marques.
Phipps, J. P. & Goodier, R. 1962. A preliminary account of the plant ecology of the Chimanimani Mountains. J. Ecol. 50:291–319.
Pilson, C. A. & Higgs, T. (MS). Geography. Climate. Geography Dept., Rhodes University, Grahamstown.
Schelpe, E. A. C. L. E. 1946. The plant ecology of the Cathedral Peak area. M.Sc. thesis, Univ. of S. Africa. (Unpubl.).
Schelpe, E. A. C. L. E. 1953. The distribution of bryophytes in the Natal Drakensberg, South Africa. Revue bryol. lichen. 22(1–2):86–90.
Shand, N. et al. 1963. Report on the hydrological investigation in the mountain area of Basutoland. 28 pp. (roneoed).
Story, R. 1952. A botanical survey of the Keiskammahoek district. Mem. bot. Surv. S. Afr. 27:1–184.
Styles, R. R. & Hudson, W. K. 1938. An ecological survey of the mountain area of Basutoland. Crown Agents for the Colonies, London.
Tansley, A. G. 1935. The use and abuse of vegetational concepts and terms. Ecology 16:284–307.
Thode, H. J. 1894. Die botanischen Höhenregionen Natals. Bot. Jb. 18; III, Beibl. 43:14–15.

Tidmarsh, C. E. M. & Havenga, C. M. 1955. The wheel-point method of survey and measurement of vegetation of semi-open grasslands and karoo vegetation in South Africa. Mem. bot. Surv. S. Afr. 29:1–49.

Thorarinsson, S. 1951. Notes on patterned ground in Iceland. Geogr. Annlr. 3–4:144–156.

Troll, C. 1944. Strukturboden und Solifluktion, und Frostklimate der Erde. Geol. Rdsch. 34:545–694.

Van der Merwe, C. R. 1941. Soil groups and subgroups of South Africa. Dep. Agr. Sci. Bull. No. 231. Pretoria.

Van der Schijff, H. P. & Schoonraad, E. 1971. The flora of the Mariepskop Complex. Bothalia 10:461–500.

Van Steenis, C. G. G. J. 1935. On the origin of the Malaysian mountain flora. II. Altitudinal zones, general considerations and renewed statement of the problem. Bull. Jard. bot. Buitenz. sér. III, 13, 3:289–417.

Van Zinderen Bakker, E. M. 1955. A preliminary survey of the peat bogs of the alpine belt of northern Basutoland. Acta geogr., Helsingf. 14:413–422.

Van Zinderen Bakker, E. M. 1962. Botanical evidence for quaternary climates in Africa. Ann. Cape Prov. Mus. 2:16–31.

Van Zinderen Bakker, E. M. 1964. Pollen analysis and its contribution to the palaeoecology of the Pleistocene in Southern Africa. In: D. H. S. Davis (ed.), Ecological studies in Southern Africa. pp. 24–34. Junk, The Hague.

Van Zinderen Bakker, E. M. 1965. Ueber Moorvegetation und den Aufbau der Moore in Süd-und Ostafrika. Bot. Jb. 84, 2:215–231.

Van Zinderen Bakker, E. M. & Werger, M. J. A. 1974. Environment, vegetation and phytogeography of the high-altitude bogs of Lesotho. Vegetatio 29:37–49.

Venter, F. A. 1938. Short general report on the geology of the extreme northern and north-eastern portion of Basutoland. In: R. R. Staples & W. K. Hudson, An ecological survey of the mountain area of Basutoland. pp. 47–56. Crown Agents for Colonies, London.

Wardle, P. 1962. Subalpine forest and scrub in the Tararua Range. Trans. R. Soc. N.Z. 1(6):78–89.

Warming, E. 1909. Oecology of plants. Oxford University Press, London.

Weimarck, H. 1941. Phytogeographical groups, centres and intervals within the Cape flora. Acta Univ. Lund. N.F. Avd. 2. 37(5):1–143.

West, O. 1951. The vegetation of Weenen County, Natal. Mem. bot. Surv. S. Afr. 23:1–180.

White, F. 1965. The savanna woodlands of the Zambezian and Sudanian domains: an ecological and phytogeographical comparison. Webbia 19:651–681.

White, F. 1971. The taxonomic and ecological basis of chorology. Mitt. bot. StSamml., Münch. 10:91–112.

Wild, H. 1965. The vegetation of Rhodesia. In: M. O. Collins, Rhodesia: its natural resources and economic development, pp. 22–23. Collins, Salisbury.

Wild, H. 1968. Phytogeography in south central Africa. Kirkia 6:197–222.

Wild, H. & Barbosa, L. A. Grandvaux. 1967. Vegetation map of the Flora Zambesiaca area, 1:2,000,000. 71 pp. Suppl. to Flora Zambesiaca. Collins, Salisbury.

13 The Indian Ocean Coastal Belt

E. J. Moll* and F. White

I   Phytogeography
    F. White & E. J. Moll

1. Introduction 563
2. The Zanzibar–Inhambane Regional Mosaic 565
2.1 The Zanzibar–Inhambane endemic flora 567
2.2 Non-Zambezian, Zanzibar–Inhambane linking elements 568
2.3 Zambezian/Zanzibar–Inhambane linking species 569
3. The Tongaland–Pondoland Regional Mosaic 570
3.1 The Tongaland–Pondoland endemic flora 572
3.2 Tongaland–Pondoland linking elements 573

II   Vegetation and Ecology
    E. J. Moll

4. The Zanzibar–Inhambane Regional Mosaic 575
4.1 Forest 576
4.2 Woodland 578
4.3 Thicket 578
4.4 Grassland 578
5. The Tongaland–Pondoland Regional Mosaic 579
5.1 Forest 579
5.2 Woodland, bushland and thicket 589
5.3 Grassland 592
5.4 Fynbos 594
5.5 Swamps 595

References 595

* Photographs are by E. J. Moll unless otherwise stated.

13 The Indian Ocean coastal belt

I. Phytogeography
F. White & E. J. Moll

1. Introduction

The vegetation of the relatively narrow coastal strip extending along the eastern seaboard of Africa from the extreme southeastern corner of Somalia to the neighbourhood of Port Elizabeth in the eastern Cape differs considerably from that occurring under more continental climates further inland. The flora has long been known to include many endemic species, and, at least in the north, to bear a close relationship to that of the Guineo–Congolian Region. No comprehensive studies, however, have previously been attempted.

In his survey of the great chorological divisions of Africa, Monod (1957) treated this narrow coastal strip as an outlying 'Oriental Domain' of the Guineo–Congolian rain forest Region. This view was provisionally accepted by White (in Chapman & White 1970), who suggested that it should be re-named the Usambara–Zululand Domain, since Lebrun had previously (1947) proposed an Oriental Domain for a different but overlapping phytochorion. Further work has shown this view to be untenable.

In a comprehensive study of the phytochoria of Africa, which stems from earlier tentative proposals (White 1965), and has not yet been published in full, but has recently been summarized, White (1976a) suggests that a hierarchical arrangement into regions, domains, sectors and districts is ill-adapted to accommodate the most interesting facts of plant distribution and provides a clumsy reference system (see Chapter 7).

The Indian Ocean Coastal Belt is both a Regional Transition Zone and a Regional Mosaic in White's (1976a) terminology. It also has a far higher proportion of endemic species than the other African transition zones and mosaics. The complexity of its chorological relationships is comparable to that of the Afromontane Region.

The flora at the northern end of the Indian Ocean Coastal Belt is almost completely different from that at the southern end. Superimposed upon this gradual change affecting almost the whole flora is a much more abrupt pattern of local change. This is because rapid local changes in climate and soil are responsible for a complex mosaic of different vegetation types. The latter include a wide range of different kinds of forest, bushland, thicket, transition woodland, woodland, wooded grassland† and edaphic grassland.

Most of the natural vegetation, however, has been destroyed and replaced by secondary, fire-maintained grassland and wooded grassland, or is under cultivation. Different vegetation types tend to show different patterns of chorological

---

† The term 'wooded grassland' (White in press) is equivalent to the term 'savanna' used in other chapters of this volume (see Preface and p. 311).

relationship, though there has been some intermingling of the various genetic elements.

For many species information on distribution is difficult to obtain. The taxonomy of others is insufficiently advanced. Hence only tentative conclusions can be presented here. They are based on detailed studies of certain families of woody plants, especially Ebenaceae, Meliaceae, Chrysobalanaceae and Combretaceae, and less detailed studies of the remaining larger woody plants, amounting to about 1000 species in all. In general, much less is known of the taxonomy and chorology of herbaceous plants, though some better-known groups have been taken into account.

Of the larger woody plants occurring in the Indian Ocean Coastal Belt c. 40 per cent are confined to it. Their distribution within it is far from even. Relatively few (c. 5 per cent) are widely distributed from north to south. The remaining endemics fall into two groups, a northern and a southern. North of the town of Moçambique c. 35 per cent of the larger woody species are endemic, whereas south of the lower reaches of the Limpopo River the figure is c. 40 per cent. The central part of the Indian Ocean Coastal Belt between the Limpopo River and the town of Moçambique has relatively few endemic species.

The non-endemic or linking species occurring in the coastal belt provide connections with several other major phytochoria, including the Somalia–Masai, Zambezian, Guineo–Congolian, Karoo–Namib, Cape, Malagasy and Afromontane. As with the endemics, so with the linking species, the representation of the different elements varies greatly in different parts of the coastal belt.

The floras of the northern and southern parts of the Coastal Belt, although part of the same continuum, are so different that it is necessary to recognize two major phytochoria – the Zanzibar–Inhambane and the Tongaland–Pondoland transitional and mosaic Regions*(Figs. 1 and 2).

The delimitation between them is far from arbitrary and is provided by the lower reaches of the Limpopo River. Very many Tongaland–Pondoland species, both endemic and linking extend northwards into Moçambique, but very few cross the lower Limpopo River. The converse is also true. Relatively few species occurring in the Zanzibar–Inhambane Regional Mosaic extend to the south beyond the Limpopo River. Those that do penetrate the Tongaland–Pondoland Regional Mosaic, reach their southern limits at different places, and some extend to its southern limits or beyond. Where a distinctive vegetation type, e.g. sand forest (p. 582) abruptly ends, several species may reach their southern limit more-or-less concurrently, but there are few species in this category and the feature is of only local significance.

The most important differences between these two regional mosaics are as follows:

Zanzibar–Inhambane Regional Mosaic:

Many endemic species belong to otherwise Guineo–Congolian genera. Guineo–Congolian linking species are numerous. Afromontane linking species are few and of very sporadic occurrence. Zambezian-linking species include several that are widely distributed in the wetter and upland parts of the Zambezian

---

* Regarding the use of the term 'Region' see p. 159.

Region, e.g. species of *Brachystegia, Julbernardia, Uapaca, Parinari*. Cape and Karoo–Namib linking species are virtually absent.

Tongaland–Pondoland Regional Mosaic:

Very few endemic species belong to otherwise Guineo–Congolian genera. Guineo–Congolian linking species are few. Afromontane linking species are numerous and relatively conspicuous, especially towards the south. Zambezian linking species are, within the Zambezian Region, mostly confined to the hotter, drier, low-lying parts or are widespread and transgress its limits and are of little diagnostic value. Cape and Karoo–Namib linking species are, at least locally, conspicuous.

2. The Zanzibar–Inhambane Regional Mosaic

The Zanzibar–Inhambane Regional Mosaic extends to the mouth of the Limpopo River, and varies in width from 50 to 200 km except where it penetrates inland along certain broad river valleys (Fig. 1). It mostly lies below 200 m, but in the northern half there are scattered hills and plateaus rising to considerably more than this. The mean annual rainfall is more than 800 mm, but rarely over 1250 mm. Nearly everywhere there is a perceptible dry season, which varies from one month with less than 40 mm to six months with less than 40 mm, of which as many as five may receive less than 20 mm. The contrast between the wet season and the dry season, however, is everywhere much less marked than further inland, and the mean monthly rainfall for each month of the dry season nearly everywhere exceeds 5 mm. Relative humidity throughout the dry season is high. Mean annual temperature is more or less 26°C north of the Zambezi; it diminishes steadily southwards (see Chapter 2).

To the interior of the coastal belt the climate is too continental in character to support the characteristic vegetation of the Zanzibar–Inhambane Region, except very locally on the lower windward slopes of mountainous massifs below 1500 m (or less depending on the distance from the sea and other factors), where the local increase in precipitation and dry season relative humidity favours the development of lowland and transitional rain forest. These islands of lowland rain forest are mostly composed of species which are also found in the coastal belt but are absent from the surrounding Zambezian vegetation except sometimes in fringing forest. For this reason they can be regarded as small outlying satellites of the Zanzibar–Inhambane Regional Mosaic.

In the extreme north, the Zanzibar–Inhambane Regional Mosaic is adjacent to the Somalia–Masai Region; further south it lies next to the Zambezian Region. The transition zones between these phytochoria have not yet been studied in detail.

No comprehensive account of the vegetation or phytogeography of the Zanzibar–Inhambane Regional Mosaic has ever been published, and very few detailed studies of individual vegetation types have been made. For the most part, there are only generalized descriptions and a wealth of scattered ecological observations and chorological records locked-up in taxonomic publications and herbarium cupboards.

For Somalia a brief account is provided by Pichi-Sermolli (1957, types 9, 14,

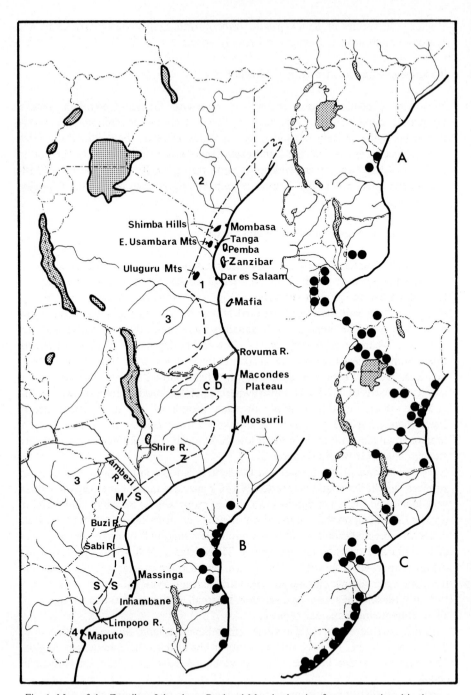

Fig. 1. Map of the Zanzibar–Inhambane Regional Mosaic showing features mentioned in the text. 1. Zanzibar–Inhambane Region. 2. Somalia–Masai Region. 3. Zambezian Region. 4. Tongaland–Pondoland Region. CD = Cabo Delgado. Z = Zambezia. MS = Manica e Sofala. SS = Sul do Save.

A, B and C. Inset maps showing distributions of *Maranthes goetzeniana*, *Cussonia zimmermannii* (after Bamps, Distr. Pl. Afr. 8:224, 1974), and *Turraea floribunda* respectively.

16). For Kenya the coastal vegetation is described and mapped by Dale (1939), and in less detail, but with more emphasis on environmental factors by Moomaw (1960). The floristic composition of the coastal forests is recorded by Lucas (1968). For Tanzania published information is sparse. The most comprehensive account is that of Polhill (1968), who only briefly describes the coastal vegetation types but gives valuable information on their floristic composition. The lowland rain forest of the East Usambara Mts. is very briefly described by Moreau (1935) and that of the Uluguru Mts. by Pócs (1976). The coastal thickets dominated by *Philippia mafiensis* on Mafia Island are described very briefly by Greenway (1973) and more fully in an unpublished manuscript deposited in the East African herbarium (1938). For the region covered by *Flora Zambesiaca* the coastal vegetation and its small inland satellites has been described in outline by Wild & Barbosa (1968, types 1 (pro parte), 2 (p.p.), 5, 6, 9, 10, 13 (p.p.), 14, 20, 25–27, 31–33, 44, 53, 54). Works of more local significance are cited in the next section. The lowland rain forests of Malawi, which are best regarded as Zanzibar–Inhambane satellites, are dealt with more fully by Chapman & White (1970). The mangrove swamps of Kenya have been described by Graham (1929) and those of East Africa generally by Walter & Steiner (1936). Brief descriptions of the coastal vegetation of East Africa are given by Lind & Morrison (1974).

In the northern part of the Zanzibar–Inhambane Regional Mosaic in Kenya and Tanzania, the prevalent vegetation is forest, bushland and thicket, in which endemic species and Guineo–Congolian linking species predominate (White unpubl.). Vegetation dominated by Somalia–Masai and Zambezian linking species is relatively restricted. In Moçambique, however, although patches of forest, bushland and thicket, not dissimilar to those of Kenya and Tanzania, occur too as far south as the Limpopo River, the area occupied by Zambezian vegetation seems to be much more extensive (Wild & Barbosa 1968), though precise information is lacking. Elements occurring in Zambezian and non-Zambezian vegetation types will be considered separately below. The analyses of the non-Zambezian elements in the Zanzibar–Inhambane flora are based on 190 tree species which normally attain 9 m or more in height. Only a few taxonomically critical species have been excluded from consideration. The total number of Zambezian tree species occurring, other than as marginal intruders, in the Zanzibar–Inhambane Regional Mosaic is unknown, but is estimated at 100.

2.1 *The Zanzibar–Inhambane endemic flora*

No family and only 3 genera of forest trees, *Cephalosphaera*, *Englerodendron* and *Lettowianthus* are endemic to the Zanzibar–Inhambane Regional Mosaic. They are confined or almost confined to the East Usambara Mts. A further 7 genera, however, including *Bivinia*, *Hirtella*, *Ludia* and *Trachylobium*, which also occur in Madagascar, are confined or almost confined on the African mainland to the Zanzibar–Inhambane Regional Mosaic.

131 species (70 per cent) are confined on the African mainland to the Indian Ocean Coastal Belt. 9 species (4.7 per cent) (see below) also occur in Madagascar and 29 species (15.3 per cent) (see below) also occur in the Tongaland–Pondoland Regional Mosaic. Of the 92 (48.4 per cent) Zanzibar–Inhambane endemic species by far the greatest concentration is in Kenya and northern Tanzania, centred on the Shimba Hills and East Usambara Mts, respectively. Impoverishment to the

south is rapid. Relatively few species extend to Moçambique. Among them are: *Berlinia orientalis, Diospyros verrucosa, Fernandoa mangifica, Guibourtia schliebenii, Manilkara sansibarensis, Nesogordonia parvifolia, Rhodognaphalon* and *Sterculia appendiculata.*

A few Zanzibar–Inhambane endemic and near-endemic species, are confined, in the coastal belt, to relatively elevated massifs, principally the Shimba Hills and East Usambara Mts. Some near-endemic species, including *Lovoa swynnertonii* and *Maranthes goetzeniana* (Fig. 1), which extend to south tropical Africa, also occur in the small outliers of moist lowland and transitional forest found on the lower slopes of certain mountains inland from the coastal belt. For the present purpose these are regarded as members of the Zanzibar–Inhambane flora.

2.2   *Non-Zambezian, Zanzibar–Inhambane linking elements*

(i) Tongaland–Pondoland linking species. Twenty-nine coastal endemic species (15.3 per cent) occur also in the Tongaland–Pondoland Regional Mosaic. They include *Albizia forbesii, A. petersiana, Balanites maughamii, Bequaertiodendron natalense, Casearia gladiiformis, Cleistanthus schlechteri, Commiphora zanzibarica, Cordyla africana, Craibia zimmermannii, Crossonephelis (Melanodiscus) oblongus, Drypetes natalensis, Inhambanella henriquesii, Newtonia hildebrandtii, Pseudobersama mossambicensis, Sideroxylon inerme, Suregada zanzibarensis, Turraea floribunda,* and *Vepris undulata.*

All these species are widely distributed in the Zanzibar–Inhambane Regional Mosaic and extend at least as far north as northern Tanzania or Kenya, but most of them extend only a short way into the Tongaland–Pondoland Regional Mosaic and reach their southern limit in Zululand. Only six extend south of Durban, viz. *Casearia gladiiformis, Turraea floribunda* (Fig. 1), *Drypetes natalensis, Vepris undulata, Bequartiodendron natalensis* (Pondoland) and *Sideroxylon inerme* (Cape Peninsula). Most are more or less confined to the coastal belt and all are most abundant there, but some have a few scattered occurrences inland, even as far as Uganda (*Crossonephelis* and *Turraea floribunda*). *Newtonia hildebrandtii* and *Cordyla africana* penetrate some distance up the Zambezi and Shire valleys.

(ii) Malagasy linking species. Nine species (4.7 per cent) are confined in Africa to the Zanzibar–Inhambane Regional Mosaic but also occur in Madagascar. They all extend as far south as Moçambique, where two, *Hirtella zanzibarica* and *Trachylobium verrucosum,* are important constituents of the vegetation.

(iii) Guineo–Congolian linking species. Forty-nine species (25.8 per cent) also occur in the rain forest of the Guineo–Congolian Region and are absent from typical vegetation in the intervening Zambezian and Somalia–Masai Regions, though about half also occur in small populations in fringing forest or in small islands of forest on other specially favoured sites. The remaining species show a wide interval between their Guineo–Congolian and Zanzibar–Inhambane occurrences. Every degree of gradation exists between species such as *Parkia filicoidea,* which occur extensively in the intervening areas and those like *Pterocarpus mildbraedii* where the interval is very wide. Nearly all those species which do not occur in the interval between the Guineo–Congolian and Zanzibar–Inhambane phytochoria, occur only in the northern part of the Zanzibar–Inhambane Regional Mosaic and are concentrated in the Shimba Hills and East Usambara Mts. An important exception is *Croton sylvaticus* which extends as far south as Transkei.

In contrast to the situation described above, the majority of species which do occur in the interval between the Guineo–Congolian Region and the Zanzibar–Inhambane Regional Mosaic, are widely distributed in the latter, and no less than 17, including *Blighia unijugata*, *Celtis mildbraedii*, *Ficus capensis*, *Rauvolfia caffra* and *Sapium ellipticum*, penetrate the Tongaland–Pondoland Regional Mosaic to various distances.

(iv) Only seven species (3.7 per cent) are common to the Afromontane and Zanzibar–Inhambane phytochoria. In the latter they are rare and localized, with most occurrences on the islands of Zanzibar, Mafia and Pemba. This element is singularly inconspicuous in the coastal plain of Moçambique.

(v) Thirteen species (7.0 per cent) are ecological and chorological transgressors. Only two of them, *Diospyros natalensis* and *Euclea natalensis*, both of which extend to the Tongaland–Pondoland Regional Mosaic, are more characteristic of the Indian Ocean Coastal Belt than elsewhere.

2.3 *Zambezian/Zanzibar–Inhambane linking species*

Until the boundary between the Zanzibar–Inhambane Regional Mosaic and its neighbouring phytochoria has been defined more precisely on the basis of detailed studies in the field, it will be impossible to compute at all accurately the number of Zambezian species occurring within the former. From available information, however, it is clear that the Zambezian element is subordinate to the non-Zambezian elements. It seems that only about 100 tree species occur other than marginally. Published information on the ecology of Zambezian species in the Zanzibar–Inhambane Regional Mosaic is extremely meagre. White (unpubl.) has made a study of the problems in Kenya and Tanzania, on which the following account is largely based. More than half the species studied in the field in Kenya and Tanzania also extend south into the coastal parts of Moçambique and there is no reason to believe that they behave differently there. Nearly all the species mentioned below are common to the northern and southern parts of the Zanzibar–Inhambane Regional Mosaic.

Vegetation in which Zambezian species are prominent is widely distributed in the Zanzibar–Inhambane Regional Mosaic today, but most of it is anthropic in origin. There is, however, impressive evidence, both ecological and taxonomic, that a substantial proportion of Zambezian tree species occurring in the Zanzibar–Inhambane Regional Mosaic do not owe their occurrence there to human activity. They are found in six main types of natural habitat. The vegetation of only the first is composed almost exclusively of Zanzibar–Inhambane endemic species. In the remaining types the Zambezian species are represented to varying degrees. They are favoured by locally occurring climatic, edaphic, or biotic factors, operating singly or in combination.

(i) Drier types of forest. *Afzelia quanzensis* occurs in a wide range of dry forest types, in which its associates are completely different from those in its Zambezian habitats. Because of its short bole and very wide-spreading crown, its appearance is also strikingly different. The baobab, *Adansonia digitata*, is a conspicuous feature of the coastal belt. Large individuals are found in forest, especially on outcrops of coral limestone. Its distribution, however, seems to have been determined largely by man, and regeneration does not occur under a closed forest canopy. On certain shallow or nutrient deficient soils *Brachystegia spiciformis* is the dominant of transition woodland in which the understorey is comprised of a mixture of heliophilous Zambezian savanna species and sciaphilous Zanzibar–Inhambane forest species. Elsewhere, *B. spiciformis* dominates the early seral stages of moist evergreen forest.

(ii) Coastal thicket. Several Zambezian species, including *Bridelia cathartica*, *Cardiogyne africana*, *Diospyros squarrosa*, *Lannea stuhlmannii*, *Strychnos spinosa* and *Turraea nilotica*, figure prominently in this unstable community, which is 'migratory' in the sense of Crampton (1912, see also Tansley 1939).

(iii) Biotically disturbed seral stages. Large mammals, especially buffalo and elephant, are sufficiently plentiful today, at least locally, to degrade climax forest so that heliophilous Zambezian species can gain a foothold. *Annona senegalensis*, *Antidesma venosum*, *Crossopteryx febrifuga* and *Sclerocarya caffra*, among others, are frequent in such habitats.

(iv) Rain shadow habitats. Small areas with low rainfall, especially where there is only a single rainy season, support bushland or stunted woodland dominated by *Combretum collinum*, *Lonchocarpus bussei*, *Sclerocarya caffra* and *Stereospermum kunthianum*.

(v) Mosaics of edaphic grassland and termite-mound thicket. Plains and depressions which are seasonally waterlogged or shallowly flooded support grassland with or without a sparse cover of trees, which include the endemic species, *Hyphaene compressa*.

The large termite mounds which are abundant in this habitat are covered by thicket with a few emergent trees. Zambezian and coastal species occur in intimate mixture. Among the former are *Commiphora africana*, *Crossopteryx febrifuga*, *Fagara chalybea*, *Gardenia jovis-tonantis*, *Maerua angolensis*, *Pterocarpus angolensis* and *Tamarindus indica*.

(vi) Saline grassland. On coasts of low relief the landward side of mangrove swamps is frequently fringed with poorly drained slightly saline grassland, in which scattered trees of a few Zambezian and Somalia–Masai species also occur, especially *Acacia nilotica*, *A. zanzibarica*, *Albizia anthelmintica*, *Dalbergia melanoxylon* and *Dichrostachys cinerea*.

Most of the features of the Zanzibar–Inhambane Regional Mosaic that are mentioned in the text are shown in Fig. 1, which also shows the distribution of *Cussonia zimmermannii*, a Zanzibar–Inhambane endemic species, *Turraea floribunda*, a Zanzibar–Inhambane near-endemic species, and *Maranthes goetzeniana* which occurs in upland areas in the Zanzibar–Inhambane Region and in the transitional rain forest on the lower slopes of mountains further inland.

3. The Tongaland–Pondoland Regional Mosaic

The Tongaland–Pondoland Regional Mosaic extends from the mouth of the Limpopo River (25°S) to about Port Elizabeth (34°S) (Fig. 2). In the north it is about 240 km wide, but in the south where mountains come close to the sea it is locally no more than 8 km wide. Elsewhere it penetrates along deep river valleys far into the interior. Because the Tongaland–Pondoland Regional Mosaic is a transition zone of complex chorological relationships it is difficult to define its limits precisely (see Chapter 7). For most of its length it lies below the Afromontane Region, but owing to the compensating effects of latitude many Afromontane species descend to sea-level and the lower limit of the Afromontane Region is somewhat

Fig. 2. Map of the Tongaland–Pondoland Regional Mosaic showing features mentioned in the text. 1 = Tongaland–Pondoland Region. 2 = Afromontane Region. 3 = Austro-afroalpine Region. 4 = Transition between Afromontane and Zambezian Regions. 5 = Transition between Afromontane Region and Highveld. 6 = Cape Region. 7 = Karoo–Namib Region.

A, B and C. Inset maps showing distributions of *Diospyros dichrophylla*, *D. glandulifera* and *D. simii* respectively.

blurred compared with the situation farther north. In general, the forests of the 'mist belt' (a local name for elevated areas that receive much fog during several months of the year, cf. Edwards 1967, Moll 1968b) are essentially Afromontane, in contrast to those at lower altitudes which include many endemic Tongaland–Pondoland species and are mostly much less rich in epiphytic lichens, bryophytes, ferns and orchids.

The vegetation comprises the following of Acocks's (1953) mapping units: 1, 2, 6, 10, 23 and 24. Types 3 and 5 are transitional between Afromontane and Tongaland–Pondoland vegetation. The latter is overwhelmingly tropical in affinity in contrast to other vegetation at comparable latitudes elsewhere in South Africa. This is due to the ameliorating effects of the warm Moçambique Current which flows a short distance offshore and causes an attenuated extension of sub-tropical conditions down the east coast. Climate however can vary rapidly over short distances, and there is often a great contrast between xerocline and mesocline vegetation and between that of deeply incised valleys and slopes, and of plains under maritime influences. The desiccating 'berg' winds, in particular, have a profound influence on valley vegetation (see Chapters 2 and 12).

The geology is also diverse and this too greatly influences the vegetation by giving rise to soils which are very different in their nutrient status and water relations. The coastal plain is composed of Cretaceous and Caenozoic marine sediments. Elsewhere the more undulating landscape is carved out of rocks of the Basement Complex, Table Mountain Sandstone and sedimentary strata of the Karoo system (Truter & Rossouw 1955 and Chapter 1).

The vegetation of the valleys shows strong Zambezian links as far south as the Kei River, and Karoo–Namib links further south from the Fish to the Sundays River. Afromontane species are particularly plentiful south of East London in the zone between the humid and subhumid zones of Poynton (1974). Cape linking species and Cape 'relatives' also occur, particularly on nutrient deficient sandy soils.

Over much of the Regional Mosaic the vegetation has been drastically altered or obliterated relatively recently by the hand of man, beginning with the arrival of the Bantu in about 1400 A.D. (Brooks & Webb 1965)* with their slash and burn agriculture, followed by European settlement of much of the area in the early nineteenth century. Fortunately for the phytogeographer and ecologist many of the forest patches, particularly in the Bantu occupied areas, are still relatively intact; though their present conservation status is causing considerable alarm (Edwards 1974). The grasslands, woodlands and shrublands have, on the other hand, been much altered and there has been a dramatic expansion of the Karoo–Namib flora in the Tongaland–Pondoland Regional Mosaic (see Acocks 1953, maps 1 and 3).

The flora of the Tongaland–Pondoland Regional Mosaic is believed to comprise about 3000 species. Of these, about 650 are trees or large shrubs or climbers. The chorological analyses summarized below are based on 523 species of larger woody plants. Taxonomically critical species have been excluded from consideration.

Most of the features of the Tongaland–Pondoland Regional Mosaic that are mentioned in the text are shown in Fig. 2, which also shows the distributions of three endemic and near-endemic species.

3.1   *The Tongaland–Pondoland endemic flora*

Endemism is more pronounced than in the Zanzibar–Inhambane Regional Mosaic. There is one almost endemic family, the Achariaceae, and 23 endemic genera including: *Anastrabe, Bachmannia, Burchellia, Duvernoya, Ephippiocarpa, Galpinia, Harpephyllum, Hippobromus, Jubaeopsis, Loxostylis, Mackaya, Pseudosalacia, Rhynchocalyx, Stangeria* and *Umtiza*.

The following genera are confined in Africa to the Tongaland–Pondoland Regional Mosaic but also have representatives elsewhere: *Atalaya* (Indochina, Malaysia, Australia), *Ernestimeyera* (*Alberta*, 2 spp. in Madagascar), *Protorhus* (20 spp. in Madagascar).

More than 200 larger woody species, about 40 per cent of the total, are endemic. Their chorological relationships are diverse. Only a few examples can be given here.

Encephalartos is predominantly South African. Twenty-six of its 35 species occur there, and 18 of them are Tongaland–Pondoland endemics. *Aloe* which has its greatest concentration of species in South Africa is generally diffused over large parts

* Backdated recently to 200 A.D. or possibly 4000 B.P. (Maggs 1977).

of southern Africa, but 12 out of the 23 tree species that occur in South Africa, including the tallest, *A. bainesii*, are Tongaland–Pondoland endemics. Similarly, 9 of the 13 South African succulent tree euphorbias occur in the Tongaland–Pondoland Regional Mosaic.

Some genera including *Bersama*, *Diospyros* (Sectio *Royena*), *Euclea* and *Rhoicissus*, which are widespread in Africa, have their centre of variation in the Tongaland–Pondoland Regional Mosaic. Others, which are more widespread both in Africa and elsewhere, have, in relation to its size, high concentrations of species in the Tongaland–Pondoland Regional Mosaic. They include *Canthium*, *Cassine*, *Cussonia*, *Dovyalis* and *Eugenia*.

Some Tongaland–Pondoland endemic species are separated by very wide intervals from the other species in their genera. *Cavacoa* consists of the Tongaland–Pondoland, *C. aurea*, and two Guineo–Congolian species. In addition to the four Tongaland–Pondoland endemic and near-endemic species of *Schotia*, there are three Guineo–Congolian species. In Africa *Wrightia* is represented by the Tongaland–Pondoland *W. natalensis* and a semi-desert Somalia–Masai species, *W. demartiniana*. The remaining species occur in Asia and Australia. The Tongaland–Pondoland *Xylotheca kraussiana* is replaced by a related species in the Zanzibar–Inhambane Regional Mosaic. The only other species in the genus is confined to a small area in southern Angola. *Mitriostigma* has two species, the Tongaland–Pondoland, *M. axillare*, and *M. barteri*, which is confined to the island of Fernando Po in the Bight of Benin.

Some of the above facts suggest that the Tongaland–Pondoland Regional Mosaic has served as a refuge for genera which were formerly more widespread on the African mainland. However, the fact that within the Tongaland–Pondoland Regional Mosaic closely related species, as in *Cussonia*, or subspecies as in *Diospyros* (Sectio *Royena*) and *Euclea*, occupy contrasting but often adjacent habitats suggests that it has also been a region of recent diversification.

3.2 *Tongaland–Pondoland linking elements*

(i) Zanzibar–Inhambane linking species. Forty species (7.6 per cent). The trees in this element have already been dealt with (p. 568).
(ii) Zambezian linking species. In the northern part of the Tongaland–Pondoland Regional Mosaic, especially in the drier more open vegetation types, there has been a considerable intermingling of the Zambezian and the Tongaland–Pondoland floras, but it is easy to exaggerate the significance of this.

Several species which are widespread and abundant in the Tongaland–Pondoland Regional Mosaic, such as *Burchellia bubalina*, *Cnestis natalensis*, *Diospyros dichrophylla* (Fig. 2), *Entada spicata* and *Schotia brachypetala*, also occur beyond its limits in the Zambezian Region in outliers of similar vegetation or as satellite populations in different vegetation types. If they penetrate only a relatively short distance into the Zambezian Region and are more abundant in the Tongaland–Pondoland Regional Mosaic they are best regarded as Tongaland–Pondoland near-endemic species and, for classification purposes, as actual endemic species. Apart from these, there are c. 100 species (20 per cent) which are widespread in the Zambezian Region and enter the Tongaland–Pondoland Regional Mosaic. Twenty-six of them, including *Bauhinia galpinii*, *Berchemia zeyheri*, *Bolusanthus speciosus*, *Cassine transvaalensis*, *Combretum erythrophyllum, Rhigozum zambeziacum* and *Sterculia rogersii,* have relatively restricted ranges in the Zambezian Region and are confined to its southern fringes.

Of the remainder no less than 30, including *Acacia senegal, A. tortilis, Cordia sinensis, Dichrostachys cinerea, Euclea divinorum, Euphorbia ingens, Flacourtia indica, Mundulea sericea,* and *Ximenia americana,* also extend beyond the Zambezian Region in other directions and are of limited value in chorological analysis. Many 'Zambezian' species, including *Entandrophragma caudatum, Pterocarpus angolensis,* and *P. rotundifolius* penetrate only a short way into the Tongaland–Pondoland Regional Mosaic and are best regarded as marginal intruders. A minority, including *Acacia gerrardii, A. nilotica, A. sieberana, Combretum molle, Ficus ingens, Sclerocarya caffra* and *Ziziphus mucronata* extend much further south.

(iii) Afromontane Tongaland–Pondoland linking species. Forty-six species (8.7 per cent). The relationships between the Afromontane and Tongaland–Pondoland floras are described in Chapter 11. Although Afromontane species occur in most types of Tongaland–Pondoland woody vegetation, their role is usually subordinate. Most types of Tongaland–Pondoland forest are quite different from typical Afromontane forest. Some Afromontane species occur in the Tongaland–Pondoland Regional Mosaic only as marginal intruders. *Diospyros whyteana,* for instance, can occur, as in the Tsitsa Valley, in valley bushland (White unpubl.) with *Euclea undulata, Aloe ferox, Encephalartos altensteinii,* succulent euphorbias and *Crassula portulacea.* It is, however, absent from the greater part of the Tongaland–Pondoland Regional Mosaic.

(iv) 'Upland' linking species. Twenty-two species (4.4 per cent). This element includes *Canthium gueinzii, Celtis gomphophylla (durandii), Chrysophyllum viridifolium, Drypetes gerrardii, Erythrina lysistemon, Heywoodia lucens, Macaranga capensis, Manilkara discolor, Rawsonia lucida, Suregada procera, Tarenna pavettoides, Scolopia stolzii, S. zeyheri,* and *Strychnos mitis.*

In South Africa these species are absent from the Afromontane Region sensu stricto but are well represented in the Tongaland–Pondoland Regional Mosaic. All are widely distributed in upland areas in tropical Africa but some are localized and show wide disjunctions, e.g. *Erythrina lysistemon, Heywoodia lucens,* and *Scolopia zeyheri.* They must have been much more widespread in the past. *Canthium gueinzii, Celtis gomphophylla* and *Tarenna pavettoides* occur in West Africa, where they are confined to the Cameroun Highlands and other upland areas. Most of these species do not occur in typical Afromontane vegetation in tropical Africa other than as marginal intruders. Only *Drypetes gerrardii* and *Suregada procera* are important constituents of Afromontane forest. This element is virtually absent from the Zanzibar–Inhambane Regional Mosaic. *Macaranga capensis* and *Rawsonia lucida,* however, occur on the East Usambara Mts. and the Shimba Hills. *Scolopia zeyheri* occurs on the Tanzania coast at Tanga, *Rawsonia lucida* and *Scolopia stolzii* occur on Pemba Island, and *Strychnos mitis* on Mafia Island.

(v) Cape/Afromontane/Tongaland–Pondoland linking species. The following eight species (1.5 per cent) are widely distributed in the Cape, the South African part of the Afromontane Region, and are also conspicuous, at least locally, especially in coastal communities, in the Tongaland–Pondoland Regional Mosaic: *Chrysanthemoides monilifera, Clutia pulchella, Erica caffra, Metalasia muricata, Phylica paniculata, Polygala myrtifolia, Passerina filiformis,* and *Psoralea pin-*

nata. These are all typical Cape genera except *Polygala* and *Psoralea*. It is perhaps noteworthy that nearly all these linking species are among the tallest members of their respective genera. Most are shrubs 3 m or more tall.

(vi) Karoo–Namib linking species. Twelve species (2.5 per cent) as follows: *Aloe speciosa, Cadaba aphylla, Carissa haematocarpa, Crassula portulacea, Euclea undulata, Euphorbia grandidens, Lycium austrinum, Montinia caryophyllacea, Maytenus linearis, Portulacaria afra, Schotia afra, S. latifolia*. The majority of these occur only towards the southern end of the Tongaland–Pondoland Regional Mosaic, especially in the Sundays River and Fish River Scrub, but some, e.g. *Euclea undulata, Euphorbia grandidens* and *Portulacaria afra*, occur scattered throughout. Within the Karoo Region they are most characteristic of 'Spekboomveld', which occurs in the extreme south, though some species, e.g. *Euclea undulata, Montinia*, and *Schotia afra*, are much more wide-ranging (see Chapter 9).

(vii) Guineo–Congolian linking species. Twenty-seven species (5.1 per cent) including *Blighia unijugata, Celtis mildbraedii, Croton sylvaticus, Ficus capensis, Morus mesozygia, Rauvolfia caffra, Tapura fischeri, Trema orientalis* and *Voacanga thouarsii*.

A few species in this element, e.g. of *Blighia* and *Morus*, extend no further south than Zululand, and many are also widely distributed in fringing forest, and, less often, in other forest and thicket types in the Zambezian Region.

(viii) Chorological and ecological transgressors. Thirty-four species (6.5 per cent). Several transgressors including *Acacia karroo, Aloe arborescens, Cassine aethiopica, Celtis africana, Diospyros lycioides, Euclea crispa, Olea africana, O. capensis, Pappea capensis, Ptaeroxylon obliquum*, and *Tarchonanthus camphoratus*, occur in a wide range of vegetation types in southern Africa. Most of the above are absent from the Zanzibar–Inhambane Regional Mosaic. *Celtis africana*, however, occurs on Zanzibar.

II. Vegetation and Ecology

E. J. Moll

4. The Zanzibar–Inhambane Regional Mosaic

The vegetation of the Indian Ocean Coastal Belt in Moçambique between the mouths of the Rovuma and Limpopo Rivers is a complex mosaic of various types of forest, woodland, bushland, thicket, wooded grassland, grassland and aquatic and semi-aquatic communities. A general account has been written by Wild & Barbosa (1968). A few local studies have also been published (Pedro & Barbosa 1955, Macedo 1970, Tinley 1971b, 1971c) but there are no detailed descriptions and only a brief and tentative outline can be offered here. Some idea of the complexity of the vegetation can be gained from the fact that 20 of the 74 mapping units of the 'Vegetation Map of the Flora Zambesiaca area' (Wild & Barbosa 1968) are represented in this narrow coastal strip, although it represents only a small proportion of the Flora Zambesiaca area. Furthermore, most of these mapping units represent mosaics, not individual vegetation types.

4.1 Forest

Six types are recognized by Wild & Barbosa.

(i) Moist evergreen forest (Type 1 of Wild & Barbosa, part only). The forest occurring on the lower windward slopes of many mountains lying just inside the Zambezian Region is best regarded as outliers of Zanzibar–Inhambane vegetation since many of its constituent species, e.g. *Lovoa swynnertonii, Maranthes goetzeneana* (distribution map in White 1976b), *Crossonephelis (Melanodiscus) oblongus*, occur in the Zanzibar–Inhambane Regional Mosaic, though they are better represented in the northern half, and many are absent from the coast of Moçambique. At higher altitudes Afromontane species become increasingly numerous and lowland forest gives way to transitional forest and ultimately, where the massifs are high enough, to montane forest.

(ii) Moist semi-deciduous forest (Type 2 of Wild & Barbosa). It occurs between 700 and 1200 m on the subplanaltic slopes of Manica e Sofala and between 750 and 860 m on the western slopes of the Macondes Plateau (Fig. 3). Characteristic species in the former are *Newtonia buchananii, Erythrophleum suaveolens, Millettia stuhlmannii, Khaya nyasica, Ekebergia capensis* and *Pachystela brevipes*, and in the latter *Erythrophleum suaveolens, Pachystela brevipes, Ekebergia capensis, Xylopia aethiopica, Albizia gummifera* and *Syzygium guineense*. These trees form a canopy some 20 m or more in height, and where disturbed a dense shrub layer, with many species, develops. Some common shrub species are *Harungana madagascariensis, Macaranga capensis, Albizia adianthifolia, Trema orientalis* and *Bauhinia* spp. (Pedro & Barbosa 1955, Wild & Barbosa 1968, Macedo 1970).

(iii) Dry semi-deciduous forest (Type 5 of Wild & Barbosa, part only). It occurs principally on the sublittoral belt of ancient dunes. There is appreciable change in floristic composition towards the north. Between Inhambane and the Sabi River *Chlorophora excelsa, Ficus* spp., *Morus mesozygia, Celtis africana, Afzelia quanzensis* and *Dialium schlechteri* are characteristic species. In Manica e Sofala and Zambezia *Trachylobium verrucosum, Afzelia quanzensis, Nesogordonia parvifolia, Cola mossambicensis* and *Cynometia* aff. *webberi* are representative, whereas in Cabo Delgado District the floristic assemblage includes *Trachylobium verrucosum, Balanites maughamii, Afzelia quanzensis*, associated with thickets of *Guibourtia schliebenii, Pseudoprosopis euryphylla* and *Grewia* spp., and woodland of *Berlinia orientalis*. The forest is of medium stature (up to 15 m tall), but only small, much damaged remnants occur (Wild & Barbosa 1968, Tinley 1971c).

(iv) Dry deciduous forest (Type 6 of Wild & Barbosa) occurs scattered along the entire length of Moçambique north of Massinga. Rainfall varies from 700 to 1400 mm p.a. and floristic composition is correspondingly varied. This type collectively is characterized by *Adansonia digitata, Afzelia quanzensis, Balanites maughamii, Chlorophora excelsa, Cordyla africana, Dialium mossambicense, Fernandoa magnifica, Inhambanella henriquesii, Khaya nyasica, Millettia stuhlmannii, Pteleopsis myrtifolia, Rhodognaphalon* and *Sterculia appendiculata*. The sub-canopy layer of this forest is usually well developed and may even form a closed, thick, almost impenetrable layer of deciduous to semi-deciduous shrubs and small trees. Some of the characteristic genera are *Fernandoa, Markhamia, Tabernaemontana, Deinbollia, Baphia, Hugonia* and *Combretum*. These forest

Fig. 3. Lowland rain forest, west of the River Shire, overlooking Port Herald, Malawi Hills, alt. 690 m. *Burttdavya nyasica* 30 m high with a clear bole for 16 m and a diameter of 1 m at 3.2 m. above the ground. This fine tree occurs in association with *Newtonia buchananii* and *Khaya nyasica*. *Burttdavya nyasica* in Malawi appears to be restricted to the extreme south (photo J. D. Chapman).

patches are usually scattered in a mosaic of grassland and woodland types which are mainly Zambezian in affinity (see Chapter 10).

(v) *Hirtella* forest. Fringing the Zambezi delta both to the north and the south, moist *Pteleopsis–Erythrophleum* semi-deciduous forest occurs in mosaic with miombo woodland (Wild & Barbosa, type 9). Rainfall is from 1200–1400 mm p.a. and the water-table is high. In wetter areas *Hirtella zanzibarica* (distribution map in White 1976b) is abundant or even locally dominant. *Pandanus* and *Barringtonia* occur in swampy glades.

(vi) Fringing forest. The fringing forest occurring on the alluvium of the deltas and lower courses of the bigger rivers is described by Wild & Barbosa (type 54). The more important tree species include *Adansonia digitata, Cordyla africana, Garcinia livingstonei, Khaya nyasica, Parkia filicoidea, Rhodognaphalon, Sterculia appendiculata* and *Treculia africana.*

4.2 Woodland

The most extensive type of woodland seems to be a floristically impoverished version of miombo woodland dominated by species of *Brachystegia* and *Julbernardia globiflora*. Miombo woodland has a scattered distribution from the Rovuma River to just south of the Limpopo River. Nearly everywhere it interdigitates with, or occurs in mosaic with, patches of forest and other vegetation types. Miombo woodland occurs in mosaic with forest and, often also, other vegetation types in the following of Wild & Barbosa's mapping units: 9, 10, 13, 25, 26, 27, 31, 32 and 33. These miombo communities are discussed in Chapter 10.

Brachystegia spiciformis woodland (Wild & Barbosa, type 20) is extensively developed on 'Sul do Save' sublittoral sand dunes between the Sabi and Limpopo Rivers. Pedro & Barbosa (1955) believe this to be secondary to semi-deciduous coastal forest of *Dialium schlechteri, Cordyla africana, Afzelia quanzensis*, etc.

4.3 Thicket

Littoral thicket on recent dunes (Wild & Barbosa, type 14b) similar to that described under the seral stages of Dune forest occurring in the Tongaland–Pondoland Regional Mosaic occurs along almost the entire length of Moçambique. In Zambesia the commonest shrub species include *Sideroxylon inerme, Mimusops caffra, Carissa bispinosa, Salacia madagascariensis, Hugonia elliptica* and *Cardiogyne africana*. Near the Rovuma River, *Grewia glandulosa, Ozoroa reticulata* subsp. *grandifolia* and *Erythroxylum platycladum* are frequent.

Another type of littoral thicket (Wild & Barbosa, type 14) occurs on Quaternary sands in northern Moçambique. Its more characteristic species include *Guibourtia schliebenii, Pseudoprosopis euryphylla, Platysepalum inopinatum, Gossypioides kirkii* and *Combretum pisoniiflorum*. Scattered, usually tall, trees of *Rhodognaphalon, Afzelia quanzensis, Sterculia schliebenii* and *Manilkara discolor* (*altissima*) occur as emergents.

4.4 Grassland

Edaphic grassland on waterlogged soils occurs in mosaic with other vegetation types.

On the coastal plain, between the R. Sabi and R. Buzi, 'tandos' (Wild & Barbosa, type 42), or seasonally flooded clayey depressions occurring in sandy Quaternary or calcareous Cretaceous deposits, are covered with *Hyparrhenia, Ischaemum, Setaria* grassland. They are bordered by wooded grassland with *Parinari curatellifolia, Uapaca nitida, Syzygium guineense*, etc. (see Chapter 10).

Badly drained grassland with scattered palms, chiefly *Hyphaene coriacea* and *Borassus aethiopum*, together with other trees, including *Garcinia livingstonei* and *Syzygium cordatum* (Wild & Barbosa, type 44), occurs at several places along the Moçambique coast.

There are also extensive areas of badly drained grassland on the deltas of the larger rivers (Wild & Barbosa, type 54). Characteristic species include: *Setaria holstii* (sometimes dominant), *S. mombassana*, *S. sphacelata*, *Ischaemum afrum* (very common), etc. In very wet places in the Zambezi delta 'Elephant grass' *Pennisetum purpureum* is common.

Scattered in these badly drained grasslands, especially in the deltas of the larger rivers, acquatic and semi-aquatic communities are found (see Chapters 10 and 33).

5. The Tongaland–Pondoland Regional Mosaic

The vegetation of the Tongaland–Pondoland Regional Mosaic is even more complex than that of the Zanzibar–Inhambane, but it has been more fully described in the literature. Owing to the greater physiographic diversity of the Tongaland–Pondoland Regional Mosaic, climatic gradients are steeper, but vegetation types influenced by flooding or a high water-table are chiefly concentrated in the coastal plain in the northern part of the region. The Tongaland–Pondoland forests are floristically very diversified. They receive a fuller treatment than other vegetation types.

5.1 Forest

There are five main types of forest. Four of them, Sand forest, Dune forest, Swamp forest and Fringing forest occupy specialized sites. The remaining type, which is more widely distributed, can be referred to as 'Undifferentiated lowland forest'. Where the rainfall is adequate (>1000 mm p.a.) forest is climax and many remnants occur, usually in small patches less than 100 ha in area, though there are a few larger tracts (Palmer & Pitman 1972).

The forests of the Tongaland–Pondoland Regional Mosaic contain many Afromontane species. Their relationships to Afromontane forests are complex and are discussed elsewhere in this work (pp. 475, 570).

(i) Undifferentiated lowland forest. The larger surviving forest patches include the Gwaleweni Forest on the Lebombo mountains near Ingwavuma, Ngomi Forest near Nongoma, Ngoye Forest near Empangeni (Huntley 1965), Nkandhla Forest near Krantzkop (Edwards 1967), the lower parts of the Karkloof Forest near Pietermaritzburg (Rycroft 1943), the forests inland of Port Shepstone and Port St. Johns, the Dwesa Forest (Moll 1974) and the Manubi Forest. Many of the forests such as the Alexandria Forest and the Karkloof and Nkandhla forests at their upper altitudinal limits are intermediate between Zululand–Pondoland and typical Afromontane forests. Table 1 gives a summarized outline of the relationships of some of the forests from which quantitative data are available. This table clearly illustrates that Alexandria has a strong Afromontane affinity, though this forest also has certain distinctive features.

These Lowland Forests (Acocks 1953, type 1) are essentially evergreen, with a varying proportion of semi-deciduous species. The height and stratification of the forest depends largely on local site factors, and canopy height ranges from 10 to

Table 1. Summary of forest data (Alexandria, Knysna, Hogsback and Kologha, Dwesa, and Oribi from unpublished data Moll & Campbell; Karkloof, Hawaan and Hlogwene from Moll 1971; Stainbank from Rogers & Moll 1975; Krantzkloof from Moll 1968a; and Umdoni Park from Guy & Jarman 1969). Values are maximum cover-abundance values in the Braun-Blanquet scale.

	Alexandria Type 1	Alexandria Type 2	Knysna	Hogsback & Kologha	Karkloof	Hawaan	Hlogwene	Umdoni park	Stainbank	Krantzkloof	Dwesa Type 1	Dwesa Type 2	Oribi Type 1	Oribi Type 2	Oribi Type 3
Number of relevés	3	2	3	2	12	12	24	2	2	12	9	9	4	3	2
Schotia latifolia	5														
Cassine crocea	5														
Eugenia woodii	4														
Sideroxylon inerme	4														
Gonioma kamassi			2												
Curtisia dentata			2												
Rinorea angustifolia					4										
Syzygium gerrardii					1										
Maytenus heterophylla	5	5	2	5	1										
Scolopia mundii	5	1	1	3	1										
Diospyros whyteana	2	3	4	5	1										
Podocarpus falcatus			3	4	5	1									
Rapanea melanophloeos			1	5	3	1									
Canthium loculus			1	4	1	1									
Linociera foveolata				1	1	1									
Ocotea bullata				3	1	1									
Kiggelaria africana				1	1	1									
Prunus africana				1		1									
Cavacoa aurea						5									
Pancovia golungensis							4								
Canthium obovatum						1	2	1							
Drypetes natalensis						1	1	1							
Sapium ellipticum										4					
Erythroxylum pictum										2					
Faurea macnaughtonii										1					
Memecylon sp.										3					
Strychnos innocua						5	4	2	3	1					
Ziziphus mucronata						1	1	1		2					
Chrysophyllum viridifolium							1	2	3	2					
Baphia racemosa						2		1		2					
Anastrabe integerrima								1	2	3					
Manilkara discolor								1	5	2					
Schrebera alata									1	1					
Millettia sutherlandii											4				
Buxus macowanii	2										4	3			
Encephalartos villosus										4	3	5	3		
Heywoodia lucens											5	1	4		
Olea woodiana					1	3	R	1	3	2	1				
Deinbollia oblongifolia					3	5	R	1	R	2	4				
Strychnos usambarensis					2	1	3	1	1	4	3	5			
Diospyros natalensis				1	3	3	1	1	1	2	4	3			
Erythroxylum emarginatum				1	2	3	1	2	1	2	5	3			
Dracaena hookerana				1	2	1	2	2	3	2	1	2			
Ochna natalitia			2	1	1	1	1	1	1	4	4	3			
Gardenia thunbergia		1	1	1	1	1	1	1	1	1	1	3			
Cryptocarya woodii		1	1	2	1	1	4	4	3	2	1	3			
Vitellariopsis emarginatum										1			4		
Buxus natalensis					1					5	5		4		
Celtis africana	3	1	5	1	3	3	1	1	2	4	2	4	2		
Peddiea spp.	1	1	R	1	3	2	1	1	1	2	3	1	5		
Calodendrum capense		1	1	3		1			1	1	1	2	2		
Croton sylvaticus	4			1	3	1	5	5	1		1	2	5		

Table 1 (contd.)

	Alexandria Type 1	Alexandria Type 2	Knysna	Hogsback & Kologha	Karkloof	Hawaan	Hlogwene	Umdoni park	Stainbank	Krantzkloof	Dwesa Type 1	Dwesa Type 2	Oribi Type 1	Oribi Type 2	Oribi Type 3
Number of relevés	3	2	3	2	12	12	24	2	2	12	9	9	4	3	2
Clerodendrum glabrum		3		1	1			2		1	2	1	2		
Suregada africana	4				1						2	2			
Cussonia sphaerocephala	1	1	1	1	1	1	1	2	1	4	3	1	5	4	
Canthium ciliatum	2		3	1	1	1	1	1	1	2	2	2	4	2	
Ptaeroxylon obliquum	5	1	1	1	1					1	1	2	2	2	
Mimusops obovata						2	1	1	1	1	1		4	3	
Pavetta natalensis															4
Protorhus longifolia							5	5	4	1	3	3	5	2	
Drypetes gerrardii							1	R	2	4	5	5	5	5	
Millettia grandis							1	2		4	4	3	3	4	
Strangeria sp.										R	2	5		5	3
Celtis gomphophylla										1		2	4	2	
Excoecaria simii					1						5	5	4	2	
Harpephyllum caffrum								2	1	1			3	3	4
Bachmannia woodii							1	1		1			2		4
Garcinia gerrardii							1			1				3	4
Cola natalensis						5		4		2			5		4
Fagara davyi				5	1					1			2		
Xymalos monospora				5	3					1					2
Nectaropetalum zuluense													3		4
Euclea natalensis		2			R	2	1	1	1	1	3	3	5	R	R
Monanthotaxis caffra			R		2	4	4	3	1	3	5	5	5	5	5
Chaetacme aristata	4				R	3	3	2	3	1	2	4	5	2	4
Strychnos henningsii					3	R	1	R	1	4	3	3	2	4	
Bequaertiodendron natalense					1	4	5	3	3	2	1	5	4	5	
Oricia bachmanii					1	1	1	1	1	3	2	2	5	5	
Trichilia dregeana					1	4	2	R	3		3	2	2	5	
Rawsonia lucida					1	R	3	1	1	1		3	5	4	
Strophanthus speciosus					R						1		4	2	
Rothmannia globosa					2	R	3	2	2	1	4	3	4	4	2
Cassipourea gerrardii					1	1	R	1	1	2	5	4	4		5
Tricalysia capensis						2	1	1	3	4	3	2		5	5
Eugenia natalitia					1		2	2	2			2	3	3	
Teclea gerrardii	2					2	2	1			5	4	3		2
Apodytes dimidiata	2	R	4	R	1	R	2	1	R	1	2	2	R	R	R
Canthium ventosum	2	R	R	5	R	1	1	2	R	3	1	3	3	4	2
Gardenia amoena	1	R	R	3	1	R	1	4	5	1	3	2	3	2	4
Olea capensis	2		5	3	1						1		2	R	2
Vepris undulata	4	5	R		1		R	2	R	1	3	3		R	
Fagara capensis	4	3	R	3		1	1	1	1	R	2	2		R	
Dovyalis rhamnoides	5	5		3		5	2	1	R	R	4	3			
Ochna arborea	5	4			1			1	R	2		5		2	
Linociera peglerae	2		R				1	2	R		4	4		5	
Podocarpus latifolius			4	3	3					2		4			2
Durvernoia adhatodoides				1		1				1	2		5	4	
Psychotria capensis					1	1	1	4	1	3	3		1		
Drypetes arguta						2		3	3	2		1	4		5
Tricalysia sonderana						2	2	1	3	2	3		2		
Maytenus mossambicensis					2					1	2		3		
Turraea floribunda							2	1	2			2	2		
Cryptocarya myrtifolia				1						1	2		4		
Cordia caffra		5	2			1	2			1					
Calpurnia aurea					1		1				2	1			
Scolopia zeyheri					2			1	1	2					
Xylotheca kraussiana					1			1	1	2					
Combretum kraussii					1					2					
Ekebergia capensis					1		1								
Cassipourea gummiflua					1		1	1							

30 m. One of the notable differences between these forests and the Afromontane forests a little further inland is that regeneration and growth is good (see Moll 1972c, Moll & Woods 1971). Another feature of these forests is that a large number of species of woody lianas occur; they are particularly common in disturbed areas. Some of the more common lianas and other climbers are *Acacia ataxacantha, A. kraussiana, Entada spicata, Dalbergia armata, D. obovata, Rhoicissus rhomboidea, R. tridentata, R. tomentosa, Cissus* spp., *Uvaria caffra, Monanthotaxis caffra, Combretum edwardsii, Quisqualis parviflora, Secamone alpinii, Oncinotis inandensis, Acridocarpus natalensis, Buddleia dysophylla, B. pulchella, Capparis sepiaria, Cnestis natalensis, Diospyros simii, D. villosa, Embellia ruminata, Canthium gueinzii, Putterlickia* sp., *Scutia myrtina*, and *Urera cameroonensis*. Exploitation of all valuable timber trees (Fourcade 1889) makes the investigation of the ecology of these forests extremely difficult, however. Characteristic canopy trees include *Drypetes gerrardii, Strychnos* spp., *Vepris undulata, Trichilia dregeana, Protorhus longifolia, Millettia grandis, Albizia adianthifolia, Brachylaena* spp., *Celtis* spp., *Chrysophyllum viridifolium, Chaetacme aristata, Ficus* spp., *Croton sylvaticus, Combretum kraussii, Cassipourea* spp. and *Mimusops obovata*. Other canopy trees that are infrequent, yet are nevertheless fairly characteristic, are *Phyllanthus* spp., *Nuxia* spp., *Scolopia zeyheri, Syzygium gerrardii, Podocarpus* spp., *Homalium dentatum, Cola natalensis* and *Cavacoa aurea*. The sub-canopy layer is usually well developed, but this could be partly due to exploitation which has opened up the upper canopy. Some of the common, widespread genera in the subcanopy are *Bequaertiodendron, Canthium, Clausena, Dracaena, Drypetes, Erythroxylum, Eugenia, Memecylon, Pavetta, Peddiea, Teclea* and *Tricalysia*, with other fairly common genera being *Allophylus, Anastrabe, Baphia, Buxus, Cryptocarya, Diospyros, Gardenia, Ochna, Oxyanthus, Pancovia, Rawsonia, Rothmannia* and *Xylotheca*. The herbaceous field layer of the forest is usually well developed; many grasses, sedges and members of the Acanthaceae are dominant and widespread.

 A feature of these lowland forests is the well developed forest margin, where this has not been continually destroyed by annual grass fires or cleared for cultivation. On such well-developed margins woody shrubs and small trees of the genera *Antidesma, Clerodendrum, Combretum, Dichrostachys, Fagara, Grewia, Heteropyxis, Hippobromus, Kraussia, Maesa, Maytenus* and *Rhus* are usually present. In addition, tall (up to 2 to 3 m) sub-woody plants such as *Clutia, Euphorbia, Euryops, Hibiscus, Leonotis, Lippia, Polygala, Pseudarthria, Solanum* and *Syncolostemon* occur, together with tall coarse grasses such as *Cymbopogon, Hyparrhenia, Miscanthidium* and *Setaria*. Many climbers occur in this marginal community and perhaps the most noteworthy, as they are usually felt if one walks through the community, are *Smilax kraussiana* with its hooked spines, and *Tragia* spp. and *Dalechampia* spp. with their stinging hairs.

(ii) Sand forest. In Tongaland a distinctive type of dry semi-deciduous to deciduous forest occurs on sandy soils and is known locally as 'Sand Forest' (Moll 1968c, 1973). Similar forest occurs in Moçambique south of Maputo (Wild & Barbosa 1968, type 5, 'Dry semi-deciduous (lowland sublittoral) forest', pro parte). The other types of forest included by Wild & Barbosa in this category, which occur further north in Moçambique, mostly on a sublittoral belt of ancient dunes, are very different in floristic composition and are composed mainly of

Zanzibar–Inhambane species. There is very little published information about them.

South of Maputo the forests are better documented (Myre 1964, Vahrmeijer 1966, Moll 1973). They occur on pale orange to grey sandy soils, where the rainfall is between 700 and 900 mm p.a.; additional moisture is derived from inversion mists in winter. This forest is typified by the trailing *Usnea* lichens growing on exposed branches, as well as by the general structure and floristic composition of the community. Structurally the Sand forest is 10 to 25 m high and forms a dense, almost impenetrable community with three principal strata: a canopy of *Newtonia hildebrandtii* (usually in localized pure stands), *Cleistanthus schlechteri, Balanites maughamii, Ptaeroxylon obliquum, Dialium schlechteri, Erythrophleum lasianthum, Afzelia quanzensis* and *Pteleopsis myrtifolia*; a well developed sub-canopy of small trees and shrubs such as *Hymenocardia ulmoides, Cola microcarpa, Monodora junodii, Toddaliopsis bremekampii, Croton* spp., *Suregada zanzibarensis, Grewia* spp., *Ochna* spp., *Salacia leptoclada* and *Cavacoa aurea*; and a poorly developed herbaceous ground layer of *Sansevieria* spp., *Gonatopus* spp., *Commelina* spp. and others (Figs. 4, 5 and 6).

Fig. 4. A part of the Iseleni Sand forest in northeastern Natal showing the longitudinal east–west relationship of the forest and the sharp forest/wooded grassland interface.

(iii) Dune forest. This type is particularly well developed in the northern part of the Tongaland–Pondoland Regional Mosaic, especially north of Cape St. Lucia. It is sufficiently distinct to warrant special attention, and has been fairly well studied floristically south of Maputo (Bayer 1938, Myre 1964, Vahrmeijer 1966, Tinley 1967, Breen 1971, Breen & Jones 1971, Venter 1976, Moll 1972a). Newly formed parabolic dunes which are initially unstable and saline, become rapidly colonized by pioneer sand-binding, gregarious species particularly *Scaevola*

Fig. 5. Sand Forest in northeastern Natal showing the height of the canopy (*Newtonia*) and the structure of the forest.

thunbergii. Some of the more common associated species are *Ipomoea pes-caprae*, *Canavalia maritima*, and *Carpobrotus* spp. (Fig. 7). These pioneers create conditions suitable for the establishment of littoral thickets or Dune forest. The canopy of this closed woody community of essentially evergreen, coriaceous-leaved, gnarled plants shows a typical wind-pruned effect (Fig. 8, 9 and 10). The individual canopy species are characteristically laced together by abundant woody lianas, the combined effect being a relatively tightly sealed, dense canopy with very little understorey. Common

Fig. 6. Schematic profile diagram running in an east to west direction through Sand Forest and marginal communities (after Moll 1968c).

Fig. 7. *Scaevola* stabilizing new dunes on the Natal coast. The first dune less than 5 years old, while the second dune is about 10 years old.

canopy species are *Mimusops caffra, Diospyros rotundifolia, D. inhacensis, Euclea natalensis* and many more. Common species of lianas are *Dalbergia armata, D. obovata, Rhoicissus* spp., *Acacia kraussiana, Entada spicata, Ficus burtt-davyi, Uvaria* spp., *Monanthotaxis caffra, Landolphia petersiana, Cynanchum* spp., *Asparagus* spp. and many more. The ground flora, back from the first forested dune, may be fairly well developed with a dense layer of the fern, *Phymatodes scolopendria*, plus *Zamioculcas, Gonatopus, Sansevieria* and others, or on drier sites a dense, almost pure stand of *Isoglossa woodii*. In many areas the forest has been cleared for cultivation and common pioneer trees on abandoned sites are *Acacia karroo, Brachylaena discolor* and *Apodytes dimidiata* with a dense herbaceous understorey of grasses and herbs. In areas where blow-outs have occurred *Casurina equisetifolia* has sometimes been planted to stabilize the shifting sands (particularly south of Maputo).

Further south Dune forest is floristically more depauperate though structurally similar. The Dune forest which formerly occurred north of Durban, before it was cleared for housing in the 1960's, was perhaps the tallest, best developed Dune forest of the whole Indian Ocean Coastal Belt. Here the canopy trees were once 30 to 35 m tall. Common species included the characteristic, *Mimusops caffra*,

Fig. 8. Wind-pruned dune forest at Mapelana.

Fig. 9. Dune forest at Mapelana just south of Lake St. Lucia, showing the high, first dune with a well pruned canopy. Note the lack of pioneer communities and the sharp beach/forest interface.

which formed an almost pure community, as well as scattered trees of *Euclea natalensis*, *Cordia caffra*, *Strychnos decussata*, *Canthium obovatum* and *Ziziphus mucronata*. South of Durban there is a rapid decline in species occurring in Dune forest, and in Transkei the common tree species are limited to *Mimusops caffra* and a few others, with *Sideroxylon inerme* becoming increasingly important. *M. caffra* does not occur quite as far south and west as Port Elizabeth, and the isolated patches of Dune forest found west of there, in Capensis, as far west as Cape Town, comprise almost pure stands of *S. inerme*.

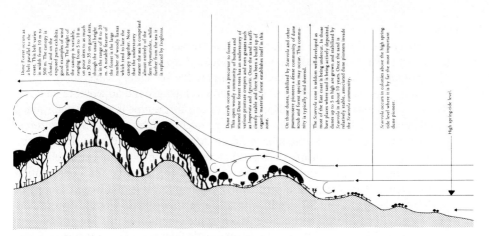

Fig. 10a. Schematic profile diagram running at right angles to the coast, through Dune forest, showing the pioneer, sand stabilizing communities and mature forest (after Moll 1972a). Examples are rare, where all the various successional stages illustrated above are clearly shown in stretched zones. The arrowed lines indicate the wind profile over the community.

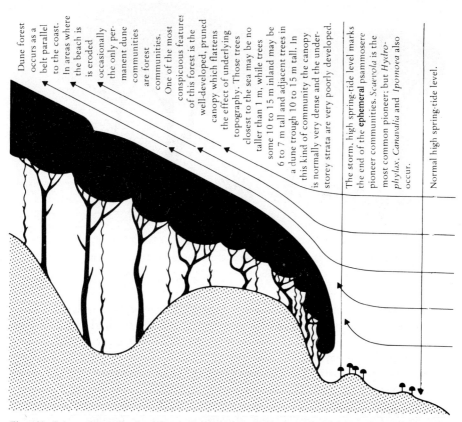

Fig. 10b. Schematic profile diagram running at right angles to the coast, through Dune forest. The example illustrated is the more usual one, showing poorly developed pioneer communities due to the undercutting of the east coast of South Africa (after Moll & Pierce 1975). The arrowed lines indicate the wind profile over the community.

In many areas the Dune forest has been disturbed by urban development along the coast. In such disturbed areas *Brachylaena discolor*, *Strelitzia nicolai* and/or *Chrysanthemoides monilifera* tend to dominate the secondary vegetation.

(iv) Swamp forest. On lowland wet sites and swampy ground adjacent to streams, Swamp forest may occur. The distribution of this forest type is mainly determined by edaphic factors and tends to be independent of rainfall. These Swamp forest patches were never extensive, and in South Africa many have been drained and cleared for agriculture (Moll 1976). What remains today is an extremely poor representation of this forest type and they rate top priority for conservation (Bayer et al. 1968). The general structure of Swamp forest is of a closed canopy of fairly even height, some 30 m tall, with a sparse subordinate woody stratum and usually a well developed herbaceous layer; some common plants are grasses and sedges up to 2.5 m tall (Fig. 11). A characteristic feature is the profusion of ferns,

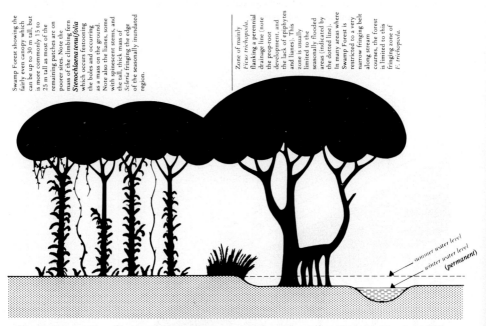

Fig. 11. Schematic profile diagram through a section of Swamp forest (modified after Huntley 1965). 1965).

the most common and widespread being *Stenochlaena tenuifolia*, which is not only a climber festooning the boles of canopy trees, but fills the herbaceous ground layer also. In Swamp forest proper there are usually few tree species that are important in the canopy, though the species diversity increases greatly on slightly drier sites. True Swamp forest canopy trees are *Ficus trichopoda*, *Syzygium cordatum*, *Raphia australis*, *Voacanga thouarsii* and *Rauvolfia caffra*. Additional species on drier sites are *Podocarpus falcatus*, *Scolopia stoltzii*, *Ficus capensis*, *Ilex mitis*, *Rapanea melanophloeos*, *Macaranga capensis*, *Schefflera umbellifera*, *Erythrina caffra*, *Cussonia sphaerocephala* and *Bridelia micrantha*. Woody lianas, such as *Dalbergia armata*, *Canthium gueinzii* and *Pisonia aculeata* can be fairly common, especially on drier sites. The literature on Swamp forest is particularly poor, and, though Tinley (1971c) gives an indication of its distribution (between 21 and 28°S), he gives no indication of the structure and composition. In South Africa the southern limit of Swamp forest relics occurs at about 31°S, with the best developed areas being in the region of Kozi Bay (Fig. 12).

(v) *Fringing forest*. Most of the rivers in the northern part of the Tongaland–Pondoland Regional Mosaic flow fairly slowly across the flat coastal plain, and, where it has not been destroyed, well developed Fringing forest occurs. Common Fringing forest species are *Ficus* spp., *Trichilia emetica*, *Syzygium* spp., *Xanthocercis zambesiaca*, *Rauvolfia caffra*, and *Ekebergia capensis*. These species are extremely important for stabilizing the river banks. In areas where this forest has been destroyed the river banks are eroded, and flooding of local, low-lying areas occurs (often destroying gardens and crops). On the river flood plains

Fig. 12. Oblique aerial view of Swamp forest flanking one of the Kosi Lakes (Manzenyama = black water, due to the humic acid concentration of the water). The forest comprises a mosaic of *Raphia* palms and *Syzygium* dominated forest. This patch of Swamp forest is by far the most extensive and best conserved in the Republic of South Africa.

Acacia xanthophloea forms localized pure stands in low lying areas, particularly fringing pans on the Pongolo, Usutu, Limpopo and Save Rivers (Figs. 13 and 14, and Chapter 10, Fig. 67). Other widespread tree species of the flood plains are *Acacia albida*, *Kigelia africana*, *Combretum imberbe*, *Tabernaemontana elegans*, *Lonchocarpus capassa* and *Trichilia emetica* (see Chapter 10).

5.2 *Woodland, bushland and thicket*

Non-forest woody vegetation in the Tongaland–Pondoland Regional Mosaic is composed mainly of large bushes or bushy trees which are less than 9 m tall and are multiple-stemmed or have extremely short boles. True trees taller than 9 m

Fig. 13. The zonation of floating aquatics (*Pistia* and *Paspalidium*), rooted aquatics (*Phragmites*) and fringing trees (*Ficus sycomorus* and *Acacia xanthophloea*) is clearly defined on Banzi Pan on the Usutu River floodplain.

with clear straight boles 2 m or more in length are scattered or absent. This vegetation fits the category of bushland in the classification of Greenway (1973) and that of the UNESCO/AETFAT 'Vegetation Map of Africa' (White, in press) when it is open, or thicket when it is closed. Patches of woodland or open woodland also occur, often dominated by *Sclerocarya caffra*, *Acacia burkei*, *A. nigrescens* or *Terminalia sericea*. They are of restricted occurrence, however, and may, in part, have been derived from bushland by selective removal of the smaller woody plants. A selection of non-forest woody vegetation is described below.

In southern Moçambique and Tongaland open to closed deciduous bushland occurs. It corresponds to types 34 and 43 of Wild & Barbosa (1968). This vegetation is fairly well documented in the literature (Myre 1964, de Moor et al. in prep.,

Fig. 14. Schematic profile diagram through a section of Fringing forest and flanking communities (partly after Tinley 1958). The left bank of the river has well-conserved communities which, once radically disturbed, result in an unstable river bank and subsequent erosion (shown diagrammatically on the right bank).

Tinley 1964, Vahrmeijer 1966). The area supports a fairly dense population, consequently the vegetation is severely damaged and many of the woody species that remain are those that bear edible fruits. The original vegetation of the area was probably a closed bushland or even low forest of a semi-deciduous type. Today common tree species, some 5 to 7 m tall, are *Albizia* spp., *Afzelia quanzensis*, *Sclerocarya caffra*, *Terminalia sericea*, *Acacia burkei*, *Dialium schlechteri*, *Garcinia livingstonei*, *Strychnos* spp., *Ficus* spp., *Trichilia emetica* and *Syzygium cordatum*. The woody understorey can be well developed in parts, common species being *Antidesma venosum*, *Tecomaria capensis*, *Dichrostachys cinerea*, *Grewia* spp., *Vangueria infausta*, *Xylotheca kraussiana*, *Euclea* spp., *Rhus* spp. and many more.

The vegetation on the rhyolitic hills of the Lebombo Range, occurring in southern Moçambique and the northern part of Natal and rising steeply from the flat coastal plain to an altitude of some 1000 m, consists of a mosaic of grassland, bushland and isolated patches of forest where the local conditions are sufficiently moist. This vegetation is strongly Zambezian in affinities and is discussed in Chapter 10.

On heavy clay soils, often of alluvial origin, which are subject to seasonal waterlogging, various types of short deciduous thicket occur. They vary greatly in species composition from site to site but in structure are fairly similar, comprising a dense, almost impenetrable thicket of tall shrubs and short trees. Occasional tall trees may emerge from the canopy. South of Maputo this type of vegetation has been described by de Moor et al. (in prep.) and Tinley (1964). The floristically diverse canopy is 2 to 5 m tall; perhaps the most characteristic species are *Acacia grandicornuta*, *Gardenia cornuta*, *Euphorbia grandicornis*, *Dichrostachys cinerea*, *Commiphora* spp. and *Pappea capensis*. Many of the species are spinescent. A large number of succulents and climbers occur, and there is a fairly dense field layer of *Panicum* spp., *Urochloa* spp., *Sansevieria* spp., *Jatropha* spp. and various species of Acanthaceae.

Much of the woodland, bushland and thicket occurring in the deep valleys of the Tongaland–Pondoland Regional Mosaic is secondary in origin or has at least been profoundly modified by human activity. As was mentioned earlier these types show strong affinities to various other regions. North of the Umtamvuna River the relationship is mainly Zambezian, especially in the deep, dry river valleys of the Umkomazi, Umgeni, Umvoti and Tugela and on the intervening ridges and spurs.

Bush Clump Mosaic (Moll 1975, 1976) which consists of scattered bush clumps, 5 to 20 m in diameter and with a density of 1 to 10 per ha, in a grassland matrix, occurs on hill-slopes from the Tugela to the Umtamvuna Rivers, where conditions are not quite sufficiently mesic for a forest climax to develop. The bush clumps have a characteristic appearance with one or more large trees (7 to 15 m tall) in the centre surrounded by smaller bushes and shrubs, and often occur on old termitaria (Fig. 15). Over much of its range this vegetation type has been drastically changed, not only have the bush clumps been cleared for firewood and patch cultivation, as the soils beneath the clumps have a better than average fertility, but the species composition of the grassland mosaic has been completely altered. In addition large areas have been cleared for cultivation, notably sugarcane. Most of the woody species of the bush clumps are either evergreen or partially deciduous, forest species or forest pioneers, though on drier sites species like

Fig. 15. Schematic profile diagram through a section of Bush Clump Mosaic, showing bush clumps on old termitaria in a tussocked grassland mosaic.

Euphorbia ingens, Acacia spp. and *Sclerocarya caffra* may be important (the last two being deciduous). Woody lianas are common and the clumbs often have an impenetrable thicket as an understorey. On some edaphically wetter sites palms, particularly *Phoenix reclinata*, but occasionally *Hyphaene natalensis*, may be important in the clumps and then the Bush Clump Mosaic resembles Palm Veld. The composition of the original grassland matrix is not well documented, but it is probable that *Themeda, Cymbopogon, Setaria, Digitaria* and *Hyparrhenia* were common genera, forming a coarse, tussocked grassland some 1.5 to 2 m tall. Today most of the remaining grassland matrix is dominated by *Aristida junciformis*, with *Eragrostis* spp., *Sporobolus* spp. and *Hyparrhenia* spp. being locally important (see also Acocks 1953, types 1 and 5).

In the dry river valleys north of the Umtamvuna River the vegetation is mixed, and its structure and composition is dependent on local site characteristics and biotic factors (see West 1951, Acocks 1953, types 6, 10 and 23, Killick 1959, Wells 1962, Morris 1967, Edwards 1967, Moll 1976). In essence it is a short (3 to 10 m tall) open or closed bushland of bushes and multiple stemmed tall shrubs which are either evergreen or deciduous though the majority of dominants are deciduous. The fieldlayer consists of grasses. On more xeric sites succulents such as *Euphorbia* spp., *Aloe* spp., *Crassula* spp. and *Portulacaria afra* are most common and the grass understorey is either absent or sparse, though herbs may be fairly common. On wetter sites an open to closed *Acacia*-dominated wooded grassland occurs on moderately deep soils, while on rocky sites *Combretum* spp., *Heteropyxis natalensis, Erythrina latissima, Cussonia spicata* and many others occur in local vegetation complexes. On valley bottoms a *Spirostachys africana* community occurs in scattered, favourable, flat, hot sites with light winter frosts. Of phytogeographic importance is the rapid fall off in species composition from north of the Tugela River where Acocks (1953) lists over 100 important large woody species, to the similar vegetation type south of the Umtamvuna River where only some 20 to 30 species occur (Comins 1962). Some of these vegetation types have Karoo–Namib affinities rather than Zambezian affinities.

5.3 Grassland

Very little literature is available on the grasslands of the Tongaland–Pondoland Regional Mosaic, and even the written descriptions are unsatisfactory because most attention has been given to the woody vegetation in the literature. The only quantitative work is that by Myre (1964, 1971) who studied the vegetation of the coastal plain of Moçambique south of Maputo. These grasslands are chiefly composed of tussocked forms, which develop into a closed grassland some 1 to 1.5 m tall, and are mainly of Zambezian affinity. The genera *Aristida, Pogonarthria, Tragus* and *Perotis* are the important pioneer grasses, and *Rhynchelytrum repens* forms a pink waving sward on abandoned cultivated sites when the inflorescences are developed. The communities are discussed in somewhat more detail in Chapter 10.

The only grassland type that requires additional discussion is that of the sublittoral. These grasslands are essentially edaphically controlled (high water-table grasslands) and are perhaps more extensive today because of the increased frequency of fires. It is seldom possible to look across the Tongaland coastal plain without seeing at least one smoke plume on the horizon, indicating a patch-burn. These coastal grasslands experience a climate which allows growth throughout the year, and are not, therefore, as distinctly seasonal as the grasslands of the hinterland. Because growth is not distinctly seasonal, and production, even on the infertile soils, is high, the same areas of grassland can be burned twice or even thrice in one year and often are. As a result the species composition is such that the most fire resistant genera occur commonly, particularly *Themeda, Tristachya, Trachypogon* and *Aristida*. In depressions *Hemarthria, Ischaemum, Acroceras* and *Paspalum* are locally common and because these areas are more moist they are difficult to burn and are, therefore, better protected. A characteristic feature of this grassland is the abundance of dwarf woody plants of distinctive habit (Henkel, Ballenden & Bayer 1936, Bayer 1938) which White (in press) refers to as 'geoxylic suffrutices' (Fig. 16). Some, e.g. *Syzygium cordatum*, are suffruticose

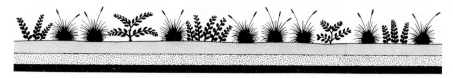

Fig. 16. Schematic profile diagram through a section of edaphically and fire-controlled tussock grassland and 'geoxylic suffrutices' of *Parinari, Diospyros, Salacia* and *Eugenia* (multi-stemmed) and *Syzygium, Parinari* and *Diospyros* (dwarf single-stemmed). The substrate basically comprises an upper layer of sandy soil which may be seasonally dry, a second layer of waterlogged sandy-clay overlying an impermeable layer.

variants of tree species, whereas other species, e.g. *Parinari capensis* and *Diospyros galpinii*, are always suffruticose. Other dwarf woody species belong to the genera *Salacia* and *Eugenia*. If these grasslands are protected from fire, especially on the drier sites where plantations of *Pinus patula* have been established, the woody plants grow taller and this community is rapidly invaded by other forest precursor species, and within 10 to 15 years scrub forest has become established. Where areas of grassland have been cultivated and the fields abandoned *Imperata cylindrica* becomes established, particularly on the deeper, better drained, sandy sites.

On more inland situations where the high water-table is of a more seasonal nature, various palms tend to characterize the community (Moll 1972b, Wild & Barbosa 1968, type 44). This Palm Veld occurs mainly south of Inhambane and north of St. Lucia in Tongaland proper (Campbell 1949) and its ecology has been well described in the literature (Henkel et al. 1936, Bayer 1938, Wild & Barbosa 1968, Moll 1972b, 1976). It does, however, occur to the north and south of this region, but more as scattered patches on environmentally suitable sites. In

Tongaland the Palm Veld consists of scattered *Hyphaene natalensis* with some *Phoenix reclinata* and *Borassus aethiopum* to the north, in a tussock grassland matrix (Figs. 17 and 18). The average number of palms per hectare is about 100, and they are used extensively by the local population for fibre and to provide 'wine' (ubu Sulu). To obtain the latter the palms are 'tapped' by a process which involves the removal of the apical growing point. Thus most of the individuals are never more than a few metres in height though they are capable of growing from 5 to 12 m tall (Moll 1972b).

Fire-maintained grassland is also widespread, especially on deep soils, where rainfall is c. 1000 mm p.a. and temperatures are cooler, particularly south of the Umzimkulu River (Acocks 1953, type 3 and part of 5). These fire-maintained grasslands are dense, tussocky, vigorous types growing to a height of 1 to 1.5 m

Fig. 17. Palm Veld with *Hyphaene natalensis*.

tall. The natural fire climax composition was one where members of the Andropogoneae dominated, and the fire régime was probably one burn at least once every three to five years in late winter when the grassland is most inflammable. Man's impact on this grassland was two-fold; firstly he began burning the grassland annually, usually in late summer and autumn (Staples 1926, 1930, Scott 1947, 1951) and secondly his domestic stock were selective grazers (Acocks 1966) which were not allowed free movement over the whole range. These changes resulted in a completely different grassland composition with *Aristida junciformis* becoming virtually dominant over most of the area.

5.4 *Fynbos*

True fynbos does not occur but species belonging to characteristic fynbos genera are conspicuous in some communities, particularly on sandy soils of low nutrient status. For example, just north of Lake St. Lucia some of the seasonal wet-lands

Fig. 18. Schematic profile diagram through a section of Palm Veld. This edaphic community (high water-table) occurs on sandy soils which are underlain by a hard impermeable layer of varying depth. Where the sand is deep (on ridges and hillocks) taller woody communities grow, while in depressions the water table lies above the soil surface forming small to large, seasonal or perennial, pans (see Tinley 1967, 1971a).

are locally dominated by *Restio* cf. *sieberi*, and a dwarf *Protea* occurs in the high water-table grassland with the suffruticose forms of *Syzygium cordatum* and *Parinari capensis* (Bayer 1938). Further south the fynbos links become stronger, and near Port Shepstone there are localized areas of vegetation with almost complete Capensis affinities. Here the dominant species belong to *Protea*, *Leucospermum*, *Passerina*, *Erica* and Restionaceae (R. G. Strey, pers. comm.). The size and frequency of these Capensis links increase until beyond the Sundays River true Capensis begins (Chapter 8).

5.5 Swamps

Before completing the description of the vegetation of the Tongaland–Pondoland Regional Mosaic mention must be made of two kinds of wet-land community. The first is the *Cyperus papyrus* swamp which occurs on permanently wet, low-land sites such as river deltas and fringing pans and lakes. Suitable sites are not common and occur at such places as the Mfolozi and Mkuzi Swamps at St. Lucia (Bayer & Tinley 1965). Where *C. papyrus* occurs it usually forms an almost pure stand some 2 to 3 m tall. A much more widespread form of this community, which occurs commonly along the coast belt on periodically flooded areas, is the characteristic hydrosere successional stages (Fig. 13) typified by the floating aquatics *Pistia*, *Eichhornia*, *Lemna* and *Wolffia*, rooted aquatics *Nymphaea* and *Nymphoides*, submerged aquatics *Potamogeton*, and fringing species of *Paspalidium*, *Echinochloa*, *Phragmites*, *Typha* and *Cyperus papyrus* (Wild & Barbosa 1968, type 53, de Moor et al. in prep., and Chapters 33 and 10).

References

Acocks, J. P. H. 1953. Veld types of South Africa. Mem. Bot. Surv. S. Afr. 28:1–192.
Acocks, J. P. H. 1966. Non-selective grazing as a means of veld reclamation. Proc. Grassld Soc. S. Afr. 1:33–39.
Bayer, A. W. 1938. An account of the plant ecology of the Coastbelt and Midlands of Zululand. Ann. Natal Mus. 8:371–454.
Bayer, A. W., Bigalke, R. C. & Crass, R. S. 1968. Conservation of vegetation in Africa south of the Sahara: Natal. Acta Phytogeogr. Suecica 54:243–247.
Bayer, A. W. & Tinley, K. L. 1965. The vegetation of the St. Lucia Lake area. Appendix 9, Report of the Commission of Inquiry into the alleged threat to animal and plant life in St. Lucia Lake. Govt. Printer, Pretoria.
Breen, C. M. 1971. An account of the plant ecology of the dune forest at Lake Sibayi. Trans. Roy. Soc. S. Afr. 39:223–234.
Breen, C. M. & Jones, I. D. 1971. A preliminary list of angiosperms collected in the vicinity of Lake Sibayi. Trans. Roy. Soc. S. Afr. 39:235–245.

Brooks, E. A. & Webb, C. de B. 1965. A history of Natal. Natal University Press, Pietermaritzburg.
Campbell, G. G. 1969. A review of scientific investigations in the Tongaland area of northern Zululand. Trans. Roy. Soc. S. Afr. 38:305–316.
Chapman, J. D. & White, F. 1970. The evergreen forests of Malawi. Commonwealth Forestry Institute, Univ. of Oxford.
Comins, D. M. 1962. The vegetation of the districts of East London and King William's Town, Cape Province. Mem. Bot. Surv. S. Afr. 33:1–32.
Crampton, C. B. 1912. The geological relations of stable and migratory plant formations. Scot. Bot. Rev. 1:1–61.
Dale, I. R. 1939. The woody vegetation of the Coast Province of Kenya. Imp. For. Inst. Paper 18:1–28. Comm. For. Inst., Oxford.
de Moor, P. P., Pooley, E., Neville, G. & Bowbrick, J. The vegetation of Ndumu Game Reserve, Natal: a quantitative physiognomic survey (in prep.).
Edwards, D. 1967. A plant ecological survey of the Tugela River Basin. Mem. Bot. Surv. S. Afr. 36:1–285. Town and Regional Planning Commission, Natal, Pietermaritzburg.
Edwards, D. 1974. Survey to determine the adequacy of existing conserved areas in relation to vegetation types. A preliminary report. Koedoe 17:2–37.
Fourcade, H. G. 1889. Report on the Natal forests. W. Watson, Pietermaritzburg.
Graham, R. M. 1929. Notes on the mangrove swamps of Kenya. J.E. Afr. Uganda nat. Hist. Soc. 38:157–164.
Greenway, P. J. 1973. A classification of the vegetation of East Africa. Kirkia 9:1–68.
Guy, P. R. & Jarman, N. G. 1969. A preliminary qualitative and quantitative account of the vegetation of Umdoni Park, Natal South Coast. B.Sc. Hons. project, Univ. of Natal, Pietermaritzburg (unpubl.).
Henkel, J. S., Ballenden, S. st.C. & Bayer, A. W. 1936. An account of the plant ecology of the Dukuduku Forest Reserve and adjoining areas of the Zululand Coast Belt. Ann. Natal. Mus. 8:95–125.
Huntley, B. J. 1965. A preliminary account of the Ngoye Forest Reserve, Zululand. J. S. Afr. Bot. 31:177–205.
Killick, D. J. B. 1959. An account of the plant ecology of the Table Mountain area of Pietermaritzburg, Natal. Mem. Bot. Surv. S. Afr. 32:1–133.
Lebrun, J. 1947. La végétation de la plaine alluviale au sud de lac Édouard. Expl. Parc. Nat. Alb. Miss. Lebrun (1937–38). I. Inst. Parcs Nat. Congo Belge, Brussels.
Lind, E. M. & Morrison, M. E. S. 1974. East African vegetation. Longman, London.
Lucas, G. Ll. 1968. Conservation of vegetation in Africa south of the Sahara: Kenya. Acta Phytogeogr. Suecica 54:152–163.
Macedo, J. M. de Aguiar. 1970. Serra da Gorongosa: Necessidade e bases da sua protecção. Inst. Inv. Agron. de Moç. Communicação 44:1–92.
Maggs, T. 1977. Some recent radio carbon dates from eastern and southern Africa. J. Afr. Hist. 38:161–191.
Moll, E. J. 1968a. A quantitative ecological investigation of the Krantzkloof Forest, Natal. J. S. Afr. Bot. 34:15–25.
Moll, E. J. 1968b. A plant ecological reconnaissance of the Upper Ungeni Catchment. J. S. Afr. Bot. 34:401–420.
Moll, E. J. 1968c. Some notes on the vegetation of Mkuzi Game Reserve. Lammergeyer 8:25–30.
Moll, E. J. 1971. Vegetation studies in the Three Rivers Region, Natal. Ph.D. Thesis, Univ. of Natal, Pietermaritzburg (Unpubl.).
Moll, E. J. 1972a. A preliminary account of the dune communities at Pennington Park, Mtunzini, Natal. Bothalia 10:615–626.
Moll, E. J. 1972b. The distribution, abundance and utilization of the Lala Palm, Hyphaene natalensis, in Tongaland, Natal. Bothalia 10:627–636.
Moll, E. J. 1972c. The current status of Mistbelt mixed Podocarpus Forest. Bothalia 10:595–598.
Moll, E. J. 1973. A preliminary report on the vegetation of Maputaland with recommendations as to future land-use. Botanical Research Unit, Durban (Unpubl.).
Moll, E. J. 1974. A preliminary report on the Dwesa Forest Reserve, Transkei. Wildlife Society of S. Afr., Johannesburg.
Moll, E. J. 1975. A report on the general ecological status of the New Germany Commonage, and some recommendations for the future development and use of the area. Commonage Trustees, New Germany.

Moll, E. J. 1976. A plant ecological survey of the Three Rivers Area, Natal. Natal Town and Regional Planning Commission, Pietermaritzburg (in press).
Moll, E. J. & Pierce, S. M. 1975. The use of the profile diagram technique as a tool for describing vegetation structure. Trees in South Africa 27:2–10.
Moll, E. J. & Woods, D. B. 1971. The rate of forest tree growth and a forest ordination at Xumeni, Natal. Bothalia 10:451–460.
Monod, T. 1957. Les grandes divisions chorologiques de l'Afrique. CCTA/CSA, Publ. No. 24, London.
Moomaw, J. C. 1960. A study of the plant ecology of the Coast Region of Kenya Colony. Govt. Printer, Nairobi.
Moreau, R. E. 1935. A synecological study of Usambara, Tanganyika Territory, with particular reference to birds. J. Ecol. 23:1–43.
Morris, J. W. 1967. Descriptive and quantitative plant ecology of Ntshongweni, Natal. M.Sc. Thesis, Univ. of Natal, Pietermaritzburg (Unpubl.).
Myre, M. 1964. A vegetação do extremo sul da provincia de Moçambique. Estudos, Ensaios e Documentos 110:1–145.
Myre, M. 1971. As pastagens da Região do Maputo. Inst. de Investigação de Moçambique 3:1–181.
Palmer, E. & Pitman, N. 1972. Trees of Southern Africa. 3 vols. A. A. Balkema, Cape Town.
Pedro, J. G. & Barbosa, L. A. Grandvaux. 1955. A Vegetação, Chapter V of Esboço do Reconhecimento Ecológico–Agricola de Moçambique. C.I.C.A. Mem. 23:67–224.
Pichi-Sermolli, R. 1957. Une carta geobotanica (1 : 5 000 000) dell'Africa orientale (Eritrea, Ethiopia, Somalia). Webbia 13:15–132.
Pócs, T. 1976. Vegetation mapping in the Uluguru Mountains (Tanzania, East Africa). Boissiera 24:477–498.
Polhill, R. M. 1968. Conservation of vegetation in Africa south of the Sahara: Tanzania. Acta Phytogeogr. Suecica 54:166–178.
Poynton, R. J. 1974. Characteristics and uses of trees and shrubs. Bull. 39, Dept. Forestry, Govt. Printer, Pretoria.
Rogers, D. J. & Moll, E. J. 1975. A quantitative description of some coast forests in Natal. Bothalia 11:523–537.
Rycroft, H. B. 1943. The plant ecology of the Karkloof Forest. M.Sc. Thesis, Univ. of Natal, Pietermaritzburg (Unpubl.).
Tansley, A. G. 1939. The British Islands and their vegetation, Cambridge University Press, Cambridge.
Tinley, K. L. 1958. A preliminary report on the ecology of the Pongolo and Mkuze flood plains. Natal Parks Board Rept. (Unpubl.).
Tinley, K. L. 1964. The vegetation of Ndumu Game Reserve. Unpublished report to Natal Game and Fish Preservation Board, Pietermaritzburg.
Tinley, K. L. 1967. The moist evergreen forest – tropical dry semi-deciduous forest tension zone in north-eastern Zululand and hypotheses on past temperate/montane rain forest connections. In: E. M. van Zinderen Bakker (ed.), Palaeoecology of Africa 2. pp. 82–85.
Tinley, K. L. 1971a. Lake St. Lucia and its peripheral sand catchment – the ecology and implications of proposals to reprieve a condemned system. Wildlife Society of S. Afr., Johannesburg.
Tinley, K. L. 1971b. A sketch of Gorongosa National Park, Moçambique. SARCCUS: 163–172.
Tinley, K. L. 1971c. Determinants of coastal conservation: dynamics and diversity of the environment as exemplified by the Moçambique coast. SARCCUS: 125–153.
Truter, F. C. & Rossouw, P. J. 1955. Geological map of the Union of South Africa. Govt. Printer, Pretoria.
Scott, J. D. 1947. Veld management in South Africa. Dept. Agric. Bull. 278; Govt. Printer, Pretoria.
Scott, J. D. 1951. A contribution to the study of the problems of the Drakensberg conservation area. Dept. Agric. Bull. 324, Govt. Printer, Pretoria.
Staples, R. R. 1926. Experiments in veld management I. Union of S. Afr. Dept. Agric. Sci. Bull. 49, Govt. Printer, Pretoria.
Staples, R. R. 1930. Experiments in veld management II. Union S. Afr. Dept. Agric. Sci. Bull. 91, Govt. Printer, Pretoria.
Vahrmeijer, J. 1966. Notes on the vegetation of northern Zululand. African Wildlife 20:151–161.
Venter, H. J. T. 1972. Die plantekologie van Richardsbaai, Natal. D.Sc. thesis, Univ. Pretoria (Unpubl.).

Venter, H. J. T. 1976. An ecological study of the dune forest at Mapelana, Cape St. Lucia, Zululand. J. S. Afr. Bot. 42.

Walter, H. & Steiner, M. 1936. Die Ökologie des ostafrikanischen Mangroven. Zeitschr. Bot. 30:65–193.

Wells, M. J. 1962. An account of the plant ecology of the Nagle Dam area of Natal. M.Sc. Thesis, Univ. of Natal, Pietermaritzburg (Unpubl.).

West, O. 1951. The vegetation of Weenen County, Natal. Mem. Bot. Surv. S. Afr. 23:1–183.

White, F. 1965. The savanna woodlands of the Zambezian and Sudanian Domains: an ecological and phytogeographical comparison. Webbia 19:651–681.

White, F. 1976a. The vegetation map of Africa: the history of a completed project. Boissiera 24:659–666.

White, F. 1976b. The taxonomy, ecology and chorology of African Chrysobalanaceae (excluding Acioa). Bull. Jard. Bot. Nat. Belg. 46:265–350.

White F. (in press). The underground forests of Africa: a preliminary review. Gardens Bull. Singapore 29:57–71.

White, F. (ed., in press). Vegetation map of Africa. UNESCO, Paris.

Wild, H. & Barbosa, L. A. Grandvaux. 1968. Vegetation map of the Flora Zambesiaca area, 1:2 000 000, supplement to Flora Zambesiaca, Collins, Salisbury.

14 The Guineo–Congolian transition to southern Africa

F. White and M. J. A. Werger

1. Introduction 601
2. The Congo–Zambezia transition zone in Angola and western Zaïre 604
2.1 Cloud Forest 606
2.2 Gallery forest, fringing forest and swamp forest 608
2.3 Mangrove and coastal thicket 609
3. The Congo–Zambezia transition zone east of Angola 610
3.1 In Kwango 610
3.2 In Kasai and lower Shaba 613
4. Guineo–Congolian vegetation and species in the Zambezian Region .. 615
4.1 Exclaves of Guineo–Congolian rain forest 615
4.2 Swamp forest and fringing forest 616
4.3 Dry evergreen forest 617
5. Phytosociological classification 618

References 619

14 The Guineo–Congolian transition to southern Africa

1. Introduction

The floras of the Guineo–Congolian and Zambezian Regions,* except for a few cosmopolites, pluri-regional linking species and chorological and ecological transgressors, are almost mutually exclusive so far as the greater part of their areas is concerned. There is, however, a transition zone between them (White 1976a, in press, A) up to 500 km wide, where an impoverished Guineo–Congolian flora and an even more impoverished Zambezian flora interdigate or occur in mosaic, and, locally, intermingle. The precise distribution of vegetation dominated by these floras is determined by local edaphic and climatic factors. Most of the Zaïre part of this Congo–Zambezia transition zone is included in the Guineo–Congolian Region by Duvigneaud (1952), who, nevertheless, admits its transitional nature. Certain authors, e.g. Devred (1957) and Liben (1958), attempt to define a precise boundary between the Guineo–Congolian and Zambezian Regions. The general needs of biogeography, however, are equally well served if transitional zones are clearly recognized as such.

The natural northern biogeographic limit to southern Africa could be drawn to coincide with either the southern limit of the Guineo–Congolian Region or of the Congo–Zambezia transition zone. The former alternative is chosen here since almost the whole of the western end of the transition zone lies within Angola which for the purpose of this book is included in southern Africa. The general features of the Congo–Zambezia transition zone, which has a voluminous literature, have been summarized by White (in press, A). No other comprehensive reviews have been written, though a few topics have been briefly mentioned by Exell (1957).

To the south of the transition zone the Guineo–Congolian flora is represented in the Zambezian Region by:

(a) a few relatively small enclaves of floristically impoverished rain forest 40–80 km from the coast in western Angola;

(b) many typical rain forest species which extend a relatively short distance into the Zambezian Region as marginal intruders in forest on the banks of or in the valleys of the major tributaries of the Zaïre River. The latter run approximately in a north/south direction;

(c) a relatively small number of Guineo–Congolian linking species and transgressors, which, in the Zambezian Region, have very restricted distributions and occur, principally in areas of higher rainfall, in fringing forest, or in dry evergreen forest ('muhulu' in Shaba; 'mateshi' in Zambia) on deep water-retaining soils, or in rain forest on the slopes of certain mountains and escarpments where the rainfall is locally augmented and better distributed throughout the year, or where the severity of the dry season is ameliorated by frequent mists. These rain forests have been described by Chapman & White (1970). In the present work

* For use of the term 'region' see p. 159.

601

they are included with the forests of the Indian Ocean Coastal Belt because of their floristic similarity.

The greater part of the Congo–Zambezia transition zone is occupied today by secondary grassland and wooded grassland dominated almost exclusively by Zambezian species. There is little doubt that the former relative extent of Guineo–Congolian and Zambezian vegetation types varied greatly from one part of the zone to another according to local climatic and edaphic features, but this is a subject which has not been studied in detail. Most of the features of the Congo–Zambezia transition zone that are mentioned in the text are shown in Fig. 1. The geographical distributions of 10 representative species occurring in the transition zone are shown in Fig. 2. Of the species shown in Fig. 2, two, *Gilbertiodendron dewevrei* and *Sacoglottis gabonensis*, are Guineo–Congolian species

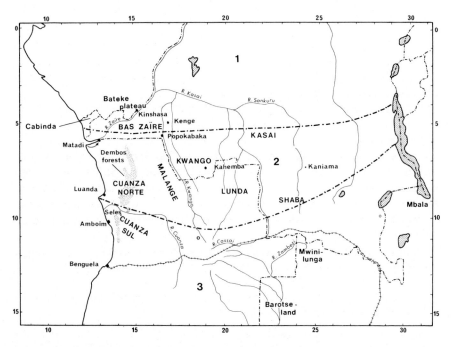

Fig. 1. Map of the Congo–Zambezia transition zone showing features mentioned in the text. The stippled area in Angola shows the 'cloud' forests of Dembos and exclaves of similar forests further south. 1. Guineo–Congolian Region. 2. Congo–Zambezia transition zone. 3. Zambezian Region.

which penetrate the Congo–Zambezia transition zone to a greater or lesser extent; one, *Pteleopsis diptera*, is almost confined to the transition zone; five, *Brachystegia wangermeeana*, *Cryptosepalum pseudotaxus*, *Daniellia alsteeniana*, *Entandrophragma delevoyi* and *Marquesia macroura*, are widespread in the transition zone and in the wetter northern half of the Zambezian Region, and two, *Brachystegia spiciformis* and *Parinari curatellifolia*, are widespread in the transition zone and occur almost throughout the Zambezian Region. Both also occur in the Indian Ocean Coastal Belt, and the latter is also widely distributed in the Sudanian Region.

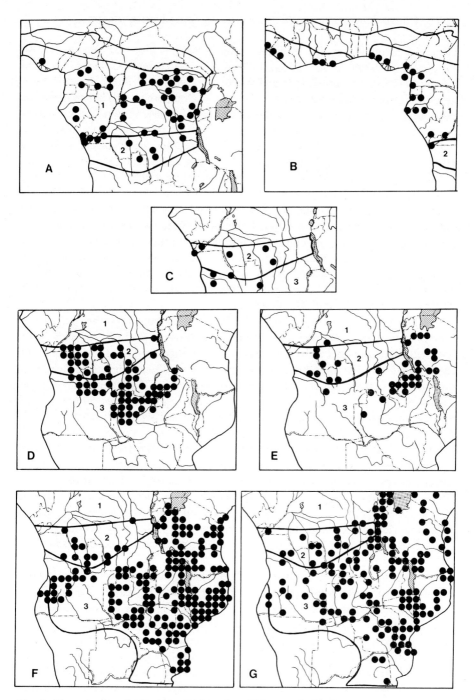

Fig. 2. Maps showing distributions of 10 species which occur in the Congo–Zambezia transition zone. A. *Gilbertiodendron dewevrei* (after Liben, Distr. Pl. Afr. 2:47, 1970). B. *Sacoglottis gabonensis* (after Liben, Distr. Pl. Afr. 3:80, 1971). C. *Pteleopsis diptera*. D. Combined distribution of *Cryptosepalum pseudotaxus, Daniellia alsteeniana, Entandrophragma delevoyi* and *Marquesia macroura*. E. *Brachystegia wangermeeana*. F. *Brachystegia spiciformis*. G. *Parinari curatellifolia* (after, White, Distr. Pl. Afr. 10:333, 1976). 1. Guineo–Congolian Region. 2. Congo–Zambezia transition zone. 3. Zambezian Region.

2. The Congo–Zambezia transition zone in Angola and western Zaïre

General information on the transition zone in Zaïre is given by Devred (1958), Duvigneaud (1949a, 1952, 1953a), Lebrun (1961), Lebrun & Gilbert (1954) and Léonard (1953). The grasslands and wooded grasslands of Bas Zaïre have been described in some detail by Duvigneaud (1953). Compère (1970) has mapped the vegetation of Bas Zaïre in considerable detail, but his descriptions of the different types are somewhat brief. More local studies of the vegetation have been published by Lebrun (1954) for the littoral zone, and by Denisoff & Devred (1954) and Devred (1956) for a small area in Bas Zaïre near Mvuazi.

The greater part of Bas Zaïre has a mean annual rainfall greater than 1200 mm and corresponds to Duvigneaud's (1952) 'District subatlantique du Bas Congo'. The climax vegetation is Guineo–Congolian rain forest but vast areas are covered with anthropic grassland and wooded grassland, composed almost exclusively of Zambezian species. In the extreme southeastern corner the forests consist of a mixture of Guineo–Congolian species such as *Gnetum africanum*, *Paramacrolobium coeruleum*, *Maranthes glabra*, *Petersianthus macrocarpus* and *Staudtia stipitata*, and Zambezian species which include *Berlinia giorgii*, *Marquesia acuminata* and *M. macroura*.

In western Bas-Zaïre the rainfall diminishes rapidly from 1200 mm p.a. east of Matadi to less than 800 mm p.a. near the mouth of the Zaïre River. This is the 'District atlantique de l'embouchure du fleuve Congo' of Duvigneaud. The dry season lasts for 5–6 months but atmospheric humidity is high. The flora is a mixture of Guineo–Congolian and Zambezian elements. The latter dominates the extensive anthropic vegetation seen today but Zambezian species probably were plentiful in some climax communities. In the extensive secondary *Heteropogon–Hyparrhenia* grasslands clumps of baobabs (*Adansonia digitata*), mangoes (*Mangifera indica*) and cashew nuts (*Anacardium occidentale*) indicate the sites of abandoned villages. Near Matadi and on the western slopes of the Monts de Cristal, patches of sclerophyllous thicket occur on quartzite. They have not been studied in detail but include *Hymenocardia ulmoides*, *Manilkara lacera*, *Strychnos henningsii* and *S. gossweileri*. Granite outcrops support a special vegetation rich in Melastomataceae, Commelinaceae, Labiatae, Cyperaceae, *Riccia*, lichens, and succulents such as *Sansevieria cylindrica*, *Aloe*, *Rhipsalis* and *Euphorbia*.

For Angola, Gossweiler & Mendonça (1939) have mapped the principal vegetation types using the physiognomic/ecological system of Rübel (1930). In the memoir which accompanies their map the vegetation types are described with abundant floristic detail. Gossweiler & Mendonça's work is summarized in English by Airy Shaw (1947). The more recent vegetation map of Angola by Barbosa (1970) largely follows the Yangambi classification (Anon. 1956). Barbosa's map is less detailed than Gossweiler & Mendonça's and fewer floristic details are given in the text, which includes, however, much more information on climate and soils. These two works are largely complementary.

Diniz & Aguiar (1969) have divided Angola into 16 natural regions, of which 8 occur in the Congo–Zambezia transition zone. Only a few characteristic species are mentioned. Diniz (1973) recognizes 36 agricultural zones in Angola. Fifteen occur, at least in part, in the transition zone. For each zone the physiography, landscape, climate, geology, soil and vegetation are described and a small vegetation map is provided. Less comprehensive studies have been published by Nolde (1938), Huntley (1974), Monteiro (1965) and Mendonça (1961).

In Cabinda, the satellite of Angola to the north of the Zaïre River, the climax vegetation in the upper part, i.e. more than 60 km inland, where the rainfall exceeds 1250 mm p.a., is typical Guineo–Congolian rain forest, which is much richer in species than is the cloud forests of Dembos in Angola proper (see below). The coastal belt is drier and the vegetation is essentially Zambezian and is an extension of that occurring in the coastal belt of Angola. Along the courses of the Rio Chiloango and its tributaries, however, about 50 km from the coast, extensive gallery or fringing forest of rain forest type and of great luxuriance occurs in bands up to 1 km in width. Its presence is due to the flooding of these valleys during the period of greatest heat, though the rainfall does not exceed 1100 mm p.a.

In Angola proper, Guineo–Congolian rain forest, which is of the semi-evergreen type, is extensively developed only between 350 and 1000 m on the western slopes of the interior plateau 100 km or more from the sea between 6°30′ and 9°30′ S in the Dembos region. Rainfall is between 1100 and 1500 mm p.a. The fertile paraferralitic and ferralitic soils in this region favour the growth of forest, but it largely depends for its existence on the constant condensation of water-vapour brought in by the moisture-laden westerly sea-breeze, and is sometimes referred to as 'cloud forest'. It is also known locally as 'coffee forest'. Two wild coffees, *Coffea canephora* and *C. welwitschii* formerly occurred in the underwood, but they have almost been exterminated by exploitation. Tall trees have been left to provide shade for present coffee plantations. Airy Shaw (1947) refers to this community as 'montane forest' but it should be pointed out that, although a few Afromontane species do occur, they are inconspicuous. Undisturbed forest is found only in the most inaccessible places. Very small patches of forest similar to the Dembos forest also occur in deep valleys on the east and south-east of the Malange plateau between 9° and 11° 30′ S, where they clothe the slopes above the fringing forest up to the level of the plateau. Here they meet the miombo woodland which is known locally as 'panda'.

Elsewhere in the Congo–Zambezia transition zone in northern Angola the vegetation is overwhelmingly Zambezian except for various types of ground-water forest, mangrove and other littoral communities. The Zambezian vegetation will not be described here in detail (see Chapter 10).

The rainfall of the coastal belt to the west of the cloud forests of Dembos is too low to support Guineo–Congolian vegetation except along watercourses. The prevalent vegetation is grassland and wooded grassland†, most of it probably secondary. *Adansonia digitata*, which is rare or absent further inland, is a conspicuous feature of the landscape. Its most important associates are *Acacia welwitschii*, *Sterculia africana* and *Euphorbia conspicua*. The two introduced trees, *Anacardium occidentale* and *Mangifera indica*, are also plentiful. In some areas there are extensive societies of *Sansevieria cylindrica*. On sandy soil, on elevated country of the littoral, where there is regular condensation of moisture from the westerly winds, *Strychnos henningsii* (*ligustroides*) forms dense, sometimes impenetrable, thickets thousands of square kilometres in extent. Its associates include *Combretum camporum, *Croton angolensis, Diospyros abyssinica, *D.

† For use of this terminology see p. 311.

heterotricha, **Hymenostegia laxiflora*, *Sansevieria cylindrica*, *Sterculia africana*, *Strychnos floribunda* (*welwitschii*) and *Tarchonanthus camphoratus*. Species marked with an asterisk are more or less endemic to the western end of the Congo–Zambezia transition zone.

To the east of the Dembos cloud forests there are extensive areas of secondary grassland and wooded grassland. The nature of the forest which they have replaced is unknown. Further east still, especially in Lunda District, typical Zambezian communities occur extensively. Among them are *Brachystegia–Julbernardia* miombo woodland, *Daniellia alsteeniana–Marquesia macroura* dry evergreen forest, and seasonally waterlogged grassland with abundant geoxylic suffrutices (see White 1976b, in press, B and Chapter 10). Their relationships to the Guineo–Congolian vegetation which also occurs in the same general area, especially in deep valleys and rejuvenated landscapes, have not been studied in detail.

2.1 Cloud forest

The Dembos 'cloud' forests vary greatly in stature, density and floristic composition from place to place, depending on the local conditions and the degree of disturbance. The canopy is frequently at 30 m and the tallest trees are up to 50 m (Fig. 3). Among the taller trees are **Afrosersalisia cerasifera*, **Afzelia bipindensis*, *Albizia adianthifolia*, **A. glaberrima*, *Amphimas ferrugineus*, **Antiaris toxicaria*, **Bombax buonopozense* subsp. *reflexum*, *Ceiba pentandra*, **Celtis gomphophylla*, *C. mildbraedii*, **C. zenkeri*, **Chlorophora excelsa*, *Diospyros dendo, **Entandrophragma angolense*, *Fagara macrophylla*, *Ficus mucuso and other *Ficus* spp., *Fillaeopsis discophora*, *Funtumia africana*, *Gilletiodendron kisantuense*, *Hannoa klaineana, *Khaya anthotheca, **Lannea welwitschii*, **Monodora angolensis*, **M. myristica*, *Morus mesozygia*, *Nauclea diderichii*, *Pachystela brevipes*, **Parkia filicoidea*, **Petersianthus macrocarpus*, **Piptadeniastrum africanum*, **Premna angolensis*, **Pseudospondias microcarpa*, **Pterocarpus soyauxii*, *P. tinctorius*, **Pycnanthus angolensis*, **Ricinodendron africanum*, *Staudtia stipitata*, *Sterculia tragacantha*, **Symphonia globulifera*, *Tetrapleura tetraptera*, **Treculia africana*, *Xylopia acutiflora* and **Zanha golungensis*. Species marked with an asterisk also occur further south in exclaves of Guineo–Congolian forest inside the Zambezian Region (see p. 615).

All species mentioned above are widespread in the Guineo–Congolian Region, but relatively few, e.g. *Diospyros dendo*, *Entandrophragma angolense* and *Gilletiodendron kisantuense*, are more or less strictly confined to it. Several, e.g. *Afrosersalisia cerasifera*, *Antiaris toxicaria*, *Ricinodendron africanum* and *Zanha golungensis*, are linking species and are widespread in fringing forest and various other types of forest in the Zambezian and Zanzibar–Inhambane Regions. Others, including *Afzelia bipindensis*, *Sterculia tragacantha* and *Symphonia gabonensis*, penetrate some distance into the Zambezian Region in fringing forest. A few species, e.g. *Pteleopsis diptera*, have distributions which more or less coincide with the Congo–Zambezia transition zone. There are also a few transgressors, e.g. *Trichilia dregeana*, which is absent from the Guineo–Congolian Region. In south tropical Africa it occurs in upland but not Afromontane areas and descends almost to sea-level in Natal and the Eastern Cape. In more open and rocky parts of the forest, the widespread

Fig. 3. Dembos 'cloud' forest northwest of Dalatando (Salazar). The undergrowth and some of the trees have been cleared to promote the cultivation of coffee (*Coffea robusta*) (photo M. J. A. Werger).

Sudano–Zambezian species *Adansonia digitata* is a conspicuous feature.

Alchornea cordifolia, Clausena anisata, Croton mubango, Harungana madagascariensis, Melia bombolo, Millettia versicolor, Musanga cecropiodes, Myrianthus arboreus, Phyllanthus discoideus, Spathodea campanulata and *Trema orientalis* occur in the early stages of forest regrowth.

Small woody plants, mostly occurring in the shrub layer which is up to 8 m high, include *Buettnera africana, Byrsocarpus coccineus, Caloncoba welwitschii, Carapa procera, Carpolobia alba, Cyathea manniana, Fernandoa superba, Leea guineensis, Lindackeria dentata, Morelia senegalensis, Napoleona vogelii, Olax viridis, Rhaphiostylis beninensis, Vismia affinis,* and species of Annonaceae, *Cola, Millettia, Ochna, Ouratea, Rinorea* and Rubiaceae, etc. Several lianes reach the canopy at 30 m. Among them are *Adenia lobata, Chasmanthera welwitschii, Cissus aralioides, Dalbergia altissima, Entada gigas, Hippocratea andongensis, H. apocynoides, H. clematoides, Illigera vespertilio, Landolphia owariensis, Raphidiocystis welwitschii, Salacia elegans, Securidaca welwitschii, Tetracera podotricha, Urera thonneri* and *Ventilago africana.* Smaller climbers include *Camoensia scandens, Cissus petiolata, Cnestis ferruginea, Combretum paniculatum* and *Combretum* spp., *Gnetum africanum, Griffonia speciosa, Hugonia platysepala, Jaundea pinnata, Mezoneuron angolense, Monanthotaxis parvifolia, Mussaenda erythrophylla, Quisqualis exannulata, Q. falcata, Triaspis lateriflora, Uncaria africana, Uvaria angolensis,* and species of *Agelaea,* Apocynaceae, Asclepiadaceae, *Canthium, Dichapetalum,* Menispermaceae and Rubiaceae. Herbaceous and subherbaceous climbers include *Gloriosa superba, Smilax kraussiana,* the sarmentose Marantaceae *Hypselodelphis violacea,* and species of *Asparagus,* Convolvulaceae, Cucurbitaceae and *Dioscorea.* Epiphytes are sometimes abundant, especially in the wettest places and in the valleys. They are represented by several species of *Polystachya* and *Angraecum, Listrostachys rhipsalisocia, L. iridifolia* and *Mystacidium xanthopollinium* among orchids; by *Asplenium hypomelas, Davallia chaerophylloides (vogelii), Drynaria laurentii, Microgramma lycopodioides* and *Platycerium elephantotis (angolense)* among ferns, and by the cactus, *Rhipsalis baccifera.* The development of the herb layer depends on the amount of light filtering through the taller strata. Characteristic species include *Acanthus montanus, Aframomum erythrocarpum, Anchomanes difformis, Asystasia gangetica, Desmodium repandum, Dorstenia psilurus, Drymaria cordata, Geophila afzelii, G. repens, Haemanthus angolensis, Palisota schweinfurthii, Renealmia africana* and *Whitfieldia elongata.* Grasses are represented by *Acroceras zizanioides, Isachne buettneri, Leptaspis cochleata, Olyra latifolia* and *Oplismenus hirtellus,* and ferns by *Adiantum philippense, Ctenitis cirrhosa, Didymochlaena truncatula, Gleichenia linearis, Microlepia speluncae* and *Thelypteris dentata* (Gossweiler & Mendonça 1939, Monteiro 1965, Barbosa 1970, Diniz 1973).

The small patches of coffee forest in Malange District mentioned above are 20–30 m tall. *Sterculia purpurea, Piptadeniastrum africanum, Funtumia latifolia, Syzygium guineense* and *Afrosersalisia cerasifera* occur in the canopy. *Tabernaemontana angolensis, Diospyros hoyleana, Monanthotaxis schweinfurthii* and *Coffea canephora* occur in the understorey.

2.2 *Gallery forest, fringing forest and swamp forest*

Fringing forest composed predominantly of Guineo–Congolian species occurs not only in regions of high (c. 1300 mm p.a.) rainfall in northeast Angola, where it is particularly well-developed in the great tributary valleys of the Zaïre River which run in an approximately north–south direction, but also in the much drier (700 mm p.a.) coastal belt.

East of longitude 19°E the trans–Angolan railway line follows the watershed separating the Cassai and Zambezi drainage systems and this line roughly represents the southern limit of fringing forest of Guineo–Congolian type, which occurs between 600 and 1200 m. Near the northern frontier it is 30 m tall. Characteristic large tree species are *Chrysophyllum magalismontanum*, *Mitragyna stipulosa*, *Nauclea? pobeguinii*, *Prunus africana*, *Pycnanthus angolensis*, *Sterculia subviolacea*, *Symphonia gabonensis*, *Treculia africana*, *Uapaca guineensis* and *Xylopia aethiopica*. The large liane, *Leptoderris goetzei*, and the climbing palm, *Eremospatha cuspidata*, are plentiful.

In the littoral zone north of the Rio Cuanza at altitudes up to 350 m the leguminous tree *Oxystigma bucholzii* (mafuto) frequently forms galleries.

Three palm genera are prominent in riverine vegetation in north-western Angola. The oil palm, *Elaeis guineensis*, forms communities on alluvium of the Zaïre estuary. In the districts of Cuanza Norte and Cuanze Sul it also forms natural societies almost at sea-level in very hot sheltered situations, where it is dependent on phreatic water, as in oases, though about 600 mm of rain fall annually. Species of *Raphia* are gregarious on alluvium at the mouth of the Zaïre River, where their roots are in running water during the whole year. The dwarf, stemless palm, *Sclerosperma mannii*, occurs in alluvial forests in Lower Cabinda.

At the mouth of the Zaïre River, between Sumba and Ponta de Quiombe, there is an alluvial area of about 250 sq.km occurring on the landward side of the mangrove. The swamp forest occurring here is known as 'taba'. Although the rainfall is no more than 700 mm p.a., the constituent species are almost exclusively Guineo–Congolian rain forest species and some, e.g. *Sacoglottis gabonensis*, are especially characteristic of the wetter types. Other important species are *Canarium schweinfurthii*, *Cathormion altissimum*, *Ctenolophon engleranus*, *Elaeis guineensis*, *Eremospatha cabrae*, *Irvingia smithii*, *Mitragyna stipulosa*, *Musanga cecropioides*, *Napoleona imperialis*, *Nauclea? pobeguinii*, *Pierrodendron africanum*, *Polyscias ferruginea*, *Pycnanthus angolensis*, *Raphia gossweileri*, *Tetracera alnifolia* and *Xylopia aethiopica*.

2.3 Mangrove and coastal thicket

Neither mangrove nor coastal thicket can be considered as Guineo–Congolian vegetation types, but on the western side of Africa for the greater part of their length they occur adjacent to Guineo–Congolian vegetation. They do extend, however, some distance further to the north and to the south but only as impoverished and depauperate variants.

Mangrove is extensively developed on alluvial mud at the mouth of the Zaïre River and is always sheltered from the direct action of the waves but always reached by the sea. The principal constituent is *Rhizophora mangle* s.l., which forms dense, almost pure, communities up to 20 m in height. It is frequently associated with *Avicennia africana*, *Conocarpus erectus*, *Laguncularia racemosa* and the halophytic fern, *Acrostichum aureum* (see Chapter 37).

Immediately behind the pure *Rhizophora* community is a mixed swamp forest community of lower stature which on the Angolan side of the Zaïre River covers 50,000 hectares and extends nearly 100 km inland. The most important constituent is *Chrysobalanus icaco* subsp. *icaco* which occurs here as a shrub of 6 m. Its associates include *Baikiaea insignis*, *Camoensia brevicalyx*, *Drepanocarpus lunatus, Heteropterys leona, Hibiscus tiliaceus, Pandanus* sp. and *Ternstroemia africana.*

Small patches of dense thicket occur on littoral sands. Its most characteristic species include *Chrysobalanus icaco, Caesalpinia bonduc, Syzygium guineense* subsp. *littorale, Ximenia americana, Dalbergia ecastaphyllum* and *Rhopalopilia marquesii*. Most of these species are widespread on the West African littoral.

3. The Congo–Zambezia transition zone east of Angola

General publications were mentioned on p. 604. More detailed but more local studies have been published by Devred (1957), Devred et al. (1958), Duvigneaud (1949b, 1950, 1958), Germain (1949) and Mullenders (1954, 1955).

3.1 In Kwango

In Kwango the pattern formed by different types of Guineo–Congolian and Zambezian vegetation is more complex than elsewhere in southern Zaïre and is dependent on the complex pattern of edaphic and meso-climatic conditions due to relatively recent partial rejuvenation of the landscape by the major rivers and their tributaries. Guineo–Congolian elements penetrate deeply to the south, favoured by the wide valleys incised into Karoo strata supporting a gallery or 'mushitu' forest, and, elsewhere, by intense rejuvenation. Zambezian elements occur principally on the Kalahari Sand-covered plateaux between the valleys and extend northwards beyond 5°S on ridges which become progressively lower until at about 4°N they merge into Guineo–Congolian vegetation. Northwest of Kenge, Zambezian vegetation spreads out on the Kalahari Sand-covered plateau between the Kwango and Nsele Rivers, but its occurrence further north, especially on the Bateke plateau north of the Zaïre River, is probably entirely secondary due to the destruction of the original Guineo–Congolian vegetation for cultivation and by fire (Devred et al. 1958).

In northern Kwango vast tongues of Guineo–Congolian forest extend towards the south on red Karoo soils in the wide valleys. The vegetation on the intervening plateaux consists of small patches of Guineo–Congolian forest, and of mixed forest ('mabwati', see below) which is comprised mostly of Zambezian species, but with an admixture of Guineo–Congolian species. These forest patches are set in an enormous expanse of secondary grassland and wooded grassland, dominated almost exclusively by Zambezian species.

The Guineo–Congolian forest in the valleys, which is 35–40 m tall, is characterized by *Hymenostegia mundungu, Scorodophloeus zenkeri, Cynometra hankei, Ongokea gore, Paramacrolobium coeruleum* and *Mammea africana*. The fragments of Guineo–Congolian forest surviving on the Kalahari sands of the plateau are more distinctly deciduous in character. They are shown as extending to almost 6°S by Devred et al. (1958). Principal species include *Albizia ferruginea, Amphimas ferruginea, Anthonotha gilletii, Aphanocalyx marginatus, Bussea gossweileri, Diospyros hoyleana, Lovoa trichilioides, Parinari excelsa, Tessmannia anomala* and *Xylopia staudtii* (Devred et al. 1958). This type of forest regenerates freely in its own shade, but is unable to colonize adjacent fire-protected grassland. It has almost completely disappeared. The few remaining vestiges have survived because they are sacred forests (Devred 1957).

Elsewhere on the high plateau of Kwango there are patches of dense, evergreen 'sclerophyllous' forest with glossy foliage which are known locally as 'mabwati'. Duvigneaud (1950) refers them to the category of Laurisilva of Rübel (1930). This type of forest, which may reach a height of 25 m, represents a transition, both floristic and physiognomic, from Guineo–Congolian rain forest to Zambezian woodland. The transition is continuous when the soil permits, or takes the form of a mosaic when edaphic conditions override the climate, as is often the case. The leaves of most mabwati species are more coriaceous than those of rain forest species and lack 'drip-tips'.

The most characteristic taller tree species of mabwati are *Marquesia macroura, M. acuminata, Berlinia giorgii, Lannea antiscorbutica, Daniellia alsteeniana, Brachystegia spiciformis, B. wangermeeana,* and *Parinari curatellifolia.* Smaller trees include *Uapaca nitida, U. sansibarica, Memecylon sapinii, Diospyros batocana, Anisophyllea gossweileri, Monotes dasyanthus* and *Diplorhynchus condylocarpon.* The two species of *Marquesia, Berlinia giorgii,* and *Daniellia alsteeniana* only occur in the Congo–Zambezia transition zone and in the wetter northern parts of the Zambezian Region. Most of the other species are very widely distributed in the Zambezian Region.

The floristic composition of mabwati forest varies from north to south. In the north of Kwango, in the zone of the Kalahari–Karoo geological contact in the Popokabaka–Kenge region, there is a strong admixture of Guineo–Congolian species, especially species which also occur on Kalahari sand in the Bateke forests of the Brazzaville–Kinshasa region. In southern Kwango the mabwati forests occur as fragmented and isolated populations in miombo woodland. Duvigneaud (1950) suggests that soil degradation due to human activity may have permitted mabwati to replace rain forest in the north, just as mabwati itself may have been largely replaced by miombo in the south. Devred (1957) records a number of species of secondary Guineo–Congolian forest, such as *Gaertnera paniculata, Millettia drastica* and *Pentaclethra eetveldeana* from the northern mabwati.

Epiphytic pteridophytes of the genera *Polypodium, Platycerium* and *Lycopodium* are extraordinarily abundant in mabwati. According to Duvigneaud (1952) the shrub layer of mabwati consists of *Hymenocardia, Maprounea* and *Psorospermum,* and the herb layer is dominated by the tall grasses, *Hyparrhenia diplandra* and *Tristachya nodiglumis,* more or less mixed with *Aframomum, Pteridium* and *Smilax.* The presence of all these species suggests that Duvigneaud is referring to a degraded or secondary community rather than a climax type of forest.

Forest of the mabwati type is widely distributed in the Lunda and Malange Provinces in Angola, but has not been described in detail. It also occurs in the Haut–Lomami District of Shaba (p. 610). The muhulu (muulu) forests of southern Shaba and the mateshi and mavunda forests of Zambia (p. 617) are closely related to mabwati. White (in press, A) refers all these types to 'Zambezian dry evergreen forest'.

When mabwati is degraded by clearing and burning it is replaced by a wooded grassland known locally as 'mikwati' in which the most frequent fire-resistant trees are: *Erythrophleum africanum, Dialium engleranum, Burkea africana, Hymenocardia acida, Diplorhynchus condylocarpon, Pterocarpus angolensis, Protea petiolaris, Combretum laxiflorum* and *Strychnos pungens.* The herb layer consists largely of the grasses *Hyparrhenia diplandra, H. familiaris, Loudetia arundinacea, Digitaria uniglumis, Brachiaria brizantha* and *Ctenium newtonii*.

In the basin of the upper Wamba and its tributaries, the Tundwala and Zamba, semi-evergreen forest, 30–40 m tall, dominated by *Marquesia acuminata* and *Pteleopsis diptera* is extensively developed on the Karoo soils of the rejuvenated land surface. *Marquesia acuminata* is a Zambezian species, and the distribution of *Pteleopsis diptera* is largely centred on the Congo–Zambezian transition zone. Most of the other species occurring in this forest, including *Antiaris toxicaria, Canarium schweinfurthii, Chlorophora excelsa, Entandrophragma utile, Petersianthus macrocarpus, Pycnanthus angolensis* and *Turreanthus africanus*, are Guineo–Congolian species or Guineo–Congolian linking species (Devred 1957, Devred et al. 1958).

In the south-east of Kwango in Kahemba district the prevalent vegetation is woodland in which Zambezian species predominate. It is known locally as 'tumbi' or 'mikondo'. The most important larger tree species include *Brachystegia spiciformis, B. longifolia, B. wangermeeana, Julbernardia globiflora, Daniellia alsteeniana, Marquesia macroura, Parinari curatellifolia, Berlinia giorgii* and *Pericopsis angolensis* (see Chapter 10). The extent to which mabwati might have been replaced by tumbi owing to human activity is not known.

The importance of parent material in determining the distribution of vegetation in Kwango is well shown in the region of Kimvula and Popokabaka, where the plateau has been reduced to scattered sugar-loaf mountains, capped by loose Kalahari sand. The latter is often separated from the underlying Karoo beds by a basal band of Kalahari sandstone. The Kalahari sand is covered with short sparse Zambezian grassland of *Loudetia simplex, Monocymbium ceresiiforme, Schizachyrium semiberbe* and *Trachypogon thollonii*. The outcrop of Kalahari sandstone supports a bushland of *Philippia* sp. and *Protea melliodora*, whilst on the lower slopes mabwati forest or Guineo–Congolian forest of the Bateke type occurs.

On the plateau north of the 7th parallel in Kwango there are enormous areas of sparse short grassland, which is usually referred to by Belgian authors as 'steppe', 'pseudosteppe' or 'savanne steppique' (Devred et al. 1958, Duvigneaud 1949a, 1952). It occurs on deep aeolian sand, more than 100 m thick. The structureless soil is without humus and is extremely deficient in nutrients. Although the rainfall is 1600–1800 mm p.a. it percolates rapidly and the water-table is at a considerable depth. Fires occur each year and initiate the flowering of the numerous geophytes and geoxylic suffrutices. The principal grasses are *Aristida vanderystii, Ctenium newtonii, Digitaria brazzae, Diheteropogon emarginatus, Elyonurus argenteus, Loudetia demeusii, L. simplicifolia, Monocymbium ceresiiforme, Rhynchelytrum amethysteum, Schizachyrium thollonii, Trachypogon thollonii* and *Tristachya eylesii*. The geoxylic suffrutices include *Anisophyllea dichostyla, Brackenridgea arenaria, Erythrina baumii, Gnidia kraussiana, Landolphia camptoloba, Ochna manikensis, Parinari capensis* and *Rauvolfia nana*. These grasslands are sometimes lightly wooded with scattered trees of *Combretum, Dialium engleranum, Erythrophleum africanum* and *Daniellia alsteeniana*, but often are treeless for considerable distances, or contain only stunted shrubby individuals of *Swartzia madagascariensis, Burkea africana, Oldfieldia dactylophylla, Hymenocardia acida*, etc. According to Devred et al. (1958) these grasslands are secondary and have relatively recently replaced forest and woodland. It is likely, however, that locally, where the water-table is high, at least for part of the year, a similar community represents an edaphic climax. Further south in the upper Zambezi basin in Barotseland in Zambia this type of vegetation is very extensively developed on seasonally waterlogged Kalahari sand as an edaphic climax (see Chapter 10). It does, however, spread onto deep, well-drained Kalahari sand as the ultimate stage of the degradation of forest and woodland (White in press, B). Duvigneaud (1949) very briefly mentions that in places the 'pseudosteppe' occurs where the water-table is high and then is characterized by the abundance of Cyperaceae, Xyridaceae and Eriocaulaceae. Hydromorphic soils supporting edaphic grassland also occur in Kwango in the upper unrejuvenated parts of shallow, 2–3 km wide, valleys near the sources of tributary streams above the waterfalls (Devred et al. 1958, Duvigneaud 1952).

3.2 In Kasai and Lower Shaba

For most of this area there is little published information. The vegetation of one small part, however, between the Lubilash and Lubishi Rivers at the latitude of Kaniama (7°31'S) has been described in considerable detail by Mullenders (1954, 1955).

The prevalent vegetation is forest of Guineo–Congolian affinity or secondary grassland and wooded grassland comprised of Zambezian species, but the scattered occurrence of Zambezian species such as *Daniellia alsteeniana* (map in Duvigneaud 1949), suggests that small pockets of primary Zambezian or mixed vegetation formerly occurred wherever edaphic conditions are suitable.

In that part of the Kaniama region studied by Mullenders the altitude varies between 750 and 1000 m. The mean altitude of the plateau is 850–900 m. The parent material consists of crystalline rocks of the Basement Complex in the form of a giant batholith of gabbro surrounded by a girdle of tonalite, itself surrounded by a girdle of granite. The relief constitutes an undulating end-Tertiary peneplain with incised V-shaped valleys of Quaternary age. The mean annual rainfall varies from 1350–1650 mm and the dry season lasts from 2–3 months. Such a climate should favour Guineo–Congolian rain forest but in the Kaniama region forest is almost restricted to narrow bands on the slopes of the valleys which have cut into the peneplain and are mostly 20–200 m wide. Elsewhere, there are only small patches of forest on the plateau, on red clay soils derived from basic rocks. The great forest of Sankuru which marks the present-day southern limit of Guineo–Congolian rain forest is situated 300 km to the north. Most of the landscape in Kaniama District is occupied by secondary wooded grassland, which frequently is only very sparsely wooded. According to Mullenders the climax vegetation on soils derived from gabbro and tonalite is Guineo–Congolian forest. There is some experimental evidence for this. An extensive area of plateau near Kaniama, which included small patches of forest surrounded by grassland showed a considerable amount of forest regrowth after 12 years of fire-protection.

For granite the situation is more equivocal. Guineo–Congolian forest appears to be the climax only in the ravines. Mullenders suggests that if Guineo–Congolian forest represents the climax on the granite plateau it would have been floristically impoverished and included an admixture of Zambezian species. It appears from Mullender's descriptions, however, that where there are extensive outcrops of granite or the soil is coarse and shallow, Zambezian or mixed communities, which are not dependent on fire for their occurrence represent the climax.

In the valley bottoms swamp forest occurs on badly drained soils. The most abundant species are *Mitragyna stipulosa, Spondianthus preussii, Nauclea diderichii, Treculia africana, Symphonia globulifera* and *Beilschmiedia hermannii*. Less common species include *Erythrina excelsa, Pseudospondias microcarpa* and *Magnistipula butayei* subsp. *transitoria*. The Afromontane species *Prunus africana* occurs locally.

Most species occurring in forest on better drained soils are widespread Guineo–Congolian species or Guineo–Congolian linking species. Among larger trees the following are characteristic: *Albizia zygia, Bosquiea angolensis, Canarium schweinfurthii, Carapa procera, Celtis wightii, C. zenkeri, Chlorophora excelsa, Cordia millenii (chrysocarpa), Cynometra alexandri, Dacryodes edulis, Funtumia africana, Khaya anthotheca, Klainedoxa gabonensis, Lovoa trichilioides, Maesopsis eminii, Monodora myristica, Myrianthus arboreus, Pachystela brevipes, Parkia filicoidea, Piptadeniastrum africanum, Pycnanthus angolensis, Ricinodendron heudelotii, Staudtia gabonensis* and *Trichilia prieuriana*. Other species include *Pterygota mildbraedii* which is otherwise almost restricted to the eastern part of the Guineo–Congolian Region, *Pteleopsis diptera*, which is more or less confined to the Congo–Zambezia transition zone, and the ecological and chorological transgressors, *Celtis africana* and *Newtonia buchananii*, which are absent from the Guineo–Congolian Region.

Where granite is the parent material, forest of Guineo–Congolian affinity is strictly confined to the river valleys (Chapter 41, Fig. 5). At its upper margin it is often replaced by a band of short mixed forest, related to the mabwati of Kwango, and dominated by *Berlinia giorgii* and *Uapaca nitida*. On the plateau above the mixed forest the prevalent vegetation, which consists almost exclusively of Zambezian species, is a sparsely wooded grassland with abundant geoxylic suffrutices, the *Loudetia arundinacea–Ochna leptoclada (debeerstii)* association of Mullenders.

The *Berlinia giorgii–Uapaca nitida* forest is 8–12 m tall. It differs strikingly from the Guineo–Congolian forest of the ravines in its low stature, and from the wooded grassland by its dense canopy of large, dark green, shining leaves. There are virtually no transition zones between these types. *Berlinia giorgii–Uapaca nitida* forest occurs on light, colluvial non-gravelly soils with good water-retention. *Berlinia giorgii* is by far the most abundant species, followed by *Uapaca nitida* and *Combretum psidioides*. Other Zambezian species such as *Albizia versicolor, Cussonia sessilis, Maprounea africana, Monotes dasyanthus, Piliostigma thonningii, Sterculia quinqueloba, Stereospermum kunthianum, Strychnos cocculoides* and *Terminalia mollis*, are very much rarer. Although fire rarely passes through the *Berlinia giorgii* forests they show little tendency to change in the direction of Guineo–Congolian forest. Mullenders suggests that the *Berlinia* forests are the degraded and opened-up relics of formerly denser forests which were a mixture of Zambezian and Guineo–Congolian species. The latter, however, are not much in evidence today. Among trees they are represented only by a few widely distributed forest-edge species such as *Antidesma membranaceum, Harungana madagascariensis* and *Phyllanthus muelleranus*, which are widespread in several phytochoria in addition to the Guineo–Congolian Region. Other Guineo–Congolian species, however, such as *Canarium schweinfurthii, Pterygota mildbraedii, Ficus vallis-choudae* and *Setaria megaphylla*, occur on termite mounds in *Berlinia giorgii* forest.

The *Loudetia arundinacea–Ochna leptoclada* wooded grassland occupies the upper slopes and ridges of the granite plateau. The soils are shallow, rarely more than 1.5 m deep, and at a depth of 20 cm consist of more than 40 per cent of gravel. They are very porous and retain little water. The 1.5 m tall grass, *Loudetia arundinacea*, dominates the herb layer with 55 per cent cover. *Ochna leptoclada*, an extensively rhizomatous geoxylic suffrutex, 5–50 cm tall, forms an understorey to *Loudetia* with 25–50 per cent cover. The scattered small trees vary greatly in abundance from place to place. They rarely exceed 7–8 m in height. *Stereospermum kunthianum, Securidaca longepedunculata, Entada abyssinica* and *Parinari curatellifolia* locally form small groves. Where there are piles of rocks *Sterculia quinqueloba, Albizia versicolor* and *Sclerocarya caffra* are found. *Terminalia mollis* and *Entada abyssinica* are often rooted in fissures in the granite pavement. Their boles and branches are thickly covered with *Usnea* and other lichens and bryophytes and their appearance is puny and ailing. *Steganotaenia araliacea* and *Ochna schweinfurthiana* find their optimum in this habitat and *Aloe 'congolensis'* is confined to it. Other Zambezian tree species include *Pericopsis angolensis, Annona senegalensis, Hymenocardia acida, Maprounea africana, Maytenus senegalensis, Psorospermum febrifungum, Pterocarpus angolensis* and *Strychnos cocculoides*.

Grasses, in addition to *Loudetia arundinacea*, include *Imperata cylindrica, Hyparrhenia filipendula, H. lecomtei, Andropogon schirensis* and *Brachiaria brizantha*. Additional suffrutices include *Annona stenophylla, Lannea edulis* ('*velutina*') and *Tetracera masuiana*.

According to Mullenders the *Loudetia arundinacea–Ochna leptoclada* community represents a fire climax, although he suggests that there is little evidence that in the absence of fire it would progress to forest. Many of the species mentioned above as occurring in the *Loudetia–Ochna* community and in the *Berlinia giorgii* forests are also abundant today in the secondary grasslands and wooded grasslands of the Congo–Zambezia transition zone. There can be little doubt that they are truly indigenous members of the flora of the transition zone, but, that before man's destructive activities began, they were confined to edaphically specialized sites, such as those, among others, described above.

4. Guineo–Congolian vegetation and species in the Zambezian Region

Although a relatively high number of Guineo–Congolian and Guineo–Congolian linking species occur in the Zambezian Region the area they occupy is very restricted. They occur in three main vegetation types:

(1) exclaves of Guineo–Congolian rain forest;
(2) various types of ground-water forest;
(3) dry evergreen forest on well-drained but usually deep and water-retaining, soils.

4.1 Exclaves of Guineo-Congolian rain forest

Forest similar to the cloud forest of the Dembos region of the Congo–Zambezia transition zone also occurs further south in Angola in a discontinuous band on the escarpment about 80 km inland from Porto Amboim to as far south as 11°30′.

These islands of Guineo–Congolian forest owe their existence to mist precipitation and are surrounded by Zambezian vegetation (Fig. 4).

The Amboim forests occur in the most favourable area for coffee cultivation and are only a shadow of their former selves. They are somewhat shorter than the Dembos forests, since few trees exceed a height of 30 m, and are less rich floristically, but nearly all the species occurring in the Amboim forests also occur in the forests of Dembos (see p. 606).

A floristically even poorer form of this forest occurs about 50 km further south on the escarpment near Vila Nova do Seles between 350 and 900 m. These forests too are used for the cultivation of coffee. Larger trees include *Antiaris toxicaria*,

Fig. 4. Northeast of Porto Amboim and Gabela exclaves of Guineo–Congolian rain forest occur patchily distributed between the mountains, as shown in the background. These forests are used to grow coffee. Part of the forest has been cleared for other crops. On a moist site *Acacia sieberana* occurs (centre right) and the rocks in the foreground carry a pioneer vegetation with *Aloe* sp. and *Plectranthus* sp. (photo M. J. A. Werger).

Albizia glabrescens, Bosquiea angolensis, Celtis gomphophylla, C. zenkeri, Croton mubango, Ficus capensis, F. mucuso, Maesopsis eminii, Monodora angolensis, Piptadeniastrum africanum and *Ricinodendron heudelotii*. The majority are Guineo–Congolian linking species which extend far beyond the boundaries of the Guineo–Congolian Region.

4.2 *Swamp forest and fringing forest*

In the wetter parts of the Zambezian Region permanent swamp forest occurs around springs at the sources of tributary streams and locally along watercourses where the water movement is sluggish. Swamp forest in the latter situation merges into other types of fringing forest in which the water-table descends some distance below the surface for at least part of the year. Swamp forest is relatively poor floristically. Nearly all its most abundant and most characteristic species are otherwise virtually confined to the Guineo–Congolian Region. Most of them also occur sporadically in various types of fringing forest. Fringing forest is rich in species, many of which, but probably a minority, also occur in the Guineo–Congolian Region. The fringing forests and swamp forest of Shaba have been described by Schmitz (1963, 1971) and those of Zambia by Fanshawe (1969) and Lawton (1967, 1969).

The most abundant dominant trees of swamp forest, *Mitragyna stipulosa*, *Syzygium owariense*, *Xylopia aethiopica*, *X. rubescens* and *Uapaca guineensis*, are widespread in the Guineo–Congolian Region, as are the most characteristic members of the understorey, *Aporrhiza nitida*, *Garcinia smeathmannii* and *Gardenia imperialis*, and of the shrub layer, *Cephaelis peduncularis*, *Craterispermum laurinum* and *Dracaena camerooniana*.

Other Guineo–Congolian tree species and Guineo–Congolian linking species which occur in swamp forest and fringing forest include *Afrosersalisia cerasifera*, *Afzelia bipindensis*, *Albizia glaberrima*, *Anthocleista schweinfurthii*, *A. vogelii*, *Antidesma vogelianum*, *Canarium schweinfurthii*, *Canthium glabriflorum*, *Cathormium altissimum*, *Dacryodes edulis*, *Erythrina excelsa*, *Erythrophleum suaveolens*, *Fagara macrophylla*, *Ficus capensis*, *F. congensis*, *Homalium africanum*, *Maesopsis eminii*, *Nauclea pobeguinii*, *Newtonia aubrevillei*, *Pachystela brevipes*, *Parkia filicoidea*, *Phoenix reclinata*, *Rauvolfia caffra*, *Samanea leptophylla* s.l., *Sapium ellipticum*, *Sterculia tragacantha*, *Treculia africana* and *Voacanga thouarsii*. Many of these species are extremely localized in the Zambezian Region.

The majority extend not further south than the northern high-rainfall belt of Zambia extending from Mwinilunga to Mbala.

4.3 Dry evergreen forest

This type occurs on deep, water-retaining soils in the wetter half of the Zambezian Region. It varies greatly in floristic composition from place to place. Although none of the eight or so dominant species occur throughout, each one overlaps considerably with most of the others. The mabwati forests already described from Kwango (p. 611) seem to be a somewhat degraded and floristically impoverished variant. The dry evergreen forests of Shaba, where they are known as 'muhulu', have been described briefly by Duvigneaud (1958) and in greater detail by Schmitz (1962). Those of Zambia, which are mostly referred to as 'mateshi', have been described by Fanshawe (1960, 1969). The *Cryptosepalum exfoliatum* subsp. *pseudotaxus* forests ('mavunda') occurring on Kalahari sands are the subject of a study by Cottrell & Loveridge (1966) (see Chapter 10). Lawton (1964) has described an interesting stand of *Marquesia acuminata* forest.

Dry evergreen forest is of much less pronounced Guineo–Congolian affinity than swamp forest and fringing forest. Of its eight most abundant dominant tree species six, namely *Berlinia giorgii*, *Cryptosepalum exfoliatum* subsp. *pseudotaxus*, *Daniellia alsteeniana*, *Entandrophragma delevoyi*, *Marquesia acuminata* and *M. macroura*, are confined to the northern half of the Zambezian Region and the Congo–Zambezia transition zone (see Chapter 10). Of the other two, one, *Parinari excelsa* occurs throughout the greater part of the Guineo–Congolian Region but also extends far beyond its limits (see White 1976b), and the other, *Syzgium guineense* subsp. *afromontanum*, is more characteristic of the Afromontane Region. Several subordinate species, however, are Guineo–Congolian linking species. They include *Ancylobotrys amoena, Craterispermum laurinum, Dictyophleba lucida, Diospyros hoyleana, D. pseudomespilus, Erythrophleum suaveolens, Erythroxylum emarginatum, Landolphia parvifolia, Monanthotaxis schweinfurthii, Opilia celtidifolia, Paropsia brazzeana, Rhaphiostylis beninensis, Rothmannia whitfieldii, Strychnos angolensis* and *Uvaria angolensis*. With the exception of *Erythrophleum* these are all small trees, shrubs or lianes.

5. Phytosociological classification

Several Belgian ecologists, notably Lebrun & Gilbert (1954), Mullenders (1954), Devred (1958) and Schmitz (1962, 1963, 1971), have attempted a hierarchical phytosociological classification system for the vegetation of Zaïre, either in whole (Lebrun & Gilbert) or in part. An attempt is made here to show how the vegetation types discussed in this chapter have been so classified. For some types this is precluded for lack of published information.

All those forests in the foregoing account which are referred to as rain forest of Guineo–Congolian affinity, namely the 'cloud' forests of Dembos (Section 2.1), the floristically impoverished but similar forest exclaves occurring further south (Section 4.1), the forests occuring in the wide river valleys of Kwango and the fragments of Guineo–Congolian forest (Section 3.1) on the plateau in northern Kwango, and the rainforest of Kaniama in Lower Shaba (Section 3.2), belong to the order Piptadeniastro– ('Piptadenio–') Celtidetalia of the class Strombosio–Parinarietea of Lebrun & Gilbert (1954). Among them the forests of the wide river valleys of northern Kwango belong to the Oxystigmo–Scorodophloeion alliance and all the others belong to the Canarion schweinfurthii (=Albizio–Chrysophyllion) although they contain several species typical of secondary forests. In the Kaniama region Mullenders (1954) recognizes two associations within the latter – an association of *Mellera lobulata* and *Canarium schweinfurthii* on gabbro and tonalite, and an association of *Klainedoxa gabonensis* and *Pterygota mildbraedii* on granite. Aubréville, however, has pointed out (1955) that, the floristic differences betweeen these associations are slight, and that most of the species recorded are extremely widespread within the Guineo–Congolian Region, where their associates change rapidly and kaleidoscopically from place to place, so that an unmanageable multitude of associations could be recognized with equal justification.

Lebrun & Gilbert (1954) divide the forests occurring in Zaïre on non-saline hydromorphic soils into 5 orders. Their classification, however, was not apparently based on detailed relevés, and is often difficult to apply in practice, especially to the Congo–Zambezia transition zone, since little floristic information was available from there at the time it was proposed. There is still only sparse published information (Compère 1970, Devred et al. 1958, Mullenders 1954) on the phytosociology of swamp and riparian forest in the Congo–Zambezia transition zone. It seems that the orders Alchorneetalia and Lanneo–Pseudospondietalia, which represent successional stages, occur chiefly in the northwest of the transition zone. The orders Mitragyno–Raphietalia and Pterygotetalia, which are based on climax forest, respectively of permanent swamp and of seasonally inundated alluvium, occur much more widely, and penetrate some distance into the

Zambezian Region. Stands of Mitragyno–Raphietalia and Pterygotetalia forest often occur in close proximity and are connected by intermediates. Many of the communities mentioned in Sections 2.2 and 3.2 and for swamp forest in Section 4.2 belong to Mitragyno–Raphietalia.

Of the dry evergreen forests mentioned in Sections 3.1, 3.2 and 4.3, the 'muhulu' forests are placed by Lebrun & Gilbert (1954) in the alliance Mabo–Parinarion. The latter has been renamed, more appropriately, by Devred (1958), followed by Schmitz (1962, 1971), who also includes in it the *Cryptosepalum* forest, as Diospyro–Entandrophragmion delevoyi of the order Piptadeniastro–Celtidetalia (but compare Chapter 10). The bateke forests, however, are attributed to the alliance Berlino–Marquesion by Lebrun & Gilbert and by Schmitz, who places them in the class Erythrophleetea africani circumguineensia. The latter includes most types of woodland occurring in the Zambezian Region.

In this chapter the following two facts have been stressed:

(1) That dry evergreen forest is transitional both floristically and physiognomically between Guineo–Congolian rain forest and Zambezian woodland.
(2) That dry evergreen forest is floristically variable but that the variation is remarkably continuous.

The mangroves (Section 2.3) belong to the class Avicennio–Rhizophoretea and the coastal thicket (Section 2.3) to the alliance Chrysobalano–Syzygion of the order Ecastaphylletalia brownei forming a separate class (Lebrun 1954).

References

Airy Shaw, H. K. 1947. The vegetation of Angola. J. Ecol. 35:23–48.
Anon. 1956. Phytogéographie. Phytogeography. CCTA/CSA Publ. No. 53:1–33.
Aubréville, A. 1955. Prospection en chambre, 52. Bois Forêts Trop. 44:62–65.
Barbosa, L. A. Grandvaux. 1970. Carta fitogeográfica de Angola. I.I.C.A., Luanda.
Chapman, J. D. & White, F. 1970. The evergreen forests of Malawi. Comm. For. Inst., Oxford.
Compère, P. 1970. Carte des sols et de la végétation du Congo, du Rwanda et du Burundi. 25, Bas Congo. Min. Belge Ed. Nat. Cult., Bruxelles.
Cottrell, C. B. & Loveridge, J. P. 1966. Observations on the Cryptosepalum forest of the Mwinilunga District of Zambia. Proc. Trans. Rhod. Sci. Ass. 51:79–120.
Denisoff, I. & Devred, R. 1954. Carte des sols et de la végétation du Congo Belge et du Ruanda-Urundi. 2, Mvuazi (Bas Congo). I.N.E.A.C., Bruxelles.
Devred, R. 1956. Les savanes herbeuses de la région de Mvuazi (Bas-Congo). Publ. I.N.E.A.C. Sér. Sc. 65:1–115.
Devred, R. 1957. Limite phytogéographique occidento-méridionale de la Région Guinéene au Kwango. Bull. Jard. Bot. Brux. 27:417–431.
Devred, R. 1958. La végétation forestière du Congo belge et du Ruanda-Urundi. Bull. Soc. Roy. For. Belg. 65:409–468.
Devred, R., Sys, C. & Berce, J. M. 1958. Carte des sols et de la végétation du Congo belge et du Ruanda-Urundi. 10. Kwango. I.N.E.A.C., Bruxelles.
Diniz, A. Castanheira. 1973. Características mesológicas de Angola. M.I.A.A., Nova Lisboa.
Diniz, A. Castanheira & Aguiar, F. Q. de Barros. 1969. Regiões naturais de Angola. I.I.A.A. Ser. Cien. 7:1–6.
Duvigneaud, P. 1949a. Voyage botanique au Congo belge à travers le Bas-Congo, le Kwango, le Kasai et le Katanga. De Banana à Kasenga. Bull. Soc. Roy. Bot. Belg. 81:15–34.
Duvigneaud, P. 1949b. Le 'Mulombe' du Kwango (Daniellia alsteeniana Duvign.) et le mode de distribution kwango-katangais au Congo-belge. Inst. Roy. Col. Belge, Bull. Séances, 20:677–689.

Duvigneaud, P. 1950. Sur la véritable identité du Parinari sp. 'Mafuca' de Gossweiler et sur l'existence d'une laurisilve de transition Guinéo-Zambézienne. Bull. Soc. Bot. Belg. 83:105–110.
Duivgneaud, P. 1952. La flore et la vegetation du Congo méridional. Lejeunia 16:95–124.
Duvigneaud, P. 1953a. Les savanes du Bas-Congo. Lejeunia 10:1–192.
Duvigneaud, P. 1953b. Les formations herbeuse (savanes et steppes) du Congo méridional. Les Nat. Belges 34:66–75.
Duvigneaud, P. 1958. La végétation du Katanga et de ses sols métallifères. Bull. Soc. Roy. Bot. Belg. 90:127–286.
Exell, A. W. 1957. La végétation de l'Afrique tropicale australe. Bull. Soc. Roy. Bot. Belg. 89:101–106.
Fanshawe, D. B. 1960. Evergreen forest relics in Northern Rhodesia. Kirkia 1:20–24.
Fanshawe, D. B. 1969. The vegetation of Zambia. Kitwe. For. Res. Bull. 7:1–67.
Germain, R. 1949. Reconnaissance geobotanique dans le nord du Kwango. Publ. I.N.E.A.C. Sér. Sc. 43:1–22.
Gossweiler, J. & Mendonça, F. A. 1939. Carta fitogeográfica de Angola. Gov. geral de Angola, Luanda.
Huntley, B. J. 1974. Vegetation and flora conservation in Angola. Serv. de Vet. Rep. no. 22, Luanda (Unpubl.)
Lawton, R. M. 1964. The ecology of Marquesia acuminata (Gilg) R.E. Fr. evergreen forests and the related chipya vegetation types of North-Eastern Rhodesia. J. Ecol. 52:467–479.
Lawton, R. M. 1967. The conservation and management of the riparian evergreen forests of Zambia. Comm. For. Rev. 46:223–232.
Lawton, R. M. 1969. A new record, Maesopsis eminii Engl., for Zambia. Kirkia 7:145–146.
Lebrun, J. 1954. Sur la végétation du Secteur Littoral du Congo belge. Vegetatio 5–6:157–160.
Lebrun, J. 1961. Les deux flores d'Afrique tropicale. Mém. Acad. Roy. Belg. Cl. Sc. 32(6):1–81.
Lebrun, J. & Gilbert, G. 1954. Une classification écologique des forêts du Congo. Publ. I.N.E.A.C. Sér. Sc. 63:1–89.
Léonard, J. 1953. Les forêts du Congo Belge. Les Nat. Belges 34:53–65.
Liben, L. 1958. Esquisse d'une limite phytogéographique Guinéo-Zambézienne au Katanga occidental. Bull. Jard. Bot. Brux. 28:299–305.
Mendonça, F. A. 1961. Indices fitocorológicos da vegetacão de Angola. Garcia de Orta 9:479–483.
Monteiro, R. F. Romero. 1965. Correlacão entre as florestas do Maiombe e dos Dembos. Bol. I.I.C.A. 1:257–265.
Mullenders, W. 1954. La végétation de Kaniama (Entre-Lubishi-Lubilash, Congo belge). Publ. I.N.E.A.C. Sér. Sc. 61:1–500.
Mullenders, W. 1955. The phytogeographical elements and groups of the Kaniama District (High Lomami, Belgian Congo) and the analysis of the vegetation. Webbia 11:497–517.
Nolde, I. 1938. Botanische Studie über das Hochland von Quela in Angola. Feddes Repert. Beih. 15:35–54.
Rübel, E. 1930. Pflanzengesellschaften der Erde. Berne.
Schmitz, A. 1962. Les muhulus du Haut-Katanga méridional. Bull. Jard. Bot. Brux. 32:221–299.
Schmitz, A. 1963. Aperçu sur les groupements végétaux du Katanga. Bull. Soc. Roy. Bot. Belg. 96:233–447.
Schmitz, A. 1971. La végétation de la plaine de Lubumbashi (Haut-Katanga). Publ. I.N.E.A.C. Sér. Sc. 113:1–388.
White, F. 1976a. The vegetation map of Africa: the history of a completed project. Boissiera 24:659–666.
White, F. 1976b. The taxonomy, ecology and chorology of African Chrysobalanaceae (excluding Acioa). Bull. Jard. Bot. Nat. Belg. 46:265–350.
White, F. (ed. in press, A). UNESCO/AETFAT Vegetation of Africa. U.N.E.S.C.O., Paris.
White, F. (in press, B). The underground forests of Africa: a preliminary review. Gardens Bull. Singapore, 29:57–71.

15 Primary production ecology in southern Africa

M. C. Rutherford*

1.	Introduction	623
1.1	Production ecology development	623
1.2	Production concepts and units	624
1.3	Delimitation of 'natural' production areas	626
2.	Biomass and production of main vegetation types	627
2.1	Fynbos	627
2.2	Karoo–Namib	629
2.3	Grassland	630
2.4	Savanna and woodland	631
2.4.1	Herbaceous layer	631
2.4.2	Woody component	634
2.5	Forest	638
2.6	Below ground production	640
3.	Some ecological production controls	641
3.1	Rainfall	643
3.2	Fire	644
3.3	Defoliation	645
3.4	Woody-herbaceous production relations	646
3.5	Soils	647
4.	Production relative to method	648
5.	Conclusions	650
	References	652

* The help of those whose data have been used to derive relationships such as in Fig. 3 and those who supplied unpublished data, is gratefully acknowledged.

15 Primary production ecology in southern Africa

1. Introduction

'Perhaps the most fundamental dimension of an ecosystem is its productivity – the rate of creation of organic material, by photosynthesis primarily, per unit area and time' (Whittaker 1970). Of importance are the structural parameter, biomass, and the functional parameter, production (Odum 1962), both essential in the process of ecosystem analysis (Reichle 1975). Economic requirements for increased food and structural material production have long drawn attention to the importance of understanding and optimizing primary production within different ecosystems.

1.1 *Production ecology development*

Production ecology is not a new field of study. Estimates of world primary production were already made before 1920 (Lieth 1972) and ecological implications of rates of plant production per unit ground area were emphasized by Weaver in 1924. The term 'production ecology' has often been employed to unify ecology and (primary) production (for example, Pearsall 1959 and Newbould 1963). According to Duvigneaud (1971) (translated, Major 1974a), it was the symposium at Hohenheim in 1960 on productivity of vegetation (edited by Lieth 1962) that resulted in a 'capture of the consciousness of western European ecologists who until then were straight-jacketed in interminable descriptions of plant associations'. This symposium, reportedly following guidelines laid down much earlier in the 1932 work of Boysen Jensen (Müller 1962), included presentation of a listing of about 190 production data from certain parts of the world. With the added stimulus of the production orientated International Biological Program (IBP) from 1964 to 1974 (Lieth 1972, Cooper 1975), newly available data on geographical variation of plant production enabled work to venture beyond detailed production comparisons such as that of Westlake (1963) and made possible attempts at mapping production on a world scale (for example, Bazilevich et al. 1971, Lieth 1972). However, the first cited writers pointed to the incompleteness of data for various zones necessitating large generalizations. One of the main areas of incomplete data is stated to be that for tropical and subtropical zones, and although Bourlière & Hadley (1970), Golley & Lieth (1972), Murphy (1975) and other subsequent work on tropical productivity has assisted in filling some gaps in this area, the amount of suitable production data for the largely tropical and subtropical southern African region remains very incomplete. It is only since about 1970 that increased interest in production ecology has been generated in some areas of southern Africa. Many new studies are in the planning stage or underway in southern Africa. This heralds a new era of giving more detailed attention to net primary production of this region. This chapter attempts to summarize existing information relating to net primary production. Although based on data where less precise techniques have often been used and immediate economic or practical considerations prevailed, it is hoped that this chapter will serve as a comparative basis from which new systems orientated studies may be viewed.

1.2 Production concept and units

There is little agreement on precise use of production related terminology. In the present work, for purposes of comparison and taking into account data condition for some categories of vegetation, production refers to the rate of formation of dry matter of plants typically expressed as an annual rate and in this chapter only applies to above ground parts unless otherwise stated. Biomass refers to the amount of dry (normally live or functional) material present per unit ground area at any stated time and also only applies to above ground parts unless otherwise stated. This usage is in basic agreement with that of Ovington (1962), Westlake (1963, 1965) and Květ et al. (1971). The term standing crop has often been equated with the term biomass (for example, Petrusewicz 1967, Whittaker 1970) but sometimes it is implied that standing crop includes dead non-functional material (Westlake 1965). Standing crop, however, may refer to, for example, the mass of loosely defined economically important parts of the total mass present and is therefore a scientifically imprecise term (Ovington 1962, Westlake 1963). For this reason the term standing crop is avoided in this chapter and where much dead material appears to be present, the term 'mass' or 'mass present' is used instead of biomass.

It is clear that the production referred to is only loosely related to above ground net primary production which is the change in biomass over a certain time interval to which is added the losses that occurred in the form of dead material, litter, graze or browse and translocation within the same interval. Field techniques have seldom measured the net primary production term directly and almost always measured the change in above ground biomass without taking into account each potential loss term. Total net primary production requires additional data on difficultly obtained below ground losses including exudates. To date, no data of net primary production appears to exist for southern African vegetation and production data that do exist are seldom very closely related to above ground net primary production.

Production and biomass may be measured in various units (Westlake 1965), for example, volume, carbon content or energy content, but practical usage is that of mass of dry material in $kg.ha^{-1}$ for biomass and $kg.ha^{-1}.year^{-1}$ for production. Conceptual difficulties may arise where biomass is expressed per m^2, for example, where large scattered savanna trees are involved, variation in biomass and possibly production of individual square metres may be of several orders of magnitude. Mass units of kg are employed to express both biomass and production to facilitate interrelation and for added simplicity of dealing with whole numbers. The time interval of a year is suitable for southern African vegetation since in most areas plant growth activities are seasonal. In vegetation with a significant woody element use of more easily measured parameters have been employed, such as basal area ($m^2.ha^{-1}$ and $m^2.ha^{-1}.year^{-1}$) and volume ($m^3.ha^{-1}$ and $m^3.ha^{-1}.year^{-1}$) which nevertheless have often been found to correlate with dry mass of woody vegetation. In an assessment of southern African biomass and production the nature of existing data unfortunately compels some use of these related but not directly comparable units. Presence of varying units for biomass and production data which have been found for various southern African vegetation types are given in Table 1.

The implications of applying conversion factors to production data in a form

Table 1. Units of available data related to total above ground biomass and production for the main vegetation types in southern Africa.

CONCEPT	DATA TYPE	UNITS	FYNBOS	NAMIB – KAROO	GRASSLAND	SAVANNA WOODLANDS		FOREST
						HERBACEOUS LAYER	TREE LAYER	
BIOMASS	BASAL AREA	$m^2 \cdot ha^{-1}$		n/a	n/a		■	■
	VOLUME	$m^3 \cdot ha^{-1}$		n/a	n/a		■	■
	MASS	$kg \cdot ha^{-1}$	■	■	■	■	■	
PRODUCTION	BASAL AREA INCREMENT	$m^2 \cdot ha^{-1} \cdot year^{-1}$		n/a	n/a	n/a	■	■
	VOLUME INCREMENT	$m^3 \cdot ha^{-1} \cdot year^{-1}$		n/a	n/a	n/a	■	■
	RADIAL INCREMENT	mm Radial $\cdot indiv \cdot year^{-1}$	■	n/a	n/a	n/a	■	■
	INDIVIDUAL BASAL AREA INCREMENT	$cm^2 \cdot indiv^{-1} \cdot year^{-1}$		n/a	n/a	n/a	■	
	MASS INCREMENT	$kg \cdot ha^{-1} \cdot year^{-1}$	■		■	■		

■ DATA PRESENT   n/a DATA TYPE NOT SUITABLE

not suitable for direct comparison with other production data, appears to have rarely been closely questioned. There is good reason to convert published values using a minimum of well based assumptions but there is probably little purpose in further applying several tenuous assumptions to obtain fully comparable 'corrected' data for which there is virtually no indication of degree of error. Some phrases that have been used include: 'appropriate adjustments', 'simplifications', 'shrewd guess' and 'predictable ratios'. Some of these derived data are then later quoted as 'reliable production measurements'! Statistical analysis of derived data can certainly create an impression of false confidence.

In some works the correction factors are explicitly stated, for example, in forests, annual leaf fall multiplied by 3 to give total annual production (Murphy 1975); in North Carolina's softwood and hardwood forests commercial dry mass yield multiplied by 2 to give total primary production (Sharp et al. 1975). In some cases where a single factor is applied, for example multiplying above ground production by 1.66 to include root production in different grassland and savanna communities (Murphy 1975) comparisons of these values are essentially comparisons of the original data, no further information having been introduced. Sharp (1975) points out how critical conversion ratios are. Unless corrections are accurate and may be validly applied to a particular situation, only relatively large differences in production between regions can be confidently accepted. For available southern African data, it is considered that conversion of production related values should be limited so that a restricted number of sets of values are obtained which are at least internally comparable within a set with a greater degree of sensitivity. When more specific information on required correction factors become available, conversion to fully comparable values will become possible.

Considerable selectivity has been exercised in the incorporation of data from available literature. Some published works with promising titles have for many different reasons been found unsuitable while in other apparently irrelevant articles, relevant and suitably defined data have been located. In a few cases data have been included from sources that do not state sufficient details of method for full interpretation of the data, but for want of any other data for the area concerned, have been incorporated. Several sources provide only the most basic, unprocessed form of data and here own computations have been required for comparison with other data, for example, in frequency distributions of woody plant basal areas which when given in histogram form have required assumptions of class midpoints and circumspection when an open ended frequency class is present. The principles involved in the propagation of errors through rounding-off, multiplication and division of values have been considered in an attempt to retain the originally indicated level of accuracy although, owing to many diverse ways of data presentation, consistency in this regard has been difficult to maintain.

Data have not been tabulated since often there are so many qualifying statements, that the unavoidable omission of some of these in a table might create an undue impression of precise comparability.

The sequence of data presentation within sections usually follows a numerical sequence according to value except when excessive geographic differences are involved. Within a particular section accurate data usually preceeds apparently less accurate data.

1.3 Delimitation of 'natural' production areas

It is intended to restrict discussion of comparative production and biomass and some major ecological factors controlling production to those areas of southern Africa that bear so-called 'natural' vegetation. The inferred 'low' production of veld of several parts of southern Africa (West 1955, Edwards 1957, Birch 1972) appears to have added to the considerable interest in determining in which areas plant production may be artificially increased. Production of vegetation may be changed in four basic ways (Grunow et al. 1970): (i) by grazing management practices including, for example, movement of grazing animals and manipulation of fire; (ii) by changing botanical composition, involving, for example, overseeding; (iii) by changing soil conditions through, for example, fertilization; and (iv) by changing climatic factors, for example, by irrigation.

Arable land, that is, where the last three types of changes may be very heavily applied, covers only 15 per cent of South Africa, and almost all of this is already under cultivation (Edwards & Booysen 1972). The remaining 85 per cent, that is uncultivitable land may be divided into two basic parts: those areas where only the first method of change may be suitably applied, that is in 'veld' (Booysen 1967) and those areas where the other three radical methods of change may still be applied to some degree. In South Africa, the main factor in many such delimitations has been rainfall (for example, Tidmarsh 1968) and where rainfall is not limiting to radical change, plant production is more often limited by economic factors (Booysen 1972). The potential area of radical veld production change has been delimited using as lower limits the mean annual isohyet lines of 300 mm in the Cape winter rainfall region, 400 mm in the all year rainfall area west of Port Elizabeth, 500 mm east of Port Elizabeth and on the Drakensberg escarpment and 600 mm west of the Drakensberg escarpment (Edwards & Booysen 1972). Thus, for example, west of the escarpment near Bloemfontein, with mean annual rainfall of only about 500 mm, Mostert & Voster (1969) report that a 12 year study indicated it uneconomical to fertilize the veld.

Similarly afforestable areas have been delimited for South Africa on the basis of mean annual rainfall and temperature. Grut (1965) gives the lower limits of potential afforestation areas as the mean annual isohyets of about 650 mm in the winter rainfall region to about 900 mm on the Drakensberg escarpment and about 1000 mm in warmer areas of the Natal coast and northern Transvaal escarpment. However, where silvicultural changes are envisaged that employ indigenous tree species (Vowinckel 1961) rainfall limits may be considerably lower. Nevertheless, it is clear that afforestable areas fall largely within areas of potential radical change for agricultural grazing production. The estimated proportion of this area of potential change is 20 per cent of South Africa which together with the 15 per cent of potential arable land, leaves 65 per cent of South Africa's vegetation which may only be suitably manipulated through grazing management practices (Edwards & Booysen 1972) and it seems probable that this last percentage is not much lower for southern Africa as a whole.

Some not uncommon exceptions to this general delimitation of production areas do occur. In areas delimited as potentially suitable for radical production change, some areas are not suited, for example, where limitations of access are present in the higher mountain regions of Lesotho (Bawden & Carroll 1968). In areas usually suited only to grazing management practices, bush and tree killing or removal constitutes a radical change in many dryer savanna areas, while future development of applied techniques such as increased rainfall through cloud seeding (Garstang et al. 1974) may also result in further radical changes in production.

Production of vegetation in areas where only grazing management practices are suited, or in areas in which production may be radically changed but has not yet been undertaken, is considered here as 'natural' production and further discussion is limited to such production. Geographical origins of biomass or production data quoted is indicated in Fig. 1.

2. Biomass and production of main vegetation types

2.1 Fynbos

Kruger (1977) reports biomass and production data for several fynbos communities at three sites between Stellenbosch and Botrivier, that is at Jonkershoek, Zachariashoek and at Jakkalsrivier. In fynbos the age of communities is most often directly related to time of previous fire and has a direct bearing on the biomass present. Biomass of communities of mesophyllous evergreen broad sclerophyll scrub 10 and 17 years old was between 13,000 and 26,000 kg.ha^{-1}; that of those from 6 to 7 years old was between 6000 and 16,000 kg.ha^{-1} (except for one six-year-old phreatic *Elegia thyrsifera-Osmitopsis asteriscoides* swamp community where biomass was 21,000 kg.ha^{-1}), while that of those from 2 to 5 years ranged between 2000 and 8000 kg.ha^{-1}. Biomass of mature 16 year old heath communities ranged from 11,000 to 15,000 kg.ha^{-1}; while that of 3.5 to 4.5 year old heath communities was between 5000 and 9000 kg.ha^{-1}. Near the summit on part of the Riviersonderend Mountains, south of Worcester, biomass of a north facing mature *Elegia-Tetraria-Restio* dominated restionaceous community was found to be about 14,000 kg.ha^{-1}, while that of a *Protea laurifolia* dominated community (with occasional *Protea* individuals up to 15 years old) on lower more xeric north facing slopes of the same mountain was

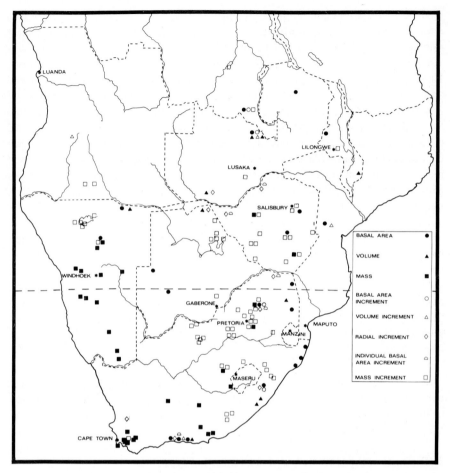

Fig. 1. Map showing the locations from which biomass or production related data have been obtained.

about 11,000 kg.ha^{-1} (Rutherford 1978a). On a xeric site on Kanonkop in the Cape Point Nature Reserve south of Cape Town, Taylor (pers. comm.) found biomass was only about 5000 kg.ha^{-1} for the Dry Hillveld variation of Upland Mixed Fynbos at least 12 years old. Moll (pers. comm.) reports a particularly low biomass of about 4500 kg.ha^{-1} for mature restioland in Bainskloof east of Wellington. Fynbos with possibly the lowest mean annual rainfall occurs in the form of *Elytropappus rhinocerotis* dominated communities often found in zones between more typical fynbos communities and Karoo vegetation (Acocks 1975, and Chapters 8 and 9). Biomass of an *Elytropappus rhinocerotis* community (Acocks 1975, Veld type 43) south of Worcester has been found to be 11,000 kg.ha^{-1} (Rutherford 1978a). It should be noted that owing to high variability in fynbos communities the 95 per cent confidence intervals found for most of the above data are seldom much below 20 per cent of the mean.

In a study of growth of 23 individuals (all less than 3 m high at 25 years old) of *Widdringtonia cedarbergensis*, 22 of which were naturally regenerated at several localities in the Cedarberg range between Clanwilliam and Citrusdal, Hubbard (1937) provides data based on growth rings which show a mean radial stem increment of 0.8 mm.year^{-1} over the first 45 years of growth. Kruger (1977) gives production for the first 2 years of growth in fynbos as varying from 1000 kg.ha^{-1}.year^{-1} for a heath community to 4000 kg.ha^{-1}.year^{-1} for the *Elegia-Osmitopsis* swamp community with the average production in fynbos in the order of 2500 kg.ha^{-1}.year^{-1}.

2.2 Karoo–Namib

Very few biomass or production measurements have been made in the Karoo–Namib Region, probably partly due to association between assumed low production and reduced economic importance. In the eastern more mesic Karoo areas (False Karoo, Acocks 1975) attention appears to have been largely diverted from mass determination methods through the evolution and application of point sampling methods for monitoring relative changes in basal area of differently treated Karoo vegetation (Tidmarsh & Havenga 1955, Roux 1963). However, in heterogeneous Karoo vegetation which may include herbaceous, woody and succulent plant forms together, basal cover of these forms does not appear additive for estimating total biomass and even in herbaceous vegetation the relation between basal cover and production is dependent upon species composition (Rutherford 1975). Also, using the Tidmarsh & Havenga (1955) point method in herbaceous vegetation, total basal cover estimates have been found to differ by as much as 20 or 25 per cent between different operators (Walker 1970). For purposes of overall comparison, therefore, basal area of much of Karoo, grassland or other herbaceous vegetation appears not to provide a reliable estimate of biomass or production.

Walter (1939) has determined plant mass data for several areas in the Karoo–Namib area in South West Africa. However, he points out that these are maximal values for localized areas and from his contemporary data gathered in savanna areas in South West Africa (see Section 2.4.1) it appears that these data be recognized as accumulated plant mass present at the time of sampling and not as annual production which might be in the order of half or less the oft-cited (for example, Walter & Volk 1954, Odum 1959, Rodin & Bazilevich 1967, Walter 1973, Lieth 1975c) stated values. Mass values obtained by Walter (1939) for mainly grass dominated communities often with *Stipagrostis uniplumis* and *Aristida* and *Eragrostis* species present were about 600 kg.ha^{-1} on a stony, steep site at Blässkranz in the Rehoboth district; 700 to 800 kg.ha^{-1} at sites near Warmbad, Grünau and at Voigtsgrund (in the Mariental district); 900 kg.ha^{-1} at Gellap Ost in the Keetmanshoop district and at Klein Spitzkop west of Usakos and between 1000 and 1700 kg.ha^{-1} at 2 sites in the Karabib district and at Ababis in the Rehoboth district. Mean annual rainfall at almost all these sites is between 100 and 200 mm.

In the Upper Karoo near Carnavon mass of the shrub vegetation has been determined at 3700 kg.ha^{-1} while that of the shrub vegetation in a higher rainfall area near Middelburg, Cape, at 4800 to 5600 kg.ha^{-1} (Department of Agricultural Technical Services 1975/76). In a localized frequently moist dense stand of *Tetrachne dregei* and other grass species near Middelburg mass accumulated over 3 years was 5100 kg.ha^{-1} (Roux 1968). In a less arid part of the Robertson Karoo, south of Worcester, the biomass of a succulent form of Broken Karoo vegetation (Acocks 1975, Veld Type 26), dominated by species of

Pteronia, *Delosperma* and *Euphorbia*, was 7500 kg.ha^{-1} (Rutherford 1978a). In more arid parts of the Karoo, for example in true Succulent Karoo (Acocks 1975, Veld Type 31), this biomass value may be expected to be considerably lower.

2.3 *Grassland*

Production studies have been carried out at several centres in the grassland area of the central high plateau of South Africa.

In the drier southwestern part of the grassland area near Bloemfontein, with an annual mean rainfall of slightly higher than 500 mm, Vorster & Mostert (1968) report a 10 year average production of 630 kg.ha^{-1}.year^{-1} for *Themeda triandra* dominated climax grassland and 700 kg.ha^{-1}.year^{-1} for *Eragrostis* and *Aristida* dominated pioneer grassland. Both these means fall within a previously stated range for this area of 600 to 1200 kg.ha^{-1}.year^{-1}(Meredith et al. 1955). In contrast a single season's production of *Themeda triandra* dominated grassland also near Bloemfontein has been found to be 2150 kg.ha^{-1} (van Schalkwyk et al. 1968). Herbst & Goosen (1973/74 and 1974/75) have demonstrated in the Bloemfontein area that mass of grassland varies greatly according to soil type and have obtained mass values as high as 4900 kg.ha^{-1} on Arcadia, 3300 kg.ha^{-1} on Glendale and 3100 kg.ha^{-1} on Bonheim soil series.

In a slightly higher rainfall area in the western grassland area in the vicinity of Potchefstroom (mean annual rainfall slightly higher than 600 mm), Grunow et al. (1970) indicate a 10 year average production of 500 to 600 kg.ha^{-1}.year^{-1} for a *Themeda triandra-Elyonurus argenteus- Eustachys mutica* community while for the same area Visser (1966) indicates a 6 year average of about 800 kg.ha^{-1}.year^{-1}. This contrasts with production of about 1200 kg.ha^{-1} given for this area by Meredith et al. (1955). In the Zwartrand bankenveld in so-called 'klipveld' (Acocks 1975, Veld Type 61a) northwest of Potchefstroom, Coetsee (1972) found a 3 year average production of 560 kg.ha^{-1}.year^{-1}. This relatively low production is possibly related to local stony conditions that typify this part of the grassland. Summing three regrowths over one season, Edwards & Nel (1973) obtained a production of 970 and 1360 kg.ha^{-1}.year^{-1} for 2 *Eragrostis-Themeda* 'klipveld' plots.

On the northern boundary of the grassland area at Pretoria, with mean annual rainfall in the region of 700 mm, Henrici, quoted by Walter (1939), states an average annual production of 1000 kg.ha^{-1}.year^{-1} while a 5 year average production of about 1200 kg.ha^{-1}.year^{-1} ('usually three cuts ... each growing season') is indicated for Hatfield in Pretoria by Hall et al. (1955). Similarly, for Rietvlei immediately south of Pretoria, an annual production of 1200 kg.ha^{-1} (ranging from 500 to 1600 kg.ha^{-1}) has been obtained (Meredith et al. 1955). Grunow et al. (1970) indicate long term average production over 13 to 20 years to be from 1500 to 1900 kg.ha^{-1}.year^{-1} for a *Themeda triandra-Heteropogon contortus* site and 2 other sites at Pretoria. Possibly comparable results are those for a site near the southern grassland limit in the eastern Cape sourveld (Du Toit & Ingpen 1970). A 10 year average annual production of 1540 kg.ha^{-1}.year^{-1} has been obtained for naturally reclaimed *Eragrostis* grassland in the eastern Orange Free State with average annual rainfall of about 700 mm (Kruger & Smit 1973).

On higher lying regions of the highveld at Frankenwald north of Johannesburg, with mean annual rainfall of about 750 mm, Weinmann (1943), for *Trachypogon-*

Tristachya grassland obtained production of 2163 kg.ha⁻¹.year⁻¹ averaged over 3 seasons and reported (Weinmann 1948a) an equal amount (2163 kg.ha⁻¹) for a later single season. Meredith (quoted by Weinmann 1955) found production of 1650 kg.ha⁻¹.year⁻¹ in camps that had been intensively grazed while in an ungrazed area production accumulated over 3 years was roughly 3500 kg.ha⁻¹ (Hall et al. 1937). Further east in the vicinity of Ermelo with mean annual rainfall between 750 and 800 mm but generally colder than Frankenwald, Rethman & Beukes (1973) indicate a production of about 1800 kg.ha⁻¹.year⁻¹ over one season and over 3 seasons production of *Themeda triandra*, *Heteropogon contortus* and *Digitaria tricholaenoides* dominated grassland as sampled in January or February ranged from 1980 kg.ha⁻¹ to 2340 kg.ha⁻¹ with an average of 2120 kg.ha⁻¹.year⁻¹ (Rethman et al 1971).

In the relatively mesic Natal grasslands production has been stated to be between 2000 and 4000 kg.ha⁻¹.year⁻¹ (Henrici, quoted by Walter 1939) and that of the highland sourveld to be between 2500 and 3800 kg.ha⁻¹.year⁻¹ (Edwards 1966). For grassland at Cedara near Pietermaritzburg, annual production has been given as between 3000 and 3600 kg.ha⁻¹.year⁻¹ (Meredith et al. 1955). At Ukulinga near Pietermaritzburg average production over 15 years has been found to be 3900 kg.ha⁻¹.year⁻¹ ranging from 2880 to 5560 kg.ha⁻¹ (Tainton et al. 1970), although in a below average rainfall year in *Themeda triandra* dominated vegetation in the same area only about 1700 kg.ha⁻¹ of living material was found at the peak of the season (Rethman & Booysen 1969). In the high rainfall region of the southeastern Free State, north of Zastron, Herbst & Goosen (1973/74 and 1974/75) have found grassland mass to vary from 5000 to 5500 kg.ha⁻¹ for an *Andropogon appendiculatus–Eragrostis plana* association on clay soil to between 1800 and 2600 kg.ha⁻¹ for grass associations on sand. Grass mass on litholitic complexes in this area were lower, for example, 1600 kg.ha⁻¹ for a warm aspect *Aristida diffusa–Heteropogon contortus* association.

2.4 *Savanna and woodland*

Savanna and woodland may be divided into two important producing components, the herbaceous layer and the woody layer, and these two components have most often been studied separately. The vegetation of this region comprises at least half the area of southern Africa and is inter-dispersed with occasional patches of open grassland, the production of which is given together with discussion of herbaceous layer production of savannas and woodlands. In some information sources, it is not entirely clear whether production has been measured in herbaceous layer of savanna or in a more extensive open grassland area within a savanna region.

2.4.1 Herbaceous layer

Areas with reported herbaceous layer production of usually less than 1000 kg.ha⁻¹.year⁻¹ are considered first. Le Roux (1972/73) determined herbaceous production over one season for several calcareous sites with mean annual rainfall between 400 and 450 mm in the Etosha National Park in northern South West Africa. A production of 230 kg.ha⁻¹ was obtained for open savanna with

Acacia spp. in the central Leeubron–Adamax area; 390 kg.ha^{-1} for very open savanna in the south central Gemsbokvlakte area and 620 kg.ha^{-1} for both the very open savanna in the eastern Charitsaubvlakte area and the open *Sporobolus spicatus* grassland of Andoni in the northeast. Other data from a subsequent year with above average rainfall indicated production to exceed 1000 kg.ha^{-1} in several areas.

Some other low herbaceous layer production values have been found along the southern boundary of the savanna zone not only where mean annual rainfall is relatively low but where so called 'bush encroachment' is prevalent (see Chapters 9 and 10). In the northwestern Cape at Vryburg, annual production of herbaceous material has been stated to be about 500 kg.ha^{-1}.year^{-1} (Henrici, quoted by Walter 1939) while further north, in the Molopo area, Donaldson (1967) indicates 'normal' production to be in the order of 750 kg.ha^{-1}.year^{-1}. Herbaceous layer production may vary greatly according to: (i) erratic rainfall where, for example, only 51 kg.ha^{-1} was produced in a season of severe drought in the Molopo area (Donaldson 1967) and (ii) density of trees and shrubs where, for example, Donaldson & Kelk (1970) show that annual production varied from 119 to 1071 kg.ha^{-1} in inverse relation to density of *Acacia mellifera* subsp. *detinens* individuals (see Fig. 3). A similar range of herbaceous layer production may be expected for similar *A. mellifera* dominated vegetation in some more arid savanna areas in South West Africa. In more open parts of Kalahari thornveld (Acocks 1975, Veld Type 16) with scattered *Acacia* and *Tarchonanthus* shrubs in the vicinity of Vryburg in a below average rainfall season, Fourie & Roberts (1976) obtained production of 650 kg.ha^{-1} for a *Digitaria pentzii, Eragrostis nindensis* and *Themeda triandra* dominated dolomite community; 950 kg.ha^{-1} for a *Eragrostis lehmanniana, Stipagrostis uniplumis, Eragrostis pallens* and *Anthephora pubescens* dominated sand community and 1150 kg.ha^{-1} for an *Eragrostis lehmanniana, Digitaria polevansii* and *Stipagrostis uniplumis* dominated limestone community.

In the Eastern Cape, a situation corresponding to that in the northwestern Cape is found in the form of grassland invaded by *Acacia karroo* trees (Acocks 1975, Veld Type 22). North of Cathcart in an area with average tree density of approximately 1000 per hectare, du Toit (1968) determined over two years (one of which had below average rainfall) an herbaceous layer production of 160 and 620 kg.ha^{-1} and later reported (du Toit 1972) production ranging from 70 to 1000 kg.ha^{-1} for a number of subsequent seasons with greatly differing rainfall. In an area with about 1600 woody plants (mostly *Acacia karroo*) per hectare, Trollope (1974) reports production of about 2500 kg.ha^{-1} over one year in a *Themeda triandra* dominated herbaceous layer.

Other production values of less than 1000 kg.ha^{-1} have been reported for diverse savanna areas. The values appear low relative to that expected for the respective regions, probably owing to, for example, grazing, drought and high woody plant density. In northeastern Botswana, Blair Rains & McKay (1968) provide herbaceous layer production values for several savanna communities, presumably in a grazed condition, ranging from 440 kg.ha^{-1} for *Schmidtia pappohoroides–Stipagrostis uniplumis* veld with scattered low *Combretum* species to 860 kg.ha^{-1} for *Andropogon gayanus–Setaria sphacelata* veld with scattered shrubs and low trees of *Combretum imberbe, C. hereroense* and *Piliostigma thonningii*. In *Acacia nigrescens–Sclerocarya caffra* savanna near Mahalapye,

Botswana, McKay (1968) found production of a *Digitaria* dominated herbaceous layer with a moderate grazing intensity history to be 700 kg.ha^{-1} in a season with below average rainfall. In an arid portion of the western Transvaal near Thabazimbi, Du Plessis & van Wyk (1969) determined an annual production of 750 and 800 kg.ha^{-1} for *Eragrostis* and *Panicum* spp. dominated herbaceous layer of 'sweet bushveld' in an open mixed *Acacia heteracantha–A. senegal* community after a severe drought. In *Brachystegia* and *Julbernardia* woodland near Fort Victoria, Rhodesia, a particularly low production of 340 kg.ha^{-1}.year^{-1} averaged over five years was obtained for *Digitaria–Bulbostylis–Rhynchelytrum* dominated herbaceous layer (Ward & Cleghorn 1964). In the same area, Ward & Cleghorn (1970) report production of 500 to 560 kg.ha^{-1}.year^{-1} under light grazing and data of Kennard & Walker (1973) point to much higher herbaceous production in more open areas with *Panicum maximum* associated with, for example, *Combretum molle* and *Terminalia sericea*.

Production falling roughly between 1000 and 2000 kg.ha^{-1}.year^{-1} has been reported for several areas. Production between 750 and 1000 kg.ha^{-1}.year^{-1} has been quoted for *Andropogon–Combretum* savanna (Bourliére & Hadley 1970) northwest of Pretoria. Herbaceous layer production values of about 1000 kg.ha^{-1}.year^{-1} have been found over 3 seasons near Potgietersrus, Transvaal (Huntley 1971) and for *Digitaria polevansii, Brachiara nigropedata* and *Andropogon gayanus* dominated vegetation in *Burkea africana* savanna in northeastern South West Africa (Rutherford 1975). At Nylsvley, near Naboomspruit in the northern Transvaal, Hirst (1975) quotes Grunow as finding a seasonal peak of green material of *Eragrostis pallens–Digitaria eriantha* dominated herbaceous layer in *Burkea africana* savanna of about 1230 kg.ha^{-1} in between trees, but 970 kg.ha^{-1} directly under trees and 640 kg.ha^{-1} under shrubs but slightly lower values were obtained in a following year (Grunow 1975/76). For several sites in unutilized *Colophosperum mopane* savanna in south eastern Rhodesia, Kelly (1973) found herbaceous layer production to range from 650 kg.ha^{-1}.year^{-1} to 2084 kg.ha^{-1}.year^{-1} but averaging 1219 kg.ha^{-1}.year^{-1} while Knapp (1965) obtained 1220 kg.ha^{-1}.year^{-1} for dry savanna in a southern part of Rhodesia. At Matopos and Nyamandhlovu near Bulawayo, West (1955) states production to be not more than about 1300 kg.ha^{-1}.year^{-1}. A 9 year average production of 1450 kg.ha^{-1}.year^{-1} is reportedly given by West for *Themeda–Heteropogon* vegetation at Matopos (Bourliére & Hadley 1970) and herbaceous production over 3 years varied from 700 to 1300 kg.ha^{-1}.year^{-1} for *Hyparrhenia filipendula* sandveld to between 1200 and 1800 kg.ha^{-1}.year^{-1} for *Heteropogon contortus–Acacia* veld (Mills 1968). In a single particularly 'good' growing season at Matopos, Mills (1964) reported higher production values with some communities exceeding 2000 kg.ha^{-1}. An average annual production over 13 years of about 1400 kg.ha^{-1}.year^{-1} has been reported for an area of 'natural veld on the Springbok flats' at Tawoomba near Warmbaths in the northern Transvaal (Louw 1968). Herbaceous layer mass data obtained for various savannas in South West Africa by Walter (1939) appear, as mentioned in Section 2.2, to be at least twice that of probable annual production. It appears likely therefore that mass values of 3970 kg.ha^{-1} averaged for an area near Windhoek, 4530 kg.ha^{-1} near Gobabis and 5530 kg.ha^{-1} near Otjiwarongo interpreted in terms of annual production probably all fall below 2000 kg.ha^{-1}.year^{-1}. On part of the actual open pan area of the Etosha pan in northern South West Africa, Le

Roux (pers. comm.) determined production of 1300 kg.ha^{-1} for a monostand of *Sporobolus salsus* after a season of exceptionally high rainfall. In Botswana a localized open grassland within the savanna zone adjoining the northern arm of the Makarikari Pan, with *Cenchrus ciliarus*, *Panicum coloratum* and *Schmidtia pappophoroides* predominant, has been found to produce 1780 kg.ha^{-1} (Blair Rains & McKay 1968).

Most of the areas with production usually more than 2000 kg.ha^{-1}.year^{-1} lie in the northerly parts of the savanna-woodland zone of southern Africa. In southern Zaïre at Luiswishi near Lubumbashi, in miombo-woodland with *Marquesia*, *Brachystegia* and *Monotes* species, Freson (1973) found production of *Tristachya* and other species dominated herbaceous layer to be 2200 kg.ha^{-1} in one season. In southern Angola, Menezes (1971) reports presumably annual production to range from 1200 to 3300 kg.ha^{-1} for woodlands of *Baikiaea plurijuga*, *Colophospermum mopane* and other species. In a riverine community in northwestern Rhodesia, Goodman (1975) determined a January biomass of 4100 kg.ha^{-1}. In an area near Salisbury where *Brachystegia–Julbernardia* had been stumped 16 years previously but had coppiced, Weinmann (1948b) reports herbaceous production of about 2250 kg.ha^{-1}. According to Thomas (pers. comm.), production of natural *Hyparrhenia* communities in several provinces of Zambia average about 3800 kg.ha^{-1}.year^{-1} while Van Rensburg (1968) also indicates the Kafue flood plain to be highly productive. For *Hyparrhenia dissoluta–Heteropogon contortus* grassland at Mazabuka in Zambia, production exceeded 2000 kg.ha^{-1} in January (Smith 1963) and attained about 4500 kg.ha^{-1} at the end of the season (Smith 1961). In contrast, Thomas (pers. comm.) indicates that in a *Hyparrhenia* community in the central region of Malawi, production in 1 year amounted to 1180 kg.ha^{-1}. Elliott & Folkertsen (1961) indicate a 3 year average production of about 3300 kg.ha^{-1}.year^{-1} for grassland near Salisbury. In southern Angola, Menezes (1971) reports production from between 3300 and 4800 kg.ha^{-1} for open grassland with some *Themeda* and *Loudetia* species while in the central region of Malawi an *Echinochloa pyramidalis* vlei had a 3 year production averaging 5370 kg.ha^{-1}.year^{-1} according to Thomas (pers. comm.). Remarkably high production values of 7400 kg.ha^{-1} to 10,940 kg.ha^{-1} (presumably for herbaceous layer vegetation) have been reported for areas near the southern and eastern Rhodesian border and a moist part of the Zambezi valley (Knapp 1965).

2.4.2 Woody component

The most commonly measured woody biomass related parameter in the savannas and woodlands of southern Africa has been basal stem area, although choice of minimum tree size and measurement height above ground differ with the result that some data can only be approximately compared. Some of the following data have been derived from frequency distributions of tree circumferences in areas of known size.

Data of Blair Rains & Yalala (1972) indicate a basal area of about 0.5 m^2.ha^{-1} for trees larger than 10 cm diameter at breast height (here only *Acacia erioloba* and *A. uncinata*) for bush savanna with *Grewia flava* and *Terminalia sericea* south of Mamuno in western Botswana and about 5 m^2.ha^{-1} basal area for savanna between Kang and Lone Tree Pan in southwestern Botswana where the dominant species were *Acacia erioloba, Acacia uncinata, Lonchocarpus nelsii* and *Acacia mellifera* subsp. *detinens*. For all trees and shrubs with stems greater than 1 cm diameter in a long protected savanna woodland of *Burkea africana, Terminalia sericea* and *Combretum psidioides*, Rutherford (1975) obtained 8 m^2.ha^{-1} basal area while in a similar *Burkea africana* woodland in the northern Transvaal, Rutherford & Kelly (1977) obtained, for stems larger than 0.5 cm diameter (excepting *Grewia flavescens* shrubs) 8.5 m^2.ha^{-1} basal area of live woody plants and an additional 2.1 m^2.ha^{-1} of dead individuals. For *Brachystegia spiciformis, Julbernardia globiflora* and *Burkea africana* just north of Fort Victoria in Rhodesia, Ward & Cleghorn (1964), for stem diameters above 2.5 cm, obtained 8.3 m^2.ha^{-1} basal area. In the Okavango region of South West Africa, using a more restrictive criterion of 10 cm and greater diameter at breast height, Geldenhuys (1976) obtained for several woodland types values which ranged from 6.96 m^2.ha^{-1} for *Burkea africana* dominated woodland with *Baikiaea plurijuga* and *Pterocarpus angolensis* to 3.53 m^2.ha^{-1} for *Burkea africana, Terminalia sericea, Combretum* species and *Ochna* vegetation severely subjected to fire. Data from Van der Schijff (1957) indicate a ground level basal area for all trees and single stemmed shrubs to be not less than* 9 m^2.ha^{-1} for *Acacia nigrescens* dominated vegetation on dolerite intrusions in the Kruger National Park, eastern Transvaal, and not less than 13 m^2.ha^{-1} for *Colophospermum mopane* woodland on colluvial-alluvial soil in the northern part of the park. For various originally cleared sites on the Rhodesian highveld with *Brachystegia spiciformis* and *Julbernardia globiflora* Strang (1974) found that basal area stabilized after 50 years at 10 to 11 m^2.ha^{-1} (From formula $y = -3.399 + 4.966x - 0.426x^2$ with $r = 0.978$ and where y = basal area of trees in m^2.ha^{-1} and x = age in decades). For all species with more than 5 cm diameter at breast height, Endean (1967), in what he considered maximum producing undisturbed miombo vegetation in Zambia with mature *Brachystegia longiflora, Julbernardia paniculata, Isoberlinia angolensis* and some *Pterocarpus angolensis* and *Burkea africana* immediately west of Ndola, initially measured in 1933 a basal area of 10.9 m^2.ha^{-1} and by 1960 measured 16.1 m^2.ha^{-1}. A basal area of all trees at 1.5 m above ground in a *Brachystegia* spp. – *Julbernardia paniculata* miombo vegetation in southern Zaïre was found to be 13.3 m^2.ha^{-1} (Malaisse et al. 1975). In the Mucheve woodland reserve in Manica, Sofala district of Moçambique, Guerreiro (1966) found for trees, larger than 15 cm in diameter, a basal area of 13.95 m^2.ha^{-1} of which more than half was *Amblygonocarpus andongensis, Pseudolachnostylis maprouneifolia* and *Pterocarpus angolensis*.

Volume of trees per hectare is a parameter which has been used in savanna and woodland mainly for timber exploitation potential and, as indicated by Endean (1967), is often a very subjective estimate depending on what is considered exploitable or merchantable. In the savanna zones, volume has been estimated in the better developed woodland portions. Miller (1939) estimated volume of merchantable trees for *Baikiaea plurijuga–Pterocarpus angolensis* and *Burkea africana* type woodland in the Chobe area in northeastern Botswana at 3.0 m^3.ha^{-1} over 280 km^2 and 1.6 m^3.ha^{-1} over 195 km^2. In southwestern Zambia, Martin (1940) reports timber volume between 2.1 and 5.6 m^3.ha^{-1} for *Baikiaea* woodlands, but refers to a particular *Baikiaea* forest from which

---

* 'not less than' due to open ended maximum stem diameter class.

31 m^3.ha^{-1} was harvested. Lawton (1967) states that remaining *Baikiaea* forests only yield 4.9 to 5.6 m^3.ha^{-1}. For trees with diameters greater than 45 cm at breast height for *Baikiaea plurijuga, Guibourtia coleosperma* and *Pterocarpus angolensis* and greater than 35 cm diameter at breast height for *Burkea africana*, Geldenhuys (1976) obtained a basal area from 0.88 to 7.44 m^3.ha^{-1} for several localities of Okavango woodland in northeastern South West Africa. Possibly illustrating the difficulty involved in volume estimation of savanna trees, Endean (1967) obtained for well-developed undisturbed *Brachystegia* woodland in northern Zambia values of 47 to 60 m^3.ha^{-1} (for all species with diameters larger than 5 cm at breast height), while King (1952) for similar *Brachystegia, Isoberlinia* and *Pterocarpus* on the copperbelt in Zambia obtained only 3.5 m^3.ha^{-1} of 'saw logs' but roughly 200 m^3.ha^{-1} of 'firewood'. For *Brachystegia–Julbernardia* woodland of the copperbelt, Lees (1962) is quoted (Lawton 1967) as reporting a volume of 3.5 m^3.ha^{-1}.

Very few biomass values have been determined for the woody component of savannas or woodlands in southern Africa. Total woody mass per hectare has been calculated for 9 sites in *Colophospermum mopane* dominated communities in the southwestern Rhodesian lowveld by Kelly (1973) with values ranging from 8700 to 30,800 kg.ha^{-1} dry mass averaging 19,700 kg.ha^{-1} in a region with mean annual rainfall of about 500 mm. In an area with similar mean annual rainfall in South West Africa, in *Burkea africana* dominated woodland savanna, a value of 22,300 kg.ha^{-1} was obtained (Rutherford 1975). By applying an assumed 750 air dry kg per green m^3 (from data of Miller 1938, Van Wyk 1972/74 and others) to Endean's (1967) volume data for well developed undisturbed *Brachystegia* woodland in northern Zambia, a biomass value of between 35,000 and 45,000 kg.ha^{-1} is obtained.

Several biomass determinations have been made in the often dense shrubby 'savanna' of the eastern Cape, but apparently not on a dry mass basis. In Valley Bushveld (Acocks 1975, Veld Type 23) where the semi-succulent *Portulacaria afra* is dominant, Aucamp (1967) obtained a mass of 50,000 kg.ha^{-1} while in *Portulacaria afra* dominated bushveld at Addo north of Port Elizabeth a mass of 184,000 kg.ha^{-1} was obtained although only 46 per cent of this value was obtained in an area stocked with about 2.5 elephant per square km (Penzhorn et al. 1974). In the 'Noorsveld' (Acocks 1975, Veld Type 24) near Jansenville in the eastern Cape a fresh mass of 74,000 kg.ha^{-1} was obtained by Louw (1964) (quoted by Van der Walt 1965) for a community dominated by the succulent *Euphorbia coerulescens*. Owing to the high degree of succulence in the communities of the eastern Cape referred to, the dry mass values for such communities is likely to be distinctly less than half the stated values.

Increments in woody species have often been given in terms of radial, diameter or circumference increment, all of which are dependent upon the size of the tree considered. Rutherford & Kelly (1977) found, for 6 species in a northern Transvaal *Burkea africana* woodland, radial increment over a 1 year period was from about 1 to 4 mm averaging 2.5 mm. Miller (1938, after main section) gives a mean annual increment of 3.6 mm.individual^{-1}.year^{-1} for *Pterocarpus angolensis* and 2.0 mm.individual^{-1}.year^{-1} for *Baikiaea plurijuga* in the Chobe area in Botswana. Boaler (1963) found that one year radial increment in a number of *Pterocarpus angolensis* individuals (in Tanzania) varied from 0.8 to 4.9 mm.individual^{-1}.year^{-1}. Miller (1952) states that his data suggest a radial increment of

2.0 mm.individual^{-1}.year^{-1} for individuals of *Baikiaea plurijuga* from 8 to 36 cm in diameter in Zambia. In an apparent refinement, Osmaston's (1956) adjusted radial increment rate for *Baikiaea plurijuga* is 2.2 mm.individual^{-1}.year^{-1}. In an area south of Messina in the northern Transvaal, Guy (1970) found that over a 36 year period *Adansonia digitata* trees with a circumference of more than 5 m decreased radially by 0.3 mm.year^{-1}; those between 2.5 and 5 m circumference increased radially 2.1 mm.year^{-1}, while those below 2.5 m circumference had a radial increase of about 1.0 mm.year^{-1}. However, for one tree of the same species with circumference of 14.4 m (diameter 4.6 m) and an estimated age of 1000 years in the Kariba valley in Rhodesia, Swart (1963) reports a positive rate of radial growth of 1.5 mm.year^{-1} (based on radio-carbon measurement) or 1.1 mm.year^{-1} (based on ring measurement). At Wankie township in Rhodesia, Guy (1970) reports for a period of 30 years an annual radial rate of growth of 7.6 mm.year^{-1} for 3 individuals with circumference below 2.5 m.

Increment in basal area per individual tree accounts for differences in tree size. In *Burkea africana* woodland in the northern Transvaal most species' basal area increment varied from about 10 to 35 cm^2.individual^{-1}.year^{-1} (Rutherford & Kelly 1977). For *Adansonia digitata* near Messina, trees with circumference of greater than 5 m decreased by 20.7 cm^2.individual^{-1}.year^{-1}, those with circumference 2.5 to 5 m increased by 64.2 cm^2.individual^{-1}.year^{-1} and those less than 2.5 m circumference by 10.3 cm^2.individual^{-1}.year^{-1} (Guy 1970). Swart's (1963) 14.4 m circumference *Adansonia* at Kariba, however, increased in the order of 200 cm^2.year^{-1}, while the 3 trees at Wankie with circumference less than 2.5 m increased between 21 and 301 cm^2.individual^{-1}.year^{-1}.

In *Burkea africana* woodland in the northern Transvaal, for all trees and shrubs (omitting *Grewia flavescens* individuals) with diameters greater than 0.5 cm, basal area of the main body of woody species increased by approximately 0.4 m^2.ha^{-1}.year^{-1} or 4.7 per cent of the total basal area (Rutherford & Kelly 1977) whereas in undisturbed woodland in northern Zambia Endean (1967) taking all woody species greater than 5 cm diameter at breast height found an increase of about 0.3 m^2.ha^{-1}.year^{-1} or roughly 2.2 per cent of the total basal area over a 14 year period. In southern Zaïre, basal area of miombo woodland with *Brachystegia* spp. and *Julbernardia paniculata* increased by about 0.4 m^2.ha^{-1}.year^{-1} or 2.9 per cent (Malaisse et al. 1975). These percentage annual increases in basal area correspond closely to the 3 per cent annual increase obtained over four years in *Isoberlinia–Monotes* unfelled woodland in Nigeria (Kemp 1963).

For timber volume increments per unit ground area, Endean (1967) reports for northern Zambian woodland an increment of about 0.5 m^3.ha^{-1}.year^{-1}, Guerreiro (1966) also found 0.5 m^3.ha^{-1}.year^{-1} for Moçambique woodland and Monteiro (1970) gives 0.6 m^3.ha^{-1}.year^{-1} for savanna in Bié, Angola. For presumably South African savanna vegetation, King (1951) estimated an increment of 'probably less than' 0.7 m^3.ha^{-1}.year^{-1}.

Data on dry mass production per unit area appears to be linked only to parts of the woody plants in savanna vegetation. For nine *Colophospermum mopane* sites in the southeastern Rhodesian lowveld, Kelly (1973) found that annual production of twigs and leaves for one season varied from 590 to 2120 kg.ha^{-1} averaging 1510 kg.ha^{-1}. In Lutope riverine vegetation in northwestern Rhodesia, Goodman (1975) estimated in January current twig and leaf mass at 2240 kg.ha^{-1}, 50 per cent of which was above 5 m height above ground level and only 33 per cent below 2.5 m height or within reach of many browsing animals. In *Burkea africana* woodland savanna in South West Africa annual mass of leaves was found to be in the order of 1000 kg.ha^{-1} or about 5 per cent of total mass (Rutherford 1975). In a 0.2 ha sample of relatively dense shrub *Colophospermum mopane* and *Grewia* vegetation in southwestern Rhodesia, Kennan (1969) reports production of 'green leaf' to be 1490 kg.ha^{-1}. In the high rainfall, mostly deciduous miombo vegetation of southern Zaïre, annual tree and shrub leaf fall averaged 2900 kg.ha^{-1}.year^{-1} over a period of 5 years (Malaisse et al. 1975).

In Rhodesia, West (1950), assuming theoretical but not unnaturally encountered densities of large trees such as *Acacia albida* and *Acacia erioloba* at 10 to 20 individuals per hectare, calculated that fruit production would lie between 1100 and 2200 kg.ha^{-1}.year^{-1} while with spacing of 25 to 50 individuals per hectare of smaller trees such as *Acacia tortilis* subsp. *heterancantha* and *Piliostigma thonningii* would lie between 280 and 560 kg.ha^{-1}.year^{-1}. Malaisse et al. (1975) found that in the southern Zaïre miombo vegetation fruit production (measured as fruit fall) varied greatly with an average of 160 kg.ha^{-1}.year^{-1} over a 4 year period, but about 2000 kg.ha^{-1} in a subsequent year.

Owing to the apparent lack of any direct means of comparing annual production rates between total above ground savanna or woodland vegetation on the one hand and, for example, grassland on the other (see Table 1), it appears justified to attempt a rough indication of the order of total savanna dry mass production, based on presently available data. The basic unknown is what annual biomass increment is found in the form of radial growth of trunk and branches.

In savanna types such as that studied by Rutherford (1975) and some of those by Kelly (1973) (both in roughly 500 mm mean annual rainfall areas), it appears that herbaceous layer growth is roughly 1000 kg.ha^{-1}.year^{-1} or more, whereas the woody species leaf and current twig component has a production in the order of 1500 kg.ha^{-1}.year^{-1} (Kelly 1973). Aportioning percentage basal area annual increment per hectare of 3 per cent (averaged from available savanna data above) to total biomass (without terminal growth) of the roughly 20,000 kg.ha^{-1} for these savanna types, radial production of trunk and branches appear to be in the order of 600 kg.ha^{-1}.year^{-1}. It may thus be concluded that in savannas such as these, average total above ground production will seldom be less than 3000 kg.ha^{-1}.year^{-1}. Since most savanna areas occur where mean annual rainfall is greater than 500 mm, one may expect total above ground production for savanna areas in southern Africa in general to be distinctly greater than 3000 kg.ha^{-1}.year^{-1} and one may conclude that savannas are probably more productive than vegetation in equivalent rainfall zones in major grassland areas such as that of the South African Highveld.

2.5 *Forest*

Basal area of all trees, including *Olea capensis*, *Podocarpus latifolius* and *P. falcatus*, with diameter at breast height greater than 10 cm at several sites in the southern Cape between Bloukrans and Grootrivier passes have been found to vary from 39.62 to 49.67 m^2.ha^{-1}, averaging 45.57 m^2.ha^{-1} (Geldenhuys pers. comm.). Using the same criterion in the Groenkop State forest near George, Van Laar & Lewark (1973) state basal area to be 39.63 m^2.ha^{-1}. Using data derived from a 150 ha sample of Geldenhuys (pers. comm.) for Deepwalls forest with similar species composition near Knysna for trees of similar dimensions, a basal area of about 41 m^2.ha^{-1} was obtained. In southern Natal in the Xumeni forest with *Podocarpus henkelii*, data derived from Moll & Woods (1971), suggest a basal area of about 30 m^2.ha^{-1} for all canopy trees larger than 16 cm diameter at breast height. In the immediate vicinity of Richardsbay on the north coast of Natal, Venter (1972) found for all trees larger than 10 cm diameter at breast height, coastal dune forests varied in basal area from 26.07 m^2.ha^{-1} for the *Celtis africana* community, through 29.54 m^2.ha^{-1} for the *Acacia karroo* community, 39.98 m^2.ha^{-1} for the *Mimusops caffra* community to 42.25 m^2.ha^{-1} for the *Strelitzia nicolai* community. On the surrounding flats at Richardsbay the *Strelitzia–Acacia* community had a basal area of only 7.38 m^2.ha^{-1} while the *Manilkara discolor* community had a basal area of 29.76 m^2.ha^{-1}. Venter (1972) reports not dissimilar basal areas for dune forest communities further north at Mapelana, north of Cape St. Lucia where the *Diospyros natalensis* community had a basal area of 39.46 m^2.ha^{-1}, while at Sibayi in northeastern Zululand, the *Mimusops–Apodytes* community was 13.15 m^2.ha^{-1}, the *Ziziphus mucronata* community 26.58 m^2.ha^{-1} and the *Ptaeroxylon obliquum* community with 30.95 m^2.ha^{-1}. For 'marsh' forests at Richardsbay Venter (1972) found a far greater variation of basal areas ranging from 20.5 m^2.ha^{-1} for the *Myrica serrata* community in dune vleis, through the *Avicennia marina* community (20.7 m^2.ha^{-1}); the *Bruguiera gymnorrhiza* community (50.7 m^2.ha^{-1}); the *Cassipourea gummiflua* community (65.0 m^2.ha^{-1}) to the *Barringtonia racemosa* community with 131.2 m^2.ha^{-1}. The highest average basal area found was for forests of the watercourses at Richardsbay with 124.4 m^2.ha^{-1} for the *Ficus hippopotami* community and 144.4 m^2.ha^{-1} for the *Barringtonia–Ficus sycomorus* community. Data derived from Chapman & White (1970) for submontane forests in Malawi near the Zambian border, more specifically in the *Entandrophragma* forest, Kasyaula, Nyika Plateau give, for all trees included, a basal area of about 50 m^2.ha^{-1}. An anomalously low basal area of 5.1 m^2.ha^{-1} has been given by Lawton (1967) for trees larger than 10 cm in diameter in *Syzygium–Uapaca* riparian forest in northeastern Zambia.

Timber volume estimations, an important parameter estimated in the past in several natural forests, have, however often made use of less precise relations for calculation than those derived more recently by Van Laar & Geldenhuys (1975). For all trees more than 15 cm in diameter at breast height in a 4 per cent sample of the whole Knysna-Humansdorp natural forest areas of the southern Cape, Laughton (1937) gives 87m^3.ha^{-1} of merchantable timber for the then 686 ha of virgin unworked forest but quotes natural but overutilized forests as containing only 35 m^3.ha^{-1}. Volume of merchantable and unmerchantable timber was 74 m^3.ha^{-1} over the whole sample area, but attained 210 m^3.ha^{-1} in the 'best virgin forest'. King (1941) states that the Tonti-forest in 1924 contained 180 m^3.ha^{-1} of merchantable timber (mainly *Podocarpus* species) whereas the Maxego forest, also in the Transkei, in 1936 contained a very high 550 m^3.ha^{-1} of which 470 m^3.ha^{-1} was made up of *Podocarpus* species. The same writer estimates that the Mariepskop forest on the eastern Transvaal escarpment contained 140 m^3.ha^{-1}. In Malawi, Chapman (1961) indicates the volume of mainly mature to over-mature *Widdringtonia whytei* forest in 1955 as in the order of 110 m^3.ha^{-1}.

No total forest mass figures per unit ground area appear to be available for southern Africa.

Rates of growth in forest trees have usually been expressed in some form of radial increment per tree individual. Stapleton (1955) states that under natural conditions at Knysna it takes between 160 and 230 years for trees such as *Podocarpus latifolius*, *Olea capensis* and *Apodytes dimidiata* to attain 'mature' diameters of 35 to 50 cm. Phillips (1931) gives mean radial annual increment for several species in the southern Cape forests, for example, *Gonioma kamassi* 0.23 mm, *Apodytes dimidiata* 0.53 mm, *Olea capensis* 0.76 mm and *Podocarpus latifolius* 0.56 mm. Average increment rates for the last 4 species together in exploited forests was 42 per cent greater than in natural forests. In exploited situations, Laughton (1937) reports annual rates of 0.92 mm for *Podocarpus latifolius*, 1.05 mm for *Olea capensis* and 2.05 mm for *Ocotea bullata* and in some indigenous trees planted in natural sites 25 years old *Cunonia capensis*, for example, averaged 1.3 mm annual radial increment over a subsequent 20 years (Geldenhuys 1975). The particularly slow rate of growth of some southern Cape forest species such as that of *Buxus macowanii* which averages only 0.4 mm per year over 30 years is in strong contrast with the corresponding rate of 7.2 mm per year for the exotic species *Pinus radiata* for the same period (King 1951). In the Xumeni forest species in Natal, Moll & Woods (1971) found similar rates to those in the Knysna area averaging 0.8 mm per year with, for example, *Podocarpus henkelii* 0.5 mm and *Kiggelaria africana* 1.1 mm. However, in the vicinity of Westfalia estate in the northeastern Transvaal escarpment, Keet (1962) reports Merensky's data on increments in 'representative natural trees' of the pioneer species *Trema orientalis* which in the period January to March alone increased by a mean of 4.4 mm and in the later period of April to September further increased by 4.1 mm so that taking into account the October to December period for which data is not given, the annual radial increment of natural *Trema* is probably equal to or greater than that of 8.4 mm reported in *Trema* plantations (Keet 1962) and values for several exotic timber species plantations.

In terms of average basal area increment per individual per year Phillips' (1931) data give *Gonioma kamassi* as 0.77 cm^2, *Apodytes dimidiata* as 3.45 cm^2, *Olea capensis* as 6.90 cm^2 and *Podocarpus latifolius* as 6.83 cm^2. For the Xumeni forest in Natal, Moll & Wood's (1971) data indicate a basal area increment per unit ground area of 0.12 m^2.ha^{-1}.year^{-1} that is about 0.4 per cent of basal area present. For annual volume increments per unit ground area Laughton (1937) quotes a 'Departmental' figure of 0.52 m^3.ha^{-1}.year^{-1} for the Knysna forest. However, mean annual timber volume increment, measured in unthinned plots of unstated size, is given as 1.20 per cent over 5 years in Deepwalls forest (Knysna) and 1.44 per cent over 7 years in Woodville forest (George) (Laughton 1937). Applying these increments to the 87 m^3.ha^{-1} of virgin unworked forest stated before, the annual volume increment per ground area of this undisturbed form of forest should lie between 1.0 and 1.3 m^3.ha^{-1}.year^{-1}.

2.6 *Below ground production*

Apart from sporadic work on root morphology and distribution in relation to ecological conditions (for example, Scott & Van Breda 1937, Savory 1963 and Kerfoot 1963 for species indigenous to southern Africa, Nänni 1960 and Haigh

1966 for introduced species) very little work has been done on production of roots in situ.

In *Trachypogon–Tristachya* grassland near Johannesburg Weinmann (1943) determined root mass to 15 cm depth and obtained 2915 kg.ha^{-1} but indicated that to a depth of 30 cm a mass of 3210 kg.ha^{-1} should be obtained which gives a root/shoot ratio of 1.41:1. In a later determination Weinmann (1948a) found a root mass of 3015 kg.ha^{-1} to 23 cm depth which gives a root/shoot ratio of 1.39:1. Since in both these ratios shoot mass is only that produced over 1 year the ratios may be regarded as maximum ratios. In the same area Coetzee et al. (1946) report root mass to a depth of 61 cm to be between 2400 and 4500 kg.ha^{-1} but this may possibly include fertilized grassland.

In a sandy *Burkea africana* savanna in the northern Transvaal, Van Wyk (Huntley 1977) obtained a mass of 2260 kg.ha^{-1} to 1 m depth for herbaceous roots in openings between trees. This together with the average total above ground herbaceous mass of 1710 kg.ha^{-1} for these areas (Grunow 1975/76) gives a root/shoot ratio of 1.32:1. For woody species, data collected (in a 50 m^3 sample) by Rutherford & Kelly (unpubl.) for 1 to 4 m high *Ochna pulchra* individuals in deep sand in *Burkea africana* woodland that had not been burned for about 50 years, indicated a root/shoot ratio for a depth to 1 m as somewhat greater than 1:1 and in sandy woodland that is subject to fire this ratio may sometimes be expected to be higher, possibly greater than 2:1.

Whittaker (1970) has noted that root/shoot ratios increase in fire adapted shrubland and forest and towards drier environments. This together with the meagre available data would indicate that in several parts of southern Africa a considerable proportion of total annual primary production accumulates below ground.

3. Some ecological production controls

Primary production of a plant community is usually subject to change, yet there may also be consistent differences between communities. The number of factors that control production are great but for regional production comparisons the number of controls must often be restricted to major factors to make extensive evaluation of the effects possible. In southern Africa many important environmental production relations have not been adequately investigated.

Some production controls have been incorporated into production models and been extensively applied. Several models that predict production from various environmental variables have been widely publicized in recent years. These include the following: The Miami Model where the lower of two production values predicted from two relations is accepted. The formulae are:

$$P = 3000 (1 - e^{-0.000664 x})$$

and

$$P = \frac{3000}{1 + e^{1.315 - 0.119 z}}$$

where P is the production in g.m^{-2}.year^{-1}, x is the mean annual precipitation in mm and Z is the mean annual temperature in °C.

The C. W. Thornthwaite Memorial (Montreal) Model where

$$P = 3000 \, [1 - e^{-0.0009695 \, (E-20)}]$$

with P as above and E the annual actual evapotranspiration (mm).
The Hague model where

$$P = -157 + 5.17s$$

with P as above and S the length of the photosynthetic season in days.

These models have been applied on a world scale and mapped by computer application (Lieth 1975a) thus providing a small scale coverage of the variation of primary production over southern Africa. Barreto & Soares in 1972 (quoted by Lieth 1975b) have gone farther and applied the Miami Model on a regional basis to Moçambique and mapped the variation of primary production at a scale of about 1:10 million (reproduced in Lieth 1975b).

Other informal 'models' in the form of handdrawn world production maps also provide a small scale production map for southern Africa (namely, the Innsbruck productivity map which with inclusion of marine production became the Seattle productivity map (Lieth 1975a), the map of Bazilevich et al. (1971) and Rodin et al. (1975)).

Other production models take additional production controlling variables into account. Ryahchikov's model (Rodin et al. 1975) requires annual effective precipitation and Czarnowski's model (Lieth 1975a) requires water vapour pressure for the growing season. Since such environmental parameters are not available in many regions, use of these models for regional comparative production purposes requires that some variables must be approximated, thus reducing the effects of improved prediction accuracy of the model.

Since small scale primary production maps have been produced which include southern Africa, their evaluation is important to this chapter. Box (1975) in his evaluation of global primary production models does not evaluate the actual basis of the models and is mainly concerned with the mapping procedure as it affects the total world production value.

To test model predicted production against measured production, a southern Transvaal grassland is selected where above ground production has been determined at about 2000 kg.ha^{-1}.year^{-1} with below ground production in the same order, average total production of the natural grassland is thus unlikely to exceed 5000 kg.ha^{-1}.year^{-1}. The Miami model and the Hague model applied to this region both predict total production of about 10,000 kg.ha^{-1}.year^{-1} or more and even applying Lieth's (1975c) restriction of actual evapotranspiration not being allowed to exceed 50 per cent of precipitation in the Thornthwaite Memorial model results in only slightly less than 10,000 kg.ha^{-1}.year^{-1}. The handdrawn production maps indicate production to lie between 10,000 and 15,000 kg.ha^{-1}.year^{-1} for the test area. Indications are that the tendency toward the consistent over-estimation of production also applies to other regions of southern Africa. For the United States of America, Sharp (1975) compares one of the above models (the Thornthwaite Memorial model) with a locally based model produced by Rosenzweig (1968) and notes consistently high values from the former model and conjectures that the model is based on data from 'well-stocked ... better sites of a region'. It is also possible that the models are based on data which includes fertilization and irrigation which Westlake (1963) calls normal agricultural practice and groups with natural conditions. Lieth (1975c) also points out that gaps in data exist between many individual regions so that further refinements of both maps of data and prediction model maps are required before a satisfactory situation is arrived at.

Having noted some shortcomings of the inclusion of some production controlling factors in several existing models, some individual production controlling factors as they occur in southern Africa are briefly discussed. Some controls which have received attention on a natural ecological basis are rainfall, fire, defoliation, soil and woody-herbaceous relations.

3.1 *Rainfall*

Rainfall often forms a central part of many primary production prediction models. That there is often a relationship between rainfall and production has long been observed although it is especially in relatively arid regions that a linear relationship between annual production and annual rainfall may be expected (Newbould 1963, Whittaker 1970). In high rainfall areas, other factors may become more important in limiting production. Since most of southern Africa classifies in this context as an arid zone an attempt has been made to identify linear relations between current season's rainfall and corresponding herbaceous production values for several grassland and savanna areas in southern Africa. The results are given in Fig. 2. It is clear that linear relationships exist but that differences in relationship depend strongly upon type of vegetation. The much quoted relationship between mean annual rainfall and primary production in South West Africa is included in Fig. 2 to illustrate its particularly steep gradient and confirm earlier views on the interpretation of the data upon which the relation is based.

Rainfall effects are often complex. For example, in a grassland area in the Orange Free State herbaceous production related significantly to the infiltration capacity of the soil (Van den Berg et al. 1976). In northern Cape thorn savanna, the effect of low rainfall reducing herbaceous production was much pronounced in areas with greater levels of bush infestation (Donaldson & Kelk 1970), an effect also found elsewhere in the world (Cable 1975).

In parts of southern Africa, rainfall is sometimes associated with the extremes of flood and drought. A long term reduction in production may result from soil erosion after floods, whereas extreme drought may reduce 'normal' herbaceous production (through increased mortality) by as much as 93 per cent (Donaldson 1967). Savanna trees are possibly more resistant to severe drought. Utilization of drought resistant shrubs as browse has sometimes been advocated and relatively low levels of tree mortality have been reported after even a major drought in the northwestern Transvaal (Van Wyk et al. 1969).

'Dendroclimatological' work is currently being conducted in South Africa to find indigenous trees whose annual rings correlate well with rainfall patterns (Gillooly 1976, and Chapter 3). In a Tanzanian study of *Pterocarpus angolensis*, a tree species indigenous in southern Africa, Bryant & Procter (1970) related radial increment in the form of ring width to seasonal rainfall. No correlation with the entire season's rainfall appeared apparent but the most significant positive correlation found was with the number of months of the season to receive between 30

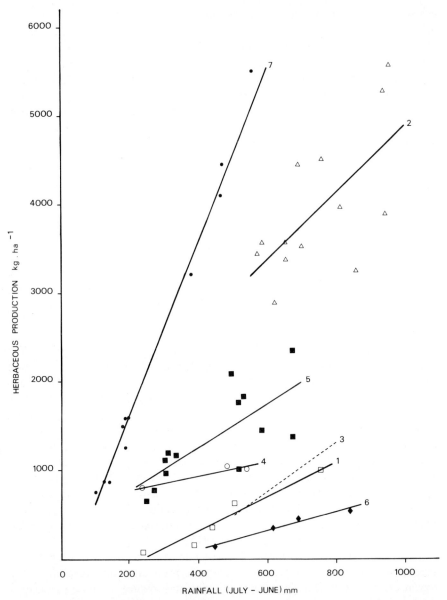

Fig. 2. Relation between specific season's (July–June) rainfall and herbaceous dry matter production for certain areas of various grassland and savanna vegetation types in southern Africa. 1. Eastern Cape *Acacia karroo* savanna (r = 0.97); 2. Natal tall grassland (r = 0.66); 3. South African western Highveld grassland; 4. Northern Transvaal savanna (r = 0.96); 5. Rhodesian *Colophospermum mopane* savanna (r = 0.73); 6. Rhodesian *Brachystegia* woodland (r = 0.98); 7. South West African arid grassland/savanna. No. 7 is for mean annual rainfall and r is the correlation coefficient.

and 100 mm rainfall. The writers also quote Boaler (1966) who found that radial increments for the same species decrease with earlier time of leaf fall.

3.2 *Fire*

Fire is a factor which is encountered in many natural production areas and is especially common in fynbos, grassland and savanna. Occurrence of fire in forest is rare and usually occurs only after prolonged drought and in Karoo and desert and even in, for example, dry *Colophospermum–Kirkia–Commiphora* savanna in southwestern Rhodesia (West 1965) fire may occur only after exceptionally high rainfall seasons. The effect of fire on biomass and production varies according to vegetation type. The immediate effect of fire on biomass may vary but after fire there is usually a successive biomass accumulation until steady state is attained (Major 1974b). Rate of accumulation of combustible material also determines potential periodicity of fire. The effects on production are often complex depending on many other factors.

In fynbos, where fire is used to maintain plant species diversity for nature conservation purposes and to improve water discharge into storage reservoirs (Wicht 1971), above ground biomass is often reduced to virtually nothing at time of the fire. Data of Kruger (1977) suggest that annual rates of production in the first few seasons after fire may not be very different to production of more mature fynbos communities.

In grassland vegetation, where burning is used as a grazing management tool for removing dead, self-smothering and unpalatable grass material and for stimulating earlier growth, the effects on biomass and production vary. Biomass is often reduced to almost nothing by burning in the dry season but by burning in other seasons some may remain in the form of stalks and smaller islands (see Van Wyk 1972). Daubenmire (1968) states that in Africa, regular burning increases production in relatively moist regions, but is generally detrimental in more arid regions. Areas cited by West (1965) in accordance with this, indicate moist regions to be areas with mean annual rainfall above 850 mm and arid areas with mean annual rainfall less than 650 mm. In medium rainfall areas no effect of fire on production may be expected and this has been indicated by, for example, work of Le Roux & Morris (1977). Yet, even in such areas regular fire may reduce production in dry years (West 1965). Total long term protection of grassland, not only from fire but from other means of grass removal results in eventual reduction of production in most areas (West 1965). Although fire most often results in higher production relative to total protection, in moister areas production after material is removed by fire has been found to be lower than production after removal of material through mowing (Edwards 1968, Scott 1972).

In savanna vegetation, burning is employed for the same purpose as in grassland, but in addition is used for decreasing presence of undesirable plant species or at least preventing an increase. That fire often decreases the amount of, especially young, woody plant individuals has often been acknowledged (West 1965), the degree of control depending upon the time of year that is burnt (Trapnell 1959, and Chapter 10). With reduction of woody plant cover through fire there is often an expected increase in herbaceous production (for example, Trollope 1972) but sometimes little change occurs (Kennan 1972). Although regular burning in savanna may markedly reduce total biomass and change plant species composition and community structure, it appears uncertain whether there is any great change in total above ground production.

3.3 *Defoliation*

A most common primary production governing factor in many 'natural' production areas of southern Africa, is defoliation through grazing. Previous grazing history and current grazing can markedly affect production. Plant reaction to defoliation is often highly complex. Measurement of plant production under different animal stocking rates has sometimes produced anomalous or unexpected results (for example, Edwards & Nel 1973). Other approaches have attempted to isolate the defoliation factor from many other grazing factors by direct clipping. Even here results are often contradictory. However, some possible trends in grassland are that seasonal production declines with increasing frequency of cutting, this being particularly marked at lower cutting heights (for example, Weinmann 1943, Tainton et al. 1970). (Clipping at different heights of between 4 and 10 cm above ground over a series of years is reported not to affect subsequent production (Tainton et al. 1970).)

Time of defoliation is important with defoliation in mid or late summer reducing production relative to that when defoliation takes place in winter or early spring. Barnes (1972) has summarized much of the conflicting data but states that there may be 'factors relating to defoliation effects of which, as yet, we have no conception'.

Defoliation effects in woody species are even less clear, particularly with regard to browse mammals. Recently, increased attention is being given to testing techniques for determining amounts of browse available and removed by animals (Barnes 1976, Barnes et al. 1976, Aucamp et al. 1973, Aucamp 1976). In some deciduous savanna tree species, for example *Burkea africana*, that are sometimes subjected to rapid and complete defoliation by insects (Van den Berg 1974), an entirely new crop of leaves is produced in the same season. In such situations, particularly where the species concerned is frequently a virtual exclusive dominant, for example, *Colophospermum mopane*, woody species seasonal leaf production may almost double through defoliation.

3.4 *Woody-herbaceous production relations*

The inverse relation between herbaceous production and amount of woody species material present has been referred to. An example of a relation (from data of Donaldson & Kelk 1970) between herbaceous production and naturally occurring tree density is given in Fig. 3. Such inverse relationships have been long identified as a key problem in effective use of savanna vegetation (Walker 1971). Considerable work has been undertaken on methods to kill trees (for example, Strang 1960, Rees 1974) which, once effected, has invariably resulted in increased herbaceous production (Rowland 1974). Artificial removal of woody plants has been earlier identified with radical forms of change, but is applied increasingly in the savanna areas of southern Africa. It is thus of interest to note in what direction this widespread plant removal is likely to affect production estimates of herbaceous and woody savanna vegetation. Percentage increases in herbaceous production after tree clearing, for example, stated increases in excess of 1000 per cent (Cleghorn 1969), are often misleading where original herbaceous production was low. For 15 savanna sites with mean annual rainfall larger than 500 or 600 mm in Rhodesia, West's (1969) data shows increases of between 200 and 1200 per cent, but actual increases were between 200 and 1500 kg.ha^{-1} averaging about 800 kg.ha^{-1}. Since woody plant production in such areas has been earlier estimated at not less than 2000 kg.ha^{-1}.year^{-1}, clearing of woody

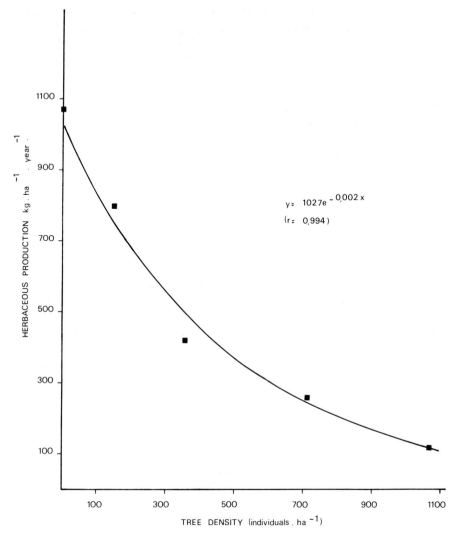

Fig. 3. Relation between density of mature *Acacia mellifera* spp. *detinens* individuals and annual herbaceous dry matter production averaged over three years for *Acacia mellifera* dominated areas of the Molopo zone in the northern Cape. Based on data from Donaldson & Kelk (1970).

plants in such savanna areas probably does not increase total above ground production, a decrease often being more likely. Work on a *Panicum maximum* – tree association has demonstrated a tendency for higher herbaceous biomass on sites partially covered with trees (opposed to totally covered or totally open sites) (Kennard & Walker 1973). Production of more palatable grass species may decrease once the trees have been removed in savanna areas, where occurrence of species such as *Panicum maximum* are strongly correlated with tree canopy cover (Bosch & Van Wyk 1970).

3.5 *Soils*

Whittaker (1970) stated that effects of soil fertility on natural vegetation had as yet been little studied. Vegetation on different soils in close proximity are suitable for determination of the effect of soil on production. In savanna at Matopos, Rhodesia, herbaceous layer production is higher on clay soils than on sandy soil (Mills 1964, Mills 1968). Herbst & Goosen (1973/74 and 1974/75) have made a study specifically directed at the effect of soil type on production in southcentral and southeastern Orange Free State grassland and report highest mass values on soils with high clay content, with decreasing mass values through medium clay content soils, sandy soils to lowest values for grassland on litholitic complexes.

4. Production relative to method

There are many physiological production measuring techniques (Šesták et al. 1971), indirect techniques (for example, Symons & Jones 1971) and 'biological' methods (for example, Von Berg 1970) in many of which difficulties in extrapolation of primary measurement, in callibration or in expressing error, occur. In southern Africa, however, primary production measurement has usually been of the difference type, for example, where difference in biomass over a certain interval is used as an estimate of production. The inadequacy of such difference methods has been referred to earlier and for further comments it is referred to, for example, Singh et al. (1975) and Whittaker & Marks (1975). It is clear that production obtained only through difference methods does not provide net primary production (NPP) but a biomass accumulation rate or allied concept. However, even accepting results from difference methods owing to their wide use in southern Africa, there is need for further consideration of potential error. In several published works assumptions of details of procedure must often be made but should these assumptions be incorrect, the stated values may contain significant error.

Some of the following considerations are important. In clipped herbaceous samples, the clipping height above ground is critical to the mass measured. Tainton et al. (1970) found that under field conditions, there was significantly lower mass when cut at 6.3 cm than that when cut at 3.8 cm above ground. Under more controlled conditions, Opperman (1973) (quoted in Roberts & Opperman 1974) found that in species such as *Themeda triandra* less than 50 per cent of the total above ground mass is sampled by clipping at 5 cm or more above ground. Clipping at about 2 cm and at 5 cm resulted in a mass difference of about 25 per cent of total above ground production in several grass species. In taller grasses, clipping height becomes less critical.

The timing of the final sample relative to the peak of seasonal growth is important. Where final herbaceous samples have been taken two months before the growth peak, various data (for example, Rethman & Booysen 1969, Grunow 1975/76) indicate that an underestimate in the order of 20 per cent is not unexpected. Since the point of maximum mass is not stable under field conditions (Rutherford 1976), sampling after the growth peak may also result in a significant underestimate.

To interpret a particular season's production as reflecting the mean annual production may incur large error. In the herbaceous layer production of some more arid savanna areas, for example, maximum variation in production between seasons may be in the order of 600 per cent (Kennan & Barnes 1973).

Error estimation is often particularly difficult where a part is apparently identified with the whole above ground production. Examples are where some rarer plant species, smallest tree size categories, and unmerchantable timber or non-grazeable material are omitted. In determinations for calculation of animal stocking capacity, recommendations that undesirable bushes or grass should be discarded before weighing (Edwards & Coetsee 1971) have sometimes been carried further with the result that, for example, 'veld production' sometimes excludes forbs. In the herbaceous layer of savanna, data of Goodman (1975) and Grunow (1975/76) show that omission of forbs alone may result in an underestimate of more than 10 per cent.

Sometimes new production is not distinguished from total mass present. If, in grassland, material formed in previous seasons is not separated from current season's growth, measurement of difference in total above ground mass is likely to underestimate production (Rutherford 1976) although the magnitude of the error depends on several factors such as plant age and community. If the only sample taken is around peak of growth, this may result in an overestimate of more than 100 per cent (Grunow 1975/76) when new material is not separated.

Taking oven dry material as standard, air dry herbaceous material, with between 10 and 15 per cent water content (Weinmann 1955), may overestimate oven dry mass by about 10 per cent depending on air humidity. In succulent vegetation, biomass expressed as dry mass is essential for comparison with biomass of other non-succulent vegetation types.

Dry mass units, although widely accepted for production are not as directly comparable as energy values (heats of combustion). Production of woody communities may be underestimated by more than 10 per cent relative to that of an herbaceous community when not based on energy units (Lieth 1968).

Natural heterogeneity occurs to some degree in all plant communities. Biomass sampling usually can only be done in part of a community with the result that the measured value may not fully represent that of the community. It is especially where there is interest in total above ground biomass of a larger demarcated region that considerable error may arise. An overestimate of biomass may arise, for example, where rock outcrops or other growth limiting surfaces are common and underestimation may result, for example, in grassland intersected by river gallery forest or other 'weighted' patches. Even in a very uniform region underestimation is possible where no account is taken of average slope. Steep slopes also tend to be avoided as sample sites since among other reasons, specially modified equipment is often needed (see, for example, Müller 1963).

Difference methods in non-herbaceous plants often require relationships to, for example, convert selected dimension measurements of the plant to dry mass of the plant or part of the plant. In southern Africa, various such relations have been established, for example, for *Serruria aemula, Elytropappus glandulosus* (Van der Pas 1972) and *Protea laurifolia* (Rutherford 1978a) in fynbos; for *Cotyledon paniculata* (Rutherford 1978b) in karoo and for *Portulacaria afra* (Aucamp et al. 1973), *Colophospermum mopane* (Kelly 1973, Barnes et al. 1976) and *Brachystegia boehmii* (Barnes et al. 1976) in 'savanna'. The confidence interval of the regression of these relations is often such that possible compounded error for total mass per unit ground area may exceed 10 per cent.

5. Conclusions

Some generalizations of the above data are possible. The mass values for karoo shrub vegetation (3700 to 7500 kg.ha^{-1}) are similar to those for very young or particularly arid fynbos communities, and similar to mass accumulated over several seasons in moister grassland areas or grassland areas with particularly fertile soil. Mass values for several older fynbos communities (up to 15,000 or 20,000 kg.ha^{-1}) correspond to the lower mass range for savanna vegetation. Greatest fynbos mass may be expected to fall within the typical range of mass for woodland. Mass of northern miombo woodland and possibly that of dense valley bushveld such as that in parts of the eastern Cape, may be exceeded only by mass of forest communities.

Production in grassland vegetation of South Africa increases from below 1000 kg.ha^{-1}.year^{-1} in the western parts, through about 1000 to 2000 kg.ha^{-1}.year^{-1} for higher rainfall but marginal grassland areas and about 2000 kg.ha^{-1}.year^{-1} for more typical Highveld vegetation, to often considerably greater than 2000 kg.ha^{-1}.year^{-1} in the often relatively mesic and eastern grasslands of Natal and extreme eastern Orange Free State. Herbaceous layer production in savanna or woodland is more variable but there appears to be a trend with lowest production (less than 1000 kg.ha^{-1}.year^{-1}) in southern lower rainfall areas with relatively high density of woody plants and highest production (more than 2000 kg.ha^{-1}.year^{-1}) in more open parts of the northern savannas. Production of fynbos vegetation probably corresponds to that of high producing grassland vegetation while production of herbaceous and woody savanna appears to be greater than that of grassland with similar rainfall. That savanna production exceeds that of 'equivalent' grassland areas has been found elsewhere by, for example, Ovington et al. (1963), and Rodin & Basilevič (1968) indicate that grassland production is not higher than dry savanna. However, Golley (1975), in a review of production of tropical vegetation that is assumed to be at or near steady state conditions, indicates that mean savanna production is less than the mean net production of grassland. Murphy (1975) also indicates savanna production to be less than grassland production but this does appear not to account for the production of the savanna woody component. There seems to be little basis for Bazilevich et al. (1971) to indicate production of the southern part of southern Africa's savanna zone as greater than that of the northern part.

The relation of biomass to production may be conveniently expressed as the biomass accumulation ratio — the ratio of dry mass to annual net primary production (Whittaker 1970) which may be viewed as the reciprocal of the relative growth rate (RGR) from classical growth analysis. Such ratios for fynbos probably fall within Whittaker's (1970) range of 3 to 12 for shrubland; near the top of his 1.5 to 3 range for undisturbed grasslands, and that for savannas at the lower part of the 10 to 30 range given for woodlands.

Production, as given for southern Africa by certain world production models, has already been indicated as a probable overestimation of production (allowing for root production). In 1940, Clarke subdivided South Africa in 17 regions and gave production of seven of these based on 'productivity of efficiently managed veld'. Although relative values are similar to more recent data, his values, if interpreted as annual production values of natural production areas as defined earlier, appear generally greater than average production.

Basal area of woody vegetation varies from less than 5 m².ha⁻¹ for open shrubby savanna of more arid areas with mean annual rainfall usually less than 500 mm to between 5 and 10 m².ha⁻¹ for several southern savanna areas with mean annual rainfall usually more than 500 mm to between 10 and 16 m².ha⁻¹ for several northern savanna woodlands. Forest communities with basal area up to about 40 m².ha⁻¹ include some of the north coast of Natal while basal area of between 40 and 50 m².ha⁻¹ include the temperate forests of the southern Cape. Communities with basal area between 50 and 130 m².ha⁻¹ include three marsh forests with highest basal area value (maximum of 144 m².ha⁻¹) for two water course forests, all on the north coast of Natal.

Annual radial increments of woody species of the southern Cape and Natal mist belt forests usually average between 0.5 and 1.0 mm.individual⁻¹.year⁻¹ whereas several savanna tree species appear to increase in the order of 2 or 3 mm.individual⁻¹.year⁻¹.

Eckardt (1975) has assessed some of the major achievements of the PT activities of the International Biological Programme. One of the achievements has been the acquisition of production parameters that provide net primary production per unit land area of the world's biomes. In this particular section of endeavour, southern Africa does not appear to have benefited fully from the programme (for example, International Biological Programme 1972). Although much new production data are being gathered in southern Africa at present, large gaps in knowledge of comparative biomass and production remain. The most important needs appear to be for measurement for biomass of forests and for dry mass production rates for both forests and whole savanna and woodland communities. The largest single gap in knowledge of primary production in southern Africa, however, remains that of below ground production. Other important unknowns include the effect of commonly reported shrub encroachment on total production and its interpretation in the light of successional changes (for example, Lieth 1974). Relationships that are sometimes assumed should be further tested under natural conditions in southern Africa. As illustration, Caldwell (1975) has tabulated for several vegetation types, dry matter production, the probable photosynthetic pathway of the dominant species for each community and the efficiency of solar radiation utilization. These data appeared not to support the sometimes accepted hypothesis that C_4-species in natural grazing ecosystems should be more productive. A further important consideration is the proportion of primary production which may be directly utilized. Thus in natural production areas of southern Africa, data of, for example, Goodman (1975), Grunow (1975/76) and Aucamp (1976), show that often only a relatively small proportion of total above ground production is available for grazing or browse.

In some past primary production work in southern Africa there appears to have been little realization that data might be required for later multi-disciplinary purposes, the form of the data presentation often precluding extraction of additional information. Some production work also appears uncritical when considered together with the many sources of potential error mentioned earlier, as well as other practical difficulties in obtaining valid production data (Olson 1975).

A final pertinent implication arising from southern African production data, is that if net primary production per year in a stable ecosystem ideally approaches zero (Wassink 1975), existing production data indicates that there are few, if any such stable ecosystems in southern Africa.

References

Acocks, J. P. H. 1975. Veld types of South Africa. Mem. Bot. Surv. S. Afr. 40:1–128. 2nd ed. Government Printer, Pretoria.

Aucamp, A. J. 1976. The role of the browser in the bushveld of the eastern Cape. Proc. Grassld. Soc. Sth. Afr. 11:135–138.

Aucamp, A. J., Smith, D. W. W. Q., Howe, L. G. & Viljoen, B. D. 1973. 'n Belowende tegniek vir die bepaling van struikbenutting. Proc. Grassld. Soc. Sth. Afr. 8:129–132.

Barnes, D. L. 1972. Defoliation effects on perennial grasses – continuing confusion. Proc. Grassld. Soc. Sth. Afr. 7:138–145.

Barnes, D. L. 1976. A review of plant-based methods of estimating food consumption, percentage utilisation, species preferences and feeding patterns of grazing and browsing animals. Proc. Grassld. Soc. Sth. Afr. 11:65–71.

Barnes, D. L., Lloyd, B. V. & McNeill, L. 1976. The use of shoot dimensions to estimate the leaf mass or leaf area of certain indigenous trees in Rhodesia. Proc. Grassld. Soc. Sth. Afr. 11:47–50.

Bawden, M. G. & Carroll, D. M. 1968. The land resources of Lesotho. Land Resource Study No. 3. Land Resources Division, Directorate of Overseas Surveys, Tolworth, Surrey, England.

Bazilevich, N. I., Drozdov, A. V. & Rodin, L E. 1971. World forest productivity, its basic regularities and relationship with climatic factors. In: Duvigneaud, P. (ed.), Productivity of forest ecosystems. pp. 345–353. UNESCO, Paris.

Birch, E. B. 1972. Why so little progress with improved pastures? Proc. Grassld. Soc. Sth. Afr. 7:56–60.

Blair Rains, A. & McKay, A. D. 1968. The Northern State Lands, Botswana. Land Resource Study No. 5. Land Resources Division. Directorate of Overseas Surveys, Tolworth, Surrey, England.

Blair Rains, A. & Yalala, A. M. 1972. The Central and Southern State Lands, Botswana. Land Resource Study No. 11. Land Resources Division, Tolworth Tower, Surbiton, Surrey, England.

Boaler, S. B. 1963. The annual cycle of stem girth increment in trees of Pterocarpus angolensis D.C., at Kabungu, Tanganyika. Commonw. For. Rev. 42:232–236.

Booysen, P. de V. 1967. Grazing and grazing management terminology in southern Africa. Proc. Grassld. Soc. Sth. Afr. 2:45–57.

Booysen, P. de V. 1972. Pastoral productivity and intensification. Proc. Grassld. Soc. Sth. Afr. 7:51–55.

Bosch, O. J. H. & Van Wyk, J. J. P. 1970. Die invloed van bosveldbome op die produktiwiteit van Panicum maximum: 'n Voorlopige verslag. Proc. Grassld. Soc. Sth. Afr. 5:69–74.

Bourlière, F. & Hadley, M. 1970. The ecology of tropical savannas. Ann. Rev. Ecol. Syst. 1:125–152.

Box, E. 1975. Quantitative evaluation of global primary productivity models generated by compunters. In: Lieth, H. & Whittaker, R. H. (eds.), Primary productivity of the biosphere. Ecological studies 14. pp. 265–283. Springer-Verlag, New York.

Bryant, C. L. & Procter, J. 1970. Rainfall distribution as a growth indicator for Pterocarpus angolensis D.C. in Tanzania. Commonw. For. Rev. 49:180—184.

Cable, D. R. 1975. Influence of precipitation on perennial grass production in the semidesert southwest. Ecology 56:981–986.

Caldwell, M. 1975. Primary production of grazing lands. In: Cooper, J. P. (ed.), Photosynthesis and productivity in different environments. pp. 41–73. Cambridge University Press, Cambridge.

Chapman, J. D. 1961. Some notes on the taxonomy, distribution, ecology and economic importance of Widdringtonia, with particular reference to W. whytei. Kirkia 1:138–154.

Chapman, J. D. & White, F. 1970. The evergreen forests of Malawi. Commonwealth Forestry Institute, Oxford.

Clarke, P. F. 1940. An evaluation of the productive capacity, potentialities and limitations of the South African veld. M.Sc. thesis, University of Pretoria, Pretoria.

Cleghorn, W. B. 1969. Some factors affecting vegetation. Rhod. Sci. News 3:327–329.

Coetsee, G. 1972. Control of Pachystigma pygmaeum with herbicide. Proc. Grassld. Soc. Sth. Afr. 7:28–31.

Coetzee, J. A., Page, M. I. & Meredith, D. 1946. Root studies in Highveld grassland communities. S. Afr. J. Sci. 42:105–118.

Cooper, J. P. (ed.) 1975. Photosynthesis and productivity in different environments. International Biological Programme 3. Cambridge University Press, Cambridge.

Daubenmire, R. 1968. Ecology of fire in grasslands. Adv. Ecol. Res. 5:209–266.

Department of Agricultural Technical Services. Karoo Region, Annual Report 1975/1976. A.T.S., Pretoria.
Donaldson, C. H. 1967. The immediate effects of the 1964/66 drought on the vegetation of specific study areas in the Vryburg district. Proc. Grassld. Soc. Sth. Afr. 2:137–141.
Donaldson, C. H. & Kelk, D. M. 1970. An investigation of the veld problems of the Molopo area 1. Early findings. Proc. Grassld. Soc. Sth. Afr. 5:50–57.
Du Plessis, G. J. & Van Wyk, J. J. P. 1969. 'n Vergelyking tussen agt verskillende insaaitegnieke ter verbetering van natuurlike veld. Proc. Grassld. Soc. Sth. Afr. 4:116–125.
Du Toit, P. F. 1968. A preliminary report on the effect of Acacia karroo competition on the composition and yield of sweet grassveld. Proc. Gassld. Soc. Sth. Afr. 3:147–149.
Du Toit, P. F. 1972. Acacia karroo intrusion: The effect of burning and sparing. Proc. Grassld. Soc. Sth. Afr. 7:23–27.
Du Toit, P. F. & Ingpen, R. A. 1970. Enkele gevolge van winterbeweiding op Dohne-suurveld. Proc. Gassld. Soc. Sth. Afr. 5:27–31.
Duvigneaud, P. 1971. Remarques préliminaires concernant la productivité des écosystèmes forestiers et le programme 'Productivité forestière' du Programme biologique international. In: Duvigneaud, P. (ed.), Productivity of forest ecosystems. pp. 13–18. UNESCO, Paris.
Eckardt, F. E. 1975. Functioning of the biosphere at the primary production level – objectives and achievements. In: Cooper, J. P. (ed.), Photosynthesis and productivity in different environments. pp. 173–185. Cambridge University Press, Cambridge.
Edwards, P. J. 1957. Curing and storage of veld hay. Fmg. S. Afr. 33(7):27–30.
Edwards, P. J. 1966. Veld replacement by improved grasslands in Natal. Proc. Grassld. Soc. Sth. Afr. 1:63–67.
Edwards, P. J. 1968. The long-term effects of burning and mowing on the basal cover of two veld types in Natal. S. Afr. J. Agric. Sci. 11:131–140.
Edwards, P. J. & Booysen, P. de V. 1972. The future for radical veld improvement in South Africa. Proc. Grassld. Soc. Sth. Afr. 7:61–66.
Edwards, P. J. & Coetsee, G. 1971. Growth vigour and carrying capacity of veld – how to determine it. Fmg. S. Afr. 47(8):38–39.
Edwards, P. J. & Nel, S. P. 1973. Short-term effects of fertilizer and stocking rates on the Bankenveld: 1. Vegetational changes. Proc. Grassld. Soc. Sth. Afr. 8:83–88.
Elliott, R. C. & Folkertsen, K. 1961. Seasonal changes in composition and yields of veld grass. Rhod. Agric. J. 58:186–187.
Endean, F. 1967. The productivity of 'Miombo' Woodland in Zambia. Forest Research Bulletin No. 14. Government Printer, Lusaka.
Fourie, J. H. & Roberts, B. R. 1976. A comparative study of three veld types of the northern Cape: species evaluation and yield. Proc. Grassld. Soc. Sth. Afr. 11:79–85.
Freson, R. 1973. Contribution a l'étude de l'écosystème forêt claire (Miombo). Note 13. Aperçu de la biomasse et de la productivité de la strate herbacée au miombo de la Luiswishi. Annales de l'université d'Abidjan. Série E. 6:265–277.
Garstang, M., Tyson, P. & Moskowitz, S. 1974. A case study of the cloud seeding potential of the southeastern coastal belt of Africa. S. Afr. J. Sci. 70:208–210.
Geldenhuys, C. J. 1975. Die kunsmatige vestiging van inheemse bos in die Suid-Kaap. Forestry in South Africa 16: 45–53.
Geldenhuys, C. J. 1976. An extensive enumeration survey and type classification of the woodlands in Kavango. In press.
Gillooly, J. F. 1976. Dendroclimatology in South Africa. S. Afr. For. J. 98:64–65.
Golley, F. B. 1975. Productivity and mineral cycling in tropical forests. In: Reichle, D. E., et al. (preface writers), Productivity of world ecosystems. Proceedings of a symposium, 1972. Seattle, Washington. pp. 106–115. National Academy of Sciences, Washington, D.C.
Golley, F. B. & Lieth, H. 1972. Bases of organic production in the tropics. Paper from a symposium on 'Tropical ecology with an emphasis on organic productivity' convened by F. B. Golley & R. Misra in New Delhi, India. 25 pp.
Goodman, P. S. 1975. The relation between vegetation structure and its use by wild herbivores in a riverine habitat. M.Sc. thesis in tropical resource ecology. University of Rhodesia.
Grunow, J. O. 1975/76. Annual report, 1975/76. Savanna Ecosystem Project, Nylsvley. Producer group: grass layer. Council for Scientific and Industrial Research, Pretoria.
Grunow, J. O., Pienaar, A. J. & Breytenbach, C. 1970. Long term nitrogen application to veld in South Africa. Proc. Grassld. Soc. Sth. Afr. 5:75–90.
Grut, M. 1965. Forestry and forest industry in South Africa. A. A. Balkema, Cape Town.

Guerreiro, M. G. 1966. A floresta africana e os factores bióticos. Primeiras observações de um ensaio em Moçambique. Instituto de Investigação Cientifica de Angola, Luanda.

Guy, G. L. 1970. Adansonia digitata and its rate of growth in relation to rainfall in south central Africa. Proc. Trans. Rhod. Sci. Ass. 54:68–84.

Haigh, H. 1966. Root development in the sandy soils of Zululand. Forestry in South Africa 7:31–36.

Hall, T. D., Meredith, D. & Altona, R. E. 1955. The role of fertilizers in pasture management. In: Meredith, D. (ed.), The grasses and pastures of South Africa. pp. 637–652. Central News Agency, South Africa.

Hall, T. D., Meredith, D. & Murray, S. M. 1937. The productivity of fertilized natural Highveld pastures. S.A. J. Sci. 34:275–285.

Herbst, S. N. & Goosen, P. C. N. 1973/74 and 1974/75. Die vasstelling van norme om drakrag van natuurlike veld van die O.V.S.-streek te bepaal: Die produksiepotensiaal van die natuurlike weiveld van die Vrystaat gedeelte van die O.V.S.-streek. Annual progress reports 1973/74 and 1974/75. Project O. Gl. 141/2. Department of Agricultural Technical Services, Pretoria.

Hirst, S. M. 1975. Savanna ecosystem project. Progress report 1974/75. South African National Scientific Programmes Report No. 3. National Scientific Programmes Unit, Council for Scientific and Industrial Research, Pretoria.

Hubbard, C. S. 1937 (sic). Observations on the distribution and rate of growth of Clanwilliam cedar Widdringtonia juniperoides Endl. S. Afr. J. Sci. 33:572–586.

Huntley, B. J. 1971. Aspects of grassland production and utilization on a northern Transvaal nature reserve. J. Sth. Afr. Wildl. Mgmt. Ass. 2:24–28.

Huntley, B. J. 1977. Savanna ecoystem project – progress report 1975/76. South African National Scientific Programmes Report No. 12. Council for Scientific and Industrial Research, Pretoria.

International Biological Programme. 1972. Summary report 1967–1972 of the South African National Committee for the I.B.P. Obtainable from C.S.I.R., Pretoria.

Keet, J. D. M. 1962. The Trema plantations of Westfalia estate. S. Afr. For. J. 41:15–27.

Kelly, R. D. 1973. A comparative study of primary productivity under different kinds of land use in southeastern Rhodesia. Ph.D. thesis, University of London.

Kemp, R. H. 1963. Growth and regeneration of open savanna woodland in northern Nigeria. Commonw. For. Rev. 42:200–206.

Kennan, T. C. D. 1969. The significance of bush in grazing land in Rhodesia. Rhod. Sci. News 3:331–336.

Kennan, T. C. D. 1972. The effects of fire on two vegetation types at Matopos, Rhodesia. Proc. Tall Timbers Fire Ecol. Conf. 11:53–98.

Kennan, T. C. D. & Barnes, D. L. 1973. The development of semi-arid bushveld to its full potential 4. Rhodesia. Proc. Grassld. Soc. Sth. Afr. 8:147–148.

Kennard, D. G. & Walker, B. H. 1973. Relationships between tree canopy cover and Panicum maximum in the vicinity of Fort Victoria. Rhod. J. Agric. Res. 11:145–153.

Kerfoot, O. 1963. The root systems of tropical forest trees. Commonw. For. Rev. 42:19–26.

King, N. L. 1941. The exploitation of the indigenous forests of South Africa. J. S. Afr. For. Ass. 6:26–48.

King, N. L. 1951. Tree-planting in South Africa. J. S. Afr. For. Ass. 21:1–102.

King, N. L. 1952. Some notes on forestry in Northern Rhodesia. J. S. Afr. For. Ass. 22:34–37.

Knapp, R. 1965. Pflanzenarten – Zusammensetzung, Entwicklung und natürliche Produktivität der Weide-Vegetation in Trockengebieten in verschiedenen Klima-Bereichen der Erde. In: Knapp, R. (ed.), Weide-Wirtschaft in Trockengebieten. pp. 71–97. Gustav Fischer Verlag, Stuttgart.

Kruger, F. J. 1977. A preliminary account of aerial plant biomass in fynbos communities of the Mediterranean-type climate zone of the Cape Province. Bothalia 12:299–305.

Kruger, J. A. & Smit, I. B. J. 1973. Herwinning van oulande in die Oos-Vrystaat deur die insaai van Digitaria smutsii, Eragrostis curvula en Themeda triandra. Agroplantae 5:101–106.

Květ, J., Ondok, J. P., Nečas, J. & Jarvis, P. G. 1971. Methods of growth analysis. In: Šesták, Z., Čatský, J. & Jarvis, P. G. (eds.), Plant photosynthetic production. Manual of methods. pp. 343–391. Junk, The Hague.

Laughton, F. S. 1937. The sylviculture of the indigenous forests of the Union of South Africa with special reference to the forests of the Knysna region. Science Bulletin No. 157. Forestry Series No. 7. Government Printer, Pretoria.

Lawton, R. M. 1967. The conservation and management of the riparian evergreen forests of Zambia. Commonw. For. Rev. 46:223–232.

Le Roux, A. & Morris, J. W. 1977. The effects of some burning and cutting treatments and fluctuation of annual rainfall on the seasonal variation of grassland basal cover. Proc. Grassld. Soc. Sth. Afr. 12: In press.

Le Roux, C. J. G. 1972/73. Grazing problems in the Etosha National Park: Carrying capacity of Andoni vlakte/Gemsbokvlakte/Charitsaubvlakte/Leeubron–Adamax Area. Progress Reports 1972/73. Department of Nature Conservation and Tourism, Windhoek.

Lieth, H. 1962. Die Stoffproduktion der Pflanzendecke. Vorträge und Diskussionsergebnisse des internationalen ökologischen Symposiums in Stuttgart – Hohenheim vom 4.–7. Mai 1960. Gustav Fischer Verlag, Stuttgart.

Lieth, H. 1968. The measurement of calorific values of biological material and the determination of ecological efficiency. In: Eckardt, F. E. (ed.), Functioning of terrestrial ecosystems at the primary production level. pp. 233–242. UNESCO, Paris.

Lieth, H. 1972. Über die Primärproduktion der Pflanzendecke der Erde. Angew. Botanik 46:1–37.

Lieth, H. 1974. Primary productivity of successional stages. In: Knapp, R. (ed.), Vegetation dynamics. pp. 187–193. Junk, The Hague.

Lieth, H. 1975a. Primary productivity in ecosystems: comparative analysis of global patterns. In: Van Dobben, W. H. & Lowe-McConnell, R. H. (eds.), Unifying concepts in ecology. pp. 67–88. Junk, The Hague.

Lieth, H. 1975b. Primary production of the major vegetation units of the world. In: Lieth, H. & Whittaker, R. H. (eds.), Primary productivity of the biosphere. Ecological studies vol. 14. pp. 203–215. Springer-Verlag, Berlin, Heidelberg, New York.

Lieth, H. 1975c. Modeling the primary productivity of the world. In: Lieth, H. & Whittaker, R. H. (eds.), Primary productivity of the biosphere. Ecological studies vol. 14. pp. 237–263. Springer-Verlag, Berlin, Heidelberg, New York.

Louw, A. J. 1968. Bemesting van natuurlike veld op rooi-leemgrond van die Springbokvlakte. 2. Invloed van ammoniumsulfaat- en superfosfaatbemesting op lugdroëmateriaalopbrengs en die minerale-inhoud daarvan. S. Afr. J. Agric. Sci. 11:629–636.

Major, J. 1974a. A review symposium on forest productivity research. Ecology 55:1167–1168.

Major, J. 1974b. Biomass accumulation in succession. In: Knapp, R. (ed.), Vegetation dynamics. pp. 197–203. Junk, The Hague.

Malaisse, F., Freson, R., Goffinet, G. & Malaisse-Mousset, M. 1975. Litter fall and litter breakdown in miombo. In: Golley, F. B. & Medina, E. (eds.), Tropical Ecological Systems. Ecological Studies vol. 11. pp. 137–152. Springer-Verlag, Berlin, Heidelberg, New York.

Martin, J. D. 1940. The Baikiaea forests of Northern Rhodesia. Emp. For. J. 19:8–18.

McKay, A. D. 1968. Rangeland productivity in Botswana. E. Afr. Agric. For. J. 34:178–193.

Menezes, O. J. A. de 1971. Estudo fitoecológico da região do Mucope e carta da vegetação. Sep. do Bol. Inst. Invest. Cient. Ang., Luanda.

Meredith, D., Scott, J. D. & Rose, C. J. 1955. The preservation and utilisation of grassland products. In: Meredith, D. (ed.), The grasses and pastures of South Africa. pp. 672–684. Central New Agency, South Africa.

Miller, O. B. 1938. A report of the forests of the Chobe district, Bechuanaland Protectorate. Unpubl. report submitted to the Administration 31.5.1938. 22 pp. and 23 pp. Typescript.

Miller, O. B. 1939. The Mukusi forests of the Bechuanaland Protectorate. Emp. For. J. 18: 193–201.

Miller, R. G. 1952. A girth increment study in Baikiaea plurijuga in Northern Rhodesia with reference to the determination of age and rotations for species without annual growth rings in irregular natural forest. Emp. For. Rev. 31:45–52.

Mills, P. F. L. 1964. Effects of fertilizer on the botanical composition of veld grassland at Matopos. Rhod. Agric. J. 61:91–93.

Mills, P. F. L. 1968. Effects of fertilizers on the yield and quality of veld grassland in the dry season. Rhod. J. Agric. Res. 6:27–39.

Moll, E. J. & Woods, D. B. 1971. The rate of forest tree growth and a forest ordination at Xumeni, Natal. Bothalia 10:451–460.

Monteiro, R. F. R. 1970. Estudo da flora e da vegetação das florestas abertas do planalto do Bié. Instituto de Investigação Científica de Angola, Luanda.

Mostert, J. W. C. & Vorster, L. F. 1969. The origin and development of the grazing section at Glen. Fmg. S. Afr. 45(8):86–88, 90.

Müller, P. J. 1963. Stave-point apparatus for vegetation survey and measurement on steep and rock-strewn hill slopes. S. Afr. J. Agric. Sci. 6:339–343.

Müller, Von D. 1962. Boysen Jensen und die Stoffproduktion der Pflanzen. In: Lieth, H. (ed.), Die Stoffproduktion der Pflanzendecke. pp. 6–10. Gustav Fischer Verlag, Stuttgart.
Murphy, P. G. 1975. Net primary productivity in tropical terrestrial ecosystems. In: Lieth, H. & Whittaker, R. H. (eds.), Primary productivity of the biosphere. Ecological studies vol. 14. pp. 217–231. Springer-Verlag, New York.
Nänni, U. W. 1960. Root distortion of young Pinus patula trees at Cathedral Peak. J. S. Afr. For. Ass. 34:13–22.
Newbould, P. J. 1963. Production ecology. Science Progress 51:91–104.
Odum, E. P. 1959. Fundamentals of ecology. 2nd ed. W. B. Saunders Company, Philadelphia and London.
Odum, E. P. 1962. Relationships between structure and function in ecosystems. Jap. J. Ecol. 12:108–118.
Olson, J. S. 1975. Productivity of forest ecosystems. In: Reichle, D. E. et al. (Preface writers), Productivity of world ecosystems. Proceedings of a symposium, 1972. Seattle, Washington. pp. 33–43. National Academy of Sciences, Washington, D.C.
Osmaston, H. A. 1956. Determination of age/girth and similar relationships in tropical forestry. Emp. For. Rev. 35:193–197.
Ovington, J. D. 1962. Quantitative ecology and the woodland ecosystem concept. Adv. Ecol. Res. 1:103–192.
Ovington, J. D., Heitkamp, D. & Lawrence, D. B. 1963. Plant biomass and productivity of prairie, savanna, oakwood and maize field ecosystems in central Minnesota. Ecology 44:52–63.
Pearsall, W. A. 1959. Production ecology. Science Progress 47:106–111.
Pentzhorn, B. L., Robbertse, P. J. & Olivier, M. C. 1974. The influence of the african elephant on the vegetation of the Addo Elephant National Park. Koedoe 17:137–158.
Petrusewicz, K. 1967. Suggested list of more important concepts in productivity studies (definitions and symbols). In: Petrusewicz, K. (ed.), Secondary productivity of terrestrial ecosystems. Vol. 1. pp. 51–57. Kraków, Warsawa.
Phillips, J. F. V. 1931. Forest-succession and ecology in the Knysna region. Mem. Bot. Serv. S. Afr. 11:1–379.
Rees, W. A. 1974. Bush sucker control in miombo woodland in Zambia. E. Afr. Agric. For. J. 40:44–49.
Reichle, D. E. 1975. Advances in ecosystem analysis. Bioscience 25:257–264.
Rethman, N. F. G. & Beukes, B. H. 1973. Overseeding of Eragrostis curvula on north-eastern sandy Highveld. Proc. Grassld. Soc. Sth. Afr. 8:57–59.
Rethman, N. F. G., Beukes, B. H. & Malherbe, C. E. 1971. Influence on a north-eastern sandy Highveld sward of winter utilization by sheep. Proc. Grassld. Soc. Sth. Afr. 6:55–62.
Rethman, N. F. G. & Booysen, P. de V. 1969. The seasonal growth patterns of a tall grassveld sward. Proc. Grassld. Soc. Sth. Afr. 4:56–60.
Roberts, B. R. & Opperman, D. P. J. 1974. Veld management recommendations – a reassessment of key species and proper use factors. Proc. Grassld. Soc. Sth. Afr. 9:149–155.
Rodin, L. E. & Basilevič, N. I. 1968. World distribution of plant biomass. In: Eckardt, F. E. (ed.), Functioning of terrestrial ecosystems at the primary production level. pp. 45–52. UNESCO, Paris.
Rodin, L. E. & Bazilevich, N. I. 1967. Production and mineral cycling in terrestrial vegetation. Translated Fogg, G. E. Oliver & Boyd, Edinburgh and London.
Rodin, L. E., Bazilevich, N. I. & Rozov, N. N. 1975. Productivity of the world's main ecosystems. In: Reichle, D. E. et al. (preface writers), Productivity of the world's ecosystems. Proceedings of a symposium, 1972, Seattle, Washington. pp. 13–26. National Academy of Sciences, Washington, D.C.
Rosenzweig, M. L. 1968. Net primary productivity of terrestrial communities: Prediction from climatological data. Amer. Natur. 102:67–74.
Roux, P. W. 1963. The descending-point method of vegetation survey. A point-sampling method for the measurement of semi-open grasslands and karoo vegetation in South Africa. S. Afr. J. Agric. Sci. 6:273–285, 287.
Roux, P. W. 1968. The autecology of Tetrachne dregei Nees. Ph.D. Thesis, University of Natal, Pietermaritzburg.
Rowland, J. W. 1974. The conservation ideal SARCUSS. P/Bag X116, Pretoria.
Rutherford, M. C. 1975. Aspects of ecosystem function in a savanna woodland in South West Africa. Ph.D. thesis, University of Stellenbosch, Stellenbosch.

Rutherford, M. C. 1976. Changes of biomass in some perennial grass species. Proc. Grassld. Soc. Sth. Afr. 11: 43–46.
Rutherford, M. C. 1978a. Karoo-fynbos biomas along an elevational gradient in the western Cape. Bothalia: in press.
Rutherford, M. C. 1978b. Allometric biomass relations applied to succulent plant forms. In press.
Rutherford, M. C. & Kelly, R. D. 1977. Woody basal area and stem increment in Burkea woodland. In press.
Savory, B. M. 1963. Site quality and tree root morphology in Northern Rhodesia. Rhod. J. Agric. Res. 1:55–64.
Scott, J. D. 1972. Veld burning in Natal. Proc. Tall Timbers Fire Ecol. Conf. 11:33–51.
Scott, J. D. & Van Breda, N. G. 1937. Preliminary studies on the root system of the Rhenosterbos (Elytropappus rhinocerotis on the Worcester Veld Reserve). S. Afr. J. Sci. 33:560–569.
Šesták, Z., Čatský, J. & Jarvis, P. G. 1971. Plant photosynthetic production. Manual of methods. Junk, The Hague.
Sharp, D. D., Lieth, H. & Whigham, D. 1975. Assessment of regional productivity in North Carolina. In: Lieth, H. & Whittaker, R. H. (eds.), Primary productivity of the biosphere. pp. 131–146. Springer-Verlag, New York.
Sharp, D. M. 1975. Methods of assessing the primary production of regions. In: Leith, H. & Whittaker, R. H. (eds.), Primary productivity of the biosphere. pp. 147–166. Springer-Verlag, New York.
Singh, J. S., Lauenroth, W. K. & Steinhorst, R. K. 1975. Review and assessment of various techniques for estimating net aerial primary production in grassland from harvest data. Bot. Rev. 41:181–232.
Smith, C. A. 1961. The utilization of Hyparrhenia veld for the nutrition of cattle in the dry season. I. The effects of nitrogen fertilizers and mowing regimes on herbage yields. J. Agric. Sci. 57:305–310.
Smith, C. A. 1963. Oversowing pasture legumes into the Hyparrhenia grassland of Northern Rhodesia. Nature 200:811–812.
Stapleton, C. C. 1955. The cultivation of indigeneous trees and shrubs. J. S. Afr. For. Ass. 26: 10–17.
Strang, R. M. 1960. Experiment in bush eradication. Rhod. Agric. J. 57:122–123.
Strang, R. M. 1974. Some man-made changes in successional trends on the Rhodesian Highveld. J. Appl. Ecol. 11:249–263.
Swart, E. R. 1963. Age of the baobab tree. Nature 198:708–709.
Symons, L. B. & Jones, R. I. 1971. An analysis of available techniques for estimating production of pastures without clipping. Proc. Grassld. Soc. Sth. Afr. 6:185–190.
Tainton, N. M., Booysen, P. de V. & Scott, J. D. 1970. Response of tall grassveld to different intensities, seasons and frequencies of clipping. Proc. Grassld. Soc. Sth. Afr. 5:32–41.
Tidmarsh, C. E. M. 1968. Bioclimatology with special reference to agriculture. S. Afr. J. Sci. 64:456–459.
Tidmarsh, C. E. M. & Havenga, C. M. 1955. The wheel-point method of survey and measurement of semi-open grasslands and Karoo vegetation in South Africa. Mem. Bot. Surv. S. Afr. 29:1–49.
Trapnell, C. G. 1959. Ecological results of woodland burning experiments in Northern Rhodesia. J. Ecol. 47:129–168.
Trollope, W. S. W. 1972. Fire as a method of eradicating macchia vegetation in the Amatole Mountains of South Africa – experimental and field scale results. Proc. Tall Timbers Fire Ecol. Conf. 11:99–120.
Trollope, W. S. W. 1974. Role of fire in preventing bush encroachment in the eastern Cape. Proc. Grassld. Soc. Sth. Afr. 9: 67–72.
Van den Berg, J. A., Roberts, B. R. & Vorster, L. F. 1976. Die uitwerking van seisoensbeweiding op die infiltrasievermoë van gronde in 'n Cymbopogon – Themedaveld. Proc. Grassld. Soc. Sth. Afr. 11:91–95.
Van den Berg, M. A. 1974. Biological studies on Cirina forda (Westw.) (Lepidoptera: Saturniidae), a pest of wild Seringa trees (Burkea africana Hook.). Phytophylactica 6:61–62.
Van der Pas, J. B. 1972. Investigation of techniques of measuring and estimating phytomass in the Cape fynbos. Progress report, project 116/17/1. Jonkershoek Forest Research Station. Department of Forestry, Pretoria.
Van der Schijff, H. P. 1957. 'n Ekologiese studie van die flora van die Nasionale Krugerwildtuin. D.Sc. thesis, University of Potchefstroom for C. H. E., Potchefstroom.

Van der Walt, P. T. 1965. A plant ecological survey of the noorsveld. M.Sc. thesis. University of Natal.
Van Laar, A. & Geldenhuys, C. J. 1975. Tariff tables for indigenous tree species in the southern Cape Province. Forestry in South Africa 17:29–36.
Van Laar, A. & Lewark, S. 1973. Sampling for forest inventories in the indigenous forests of the southern Cape Province. Forestry in South Africa 14:35–43.
Van Rensburg, H. J. 1968. Multipurpose survey of the Kafue river basin Zambia. Vol. 4. Ecology of the Kafue flats. Part 1. Ecology and development. FAO, Rome.
Van Schalkwyk, A., Lombard, P. E. & Vorster, L. F. 1968. Evaluation of the nutritive value of a Themeda triandra pasture in the central Orange Free State. 2. Comparison between quadrat and simulated grazing samples. S. Afr. J. Agric. Sci. 11:249–257.
Van Wyk, J. J. P., Bosch, O. J. H. & Kruger, J. A. 1969. Droogtebeskadiging van bosveldbome en groot struike. Proc. Grassld. Soc. Sth. Afr. 4:61–65.
Van Wyk, P. 1972. Veld burnings in the Kruger National Park, an interim report of some aspects of research. Proc. Tall Timbers Fire Ecol. Conf. 11:9–31.
Van Wyk, P. 1972/74. Trees of the Kruger National Park. Purnell, Cape Town.
Venter, H. J. T. 1972. Die plantekologie van Richardsbaai, Natal. D.Sc. thesis, University of Pretoria, Pretoria.
Visser, J. H. 1966. Bemesting van die veld. Proc. Grassld. Soc. Sth. Afr. 1:41–48.
Von Berg, D. A. 1970. Measuring the productive capacity of pastures ... a biological method. Fmg. S. Afr. 46(2):81–84.
Vorster, L. F. & Mostert, J. W. C. 1968. Veldbemestingstendense oor 'n dekade in die sentrale Oranje-Vrystaat. Proc. Grassld. Soc. Sth. Afr. 3:111–119.
Vowinckel, E. 1961. Potential growth areas for introduced tree species. Forestry in South Africa 1:91–104.
Walker, B. H. 1970. An evaluation of eight methods of botanical analysis on grasslands in Rhodesia. J. Appl. Ecol. 7:403–416.
Walker, B. H. 1971. Ecological problems in the management of Rhodesian veld. Rhod. Sci. News 5:20–23, 36.
Walter, H. 1939. Grassland, Savanne und Busch der arideren Teile Afrikas in ihrer ökologischen Bedingtheit. Jb. wiss. Bot. 87:750–860.
Walter, H. 1973. Die Vegetation der Erde. Band 1: Die tropischen und subtropischen Zonen. (3rd ed.). Gustav Fischer Verlag, Stuttgart.
Walter, H. & Volk, O. H. 1954. Grundlagen der Weidewirtschaft in Südwestafrika. Eugen Ulmer, Stuttgart.
Ward, H. K. & Cleghorn, W. B. 1964. The effect of ring-barking trees in Brachystegia woodland on the yield of veld grasses. Rhod. Agric. J. 61:98–105, 107.
Ward, H. K. & Cleghorn, W. B. 1970. The effects of grazing practices on tree regrowth after clearing indigenous woodland. Rhod. J. Agric. Res. 8:57–65.
Wassink, E. C. 1975. Photosynthesis and productivity in different environments – conclusions. In: Cooper, J. P. (ed.), Photosynthesis and productivity in different environments. pp. 675–687. Cambridge University Press, Cambridge.
Weaver, J. E. 1924. Plant production as a measure of environments. J. Ecol. 12:205–237.
Weinmann, H. 1943. Effects of defoliation intensity and fertilizer treatment on Transvaal Highveld. Emp. J. Exp. Agric. 11:113–124.
Weinmann, H. 1948a. Effects of defoliation intensity and fertilizer treatment on Transvaal Highveld. Emp. J. Exp. Agric. 16:111–118.
Weinmann, H. 1948b. Seasonal growth and changes in chemical composition of the herbage on Marandellas sandveld. Rhod. Agric. J. 45:119–131.
Weinmann, H. 1955. The chemistry and physiology of grasses. In: Meredith, D. (ed.), The grasses and pastures of South Africa. pp. 571–600. Central News Agency, South Africa.
West, O. 1950. Indigenous tree crops for Southern Rhodesia. Rhod. Agric. J. 47:204–217.
West, O. 1955. Veld management in the dry, summer-rainfall bushveld. In: Meredith, D. (ed.), The grasses and pastures of South Africa. pp. 624–636. Central News Agency, South Africa.
West, O. 1965. Fire in vegetation and its use in pasture management with special reference to tropical and subtropical Africa. Mimeographed publication No. 1. Commonwealth Bureau of Pastures and Field Crops, Commonwealth Agricultural Bureaux, Hurley, Berkshire.
West, O. 1969. Fire, its effect on the ecology of vegetation in Rhodesia and its application in grazing management. Rhod. Sci. News 3:321–326.

Westlake, D. F. 1963. Comparisons of plant productivity. Biol. Rev. 38:385–425.
Westlake, D. F. 1965. Theoretical aspects of the comparability of productivity data. Mem. Ist. Ital. Idrobiol. 18 Suppl.: 313, 315–322.
Whittaker, R. H. 1970. Communities and ecosystems. The Macmillan Company, London.
Whittaker, R. H. & Marks, P. L. 1975. Methods of assessing terrestrial productivity. In: Lieth, H. & Whittaker, R. H. (eds.), Primary productivity of the biosphere. pp. 55–118. Springer-Verlag, New York.
Wicht, C. L. 1971. The task of forestry in the mountains of the western and southern Cape Province. Proc. Grassld. Soc. Sth. Afr. 6:20–27.